Principles of Mucosal Immunology

Principles of
Mucosal
Immunology

Society for Mucosal Immunology

Phillip D. Smith, Thomas T. MacDonald, and Richard S. Blumberg, Editors

GS Garland Science
Taylor & Francis Group
LONDON AND NEW YORK

Vice President: Denise Schanck
Senior Editor: Janet Foltin
Associate Editor: Monica Toledo
Production Editor: Ioana Moldovan
Typesetter and Senior Production Editor: Georgina Lucas
Copy Editors: Sally Huish and Bruce Goatly
Proofreader: Jo Clayton
Illustrations: Matthew McClements, Blink Studio, Ltd
Indexer: Medical Indexing Ltd

ISBN 978-0-8153-4443-8

Library of Congress Cataloging-in-Publication Data

Principles of mucosal immunology / Society for Mucosal Immunology ; Phillip
D. Smith, Thomas T. MacDonald, and Richard S. Blumberg, editors.
      p. ; cm.
 Includes bibliographical references and index.
 ISBN 978-0-8153-4443-8 (alk. paper)
 I. Smith, Phillip D., M.D. II. MacDonald, Thomas T., 1951- III. Blumberg,
R. S. (Richard Steven) IV. Society for Mucosal Immunology.
 [DNLM: 1. Immunity, Mucosal--physiology. 2. Mucous Membrane--immunology.
QW 563]

 611'.0187--dc23

                                         2012005355

Published by Garland Science, Taylor & Francis Group, LLC, an informa business,
711 Third Avenue, 8th floor, New York, NY 10017, USA,
and 3 Park Square, Milton Park, Abingdon, OX14 4RN, UK.

Printed in the United States of America

15 14 13 12 11 10 9 8 7 6 5 4 3 2 1

Garland Science
Taylor & Francis Group

Visit our website at http://www.garlandscience.com

Preface

In 1988, Dr. Charles Janeway coined the term 'frontiers of the immune system' to describe the mysterious attributes of immunology at mucosal sites. Since then, an explosion in the study of mucosal surfaces has yielded remarkable new insights into the immunology, inflammatory processes, infectious diseases, and microbiota at these sites as well as the role of the mucosa in obesity and systemic autoimmune diseases, including diabetes mellitus, rheumatoid arthritis, and multiple sclerosis.

The Society for Mucosal Immunology (SMI) was established in 1987 to promote excellence in research and education in mucosal immunology and to foster communication among investigators examining the unique aspects of the mucosal immune system. In pursuing this mandate, the Society realized that the absence of an accessible book, which focuses on principles and concepts in a broad yet concise manner, was an impediment to its mission. Therefore, in 2007, the Society decided to undertake the development of this book to accomplish these objectives.

Principles of Mucosal Immunology is intended for graduate students receiving advanced training in mucosal immunology, doctoral students and postdoctoral fellows investigating topics in mucosal immunology, medical and dental students, and immunologists and clinicians seeking a broad perspective of the field. The book also provides a framework for specialized courses in mucosal immunology.

To achieve the goals established by the Society, we obtained input from the Board of Councilors in the development of the book's content and sought the scientific expertise of internationally recognized leaders in the field of mucosal immunology, many of whom contributed chapters to the book related to their research and practice.

Principles of Mucosal Immunology examines the commonalities of the mucosal immune system, focusing on all compartments, but especially those for which the largest body of information is available, such as the small and large intestine, lung and upper airways, genital and urinary systems, and the ocular system. We recognize that, in this first edition, some areas of the mucosal immune system are not discussed at length, including those associated with the pancreaticobiliary tree, stomach, inner ear, and mammary glands. The book introduces readers to the topic in a way that provides a launching point for considering these and related areas in greater depth.

The book is organized into seven parts. Chapters in Parts I through III present the development and structure of mucosal tissues, the cellular constituents of the mucosal immune system and their function in mucosal homeostasis, and the relationship of each of these with commensal microbes. This content relies heavily on information from the intestinal immune system, the study of which has resulted in major findings in mucosal immunology. Where possible, we make comparisons to other mucosal tissues so as to provide a more complete understanding of mucosal immune function. Chapters in Parts IV through VII focus on the genitourinary system, the airways (nose, upper airways, oral cavity, and lung) and ocular surfaces, specific infectious diseases of mucosal tissues, and immune-mediated diseases of mucosal surfaces that provide a paradigm for understanding immunopathology at these sites. Major themes relating to the unique functions that characterize mucosal immunity, such as the specialized anatomic organization and its relationship to the microbiota, secretory immunity, peripheral immune regulation, and antigen handling and leukocyte homing, are highlighted throughout the book.

Key features of the book include elegant and demonstrative illustrations and the Further Reading sections in each chapter, both of which will be valuable resources for instructors who use the book to guide their courses or training. Consequently, we believe that the book is a unique asset for teaching mucosal immunology courses, a topic of increased popularity across academic campuses.

We and the SMI are delighted to team up with Garland Science, the publisher of the acclaimed *Janeway's Immunobiology*. Accordingly, we envision that the book will serve as a companion to *Immunobiology* and hope that the opportunities that are derived from this partnership will benefit readers and instructors alike. We hope that *Principles of Mucosal Immunology* expands your knowledge of this fascinating field. We thank our contributors, as well as our students and colleagues, who share our passion for this topic and who inspire us on a daily basis with their discoveries.

Phillip D. Smith
Birmingham, Alabama

Thomas T. MacDonald
London, United Kingdom

Richard S. Blumberg
Boston, Massachusetts

Instructor Resources Website

Accessible from www.garlandscience.com, the Instructor Resource Site requires registration and access is available only to qualified instructors. To access the Instructor Resource Site, please contact your local sales representative or email science@garland.com.

The images in *Principles of Mucosal Immunology* are available on the Instructor Site in two convenient formats:

PowerPoint® and JPEG, which have been optimized for display. The resources may be browsed by individual chapter or a search engine. Figures are searchable by figure number, figure name, or by keywords used in the figure legend from the book.

Resources available for other Garland Science titles can be accessed via the Garland Science website.

PowerPoint is a registered trademark of Microsoft Corporation in the United States and/or other countries.

Contributors

William W. Agace, PhD, Professor of Experimental Medical Research and Head of the Immunology Section, Department of Experimental Medical Science, Lund University, Lund, Sweden. *Chapter 14: Lymphocyte Trafficking from Inductive Sites to Effector Sites.*

David Artis, MD, Associate Professor of Microbiology, Department of Pathobiology, University of Pennsylvania School of Veterinary Medicine, Philadelphia, Pennsylvania. *Chapter 12: Mucosal Basophils, Eosinophils, and Mast Cells.*

Kenneth W. Beagley, PhD, Professor of Immunology, Faculty of Science and Technology, Queensland University of Technology; Institute of Health and Biomedical Innovation, Brisbane, Australia. *Chapter 18: The Immune System of the Genitourinary Tract.*

Diane Bimczok, DVM, PhD, Instructor of Medicine, Department of Medicine (Gastroenterology/Hepatology), University of Alabama at Birmingham, Alabama. *Chapter 24: Helicobacter pylori Infection.*

Stephan C. Bischoff, MD, Professor of Medicine, Institute of Nutritional Medicine and Immunology, University of Hohenheim, Stuttgart, Germany. *Chapter 12: Mucosal Basophils, Eosinophils, and Mast Cells.*

Richard S. Blumberg, MD, Professor of Medicine, Harvard Medical School, Chief, Gastroenterology and Hepatology Division, Brigham and Women's Hospital, Boston, Massachusetts. *Chapter 5: Immune Function of Epithelial Cells.*

Per Brandtzaeg, MD, Professor of Medicine, University of Oslo; Laboratory for Immunohistochemistry and Immunopathology (LIIPAT), Oslo, Norway. *Chapter 1: Overview of the Mucosal Immune System Structure; Chapter 8: Mucosal B cells and Their Function.*

Jonathan Braun, MD, PhD, Professor and Chair, Department of Pathology & Lab Medicine, UCLA David Geffen School of Medicine, Los Angeles, California. *Chapter 17: The Commensal Microbiota and Its Relationship to Homeostasis and Disease.*

Elke Cario, MD, Professor of Medicine, Division of Gastroenterology & Hepatology, University Hospital of Essen, Essen, Germany. *Chapter 16: Recognition of Microbe-Associated Molecular Patterns by Pattern Recognition Receptors.*

Rachel R. Caspi, PhD, Adjunct Professor, Department of Pathology, University of Pennsylvania; Chief of the Immunoregulation Section and Deputy Chief of the Laboratory of Immunology, National Eye Institute, National Institutes of Health, Bethesda, Maryland. *Chapter 22: The Ocular Surface as a Mucosal Immune Site.*

Nadine Cerf-Bensussan, MD, PhD, Research Director, INSERM UMR989, Université Paris Descartes, Paris, France. *Chapter 23: Mucosal Interactions with Enteropathogenic Bacteria.*

Hilde Cheroutre, PhD, Professor and Division Head, Division of Developmental Immunology, La Jolla Institute for Allergy and Immunology; Adjunct Professor, Department of Gastroenterology, School of Medicine, University of California, San Diego, California. *Chapter 6: Intraepithelial Lymphocytes: Unusual T Cells at Epithelial Surfaces.*

Satya Dandekar, PhD, Professor and Chair, Department of Medical Microbiology and Immunology, School of Medicine, University of California, Davis, California. *Chapter 25: Viral Infections.*

Arlette Darfeuille-Michaud, PhD, Professor, Unité M2iSH "Microbes, Intestin, Inflammation et Susceptibilité de l'Hôte", Université d'Auvergne, Clermont-Ferrand, France. *Chapter 23: Mucosal Interactions with Enteropathogenic Bacteria.*

Charles O. Elson III, MD, Professor of Medicine and Microbiology and Basil I. Hirschowitz Chair in Gastroenterology, Department of Medicine (Gastroenterology/Hepatology), University of Alabama at Birmingham, Alabama. *Chapter 15: Mucosal Tolerance.*

Mary K. Estes, PhD, Professor, Department of Molecular Virology and Microbiology, Baylor College of Medicine, Houston, Texas. *Chapter 25: Viral Infections.*

Kohtaro Fujihashi, DDS, PhD, Professor of School of Dentistry, Department of Pediatric Dentistry and Microbiology, University of Alabama at Birmingham, Alabama. *Chapter 3: Immunological and Functional Differences Between Individual Compartments of the Mucosal Immune System; Chapter 20: The Nasopharyngeal and Oral Immune System.*

Glenn T. Furuta, MD, Professor of Pediatrics, Gastrointestinal Eosinophilic Diseases Program, Mucosal Inflammation Program, National Jewish Health, University of Colorado School of Medicine; Digestive Health Institute, Section of Pediatric Gastroenterology, Hepatology and Nutrition, Children's Hospital Colorado, Denver, Colorado. *Chapter 32: Food Sensitive and Eosinophilic Enteropathies.*

Harry B. Greenberg, MD, Joseph D. Grant Professor, Departments of Medicine and Microbiology and Immunology, Stanford University School of Medicine, Stanford, California. *Chapter 27: Principles of Mucosal Vaccine Strategies.*

Jan Holmgren, MD, PhD, Professor of Microbiology and Immunology, University of Göteborg Vaccine Research Institute, University of Göteborg, Gothenburg, Sweden. *Chapter 27: Principles of Mucosal Vaccine Strategies.*

Patrick G. Holt, DSc, Professor, The University of Western Australia; Deputy Director, Telethon Institute for Child Health Research, Perth, Australia. *Chapter 21: Bronchus-Associated Lymphoid Tissue and Immune-Mediated Respiratory Diseases.*

Akiko Iwasaki, PhD, Professor of Immunobiology, Department of Immunobiology, Yale University School of Medicine, New Haven, Connecticut. *Chapter 19: Mucosal Immune Responses to Microbes in the Genital Tract.*

Bana Jabri, MD, PhD, Professor, Department of Medicine, University of Chicago, Chicago, Illinois. *Chapter 28: Celiac Disease.*

Bruce A. Julian, MD, Professor of Medicine, Department of Medicine (Nephrology), University of Alabama at Birmingham, Alabama. *Chapter 29: IgA Nephropathy.*

Charlotte S. Kaetzel, PhD, Professor, Department of Microbiology, Immunology, and Molecular Genetics, University of Kentucky, College of Medicine, Lexington, Kentucky. *Chapter 9: Secretory Immunoglobulins and Their Transport.*

Arthur Kaser, MD, Professor of Gastroenterology, Department of Medicine II (Gastroenterology/Hepatology), University of Cambridge, Addenbrooke's Hospital, Cambridge, United Kingdom. *Chapter 5: Immune Function of Epithelial Cells.*

Brian L. Kelsall, MD, Chief and Senior Investigator, Mucosal Immunobiology Section, National Institute of Allergy and Infectious Diseases, National Institutes of Health, Bethesda, Maryland. *Chapter 10: Role of Dendritic Cells in Integrating Immune Responses to Luminal Antigens.*

Hiroshi Kiyono, DDS, PhD, Dean and Professor, Department of Microbiology and Immunology, The Institute of Medical Science, University of Tokyo, Tokyo, Japan. *Chapter 3: Immunological and Functional Differences Between Individual Compartments of the Mucosal Immune System; Chapter 20: The Nasopharyngeal and Oral Immune System.*

Leo Lefrancois, PhD, Professor, Department of Immunology, University of Connecticut Health Center, Farmington,

Connecticut. *Chapter 6: Intraepithelial Lymphocytes: Unusual T Cells at Epithelial Surfaces.*

Wayne I. Lencer, MD, Professor of Pediatrics, Harvard Medical School; Chief, Division of Gastroenterology, Children's Hospital, Boston, Massachusetts. *Chapter 5: Immune Function of Epithelial Cells.*

Graham M. Lord, MD, FRCP, PhD, Professor of Medicine and Director, Translational Research Development, Department of Experimental Immunobiology, King's College London, United Kingdom. *Chapter 17: The Commensal Microbiota and Its Relationship to Homeostasis and Disease.*

Nils Lycke, MD, PhD, Professor, Sahlgrenska Academy; Director, Mucosal Immunobiology and Vaccine Research Center, University of Göteborg, Gothenburg, Sweden. *Chapter 27: Principles of Mucosal Vaccine Strategies.*

Thomas T. MacDonald, PhD, FRCPath, FMedSci, Dean for Research and Professor of Immunology, Centre for Immunology and Infectious Disease, Barts and the London School of Medicine and Dentistry, United Kingdom. *Chapter 2: Phylogeny of the Mucosal Immune System; Chapter 7: Lymphocyte Populations Within the Lamina Propria; Chapter 31: Inflammatory Bowel Disease.*

Lloyd Mayer, MD, Dorothy and David Merksamer Professor of Medicine, Professor and Chairman, Immunology Institute, Mount Sinai Medical Center, New York, New York. *Chapter 15: Mucosal Tolerance.*

Sarkis K. Mazmanian, PhD, Assistant Professor of Biology, Division of Biology, California Institute of Technology, Pasadena, California. *Chapter 17: The Commensal Microbiota and Its Relationship to Homeostasis and Disease.*

Michael A. McGuckin, PhD, Principal Research Fellow, Mater Medical Research Institute, South Brisbane, Australia. *Chapter 4: Secreted Effectors of the Innate Mucosal Barrier.*

Jiri Mestecky, MD, PhD, Professor of Microbiology and Medicine, Department of Microbiology, University of Alabama at Birmingham, Alabama. *Chapter 3: Immunological and Functional Differences Between Individual Compartments of the Mucosal Immune System; Chapter 9: Secretory Immunoglobulins and Their Transport; Chapter 29: IgA Nephropathy.*

Suzanne M. Michalek, PhD, Professor of Microbiology, Department of Microbiology, University of Alabama at Birmingham, Alabama. *Chapter 26: Infection-Driven Periodontal Disease.*

Robert D. Miller, PhD, Professor, Center for Evolutionary and Theoretical Immunology, Department of Biology, University of New Mexico, Albuquerque, New Mexico. *Chapter 2: Phylogeny of the Mucosal Immune System.*

Giovanni Monteleone, MD, Professor of Gastroenterology, Department of Internal Medicine, University of Rome Tor Vergata, Rome, Italy. *Chapter 31: Inflammatory Bowel Disease.*

Cathryn R. Nagler, PhD, Bunning Food Allergy Professor, Department of Pathology, Committee on Immunology, The University of Chicago, Illinois. *Chapter 32: Food Sensitive and Eosinophilic Enteropathies.*

Markus F. Neurath, MD, Direktor der Gastroenterologie, Pneumologie und Endokrinologie, Department of Medicine,

University of Erlangen-Nürnberg, Erlangen, Germany. *Chapter 31: Inflammatory Bowel Disease.*

Marian R. Neutra, PhD, Professor of Pediatrics, Departments of Medicine and Pediatrics, Harvard Medical School and Children's Hospital, Boston, Massachusetts. *Chapter 13: M Cells and the Follicle-Associated Epithelium.*

Luigi D. Notarangelo, MD, Professor of Pediatrics and Pathology at the Harvard Medical School and Jeffrey Modell Chair of Pediatric Immunology Research at the Division of Immunology, Children's Hospital, Boston, Massachusetts. *Chapter 30: Mucosal Manifestations of Immunodeficiencies.*

Jan Novak, PhD, Associate Professor of Microbiology, Department of Microbiology, University of Alabama at Birmingham, Alabama. *Chapter 29: IgA Nephropathy.*

André J. Ouellette, PhD, Professor of Pathology, Department of Pathology & Laboratory Medicine, Keck School of Medicine, University of Southern California, Los Angeles, California. *Chapter 4: Secreted Effectors of the Innate Mucosal Barrier.*

Oliver Pabst, PhD, Professor of Mucosal Immunology, Institute of Immunology, Hannover Medical School, Hannover, Germany. *Chapter 14: Lymphocyte Trafficking from Inductive Sites to Effector Sites.*

Reinhard Pabst, MD, Professor of Medicine, Immunomorphology, Hannover Medical School, Hannover, Germany. *Chapter 1: Overview of the Mucosal Immune System Structure.*

Richard M. Peek, Jr., MD, Professor of Medicine and Cancer Biology and Director, Department of Medicine (Gastroenterology/Hepatology/Nutrition), Vanderbilt University School of Medicine, Nashville, Tennessee. *Chapter 24: Helicobacter pylori Infection.*

Daniel K. Podolsky, MD, Doris and Bryan Wildenthal Distinguished Chair in Medical Science and President, UT Southwestern Medical Center at Dallas, Texas. *Chapter 16: Recognition of Microbe-Associated Molecular Patterns by Pattern Recognition Receptors.*

Nick Powell, MBChB, Department of Experimental Immunobiology, Division of Transplantation Immunology and Mucosal Biology, King's College London, United Kingdom. *Chapter 17: The Commensal Microbiota and Its Relationship to Homeostasis and Disease.*

Maria Rescigno, PhD, Group Leader, Department of Experimental Oncology, European Institute of Oncology, Milan, Italy. *Chapter 10: Role of Dendritic Cells in Integrating Immune Responses to Luminal Antigens.*

Mark Siracusa, PhD, Department of Pathobiology, University of Pennsylvania School of Veterinary Medicine, Philadelphia, Pennsylvania. *Chapter 12: Mucosal Basophils, Eosinophils, and Mast Cells.*

Phillip D. Smith, MD, Professor of Medicine and Microbiology and Mary J. Bradford Professor in Gastroenterology, Department of Medicine (Gastroenterology/Hepatology), University of Alabama at Birmingham, Alabama. *Chapter 11: Intestinal Macrophages in Defense of the Mucosa; Chapter 25: Viral Infections.*

Lesley E. Smythies, PhD, Associate Professor of Medicine, Department of Medicine (Gastroenterology/Hepatology), University of Alabama at Birmingham, Alabama. *Chapter 11: Intestinal Macrophages in Defense of the Mucosa.*

Scott B. Snapper, MD, PhD, Wolpow Family Chair in IBD Treatment and Research; Director, Inflammatory Bowel Disease Center and Basic & Translational Research (Gastroenterology, Children's Hospital); Director, Inflammatory Bowel Disease Research (Gastroenterology, Brigham and Women's Hospital); and Associate Professor of Medicine, Harvard Medical School, Boston, Massachusetts. *Chapter 30: Mucosal Manifestations of Immunodeficiencies.*

Ludvig M. Sollid, MD, PhD, Professor of Immunology, Faculty of Medicine, Institute of Immunology, University of Oslo, Norway. *Chapter 28: Celiac Disease.*

Jo Spencer, PhD, FRCPath, Professor in Immunobiology, Peter Gorer Department of Immunobiology, King's College London School of Medicine at Guy's Hospital, London, United Kingdom. *Chapter 8: Mucosal B Cells and Their Function.*

Dale T. Umetsu, MD, PhD, Professor of Pediatrics, Harvard Medical School, Division of Immunology, Children's Hospital, Boston, Massachusetts. *Chapter 21: Bronchus-Associated Lymphoid Tissue and Immune-Mediated Respiratory Diseases.*

Thomas E. Van Dyke, DDS, PhD, Vice President for Clinical and Translational Research; Chair, Department of Periodontology, The Forsyth Institute, Cambridge, Massachusetts. *Chapter 26: Infection-Driven Periodontal Disease.*

Femke van Wijk, MSc, PhD, Assistant Professor, UMC Utrecht, Centre for Molecular and Cellular Intervention, Utrecht, The Netherlands. *Chapter 6: Intraepithelial Lymphocytes: Unusual T Cells at Epithelial Surfaces.*

Casey T. Weaver, MD, Professor of Pathology, Department of Pathology, University of Alabama at Birmingham, Alabama. *Chapter 7: Lymphocyte Populations Within the Lamina Propria.*

Keith T. Wilson, MD, Professor of Medicine, Department of Medicine (Gastroenterology/Hepatology/Nutrition), Vanderbilt University School of Medicine, Nashville, Tennessee. *Chapter 24: Helicobacter pylori Infection.*

Charles R. Wira, PhD, Professor of Physiology and Neurobiology, Department of Physiology and Neurobiology and Department of Microbiology and Immunology, Dartmouth Medical School, Hanover, New Hampshire. *Chapter 18: The Immune System of the Genitourinary Tract.*

Jenny M. Woof, PhD, Reader in Immunology, Division of Pathology and Neuroscience, Medical Research Institute, University of Dundee Medical School, Ninewells Hospital, Dundee, Scotland, United Kingdom. *Chapter 9: Secretory Immunoglobulins and Their Transport.*

Gary D. Wu, MD, Professor of Medicine, Division of Gastroenterology, University of Pennsylvania School of Medicine, Philadelphia, Pennsylvania. *Chapter 4: Secreted Effectors of the Innate Mucosal Barrier.*

Contents

Detailed Contents

PART VI INFECTIOUS DISEASES OF MUCOSAL SURFACES 342

Chapter 23 Mucosal Interactions with Enteropathogenic Bacteria 343

Chapter 24 *Helicobacter pylori* Infection 363

Chapter 25 Viral Infections 377

PART I

DEVELOPMENT AND STRUCTURE OF MUCOSAL TISSUE

Overview of the Mucosal Immune System Structure

Pioneering experiments in the early 1960s showed that the large lymphocytes (lymphoblasts) which enter the bloodstream from the thoracic duct migrate into the intestinal lamina propria and undergo terminal differentiation into plasmablasts and plasma cells. Many of the circulating lymphoblasts express surface IgA (sIgA), and in the gut contain cytoplasmic IgA. These lymphoid cells were thought to be derived mainly from Peyer's patches (PPs), because transfer studies demonstrated that PPs and the draining mesenteric lymph nodes (MLNs), in contrast to peripheral lymph nodes and the spleen, were enriched precursor sources for IgA-producing plasma cells in the gut mucosa. It was also shown that plasma-cell differentiation occurs during dissemination of the mucosal B cells. Thus, the fraction of cells with cytoplasmic IgA increased from an initial 2% in PPs to 50% in MLNs and 75% in thoracic duct lymph, and finally 90% in the intestinal lamina propria.

These seminal studies led to the introduction of the term 'IgA cell cycle' and subsequent research showed that B cells bearing other sIg classes than IgA, as well as T cells, when activated in PPs, also exhibit gut-seeking properties. It later became evident that different secretory effector sites can receive activated memory/effector B cells from a variety of mucosa-associated lymphoid tissue (MALT).

This work gave rise to the notion that the mucosal immune system is divided into distinct inductive sites and effector sites. The inductive sites are the organized MALT structures together with mucosa-draining lymph nodes, whereas the effector sites are the mucosal epithelia and the underlying lamina propria, which contains stromal cells and associated connective tissue stroma. The mucosae and related exocrine glands harbor by far the largest activated B-cell system of the body, and the major product—J chain-containing dimeric IgA together with some pentameric IgM—is immediately ready for external transport by the polymeric immunoglobulin receptor (pIgR) across the secretory epithelium and into the mucus layers at mucosal surfaces, to provide antibody-mediated immunity (see Chapter 9).

Immune inductive lymphoid tissue.

The MALT concept was introduced to emphasize that solitary organized mucosa-associated lymphoid follicles and larger follicle aggregates have common features and are the origin of T and B cells which traffic to secretory effector sites. This functional distinction is important because while the different tissues can be identified and discriminated by histology, single cell suspensions prepared from mucosal surfaces contain a mixture of cells from small MALT structures that cannot be dissected out, and connective tissue. This is particularly a problem in man, because solitary follicles cannot be seen in resected bowel.

1-1   MALT is different from lamina propria or glandular stroma.

MALT is subdivided according to anatomical regions (Table 1.1), and the cellular content of these lymphoid structures depends on whether the tissue is normal or chronically inflamed, superimposed upon striking age and species differences (Figure 1.1). The PPs in the distal small intestine of humans, rodents, and rabbits are classical MALT structures. PPs are inductive compartments generating conventional (B-2) sIgA-expressing memory/effector B cells and T cells, which, after a journey through lymph and peripheral blood, enter the gut mucosa. This 'homing' appears to be antigen independent, but local antigens penetrating into the lamina propria contribute to local retention, proliferation, and differentiation of the extravasated lymphoid cells. Thus, although the lamina propria is considered an effector site, it is clearly important for the expansion and terminal differentiation of T and B cells. Some expansion of mucosal memory/effector T cells can also occur in the mucosal surface epithelium, and there is considerable 'cross-talk' between these two effector compartments.

The major component of human MALT comprises the gut-associated lymphoid tissue (GALT), including the PPs, the appendix, and numerous solitary follicles (Figure 1.1) now termed isolated lymphoid follicles. Induction of mucosal immune responses can also take place in nasopharynx-associated lymphoid tissue (NALT) and bronchus-associated lymphoid tissue (BALT) as described below. Moreover, small MALT-like lymphoid aggregates are present in the conjunctiva, and are associated with the larynx and various ducts such as those connecting the ocular and nasal compartments.

MALT resembles lymph nodes with B-cell follicles, interfollicular T-cell areas, and a variety of antigen-presenting cells (APCs), but lacks afferent lymphatics and a capsule. MALT therefore samples exogenous antigens directly from the mucosal surfaces through a follicle-associated epithelium (FAE), which contains 'membrane' or 'microfold' (M) cells (see Chapter 13). These specialized thin epithelial cells do not act as APCs but are effective in the uptake of microorganisms and other particulate antigens, but they are also vulnerable 'gaps' in the mucosal barrier. Studies in the mouse have, in addition, shown that dendritic cells (DCs) in the dome region of MALT send processes through the FAE to sample gut antigens.

Table 1.1 The different regions of MALT and their components

Region	Components
GALT (Gut-associated lymphoid tissue)	Peyer's patches (PPs) and isolated lymphoid follicles constitute the major part of GALT, but also the appendix is included
NALT (Nasopharynx-associated lymphoid tissue)	In humans, NALT consists of the lymphoid tissue of Waldeyer's pharyngeal ring, including the adenoids (the unpaired nasopharyngeal tonsil) and the paired palatine tonsils. Scattered isolated lymphoid follicles may also occur in nasal mucosa. Rodents lack tonsils, but do have paired NALT structures dorsally in the floor of the nasal cavity
BALT (Bronchus-associated lymphoid tissue)	Not generally detectable in normal lungs of adult humans

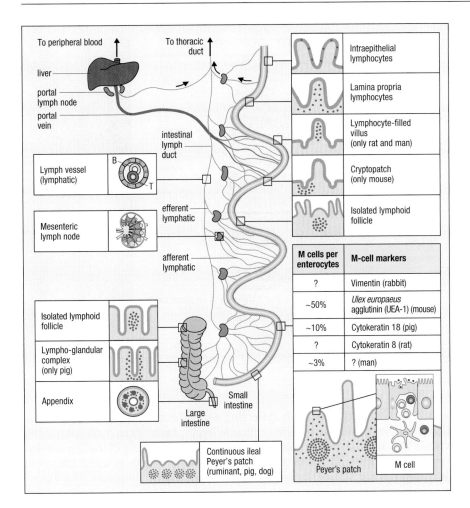

Figure 1.1 Distribution of lymphoid cells in various tissue compartments of the gut wall with species differences indicated. Lymphocytes leave the gut wall via draining lymphatics afferent to mesenteric lymph nodes, or via portal blood reaching the liver. Important regulation of immunity takes place in these organs, particularly induction of tolerance. The frequency of M cells in the follicle-associated epithelium of Peyer's patches is highly variable, and a reliable marker for this specialized cell has not been identified in humans. In contrast to the antigen-dependent activation of B cells that takes place in Peyer's patches of most species, the continuous ileal Peyer's patch present in ruminants, pigs, and dogs appears to be a primary lymphoid organ responsible for antigen-independent B-cell development, similar to the bursa of Fabricius in chicken (not shown). This Peyer's patch can be up to 2 meters long and constitute 80–90% of the intestinal lymphoid tissue. (Adapted from P. Brandtzaeg and R. Pabst, *Trends Immunol.* 25:570–577, 2004. With permission from Elsevier Ltd.)

The distinction between mucosal inductive and effector sites is not absolute, as the signals for extravasation and accumulation of naive versus memory/effector B and T cells are different. It is therefore confusing when MALT is used to refer to the mucosal effector compartments (e.g., the lamina propria and the surface epithelium of the gut and its diffusely distributed immune cells, Figure 1.2). This is in conflict with the classical definition of a lymphoid tissue, and so the Society for Mucosal Immunology and the International Union of Immunological Societies have agreed on the terminology listed in Table 1.1. Accordingly, the mucosa-draining local/regional lymph nodes should not be referred to as MALT because they do not sample antigens directly from mucosal surfaces. Thus, although being part of the intestinal immune system, MLNs should not be considered as part of GALT. Likewise, it is appropriate to refer to head-and-neck-draining cervical lymph nodes as part of the upper-airway immune system, but these are not MALT. Finally, because of the variety of homing molecules imprinted in the different MALT sites, the old term 'a common mucosal immune system' should be avoided, although some integration of the migration of activated lymphoid cells to the various effector sites does occur.

1-2   The mucosal immune system contains different types of GALT.

In humans and rodents, PPs occur mainly in the ileum and less frequently in the jejunum. By the original definition coined by Cornes in the 1960s, a PP consists of at least five aggregated lymphoid follicles but can contain up to as many as 200 (Figure 1.3). Human PPs begin to form at 11 weeks of gestation, and discrete T- and B-cell areas occur at 19 weeks; but no germinal centers

Figure 1.2 The cellular elements within the two effector compartments of the gut mucosa (lamina propria and epithelium). Panel a shows the surface mucus, the monolayered epithelium with enterocytes and other cell types, the basal lamina (basement membrane), and the diffusely distributed cellular components, as well as mediators, within the lamina propria. IgA⁺ plasma cells (green) are normally dominating. The different cell types are in reality much closer to each other and can modulate the microenvironment by secreted chemokines (e.g., derived from the epithelium) and cytokines (e.g., derived from mast cells), as well as by the interaction with juxtaposed nerve fibers. Note that dendritic cells can penetrate the basal lamina and epithelial tight junctions to actively sample antigens from the mucosal surface, while maintaining barrier integrity by expressing tight junction proteins. Panel b shows the histology (H&E staining) of normal human duodenal mucosa. Panel c shows the histology (H&E staining) of normal human colonic mucosa. (Adapted from P. Brandtzaeg and R. Pabst, *Trends Immunol.* 25:570–577, 2004. With permission from Elsevier Ltd.)

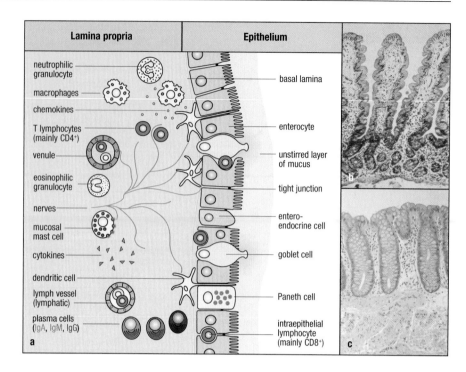

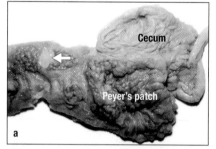

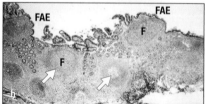

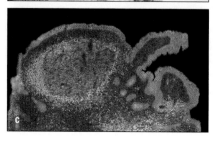

appear until shortly after birth, reflecting dependency on exogenous stimulation—a process that also induces follicular hyperplasia. Thus, macroscopically visible PPs increase from approximately 50 at the beginning of the last trimester to 100 at birth and 250 in the mid-teens, then diminish to some 100 at old age. Slightly over 50% of human PPs occur in the ileum.

Human small intestinal mucosa may harbor at least 30,000 isolated lymphoid follicles, increasing in density distally. Thus, the normal small intestine reportedly contains on average only one follicle per 269 villi in the jejunum but one per 28 villi in the ileum. In the normal large bowel the density of isolated lymphoid follicles also increases distally. Isolated lymphoid follicles have been characterized immunologically in mice and show features compatible with induction of B cells for intestinal IgA responses.

Interestingly, the organogenesis of murine isolated lymphoid follicles commences after birth, in contrast to the PPs. The isolated lymphoid follicles may actually play a special role in the induction of secretory IgA (SIgA) responses controlling the commensal microbiota, and their number and size depend on the intestinal bacterial level. These structures can function in a T-cell independent manner, and cannot fully compensate for the absence of PPs in T-cell dependent antibody responses. Isolated lymphoid follicles may indeed reflect the bridge between the ancient innate and the modern adaptive immune systems.

Figure 1.3 Peyer's patches in the terminal ileum of a young child. Panel a shows a patch where a tissue specimen has been excised (arrow). In panel b, histology of the excised specimen (H&E staining) displays several lymphoid follicles (F), and some are sectioned through the germinal center (arrows). The domes are covered with specialized follicle-associated epithelium (FAE) that lacks villi. Original ×25. Panel c shows two-color immunofluorescence staining for B cells (CD20, green) and IgD (red) in a cryosection from human Peyer's patch. Note that surface IgD (traditional marker of naive B cells) is mainly expressed on B cells in the mantle zones of the lymphoid follicle and to some extent on B cells scattered in the extrafollicular areas, whereas CD20 also decorates the germinal center. Epithelium is stained blue for cytokeratin. Original ×25.

1-3   Organogenesis of murine PPs, isolated lymphoid follicles, and NALT is not synchronous.

Experiments in knockout mice have largely helped to understand how inductive lymphoid tissue develops, and there are remarkable differences among sites. The fetal organogenesis of PPs and MLNs depends on multiple interactions between hematopoietic lymphoid tissue inducer (LTi) cells and mesenchymal stromal cells. The development of LTi cells requires two transcription factors: the inhibitory basic helix–loop–helix transcription factor Id2 and the retinoic acid-related orphan nuclear receptor RORγt.

The cellular interactions involve lymphotoxins (LTs), which are members of the tumor necrosis factor (TNF) superfamily, as well as the two TNF receptors (TNF-Rs) I and II (55 and 75 kDa). A crucial differentiating event is membrane expression of $LT\alpha_1\beta_2$ on the $CD3^-CD4^+CD45^+IL-7R^+$ LTi cells after stimulation of the receptor activator of nuclear factor κB (RANK)–TNF receptor associated factor family 6 (TRAF6) pathway and—especially in PPs—alternative cytokine signaling through the interleukin-7 receptor (IL-7R). This event is followed by LTi cell interaction with the LTβ receptor (LTβR)-positive stromal cells, which are positive for vascular cell adhesion molecule (VCAM)-1 and are called lymphoid tissue organizer cells, but display a different activation profile in the organogenesis of lymph nodes and PPs. Knockout mice deficient in LTα or LTβ virtually lack lymph nodes and have no detectable PPs. The organ development also involves several feedback loops with chemokines and adhesion molecules such as CXC chemokine ligand 13 (CXCL13)–CXCR5 and CXCR5-induced $\alpha_4\beta_1$ integrin–VCAM-1 interactions.

Specifically, at 15 days of gestation, VCAM-1⁺ stromal cells accumulate in fetal mouse small intestine. At 17.5 days, LTi cells are present and mucosal addressin cell adhesion molecule-1 (MAdCAM-1) is expressed on vessels. T and B cells migrate into PPs from 18.5 days to form T- and B-cell zones. PPs and MLNs are, by unknown mechanisms, programmed to develop in the sterile environment of the fetus. Conversely, murine isolated lymphoid follicles develop only after birth in response to innate signaling from the gut microbiota, but they nevertheless require $LT\alpha_1\beta_2$–LTβR interactions and TNF-RI function, as well as typical LTi cells. Also, CXCL13 has been shown to be important for isolated lymphoid follicle formation by promoting accumulation of LTi cells.

Interestingly, the kinetics of the appearance of lymphoid tissue in the mouse nasal mucosa are different, as development is completely postnatal. Unlike GALT structures, murine NALT organogenesis does not depend on the $LT\alpha_1\beta_2$–LTβR signaling pathway, although NALT anlagen apparently require the presence of $CD3^-CD4^+CD45^+$ cells and the transcription factor Id2. Nevertheless, NALT occurs in RORγt-deficient mice which lack the $CD3^-CD4^+IL-7R^+$ PP-type LTi cells. However, to become fully developed with follicular dendritic cells (FDCs) and optimal function, murine NALT as well as inducible murine BALT structures depend on the LT–LTβR pathway.

Detailed ontological studies linked to function are only available for mouse and man but reveal huge differences. Mice possess unique structures, termed cryptopatches, which are clusters of c-kit⁺, lineage-negative common lymphoid precursor cells. Cryptopatches can, however, mature into isolated lymphoid follicles, where primary and secondary adaptive mucosal IgA responses can be generated. Cryptopatches and isolated lymphoid follicles are dependent on LTi cells and are absent in RORγt null mice. Cryptopatches, but not isolated lymphoid follicles, are present in the fetus and in germ-free mice. Evidence indicates that isolated lymphoid follicle development requires signaling from the microbial flora via nucleotide-binding oligomerization domain 1 (NOD1) in epithelial cells to attract LTi cells which then form a lymphoid follicle; therefore, in terms of classical immunology, cryptopatches may be considered primary lymphoid organs and isolated lymphoid follicles secondary lymphoid structures. Cryptopatches and isolated lymphoid follicles appear to form a continuum, but in the mouse they may be more like the rabbit, where GALT functions as a primary lymphoid organ before puberty and matures into a secondary organ later in life.

Whether the same sequence of events occurs in humans is difficult to determine. A LTi cell, which expresses CD127 and RORC but which is CD4⁻, has been described in fetal human tissues. The lack of CD4 on the human LTi cell is somewhat surprising because many years ago human fetal gut was shown to contain clusters of CD4⁺ CD3⁻ cells, which were presumed to be the earliest PP anlagen. However, immunostaining for VCAM-1 is extremely strong in putative PPs beginning at 11 weeks of gestation, suggesting that even in man, VCAM-1 is involved in mucosal lymphoid tissue organogenesis. Beginning at 14–15 weeks of gestation, human small intestine contains loose accumulations of T cells and B cells and strongly major histocompatibility complex (MHC) class II⁺ cells, but the mucosa lacks follicular structure. By 19 weeks, organized small PPs with a FAE, primary B-cell follicles, and T-cell zones can be seen. A population of CD11c⁺ DCs clusters below the FAE, as shown in adult PPs. Although little information on the development between 20 weeks and birth is available, analysis of available specimens shows large well-organized Peyer's patches in 1- to 2-day-old tissues, containing the what appear to be early secondary follicles with IgM⁺ blasts in the developing germinal centers, presumably in response to intestinal colonization with microbes.

Cryptopatches are not present in humans and should not be confused with lymphocyte-filled villi (Figure 1.1). Whether human isolated lymphoid follicles have properties similar to the murine counterparts is not known. However, irregular lymphoid aggregates, which may have germinal centers with follicular dendritic cells, are induced in chronic inflammatory bowel disease lesions, although the function of these isolated-lymphoid-follicle-like structures remains elusive.

1-4   Lymphoid tissue is present in the nasopharynx and bronchi.

GALT is the largest and best defined part of MALT, but other potentially important inductive sites for mucosal B-cell responses are BALT and NALT—in humans particularly the unpaired nasopharyngeal tonsil (often referred to as adenoids) and the paired palatine tonsils. These structures make up most of Waldeyer's pharyngeal lymphoid ring and may play a more important role in mucosal immunity in human airways than BALT (Figure 1.4a) because the

Figure 1.4 Human BALT and NALT. Panel a: histology (H&E staining) of human bronchus-associated lymphoid tissue (BALT) from diseased lung of an adult; note submucosal glands. Panel b: human nasopharynx-associated lymphoid tissue (NALT) represented by section through a crypt region of palatine tonsil; note abundant germinal centers (GC) in lymphoid follicles. Panel c: parallel section immunostained for cytokeratin to demonstrate loose structure of reticular crypt epithelium.

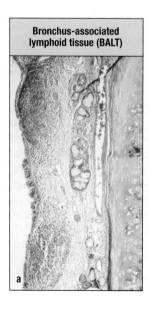

Bronchus-associated lymphoid tissue (BALT)

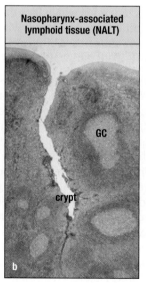

Nasopharynx-associated lymphoid tissue (NALT)

GC

crypt

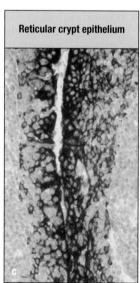

Reticular crypt epithelium

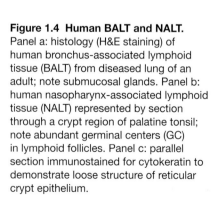

latter is rare in normal lungs of adults, and detectable in only 40–50% of small bronchial biopsy samples from adolescents and children, but is regularly encountered in large autopsy specimens. Similarly, human nasal mucosa contains only occasional isolated lymphoid follicles and they are difficult to detect in infants.

Rodents lack tonsils but possess two paired bell-shaped NALT structures laterally to the nasopharyngeal duct in the floor of the nasal cavity, dorsal to the cartilaginous soft palate. A regionalized protective IgA response can be induced by nasal vaccination in mice. Indeed, murine NALT can drive an IgA-specific enrichment of high-affinity memory B cells, but also gives rise to a major germinal center population of IgG-producing cells—quite similar to that seen in human tonsils. In contrast to tonsils, however, the anlagen of which appear at the same fetal age as those of PPs, the organogenesis of murine NALT begins after birth, similar to murine isolated lymphoid follicles. Murine BALT is also inducible, as are human BALT and nasal isolated lymphoid follicles.

Like GALT in all mammalian species, rodent NALT has a smooth surface covered by a FAE. On the other hand, human tonsils have deep and branched antigen-retaining crypts with a reticular epithelium containing M cells (Figure 1.4b,c). This difference probably explains why germinal centers develop shortly after birth in human NALT, similar to the heavily microbe-exposed GALT, whereas the absence of an abundant nasal microflora means that rodent NALT requires infection or a danger signal such as cholera toxin to drive germinal center formation. This is an intriguing species difference important in the induction of B-cell diversity and memory by vaccination via the nasal route.

1-5   Mucosal B cells may be derived from tissues other than MALT.

In mice, activated T cells rapidly obtain gut-homing properties during antigen priming in MLNs. Most likely, regional mucosa-draining lymph nodes generally share immune inductive properties with related MALT structures from which they receive antigens via afferent lymph and antigen-transporting DCs. Numerous DCs are present at epithelial surfaces where they can pick up luminal antigens by penetrating tight junctions with their processes (Figure 1.2). In the murine small intestine, a CX3CR1+ DC subset has been shown to actively sample luminal bacteria in this manner. Importantly, the human nasal mucosa is extremely rich in DCs, both within and beneath the epithelium, and a subepithelial band of putative APCs is present below the surface epithelium, as well as in the FAE of PPs in the human gut. Importantly, murine DCs that have sampled commensal bacteria in PPs do not normally migrate beyond the draining MLNs, thereby restricting the antibody response to the mucosal immune system of clean pathogen-free mice. This is clearly different in human beings where systemic antibodies to commensals are normally present in low titers.

The peritoneal cavity is recognized as yet another source of mucosal B cells in mice, perhaps providing up to 50% of the intestinal IgA+ plasma cells. The precursors are self-renewing sIgM+ B-1 (CD5+) cells, which give rise to T-cell independent polyreactive ('natural') SIgA antibodies, particularly directed against polysaccharide antigens from commensal bacteria. However, the relationship of murine B-1 cells to the conventional bone marrow derived B-2 cells remains elusive. Notably, rather than being encoded only in the germline, B-1 cells from mice may show hypermutation of Ig heavy-chain V-region (V_H) genes as a sign of selection.

Location of the B-1 subset differentiation of IgA plasma cells is controversial, although the murine intestinal lamina propria has been suggested as a site for class switching. Notably, however, evidence that peritoneal B-1 cells contribute to mucosal IgA production in humans is not available. On the other hand, considerable levels of cross-reactive SIgA antibodies directed against self as well as microbial antigens are present in human external secretions (see Chapter 9). The reason for this could be microbial polyclonal activation of B cells in GALT, independent of B-cell receptor (BCR)-mediated antigen recognition.

B-cell activation in MALT.

MALTs are on the forefront of the recognition of, and responses to the many pathogens which colonize mucosal surfaces or which use mucosal surfaces to enter the body. They also are probably important in controlling the intensity and texture of the response to harmless environmental antigens and foods. While there is a large literature on MALT in experimental animals, studies in human are relatively uncommon.

1-6   FAE is an important site of antigen uptake in the gut.

In the gut, M cells of FAE sample luminal antigens and, notably, enteropathogenic bacteria such as *Salmonella* spp. and *Yersinia* spp., and viruses such as poliovirus and reovirus, use M cells as portals of entry into the lymphoid tissue of GALT and may spread further throughout the body. Also, M cells of murine NALT have been shown to serve as a gateway for the entry of pathogens (Group A streptococci), particularly relevant to airway infections.

M-cell pockets clearly represent an intimate interface between the external environment and the mucosal immune system (Figure 1.5). In mice, the FAE expresses chemokines (e.g., CCL9 and CCL20) that may attract the professional APCs that accumulate in the PP domes. Numerous putative APCs also

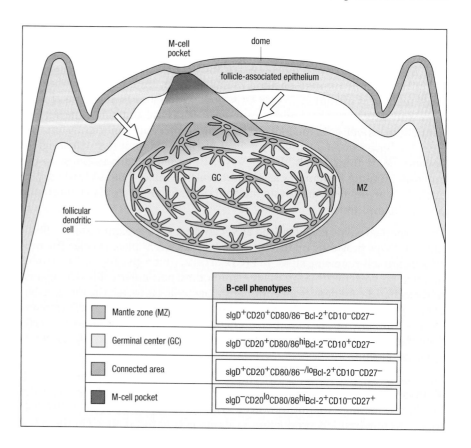

Figure 1.5 Relationship between elements of secondary (activated) B-cell follicle and M-cell pocket in human Peyer's patch. The germinal center (GC) is mostly surrounded by the mantle zone (MZ) and filled with surface (s)IgD⁻CD20⁺CD80/86ʰⁱBcl-2⁻CD10⁺CD27⁻ B cells. The mantle zone consists of naive sIgD⁺CD20⁺CD80/86⁻Bcl-2⁺CD10⁻ CD27⁻ B cells but is broken (open arrows) beneath the M-cell pocket. This connected area is only seen in a restricted part of the follicle as defined by reduced immunostaining for sIgD toward the M cell. A follicular dendritic cell network defined as CD21⁺CD20⁻ phenotype, shows a topographically similar extension but does not reach inside the M-cell pocket. This pocket contains both naive sIgD⁺CD20⁺CD80/86⁻/ˡᵒBcl-2⁺CD10⁻CD27⁻ and memory (or recently stimulated) sIgD⁻ CD20ˡᵒCD80/86ʰⁱBcl-2⁺CD10⁻CD27⁺ B-cell phenotypes, the latter being predominant. (Adapted from P. Brandtzaeg, *Immunol. Invest.* 39:303–355, 2010. With permission from Taylor & Francis.)

	B-cell phenotypes
Mantle zone (MZ)	sIgD⁺CD20⁺CD80/86⁻Bcl-2⁺CD10⁻CD27⁻
Germinal center (GC)	sIgD⁻CD20⁺CD80/86ʰⁱBcl-2⁻CD10⁺CD27⁻
Connected area	sIgD⁺CD20⁺CD80/86⁻/ˡᵒBcl-2⁺CD10⁻CD27⁻
M-cell pocket	sIgD⁻CD20ˡᵒCD80/86ʰⁱBcl-2⁺CD10⁻CD27⁺

occur immediately underneath the human FAE, while the M-cell pockets are dominated by memory T and B lymphocytes in approximately equal proportions. Interestingly, activated B cells may induce the epithelial M-cell phenotype from classical gut enterocytes. Furthermore, studies in germ-free and conventionalized rats have demonstrated that bacterial colonization drives the accumulation and differentiation of T and B cells in the M-cell pockets, apparently with an initial involvement of antigen-transporting DCs, followed by germinal center formation.

M-cell pockets may represent specialized germinal center extensions designed for rapid recall responses. The most likely cell type to mediate MHC class II interaction with cognate T cells in these microcompartments is the long-lived sIgD⁻IgM⁺Bcl-2⁺CD27⁺ memory B cell (Figure 1.5). In human tonsils, memory B cells colonize the antigen-transporting reticular crypt epithelium, and, by rapid upregulation of co-stimulatory B7 molecules, they acquire potent antigen-presenting properties. Likewise, memory B cells present in the M-cell areas of human PPs relatively often express B7.2 (CD86) and sometimes B7.1 (CD80), perhaps a prerequisite for stimulation of productive immunity (Figure 1.5). These memory cells are phenotypically similar to the so-called extrasplenic 'marginal-zone B cells,' which generally show a high frequency of BCR IgV$_\mathrm{H}$ mutations, suggesting extensive antigen stimulation.

The FAE does not express pIgR, but an uncharacterized apical receptor for IgA has been identified on M cells. SIgA may exploit this receptor for targeting of antigens to DCs in PPs. Many commensal bacteria in human saliva and feces are coated with SIgA, which may facilitate their M cell-mediated uptake. Most of the bacteria that cross the epithelium are destroyed by subepithelial macrophages but some are taken up by DCs in tiny amounts (0.001%), sufficient to induce a murine immune response restricted to GALT and MLNs. On the whole, however, the mucosal antibody response limits the penetration of bacteria into the body, with GALT 'shielded' by the SIgA antibodies. Nevertheless, the positive-feedback mechanism implied by the apical M cell IgA receptor may target relevant luminal antigens complexed with cognate maternal SIgA antibodies to the GALT of suckling newborns. Indeed many reports show that breast-fed babies over time develop enhanced secretory immunity.

Considerable effort has been directed to the characterization of M-cell receptors that may be exploited for vaccine and adjuvant targeting to GALT. Murine M cells express Toll-like receptors (TLRs), thus making TLR agonists potential mucosal vaccine adjuvants. For instance, outer membrane proteins of *Neisseria meningitidis* mixed with the TLR2 agonist PorB enhanced the microparticle transport capacity of M cells and induced migration of DCs into FAE. Also glycoprotein 2 expressed apically on murine M cells has been shown to serve as a transcytotic receptor for commensal and pathogenic bacteria carrying FimH, a component of type I pili.

1-7   Gut bacteria are important for the molecular interactions needed for germinal center formation.

Primary lymphoid follicles contain recirculating naive B lymphocytes (sIgD⁺IgM⁺) which pass into the network formed by antigen-capturing FDCs. The origin of FDCs remains obscure but both their development and the clustering that allows follicle formation depend on LT signaling, and the B cells that are one important source of LT. Among the actions of the soluble homotrimer LTα, are augmentation of B-cell proliferation and expression of adhesion molecules; mice deficient in transmembrane LTβ have no detectable FDCs.

The germinal center reaction causes the transformation of primary follicles into secondary follicles. In humans, this process has been extensively studied in tonsils, but relevant mechanistic information relies on observations of lymph nodes and spleen from immunized animals. In general, germinal centers are of vital importance for T-cell dependent stimulation of conventional (B-2) memory/effector B cells, affinity maturation of BCR, and Ig-isotype switching (see Chapter 8). Naive B cells are first stimulated in the T-cell zone, just outside the primary follicle, by cognate interactions with activated CD4+ T cells, and exposed to processed antigen complexed with MHC class II molecules on interdigitating DCs. The B cells then reenter the follicle to become proliferating sIgD+IgM+CD38+ germinal center 'founder cells' as observed in human tonsils.

The initially activated B cells produce unmutated IgM (and some IgG) antibody which binds circulating antigen with low affinity. B cells may then use their complement receptors to carry opsonized antigen or immune complexes into the follicles to become deposited on FDCs. Antigen is retained in this network for prolonged periods to maintain B-cell memory. A role for IgM in the induction of secondary immune responses with antibody affinity maturation has been strongly supported by observations in knockout mice lacking natural ('nonspecific') background IgM antibodies.

The complement receptors CR1/CR2 (CD35/CD21) are crucial in the germinal center reaction. CD21 is expressed abundantly on both B cells and FDCs and may thus function not only by localizing antigen to the FDCs but also by lowering the threshold of B-cell activation via recruitment of CD19 into the BCR. Activation of complement on FDCs bearing immune complexes is controlled by regulatory factors, but some release of inflammatory mediators may induce edema that facilitates dispersion of FDC-derived 'immune complex-coated bodies,' or iccosomes, thereby enhancing the BCR-mediated uptake of their contained antigens by the B cells.

Several other components of the innate immune system may be involved in the germinal center formation. Under certain experimental conditions, IgA differentiation driven by gut bacteria may even bypass the usual sIgM (or sIgD) BCR requirement. Nevertheless, there always appears to be a dependency on some follicle-like aggregates of B cells, which interestingly may lack antigen-retaining FDCs and germinal centers. Normally, the germinal-center reaction is driven by competition for a limited amount of antigen, but there may also be an innate drive as discussed above for isolated lymphoid follicles. The resulting B cells will then survive with a restricted repertoire and rather low BCR affinity.

In germ-free rabbit appendix, only certain commensal bacteria efficiently promote GALT development, which depends on stress responses in the same bacteria, suggesting a nonspecific impact on GALT. Other experiments have shown that GALT development can occur independently of BCR engagement, apparently because commensal bacteria promote the germinal center reaction in PPs and MLNs by interacting with innate immune receptors. This information helps explain the enormous IgA drive provided by the gut microbiota in the absence of high-affinity BCR development, thereby leading to the production of large amounts of IgA with restricted repertoire and capacity to bind with low affinity to redundant epitopes on commensal bacteria (see Chapter 8).

Thus, the gut microbiota is required and plays a critical role in the activation of GALT and normal intestinal plasma-cell development, although food proteins may also contribute to this development. Notably, commensal bacteria shape the BCR repertoire of the host. Perhaps the germinal center reaction originated evolutionarily in GALT to generate a protective antibody repertoire that was not antigen specific, but cross-reactive. The indigenous microbiota might in this context act as polyclonal B-cell activators through several mechanisms, including TLR signaling. The induction of IgA by commensal bacteria seems to represent a primitive mechanism for limiting bacterial colonization and penetration through the epithelial barrier without eliminating them from the gut.

Superimposed on this 'innate-like' defense, the B-2 system has the capacity to undergo germinal center-driven high-affinity BCR selection for particular antigens of pathogens to clear infections (see Chapter 8).

B-cell differentiation in MALT germinal centers.

The most obvious morphological features of MALT in tissue sections are the numerous large, activated germinal centers. This is especially the case in the ileum. In mice, the expansion of the germinal centers allows Peyer's patches to be identified from the serosal surface so they can be dissected out for functional studies. Typically, below each FAE there is a large germinal center.

1-8   IgA expression and class switching are dependent on activation-induced cytidine deaminase (AID).

As discussed above, the IgA switch in murine B-1 cells may circumvent the need for T-cell help which provides CD40–CD154/CD40L interaction. Instead, the B-1 cells may rely on proliferation- and survival-inducing cytokines of the TNF family such as BAFF/BLyS (B cell-activating factor of the TNF family/B lymphocyte stimulator) and APRIL (a proliferation inducing ligand) secreted by activated DCs or macrophages, for instance after exposure to lipopolysaccharide (LPS). Significantly reduced IgA antibody responses are present in APRIL-deficient mice, and it is possible that BAFF and APRIL also contribute to T-cell dependent IgA switching. In the human gut where class switch appears principally to be part of a germinal center reaction—but to some extent may also be T-cell independent—the role of BAFF and APRIL remains unclear (see Chapter 8).

Class switch in germinal centers is initiated after activation of the germinal-center founder cells, which change their BCR composition from sIgD⁺IgM⁺ to become sIgD⁻IgM⁺ memory/effector cells following affinity maturation. They may then switch to another class such as IgG or IgA by recombination of the intronic Cμ-switch (S) region with the S region of a downstream isotype, either directly or sequentially (Figure 1.6). During plasma-cell differentiation, the BCR is gradually lost with several other B-cell markers, particularly CD20 and then CD19 (in mice also B220).

Figure 1.6 Class-switch recombination leading to differentiation of B cells in mucosa-associated lymphoid tissue. This process takes place after activation of germinal center founder cells, resulting in elevated levels of the enzyme activation-induced cytidine deaminase (AID). In the classical pathway, the centroblast progeny change their Ig heavy-chain constant-region (C_H) gene expression from $C\mu$ to one of the downstream C_H genes, either sequentially or directly, by recombination between switch (S) regions (comprising repetitive sequences of palindrome-rich motifs), followed by looping-out of the intervening DNA fragment containing $C\mu$ and other C_H genes from the chromosome. This pathway gives rise to plasma cells of various IgG and IgA isotypes (and IgE, not shown), with or without concurrent J-chain expression (±J). A minor fraction of tonsillar centroblasts undergo nonclassical switching by deleting a variable part of $S\mu$ and the complete $C\mu$ region. This pathway selectively gives rise to IgD-producing plasma cells and little or no further switching because the $C\delta$ lacks a separate S region. (Adapted from P. Brandtzaeg, *Immunol. Invest.* 39: 303–355, 2010. With permission from Taylor & Francis.)

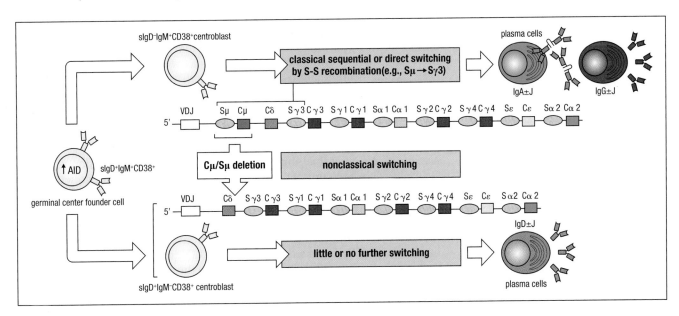

The switch process requires a variety of transcription factors and enzymatic activity expressed by cell type-specific and more generalized DNA repair enzymes, particularly activation-induced cytidine deaminase (AID, Figure 1.6). AID is present during class-switch recombination and may link class-switch recombination to somatic hypermutation of Ig V-gene segments, which takes place during the germinal center reaction (see Chapter 8). AID-deficient mice show dramatic hyperplasia of isolated lymphoid follicles in response to intestinal overgrowth of the indigenous microbiota, similar to the lymphoid hyperplasia that occurs in people with immunoglobulin deficiency who often are infected with the parasite *Giardia lamblia*. This could reflect an inadequate compensatory antibody repertoire in the gut due to lack of somatic hypermutation in the SIgM that partially replaces SIgA in these mice.

Specific alterations at any germline C_H locus prior to class-switch recombination include opening of chromatin structures—a key event in transcriptional activation. Thus, the switch to IgA1 and IgA2 in humans requires activation of the Iα1 and Iα2 promoters located upstream of the respective S regions of the C_H-gene segments of the two subclasses. DNA between the S sites then is looped out and excised, thereby deleting Cμ and other intervening C_H genes. After direct switch to IgA1, the germline Iα1–Cμ circular transcripts derived from the excised recombinant DNA, are gradually lost through dilution in progeny during proliferation.

1-9   Different class-switching pathways operate in different mucosal sites.

Human IgA responses are dominated by IgA1 both in tonsils and the regional secretory effector sites, which suggests that mucosal B-cell differentiation in these sites mainly takes place from sIgD⁻IgM⁺CD38⁺ centrocytes by sequential downstream C_H-gene switching (Figure 1.6). Conversely, the relatively enhanced IgA2 expression in PPs and the distal human gut, including the MLNs (Figure 1.7), could reflect direct switch from Cμ to Cα2. Murine PP B cells are able to switch directly from Cμ to Cα, and in human B cells this direct pathway may preferentially lead to IgA2 production.

The germinal center reaction generates relatively more intrafollicular J chain-expressing IgA⁺ plasmablasts in human PPs and appendix than in tonsils. Furthermore, immediately outside of GALT follicles, IgA⁺ plasma cells are equal to, or exceed numerically, their IgG⁺ counterparts, whereas in tonsils there is a more than twofold dominance of extrafollicular IgG-producing plasma cells. Therefore, the drive for IgA switch and J-chain expression is clearly much more pronounced in GALT than in tonsils (Figure 1.7). The reason for this disparity remains elusive, as the tonsillar crypt epithelium can secrete BAFF and thymic stromal lymphopoietin—a cytokine that further promotes class-switch recombination and a broad reactivity of local B cells by activating BAFF-producing DCs. One possibility is that the continuous superimposition of novel stimuli from the gut microbiota in PPs enhances the development of early memory/effector B-cell clones with an increased potential for IgA and J-chain production.

A regionalized microbial impact on mucosal B-cell differentiation is further exemplified by the unique sIgD⁺IgM⁻CD38⁺ centroblasts identified in the dark zone of tonsillar germinal centers. This subset shows deletion of the Sμ and Cμ gene segments, therefore selectively giving rise to IgD⁺ plasma cells by so-called nonclassical C_H-gene switching (Figure 1.6). Molecular evidence has been provided for a preferential occurrence of B cells with Cμ deletion in the normal Waldeyer's ring (human NALT) and secretory effector sites of the upper aerodigestive tract, but homing of these cells to the small intestinal mucosa appears to be quite rare. Such compartmentalized B-cell responses explain the relatively high frequency of IgD⁺ plasma cells in the respiratory

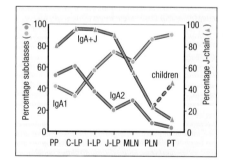

Figure 1.7 Subclass distribution and J-chain expression of IgA-producing plasma cells in various normal extrafollicular tissue compartments from adults. PP, Peyer's patch; C-LP, I-LP, and J-LP, distant colonic, ileal, and jejunal lamina propria respectively; MLN, mesenteric lymph node; PLN, peripheral lymph node; and PT, palatine tonsil. For the latter organ, J-chain data from healthy children (who have had no inflammation in their tonsils) are also indicated (connected by broken line). Based on published immunohistochemical data from Brandtzaeg's laboratory. (Adapted from P. Brandtzaeg, *Immunol. Invest.* 39:303–355, 2010. With permission from Taylor & Francis.)

tract, and particularly that IgD⁺ plasma cells are often seen in IgA deficiency. Most strains of *Haemophilus influenzae* and *Moraxella catarrhalis*, which are frequent colonizers of the nasopharynx, express outer-membrane IgD-binding factors that can activate sIgD⁺ B cells by cross-linking sIgD/BCR. In this manner it seems likely that sIgD⁺ tonsillar centrocytes are stimulated to proliferate and differentiate polyclonally, thereby driving V-gene hypermutation and Sμ/Cμ deletion. Conversely, LPS that is abundantly present in the distal gut may inhibit IgD expression which—in addition to compartmentalized homing mechanisms—might contribute to the virtual lack of IgD-producing plasma cells in the gastrointestinal tract.

Microbial influence on B-cell differentiation is supported by the observation that Sμ/Cμ deletion is more frequently detected in diseased than in clinically normal tonsils and adenoids, and extrafollicular IgD⁺ plasma cells are relatively numerous in recurrent tonsillitis and adenoid hyperplasia. But there are large individual variations, and the mean proportion of IgD-producing plasma cells is well below 5% of all isotypes in the extrafollicular tonsillar compartment.

Interestingly, sIgD⁺IgM⁻ B cells generated by nonclassical switching appear to express V-gene repertoires that may allow considerable cross-reactivity, including autoimmunity, but the biological significance of this observation remains unknown. Although numerous antimicrobial and other IgD antibody activities have been detected in mouse and human serum, the protective or pathogenic role of circulating IgD has only recently been explored. Because IgD does not activate the classical complement pathway, it likely blocks other antibody defense functions within the mucosae and reduces the immune exclusion efficiency of SIgA and SIgM antibodies in the upper airways in the face of bacterial infections that drive local IgD production. The ability of IgD to bind to monocytes/macrophages and basophils and induce release of pro-inflammatory cytokines, including interleukin (IL)-1, IL-6, IL-8, and TNF-α, may add to the pathogenic potential of IgD antibodies. Of note, IgA-deficient individuals who exhibit substantial replacement with only IgM⁺ and IgG⁺ plasma cells in their nasal mucosa have fewer clinical problems in their airways than those with an abundance of local IgD⁺ plasma cells.

How mucosal immune cells home.

Chapter 14 contains extensive information on lymphocyte homing to mucosal surfaces; however, there are some areas that are more appropriately covered here. In general, immune cells exit from inductive sites through draining lymphatics. These vessels are believed to start blindly with a fenestrated endothelium, and the egress of lymphoid cells depends on a signal mediated by sphingosine-1-phosphate. In addition, lymphatic endothelium shares with high endothelial venules (HEVs) the expression of both CCL21/SLC and other adhesion molecules such as intercellular adhesion molecule (ICAM). For example, in the gut, activated T and B cells leave PPs, enter the afferent lymphatics, and drain into the MLN. From there the cells leave in efferent lymphatics, eventually draining into the thoracic duct which empties into the blood at the junction of the left subclavian vein and left jugular vein. In humans, approximately 4 l of lymph drains into the blood every day, much of it from the gut.

1-10 Naive and activated immune cells occupy different microenvironments in GALT.

In human GALT, memory B (sIgD⁻) and T (CD45RO⁺) lymphocytes with a high level of $\alpha_4\beta_7$ integrin are often located near and within the microlymphatics, together with some CD19⁺CD38^hi$\alpha_4\beta_7$^hi B-cell blasts. However, the lymph vessels contain mainly naive $\alpha_4\beta_7^{lo}$ lymphocytes. Cytochemical and flow-cytometric analyses of human mesenteric lymph have provided similar marker profiles; also notably, the small fraction of identified B-cell blasts (2–6%) reportedly contain cytoplasmic IgA, IgM, and IgG in the proportions 5:1:<0.5.

The $\alpha_4\beta_7^{hi}$ subsets exiting through lymphatics in human GALT probably represent the first homing step to seed particularly the intestinal lamina propria with activated lymphoid cells. Relatively few memory cells express high levels of CD62L/L-selectin in intestinal and mesenteric lymph; those that do either reenter GALT or extravasate in MLNs, peripheral lymph nodes, or Waldeyer's ring together with naive cells by binding to peripheral lymph node addressin (PNAd) on HEVs.

The homing of activated lymphoid cells to the intestinal lamina propria clearly depends on their high surface level of $\alpha_4\beta_7$ in the absence of CD62L/L-selectin. The expression of $\alpha_4\beta_7$ allows binding of the circulating cells to unmodified MAdCAM-1 expressed apically on endothelial cells of the lamina propria microvasculature. The $\alpha_4\beta_7^{hi}$ phenotype is predominantly induced on antigen-specific B cells appearing in peripheral blood after enteric (peroral) immunization in humans, whereas the counterparts elicited by systemic immunization express preferentially CD62L/L-selectin but relatively little $\alpha_4\beta_7$. Although interaction of MAdCAM-1 with CD62L has been explored mainly in mice, the virtual absence of CD62L-bearing cells in human intestinal lamina propria strongly suggests that this molecule does not bind to MAdCAM-1 outside of GALT. Many large B cells retain high levels of $\alpha_4\beta_7$ after migration into the intestinal lamina propria, despite abundant co-expression of CD38 and cytoplasmic IgA. Therefore, it is possible that $\alpha_4\beta_7$ may both mediate extravasation and, together with CD44, contribute to local retention of effector cells.

In addition, the migration of both T and B cells into, and/or retention within, the small intestinal lamina propria appears to be mediated by the thymus-expressed chemokine (CCL25/TECK). Notably, this chemokine that interacts with CCR9 is selectively produced by the crypt epithelium in this part of the normal gut in humans as well as mice (Figure 1.8). Several mouse studies have suggested that small-intestinal tropism directed by high $\alpha_4\beta_7$ and CCR9 expression combined with minimal CD62L, is imprinted on T cells by DCs in PPs and MLNs.

In the healthy small intestine $\alpha_4\beta_7$ and CCR9 homing molecules determine the extravasation of memory/effector B cells. Acquisition of such gut-homing properties apparently depends at least partly on retinoic acid derived by oxidative conversion from vitamin A. Macrophages and DCs in GALT, MLNs, and intestinal mucosa express retinal dehydrogenase which drives this conversion. In the large intestine, $\alpha_4\beta_7$ appears, instead, to be assisted by the mucosae-associated chemokine (CCL28/MEC) as a decisive cue for attracting IgA$^+$ lymphoblasts with high levels of CCR10. Because the epithelial expression of CCL28 is higher in the colon than in the small intestine and appendix, this chemokine probably plays a compartmentalized role in intestinal B-cell homing. Also, interestingly, both CCL25 and CCL28 enhance integrin α_4-dependent adhesion of IgA$^+$ plasmablasts to MAdCAM-1.

The cues for gut homing of GALT- and MLN-derived B cells from the circulation may also contribute to lateral migration from GALT structures directly into the intestinal lamina propria. Vascular connections between PPs and surrounding lamina propria may be used for trafficking B cells, possibly explaining why the first IgA$^+$ plasma cells accumulate around PPs when germ-free mice are transferred to conventional conditions. Similar pathways between isolated lymphoid follicles and the surroundings are suggested by the observation that the number of intestinal IgA$^+$ plasma cells is maintained by local proliferation when the B-cell traffic via the thoracic duct is diverted by lymph

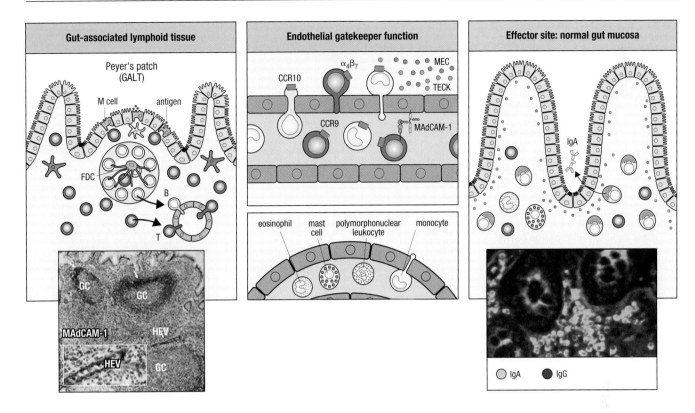

Figure 1.8 Homing mechanisms that attract gut-associated lymphoid tissue (GALT)-derived B and T memory/effector cells to the small intestinal lamina propria. A GALT structure with its epithelial M cells and antigen-presenting cells such as follicular dendritic cells (FDC), is depicted in the top left panel. Once activated, T and B cells in Peyer's patch migrate into draining lymph (solid arrows). Interactions between the multidomain unmodified (containing no L-selectin-binding O-linked carbohydrates) mucosal addressin cell adhesion molecule (MAdCAM)-1 on flat lamina propria venules preferentially targets mucosal $\alpha_4\beta_7$-bearing memory/effector B and T cells to the normal gut mucosa (middle panel). Selectively produced by the epithelium of the small intestine, the chemokine TECK (CCL25) attracts GALT-derived B and T cells that express CCR9 to this segment of the gut, whereas MEC (CCL28) is a more generalized chemokine interacting with CCR10 on mucosal B cells. The access to the mucosal effector site is normally very limited for circulating pro-inflammatory cells such as monocytes, polymorphonuclear leukocytes, mast cells, and eosinophils, whereas the favored GALT-derived cells promote mucosal immunity including polymeric Ig receptor (pIgR)-dependent secretory IgA (SIgA) generation (top right panel). The bottom left panel shows histology of a Peyer's patch with secondary lymphoid follicles containing germinal centers (GC); the insert shows that MAdCAM-1 (with L-selectin-binding capacity) is expressed on a high endothelial venule (HEV) to attract naive lymphocytes for priming. Right bottom panel shows paired immunofluorescence staining for IgA- and IgG-producing plasma cells in normal colonic human mucosa and crypts with selective transport of IgA to the lumen. (Adapted from P. Brandtzaeg, *Scand. J. Immunol.* 70:505–515, 2009. With permission from John Wiley & Sons.)

cannulation in rats. Similar observations of highly localized B-cell responses have been made in the multiple-intestinal-loop model in lambs. However, some of the plasmablasts that normally settle immediately adjacent to GALT follicles apparently belong to exhausted B-cell clones with decreased J chain-expressing potential and disproportionately increased class switch to IgG.

1-11 Homing molecules are important in homing of cells to extraintestinal sites.

Dissemination of GALT-derived B cells to extraintestinal secretory effector sites is well documented, but only limited homing from NALT or BALT to the gut has been demonstrated by immunization or infection experiments in rodents and pigs. However, indirect evidence suggests that IgA⁺ plasmablasts disperse from Waldeyer's ring (human NALT) to regional secretory effector

Figure 1.9 Model for homing of activated B cells from mucosal inductive sites with secondary lymphoid follicles, to secretory effector sites in the integrated human mucosal immune system. The specialized follicle-associated epithelium contains M cells with antigen-transporting properties. Putative compartmentalization in the trafficking is indicated—the heavier arrows representing preferential B-cell migration pathways. In the small intestinal lamina propria, attraction and/or retention of CCR9-expressing cells is mediated by the chemokine CCL25/TECK (see Figure 1.8), while CCL28/MEC interacting with CCR10 is important in the large bowel. Other adhesion molecules, such as CD62L/L-selectin and $\alpha_4\beta_1$ that bind to endothelial PNAd and VCAM-1 respectively, may be employed mainly by B cells primed in nasopharynx-associated (NALT) and bronchus-associated (BALT) lymphoid tissue. Human NALT is comprised of the various lymphoepithelial structures of Waldeyer's ring, including the nasopharyngeal tonsil (adenoids) and palatine tonsils. In this region, abundantly produced epithelial CCL28/MEC attracts activated B cells via CCR10. The female genital tract (cervix mucosa) may employ similar molecular homing mechanisms to the upper aerodigestive tract and the large bowel—therefore probably receiving activated B cells from inductive sites in both these regions. Also, lactating mammary glands appear, by shared homing mechanisms, to receive primed cells from NALT as well as gut-associated lymphoid tissue (GALT)—and much more efficiently than the female genital tract. Homing molecules integrating airway immunity with systemic immunity are shown. (Adapted from P. Brandtzaeg, *Scand. J. Immunol.* 70:505–515, 2009. With permission from John Wiley & Sons.)

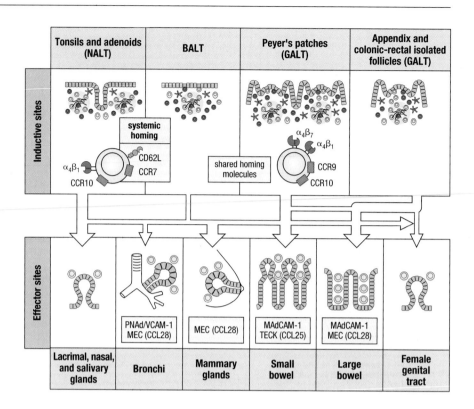

sites. Direct evidence for such regional homing has been obtained by immunization of murine NALT and rabbit palatine tonsils. A dichotomy between the upper and lower body regions (Figure 1.9) is further supported by the dispersion of human tonsillar sIgD+IgM−CD38+ plasmablasts. In the circulation these cells are variably CCR10+ but show a homing-receptor profile that does not favor the small intestinal mucosa ($\alpha_4\beta_7{}^{int/lo}$CCR9^{lo}CCR7^{hi}CD62L^{hi}) and they are virtually excluded from this effector site. Conversely, such 'marker cells' appear in lymph nodes and the bone marrow. The tissue distribution of these cells likely reflects the migratory properties of all plasma-cell precursors with a mucosal phenotype (J chain+) primed in Waldeyer's ring. In keeping with this notion, activated human tonsillar B cells transferred intraperitoneally to SCID (severe combined immunodeficiency) mice migrate to the lung but not to the gut mucosa.

Earlier studies indicate that $\alpha_4\beta_7$ plays little or no role in homing of lymphoid cells to the airways in humans, mice, or sheep. In this connection, intranasal immunization induces an insufficient level of $\alpha_4\beta_7$ on antigen-specific human B cells to make them gut seeking, whereas antibody production is evoked in both adenoids and nasal mucosa. The circulating specific B cells show substantial co-expression of CD62L and $\alpha_4\beta_7$, in contrast to the high surface levels of $\alpha_4\beta_7$ induced by peroral (enteric) immunization. Mouse experiments have indeed suggested that CD62L is involved in B-cell homing to extraintestinal mucosal effector sites, and a similar role has been implied for CCR7 in a study of murine airways.

A nonintestinal homing receptor profile might explain migration of B cells from NALT to the urogenital tract. This is reflected by the titers of IgA and IgG antibodies in cervico-vaginal secretions of mice, monkeys, and humans after intranasal immunization with a variety of antigens. The relatively consistent level of CD62L and CCR7 on NALT-derived B cells, allowing them to bind to PNAd, CCL21/SLC, and CCL19/ELC on HEVs in lymph nodes, could likewise explain the striking integration between mucosal immunity in the upper aerodigestive tract and systemic immunity. This is an attractive feature for protection against many pathogens and has been documented by immunization in rodents, humans, and nonhuman primates. Migration to secretory tissues beyond the gut might also involve $\alpha_4\beta_1$ (CD49d/CD29) expressed by NALT-derived B cells. The chief counter-receptor for this integrin is VCAM-1, which may be present on the microvascular endothelium in human bronchial, nasal, and uterine cervix mucosa.

In addition to directing activated mucosal B cells to the intestinal lamina propria, CCR10 appears to be a unifying chemokine receptor contributing to homing of plasmablasts to extraintestinal secretory effector sites. Several recent studies have shown CCR10 is expressed by IgA$^+$ plasmablasts (and less so by IgA$^+$ plasma cells) at every mucosal effector site in humans and mice. This expression pattern also characterizes IgD$^+$ plasmablasts/plasma cells in the upper airways, as well as IgM$^+$ and IgG$^+$ plasmablasts/plasma cells replacing the IgA$^+$ plasma cells in the gut of IgA-deficient subjects. As mentioned above, CCL28/MEC, the CCR10 ligand, is constitutively produced by gut epithelium, especially in the large bowel, and is also expressed at relatively high levels by secretory epithelia in the upper aerodigestive tract and lactating mammary glands. Interestingly, CCL28 (but not CCL25) was shown to attract tonsillar IgA$^+$ plasmablasts *in vitro*. Therefore, graded tissue site-dependent CCR10–CCL28 interactions, together with insufficient levels of classical gut-homing molecules, most likely explain the observed dispersion dichotomy for memory/effector B cells derived from Waldeyer's ring. Because bone marrow stromal cells reportedly produce CCL28, interactions of this chemokine with CCR10$^+$ B cells may furthermore contribute to integration between mucosal and systemic immunity.

Summary.

The mucosal immune system can be divided into the inductive MALT structures and the effector sites with diffusely dispersed immune cells such as the lamina propria and epithelium of gut and airway mucosae as well as the stroma of exocrine glands. Abundant evidence supports the notion that human intestinal plasma cells are largely derived from B cells initially activated in GALT. However, the uptake, processing, and presentation of luminal antigens which accomplish priming and sustained expansion of mucosal B and T cells are incompletely understood. Also unclear is how the germinal center reaction in GALT so prominently, compared with other MALT structures, promotes class switch to IgA and expression of J chain, although the commensal microbiota appears to contribute to both diversification and memory. B- and T-cell migration from GALT to the intestinal lamina propria is guided by rather well-defined adhesion molecules and chemokines/chemokine receptors, but the signals directing homing to secretory effector sites beyond the gut require better definition. In this respect, the role of Waldeyer's ring (including adenoids and palatine tonsils) as regional human MALT must be better defined, although the balance of evidence suggests that it functions as nasopharynx-associated lymphoid tissue like the characteristic NALT structures in rodents. Thus, the remarkable compartmentalization of the mucosal immune system must be taken into account in the development of effective local vaccines to protect the small and large intestine, the airways, and the female genital tract.

Further Reading.

Brandtzaeg, P.: **Function of mucosa-associated lymphoid tissue in antibody formation.** *Immunol. Invest.* 2010, **39**:303–355.

Brandtzaeg, P., Farstad, I.N., Johansen, F.-E., *et al.*: **The B-cell system of human mucosae and exocrine glands.** *Immunol. Rev.* 1999, **171**:45–87.

Brandtzaeg, P., and Johansen, F.-E.: **Mucosal B cells: phenotypic characteristics, transcriptional regulation, and homing properties.** *Immunol. Rev.* 2005, **206**:32–63.

Brandtzaeg, P., and Pabst, R.: **Let's go mucosal: communication on slippery ground.** *Trends Immunol.* 2004, **25**:570–577.

Heier, I., Malmström, K., Sajantila, A., *et al.*: **Characterisation of bronchus-associated lymphoid tissue and antigen-presenting cells in central airway mucosa of children.** *Thorax* 2011, **66**:151–156.

Kiyono, H., and Fukuyama, S.: **NALT- versus Peyer's-patch-mediated mucosal immunity.** *Nat. Rev. Immunol.* 2004, **4**:699–710.

Kunkel, E.J., and Butcher, E.C.: **Chemokines and the tissue-specific migration of lymphocytes.** *Immunity* 2002, **16**:1–4.

Liu, Y.J., and Arpin, C.: **Germinal center development.** *Immunol. Rev.* 1997, **156**:111–126.

Macpherson, A.J., McCoy, K.D., Johansen, F.-E., *et al.*: **The immune geography of IgA induction and function.** *Mucosal Immunol.* 2008, **1**:11–22.

Neutra, M.R., Mantis, N.J., and Kraehenbuhl, J.P.: **Collaboration of epithelial cells with organized mucosal lymphoid tissues.** *Nat. Immunol.* 2001, **2**:1004–1009.

Pabst, R., and Tschernig, T.: **Bronchus-associated lymphoid tissue. An entry site for antigens for successful mucosal vaccination?** *Amer. J. Respir. Cell Mol. Biol.* 2010, **43**:137–140.

Quiding-Järbrink, M., Nordström, I., Granström, G., *et al.*: **Differential expression of tissue-specific adhesion molecules on human circulating antibody-forming cells after systemic, enteric, and nasal immunizations. A molecular basis for the compartmentalization of effector B cell responses.** *J. Clin. Invest.* 1997, **99**:1281–1286.

Randall, T.D.: **Bronchus-associated lymphoid tissue (BALT) structure and function.** *Adv. Immunol.* 2010, **107**:187–241.

Suzuki, K., and Fagarasan, S.: **How host-bacterial interactions lead to IgA synthesis in the gut.** *Trends Immunol.* 2008, **29**:523–531.

Phylogeny of the Mucosal Immune System

2

All animals, vertebrate or invertebrate, face the challenge of combating pathogens while maintaining a tolerant relationship with symbiotic microorganisms. Tolerance to symbiotes is not a static or inert interaction but, rather, requires continuous active regulation; the front lines for these interactions are the mucosal surfaces. A fundamental question is whether or not specialized immune cells or organs have evolved in all animals to cope with the unique problems of mucosal defense, or whether specialized mucosal immunity is unique to vertebrates such as mammals. This question can be answered by investigating the phylogeny of the mucosal and systemic immune system, from invertebrates, to fish, amphibians, reptiles, and birds, to prototherians (egg-laying mammals), the metatherians (marsupials), and the eutherians (placental mammals). There is also an extremely practical reason for studying phylogeny, namely that many species are important human foods and also suffer from diseases of mucosal surfaces, so that producing mucosal vaccines against fish pathogens is arguably as important to humankind as producing vaccines against the mucosal pathogens of mammals.

Most of the paradigms for immune defense and symbiosis were established from studies of humans and a few other mammalian species such as rodents. Although these studies have provided more than a century of discovery and progress, much has also been learned from non-mammalian vertebrates (such as amphibians and fish) as well as invertebrates (flies, snails, and worms), particularly over the past two decades. Although most metazoan species have a gut and some specialized organs for performing gas exchange, it does not seem that a specialized mucosal immune system is universal. Crustaceans and worms often have a higher percentage of immune cells, such as hemocytes, surrounding the gut, but these are not organized into discrete tissues as in mammals or other higher vertebrates.

Studies in invertebrates, however, have been extremely informative. It was through studies in the fruitfly *Drosophila melanogaster*, for example, that Toll-like receptors and their human homologs were identified, among the more significant discoveries in immunology in the past decade. Similarly, studies of the sea urchin, *Strongylocentrotus purpuratus*, have revealed key steps in the evolution of the complement system, in particular the alternative pathway involving C3 and factor B.

Phylogeny of receptor diversity.

The unique aspect of the adaptive immune system is the generation of diverse receptors from germline DNA. Among the animal phyla, the chordates (Phylum Chordata) arguably exhibit the greatest diversity in immune systems.

Vertebrates are divided into two major living lineages: the jawed (gnathostomes) and the jawless (agnathans) (Figure 2.1). These two lineages have adaptive immune systems that followed very different evolutionary paths while maintaining some common features. Both lineages use lymphocytes as

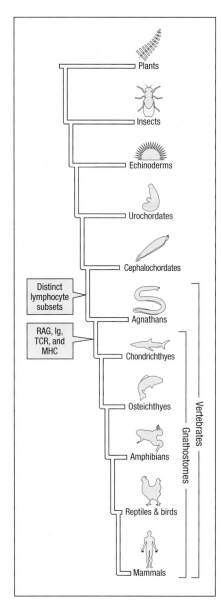

Figure 2.1 Evolutionary relationship of the animal lineages mentioned in this chapter. The branching pattern of the evolutionary 'tree' shown represents the relative order of divergence of the different lineages. For example, the plants and animals diverged from a common ancestor before the separation of the animals into the protostomes (represented by the insects) and the deuterostomes (echinoderms and chordates). The evolution of different lymphocyte subsets seems to have occurred before the divergence of the vertebrates into the jawless (agnathans) and jawed (gnathostome) lineages. However, an adaptive immune system based on the recombination-activating gene (RAG) recombinase enzymes, including the immunoglobulins (Ig), T-cell receptors (TCR), and major histocompatibility complex (MHC), is unique and common to all the gnathostomes.

the major cell type mediating adaptive immunity. These lymphocytes express cell-surface, antigen-specific, and clonally unique receptors.

2-1   T-cell receptors are similar in all jawed vertebrates.

In all jawed vertebrates the T-cell antigen receptors (TCRs) are composed of a heterodimer of either paired αβ or γδ chains (Table 2.1). The genes encoding these four chains are highly conserved both in sequence and organization from sharks to mammals, leading to the conclusion that the TCR performs common functions across all the gnathostome lineages, principally as recognition and signaling receptors and the unique way in which they recognize antigens. Unlike antibodies that serve both as signaling receptors and are also secreted, the TCR serves solely to recognize antigenic epitopes when they meet a threshold of binding affinity. The requirement of recognizing antigenic peptides and glycolipids presented on molecules encoded by the major histocompatibility complex (MHC) has further constrained the evolution of the TCR, the αβ TCR in particular. This relationship between TCR and MHC is also ancient in the gnathostomes and conserved in all living species (see Table 2.1).

2-2   Immunoglobulins have evolved in different ways in different lineages.

The evolution of different classes of immunoglobulins has occurred through both gene duplication and gene re-organization. In the bony fish and the tetrapods (amphibians, birds, reptiles, and mammals), the immunoglobulin (Ig) genes are organized in what is termed a translocon arrangement. The translocon arrangement is generally a single heavy-chain locus with clusters of separate variable (V), diversity (D), and joining (J) segments upstream of the exons encoding the constant domains. The significance of this translocon arrangement is that it facilitated the specialization of different antibody heavy-chain isotypes through gene duplication, ultimately leading to the evolution of different isotypes, such as IgG and the mucosa-associated antibodies IgT, IgX, and IgA (see Table 2.1). The evolution of class switching, however, did not drive isotype diversification, because multiple antibody classes are also found in cartilaginous fishes (sharks, rays, and skates) as well as bony fishes. However, class switching was a significant evolutionary innovation because it allowed B cells to change the functional attributes of the secreted antibody without changing antigen specificity.

2-3   T and B cells are present in lower vertebrates.

In jawless vertebrates such as lampreys and hagfish, and the jawed vertebrates such as sharks, bony fish, birds, and mammals, there are discrete lymphocyte subsets. One subset does not secrete its antigen receptor but is capable of producing cytokines that regulate other cell types. In gnathostomes these are the T cells, whereas in agnathans they are the VLRA+ lymphocytes. Also present is a subset that is capable of producing its antigen receptor both as a cell-surface receptor and as a secreted molecule, often in a polymeric form (Figure 2.2). These are the B cells in gnathostomes, and VLRB+ lymphocytes in agnathans. Therefore, the paradigm of the dichotomy between B and T cells in mice and humans probably arose in a common vertebrate ancestor more than 500 million years ago, before the evolution of the thymus, spleen, TCRs, Igs, or other structures with which most (conventional) immunologists are familiar. The strategy of having two distinct lymphocyte lineages, with one playing a regulatory role while the other produces antibodies, is an ancient defense strategy.

In the hagfish and lamprey, diversity occurs at a gene cluster encoding a class of receptors called the variable lymphocyte receptors (VLRs). The VLRs are the

Table 2.1 The diversity of cells, receptors, and immunological machinery used by vertebrates

Lineage	Age*	VLR	Ig types	Ig class switch	Specialized mucosal Ig	TCR	MHC	AID†	RAG‡
Agnathan									
Lamprey, hagfish	~500	VLRA, VLRB	no	no	Undetermined	no	no	yes	no
Gnathostomes									
Chondrichthyes (sharks, rays, skates)	~450	no	IgM, IgW, IgNAR	no	Undetermined	$\alpha\beta$, $\gamma\delta$	Class I, II	yes	yes
Osteichthyes (bony fish)	~400	no	IgM, IgD, IgT	no	IgT	$\alpha\beta$, $\gamma\delta$	Class I, II	yes	yes
Amphibians (frogs, toads, salamanders)	~360	no	IgM, IgD, IgY, IgX	yes	IgX	$\alpha\beta$, $\gamma\delta$	Class I, II	yes	yes
Lepidosaurs (snakes, lizards)	~300	no	IgM, IgD, IgY, IgA	yes	IgA	$\alpha\beta$, $\gamma\delta$	Class I, II	yes	yes
Archosaurs (birds, crocodilians)	~300	no	IgM, IgY, IgA	yes	IgA	$\alpha\beta$, $\gamma\delta$	Class I, II	yes	yes
Mammals									
Monotremes (platypus, echidna)	~170	no	IgM, IgD, IgG, IgE, IgA	yes	IgA	$\alpha\beta$, $\gamma\delta$, μ§	Class I, II	yes	yes
Marsupials (opossum, kangaroo)	~135	no	IgM, IgG, IgE, IgA	yes	IgA	$\alpha\beta$, $\gamma\delta$, μ§	Class I, II	yes	yes
Eutherians (human, mouse, cow)	0	no	IgM, IgD, IgG, IgE, IgA	yes	IgA	$\alpha\beta$, $\gamma\delta$	Class I, II	yes	yes

*Time in millions of years since last common ancestor with eutherian (placental) mammals such as humans and mice. †Activation-induced cytidine deaminase or related cytidine deaminases used to diversify antigen receptors. ‡Recombination-activating genes encoding the V(D)J recombinase. §TCRμ is a unique TCR chain related to TCRδ.

VLR, variable lymphocyte receptor; Ig, immunoglobulin; TCR, T-cell receptor; MHC, major histocompatibility complex.

functional equivalent to the antibodies in jawed vertebrates. However, instead of being based on immunoglobulin protein domains, they are composed of leucine-rich repeat (LRR) domains similar to Toll-like receptors. Both hagfish and lampreys have two sets of VLR genes, VLRA and VLRB, which define the distinct VLRA$^+$ and VLRB$^+$ lymphocyte lineages. VLRA$^+$ cells express a non-secreted cell-surface receptor and produce cytokines and chemokines. VLRB$^+$ cells express a cell-surface receptor that, upon cell activation, is secreted either as tetramers or pentamers, much like IgM in gnathostomes, and functions as an antibody. VLRs differ from Igs and TCRs not only in the protein structure that is being used but also in the genetic mechanisms used to generate diversity. Ig and TCR diversity is generated via the assembly of an exon encoding

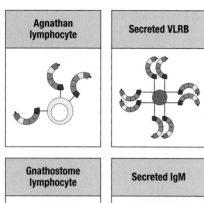

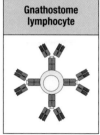 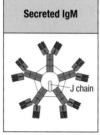

Figure 2.2 Comparison of the lymphocyte and secreted antibody forms from jawless and jawed vertebrates. Both jawless (agnathan) and jawed (gnathostome) vertebrates have lymphocytes that express a cell-surface receptor that is diversified through some process of somatic gene rearrangement. In the agnathans, these are the variable lymphocyte receptors (VLRs) containing leucine-rich repeat (LRR) domains; in the gnathostomes, they are the immunoglobulins and T-cell receptors containing Ig domains. Gene rearrangement in the VLRs occurs by gene conversion, resulting in variable numbers of LRR units being present, whereas in the gnathostomes it is mediated by the cutting and splicing of gene segments, the so-called V(D)J recombination. Similarly to IgM, VLRB can be secreted in either tetrameric or pentameric forms.

the variable domain from the V, D, and J gene segments by means of the cutting and splicing of DNA using site-specific endonucleases and DNA repair mechanisms. In contrast, the VLR gene system in lampreys and hagfish generates diversity using a gene conversion mechanism that is dependent on DNA damage induced by cytidine deaminase. The cytidine deaminases expressed in agnathan lymphocytes are related to the activation-induced cytidine deaminases (AIDs) that participate in primary antibody diversification, affinity maturation, and isotype switching in birds, mammals, and other gnathostomes. Therefore, the involvement of such deaminases in antigen receptor genetics is also ancient in the vertebrate lineage (see Table 2.1). Through gene conversion, a variable number and combination of LRR domains are added to the expressed locus. As with V(D)J recombination, the VLR system is capable of generating many (in the order of 10^{14}) possible receptors.

The phylogeny of lymphoid tissue.

It is generally agreed that primary lymphoid tissues appeared earlier in vertebrate evolution than the secondary lymphoid tissues. Indeed, many of the familiar tissues labeled as secondary lymphoid organs in mammals, such as lymph nodes, are only found in so-called higher vertebrates such as birds and mammals. Defined mucosal lymphoid structures akin to Peyer's patches in eutherian mammals are found only in birds, monotremes (echidna and platypus), and marsupials (Figure 2.3).

2-4   The lymphoid tissues in agnathans seem to have evolved from MALT.

Studies in jawless vertebrates support the evolution of early primary lymphoid tissues from mucosa-associated lymphoid tissues (MALTs). The primary hematopoietic organ in lampreys is the typhlosole, an organ unique to the agnathans. The typhlosole develops as an invagination in the larval intestine and remains associated with the intestinal wall throughout the life of the organism. In addition, the lamprey pharyngeal gill epithelium also contains accumulations of lymphocytes that express VLR genes. Furthermore, for decades it has been suggested that the pharyngeal or gill lymphoid tissue in the lamprey may represent a thymus equivalent. It is worth noting that this hypothesis is consistent with the development of the thymus in jawed vertebrates from the pharyngeal pouches. The discovery of an adaptive immune system in agnathans makes this question all the more interesting, because the generation of highly diverse VLR antigen receptors in lampreys and hagfish should create the need for some mechanism of self-tolerance, presumably via negative selection. The paradigm of the gnathostome thymus suggests the need for a specific organ in which cells undergo selection. Consistent with this idea, the pharyngeal/gill-associated lymphoid tissue in lampreys has the highest percentage of lymphocytes that are VLRA+, those lymphocytes that share the greatest number of characteristics with gnathostome T cells. Throughout the rest of the lamprey tissues, VLRB+ cells are far more prevalent than VLRA+ cells. Furthermore, there is recent evidence that the lamprey *Foxn4L* gene expressed in developing gill epithelium is the agnathan ortholog of *Foxn1*, which is expressed in the developing thymus of jawed vertebrates.

2-5   Gnathostome lymphoid tissues differ between lineages.

All jawed vertebrates, from sharks to mammals, use the same basic components for their adaptive immune system, including immunoglobulin domain-based antibodies and T-cell receptors, and an antigen presentation system

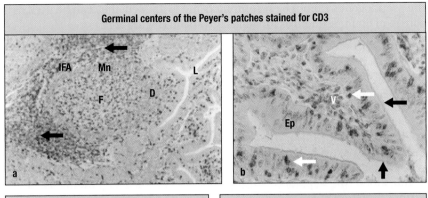

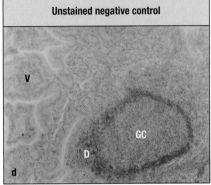

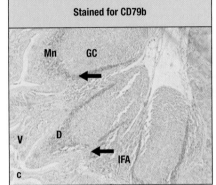

Figure 2.3 Juvenile eastern gray kangaroo gut stained for T cells and B cells. Panel a: anti-CD3 stained cells were apparent in the germinal centers of the Peyer's patches and (arrows) in the interfollicular areas (IFA). A rare cell was stained in the mantle (Mn) of the follicles (F) and in the dome (D). The lumen (L) is also observed (×100). Panel b: higher magnification (×400) of panel a, showing clearly defined T cells (white arrows) throughout the villi (V); some were in close association with epithelial cells (Ep) (black arrows). Panel c: the germinal centers (GC), mantles (Mn), and dome regions (D) heavily stained by anti-CD79b (arrows). The interfollicular area (IFA) was not stained. There seems to be no staining of B cells in the villi (V) (×100). Panel d: representative negative control slide with no staining observed. The dome (D) of the germinal center (GC) is easily seen, as are the villi (V) (×100). (Photographs courtesy of Julie Old.)

using the molecules of the MHC. Similarly, all jawed vertebrates have a thymus with clearly delineated cortical and medullary regions. The thymus is always associated with the pharyngeal regions or upper thoracic regions and develops from embryonic pharyngeal pouches. The conservation of the thymus is probably due to its critical role in the positive and negative selection of developing T cells. All jawed vertebrates, including cartilaginous fish, also have a spleen with clearly defined red and white pulp areas. However, fish do not have MALT, but the intestine does contain T cells and B cells in the epithelium and lamina propria.

In contrast to the conserved nature of the thymus, the primary lymphoid organs in gnathostomes vary between lineages. In cartilaginous fish, hematopoiesis takes place in the epigonal organ, whereas in bony fish it is in the head kidney. The role of the bone marrow as the primary hematopoietic tissue evolved in tetrapods and is found in amphibians, birds, and mammals. In well-known species such as mice and humans, B-cell development occurs primarily in the bone marrow. However, in a few mammals and all birds, the gut-associated lymphoid tissue (GALT) has an essential role in B-cell ontogeny. An extreme example of this is birds in which there is a unique avian GALT, the bursa of Fabricius. In fact, B cells acquired their name from the discovery that bursectomized chicks failed to develop antibody responses. Although, phylogenetically speaking, birds and crocodilians (crocodiles and alligators) are closely related, the latter do not have a bursa, and unfortunately little is known regarding B-cell development or the role of the GALT in this lineage.

In birds, B cells that have undergone V(D)J recombination in the bone marrow must migrate to the bursa to complete their development. In the bursa of Fabricius, avian B cells complete their development by introducing mutations into their recombined V genes using AID-mediated gene conversion. This is an essential step because of the lack of repertoire diversity generated by V(D)J recombination alone. The chicken IgH locus contains only a single functional V gene and a large array of nonfunctional V pseudogenes. Although the pseudogenes cannot be recombined and expressed, they can be used as a source of donor sequences for gene conversion (Figure 2.4).

There is also enormous species variation in the functional role of GALT in mammals. Sheep and rabbits increase diversity by somatic hypermutation in the GALT (Peyer's patches and appendix in sheep; Peyer's patch, appendix, and sacculus rotundus in rabbits); GALT involutes in sheep at sexual maturity, but becomes secondary lymphoid tissue in the rabbit. Why do some species use a GALT site to complete B-cell development whereas others do not? Species that use GALT for B-cell development all seem to have another common characteristic in that their immunoglobulin heavy-chain and light-chain genes are unable to generate sufficient antibody diversity from V(D)J recombination alone. The migration of the B cells to a GALT site apparently facilitates further mutation of the expressed V genes, increasing the diversity of the primary antibody repertoire. The mutational processes seem to be dependent on both the presence of the GALT tissue and colonization of the gut flora. Germ-free rabbits, for example, do not generate as diverse an antibody repertoire as normal rabbits. In contrast, species such as humans and mice, which have diverse germline V gene repertoires, do not require such additional mutation to generate sufficient diversity and have not evolved the need for using GALT sites for primary B-cell development.

The phylogeny of mucosal antibodies.

Several different classes or isotypes of Igs have evolved in the jawed vertebrates (see Table 2.1). This has involved both the evolution of novel heavy-chain and light-chain isotypes; however, it is the appearance of new classes of heavy chains that has had the greatest functional impact on immune responses.

2-6   Immunoglobulin A is the secretory antibody in amniotes.

Immunoglobulin A, the antibody isotype primarily associated with mucosal secretions, was discovered in mammals in the 1960s. IgA is secreted in a polymeric form as a dimer associated with a conserved J chain, and is transported across mucosal membranes by the polymeric Ig receptor (pIgR). Clear homologs of mammalian IgA have been identified in birds and reptiles, but not in amphibians and fish (see Table 2.1). Similarly, birds have a pIgR homolog

Figure 2.4 The organization of immunoglobulin genes in cartilaginous fish and other gnathostomes is different from that in mammals. The organization of the Ig heavy-chain genes in mammals, in which there are separated clusters of repeated V, D, and J gene segments, is similar to the way in which these genes are also organized in birds, reptiles, amphibians, and bony fish. This is referred to as the translocon organization. In sharks, the Ig loci consist of multiple clusters of a V gene, one or two D genes, a J gene, and the exons encoding the constant domains. These clusters can be in tandem arrays as shown or unlinked on separate chromosomes, depending on the species. In chickens, there is a single functional V gene that is rearranged in all B cells, limiting the amount of antibody diversity that V(D)J recombination alone can generate. To further diversify the antibody repertoire, B cells complete their development in the bursa of Fabricius, where the rearranged V is further mutated by gene conversion using an array of V pseudogenes upstream of the functional one as the donor sequences. (Adapted from Kenneth Murphy, Janeway's Immunobiology, 8th edition, New York, Garland Science, 2012.)

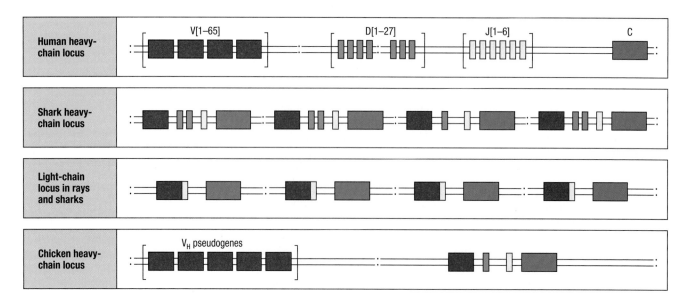

that, in chickens, is expressed in the liver, gut, and bursa of Fabricius. As in mammals, the pIgR interacts with the avian J chain and is cleaved during transcytosis across the epithelial cells. Therefore, IgA and its associated receptor appeared before the divergence of birds, reptiles, and mammals more than 300 million years ago.

2-7   Amphibians have a unique antibody, immunoglobulin X.

Although a direct homolog of IgA has not been found in amphibians, that does not mean they lack specialized mucosal antibodies. Amphibians such as *Xenopus laevis* contain at least three Ig heavy-chain isotypes: IgM, IgY, and IgX (see Table 2.1). As in birds, IgY is functionally analogous to IgG in mammals. Amphibian IgX is believed to be the functional analog of the IgA found in birds, reptiles, and mammals. Like avian and mammalian IgA, frog IgX is mainly expressed at mucosal sites such as the intestine and is secreted in a polymerized form, in this case as a hexamer. However, IgX is not related to IgA, and is more similar in sequence to IgM. Amphibian species such as *Xenopus* also express J chain and use it to form polymerized secretory IgM. They also have a pIgR that binds J chain. Curiously, however, *Xenopus* IgX is unable to interact with J chains owing to a stop codon that truncates the last constant domain. IgX therefore polymerizes in the absence of J chains; the mechanism that is used to transport frog IgX across epithelial membranes is not currently known.

2-8   Teleosts have a unique secretory antibody, IgT.

Until recently, teleost fish were thought to have only IgM and IgD and to lack an antibody class specialized for mucosal sites. However, a new Ig heavy-chain isotype called IgT (T or tau (τ) for teleost, and also referred to as IgZ in the literature as a result of its previous discovery in the zebrafish, *Danio rerio*) that may have an important role in mucosal immunity was recently discovered in teleost fish. Serum IgT exists as a monomer, whereas IgT in mucosal secretions is a polymer, similar to IgA and IgX.

In teleost fish, both IgM- and IgT-producing B cells are phagocytic and bactericidal and are critical to defense against infections. Recently both IgM- and IgT-secreting B cells have been found in the GALT of trout, a common teleost model species of economic importance (Figure 2.5). However, during infection with gut parasites, IgT-producing B cells undergo the greatest expansion, far outnumbering the IgM-producing B cells. IgT is also the most abundant antibody excreted into the lumen of the gut and has been found to bind both intestinal pathogens and bacteria that comprise the normal gut flora. This is analogous to what has been shown for IgA in mammals and further demonstrates the ancient role that the mucosal immune system has played in maintaining homeostatic relationships with microbial flora.

Bony fish do not undergo class switching in the way that it is understood from tetrapods. As in mammals, IgD in teleosts is produced through alternative mRNA splicing with IgM transcripts rather than a true class switch involving DNA deletion. The exons encoding the IgT constant-region domains are unusual in that they are upstream of those encoding IgM/D and seem to be expressed with their own unique set of D and J gene segments that are also upstream of the D and J genes used by IgM/D. Because of the organization of the genes and the lack of the class-switch mechanisms, B cells expressing IgT are a separate subset from those expressing IgM/D. In other words, during ontogeny, B cells must make a decision to produce either only IgT or only IgM/D. What determines this cell fate is unknown, and it is possible that it depends on which D gene and J gene are productively recombined. If the D and J upstream of the IgM/D are used, the IgT constant regions are deleted. If the D and J upstream of IgT are used, the B cell becomes an IgT-producing cell.

Figure 2.5 IgT-producing B cells in the gut of trout expand in numbers during parasite infection. Immunofluorescence staining of trout gut from uninfected fish (left) and fish infected with the parasite *Ceratomyxa shasta* (right). Cell nuclei are stained with DAPI (blue). IgT[+] B cells (stained green) have expanded in number in the lamina propria (LP) in infected individuals relative to IgM[+] B cells (stained red). The arrow points to a parasite located in the gut lumen (Lu). (Reprinted from Z. Yong-An et al., *Nature Immunology* 11: 827–835, 2010. With permission from Macmillan Publishers Ltd.)

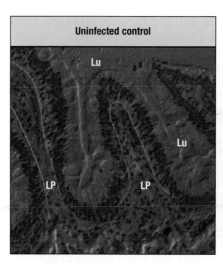
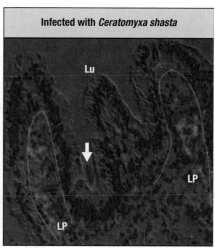

It is worth emphasizing here that the mucosal antibodies found in bony fish, amphibians, and amniotes (birds, reptiles, and mammals) are not closely related. Analyses of their gene sequences suggest origins for IgT, IgX, and IgA from independent gene duplications. Therefore, the role of these antibodies at the mucosa is presumably the product of convergent evolution due to selective pressures for each species to have at least one class of antibody specialized for mucosal sites and secretions. This is an ancient evolutionary pressure dating back at least 400 million years.

Summary.

Not long ago, immunologists were convinced that the so-called 'higher' vertebrates such as birds and mammals had a monopoly on adaptive immunity. It is clear that many of the so-called 'lower' vertebrates not only are capable of developing diverse antigen receptors and specialized cell types, but have also evolved specialized mucosal immune systems. Analyses of the immune systems of these lower vertebrates have supported both the ancient origins of the mucosal immune system as well as its role in the evolution of the primary lymphoid tissues known from mice and humans. The ancient origin of GALT in vertebrates probably illustrates how important a role that defense against pathogens at mucosal sites played in the evolution of the immune system.

Further Reading.

Amemiya, C.T., Saha, N.R., and Zapata, A.: **Evolution and development of immunological structures in the lamprey.** *Curr. Opin. Immunol.* 2007, **19**:535–541.

Flajnik, M.F.: **Comparative analyses of immunoglobulin genes: surprises and portents.** *Nat. Rev. Immunol.* 2002, **2**:688–698.

Flajnik, M.F., and Kasahara, M.: **Origin and evolution of the adaptive immune system: genetic events and selective pressures.** *Nat. Rev. Genet.* 2010, **11**:47–59.

Hofmann, J., Greter, M., Du Pasquier, L., et al.: **B-cells need a proper house, whereas T-cells are happy in a cave: the dependence of lymphocytes on secondary lymphoid tissues during evolution.** *Trends Immunol.* 2010, **31**:144–153.

Pancer, Z., and Cooper, M.: **The evolution of adaptive immunity.** *Annu. Rev. Immunol.* 2006, **24**:497–518.

Stavnezer, J., and Amemiya, C.T.: **Evolution of isotype switching.** *Semin. Immunol.* 2004, **16**:257–275.

Woof, J.M., and Mestecky, J.: **Mucosal immunoglobulins.** *Immunol. Rev.* 2005, **206**:64–82.

Immunological and Functional Differences Between Individual Compartments of the Mucosal Immune System

3

The mucosal immune system is present in anatomically and physiologically diverse tissues, including the gastrointestinal tract, nasopharynx, oral cavity, lung, eye, and urogenital tract. Although these compartments share many features, the mucosal immune systems of the tissues also display distinct characteristics, probably reflecting the anatomical and functional requirements at diverse mucosal sites. This chapter will discuss the common and unique features of the mucosal immune systems in these organ tissues, focusing on the distribution of mucosal antibodies (Figure 3.1) and immune inductive sites.

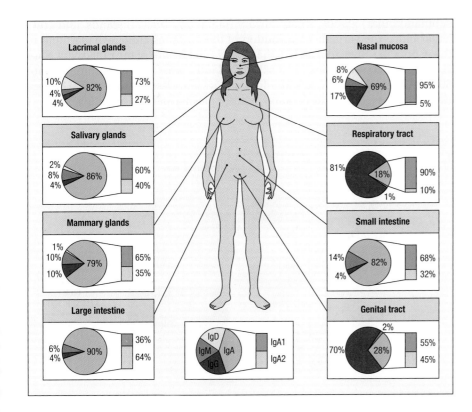

Figure 3.1 Immunoglobulin (Ig) and IgA subclass antibody distribution in mucosal compartments. IgA is the dominant Ig isotype in all mucosal secretions, except those of the lower respiratory and genital tracts, where IgG is the major Ig. In humans, two subclasses of IgA—IgA1 and IgA2—are present in different proportions in compartment secretions. IgA1 exceeds IgA2 in all secretions except those of the large intestine.

Common and distinct features of the mucosal immune system in different tissues.

3-1   The ocular immune system contains inductive sites.

Tears contain relatively high levels of secretory IgA (SIgA), the dominant Ig in ocular fluid, with small amounts of monomeric IgA and IgG antibodies. Reflecting the dominance of IgA antibodies in tears, lacrimal glands contain large numbers of IgA-producing cells. The proportion of IgA1-producing cells (53–81%) relative to IgA2-producing cells (19–47%) corresponds to the proportion of IgA1 to IgA2 in tears. Approximately 10% of the Ig-secreting cells produce IgD of unknown functional importance. Importantly, lacrimal gland acini and ducts express the polymeric Ig receptor (pIgR), a key element in the formation of SIgA and its transport into tears.

The administration of antigen onto ocular mucosa can induce antigen-specific SIgA responses in the ocular and nasal cavities, as well as systemic IgG antibody responses. The tear-duct-associated lymphoid tissues (TALT) in the conjunctival sac are connected via the tear duct to the nasal cavity (Figure 3.2) and are important in the generation of these responses. Locally administered antigens enter the tear duct and then the nasal cavity, with subsequent antigen sampling by M cells in the follicle-associated epithelium (FAE) covering the apical surface of TALT and nasopharynx-associated lymphoid tissue (NALT). Thus, the ocular mucosal immune system is involved in the induction of local and systemic immune responses, in a similar manner to ingestion- and inhalation-induced responses.

3-2   In the oral cavity, SIgA dominates and sublingual tissue is a potential inductive site.

Saliva consists of fluids derived from large salivary (parotid, sublingual, and submandibular) glands, small salivary (labial and buccal) glands, and crevicular fluid. The variable contribution of these tissues and crevicular fluid to the Ig pool in saliva depends on the periodontal health of the oral cavity. SIgA is dominant in secretions of all salivary glands, with a composition of about 60% IgA1 and 40% IgA2. IgG and IgM are present in small quantities. In contrast, the crevicular fluid contains mainly plasma-derived proteins, and IgG

Figure 3.2 Lymphoid structures of mucosa-associated lymphoid tissue (MALT) involved in the induction of antigen-specific immune responses. Tear-duct-associated lymphoid tissues (TALT), nasopharynx-associated lymphoid tissue (NALT), and gut-associated lymphoid tissue (GALT), specifically Peyer's patches (red arrow), induce antigen and vaccine-stimulated responses, leading to antigen-specific mucosal SIgA and systemic IgG production. (Immunofluorescence panel courtesy of T. Yamanaka and P. Brandtzaeg.)

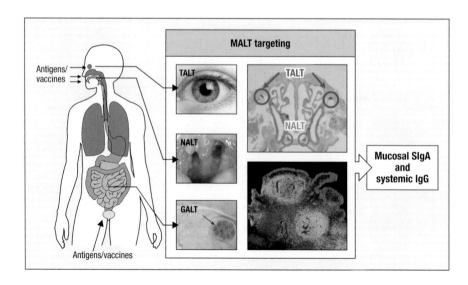

is the dominant Ig isotype. In the oral cavity, mucosal and systemic Ig contributions depend on the state of oral health. In advanced periodontal disease, the proportion of plasma-derived IgG antibodies in the Ig pool in whole saliva increases substantially.

The local application of antigen to the buccal mucosa, labial mucosa, or gingiva stimulates very low antigen-specific immune responses. In contrast, antigen ingestion, intranasal immunization, or even rectal immunization may induce antigen-specific SIgA-associated responses in saliva. Several studies have implicated the tonsils as a source of the B cells that populate salivary glands. Recent studies also indicate that the sublingual application of antigen stimulates immune responses in both systemic and mucosal compartments, including the female genital tract (see Figure 3.2). Thus, among oral cavity tissues, the sublingual immune apparatus is a potential inductive site, reflected in the ability of vaccine antigen deposited at this site to elicit antigen-specific immune responses in both mucosal and systemic compartments.

3-3 The upper and lower respiratory tract display discordant immunological features.

In nasal secretions, which are a major component of the surface barrier in the upper respiratory tract, IgA constitutes about 70% of the Ig pool. Reflecting the dominance of IgA, nasal mucosa contains large numbers of IgA-producing cells, particularly of the IgA1 isotype. Approximately 20% of Ig in nasal secretions is represented by IgG, derived mostly from the circulation with limited local production. Thus, antigen-specific IgG antibodies appear in nasal secretions after systemic immunization. In sharp contrast to the predominance of IgA in nasal secretions, IgG is the major Ig isotype in bronchoalveolar fluid in the lower respiratory tract, contributing about 80% of the Igs to lung fluid. The IgG seems to be derived predominantly from the circulation. The dominance of IgA in the upper respiratory tract and IgG in the lower tract is consistent with the mucosal immune system having the major role in protection of the upper respiratory tract and the systemic immune system having the main role in the lower tract.

The respiratory tract is equipped with several organized lymphoid structures that may serve as inductive sites for the initiation of antigen-specific immune responses against inhaled antigens. Intranasal immunization results in the induction of humoral and cellular immune responses in the systemic compartment as well as in mucosal secretions at anatomically remote sites such as saliva, intestinal fluid, and female and male genital tract fluids. Lymphoid tissues in Waldeyer's ring in the nasopharynx seem to participate in local and distant immune responses in several species, including humans. In experimental animals, including mice, rats, and rabbits, NALT and bronchus-associated lymphoid tissue (BALT) serve as mucosal inductive sites. BALT was originally identified in rodents, and its presence in humans is controversial; in normal conditions the tissue is usually not present. However, inducible BALT was recently described as an ectopic lymphoid tissue that can develop throughout the lung in both humans and rodents after infection and/or inflammation. These tissues can take up inhaled antigens to induce local immunity and are involved in the maintenance of memory-type cells. Although their anatomical location in humans may differ from that in rodents, such tissues are clearly inductive sites, because they contain antigen-sampling M cells, dendritic cells (DCs), helper T (T_H) cell subsets, IgA-committed cells, and cytotoxic T lymphocytes (CTLs) for the induction of antigen-specific humoral and/or cellular immunity.

Unlike Peyer's patch high endothelial venules (HEVs), which are present in T-cell zones, murine NALT HEVs are present in B-cell zones and express peripheral lymph node addressin (PNAd), either alone or associated with mucosal addressin cell adhesion molecule-1 (MAdCAM-1). Antibodies against L-selectin, but not against MAdCAM-1, block the binding of naive lymphocytes to NALT HEVs, suggesting a role for L-selectin and PNAd in the binding of naive lymphocytes to these HEVs. Early induction of vascular cell adhesion molecule-1 (VCAM-1), E-selectin, and P-selectin in the pulmonary vasculature was reported during pulmonary immune responses, with initial increased expression of P-selectin ligand by peripheral blood CD4$^+$ and CD8$^+$ T cells. The number of cells expressing P-selectin ligand subsequently declined in the blood as the cells accumulated in the bronchoalveolar lavage fluid. Very late antigen-4 (VLA-4) could be an important adhesion molecule involved in the migration of activated T cells into the lung, because migration of VLA-4$^+$ cells into bronchoalveolar fluids is impaired after treatment with anti-α_4 antibody. Antigen-specific L-selectin^{low} CTL effectors also have been shown to accumulate rapidly in the lung after adoptive transfer to naive mice with reduced pulmonary viral titers early during infection. Analysis of tissue-specific adhesion molecules showed that after systemic immunization, most effector B cells expressed L-selectin and only a few cells expressed the gut-homing molecule $\alpha_4\beta_7$, whereas after enteric (oral or rectal) immunization, the opposite occurred. Interestingly, effector B cells induced by nasal immunization displayed a more promiscuous pattern of adhesion molecules, with a large majority expressing both L-selectin and $\alpha_4\beta_7$. Thus, the upper respiratory immune system is an effective site for the delivery of vaccine antigen, such as through nasal and/or oral spray that targets the NALT.

3-4 The upper and lower intestinal tract are the major source of Ig and the major site for the induction of immune responses.

The largest proportion of Ig-producing cells in the body (70–80%), especially those secreting IgA, is present in the gastrointestinal tract mucosa. Most of these cells are J-chain positive and their product, polymeric IgA (pIgA), is transported into the lumen as SIgA by a pIgR-mediated mechanism through the intestinal epithelium. Approximately 3 g of SIgA enters the intestinal lumen each day (the total production of IgA is about 5 g per day). In some species, including rodents and rabbits, but not humans, the pIgR expressed on surfaces of hepatocytes is involved in the efficient transport of pIgA from the circulation into bile and thus contributes to intestinal SIgA. In humans, the dominant Ig isotype in bile is IgG, and human hepatocytes do not express pIgR. IgA1-producing cells are dominant in human gastric and small intestinal mucosa. However, in contrast to the predominance of IgA1 in all other mucosal secretions, the proportion of IgA2 in the large intestine exceeds that of IgA1. IgA2-producing cells are dominant in the large intestine, possibly related to the large numbers of colonizing Gram-negative bacteria, which seem to stimulate mainly IgA2. Unlike IgA1, IgA2 is resistant to most bacterial IgA proteases, and the IgA2 subclass may be a critical component of antibody-mediated humoral immunity in the lower digestive tract.

Oral ingestion and rectal administration of antigen stimulate immune responses at the site of antigen uptake as well as in extraintestinal mucosal sites, including the lacrimal, salivary, and lactating mammary glands, and the female genital tract (see Figure 3.2). Tissues that participate in immune response induction include Peyer's patches in the small intestine, scattered solitary lymphoid follicles, crypt patches, and accumulations of Peyer's patch-like structures in the colon and terminal large intestine called colonic patches and rectal tonsils respectively. Among the gut-associated lymphoid tissue (GALT) in both small and large intestines, Peyer's patches have been extensively characterized as an important inductive site for the initiation of antigen-specific humoral (for example SIgA) and cell-mediated (for example CTL) immunity. The dome-shaped Peyer's patches are covered by FAE containing M cells that participate in sampling mucosally encountered antigens. Beneath the FAE, antigen-processing and antigen-presenting DCs are located for the immediate capture, processing, and presentation of antigen taken up by M cells for the initiation of antigen-specific immune responses. The Peyer's patches also exhibit several B-cell follicles containing germinal centers in which isotype switching from IgM to IgA occurs, leading to the generation of IgA-committed B (Bα) cells. The follicles are surrounded by the T cell-enriched areas for the generation of T_H1, T_H2, T_H17, regulatory T cell, and CTL responses. In addition to the induction of antigen-specific helper T cells and Bα cells initiated by the M cell–DC antigen capture and presentation axis, DCs in Peyer's patches educate the antigen-specific lymphocytes to acquire gut-imprinting molecules (for example CCR9 and $\alpha_4\beta_7$) via vitamin A retinoic acids. Thus, Peyer's patches initiate antigen-specific immune responses to orally administered vaccines, with the resultant B cells being dispatched to distant effector sites, including the lamina propria of the intestine for the formation and production of SIgA antibodies. Importantly, oral and rectal immunization induce immune responses of variable strength at other mucosal sites, suggesting differences in intestinal versus rectal induction of immune cell homing to various effector sites.

3-5   Mammary glands are an important source of SIgA.

Early milk, called colostrum, and milk collected at later stages of lactation contain high levels of SIgA and small amounts of monomeric IgA, SIgM, and IgG. The subclass distribution varies between donors, and on average the proportion of IgA1 slightly exceeds that of IgA2, resembling the distribution of IgA subclasses in the adult small intestine. Humoral immune responses can be induced by the oral administration of antigens; the effectiveness of intranasal, rectal, or sublingual immunization routes in the induction of SIgA antibodies in milk has not yet been evaluated in humans. Interestingly, the injection of antigen into lactating mammary glands in experimental animals induces weak local IgA and usually strong IgG responses. The spectrum of SIgA antibodies in human milk reflects the maternal exposure to orally and intestinally encountered antigens and thus provides an appropriate passive protection for breast-feeding infants.

3-6   IgG is the major Ig in urogenital tract secretions.

The dominant Ig isotype in secretions of the female and male genital tract and in urine is IgG, which is similar to that in secretions of the lower respiratory tract (bronchoalveolar fluid). Thus, IgA is the major Ig isotype in all mucosal secretions except those of the lower respiratory and genital tracts. In females, the uterus is the most important source of Ig in cervico-vaginal secretions, which explains the profound decrease in genital Igs after hysterectomy. IgG is derived locally from numerous antibody-forming cells in the uterine endocervix and from the circulation. Systemic immunization induces IgG in genital fluids; this is consistent with the systemic source of the IgG. In this regard, the mucosal immune system of the reproductive tract is distinct from that of the digestive tract. Genital tract IgA, represented by SIgA, pIgA devoid of secretory component, and monomeric IgA, also originates from both local and systemic sources. IgA2 levels slightly exceed those of IgA1 in these fluids. The levels of Igs of all isotypes vary with the stage of the menstrual cycle, with the highest levels being present during menstruation and shortly before ovulation and the lowest levels during and a few days after ovulation. Consequently, for the induction of protective humoral immunity in female reproductive tissues, the immunological effect of the menstrual cycle must be considered in vaccine development.

In the female genital tract, humoral immune responses to sexually transmitted pathogens or intravaginally inoculated antigens are generally low. However, the introduction of antigens into the upper female genital tract, as demonstrated in animal experiments, induces measurable Ig responses. Importantly, immunization through the intranasal, sublingual, rectal, or oral routes also stimulates IgA responses in female genital secretions. An important difference between the genital tract and other mucosal tissues such as those of the gastrointestinal tract is the striking absence of M cells and inductive sites in the genital tract. Consequently, a considerable proportion of local IgG is derived from the circulation, and systemic immunization leads to the local appearance of antigen-specific IgG antibodies in the genital secretions. An, Fc receptor, FcγRn, expressed on epithelial cells in the mucosa of the female genital tract is involved in the transport of this IgG.

In the male genital tract, IgG is the dominant isotype in semen. The IgA is represented by approximately equal amounts of SIgA, pIgA devoid of secretory component, and monomeric IgA. The ratio of IgA1 to IgA2 in semen resembles that in serum. Systemic and oral immunizations induce easily detectable IgG and IgA responses in semen; other immunization routes, including the intranasal and rectal routes, have not been explored.

Human urine contains low levels of IgG and even lower levels of IgA as SIgA, pIgA, or monomeric IgA. The origin of urinary SIgA has not been firmly established. Ascendent bacterial infections induce urinary and systemic humoral responses, suggesting that urine may contain both local and systemic antibodies.

Antigen sampling in mucosal surfaces.

One of the unique characteristics of the mucosal immune system is the presence of antigen-sampling mechanisms in certain mucosal tissues. The initiation of antigen-specific immune responses in inductive sites such as GALT (Peyer's patches) and NALT begins after the sampling of ingested or inhaled antigens by M cells located in the FAE covering both inductive tissues. In addition to the M cells in Peyer's patches, similar antigen-sampling M cells, termed villous M cells, have been identified in the villous epithelium of mice without underlying organized lymphoid structures, suggesting an alternative antigen-sampling pathway (Figure 3.3). These villous M cells generally share the morphological, immunological, and functional characteristics of M cells in Peyer's patches. It is not known whether villous M cells contribute to the induction of SIgA or systemic IgG. Although clearly identified in mice and nonhuman primates, villous M cells have not yet been detected in human small intestinal mucosa.

In addition to M cells, intestinal DCs sample luminal antigen encountered at the mucosal surface. Studies performed in the mouse have shown that lamina propria DCs extend their dendrites between epithelial cell junctions into the lumen to take up antigenic material. These DCs are capable of initiating antigen-specific, systemic IgG responses, whereas antigen uptake and transport by M cells in Peyer's patches initiate local IgA production.

M cells also have been identified in the respiratory tract. In the upper respiratory tract, M cells present in the nasal mucosa (NALT M cells) are capable of taking up both particulate and soluble antigens from the nasal mucosa. M cells are also present in the airway mucosa from the bronchial bifurcation to the bronchioles, as shown in mice. These M cells serve as an alternative to alveolar macrophages as a route by which pathogens such as *Mycobacterium tuberculosis* gain entry to the mucosa. Analogous to intestinal intraepithelial DCs, some airway DCs extend their dendritic processes into the lumen of the respiratory tract for antigen sampling. Taken together, the nasal and digestive mucosal immune systems have an array of antigen uptake and presenting

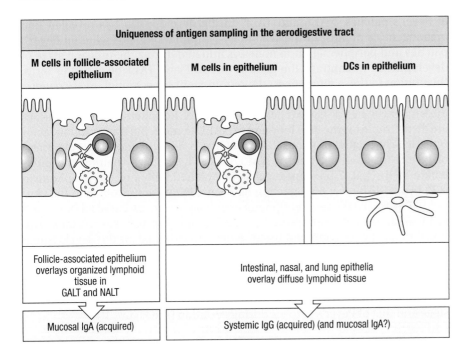

Uniqueness of antigen sampling in the aerodigestive tract		
M cells in follicle-associated epithelium	**M cells in epithelium**	**DCs in epithelium**
Follicle-associated epithelium overlays organized lymphoid tissue in GALT and NALT	Intestinal, nasal, and lung epithelia overlay diffuse lymphoid tissue	
Mucosal IgA (acquired)	Systemic IgG (acquired) (and mucosal IgA?)	

Figure 3.3 Antigen sampling in the digestive, nasal, and respiratory tracts. Mucosa-associated lymphoid tissue (MALT)-dependent M cells are located in the follicle-associated epithelium and overlay organized lymphoid structures in GALT and NALT. MALT-independent M cells (no underlying organized lymphoid structures) are present in villous epithelium, nasal cavity, and lung. Dendritic cells (DCs) are present in the intestinal and respiratory mucosa, where they extend dendrites between epithelial cells into the lumen to sample antigens. MALT-dependent antigen sampling is thought to induce antigen-specific local IgA antibody responses, whereas MALT-independent antigen sampling is thought to induce antigen-specific serum IgG responses.

systems that, depending on the tissue site and species, include Peyer's patch and NALT M cells, villous M cells, respiratory M cells, and intraepithelial DCs (see Figure 3.3).

B-cell subsets in nasal and gut-associated tissues.

At least two B-cell subsets, B-1 and B-2, are involved in the mucosal immune system. IgA-producing cells, which in mice originate from both B-1 and B-2 B cells, are present in the nasal and intestinal lamina propria. B-1 B cells in the murine peritoneal cavity and intestinal lamina propria develop from a common B-cell pool and may represent a lineage separate from that of conventional Peyer's patch B cells. Most Peyer's patch B cells belong to the B-2 cell family; similarly, NALT consists of B-2 B cells. Murine intestinal IgA antibodies with specificity for commensal bacteria are produced by B-1 B cells in a T-cell independent manner. On the basis of the murine system, B-1 B cells seem to be an important source of IgA-producing cells in the intestinal mucosa, in addition to the GALT B-2 lineage of IgA plasma cells. However, in humans, the frequency of B-1 B cells is very low in the intestinal mucosa. Whether human B-1 B cells contribute to IgA1 and/or IgA2 subclass antibody production is currently under investigation.

MALT organogenesis.

Mucosa-associated lymphoid tissue (MALT) consists of the major organized lymphoid structures and inductive sites in the lacrimal, respiratory, and digestive tracts. MALT shares common immunological characteristics and functions as the initiation site for antigen-specific humoral and cell-mediated immune responses. However, the organogenesis of the three well-characterized MALTs seems to be different (Figure 3.4). Chronological examination of Peyer's patches, NALT, and TALT during development has revealed that organogenesis of most of the secondary lymphoid tissues, including Peyer's patches and peripheral lymph nodes, is initiated during the prenatal stage of development. In contrast, the genesis of NALT and TALT begins after birth. In addition to these differences, the requirement for essential organogenesis molecules also differs in the three MALTs. The critical molecules required for the organogenesis of Peyer's patches and other peripheral lymphoid tissues are dispensable for the development of NALT. In mice, the transcription factor retinoic acid-related orphan receptor RORγt is required for the development of Peyer's patches but not for NALT or TALT. Similarly, neither the interleukin-7 (IL-7)/IL-7 receptor α subunit cytokine family nor the lymphotoxin (LT) $LT\alpha_1\beta_2$–LTβ receptor signaling pathways, which are involved in the genesis of Peyer's patches and peripheral lymph nodes, are essential for NALT and TALT organogenesis. Interestingly, even Id2, another well-known transcriptional molecule for lymphoid tissue genesis, is essential for the development of Peyer's patches, peripheral lymph nodes, and NALT but is dispensable for TALT genesis. Thus, at least three distinct tissue genesis programs operate in the development of MALTs associated with the intestine, respiratory tract, and tear duct: the organogenesis of NALT and TALT begins during the postnatal stage and that of Peyer's patches during gestation (see Figure 3.4).

Influence of aging on mucosal immunity.

In addition to the developmental differences in NALT, TALT, and GALT (Peyer's patches), it is not surprising that age-associated differences in these mucosal tissues also occur (Figure 3.5). In this connection, nasal, but not oral,

Figure 3.4 Organogenesis programs for Peyer's patches (PP), NALT, and TALT. Despite their common features, PP, NALT, and TALT undergo separate organogenesis pathways regulated by the indicated transcriptional factors, integrins, and receptors. PPi, PP inducer; NALTi, NALT inducer; TALTi, TALT inducer.

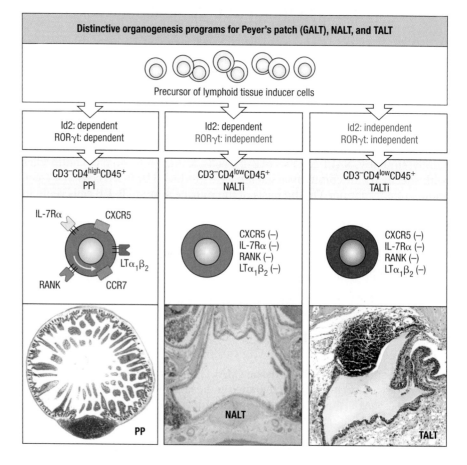

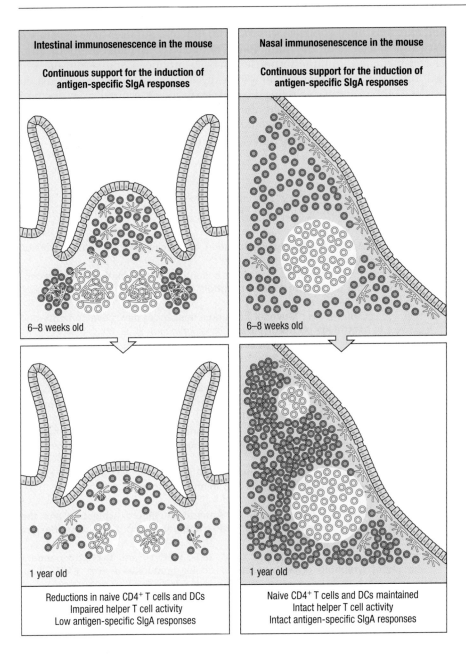

Figure 3.5 Effect of aging on NALT and GALT. In the intestine, SIgA responses decrease during aging, at least in mice; this is reflected in the reduced size of Peyer's patches. Smaller numbers of naive CD4+ T cells and follicular DCs also occur in the Peyer's patches in aging mice. In contrast, in the nasal cavity, specific SIgA (and systemic IgG) antibody responses in 1-year-old (aging) mice are similar to those in young adult (6–8-week-old) mice. The absolute numbers of naive CD4+ T cells in the NALT are maintained in aging mice.

immunization induced normal levels of antigen-specific immune responses in aging mice, supporting the notion that age influences GALT more than it does NALT. Reduced frequencies of CD4+CD45RB+ T cells are present in aging mice; however, the number of naive CD4+ T cells in the NALT of aging mice is higher than that in young adult mice. These findings suggest that a distinct immune aging process occurs in GALT in comparison with NALT, possibly contributing to differences in the induction of antigen-specific mucosal IgA and parenteral IgG responses in advanced age. Thus, nasopharyngeal administration may be a more effective route for the induction of antigen-specific mucosal and systemic immune responses in the elderly.

Summary.

The immune systems in the different mucosal tissues share a common mission: first-line defense against the entry of pathogens and undesired antigens and allergens. To accomplish this mission, mucosal surfaces have developed tissue-specific sites for the induction of antigen-specific SIgA antibody and/or cell-mediated immune responses. Locally produced and serum-derived IgG antibodies contribute immunologically to mucosal immune defense in certain compartments, including the oral cavity, lower respiratory tissues, and urogenital tissues. Distinct microbiota in the various mucosal compartments may influence the development and maintenance of the common and unique immunological features of the mucosal immune system.

Further Reading.

Brandtzaeg, P.: **Regionalized immune function of tonsils and adenoids.** *Immunol. Today* 1999, **20**:383–384.

Eberl, G., Marmon, S., Sunshine, M.J., *et al.*: **An essential function for the nuclear receptor RORγt in the generation of fetal lymphoid tissue inducer cells.** *Nat. Immunol.* 2004, **5**:64–73.

Fujihashi, K., and Kiyono, H.: **Mucosal immunosenescence: new developments and vaccines to control infectious diseases.** *Trends Immunol.* 2009, **30**:334–343.

Iwata, M., Hirakiyama, A., Eshima, Y., *et al.*: **Retinoic acid imprints gut-homing specificity on T cells.** *Immunity* 2004, **21**:527–538.

Jang, M.H., Kweon, M.N., Iwatani, K., *et al.*: **Intestinal villous M cells: an antigen entry site in the mucosal epithelium.** *Proc. Natl Acad. Sci. USA* 2004, **101**:6110–6115.

Kiyono, H., and Fukuyama, S.: **NALT- versus Peyer's-patch-mediated mucosal immunity.** *Nat. Rev. Immunol.* 2004, **4**:699–710.

Macpherson, A.J., Gatto, D., Sainsbury, E., *et al.*: **A primitive T cell-independent mechanism of intestinal mucosal IgA responses to commensal bacteria.** *Science* 2000, **288**:2222–2226.

Mestecky, J., and McGhee, J.R.: **Immunoglobulin A (IgA): molecular and cellular interactions involved in IgA biosynthesis and immune response.** *Adv. Immunol.* 1987, **40**:153–245.

Mestecky, J., Raska, M., Novak, J., *et al.*: **Antibody-mediated protection and the mucosal immune system of the genital tract: relevance to vaccine design.** *J. Reprod. Immunol.* 2010, **85**:81–85.

Nagatake, T., Fukuyama, S., Kim, D.Y., *et al.*: **Id2-, RORγt-, and LTβR-independent initiation of lymphoid organogenesis in ocular immunity.** *J. Exp. Med.* 2009, **206**:2351–2364.

Randall, T.D.: **Bronchus-associated lymphoid tissue (BALT) structure and function.** *Adv. Immunol.* 2010, **107**:187–241.

Rescigno, M., Urbano, M., Valzasina, B., *et al.*: **Dendritic cells express tight junction proteins and penetrate gut epithelial monolayers to sample bacteria.** *Nat. Immunol.* 2001, **2**:361–367.

Song, J.H., Nguyen, H.H., Cuburu, N., *et al.*: **Sublingual vaccination with influenza virus protects mice against lethal viral infection.** *Proc. Natl Acad. Sci. USA* 2008, **105**:1644–1649.

van de Pavert, S.A., and Mebius, R.E.: **New insights into the development of lymphoid tissues.** *Nat. Rev. Immunol.* 2010, **10**:664–674.

Yamamoto, S., Kiyono, H., Yamamoto, M., *et al.*: **A nontoxic mutant of cholera toxin elicits Th2-type responses for enhanced mucosal immunity.** *Proc. Natl Acad. Sci. USA* 1997, **94**:5267–5272.

Secreted Effectors of the Innate Mucosal Barrier

4

The mucosal surfaces, including the surface of the eye and the linings of the gastrointestinal, respiratory, reproductive, and urinary tracts, are the major interface between mammalian tissues and the potentially hostile external environment. Mucosal surfaces have evolved a well-regulated barrier to protect against chemical, physical, and microbial insults. In the intestinal tract, this barrier consists of epithelial cells, a secreted layer of mucus produced by goblet cells, and antimicrobial peptides released by the gastric and colonic epithelium and by small intestinal Paneth cells (Figure 4.1). In this chapter we will describe the components of this barrier and their regulation by innate and adaptive immunity, focusing mainly on the intestinal tract because it is the most thoroughly studied mucosal tissue and is continuously exposed to potential infection.

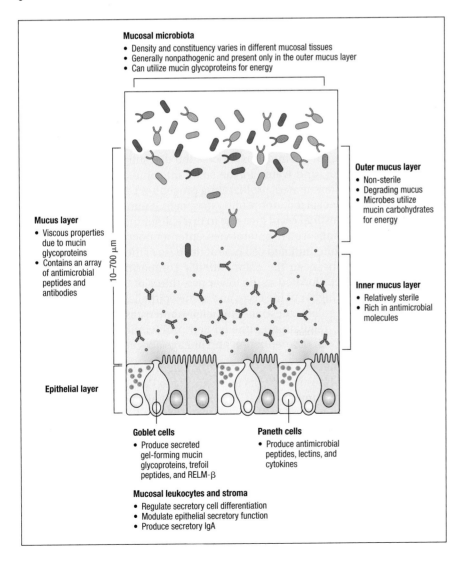

Mucosal microbiota
- Density and constituency varies in different mucosal tissues
- Generally nonpathogenic and present only in the outer mucus layer
- Can utilize mucin glycoproteins for energy

Mucus layer
- Viscous properties due to mucin glycoproteins
- Contains an array of antimicrobial peptides and antibodies

10–700 μm

Outer mucus layer
- Non-sterile
- Degrading mucus
- Microbes utilize mucin carbohydrates for energy

Inner mucus layer
- Relatively sterile
- Rich in antimicrobial molecules

Epithelial layer

Goblet cells
- Produce secreted gel-forming mucin glycoproteins, trefoil peptides, and RELM-β

Paneth cells
- Produce antimicrobial peptides, lectins, and cytokines

Mucosal leukocytes and stroma
- Regulate secretory cell differentiation
- Modulate epithelial secretory function
- Produce secretory IgA

Figure 4.1 The mucosal barrier. Diagrammatic representation of the secreted mucosal barrier showing the specialized secretory epithelial cells and the spatial relationships between the epithelial cells, the secreted mucus layer, and the microbes in the external environment.

The small intestinal epithelium absorbs nutrients, regulates water and electrolytes, and helps prevent most luminal microbes from becoming significant populations. To facilitate nutrient absorption in the small intestine, mammals maximized absorptive membrane surface area by evolving villi and an apical brush border. However, the lumen from which nutrients are absorbed is also colonized by a resident microflora and transiently exposed to microbes and foreign antigens from the diet. Although the large absorptive area of the small intestine appears to favor mucosal colonization by microorganisms, few bacteria populate the small intestine relative to populations in the cecum and colon, suggesting that the small intestine has mechanisms that counter microbial colonization of the epithelium.

Many factors influence the growth of bacteria in the small intestine. For example, extensive energy is invested by villus enterocytes in the transcytosis of antigen-specific secretory IgA from the circulation to the small intestinal lumen. Other factors including gastric acidity, digestive enzymes, bile salts, peristalsis, the resident commensal microflora, exfoliation of enterocytes during epithelial renewal, and CD8+ intraepithelial T lymphocytes contribute to mucosal health. In addition to adaptive immune responses and the potential bystander effects resulting from gastrointestinal physiologic processes, innate host defense mechanisms confer immediate biochemical protection against infection and colonization by pathogenic or opportunistic microorganisms. The secretory products of specific epithelial lineages, including mucins and antimicrobial peptides, are primary mediators of this biochemical barrier.

The intestinal epithelium: stem cells, self-renewal, and cell lineage allocation.

The intestinal epithelial barrier, along with specialized cells that secrete mucus, antimicrobial peptides, and enteric hormones, is renewed every 4–5 days. The primary function of the intestinal tract is the digestion and absorption of nutrients, electrolytes, and water. In the small intestine, the crypt compartment contains undifferentiated, proliferating progenitor cells, whereas the villi are populated by specialized, nonproliferating cells. Stem cells located in the base of the small intestinal crypt give rise to three differentiated cell types that populate the villus—absorptive enterocytes, enteroendocrine cells, and goblet cells—and a single differentiated cell type at the base of the crypt, Paneth cells. Although villi are absent in the colon, similar compartmentalization exists with proliferating cells located in the lower two-thirds of the crypt and differentiated cells found higher in the crypt and at the epithelial surface. Molecular mechanisms that regulate intestinal epithelial homeostasis include defining the stem cell niche, regulating epithelial cell proliferation in the intestinal crypt, and epithelial cell differentiation and lineage allocation (Figure 4.2).

All intestinal epithelial cell lineages arise from stem cell progenitors located near the base of the intestinal crypts in proximity to Paneth cells. These cells, termed crypt-based columnar cells (CBCs), were initially proposed to be intestinal stem cells by Cheng and Leblond in 1974. Lineage tracing of *Lgr5/Gpr49* expressing cells has shown that CBCs are (1) multipotent for all terminally differentiated intestinal epithelial cell types, (2) long lived, (3) self-renewing, and (4) radiation resistant, all features characteristic of stem cells. The notion that CBCs represent the true intestinal stem cells is contrary to the classic model of the stem cells being located at position +4 relative to the crypt bottom where Paneth cells occupy the first three positions. These cells were characterized as DNA label-retaining cells (+4 label-retaining cells). Although additional investigation will be required to fully understand the relationship between

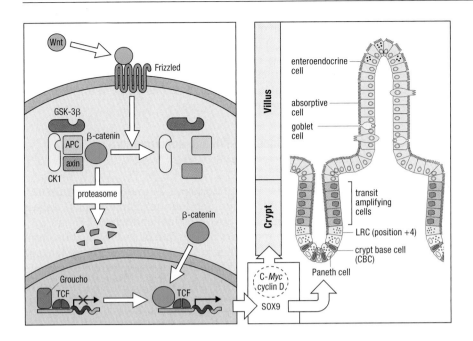

Figure 4.2 The role of the Wnt signaling pathway in the regulation of epithelial proliferation and differentiation in the small intestine. Components of the canonical Wnt signaling pathway are shown on the left. Activation of Wnt signaling by the binding of Wnt ligands to Frizzled receptors ultimately activates genes that drive cellular proliferation that is important for the maintenance of the intestinal stem cell and crypt progenitor compartment. Wnt signaling also activates the expression of SOX9, a transcription factor critical for the development of Paneth cells. CK1, casein kinase 1; GSK-3β, glycogen synthase kinase 3β; LRC, label-retaining cell. (Reprinted from G. Pitari, et al., *Clinical Pharmacology & Therapeutics* 82: 441–447, 2007. With permission from Macmillan Publishers Ltd.)

CBCs and +4 label-retaining cells, current evidence strongly favors the notion that CBCs represent the intestinal stem cell compartment. Indeed, single *Lgr5* positive CBCs are capable of building self-organizing crypt-villus organoids *in vitro* in the absence of a nonepithelial cellular niche.

4-1 Wnt signaling regulates intestinal epithelial cell positioning and differentiation.

Intestinal stem cells give rise to rapidly proliferating 'transit amplifying' cells that generate progeny that then migrate up the crypt compartment, producing the terminally differentiated cells that constitute the villus epithelium. The Wnt pathway is the primary driving force regulating the proliferation of the intestinal epithelium (Figure 4.2). Indeed, *Lgr5* was initially identified as a potential intestinal stem cell marker because of its inclusion as a target of the Wnt signaling pathway. In the absence of Wnt activation, β-catenin is targeted to undergo proteasomal degradation by sequential phosphorylation at the N-terminus through its interaction with a degradation complex consisting of axin, adenomatous polyposis coli, glycogen synthase kinase 3β, and casein kinase 1. The canonical Wnt pathway, activated by interactions between Frizzled and low-density lipoprotein receptor-related protein receptors with one of the many different Wnt ligands, results in disruption of the β-catenin destruction complex. The subsequently accumulating cytosolic β-catenin translocates to the nucleus and binds to TCF/LEF transcription factors. The newly formed active transcriptional complexes displace transcriptional repressors such as Groucho, enabling the transcription of Wnt target genes, many of which have roles in cellular proliferation, such as c-*myc* and *CCND1*.

The Wnt pathway regulates cellular differentiation of the intestinal epithelium in several ways. First, the β-catenin/TCF transcriptional complex activates genes expressing Ephrin-B ligands and their receptors EphB, which have a role in establishing migratory pathways as well as maintaining cellular boundaries. The gradient of EphB receptor expression within the crypt maintains the correct cellular architecture with CBCs and Paneth cells located at the base of the crypt compartment. Deletion of EphB3 results in aberrant Paneth cell localization. Second, signaling through the Wnt pathway is required for Paneth cell maturation and their expression of antimicrobial peptides. Furthermore, the Wnt-dependent *SOX9* gene is critical for Paneth cell development, because mice with a conditional deletion of *Sox9* lack Paneth cells completely. Finally, Wnt signaling may have an effect on cell fate determination by impeding terminal differentiation of secretory cell lineages. TCF4$^{-/-}$ mice have goblet cells and enterocytes but lack enteroendocrine cells whereas transgenic overexpression of Dkk1, an inhibitor of Wnt signaling, ablates the secretory cell lineages while absorptive enterocytes remain normal.

4-2   Notch signaling determines cell lineage specification in the intestine.

Notch signaling, which controls cell fate decisions in many different tissues, also determines secretory (goblet, enteroendocrine, Paneth cells) versus absorptive cell lineage development in the intestinal epithelium (Figure 4.3). Activation of one of four Notch receptors by any one of several ligands, from either the Delta or Jagged/Serrate families, results in the proteolytic cleavage and liberation of the Notch intracellular domain from the plasma membrane by gamma-secretase. The Notch intracellular domain subsequently translocates into the nucleus where it forms a transcriptional activation complex with RBP-jk (also known as CSL), displacing histone deacetylase co-repressors and recruiting histone acetyltransferases, leading to transcriptional activation of Notch target genes such as hairy/enhancer of split (*HES*). HES1, a basic helix–loop–helix (bHLH) transcriptional repressor, inhibits the expression of another bHLH factor, ATOH1 (Math1 for mouse, Hath1 for human), suppressing secretory cell lineage differentiation in the intestinal epithelium and leading to disproportionate numbers of absorptive enterocytes. Indeed, the intestinal epithelium of Math1 knockout mice is populated only by absorptive enterocytes. By contrast, the inhibition of Notch signaling by CSL gene knockout, or with gamma-secretase inhibitors, induces Math1 expression, leading to the conversion of all epithelial cells into goblet cells. HES1 knockout mice, although embryonic lethal, show an increase in goblet, enteroendocrine, and Paneth cells, with decreased numbers of absorptive enterocytes. Likely downstream of Math1 is the zinc finger transcriptional repressor, Gfi1, which is involved in the generation of Paneth and goblet, but not enteroendocrine, cells. Interestingly, the loss of Kruppel-like factor 4 (Klf4), a zinc finger transcription factor that is repressed by Notch signaling, leads to reduced goblet cell numbers.

Figure 4.3 The Notch signaling pathway determines cell fate decisions in the intestinal epithelium. Activation of Delta, Jagged, or Serrate Notch receptors results in the release of the Notch intracellular domain by γ-secretase. Subsequent induction of HES1 expression, a basic helix–loop–helix (bHLH) transcriptional repressor, inhibits ATOH1, another bHLH factor, resulting in the suppression of secretory cell lineages and enhancement of absorptive enterocyte development. By contrast, inhibition of Notch signaling enhances the expression of ATOH1, resulting in the development of secretory cell lineages. Gfi1, a zinc finger transcriptional repressor, induces the expression of Spdef, an Ets-domain transcription factor that is important in the development of both goblet and Paneth cells. Similarly, inhibition of Notch leads to the induction of neurogenin 3, a bHLH transcription factor, promoting the development of enteroendocrine cells. HAT, histone acetyltransferase; HDAC, histone deacetylase; Ngn3, Neurogenin 3.

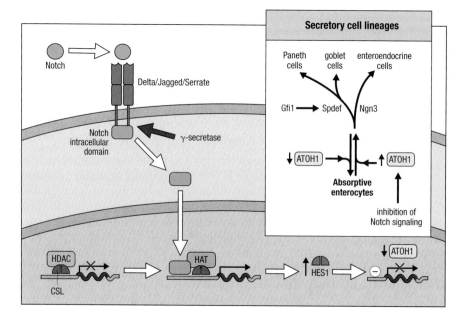

Spdef, an Ets-domain transcription factor and Notch pathway target gene downstream of Gfi1, plays an important role in the maturation of both goblet and Paneth cell lineages. Spdef is also the major driver of goblet cell phenotype in the respiratory tract. Spdef knockout mice show not only defects in the terminal differentiation of both goblet and Paneth cells but also alterations in the expression of a specific subset of Paneth cell genes including α-defensins, matrix metalloproteinase 7 (Mmp7), angiogenin-4 (Ang4), and kallikreins. In addition, Spdef induces the transcription of a suite of genes encoding proteins required for synthesis of the major components of the mucus barrier. Spdef regulates production of transcription factors such as Fox3A, mucin proteins, the endoplasmic reticulum (ER)-resident protein disulfide isomerase AGR2 required for correct mucin folding, multiple Golgi-resident glycosyltransferases required for complex mucin *O*-glycan assembly, proteins involved in mucin packaging and secretion, and several nonmucin proteins that are co-secreted with mucins.

Enteroendocrine cells account for approximately 1% of intestinal epithelial cells with about 15 different subtypes that express various enteric hormones. Downstream of the Notch–Math1 pathway is the bHLH transcription factor Neurogenin 3, which is required for development of the enteroendocrine cell lineage. Downstream of Neurogenin 3, additional bHLH and homeodomain transcription factors have been shown to be important for the development of specific enteroendocrine subtypes.

Secretory cells in mucosal epithelia.

Although the small intestinal epithelium contains differentiated lineages of specialized cells, virtually all mucosal epithelial cells are able to secrete water, ions, or macromolecules that contribute to luminal fluid and are relevant to barrier function. These epithelial cell products, as well as antimicrobial peptides produced by specialized cells in the epithelium, and mucin glycoproteins produced by goblet cells, form a barrier that limits contact between luminal microorganisms and the epithelium. Thus, nonimmune secretory cells play a critical role in host protection against the external environment.

4-3   Nonspecialized epithelial cells display remarkable plasticity.

The plasticity of certain mucosal cells enables such cells to markedly increase their secretory capabilities in response to specific stimulation. In the respiratory tract, for example, Clara cells were not recognized to be mucin-producing cells because they lack mucin storage granules. However, Clara cells continuously produce small amounts of respiratory mucins under steady-state conditions. During lung inflammation, particularly helper T cell (T_H2)-dominated inflammation, inflammatory cytokines induce goblet cell hyperplasia, that is, increased numbers of mucin-producing cells via proliferation, and transdifferentiation of Clara cells into a classical goblet cell phenotype that secretes large amounts of respiratory mucins characteristic of T_H2 lung inflammation. Thus, although we describe here the classical differentiated cells in mucosal tissues, it is important to recognize that secretory phenotypes of mucosal epithelial lineages can be modified by infection or inflammation.

4-4   Mucus-producing cells are abundant in the gastrointestinal epithelium.

The differentiated cells that produce large amounts of mucin glycoproteins are referred to as goblet cells (due to the goblet shape formed by stored mucin granules in a structure referred to as a theca) or mucus cells, depending on the location of the mucosal tissue. Ultrastructurally, these cells are characterized by copious peri-nuclear rough endoplasmic reticulum (ER) in the base of the cell, a well-developed supranuclear Golgi apparatus, large numbers of granules packed together occupying a large volume in the apical region of the cell, and reduced numbers of apical microvilli compared with adjacent non-goblet cells (Figure 4.4). In standard hematoxylin and eosin stained sections the stored mucin granules are unstained but they are revealed using carbohydrate stains such as periodic acid-Schiff or Alcian blue. In different tissues these cells are found within the surface epithelium, crypt structures, and/or submucosal glands. Mucin-producing cells are most abundant in the gastrointestinal tract, particularly in the stomach and colon where the mucus layer is thickest, but can be increased in abundance in all mucosal tissues, particularly in response to infection and inflammation.

4-5   Antimicrobial peptides are produced by specialized cells throughout the gastrointestinal tract.

All regions of the alimentary tract express antimicrobial peptides under certain conditions. Saliva delivers several antimicrobial peptides and proteins to the oral cavity, and the oral mucosa produces constitutive and inducible β-defensins. In addition, β-defensin expression occurs in the epithelium of the oropharynx, tongue, esophagus, stomach, and colon during inflammation and infection. In the small intestine, α-defensins secreted by Paneth cells constitute the major antimicrobial peptides. Antimicrobial peptides accumulate in Paneth cells within characteristic dense core secretory granules for subsequent release as components of regulated exocytotic pathways.

Biophysical features of the secreted mucus barrier.

The biophysical properties of mucus are central to its protection of mucosal surfaces. Mucus is a highly hydrated viscous secretion with a complex macromolecular constituency. The rheological properties of mucus vary between tissues and within tissues in response to different environmental and host-driven

Figure 4.4 Intestinal secretory cells. Crypt (panel a) and villous (panels b and c) epithelium of the mouse small intestine shows Paneth cells (Pc), goblet cells (gc), and intermediate cells (ic) containing granules of antimicrobial peptides (AMP), pronounced rough endoplasmic reticulum (RER), and stored mucin granules (mg) in the goblet cell theca (th). Toluidine blue stained semi-thin resin embedded sections (panels a and b) and transmission electron micrograph (panel c).

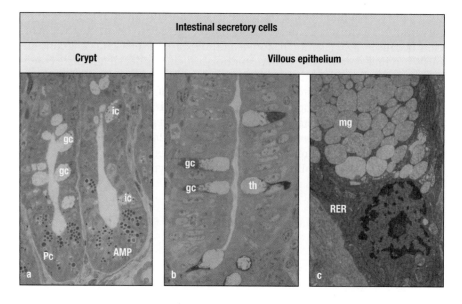

conditions. The major contributors to mucin viscosity are polymeric mucin glycoproteins. Although a barrier, mucus is permeable to macromolecules, as bidirectional diffusion of molecules through the mucus is frequently required. Nevertheless, thick mucus layers can establish concentration gradients that are functionally important, for example, in protecting the gastric mucosa from the low luminal pH. Similarly, mucus provides functional retention of some molecules, such as antimicrobial molecules and immunoglobulins. Retention in mucus is likely to increase the efficacy of these molecules in maintaining the barrier to infection. Many of these molecules may be retained in mucus due to noncovalent interactions with mucin glycoproteins, but this is poorly understood at present.

Molecules, and even viruses, can readily diffuse through mucus, although mucus is highly efficient at trapping particulate matter and cellular microbes. The surface properties of small particles determine their ability to diffuse through mucus, and binding to any component in mucus leads to highly efficient trapping. Integral to mucus physiology is the constant movement of mucus and its replacement by continuous secretion. In the gastrointestinal tract, the outer layer of mucus is moved by the flow of luminal contents, whereas in the lung, mucus is moved by the action of cilia, ensuring continuous clearing of the airway. In the lung, this system can be very rapidly increased in capacity by secretion of mucins by submucosal glands, such as after inhalation of particulate matter. Similarly, in the gut, the large number of goblet cells with stored mucin granules can be rapidly induced to secrete mucin granules in response to microbial, inflammatory, or neural signals.

4-6   Mucin glycoproteins have common structural features.

The secreted polymeric mucins that form mucus share structural similarities and most (MUC2, MUC5AC, MUC5B, and MUC6) are encoded by a family of genes in a cluster on chromosome 11, whereas the *MUC19* gene is on chromosome 12. These mucins have N- and C-terminal cysteine-rich domains that share homology with von Willebrand factor D domains and are involved in homo-oligomerization. Between these terminal domains are very large proline-, serine-, and threonine-rich domains that usually contain tandemly repeated sequences and are the scaffold for *O*-linked oligosaccharides.

The large mucin polypeptide chains are formed in the ER where the D domains are *N*-glycosylated and folded, and the C-termini are cross-linked via disulfide bonds to form dimers. The mucin dimers then move into the Golgi where the *O*-glycans are progressively assembled due to the action of Golgi-resident glycosyltransferases. Mass spectrometric analysis of mucin oligosaccharides shows that mucins are decorated by a diverse array of complex *O*-linked oligosaccharides and that these structures vary between species, within species between tissues/mucins, and within tissues between different conditions.

Following *O*-glycosylation, the dimers are further homo-oligomerized via N-terminal disulfide bond formation and packaged into granules for secretion. The true nature of the fully oligomerized mucin polymers remains clouded with conflicting data supporting the formation of linear polymers or more complex lattice-like structures. Although the details of packaging of the oligomerized mucin macromolecules into granules are not well understood, packaging involves association of Ca^{2+} ions with the negatively charged mucin oligosaccharides which allows H_2O to be excluded and the dense granules to form.

4-7   Mucin release is a tightly regulated biological process.

Secretion of mucin granules is a complex process involving movement of the granule to the apical cell surface, fusion with the cell membrane, and release of the granule contents. The process involves actin disruption and remodeling, Rab GTPases, and Munc proteins, which are essential mediators of exocytosis. In most mucosal tissues, both constitutive and inducible release of mucin granules occurs. Many different factors can promote mucin secretion, including the nucleotides ATP and UTP, which are potent mucin secretagogues via the $P2Y_2$-R purinoreceptor in the airway. The nervous system can also drive differential mucin release and, in the respiratory tract, differential innervation of submucosal glands versus the surface epithelium adds further regulation of mucin release. Once released into the luminal space, the granule contents are exposed to H_2O and the mucin complex rapidly expands as it is hydrated, increasing in volume by 100- to 1000-fold, indicating how such a large volume of mucus is maintained despite its continuous movement and degradation. Many other molecules are co-secreted with mucins into mucus, and proteomic approaches have revealed other host proteins present in mucus, some of which may bind to mucins and/or modulate the rheological properties or other functions of mucus.

The secreted mucins are expressed in a tightly regulated tissue-specific pattern. For example, MUC5B, MUC19, and the small nonpolymeric mucin MUC7 are the major constituents in saliva, and the stomach produces MUC6 deep in the glands and MUC5AC toward the surface. MUC5AC and MUC5B are the dominant respiratory tract mucins. In the duodenum, Brunner's glands produce MUC5AC, and the remainder of the goblet cells in the intestine produce mainly MUC2, although MUC6 is produced in the colon and MUC5AC can be produced during inflammation.

4-8   Antimicrobial peptides play a key role in mucosal defense.

Antimicrobial peptides collectively form a diverse population of gene-encoded protein effectors with selective microbicidal effects against bacteria and fungi. Varied epithelia release antimicrobial peptides onto mucosal surfaces, and evidence increasingly implicates them as components of a biochemical barrier against microbial challenges. Although antimicrobial peptide primary structures vary greatly (Figure 4.5), they have broad-spectrum microbicidal activities *in vitro*, generally at low micromolar concentrations, and most are ≤ 5 kDa, cationic at neutral pH, and amphipathic. Mammalian antimicrobial peptide structures range from linear, disordered peptides that form α-helices in membrane-mimetic hydrophobic environments to molecules such as defensins which consist of antiparallel β-sheet-containing peptides that are constrained by up to four disulfide bonds. Certain antimicrobial peptides may be expressed constitutively or they may be inducible by exposure to bacteria or microbial antigens. Most antimicrobial peptides kill their target cells by peptide-mediated membrane disruption, creating defects that dissipate cellular electrochemical gradients, leading to microbial cell death. Despite their diverse primary, secondary, and tertiary structures, their amphipathicity enables the peptides to interact with and disrupt microbial cell membranes. There are two major families of antimicrobial peptides in mammals: cathelicidins and defensins (Figure 4.5).

The cathelicidins are a family of highly diverse antimicrobial peptides. These peptides vary from cysteine-stabilized molecules (in swine and cattle), to linear peptides with unusually high arginine and proline content such as PR-39, to the tryptophan-rich tridecapeptide indolicidin (Figure 4.5). These highly varied primary and secondary structures derive from a C-terminal cationic antimicrobial peptide domain. Although cathelicidins exhibit remarkable diversity across phylogenetic lines, the evolutionary relationship among cathelicidins is evident in the cathelin domain, which is conserved in the proregions of cathelicidin precursors. Although swine and cattle neutrophils contain numerous divergent cathelicidins, humans express a single cathelicidin, hCAP-18, which

Defensins		
α-defensins	HNP-2	CYCRIPACIAGERRYGTCIYQGRLWAFCC
β-defensins	BNBD-4	QRVRNPQSCRWNMGVCIPFLCRVGMRQIGTCFGPRVPCCRR
θ-defensins	RTD-1	GFCRCLCRRGVCRCICTR

Cathelicidins		
human	LL-37	GKEFKRIVQRIKDFLRNLVPRTES
human	hCAP-18	GLRKRLRKFRNKIKEKLKKIGQKIQGLLPKLAPRTDY
mouse	mCRAMP	GLLRKGGEKIGEKLKKIGQKIKNFFQKLVPQPE
rat	rCRAMP	GLVRKGGEKFGEKLRKIGQKIKEFFQKLALEIEQ
cattle	Bac5b	RFRPPIRRPPIRPPFYPPFRPPIRPPIFPPIRPPFRPPLGPFP-NH2
cattle	Bac7	RRIRPRPPRLPRPRPLPFPRPGPRPIPRPLPFPRPGPRPLPFPRPGPRPIPRPL
cattle	Indolicidin	ILPWKWPWWPWRR-NH2
cattle	Dodecapeptide	RLCRIVVIRVCR
sheep	SMAP-29	RGLRRLGRKIAHGVKKYGPTVLRIIRIA-NH2
sheep	SMAP-34	GLFGRLRDSLQRGGQKILEKAERIWCKIKDIFR-NH2
horse	eCATH-2	KRRHWFPLSFQEFLEQLRRFRDQLPFP
horse	eCATH-3	KRFHSVGSLIQRHQQMIRDKSEATRHGIRIITRPKLLLAS
pig	PR-39	RRRPRPPYLPRPRPPPFFPPRLPPRIPPGFPPRFPPRFP-NH2
pig	Protegrin-1	RGGRLCYCRRRFCVCVGR-NH2
pig	PMAP-23	RIIDLLWRVRRPQKPKFVTVWVR
pig	PMAP-36	GRFRRLRKKTRKRLKKIGKVLKWIPPIVGSIPLGC-NH2

Figure 4.5 Representative members of the defensin and cathelicidin antimicrobial peptide families. To illustrate characteristic differences in the cysteine spacing in primary structures of the defensin subfamilies, HNP-2, a human neutrophil α-defensin, BNBD-4, a β-defensin from cattle neutrophils, and RTD-1, a θ-defensin from rhesus macaque neutrophils, are aligned in single-letter notation with Cys residues highlighted with red text. Because RTD-1 exists as a covalently closed macrocyclic molecule, assignment of its first residue position is arbitrary. The lower alignments are of cathelicidin peptides selected to illustrate the diversity of primary structures in the peptide family. Sequences correspond to characterized peptide products of exon 4 in the conserved peptide family genes.

is processed to functional LL-37; mice also produce only a single cathelicidin, CRAMP. LL-37 and CRAMP are expressed and inducible at mucosal surfaces. hCAP-18 is expressed both in neutrophils and in epithelial cells, and it is processed to the antimicrobial peptide LL-37 by neutrophil proteinase 3 after release of secondary granules. hCAP-18 is reported to be activated in female reproductive tract mucosa by seminal pepsin C following intercourse. The low vaginal pH activates pepsin C, enabling the processing of hCAP-18 molecule to ALL-38, the functional equivalent of LL-37. In mice, CRAMP is present in skin, eosinophils, salivary glands, and neutrophils. Disruption of mouse CRAMP results in susceptibility of mice to necrotic skin infection by Group A streptococci. In addition to its antimicrobial activities, human LL-37 also is chemotactic for neutrophils, monocytes, mast cells, and T cells; induces degranulation of mast cells; alters transcriptional responses in macrophages; and stimulates wound vascularization and reepithelialization of healing skin.

Defensins comprise three families of cationic, cysteine-rich antimicrobial peptides, the α-, β-, and θ-defensins (Figure 4.6). The α-defensins are major granule constituents of mammalian phagocytic leukocytes and secretory granules of intestinal Paneth cells. α-Defensins are microbicidal against Gram-positive and Gram-negative bacteria, certain fungi, spirochetes, protozoa, and enveloped viruses in *in vitro* assays. Accounting for 5–18% of total cellular protein in neutrophils, α-defensins are estimated to reach concentrations of ~10 mg ml^{-1} as granules dissolve in phagolysosomes following ingestion of microorganisms; Paneth cells secrete α-defensins at concentrations of 25–100 mg ml^{-1} at the point of release. The biosynthesis of α-defensins requires post-translational activation by lineage-specific proteinases. α-Defensins derive from inactive 10 kDa precursors that contain canonical signal sequences, electronegative proregions, and a 3.5–4 kDa mature α-defensin peptide in the C-terminal portion of the precursor, and α-defensin bactericidal activity is dependent on proteolytic conversion of inactive 8.4 kDa pro-α-defensins to active forms.

Figure 4.6 Structural features of defensin peptides. The general features of the three defensin peptide subfamilies are depicted. In each panel, the cysteine connectivities of the tridisulfide arrays are noted above the conserved cysteine residue positions of each subfamily displayed above ribbon diagrams of peptide solution structures. Panel a: rabbit kidney α-defensin RK-1. Panel b: human β-defensin-1. Panel c: rhesus macaque neutrophil θ-defensin RTD-1. (Reprinted from M. Selsted and A. Ouellette, *Nature Immunology* 6: 551–557, 2005. With permission from Macmillan Publishers Ltd.)

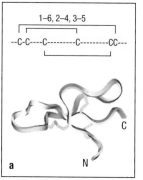
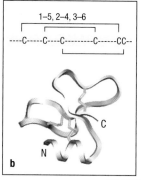
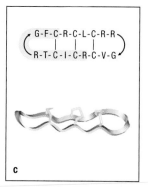

The details of human and mouse Paneth cell α-defensin processing differ in that human Paneth cells store unprocessed pro-α-defensin precursors that are processed coincident with or after secretion by trypsin, whereas mouse pro-α-defensins are activated intracellularly by Mmp7.

In humans, intestinal α-defensin-5 (HD5) is also expressed by female genital tract epithelia. HD5 mRNA is expressed in normal vagina, ectocervix, and by inflamed fallopian tube and in approximately 50% of normal endocervix, endometrium, and fallopian tube specimens. The HD5 peptide occurs in the upper half of the stratified squamous epithelium of the vagina and ectocervix, and immunoreactivity is more intense closer to the lumen. In positive endocervix, endometrium, and fallopian tube specimens, HD5 localizes to apically oriented granules and on the apical surface of some columnar epithelial cells, and HD5 peptide has been detected in cervico-vaginal lavages, and levels peak during the secretory phase of the menstrual cycle. Endometrial HD5 occurs at highest levels during the early secretory phase of the menstrual cycle.

The β-defensins, first identified in neutrophils and in airway and lingual epithelial cells of cattle, are expressed by a greater variety of epithelial cell types than are α-defensins, and the gene family extends evolutionarily to birds and reptiles. For example, β-defensin peptides or transcripts have been detected in human kidney, skin, pancreas, gingiva, tongue, esophagus, salivary gland, cornea, and airway epithelium and in epithelial cells of a number of species. Unlike α-defensins, which are components of dense core granules in a regulated pathway, β-defensins expressed by epithelial cells are released via the constitutive pathway.

Human β-defensin-1 is expressed in genital and reproductive epithelium. Constitutively high human β-defensin-1 mRNA has been localized to epithelial layers of the vagina, ectocervix, endocervix, uterus, and fallopian tubes. Intracellular storage of human β-defensins has not been detected, suggesting that peptide release occurs via the constitutive pathway. Thus, the human female genital tract expresses both α- and β-defensins that could contribute to mucosal innate immunity. Also, inducible β-defensin human β-defensin-3 and human β-defensin-4 mRNAs have been detected in human endometrium, and exogenous steroid hormones alter levels of both β-defensins *in vivo*. In primary endometrial epithelial cells and in two endometrial cell lines, mRNA for human β-defensin-1 and human β-defensin-2 was detected in primary endometrial epithelium. In mice, the homolog of human β-defensin-1, mouse β-defensin-1, is expressed in female reproductive epithelium, and mouse β-defensin-1 transcripts have been detected in kidney, liver, and female reproductive organ tissues, and mRNA for the inducible peptide mouse β-defensin-3 was detected in ovary.

The θ-defensins are unusual 2 kDa macrocyclic peptides found only in the neutrophils and monocytes of Old World monkeys. Like all defensins, they are stabilized by three disulfide bonds. θ-defensins assemble from two hemi-precursors that derive from α-defensin genes that have stop codons that terminate the peptide at residue position 12. The ligation mechanisms that circularize the closed θ-defensin polypeptide chain remain unknown.

Role of mucosal cell products in mucosal microbe homeostasis.

4-9   Microbes influence the composition and structure of the secreted mucosal barrier.

Intestinal goblet cells continue to be produced in germ-free-conditions. However, comparisons between normally exposed, germ-free, and germ-free conventionalized animals demonstrate decreased goblet cell number and reduced size of goblet cell thecae, together with a paradoxically thicker mucus layer, in the absence of microbes. The most likely explanations for this phenotype are that the microbial flora induces greater goblet cell production and greater mucin synthesis and that the mucus layer is thicker in germ-free animals because of substantially reduced mucin degradation. Mucosal epithelial cells increase production of cell-surface and secreted mucins after exposure to microbial products. Co-culture with probiotic bacteria increases expression of cell-surface mucins demonstrating how the host response to nonpathogenic bacteria may aid defense against pathogens. However, there are contrary examples where products of pathogens appear to reduce mucin synthesis, which is likely to facilitate infection. In addition to influencing mucin production, the microbial flora in the gut influences the glycosylation of mucins: conventionalization of gnotobiotic rodents and pigs results in substantial changes in mucin glycosylation, suggesting that the microbial flora either directly or indirectly influences expression of specific goblet cell glycosyltransferases. Nonmicrobial environmental factors can also influence expression of intestinal mucins. For example, dietary fructan (a prebiotic) alters goblet cell density, thecal size, and mucin glycosylation in rodents.

Two layers of mucus line the intestine: an inner layer that is resistant to removal and largely sterile, and an outer layer that can be readily removed and contains microbes (Figure 4.1). The presence of two layers is likely due to progressive degradation of mucin polymers; importantly, microbes colonize only the degraded mucus. Certain bacteria in the intestinal environment are mucolytic and can use mucin glycoproteins as an energy source. These bacteria are typically strict anaerobes, nonpathogenic, and do not encroach into the inner mucus layer. Combinations of multiple mucolytic bacteria result in more complete degradation of human MUC2 and more bacterial growth *in vitro*, suggesting that these bacteria collectively degrade mucins and utilize them as an energy source. Factors that prevent bacteria from penetrating the inner mucus layer likely include the higher pO_2, a higher concentration of antimicrobial molecules, and the continuous secretion of mucus by the epithelium. The thickness of the mucus layer varies throughout the gastrointestinal tract with a thinner layer in the small intestine than in the stomach and colon. The thinner mucus layer in the intestine may facilitate diffusion. In the respiratory tract the mucus appears to lie over a thin aqueous layer covering the apical surface of the epithelial cells. Cilia beat back and forth in this layer, propelling the overlying mucus and any trapped microbes out of the airways.

In contrast to the strong influence of the microbiota on goblet cell and mucin biology, Paneth cell ontogeny is not dependent on luminal bacteria, reflected in the development of the lineage in germ-free mice and under aseptic conditions. Human Paneth cells appear in the first trimester of gestation and begin to express α-defensins coincident with the appearance of their differentiated morphology as early as 13.5 weeks of gestation. During postnatal crypt ontogeny in mice, newly induced Paneth cell gene products influence or respond to the microflora. In adult germ-free and conventionally reared mice, α-defensins, *Pla2g2a*, Mmp7, and lysozyme levels are similar. However, some Paneth cell granule constituents are inducible in response to microbial colonization or to changes in the microflora, at least in mice monocolonized with the anaerobe *Bacteroides thetaiotaomicron*, which have increased abundance of bactericidal ribonuclease Ang4 and the C-type lectin RegIIIγ compared with germ-free animals.

4-10 The secreted mucosal barrier influences the composition of the gut microbiome.

Intestinal microbes can utilize mucins as an energy source, thereby influencing colonization. This was recently explored experimentally in germ-free mice with mono-association studies using *B. thetaiotaomicron*. These bacteria can feed on complex carbohydrates supplied in the diet or utilize mucin *O*-glycans for energy, switching to greater utilization of mucins when dietary sources are low. The bacterial genes involved in utilizing mucin *O*-glycans have been identified, and when disrupted the bacteria have diminished capacity to colonize the gut. Thus, some gut microbes depend on both host-derived and dietary factors. Further study will elucidate the relationship between the microbial community of the gut and host mucin glycosylation and production.

α-Defensins secreted by Paneth cells influence the composition of the small intestinal microbiome in mice, apparently by selecting peptide-tolerant microbial species in the microbial ecosystem. Such selection has been demonstrated in two mouse strains genetically modified to alter Paneth cell α-defensin production: the Tg-HD5 mouse, which expresses a human minigene in Paneth cells at levels comparable to those of endogenous mouse peptides, and the *Mmp7* null mouse, which is defective in the intracellular activation of pro-α-defensins in Paneth cells. Analysis of bacterial phyla and groups in the distal small intestine of *Mmp7*$^{-/-}$ mice and Tg-HD5 mice (which express HD5 in addition to murine Paneth cell α-defensins) showed that Firmicutes represented approximately 63% of the microbial composition in *Mmp7*$^{-/-}$ mice but only 26% in Tg-HD5 transgenic mice. In contrast, the percentage of Bacteroidetes was only 18% in the *Mmp7*$^{-/-}$ mouse but 69% in the Tg-HD5 mouse distal small bowel. Thus, the small bowel microflora of defensin-deficient mice compared with that of defensin-complemented mice was genotype dependent and showed reciprocal differences. Because a resident microflora is essential in inflammatory bowel disease pathogenesis, the quantity and/or the repertoire of Paneth cell α-defensins and other gene products may influence gastrointestinal inflammation by shaping a healthy microbiome or one that promotes disease in genetically predisposed individuals.

4-11 Mucins and Paneth cell products contribute to protection against infectious pathogens.

The central role of mucins during infection is reflected in changes in expression of multiple mucins during infection. In addition to forming a viscous biophysical barrier, mucin glycoproteins act as a barrier to microbes by providing carbohydrate decoy ligands, by having direct antimicrobial activity, and by facilitating retention of other antimicrobial molecules. Mucins display a complex array of *O*-linked glycans that reproduce many of the glycans present in the glycocalyx on the apical surface of mucosal epithelia. Pathogenic microbes that bind to the mucosal surface have evolved ligands for molecules in the glycocalyx and thus the secreted mucins act as decoys for these microbial adhesins, facilitating entrapment of microbes within mucus. Mice lacking the Muc2 intestinal mucin develop spontaneous inflammation and intestinal cancers in the absence of pathogens, indicating that the mucus barrier is vital for preventing normal microbial flora from causing pathology. In addition, deficient $Muc2^{-/-}$ mice develop severe, usually fatal, pathology when infected with the attaching and effacing pathogen *Citrobacter rodentium*.

A family of transmembrane mucins highly expressed on the apical surface of mucosal epithelial cells have large, filamentous, heavily glycosylated, extracellular domains that can be released from the cell after engagement by bacteria. These mucins may represent a 'final' defense against microbes that have penetrated the mucus barrier and reached the mucosal epithelial cells. Mice lacking individual cell-surface mucins appear more susceptible to infection by bacterial pathogens such as *Helicobacter pylori* and *Campylobacter jejuni*. Consistent with these findings, polymorphisms in the *MUC1* cell-surface mucin gene have been linked with susceptibility to *H. pylori* infection and the development of gastritis and gastric cancer.

In addition to binding microbes, some mucin oligosaccharides have direct antimicrobial activity or can bind to other antimicrobial molecules. The gastric mucin oligosaccharide, α1-4-linked *N*-acetylglucosamine, inhibits synthesis of *H. pylori* cell-wall components and may limit *H. pylori* expansion in gastric mucus. The MUC7 mucin, a major component of saliva, has an N-terminal histatin domain with direct candidacidal activity and thus acts to limit growth of yeast in saliva. Mucins also directly bind antimicrobial molecules. For example, MUC7 binds statherin and histatin-1, and the other major mucin in saliva, MUC5B, binds histatin-1, -3, and -5 and statherin, retaining the antimicrobial molecules in the mucus.

α-Defensins contribute to innate mucosal protection in the small intestine. MMP7 gene disruption ablates processing of pro-α-defensins to their microbicidal form, resulting in impaired clearance of orally administered *Escherichia coli* and *Salmonella enterica* serovar Typhimurium (*Salmonella* Typhimurium). Conversely, the addition of human α-defensin to Paneth cell secretions augments innate defense in mice reflected in the ability of the Tg-HD5 mice to resist infection when challenged by *Salmonella* Typhimurium. Resistance is likely due to increased killing of the bacteria, or possibly more complex HD5-related interactions in the intestinal lumen.

In mice that lack adaptor MyD88, small intestinal expression of RegIIIγ, RegIIIβ, CRP-ductin, and resistin-like molecule-β (RELM-β) fails to be induced by the normal flora or by oral bacterial infection with *Listeria monocytogenes*. In this model, cell-autonomous Paneth cell MyD88 activation, not activation in cells of myeloid origin, mediates signaling to maintain intestinal homeostasis to the infectious challenge. Also, oral infection of C57BL/6 mice with *Toxoplasma gondii* induced Toll-like receptor 9 (Tlr9) and type I interferon (IFN) mRNA, and elevated levels of selected Paneth cell α-defensin mRNAs. In $Tlr9^{-/-}$ mice, the responses to *T. gondii* were replicated with IFN-β administration, and *T. gondii*-induced effects were eliminated in mice null for type I interferon receptors. The *T. gondii* effects are mediated by Tlr9-dependent production of type I IFNs, but whether the induction of certain Paneth cell α-defensin mRNAs is a direct effect on Paneth cells or a paracrine response to IFN is not known.

Regulation of the secreted mucosal barrier by innate and adaptive immunity.

Goblet cell differentiation, mucin glycosylation, and production rates of both secreted and cell-surface mucins are regulated in response to infection and inflammation. For example, release of T_H2 cytokines in response to parasitic infections is associated with goblet cell hyperplasia in both the intestine and lung. Responses are not restricted to T_H2 cytokines, however, as T_H1 and T_H17 cytokines also upregulate goblet cell mucin production. Indeed, the ability of an array of cytokines and cell products including interleukin (IL)-1β, IL-4, IL-6, IL-9, IL-13, interferons, tumor necrosis factor-α, nitric oxide, and granulocyte proteases to upregulate mucin secretion indicates that goblet cells are an integral component of innate immunity and inflammatory responses in the mucosa.

Paneth cells release secretory granules in a dose-dependent manner in response to bacterial antigens and pharmacologic agents. Paneth cell secretion induced by bacterial and carbamyl choline antigens is regulated by cytosolic Ca^{2+} mobilized from intracellular stores and an influx of extracellular Ca^{2+}. Inhibiting influx of Ca^{2+} by selective blockers of the Ca^{2+}-activated intermediate conductance K^+ channel KCa3.1 (also known as Kcnn4 and mIKCa1) attenuates Paneth cell secretion. KCa3.1 is expressed in Paneth cells and T lymphocytes, and enables Ca^{2+} influx to sustain the Paneth cell secretory response. However, whether defects in KCa3.1 impair Paneth cell secretion sufficiently to diminish levels of luminal α-defensins and other secreted components *in vivo* is not known.

Defects in mucosal barrier secretion and the pathogenesis of disease.

Goblet cell pathology, particularly ER stress-related pathology, has been implicated in intestinal disease in murine models. $Muc2^{-/-}$ - and $Agr2^{-/-}$ -deficient mice switch off intestinal goblet cell mucin biosynthesis. Aberrant mucin biosynthesis results in accumulation of misfolded Muc2, ER stress, reduced Muc2 biosynthesis, premature goblet cell death, and spontaneous T_H17-dominated inflammation, leading to intestinal inflammation. Defects in elements of the unfolded protein response genes that are activated following misfolding, including *Xbp1*, *Ire1β*, *Mbpts1*, and *Chop*, also result in spontaneous inflammation or altered susceptibility to environmentally triggered inflammation. Information regarding goblet cell pathology in human disease is limited even though loss of goblet cells, smaller goblet cell thecae, and reduced thickness of the mucus barrier are hallmarks of ulcerative colitis.

Normally, Paneth cells are anatomically restricted to the small intestine. During conditions of inflammation, as in Barrett's esophagus, *H. pylori* gastritis, Crohn's disease, and ulcerative colitis, Paneth cells may appear ectopically. The so-called intermediate or granulo-goblet cells (Figure 4.4) are rare in healthy gut mucosa but may appear during episodes of inflammation. Reflecting the common progenitor of goblet cell and Paneth cell lineages, these cells produce mucins and contain small electron-dense cytoplasmic inclusions that are positive for Paneth cell-specific markers. Thus, pro-inflammatory conditions of diverse origin may induce the appearance of numerous intermediate cells that produce secretory proteins characteristic of both lineages.

A reduction in luminal α-defensins has been identified in cohorts of Crohn's disease patients. Such a reduction could impact the composition of resident microorganisms, promoting a more pro-inflammatory microbial population. In this connection, mouse models of cystic fibrosis have demonstrated undissolved Paneth cell secretory granules in mucus-occluded crypts, possibly contributing to the associated overgrowth of Enterobacteriaceae.

In premature infants with necrotizing enterocolitis, a leading cause of morbidity and mortality in neonates, Paneth cells appear to have reduced MD-2 expression. Premature infants appear to have a similar impaired expression of MD-2. Thus, the absence of a functional Paneth cell lipopolysaccharide-sensing apparatus in immature neonatal gut may predispose preterm newborns to necrotizing enterocolitis as the immature intestine is challenged by enteric bacteria.

Genetic defects in secretory cell pathology may become apparent under specific environmental conditions. For example, the Paneth cell phenotype in Atg16l1 hypomorphic mice is co-dependent on the genetic defect and infection by a particular strain of murine norovirus. Also, severe disruption of the unfolded protein response by deletion of *Xbp1* in epithelial cells induces massive Paneth cell apoptosis and fulminant ileitis. Germline and inducible gene deletions of *Agr2* result in goblet cell Muc2 deficiency, expansion of the Paneth cell compartment, and accumulation of intermediate cells (discussed above). Severe terminal ileitis and colitis are associated with ER stress induced by loss of Agr2. Thus, a diversity of genetic defects can disrupt Paneth cell function in mice, resulting in an associated ileitis that mimics adult human disease. In humans, familial and sporadic interstitial lung disease has been linked to misfolding mutations in the *SFTPC* gene that encodes the SP-C surfactant protein produced by lung epithelial cells. Interestingly, viral infection appears to precipitate the emergence of the phenotype in this disease, possibly through increased SP-C production and ER stress.

Summary.

Nonstructural components of the mucosal barrier, including mucus produced by goblet cells and antimicrobial peptides produced by Paneth cells, play a fundamental role in mucosal protection. Goblet and Paneth cells arise from stem-cell progenitors through intricate developmental pathways involving both Wnt and Notch signaling. Mucus, produced by goblet cells, protects the intestinal mucosal surface through its biophysical properties as well as its effects on luminal bacteria. The composition of mucus includes various mucins that are complex glycoproteins. Paneth cell products include antimicrobial peptides such as defensins, which influence the composition of the gut microbiota. Together, mucins and antimicrobial peptides form a secreted mucosal barrier that provides protection against an array of microbes. Components of both the innate and adaptive mucosal immune systems contribute to goblet cell and Paneth cell regulation in response to specific environmental challenges. Defects in either goblet or Paneth cells can lead to a disruption of the secreted mucosal barrier and thus contribute to the pathogenesis of mucosal diseases such as inflammatory bowel disease and necrotizing enterocolitis.

Further Reading.

Barker, N., van Es, J.H., Kuipers, J., *et al.*: **Identification of stem cells in small intestine and colon by marker gene Lgr5.** *Nature* 2007, **449**:1003–1007.

Davis, C.W., and Dickey, B.F.: **Regulated airway goblet cell mucin secretion.** *Annu. Rev. Physiol.* 2008, **70**:487–512.

Linden, S.K., Sutton, P., Karlsson, N.G., *et al.*: **Mucins in the mucosal barrier to infection.** *Mucosal Immunol.* 2008, **1**:183–197.

McGuckin, M.A., Eri, R., Simms, L.A., *et al.*: **Intestinal barrier dysfunction in inflammatory bowel diseases.** *Inflamm. Bowel Dis.* 2008, **15**:100–113.

Mukherjee, S., Vaishnava, S., and Hooper, L.V.: **Multi-layered regulation of intestinal antimicrobial defense.** *Cell. Mol. Life Sci.* 2008, **65**:3019–3027.

Porter, E.M., Bevins, C.L., Ghosh, D., *et al.*: **The multifaceted Paneth cell.** *Cell. Mol. Life Sci.* 2002, **59**:156–170.

Salzman, N.H., Hung, K., Haribhai, D., *et al.*: **Enteric defensins are essential regulators of intestinal microbial ecology.** *Nat. Immunol.* 2010, **11**:76–83.

Selsted, M.E., Brown, D.M., DeLange, R.J., *et al.*: **Primary structures of MCP-1 and MCP-2, natural peptide antibiotics of rabbit lung macrophages.** *J. Biol. Chem.* 1983, **258**:14485–14489.

Selsted, M.E., and Ouellette, A.J.: **Mammalian defensins in the antimicrobial immune response.** *Nat. Immunol.* 2005, **6**:551–557.

Thornton, D.J., Rousseau, K., and McGuckin, M.A.: **Structure and function of the polymeric mucins in airways mucus.** *Annu. Rev. Physiol.* 2008, **50**:5.1–5.28.

Tomasinsig, L., and Zanetti, M.: **The cathelicidins—structure, function and evolution.** *Curr. Protein Pept. Sci.* 2005, **6**:23–34.

van der Flier, L.G., and Clevers, H.: **Stem cells, self-renewal, and differentiation in the intestinal epithelium.** *Annu. Rev. Physiol.* 2009, **71**:241–260.

van Es, J.H., van Gijn, M.E., Riccio, O., *et al.*: **Notch/gamma-secretase inhibition turns proliferative cells in intestinal crypts and adenomas into goblet cells.** *Nature* 2005, **435**:959–963.

Yang, Q., Bermingham, N.A., Finegold, M.J., *et al.*: **Requirement of Math1 for secretory cell lineage commitment in the mouse intestine.** *Science* 2001, **294**:2155–2158.

PART II

CELLULAR CONSTITUENTS OF MUCOSAL IMMUNE SYSTEMS AND THEIR FUNCTION IN MUCOSAL HOMEOSTASIS

Immune Function of Epithelial Cells

5

Epithelial cells lining mucosal surfaces essentially define the border between the environment and ourselves. At some mucosal surfaces, the epithelial border is only one cell thick, interfacing with the environment over huge surface areas (200 m² for intestine and lung). These vast and delicate epithelial barriers, typified by the intestinal, respiratory, and urogenital mucosa, have to distinguish between components of the outside world and selectively transport huge quantities of those essential for life to the lamina propria or systemic circulation. At the same time, the epithelial cell must also defend against invasion and absorption of unwanted, toxic, and pathogenic molecules and microbes, as well as producing and secreting some factors required to provide for such host defense. From this pivotal position, the epithelial cell communicates dynamically with a multitude of targets: cell types immediately underlying the epithelial barrier, including immunocompetent cells of the lamina propria and those comprising the mucosal immune system; cell types at a distance that affect systemic immunity, metabolism, and organ function; and molecules and microbes, in the luminal space of the outside world. Therefore the mucosal immune system cannot be understood without first understanding the epithelial cell and its developmental and molecular biology that underlie mucosal physiology and immunology. This chapter provides a brief introduction to these fields.

Barrier function.

The epithelial barrier of mucosal surfaces separates a wealth of foreign antigen that consists of the commensal flora and ingested food, on the luminal side, and the mostly sterile environment harboring the mucosal immune system on the other side. This barrier function, which is as much physiologic as it is structural, is set up to efficiently defer any single incoming pathogen. Remarkably, the as yet only rudimentarily characterized commensal microbial communities associated with each mucosal surface and molecular components of the luminal (outside) space are required for the development of both the mucosal and systemic immune systems, the establishment of the epithelial barrier itself, and repair after the injury associated with inflammation.

5-1 Stratified squamous and simple columnar epithelia form the major types of epithelial structures.

Each of the 'wet' epithelial surfaces that line mucosal tissues has the common feature of separating the internal sterile milieus of the host from the external environment. Because each of the different mucosal surfaces serves different functions (such as gas exchange in the lungs, or absorption of nutrient solutes in the intestine) and is uniquely colonized with microbes, the associated epithelium is highly adapted and differentiated by and within organ systems. For clarity in this chapter, we will use the small intestinal mucosa as a paradigm for understanding all mucosal surfaces, because the intestine is perhaps the best-characterized and interfaces most intensely with the microbial world.

In general, there are two broad histological types of epithelia that line mucosal surfaces: stratified squamous epithelia and simple columnar epithelia (Figure 5.1). Stratified squamous epithelia line mucosal surfaces associated with the vagina, urethra, rectal canal, mouth, esophagus, pharynx, and the nasal surfaces. Most often, the stratified squamous mucosa immediately abuts the skin, or is located close to it, and these surfaces do not participate in large-scale selective transport of materials necessary for life. The stratified squamous epithelium that is associated with mucosal tissues is distinguished from the squamous epithelium of the skin by the fact that the former is non-cornified. The mucosal surfaces that do provide for solute, ion, water, and gas transport are typically lined by a delicate single layer of simple columnar or cuboidal epithelial cells. All epithelial cells lining mucosal surfaces are variously colonized with commensal bacteria of different types and densities depending on the organ system and the location within that system. They have the common properties of both resisting the invasion of pathogens and the promiscuous uptake of antigen, regulating the composition of luminal microbes and associated immune cells, and permitting the regulated sampling of antigenic materials for interpretation by the immune system.

5-2   A polarized simple epithelium permits vectorial transport.

Simple columnar and cuboidal epithelial cells forming the intestine are structurally and functionally polarized, with a specialized apical membrane facing the mucosal surface that abuts the outside environment and a specialized basolateral membrane facing the lamina propria and the internal environment of the host. The two membranes are composed of different protein and lipid compositions and structural features; the visually distinct microvilli of the apical membrane exemplify this best. Such polarity of structure and function is required for vectorial transport of solutes and water. The two membranes are continuous but are physically separated by a 'fence,' the apical tight junction that prevents mixing of apical and basolateral lipids and proteins.

Figure 5.1 Simple and stratified epithelia form the barrier between the outside world and self. The architecture of epithelial cells has evolved according to the specific requirements of the body locale. Similarly, dendritic cells localize to places specific for the relevant locales and allow one specific route of antigen sampling, a process that is controlled by various factors, in particular chemokine receptor signaling. O-MALT, organized mucosa-associated lymphoid tissue. (Adapted from P. Ogra et al. (eds): Mucosal Immunology, 2nd ed, New York: Academic Press, 1999.)

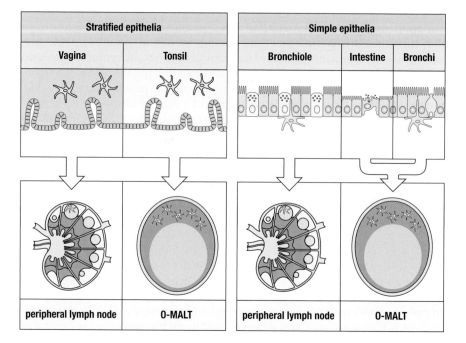

Transport of membranes and cargo between the two surfaces occurs by vesicular traffic in a process termed transcytosis. Such transcellular transport of solutes, membranes, and cargo from one side, through the cell to the other side, is termed the *transcellular pathway*. Movement of solutes and proteins via the transcellular pathway can be harnessed to cellular energy; this is how the intestine efficiently absorbs most nutrients against often very steep electrochemical gradients.

As with most eukaryotic cells, the epithelial cells lining mucosal surfaces maintain a dynamic endocytic network that continuously internalizes membranes and cargo from both sides of the cell, often in a regulated manner. The two apical and basolateral endosomal networks are distinct, and can recycle membranes and proteins back to the cell surface of origin; however, some cargoes are sorted to a common endosome that receives and processes membranes from both sides of the cell. The common endosome can then sort and deliver cargo to the degradative lysosome (relevant to antigen processing), to the Golgi apparatus and endoplasmic reticulum (ER) via the retrograde pathways (relevant to autophagy, innate sensing by intracellular Toll-like receptors (TLRs), and ER stress), or to the contralateral cell surface via transcytosis (relevant to the transport of intact proteins and particles for sensing by the mucosal immune system, antigen presentation, or innate pattern recognition). The transcytotic pathway is one way in which luminal contents can breach the epithelial barrier to enter the host, and it is the pathway for the secretion of some intact proteins, such as secretory IgA and IgG going in the opposite direction from basal to apical.

5-3   Epithelial cells form a paracellular barrier via intercellular junctions.

The other pathway for movement across epithelial barriers is to go around the cell, through the intercellular junctions that join adjacent cells together. This is termed the *paracellular pathway*. Movement of solutes and water through the paracellular pathway can only be passive. There is no way to harness cell energy directly to this process, and paracellular transport is completely dependent on concentration gradients across the barrier epithelium, or indirectly harnessed to osmotic or electrical gradients established by other activities of the epithelial cell.

Intercellular junctions between adjacent epithelial cells are of course required for barrier function. Intercellular junctions are located along the lateral surfaces of the epithelial cells. They are composed of the most apically oriented tight junctions, and the subjacent adherens junction and desmosome (Figure 5.2). The tight junctions are the rate-limiting barrier for solute and particle transport. Tight junctions are physiologically dynamic and regulate the access of small solutes, water, and macromolecules into the paracellular space. They typically exclude molecules with a Stokes radius of greater than 9–11 Å, and they are also charge (anion) selective. Some tight junctions can exclude molecules the size of mannitol, and some can exclude water and even protons. The structure and function of the tight junctions are established by the claudin, junctional adhesion molecule, and occludin families of proteins, together with their molecular connections to signaling molecules, the cytoskeleton, and the endocytic/vesicular trafficking machinery. The claudins and occludins are transmembrane proteins that essentially form the 'molecular gasket' sealing the space between cells. The permeability of the tight junction is dynamic and determined by the isoforms and quantity of claudins and occludins structurally composing the junctional complex, and by the large number of associated intracellular structural and signaling molecules (such as zonula occludens-1 and -2), which link the junction to the actin cytoskeleton. The tight junctions depend on the establishment and maintenance of the adherens junctions (the cadherins are the primary component) and desmosomes (the cadherin-like

Figure 5.2 Intercellular junctions between adjacent epithelial cells. Panel a: a transmission electron micrograph showing junctional complexes between two villous enterocytes. The tight junction (TJ) is just below the microvilli (Mv), followed by the adherens junction (AJ). The desmosomes (D) are located basolaterally. Panel b: a freeze–fracture electron micrograph showing apical microvilli (Mv) and tight junction strands (TJ) in a cultured intestinal epithelial cell. Panel c: interactions between F-actin, myosin II, zonula occludens-1 (ZO-1), claudins, tight-junction-associated MARVEL proteins (TAMPs), and immunoglobulin superfamily members such as junctional adhesion molecules (JAM) and the Coxsackie–adenovirus receptor (CAR). (From L. Shen et al.: *Annu. Rev. Physiol.*, 73:11.1–11.27, 2011.)

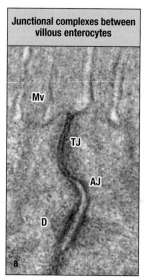
Junctional complexes between villous enterocytes

Apical microvilli and tight junction strands

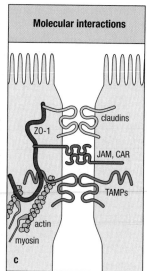
Molecular interactions

desmoglein and desmocollin proteins are the primary component). Both are also linked and regulated by the cytoskeleton (actin and cytokeratin respectively), but neither forms a significant barrier to diffusion.

Intercellular junctions are under the control of a variety of different host, dietary, and microbial factors. Cytokines that are commonly considered to have regulatory properties such as interleukin (IL)-10 and transforming growth factor-β enhance tight junction function and hence diminish paracellular permeability. In contrast, inflammatory cytokines such as IL-4, IL-13, tumor necrosis factor (TNF)-α, and interferon-γ degrade tight junctions and enhance paracellular permeability, probably through changes in claudin composition. IL-13, for example, can change the composition of claudin isoforms of the tight junction complex, thus altering the functional characteristics and barrier specificities. Dietary factors such as glucose can increase permeability by activating the actin cytoskeleton through effects on myosin light chain kinase. Microbial factors, such as repeats in toxin (RTX) from *Vibrio cholerae*, or the *Bacteroides fragilis* toxin fragilysin, may also increase paracellular permeability by affecting the regulation or structure of tight junction components respectively. Desmosomes are similarly regulated by a variety of host factors, and all junctions are affected by disruption during the process of neutrophil transmigration during inflammation. Similar factors can also affect cell polarity and the endocytic and transcellular pathways of membrane and cargo transport.

5-4 Various types of differentiated epithelial cells are derived from an epithelial stem cell.

Epithelial surfaces are rapidly self-renewing and are shed on a regular basis. This is potentially beneficial both in terms of defense against invading pathogens and for the avoidance of neoplasia given the harsh environment in which they exist. Epithelial cell renewal depends on long-lived stem cells, from which all lineages of epithelial cells are derived. In the intestines, the common stem cell resides at the base of the crypt and is characterized by the expression of specific markers such as Lgr5 and prominin-1 (Figure 5.3). Under the control of specific transcription factors, individual epithelial cell lineages differentiate from the common stem cell. In the small intestine, the common stem cell develops into four different lineages that migrate along the crypt villus axis; these are epithelial cells, goblet cells, Paneth cells, and enteroendocrine cells (see Figure 5.3). Epithelial cells are a major component of innate and adaptive

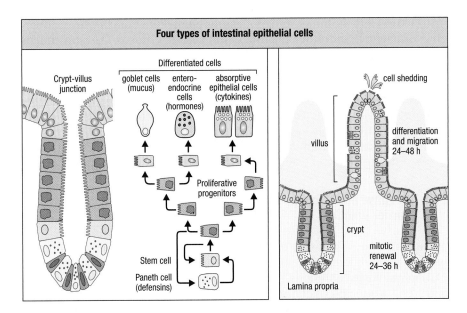

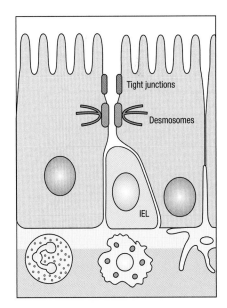

Figure 5.3 Intestinal epithelial cells arise from a common intestinal stem cell. The common stem cell is located at the base of small intestinal crypts, and gives rise to epithelial cells, goblet cells, neuroendocrine cells, and Paneth cells. The last of these form the niche for intestinal epithelial stem cells. (From Freddy Radtke and Hans Clevers, *Science* 307:1904–1909, 2005. With permission from AAAS.)

cellular immunity. Goblet cells and Paneth cells contribute significantly to innate immunity through the secretion of mucins and antimicrobial peptides respectively. Enteroendocrine cells secrete neurohumoral factors such as substance P, which are immunologically active. As they are highly secretory cells, goblet cells and Paneth cells are highly dependent on the interrelated processes of autophagy (self-eating), which arises as a consequence of intracellular nutrient deprivation and the accumulation of abnormal intracellular organelles, and the unfolded protein response, which arises secondarily to the occurrence of misfolded proteins and the consequent elicitation of ER stress.

5-5 Nonimmunologic and innate factors contribute to epithelial barrier function.

The function of mucosal barriers is also provided by the factors that they secrete into the lumen, as well as by cell types that are associated with the epithelium, such as intraepithelial lymphocytes (IELs), macrophages, neutrophils, and dendritic cells (Figure 5.4). This section discusses each of these and emphasizes the barrier as a physiologic property rather than simply a structural function.

The intestinal epithelium constitutively secretes, or is induced to secrete, inflammatory mediators and, upon recognition of microbial components such as microbe-associated molecular patterns, a variety of host-protective soluble factors. These include those with antimicrobial activity, such as lactoferrin, lysozyme, peroxidase, cathelicidin, α-defensins, and β-defensins, which are largely provided by epithelial cells (β-defensins) and Paneth cells (α-defensins). These factors serve to regulate the composition of the commensal microbiota and protect the host from pathogenic invasion. The secretion of antimicrobial peptides is a major common property of all body surfaces. Epithelial cells also secrete a variety of different complement components into the lumen; consistent with this, epithelial cells express CD55 (decay accelerating factor), which protects the epithelium from inappropriate complement deposition. In other regions of the intestinal tract, a variety of other nonspecific factors serve a role in host defense, such as gastric acid from the stomach, bile salts from the biliary system, and urea in the urogenital system.

Another major component of the nonspecific barrier is the mucus layer, which is demonstrable in all mucosal surfaces. The mucosal barrier has both cell-associated (glycocalyx) and secreted components. The glycocalyx can extend

Figure 5.4 Myeloid and lymphoid cell types contribute to epithelial barrier function. Epithelial cells directly and indirectly interact closely with intraepithelial lymphocytes (IEL), neutrophils, macrophages, and dendritic cells.

up to 500 nm into the lumen and protects the epithelium from the direct adherence of bacteria and other microbes. The secreted mucus layer (or mucus blanket) exhibits variable depths along the cephalo-caudal axis of mucosal tissues. Its structure is formed primarily by goblet cells and further modified by the enzymatic activity of the commensal microbiota. Through peristaltic activity (intestines) and ciliary function (respiratory tract), the mucus blanket is continuously renewed. The mucus layer includes not only a variety of mucin protein isoforms but also associated molecules such as intestinal trefoil factors. Mucins are important to homeostasis, because a loss of mucin-2 (*Muc2)* or intestinal trefoil factor leads to susceptibility to intestinal inflammation.

5-6   Adaptive immunologic factors contribute to epithelial barrier function.

All human and nonhuman mucosal secretions contain variable quantities of a variety of different immunoglobulin (Ig) isotypes under physiologic and pathophysiologic conditions. As this topic is discussed in detail in Chapters 8 and 9, only a brief discussion will be provided here. All Ig classes are potentially secretory, derive from mucosal lamina propria plasma cells, and are delivered into the secretions by either nonspecific or specific mechanisms. Nonspecific mechanisms can occur under homeostatic conditions, but are more common during inflammation with disruption of the intestinal barrier. Specific mechanisms are those associated with transporting receptors that are capable of specific binding and transcytosis across the polarized simple epithelium; these are necessary given the exclusivity of the paracellular barrier (see above). Whereas the polymeric Ig receptor transports secretory IgA and IgM, with IgA being the dominant Ig in mucosal secretions in the basal-to-apical direction, the neonatal Fc receptor (FcRn) is responsible for the bidirectional transport of IgG. IgD is commonly found in association with tonsillar secretions, although the transporting receptor has not yet been defined. Secretory IgE is commonly observed during parasitic and allergic inflammation, and is transported by CD23. Mucosal Igs serve to prevent microbial interactions with the immune system and/or regulate the uptake of cognate antigens through association as immune complexes and transport by their transcytosing receptors (see below).

5-7   Inappropriate barrier function leads to mucosal pathology.

Regulation of the physical epithelial barrier is necessary for the maintenance of homeostasis, the normal composition of the mucosa-associated lymphoid tissue, and the avoidance of inflammation. In animal models, disruption of junctional adhesion molecules or cadherins is associated with changes in the morphology of the lamina propria and inflammation, demonstrating the importance of these cellular structures. $\alpha_E\beta_7$ (expressed on IELs) is the receptor for cadherins on epithelial cells, which may link the latter to β-catenin and potentially to adenoma formation. Such disruptions of the physical barrier, which may be genetically determined, are possibly predisposing factors for the development of chronic intestinal inflammation as observed in inflammatory bowel disease. Proper nuclear factor κB (NFκB) signaling is required for maintaining the integrity of the epithelium and the functional linkage of the epithelium with subjacent immune cells (see below).

Normal physiologic epithelial barrier function is also necessary for intestinal homeostasis. Signals emanating from the microbiota and detected by cell-surface (TLRs) and intracellular (nucleotide-binding oligomerization domain; NOD) pattern recognition receptors regulate antimicrobial peptide secretion. Consequently, loss of this activity may lead to alterations in the commensal microbiota and susceptibility to pathogens and pathobionts. A similar impairment in the ability to respond to ER stress through elicitation of a proper unfolded protein response may also lead to alterations in antimicrobial peptide production and susceptibility to intestinal inflammation.

Antigen uptake and presentation.

The functional aspect of epithelial cell barrier activity occurs at many different levels. Epithelial cells function as the sentinel cells to detect the presence of foreign antigen and, although still insufficiently understood, contribute to the immune hyporesponsiveness of the gut immune system. Epithelial cells interact with lymphocytes that abut against their basolateral cell surface, the intestinal intraepithelial lymphocyte (IEL), as well as lymphocytes beneath the basement membrane—lamina propria lymphocytes—which stay in contact via soluble mediators as well as physically via basolateral projections of the epithelium through pores in the basement membrane. In a similar manner, epithelial cells are in contact with lamina propria macrophages, dendritic cells (DCs), and stromal cells. Notably, these interactions are two-way interactions with the epithelial cells instructing genuine immune cells and vice versa, as will be outlined in this chapter (Figure 5.5).

5-8   Three major mechanisms account for regulated antigen uptake across the epithelium.

As discussed above, the physicality of the epithelial cell barrier is formed by the polarity of a single layer of epithelial cells and their intercellular junctions, both of which are subject to a variety of homeostatic and inflammatory signals that alter composition and thus function. Inappropriate or dysregulated entry of antigens into the lamina propria may be associated with intestinal pathology (breakdown of barrier function), but intact macromolecules are also often detected within the bloodstream of a seemingly normal epithelium.

There are three known regulated pathways for antigen uptake by the epithelium (see Figure 5.5). The first is that which is associated with M cells and the lymphoepithelial structures of Peyer's patches and isolated lymphoid follicles (covered in Chapter 13). A second pathway is by transcytosis and is associated

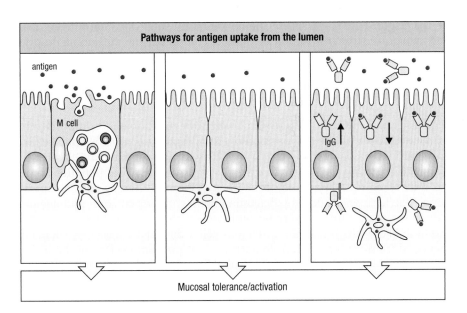

Figure 5.5 Epithelial cells sample luminal antigens through three general mechanisms. Specialized epithelial cells (M cells) sample antigen and present them to antigen-presenting cells localized immediately underneath (left panel). Dendritic cells extend dendrites in between adjacent epithelial cells and thereby gain direct access to luminal contents (center panel). The neonatal Fc receptor (FcRn) permits the bidirectional transport of IgG and thereby permits the retrieval of immune complexes from the lumen (right panel).

with the ability of transporting Ig receptors within epithelial cells such as FcRn to retrieve and deliver IgG:antigen complexes to professional antigen-presenting cells (APCs) in the lamina propria. Such antigen-laden APCs can migrate to regional lymphoid structures such as mesenteric lymph nodes and distant lymphoid tissues such as the liver and spleen. Through this mechanism, antigen exposures in the lumen can be linked to adaptive immune responses in systemic compartments. Finally, DCs that reside adjacent to the simple epithelium are capable of extending pseudopods containing tight junction proteins between epithelial cells in a chemokine-regulated pathway (such as CXCR3) for the purpose of retrieving luminal microbes and antigen.

As examples of one of the processes of antigen retrieval over mucosal barriers discussed above, IgG:antigen complexes can be efficiently and bidirectionally transported from the apical to the basolateral side of epithelial cells, and, more importantly, basolateral IgG can retrieve antigen from the apical side as an immune complex that leaves the antigen intact, in an FcRn-dependent mechanism. Experimentally translating this observation *in vivo*, it was demonstrated that the basolateral to apical transport of IgG into the lumen is dependent on FcRn, and that antigen:IgG complexes form in the lumen as a consequence of such FcRn-dependent IgG transport. In turn, the transepithelially ported antigen:IgG complexes stimulate specific CD4+ T cells in regional lymphoid structures such as the mesenteric lymph node. Further dissection of this mechanism revealed that intestinal lumen-derived antigen:IgG complexes can be transported into the lamina propria by FcRn and can be taken up by DCs, which in turn present antigen to CD4+ T cells after they migrate to the regional lymphoid structures. Hence, luminal antigen can specifically be transported across the epithelial surface in an FcRn-dependent manner; this has been applied experimentally to the development of vaccines such as for herpesvirus infection (see Chapter 9).

The relative contributions of the various transepithelial antigen transport mechanisms (FcRn, M cells, DC dendrite extension) have not yet been quantitatively assessed, and each of the mechanisms has been shown *in vivo* to have a role in the steady state as well as in experimental mucosal infection. With regard to the latter, the role of FcRn was addressed in an infectious model involving *Citrobacter rodentium*. *C. rodentium* is normally restricted in its localization to the epithelium, and its eradication is highly dependent on CD4+ T-cell responses and IgG. FcRn selectively expressed in the intestinal epithelium renders mice less susceptible to *C. rodentium* infection in comparison with FcRn$^{-/-}$ mice. Anti-*C. rodentium* IgG protected from infection via FcRn expressed selectively on epithelial cells, and affected antigen-specific CD4+ T-cell responses to *C. rodentium*. Remarkably, intestinal bacterial antigens can be transported from the lumen as immune complexes into the lamina propria by FcRn and received by CD11c+ DCs.

5-9   The epithelium exerts important innate immune functions.

The intestinal epithelium exerts innate immune function and is important in directly controlling the composition of the intestinal microbiota. Specifically, recognition of bacterial patterns via TLRs and downstream MyD88 signaling within Paneth cells regulates the release of bactericidal peptides (such as α-defensins and cryptdins) from Paneth cells and thereby controls intestinal barrier penetration by bacteria, both commensal and pathogenic. Moreover, intracellular pattern recognition receptors such as NOD2 have also been linked to the regulation of Paneth-cell α-defensin release. α-Defensins released from Paneth cells profoundly control the composition of the intestinal microbiota. It is further noteworthy that genetic deficiency in CD1d (a non-polymorphic major histocompatibility complex (MHC) class I homolog presenting lipid antigens to CD1d-restricted natural killer T (NKT) cells; see below) alters the ultrastructure of Paneth-cell granules in the germ-free state, and these Paneth cells exhibit a degranulation defect after bacterial colonization of these germ-free hosts. Through this mechanism involving Paneth cells, CD1d lipid antigen presentation to NKT cells controls the composition of the intestinal microbiota, which is particularly exemplified by an up to 10^9 difference in colony-forming units after the colonization of germ-free mice with specific commensal bacteria. Two further interrelated mechanisms that also control Paneth-cell granule morphogenesis and antibacterial function, and hence the innate immune functions of the epithelium, are autophagy and ER stress. However, innate immune functions of the epithelium are not confined to the Paneth cells. Goblet cells release resistin-like molecule-β, a molecule structurally linked to the adipocytokine resistin, which protects against worm infection. Specifically, resistin-like molecule-β not only propagates worm expulsion by acting on the host side via an IL-4-dependent pathway, but also by direct toxicity toward the worm through inhibition of the worm's ability to ingest nutrients.

5-10 The epithelium can present antigen via classical and nonclassical antigen-processing and antigen-presentation pathways.

Epithelial cells constitutively express complex (MHC) class I molecules and are capable of presenting endogenous antigens, in the context of MHC class I, for defense against infections and neoplasia. Endogenous antigens from epithelial cells can also be captured by local dendritic cells and cross-presented on MHC class I for priming of CD8⁺ T cells in draining lymph nodes. Epithelial cells also express MHC class II molecules, and do so preferentially in the small intestine. In response to pro-inflammatory stimuli, in particular interferon-γ (IFN-γ), MHC II molecules are upregulated on epithelial cells. It has also been demonstrated that IFN-γ causes antigens to be sorted into MHC class II-positive late endosomes with presentation basolaterally in epithelial cell model systems.

Epithelial cells express not only classical MHC class I and II molecules, but notably also non-polymorphic MHC class I related molecules such as CD1d, MR1, MHC class I chain-related gene A/B (MICA/B), UL-16 binding protein, and human leukocyte antigen E (HLA-E). CD1d expression on epithelial cells is regulated by IFN-γ and the heat-shock protein HSP-110 (in humans; Hsp105 in mice), which is derived in large quantities by the epithelium itself and found in rather high concentrations in the luminal content of the proximal small intestine and hence the feces of humans and mice. Antibody cross-linking of CD1d on human epithelial cell CD1d transfectants induces immediate phosphorylation of the CD1d tail and subsequent secretion of IL-10, which in turn protects epithelial cells from IFN-γ-induced leakage of the epithelial cell barrier function. Although the physiologic ligand of such retrograde CD1d signaling has not yet been established experimentally, one might predict that the T-cell receptor (TCR) on NKT cells recognizing its cognate glycolipid antigen on epithelial-cell-expressed CD1d on the basal surface might serve such a barrier protective function through the regulation of IL-10 and HSP-110 secretion. CD1d-restricted lipid antigen presentation to NKT cells serves important physiological functions, but the actual lipid antigens responsible for these functions have not yet been identified. CD1d-restricted NKT cells may either express an invariant TCR-α chain (these are therefore termed invariant NKT cells), which pairs with multiple TCR-β chains, and respond to an exogenous model glycolipid derived from marine sponge (α-galactosylceramide), or they may express a diverse set of TCR-α chains that respond to a less well characterized group of lipid antigens.

In a similar manner to the NKT cell population that is selected by CD1d, another non-polymorphic MHC class I-related molecule, MR1, selects another T-cell population that bears the canonical TCR-α chains hVα7.2-Jα33 or mVα19-Jα33. These T cells are preferentially localized to the gut lamina propria of humans and mice respectively, and are therefore genuine mucosa-associated invariant T cells. Although the selection and expansion of these cells is dependent on MR1 expressed on B cells, it is notable that MR1 expression is found on epithelial cells as well.

5-11   The epithelium can provide co-stimulatory and tertiary cytokine signals to lymphocytes.

However, in contrast to professional APCs, epithelial cells lack the expression of the classical co-stimulatory molecules CD80 and CD86, but do express novel members of the B7 family, including inducible co-stimulatory ligand (ICOS) and programmed death-1 ligand (B7-H1). Further molecules that are relevant for epithelial cell–lymphoid cross-talk and are constitutively expressed on epithelial cells are leukocyte function-associated antigen-3 (LFA-3; CD58), E-cadherin, the IL-7 receptor (IL-7R), carcinoembryonic antigen-related adhesion molecule 1 (CEACAM1), and CEACAM5 (CEA) as well as the common γ chain, which is required for signaling through the IL-2R, IL-4R, IL-7R, IL-9R, and IL-15R. CD86 may be expressed on epithelial cells during the inflammation associated with inflammatory bowel disease.

Furthermore, either constitutively or under certain conditions of specific stimulation, epithelial cells secrete multiple chemokines, cytokines, and other mediators relevant for orchestrating and regulating mucosal immune responses. Chemokines have a critical role in both homeostatic and inflammatory recruitment of myeloid cells, including dendritic cells, lymphocytes, and neutrophils. One excellent example of such chemokine-mediated recruitment of DCs to epithelial cells is CCL20 (MIP-3α) originating from epithelial cells, which recruits CCR6-expressing DCs to the subepithelial dome region of Peyer's patches just underneath M cells. In a *Salmonella* Typhimurium infection model, it has been demonstrated that pathogen-specific T-cell responses are first activated in the Peyer's patch, and that DCs are required for this activation and expansion. CCR6+ DCs are restricted to the Peyer's patches, and are rapidly recruited into the follicle-associated epithelium in this infection model. CCR6 expression on these DCs is required for the activation and expansion of *Salmonella*-specific T cells. Similarly, the CCR6 ligand CCL20 is expressed by the intestinal epithelium on exposure to *Salmonella*, indicating that a chemokine signal from the epithelium orchestrates the adaptive immune response via the recruitment and activation of DCs at mucosal surfaces and hence the activation of pathogen-specific T-cell responses. Another chemokine ligand–receptor pair implicated in the homeostatic recruitment of another subset of DCs is CX3CR1 and its ligand CX3CL1, expressed in epithelial cells. The chemokine receptor CX3CR1 on DCs is required for the extension of their processes between epithelial cells, allowing these DCs direct access to luminal contents and hence antigen uptake that circumvents the requirement for transepithelial transport of antigen (see above). The CX3CR1 ligand CX3CL1/fractalkine is a transmembrane chemokine expressed at the surface of epithelial cells and endothelial cells in the intestine. Although these CX3CR1+ DCs are not required for T-cell activation in the aforementioned *Salmonella* infection model, it has been speculated that these resident CX3CR1+ DCs may be involved in the steady-state antigen acquisition of luminal contents that contributes to tolerance induction under steady-state conditions. In line with this is the fact that orally administered labeled *Escherichia coli* can be cultured from mesenteric lymph nodes in wild-type mice but not in CX3CR1$^{-/-}$ mice, implicating CX3CR1 in the transport of commensals from the lumen to mesenteric lymph nodes via DCs. At the same time, similar numbers of *E. coli* can be recovered from Peyer's patches, implicating an unimpaired function of M cell-dependent transport mechanisms. Additional chemokines secreted by epithelial cells are IL-8 and the related molecules ENA-78, MCP-1, GRO-α, and GRO-β, but also RANTES among others.

Regulation of immune responses.

As we have seen, epithelial cells are capable of exhibiting innate and adaptive immune functions. In this section we discuss how the epithelium can integrate these functions into the fabric of the mucosa-associated lymphoid tissue (MALT). Specifically, the epithelium is strategically located as a sentry between the lumen with its associated commensal microbiota and the abluminal compartments, which are filled with the cellular components of the MALT. Thus, the polarized and stratified squamous epithelium can function as first responders to the environment (microbial and nonmicrobial) and consequently direct subsequent immunologic events and/or participate in responding to signals from abluminal immune and nonimmune cells. This section discusses these concepts.

5-12 Epithelial cells act as central organizers of immune responses.

Insight obtained over the past few years has revealed a central role of epithelial cells in orchestrating the mucosal immune response (Figures 5.6 and 5.7). Mediators derived from epithelial cells can direct the type of immune response by acting distally on DCs, with major consequences for T-cell and B-cell differentiation. Thymic stromal lymphopoietin (TSLP) derived from epithelial cells is an instructive example of such a mediator. These concepts were initially envisioned on the basis of data derived from *in vitro* experiments involving epithelial cell lines and DCs, and have been proven and further developed in mice with genetic targeting of specific genes in the intestinal epithelium by using Cre/*loxP* technology. Specifically, TSLP expression in epithelial cells is induced by the microbial flora of the intestine in a pathway that depends on NFκB signaling (in particular IKK2). TSLP binds to the TSLP receptor (TSLP-R) on DCs and renders them 'tolerogenic' in that they are induced to produce high levels of IL-10 and IL-6, but not IL-12. Remarkably, epithelial cells genetically lacking *Ikbkb* (encoding IKK2) exhibit decreased TSLP expression, and such mice are unable to eradicate the worm *Trichuris muris* as a consequence of a failure to develop T$_H$2 immunity in a TSLP-dependent pathway. Instead, these IKK2- and TSLP-R-deficient mice develop a pathological helper T cell response characterized by increased IFN-γ and IL-17 secretion, accompanied by increased TNF-α and IL-12/23p40 expression from DCs. Instead of eradicating *Trichuris muris*, these mice develop severe intestinal inflammation. Further underscoring the intricate network dictated by TSLP secretion from epithelial cells is the fact that blockade of IFN-γ and IL-12/23p40 signaling results in the restoration of resistin-like molecule-β expression from epithelial cells and the consequent expulsion of the worm. Further evidence for a key role of NFκB signaling in epithelial cells and the organization of mucosal immune responses comes from mice with a genetic deletion of NFκB essential modulator (NEMO), which acts upstream of IKK1 and IKK2. In those mice, epithelial NEMO deletion results in a MyD88- and TNF-α-dependent induction of severe inflammation of the large intestine. Interestingly, during the early stages of colonic inflammation in this model, innate immune cells dominate, whereas at later stages lymphocytes are predominantly involved.

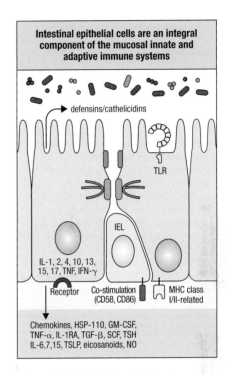

Intestinal epithelial cells are an integral component of the mucosal innate and adaptive immune systems

defensins/cathelicidins

TLR

IEL

IL-1, 2, 4, 10, 13, 15, 17, TNF, IFN-γ

Receptor   Co-stimulation (CD58, CD86)   MHC class I/II-related

Chemokines, HSP-110, GM-CSF, TNF-α, IL-1RA, TGF-β, SCF, TSH IL-6,7,15, TSLP, eicosanoids, NO

Figure 5.6 Epithelial cells direct numerous components of the innate and adaptive immune systems at various levels. Various mediators derived from myeloid, lymphoid, and stromal compartments act on epithelial cells and affect diverse biological processes including barrier function, the processing and presentation of antigens, the expression of MHC and co-stimulatory molecules, and the secretion of antimicrobial peptides. At the same time, epithelial cells secrete, and respond to, various mediators that profoundly affect the function of surrounding mucosal cell types. IFN-γ, interferon-γ; NO, nitric oxide; HSP-110, heat-shock protein 110; GM-CSF, granulocyte-macrophage colony-stimulating factor; TNF-α, tumor necrosis factor-α; IL-1RA, interleukin-1 receptor antagonist; TGF-β, transforming growth factor-beta; SCF, stem cell factor; TSH, thyroid-stimulating hormone; IL-6, interleukin-6; TSLP, thymic stromal lymphopoietin; TLR, Toll-like receptor; IEL, intraepithelial lymphocyte.

Epithelial cells not only determine cell fate decisions in mucosal T cells via their actions on DCs; they also regulate IgG/IgA class switching in the B-cell compartment at mucosal surfaces. Epithelial cells lining tonsillar crypts form pockets that contain B cells actively performing class switching at that locale. When these epithelial cells sense microbial components via TLRs, they release soluble mediators such as B-cell activating factor (BAFF, encoded by *TNFSF13B*) that induce class switching through the induction of activation-induced cytidine deaminase (AID). This epithelial-cell-induced class switching is amplified by TSLP by means of a route that involves the 'licensing' function of DCs. Remarkably, epithelial cells also secrete an inhibitor of this process, secretory leukocyte protease inhibitor, which inhibits AID function in B cells. A similar process is operative in the intestine, where bacteria induce the cytokine APRIL (a proliferation inducing ligand, encoded by *TNFSF13*) in the epithelium and thereby induce T-cell independent IgA and IgG class switching.

5-13 Epithelial cells act as bystanders and participants in immune responses and inflammation.

Although the functions described in the previous paragraphs have focused on mechanisms in which epithelial cells seem to be the central coordinators of mucosal immune functions, these central epithelial cell functions are influenced by various stimuli from outside compartments, be it from the microbial milieu or host-derived factors. In this section we discuss factors for which this outside influence in regulating epithelial cell function seems more dominant, and in which the epithelial cells therefore fulfill a contributory role in mucosal physiology. However, such bystander or participant functions of epithelial cells can nevertheless become a central feature of pathophysiological processes in the mucosa.

As discussed in other chapters, pro-inflammatory cytokines such as IFN-γ can upregulate MHC class II expression, and also upregulate the expression of co-stimulatory molecules such as CD86, which are not expressed in the baseline state. Epithelial cells express a wide variety of receptors for pro-inflammatory and anti-inflammatory cytokines, including IL-1, IL-2, IL-4, IL-10, IL-13, IL-15, IL-17, TNF-α, and IFN-γ, all of which may be expressed by various myeloid, lymphoid, and nonimmune cells in the mucosal lamina propria. As examples, IL-13, TNF-α, and IFN-γ can adversely affect epithelial cell barrier function via actions on tight junction expression, whereas anti-inflammatory IL-10 can counteract this. Although IELs may contribute to the maintenance of epithelial cell barrier function in various ways, they can also secrete IFN-γ and thereby render the epithelial cell barrier leaky by means of direct effects on

Figure 5.7 Multiple inputs at the various levels of the epithelial barrier affect the mucosal immune compartment. Epithelial cells relay various signals emanating from the mucosal environment, including signals from the microbiota and immune cells. At the same time, epithelial cells act as central organizers or participants for multiple local and systemic immune functions and the regulation of microbial composition. IFN-γ, interferon-γ; TNF-α, tumor necrosis factor-α; APRIL, a proliferation inducing ligand; BAFF, B cell-activating factor; EGF, epidermal growth factor; FcRn, neonatal Fc receptor; KGF, keratinocyte growth factor; MCP-1, mast-cell precursor 1; PAMPs, pathogen-associated molecular patterns; TSLP, thymic stromal lymphopoietin; TLR, Toll-like receptor; PRR, pattern recognition receptor; NOD, nucleotide-binding oligomerization domain. (Adapted from A. Kaser, S. Zeissig, and R. Blumberg: *Annu. Rev. Immunol.*, 28:573–621, 2010. With permission from Annual Reviews.)

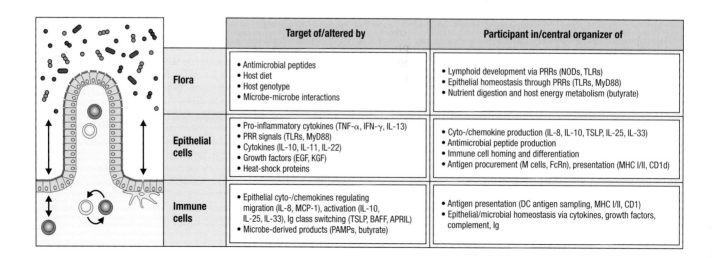

		Target of/altered by	Participant in/central organizer of
	Flora	• Antimicrobial peptides • Host diet • Host genotype • Microbe–microbe interactions	• Lymphoid development via PRRs (NODs, TLRs) • Epithelial homeostasis through PRRs (TLRs, MyD88) • Nutrient digestion and host energy metabolism (butyrate)
	Epithelial cells	• Pro-inflammatory cytokines (TNF-α, IFN-γ, IL-13) • PRR signals (TLRs, MyD88) • Cytokines (IL-10, IL-11, IL-22) • Growth factors (EGF, KGF) • Heat-shock proteins	• Cyto-/chemokine production (IL-8, IL-10, TSLP, IL-25, IL-33) • Antimicrobial peptide production • Immune cell homing and differentiation • Antigen procurement (M cells, FcRn), presentation (MHC I/II, CD1d)
	Immune cells	• Epithelial cyto-/chemokines regulating migration (IL-8, MCP-1), activation (IL-10, IL-25, IL-33), Ig class switching (TSLP, BAFF, APRIL) • Microbe-derived products (PAMPs, butyrate)	• Antigen presentation (DC antigen sampling, MHC I/II, CD1) • Epithelial/microbial homeostasis via cytokines, growth factors, complement, Ig

epithelial cells, a function of IELs that may be enhanced even further via their granzyme and FasL expression.

Moreover, epithelial cells might themselves secrete a large panel of inflammatory mediators after receiving inductive signals from other immune cells, and thereby can amplify an evolving mucosal immune response. A particularly relevant mechanism involves the induction of neutrophil- and lymphocyte-recruiting chemokines in epithelial cells.

In addition to the contribution of epithelial cells to the recruitment of the inflammatory infiltrate, epithelial cells also respond to inflammatory mediators with increased chloride and mucus secretion, as well as with an increase in paracellular permeability. The clinical correlate of these alterations is the occurrence of diarrhea. Immune cell-secreted mediators that increase chloride secretion and mucus production are serotonin, prostaglandin E_2, leukotriene C_4, histamine, and vasoactive intestinal peptide, whereas IFN-γ and TNF-α have an important role in increasing paracellular permeability.

$\gamma\delta$ T cells constitute an important subset of IELs, which have a role in IEL regeneration as reported in elegant experiments using TCR$\delta^{-/-}$ mice that were exposed to barrier-disrupting dextran sulfate sodium (DSS). TCR$\delta^{-/-}$ mice were more susceptible to DSS colitis, and it was suggested that this increase was due to a lack of keratinocyte growth factor secretion from $\gamma\delta$ T cells, because keratinocyte growth factor promotes compensatory proliferation of the injured epithelium.

Cross-talk between epithelial cell and nonepithelial cell compartments further contributes to the regenerative response of the epithelium. Various pattern recognition receptors, prominently TLRs and NOD proteins, are expressed in the mucosa in both epithelial cells and professional APCs. The microbial ligands recognized by TLRs are not unique to pathogens, and are produced by both pathogenic and commensal microorganisms. Commensal bacteria are recognized by TLRs under normal steady-state conditions; this interaction has a crucial role in the maintenance of intestinal epithelial homeostasis, and is also important for protection against gut injury during DSS colitis. Remarkably, this colonic epithelial regenerative response involves colonic epithelial progenitors, and requires the gut microbiota because it is markedly reduced in germ-free animals. It seems that it is not TLR signaling in epithelial cells itself that mediates this regenerative response; instead, TLR ligation on mucosal macrophages relays a signal to epithelial cells that is required for this regenerative response of the epithelial compartment. During the regenerative response, macrophages in the pericryptal stem cell niche extend processes to directly contact colonic epithelial progenitors near the crypt base. Cyclooxygenase-2 (COX2) expression and function are required for the aforementioned regenerative response after DSS colitis, and these function downstream of MyD88 signaling. Notably, COX2 expression is not regulated by injury and is present in MyD88$^{+/+}$ and MyD88$^{-/-}$ mesenchymal cells. However, in MyD88$^{+/+}$ mice, DSS injury leads to the repositioning of COX2-expressing stromal cells from the mesenchyme surrounding the middle and upper part of crypts to an area surrounding the crypt base adjacent to colonic epithelial progenitors.

A further factor critically involved in the regenerative response and wound healing of the epithelium is IL-22. IL-22 expression is induced in innate and adaptive immune cells, specifically a subtype of natural killer (NK)-like cells, as well as T_H17 helper T cells. IL-22-expressing NK-like cells, which are part of an increasingly recognized class of innate lymphocytes, are specifically found in tonsils, cryptopatches, and Peyer's patches; and the microbial flora is important for their induction. Remarkably, the transcription factor RORγt has a central role in the differentiation of both this subtype of mucosal NK cells and T_H17 cells. IL-22 induces the activation of STAT3 in epithelial cells and orchestrates a pro-regenerative, wound-healing program.

Intriguing insight into the importance of epithelial cell–IEL cross-talk also comes from mechanisms established in celiac disease, a disease precipitated by the exogenous antigen gluten. In the presence of gluten, epithelial cells secrete IL-15 and express nonclassical MHC class I molecules such as HLA-E and MICA. As a consequence of IL-15 secretion, natural killer receptors are upregulated on IELs such as the activating receptor NKG2D, which in turn leads to IEL activation through recognition of nonclassical MHC class I molecules (MICA) on the epithelial cell. Hence, low-affinity self-antigens lead to IEL activation as a result of a reduced activation threshold, and exhibit NK-cell-like activity, thereby destroying epithelial cells via cytolysis and IFN-γ secretion.

Summary.

Epithelial cells not only function as a physical barrier between the sterile environment of the host and the accumulation of foreign material including microbial constituents, but in fact they are also at the center of manifold immune functions of the mucosa. These involve antigen uptake and presentation, T-cell and NKT-cell interaction and reciprocal instruction, dendritic cell recruitment, and determination of the DC and T-cell response and the overall inflammatory milieu of the mucosa. Finally, epithelial cells have a central role in regulating the composition of the commensal microbiota and resisting pathogenic invasions by microbes.

Further Reading.

Baker, K., Qiao, S.W., Kuo, T., *et al.*: **Immune and non-immune functions of the (not so) neonatal Fc receptor, FcRn.** *Semin. Immunopathol.* 2009, **31**:223–236.

Cerutti, A.: **The regulation of IgA class switching.** *Nat. Rev. Immunol.* 2008, **8**:421–434.

Hansen, J., Gulati, A., and Sartor, R.B.: **The role of mucosal immunity and host genetics in defining intestinal commensal bacteria.** *Curr. Opin. Gastroenterol.* 2010, **26**:564–571.

Hill, D.A., and Artis, D.: **Intestinal bacteria and the regulation of immune cell homeostasis.** *Annu. Rev. Immunol.* 2010, **28**:623–667.

Hooper, L.V., and Macpherson, A.J. **Immune adaptations that maintain homeostasis with the intestinal microbiota.** *Nat. Rev. Immunol.* 2010, **10**:159-169.

Johansson, M.E., Larsson, J.M., and Hansson, G.C. **The two mucus layers of colon are organized by the MUC2 mucin, whereas the outer layer is a legislator of host-microbial interactions.** *Proc. Natl. Acad. Sci. USA.* 2011, **108**:Suppl 1, 4659-4665.

Kaser, A., Zeissig, S., and Blumberg, R.S.: **Inflammatory bowel disease.** *Annu. Rev. Immunol.* 2010, **28**:573–621.

Kim, M., Ashida, H., Ogawa, M., *et al.*: **Bacterial interactions with the host epithelium.** *Cell Host Microbe* 2010, **8**:20–35.

Marchiando, A.M., Graham, W.V., and Turner, J.R.: **Epithelial barriers in homeostasis and disease.** *Annu. Rev. Pathol.* 2010, **5**:119–144.

Pasparakis, M.: **Regulation of tissue homeostasis by NF-κB signalling: implications for inflammatory diseases.** *Nat. Rev. Immunol.* 2009, **9**:778–788.

Pitman, R.S., and Blumberg, R.S.: **First line of defense: the role of the intestinal epithelium as an active component of the mucosal immune system.** *J. Gastroenterol.* 2000, **35**:805–814.

Proud, D., and Leigh, R.. **Epithelial cells and airway diseases.** *Immunol. Rev.* 2011 **242**:186-204.

Rescigno, M. **The intestinal epithelial barrier in the control of homeostasis and immunity.** *Trends Immunol.* **32**:256-264.

Turner, J.R.: **Intestinal mucosal barrier function in health and disease.** *Nat. Rev. Immunol.* 2009, **9**:799–809.

van der Flier, L.G., and Clevers, H.: **Stem cells, self-renewal, and differentiation in the intestinal epithelium.** *Annu. Rev. Physiol.* 2009, **71**:241–260.

Wells, J.M., Rossi, O., Meijerink, M., and van Baarlen, P. **Epithelial crosstalk at the microbiota-mucosal interface.** *Proc. Natl. Acad. Sci. USA.* 2011, **108**:Suppl. 1, 4607-4614.

Intraepithelial Lymphocytes: Unusual T Cells at Epithelial Surfaces

6

Intraepithelial lymphocytes (IELs) are small mononuclear cells interspersed between epithelial cells. They are found in various epithelial tissues such as the skin and the epithelium of the small and large intestine, the biliary tract, the oral cavity, the lung, the upper respiratory tract, and the reproductive tract. The largest population of IELs resides within the columnar epithelial layer of the intestine (Figure 6.1), the single layer of cells that lines the lumen of the intestine and which forms the physical barrier between the core of the body and the environment. Complex innate and adaptive immune networks have developed at this prominent interface and IELs are critical in maintaining intestinal homeostasis (Figure 6.1). Although protection against pathogens is one of the major tasks of the mucosal immune system, it also needs to carry out these functions without excessive or unnecessary inflammation and tissue damage in order to preserve the integrity of the mucosal barrier. Furthermore, the intestine harbors trillions of microorganisms that establish a symbiotic relationship with their host. The fundamental role of the intestine as a digestive organ adds food to the antigenic load in the gut. Because of their location at this critical interface, IELs need to balance protective immunity while safeguarding the integrity of this major barrier, a *sine qua non* for the homeostasis of the organism.

Uniqueness and heterogeneity of mucosal IELs.

Gut IELs are notably heterogeneous in phenotype and function and are distributed differently in the epithelium of the small and large intestine. The latter is probably a reflection of the distinct digestive functions and the unique adaptations to fight infections as well as maintain tolerance toward innocuous antigens derived from the diet (mainly in the small intestine) or resident non-invasive commensals (mainly in the large intestine).

Despite this, intestinal IELs share unique phenotypical and functional characteristics that distinguish them from systemic lymphocytes in the periphery. First, gut IELs are almost exclusively T cells and there are more T cells residing within the gut epithelium than there are in other tissues. B cells are rare and only present in specialized epithelium overlying Peyer's patches (PP). Second, IELs include an abundance of γδ T-cell receptor (TCR)-expressing cells. Third, IELs are antigen-experienced cells; however, they do not express markers of

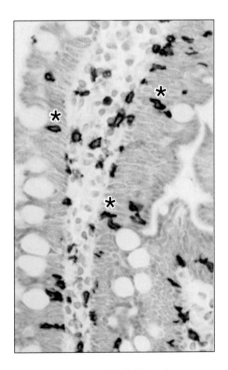

Figure 6.1 Intraepithelial lymphocytes (IELs) reside interspersed between the columnar epithelial layer of the intestine that forms the interface between the inner core of the body and intestinal lumen. Shown is small intestine villus epithelium. Brown cells are CD8+ T cells in epithelium (*) and lamina propria.

recently activated cells, such as CD25. Fourth, the majority of IELs display a typical cytotoxic phenotype, but they secrete only low levels of effector cytokines. Fifth, and typical of stress-sensing cells, IELs also express innate activating and inhibitory natural killer (NK) cell-like receptors. Sixth, IELs express the $\alpha_E\beta_7$ integrin (CD103 indicates the α_E chain), which interacts with E-cadherin on the epithelial cells. Seventh, IELs, especially in the small intestine, express CD8$\alpha\alpha$ homodimers, a hallmark of activated cells (the ligand for CD8$\alpha\alpha$, the nonclassical major histocompatibility complex (MHC) class I molecule thymus leukemia antigen, is also abundantly and constitutively expressed on gut epithelial cells). Eighth, a large fraction of TCR$\alpha\beta$ IELs are CD4$^-$CD8$^-$ double negative, typical of high affinity T cells. Ninth and finally, under steady-state conditions and in contrast to systemic and lamina propria lymphocytes, CD4$^+$ IELs are greatly underrepresented, especially in the small intestine.

Although the antigen-experienced phenotype is a hallmark of all IELs, the activation pathway they follow and the cognate antigens they recognize are very different. Consequently, IELs are divided into two major subtypes. The 'natural' IELs (nIELs) acquire their activated phenotype during a self-antigen-induced differentiation process in the thymus; whereas 'induced' IELs (iIELs) acquire their activated phenotype in response to foreign antigens (Figure 6.2).

6-1 Natural IELs are a unique cell type absent from the rest of the body.

Natural IEL-type cells are virtually absent from the human and rodent systemic lymphoid tissues. The frequency of nIELs is much higher in mice than in adult humans, and most of the data concerning the phenotype, ontogeny, and function of nIELs have been obtained from studies using mouse models.

Natural IELs, previously referred to as 'unconventional' T cells, can be either TCR$\gamma\delta$ or TCR$\alpha\beta$ and are often CD8$\alpha\alpha^+$. They are typically CD2$^-$, CD5$^-$, CD28$^-$, LFA-1$^-$, and B220$^+$ and mostly Thy1$^-$ (CD90$^-$). In addition, most TCR$\alpha\beta$ nIELs are CD69$^+$B220$^+$CD44$^{low/int}$, a phenotype seldom found among peripheral T cells. Characteristically, nIELs also share molecules expressed by natural killer (NK) type cells, such as CD16, Ly49A, Ly49G2, Ly49E, NK1.1, DAP12, and

Figure 6.2 Differentiation of antigen-experienced IEL subsets. TCR$\gamma\delta$ and TCR$\alpha\beta$ nIEL precursors acquire an antigen-experienced phenotype in the thymus in response to strong TCR signals and migrate directly to the gut epithelium where they further adapt and acquire CD8$\alpha\alpha$ expression. TCR$\alpha\beta$ iIEL precursors derive from conventional selected naive CD8$\alpha\beta$ or CD4 single-positive T cells that encounter their cognate antigen in the periphery and subsequently home to the gut mucosae where they further adapt including expression of CD8$\alpha\alpha$. MLN, messenteric lymph node.

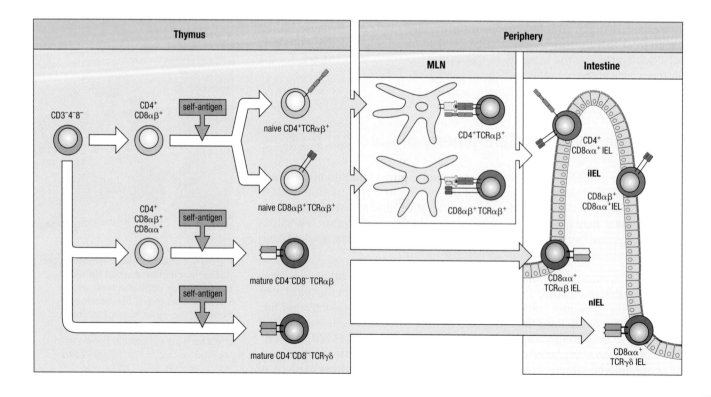

CD122, and they commonly express CD3ζ–FcεRIγ heterodimers or FcεRIγ–FcεRIγ homodimers rather than CD3ζ–CD3ζ homodimers as part of their CD3 complex. nIELs expressing the γδ TCR make up ~50% of mouse epithelial lymphocytes and only ~10% of the human epithelial lymphocytes, but this number can be quite variable and increases in various conditions, including celiac disease. The factors controlling the numbers and composition of nIELs remain poorly defined.

6-2   Induced IELs are the mucosal counterpart of T cells found in the periphery.

Induced IELs are the progeny of CD4$^+$ or CD8αβ$^+$ TCRαβ MHC class II- or class I-restricted T cells. These TCRαβ$^+$ T cells initially develop as naive cells in the thymus along the conventional selection pathway and then undergo further differentiation in response to their cognate antigens in the periphery (Figure 6.2). Consistent with this, they express an 'effector/memory-like phenotype,' (CD2$^+$, CD5$^+$, Thy1$^+$, LFA-1$^+$, CD44$^+$), often together with the activation marker, CD8αα. Most iIELs are CD8$^+$ TCRαβ cells that exhibit a typical cytotoxic effector memory phenotype, whereas CD4$^+$ IELs, although few, encompass all classical CD4$^+$ helper T cell (T$_H$) subsets (T$_H$1, T$_H$2, T$_H$17, and regulatory Foxp3$^+$ T cells) as well as newly defined CD4 cytotoxic T lymphocytes (CTL) (discussed below).

Induced IELs have much in common with other memory T cells, but they also have distinct features shared with the nIELs. Most iIELs are constitutively granzyme B^{high} and they typically express CD103 and CD69, but they lack Ly6C and CD62L. Although iIELs are all activated cells, they divide poorly *in vitro* and produce only low levels of inflammatory cytokines. In addition, both inhibitory and activating NK receptors are found on most iIELs, giving them a phenotype typical of 'stress-sensing' yet 'highly regulated' T cells. Similar to the nIELs, the unique characteristics of iIELs might also have evolved in order to deal with the gut dilemma: to provide effective immediate immune protection without compromising barrier function.

6-3   Luminal factors control the numbers of IELs.

Under normal physiological conditions there is one IEL for every 3–10 epithelial cells in the small bowel, and about one IEL per 40 epithelial cells in the colon. The different IEL subtypes are present to varying extents in the small and large intestines and the distribution depends in part on age, species, and environmental conditions. In contrast to the lamina propria, the 'natural' type T cells are far more prevalent in the epithelium, especially of the small intestine.

In rodents, small intestine nIELs are the principal population at birth, and there is only a modest expansion until weaning (Figure 6.3).

Two important events shortly after birth have a profound impact on IELs: first, the microbial colonization of the gut and second, the changes in the diet-related antigen load and composition. The increase in TCRαβ iIELs is not seen under germ-free conditions, indicating that their accumulation depends on the microbiota. The early migration of nIELs to the intestine is independent of exogenous antigens but crucially depends on CCR9/CCL25 and α$_4$β$_7$. The proportion of TCRαβ versus TCRγδ T cells varies between mouse strains and the environment, and in barrier-maintained specific pathogen-free mice TCRγδ nIELs predominate, whereas in mice kept in non-barrier facilities, TCRαβ iIELs prevail. In human gut epithelium, there is extensive infiltration of T cells from around 12 to 14 weeks of gestation and the number continues to increase, reaching about half the density of IELs found in young children by 28 weeks of

Figure 6.3 A schematic representation of IEL subsets over a lifetime. nIELs are present during gestation and at birth and decrease with age whereas iIELs are few at birth but increase with antigen exposure and with age.

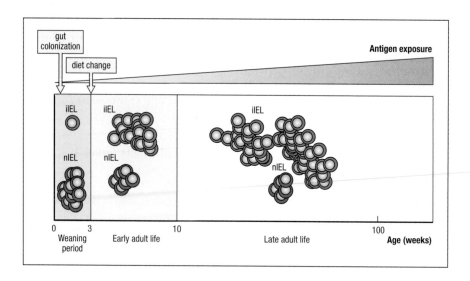

gestation. The absolute number increases dramatically during that time. The majority of IELs in humans, both at fetal and adult stages, express αβ TCRs. Very few (0.1%) human IELs are actively dividing, and in rodents, where survival can be measured, it appears that the majority of IELs are long lived. In addition, during homeostasis, there is only a limited contribution of circulating lymphocytes to the IEL compartment, as shown in parabiotic mice.

The TCR repertoire and specificity of IELs.

In young and adult animals, apart from a few exceptional cases such as γδ T cells in mouse skin, the expressed T-cell repertoire is diverse, presumably to maximize the number of peptides capable of being recognized by T cells. One of the most striking features of IELs, however, is that in mice and humans, and in the αβ and γδ subsets, TCR usage is highly skewed toward a small number of clones.

6-4   The TCR repertoire and antigen specificity of TCRγδ nIELs are different from T cells in the rest of the body.

Regardless of species, the γδ TCRs expressed by nIELs are largely distinct from their peripheral counterparts. Mouse TCRγδ IELs predominantly express Vγ5 and Vγ1.1, and a few use Vγ2. With respect to TCRγδ, depending on the strain, 15–60% of the TCRγδ IELs express Vδ4 and some express Vδ5 or Vδ6. For humans, most (~70%) colonic TCRγδ IELs utilize Vδ1. Furthermore, human IELs express a non-disulfide-linked form of the γδ TCR unlike peripheral γδ T cells. The complexity of the γδ TCR is largely contributed by the complementarity-determining region 3 (CDR3) of the δ chain and by random joining of the variable (V), diversity (D), and joining (J) segments, random N-region nontemplate deletions, and additions of nucleotide palindromes at the coding ends of the V, D, and J segments (P additions). The composition of the TCRγδ IELs changes with development with changes in V region usage and junctional diversity. The TCRδ of fetal intestinal T cells is polyclonal, with different CDR3 lengths and several template-encoded P-region, but no N-region, additions. This polyclonal nature is maintained until birth but at that time the junctional regions become more complicated with numerous N-region insertions. With age the TCRδ repertoire becomes increasingly restricted and in young adults the TCRδ usage is comparable with that in older adults. Despite the diversity of nIEL γδ TCRs, the oligoclonal expansions are distinct between individuals but appear to be stable over time.

The substantial junctional and combinatorial diversity implies that TCRγδ nIELs might recognize a diverse set of antigens. Identification of such ligands, however, remains a major deficit in our knowledge of TCRγδ nIEL biology.

6-5 TCRαβ IELs are highly unusual in that they are made up of a few dominant clones.

In the steady state, the TCRαβ repertoire of IELs is also oligoclonal and comprises a few hundred clones that are quantitatively and qualitatively unique to the individual. In the case of iIELs, which increase with age and in response to pathogen challenges (Figure 6.3), the specificity and diversity of the TCRαβ repertoire is probably a direct reflection of the environmental antigen platform encountered by that particular individual. Although also oligoclonal and unique, the repertoire of nIELs shows very little overlap with that of the iIELs. Furthermore, αβ TCRs from nIELs of littermates in the same cage or from germ-free mice, show the same degree of oligoclonality. The oligoclonality of TCRαβ IELs is also seen in humans with, similarly to the γδ TCR nIELs, striking changes throughout life. Fetal intestinal lamina propria lymphocytes and IELs contain transcripts of only a few TCRβ members as early as 13 weeks. Diversity increases with age, and at birth all 24 Vβ families are detectable. Analysis of the CDR3 regions reveals variability with numerous nontemplated N-region additions. In contrast, rearranged TCRα chains are undetectable until 16 weeks of gestation and only few can be detected at later fetal stages. A large fraction of the fetal TCRα transcripts contain an immature precursor segment, termed T early α, which is also expressed in thymocytes prior to the onset of TCRVα recombinations. By 23 weeks of gestation TCRα transcripts with productively rearranged TCRV-J-C segments can be detected in the gut epithelium. IELs at the adult stage express an oligoclonal repertoire with dominant expression of the same αβ TCRs throughout the intestine. In contrast to mice, some of the human TCRαβ IELs can be restricted to the MHC I-like CD1 molecules, CD1a, CD1b, CD1c, and CD1d.

In addition to oligoclonality, some Vβ segments, which are deleted from the TCRαβ repertoire of iIELs, are highly enriched in the repertoire of the CD8ααTCRαβ nIELs, including so-called forbidden Vβ segments. For example, in mouse strains expressing I-E and Mls-1ᵃ antigen or mouse mammary tumor virus, forbidden Vβ6, Vβ8.1, or Vβ11 are over-represented in the TCR repertoire of CD8ααTCRαβ nIELs. The presence of self-reactive TCRs among CD8ααTCRαβ⁺ nIELs indicates that their precursors do not follow conventional selection in the thymus, where self-reactive TCRs are usually purged. Nevertheless, CD8ααTCRαβ nIELs are drastically reduced in β2-microglobulin (β2m)-deficient mice, but relatively normal in mice lacking transporter associated with antigen processing, indicating that a β2m-dependent nonclassical (other than CD1d or thymus leukemia antigen) and/or transporter associated with antigen-processing-independent MHC I molecule is required for the selection and/or maintenance of CD8ααTCRαβ⁺ nIELs. Although thymus leukemia antigen does not serve as a restriction element or antigen-presenting molecule, thymus leukemia antigen does bind with high affinity to CD8αα; this interaction does not lead to activation, but to modulation of TCR reactivity and survival. Mechanisms of antigen recognition by TCRαβ nIELs are still largely unknown and research is complicated by the fact that some nIELs are reactive to MHC class I, nonclassical, or MHC class II molecules.

Development and differentiation of IELs.

An idea dating back to the 1960s is that some IELs are T cells that are generated locally in the gut from bone marrow derived precursors. However, the relative contribution of the thymus-dependent versus thymus-independent pathway has been controversial for many years. Different model systems have given different results, often depending on the subset of IEL under investigation.

6-6   Thymic versus extrathymic development of IELs is controversial.

Substantial efforts have been brought to bear on understanding the origin of IELs, with special emphasis on the CD8αα$^+$ nIELs, in part due to their unusual phenotype, and on TCRαβ nIELs, for their restricted TCR repertoire highly enriched for self-reactive TCRs. Moreover, the intestine has been described as a primary lymphoid organ based on the demonstration that certain lymphocytes are present in the intestinal epithelium of neonatal-thymectomized mice and that IELs are present in primitive vertebrates. This concept appears reasonable considering that gut-associated lymphoid tissue arose prior to the thymus, and in fact may be the evolutionary antecedent. Nevertheless, in the absence of a functional thymus, as in nude mice, IELs are drastically reduced and the residual lymphocytes chiefly express TCRγδ. Additionally, extrathymic development in the absence of irradiation results mainly in TCRγδ nIELs, suggesting that radiation promotes extrathymic T-cell development. Neonatal or fetal thymus grafting studies have demonstrated that all IELs are thymus derived and, consistent with this, neonatal thymectomy largely ablates all IEL populations. These findings indicate that under normal circumstances IELs are thymus derived but that further maturation may occur in the intestine. Lymphoid structures at the base of some small intestinal crypts in the mouse, termed cryptopatches, contain hematopoietic c-Kit$^+$, interleukin-7 receptor α (IL-7Rα$^+$), Thy1$^+$, CD44$^+$, Lin$^-$ precursors, and were proposed as a potential site where such 'extrathymic' development could occur, although cryptopatches have not been described in humans or other species. The formation of mouse cryptopatches is interleukin-7 (IL-7) and IL-15 dependent and enterocyte-restricted IL-7 is crucial for cryptopatches and sufficient for extrathymic development of TCRγδ T cells. Cryptopatches, however, are not absolutely required for TCRγδ nIELs. Moreover, athymic RAG reporter mice show RAG expression in mesenteric lymph nodes (MLN) and PP, but not in cryptopatches or IELs. Nevertheless, MLN and PP are also not required for extrathymic IELs, and RAG expression in mucosal tissues is negligible in euthymic animals, suggesting that the extrathymic pathway is a secondary consequence of a lack of thymus-derived cells and signals.

There is very little evidence for extrathymic T-cell differentiation in postnatal human gut. Although, some evidence indicates that some IELs and lamina propria lymphocytes in human fetal intestine may be maturing *in situ* from CD3$^-$CD7$^+$ cells.

6-7   Under euthymic normal conditions, all IEL subsets are progeny of bone marrow precursor cells that initially develop in the thymus.

Thymocyte differentiation and maturation is not a single streamlined process and multiple pathways exist that ultimately lie at the basis of phenotypic and functional diversity of all mature T cells, including the IEL subtypes. The γδ/αβ T cell-lineage decision occurs early during thymic development, and both TCRαβ and TCRγδ T cells arise from a common precursor, but diverge into separate lineages at the immature DN3 stage, just after T-cell commitment

(Figure 6.4). Although the process of lineage differentiation is not fully understood, several underlying mechanisms have been proposed, including differential IL-7R expression, and differences in signal strength propagated by pre-TCRα versus γδ TCR. Several observations indicate that strong signals at the immature double-negative stages favor the γδ lineage. Nevertheless, the constant association of pre-TCR with the αβ lineage choice points to unique qualities of the pre-TCR for αβ lineage commitment. The fact that nontransgenic CD8αα+TCRαβ+ nIELs are strictly dependent on pre-TCR expression indicates that these nIELs are not part of the γδ lineage but genuine pre-TCR-dependent TCRαβ-lineage cells.

6-8   Natural IEL precursor cells migrate directly from the thymus to the epithelium.

The term 'natural' refers to the lineage development of nIEL precursor cells, which go through an 'alternative' self (agonist) antigen-based thymic maturation process that results in the functional differentiation of mature double-negative, TCRγδ- or TCRαβ-expressing antigen-experienced T cells that directly migrate to the epithelium of the intestine (Figure 6.2). In contrast to conventional thymocytes, nIEL precursor cells are neither inhibited nor deleted by agonist cognate self-antigens encountered in the thymus. Instead, such agonist signals drive their antigen-experienced phenotype and functional differentiation (Figure 6.4).

6-9   Cytokines and transcription factors are involved in the development and differentiation of TCRγδ nIELs.

In addition to strong and/or sustained TCR signals for γδ-lineage commitment, specific transcription factors promote and bias specific differentiation. In that aspect, mice deficient for the Wnt-induced β-catenin-controlled T cell factor-1 showed drastic reduction in all TCRαβ+ subsets and also in the TCRγδ+ nIELs but not in TCRγδ+ splenocytes. Also signals transduced through IL-7R play nonredundant roles during survival and expansion of T cell progenitors and the rearrangement of the TCRγ locus. Consequently, IL-7 deficient (IL-7−/−) mice lack TCRγδ cells in all tissues. In addition, IL-7−/− thymus grafts do not support TCRγδ nIEL development, and neonatal thymectomy results in the appearance of TCR− IELs. Furthermore, transfer of c-Kit+IL-7R+ precursors reconstitutes the TCRγδ IELs, whereas continuous IL-7R signaling in IL-7Rα transgenic thymocytes suppresses the expression of T cell factor-1 and retinoic acid orphan nuclear receptors RORγ/RORγt, and prevents survival and differentiation of TCRαβ progenitor cells. Transgenic expression of IL-7 exclusively in the intestinal epithelium drives TCRγδ nIEL development but not the generation of TCRγδ T cells elsewhere, indicating extrathymic development for this IEL subset.

Because Vγ5 is rarely expressed by DN4 thymocytes and productive *Vγ5* gene segments are underrepresented at the DN3 stage, TCRγδ+ IEL precursors likely egress at or before the DN3 stage (Figure 6.4). Although genetic programming is involved in γδ gene rearrangements, selective TCRγδ-associated signals may also control the specific gut-homing ability of the TCRγδ+ precursors. For example, TCRγδ transgenic G8 thymocytes specific for the MHC class Ib molecules, T10b/T22b, differentiate in the presence of their cognate antigen, and directly home to the gut but nowhere else. Consistent with this, functional sphingolipid receptor, S1P1, is not required for egress of TCRγδ+ nIEL precursor thymocytes. Thymic programming of TCRγδ progenitors is imprinted at the DN3 stage and requires a full γδ TCR together with CD3 components and functional linker for activation of T cells. Gene-expression analysis of TCRγδ nIELs identified an antigen-experienced profile that distinguishes TCRγδ+

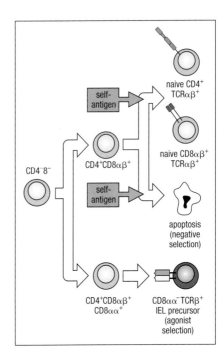

Figure 6.4 A schematic representation of thymic selection of TCRαβ precursor cells. Immature double-positive thymocytes that receive a strong selection signal undergo activation-induced death during a process of conventional negative selection. Double-positive thymocytes that receive intermediate signals are positively selected and exit the thymus as conventional naive CD4+ or CD8αβ+ T cells. Triple-positive thymocytes that receive a strong (agonist) selection signal are positively selected and functionally differentiate during an agonist selection process.

nIELs from conventional selected T cells. However, rather than being unique for TCRγδ nIELs, this profile is shared with the CD8αα⁺TCRαβ⁺ nIELs, which also require strong selection signals, and which also directly home to the intestine in a S1P1-independent manner.

The specific loss of TCRVγ5⁺ IELs in mice with defects in IL-15R-mediated signals and the increase of Vγ5⁺ IELs in IL-15 transgenic mice suggest a central role for IL-15 in the generation and/or maintenance of TCRγδ⁺ nIELs. In addition, the selective reduction of Vγ5⁺ thymocytes in IL-15$^{-/-}$ or IL-15 signal transducer Stat5$^{-/-}$ mice indicates that IL-15 directly influences development and/or survival of Vγ5⁺ thymocytes. Consistent with this, IL-15 is able to bias *Vγ5* transcription by directing chromatin domain modifications and promoting the accessibility in the exclusive vicinity of the *Vγ5* gene. Nevertheless, the absence of Vγ5 transcripts in IL-7R$^{-/-}$ thymocytes, in contrast to their presence in the gut, suggests that the IL-15 pathway is not a major mechanism during thymic differentiation, but it might operate in the gut when thymically selected IELs are limiting.

6-10   CD8ααTCRαβ nIELs with self-reactivity may be positively selected.

The accumulation of high-affinity self-reactive TCRs among the TCR repertoire of CD8ααTCRαβ nIELs is inconsistent with negative selection and more recently an alternative pathway was described where autoreactive precursor cells may be positively selected. The thymic progenitors of CD8ααTCRαβ nIELs may be positively selected and functionally differentiate under conditions that lead to the deletion of conventional TCRαβ⁺ cells (Figure 6.4). Consequently, agonist selection drives the accumulation of 'forbidden' TCRs expressed by mature CD8ααTCRαβ nIELs but purges these TCRs from the TCR repertoire of conventional naive T cells in the periphery.

In contrast to the co-receptor-expressing TCRαβ⁺ iIELs, CD8ααTCRαβ nIEL precursors acquire their antigen-experienced phenotype during this self-agonist-based selection process. Nevertheless, although both the CD8ααTCRαβ⁺ and the γδTCR⁺ nIELs depend on strong signals and acquire their activated phenotype 'naturally' in the thymus, they follow different pathways. Precursors of the CD8ααTCRαβ nIELs pass through an additional checkpoint that involves the expression of the pre-TCR and they transition to a CD4 and CD8αβ double-positive stage before they undergo the full TCR-based agonist selection process (Figure 6.4). Also, *in vitro* differentiation of double-positive thymocytes exposed to their cognate antigen leads to CD8ααTCRαβ⁺ T cells that express the typical gene transcription profile of CD8ααTCRαβ⁺ nIELs. Fate mapping studies, using expression of green fluorescent protein driven by the CD4 promoter, further confirmed that CD8ααTCRαβ⁺ nIEL progenitors transition through the double-positive checkpoint, implying that double-positive thymocytes are heterogeneous and contain precursors for conventional and CD8ααTCRαβ⁺ T cells. This led to the identification of a new subset of immature thymocytes that co-express CD8αα together with CD4 and CD8αβ, called triple positive (TP) cells (Figure 6.4). Intrathymic injection of TP, but not double-positive, thymocytes led to the generation of CD8ααTCRαβ⁺ nIELs *in vivo*, and when exposed to their cognate antigen *in vitro*, TP thymocytes survived, expanded, and acquired an antigen-experienced phenotype as well as innate-like features and gut-specific homing receptors. These observations further underscore the importance of the thymic agonist selection process as a central mechanism for the unique differentiation and migration of CD8ααTCRαβ⁺ nIEL thymic precursor cells. Agonist-selected TP thymocytes mature to CD5⁺TCRαβ⁺ double-negative thymocytes that egress the thymus and directly migrate to the intestinal epithelium (Figures 6.2 and 6.4). The reexpression of CD8αα and the downregulation of CD5 are post-thymic events that occur locally in the intestine and that are actively promoted by IL-15.

6-11 iIELs are peripheral-antigen-driven progeny of conventional thymically selected CD4$^+$ or CD8$\alpha\beta^+$TCR$\alpha\beta$ T lymphocytes.

The intestinal epithelium is particularly enriched for CD8$\alpha\beta^+$TCR$\alpha\beta^+$ iIELs, especially in man, and *in vivo* priming experiments in mice demonstrate that intestinal primary and memory CD8$\alpha\beta^+$ T cells are readily generated in response to infections. For example, oral reovirus or rotavirus infection results in the accumulation of virus-specific cytolytic iIELs. However, CTLs are present in the lamina propria but not in the intestinal epithelium following footpad immunization, indicating discordance of the response within mucosal effector sites. Analysis of CTL induction in response to a human immunodeficiency virus-1 (HIV-1) combinatorial peptide indicated that intrarectal immunization leads to primary and memory CTL within the lamina propria, intestinal epithelium, and spleen. In contrast, subcutaneous immunization with the peptide leads to CTL activity in spleen but not mucosa, implying that induction of mucosal primary and memory CTL iIELs depends on the route of infection and the form of the immunogen.

Technological advances now allow the identification of antigen-specific T cells directly *ex vivo* and even *in vivo*. In particular, the use of MHC class I tetramers and the transfer of small numbers of antigen-specific TCR-transgenic T cells to normal hosts followed by immunization, has provided a detailed description of the pathway that effector CD8$\alpha\beta^+$ T cells follow to the epithelium. Results from such studies show that few naive CD8$\alpha\beta^+$ T cells are present in the intestinal mucosa prior to immunization. Thus, it is unlikely that the epithelium is a site for primary activation of naive T cells.

Migration to and local adaptation in the gut.

IELs are part of the effector limb of mucosal immune responses, but they differ markedly from many of the paradigms that control the migration of CD4 cells into the gut. This is most dramatically seen in parabiotic mice, where there is rapid mixing of lamina propria CD4 T cells from either parabiont, but where IEL cells turn over at a much slower rate. This is also consistent with oligoclonality, because if new T cells entered the epithelial compartment at a high rate, dominant clones would be diluted with time, whereas they are stable with time.

6-12 Natural IELs are induced to home to the gut in the thymus.

In addition to their antigen-experienced phenotype, nIEL precursors also acquire gut-homing receptors in the thymus, including $\alpha_E\beta_7$ and CCR9. The specific expression of their ligands, E-cadherin and CCL25 respectively, by the small intestine enterocytes, recruits the mature nIEL precursors directly to the small intestine epithelium. Consistent with this, the seeding of these cells is independent of S1P1. The direct and early population of the mucosal barrier with these stress-sensing self-reactive effector cells provides a layer of preexisting immunity, even before birth and during the weaning period, prior to the formation of an exogenous antigen-induced defense line and before the potential for exposure to pathogens becomes significant (Figure 6.3).

6-13 Natural IELs undergo local adaptation in the gut.

Additional local factors produced by the epithelial cells, such as IL-7 and IL-15, further expand and adapt the nIELs. In the absence of IL-15 or IL-15Rα, there is a preferential loss of CD8αα⁺ nIELs. IL-15 operates by an unusual mechanism of transpresentation in which IL-15 is bound to IL-15Rα intracellularly, transported to the cell surface, and presented to responding cells expressing the IL-15β and common γ receptor chains. This mechanism is important for IEL development since both IL-15 and IL-15Rα expression by the intestinal epithelium is essential for nIEL development and maintenance. Stem cell factor (SCF), produced by epithelial cells, also plays an important role in their homeostasis and most nIELs, but not iIELs, express c-Kit. While early TCRγδ nIEL development is normal in SCF or SCF-receptor (c-Kit) mutant mice, this population is gradually lost with time.

6-14   Conventional thymically selected naive T cells are not found in the gut epithelium.

Although some conventional CD8αβ⁺ recent thymic emigrants have gut-homing receptors, in general, conventional thymically selected precursor cells and naive CD8 or CD4 T cells are not detected within the intestinal epithelium. However, organized lymphoid tissue, such as the PP and the MLN, form the main priming sites for peripheral naive T cells that acquire gut-homing capacity upon activation with their cognate antigens. In that aspect, specialized epithelial cells or M cells (discussed in Chapter 13), resident CX3CR1⁺ macrophages, and migratory CD103⁺ dendritic cells (DC) (discussed in Chapter 10) all have the unique capacity to directly or indirectly sample the antigen content of the gut lumen, allowing for constant reporting on the condition of the intestinal compartment. Upon priming in these mucosal sites, naive T cells are induced to express gut-homing receptors that drive migration to the intestine (Figure 6.5). The ability of the mucosal DC to confer these properties is mediated at least in part by the release of a diet-derived factor, the vitamin-A metabolite retinoic acid (RA), which preferentially promotes migration to the small intestine, through CCR9. The imprinting process greatly promotes migration to the intestine but it is not absolutely required and TCRαβ

Figure 6.5 Adaptive and innate responses in the intestinal epithelium. Left panel: Conventional CD8αβ TCRαβ T cells are activated in the mesenteric lymph nodes (MLN) or Peyer's patches (PP) by dendritic cells that have acquired antigen following infection of the epithelium. This process may take days. Homing molecule expression is 'imprinted' in the PP and MLN resulting in migration of activated CD8 T cells into the epithelium and destruction of infected cells through cytotoxic mechanisms. Right panel: Environmental stress or tumors induce expression of stress ligands (e.g., MICA, RAE-1) by intestinal epithelial cells. NKG2D⁺ TCRγδ IELs interact with these ligands resulting in destruction of infected cells through cytotoxic mechanisms. This process can occur in minutes to hours since the responding cells are resident in the epithelium.

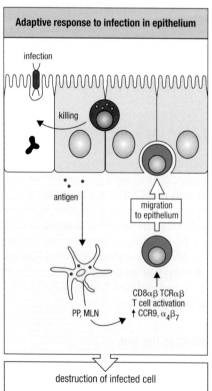
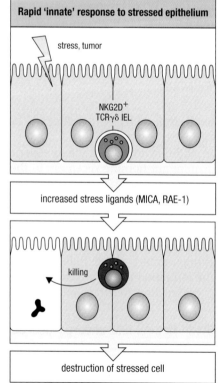

T cells activated in the periphery can also migrate to the gut, although under conditions of an intestinal challenge gut-specific homing may be significantly enhanced through imprinting.

The multi-step paradigm of lymphocyte migration applies to T-cell movement into both the inductive and effector sites of the intestinal mucosa (discussed in Chapters 1 and 14). Integrins are also important players in lymphocyte migration and retention in the mucosa and the β_2 integrin $\alpha_L\beta_2$, whose primary ligand is intercellular adhesion molecule 1 (ICAM-1), is required for establishment of IELs as is the $\alpha_1\beta_1$ integrin (very late antigen-1; VLA-1) that binds to collagen. In the case of $\alpha_L\beta_2$, it is not clear whether integrin activity is needed for initial T-cell activation, migration to the intestine, or both. Another β_2 integrin CD11c ($\alpha_X\beta_2$), expressed by most IELs and induced by activation, is linked to lytic activity but the function of CD11c *in vivo* is unknown.

The binding of $\alpha_4\beta_7$ to the mucosal addressin cell adhesion molecule-1 (MAdCAM-1), expressed by vascular endothelium in lamina propria, MLN, and PP blood vessels, promotes translocation into the respective tissues. After local or systemic infection, $\alpha_4\beta_7$ is transiently induced which provides a finite window of time to migrate to the intestinal epithelium. Following entry into the mucosa $\alpha_4\beta_7$ is lost; concomitant with this $\alpha_E\beta_7$ is induced and appears to be important for retention in the epithelium via interaction with its ligand, E-cadherin, expressed by epithelial cells. Induction of $\alpha_E\beta_7$ is mediated by transforming growth factor-beta (TGF-β) and the transcription factor Runx3 and is enhanced by signaling through CCR9. Similarly, induction of CD69 occurs when T cells enter the mucosa; however, its function on these mucosal T cells remains unknown.

6-15 Induced IELs undergo adaptation in the gut.

Activation of transferred OT-I transgenic cells by feeding soluble ovalbumin does not induce lytic activity in nonintestinal OT-I cells, but upon migration into the intestinal mucosa these cells become highly lytic. Moreover, optimal activation of mucosal T cells requires B7.1, and CD40L expressed by CD8$\alpha\beta^+$ T cells is important for optimal expansion of mucosal CD8$\alpha\beta^+$ T cells in response to virus infection. Similarly, CD70–CD27 and CD40–CD40L co-stimulation appears to be important for expansion of CD8$\alpha\beta^+$TCR$\alpha\beta^+$ iELs after oral *Listeria monocytogenes* infection, presumably for 'secondary' co-stimulation, since lymphoid tissue is required for initial priming. These observations indicate that tissue-specific cellular interactions further condition the iIELs resulting in tuning of their phenotype and function. In this light, and similarly to the nIELs, a large fraction of iIELs express CD8$\alpha\alpha$ as part of their specific adaptation to the epithelial compartment. Although most memory CD8$\alpha\beta$ iIELs retain an activated phenotype (e.g., CD69 expression and a heightened cytolytic state) that enables them to rapidly respond to antigen re-exposure, they also display reduced ability to proliferate and to produce inflammatory cytokines. These unique features might have evolved in order to maintain the balance of immunological protection without compromising organ integrity. In this respect intestinal CD8$\alpha\beta$ and also CD8$\alpha\alpha$ T cells express high levels of the P2X$_7$ purinoreceptor, which may be induced by retinoic acid in the gut environment. Expression of P2X$_7$ is associated with increased ATP-induced apoptosis and in P2X$_7$-deficient mice mucosal, but not spleen, CD8$\alpha\beta$ T-cell responses are enhanced following *Listeria monocytogenes* infection.

Overall, these results illustrate that: (1) conventional naive T-cell activation results in migration to the gut epithelium; (2) secondary activation in the mucosa occurs upon secondary encounter with antigen; (3) environmental cues in the mucosa drive iIEL activation even in response to tolerogenic forms of antigen; and (4) certain co-stimulatory elements preferentially regulate intestinal T-cell responses, including iIELs.

The beneficial functions of IELs.

The location of IELs at the intersection between the outside milieu and the core of the body prompts teleological predictions about their function. They are geared to provide immediate immune protection in order to avoid the penetration of pathogens. Consistent with this, most IELs display constitutive cytolytic activity driven by various pathways, including serine proteases, perforin release, and Fas–FasL-mediated programmed cell death and natural-killer receptor-dependent killing.

6-16   The function of nIELs appears to be to maintain epithelial homeostasis.

Although the exact functions and behavior of nIELs remain enigmatic, their primary role seems to be directed toward insuring the integrity of the epithelium and maintaining a local immune balance. Consistent with this, the absence of nIELs does not seem to significantly compromise protective immunity against a wide range of pathogens. The best known TCRγδ nIEL function is their participation in maintaining the intestinal epithelium. TCRγδ nIELs may produce keratinocyte growth factor and express the junctional adhesion molecule known to induce keratinocyte growth factor upon ligation to its receptor, the Coxsackie virus and adenovirus receptor, thereby linking this co-stimulatory pathway directly to epithelial repair and maintenance function. In mice lacking all TCRγδ T cells, there are fewer enterocytes and crypt cells and the enterocytes express less MHC class II. Intestinal infection of TCRδ$^{-/-}$ mice with the protozoan parasite *Eimeria vermiformis* results in more severe pathology and overt intestinal bleeding. This effect appears to be due to uncontrolled interferon-γ (IFN-γ) production by TCRαβ iIELs. The absence of TCRγδ cells also results in more severe injury in mouse models of colitis, and additionally, keratinocyte growth factor null mice are more susceptible to dextran sulfate sodium colitis suggesting that, in mice, TCRγδ cells control intestinal homeostasis in part by the production of keratinocyte growth factor. TCRγδ nIELs express other regulatory factors such as TGF-β1, TGF-β3, and prothymosin β4, which may aid in epithelial maintenance or repair. TCRγδ nIELs are also equipped with innate and adaptive cytotoxic mechanisms that could contribute to protective immunity. In a model of *Toxoplasma gondii* infection, TCRγδ nIELs greatly enhanced the protective function of pathogen-specific CD8αβ$^+$ T cells. Furthermore, γδ TCRs may recognize unprocessed antigens such as stress antigens expressed by enterocytes or molecular patterns generated by bacterial nonpeptide antigens, such as phosphorylated nucleotides and prenyl pyrophosphates. In humans, TCRγδ nIELs recognize the highly polymorphic MHC class I chain-related genes MICA and MICB, induced on stressed epithelium, which leads to death of damaged or infected epithelial cells. The invariant NKG2D receptors expressed by TCRγδ nIELs also bind to MICA and MICB, thereby inducing killing in a TCR-independent fashion (Figure 6.5). Together with their cytotoxic capacity, TCRγδ nIELs produce protective cytokines, such as tumor necrosis factor-α and IFN-γ. Therefore, TCRγδ nIELs not only control the production of IFN-γ by other inflammatory cells but they can also make IFN-γ that contributes to host protection. Nevertheless, much remains to be learned regarding the function of TCRγδ nIELs, not the least of which is the gap in our knowledge of their ligands for γδ TCRs.

6-17   The functions of CD8ααTCRαβ nIELs remain largely unknown.

The CD8ααTCRαβ nIELs are the most enigmatic subset of IELs. Similar to all other IEL subsets, they express an activated phenotype and show high cytolytic potential, but they do not seem to behave as activated T cells by other criteria. They have a limited capacity to proliferate and their presence often correlates with immune quiescence rather than immunity. In a lymphocytic choriomeningitis virus (LCMV) infection model using mice double-transgenic for an LCMV-epitope-specific TCR and an LCMV-specific antigen, LCMV-reactive CD8ααTCRαβ nIELs displayed signs of virus-induced activation; however, unlike the conventional virus-specific CD8αβ T cells, they did not induce LCMV-specific cytotoxicity nor did they promote an inflammatory response.

This dual nature of CD8ααTCRαβ nIELs was confirmed by gene array analysis, which showed that they (1) express high levels of regulatory or inhibitory molecules including LAG-3, CTLA-4, PD-1, TGF-β, and the NK receptors 2B4 and Ly49 A, E, and G; and (2) express molecules associated with protective immunity, including granzymes, Fas ligand, RANTES, and CD69. Effector cytokines such as IFN-γ, IL-2, and IL-4, and cytokine receptors (IL-2Ra, IL-12-R) are not abundantly expressed.

Therefore, similarly to their TCRγδ counterparts, CD8ααTCRαβ nIELs may also fulfill dual roles in the epithelium, with the emphasis on promoting homeostasis and maintaining barrier integrity; in addition, upon the appropriate signals, they may gain a full effector status, but they do not mediate tissue injury.

Potential self-reactivity together with inherent cytotoxic capability and their close proximity to enterocytes also suggest important roles for nIELs in sensing and eliminating transformed epithelial cells. Nevertheless, although self-reactive, nIELs are not self-destructive and, as discussed above, self-tolerance features seem to be in part preprogrammed during agonist selection in the thymus. These observations underscore the importance of the thymic agonist selection process as a central mechanism to drive the antigen-experienced effector yet regulate differentiation of these unique nIELs. The agonist-selected double-negative thymic precursors that migrate to the intestine induce CD8αα expression driven by IL-15 from enterocytes. In contrast to the conventional TCR co-receptors, CD8αα does not function to enhance TCR activation, rather it suppresses TCR activation. This may be due to its ability to bind signaling molecules such as p56[lck] and linker for activation of T cells, but its inability to join lipid rafts that contain the TCR activation complex (Figure 6.6). Therefore,

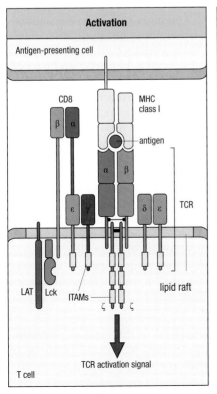
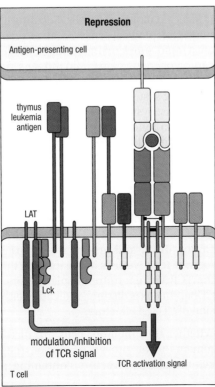

Figure 6.6 CD8αβ functions as a TCR co-receptor. CD8β moves CD8α into the lipid raft adjacent to the TCRαβ whereas CD8α recruits signaling molecules, such as lymphocyte-specific protein tyrosine kinase (Lck) and linker for activation of T cells (LAT) that promote downstream signaling. CD8αα functions as a repressor. CD8αα does not move into the lipid raft and negatively influences TCR activation by sequestering signaling molecules such as LAT and Lck away from the TCR activation complex.

CD8αα expression on these high-affinity self-reactive nIELs may reset the threshold for activation, reducing the potential for unnecessary, or excessive, self-reactivity. CD8αα-mediated inhibition is not irreversible and increased antigen stimulation or cross-ligation of CD8 and TCR can readily take precedence over the CD8αα repressor function.

Other molecules associated with immune regulation, such as LAG-3, CTLA-4, PD-1, and TGF-β, together with the SLAM-related receptor 2B4 and several members of the killer-cell immunoglobulin-like receptors, KIR or Ly49 (Ly49 A, E, and G), may further regulate the activation status of high-affinity autoreactive nIELs. The NK-inhibitory receptors suppress cytotoxicity and cytokine secretion; however, 2B4 can also function as an activation receptor and under certain conditions even promote cytotoxicity. The dual expression of activating and inhibitory receptors is consistent with their 'activated yet quiescent' state, which indicates that nIELs may gain full activation status for self-based protective and/or regulatory functions in response to self-antigen-based induction signals. In support of this, in a model of induced colitis, TCRαβ⁺CD8αα⁺ nIELs were able to prevent inflammation mediated by non-self specific naive CD4⁺CD45RB^hi donor T cells. In addition, their agonist self-antigen-based thymic functional differentiation, their cytolytic phenotype, and their early seeding and strategic location in the gut epithelium all suggest that they might serve an important role in maintaining and protecting the mucosal barrier at the time of gut colonization and before an exogenous-antigen-induced defense layer of iIELs has been formed (Figure 6.3).

6-18 TCRαβCD8αβ iIELs appear to have a protective function.

Induced IELs line the intestinal interface with a protective immune layer that gradually increases with age and which provides preexisting immunity adapted toward those pathogens that are most likely to rechallenge (Figures 6.5 and 6.7). In the small intestine, iIELs are mostly CD8αβ⁺TCRαβ⁺ cytotoxic effector memory cells. Compared with splenic memory T cells, they can be rapidly activated and provide immediate cytotoxic function. Their TCR repertoire is more restricted than that of peripheral memory CD8αβ T cells suggesting that repeated restimulation in the intestine may lead to survival and expansion of selected clones of T cells. Locating effector memory cells in mucosal tissues makes teleological sense because a rapid response may be important to mediate a rapid attack on incoming pathogens. Early work showed that TCRαβ IELs can be primed against alloantigens, and against viral, bacterial, and parasite antigens. In mice the intestinal CD8αβ T-cell response has been characterized following systemic vesicular stomatitis virus, *L. monocytogenes*, and rotavirus infections. Systemic vesicular stomatitis virus infection results in CD8αβ T-cell activation followed by migration of effector cells to all nonlymphoid tissues. The magnitude of the response in the IEL compartment is substantially larger and more sustained, and the contraction phase is more prolonged, compared with the spleen. Repeated challenge infections result in enhanced and sustained granzyme B expression in secondary and tertiary antigen-specific IELs. The CD8αβ T-cell response to enteric infection with *L. monocytogenes* largely parallels that observed after systemic infection, but with some significant differences. Nevertheless, either route of infection generates antigen-specific CD8αβ iIELs, and enteric infections induce larger numbers of *L. monocytogenes*-reactive cells in the mucosa than do intravenous infections. Moreover, the TCR Vβ repertoire of splenic versus intestinal *L. monocytogenes*-specific CD8αβ T cells is distinct, indicating either differential migration and/or expansion of CD8αβ effectors.

Whether iIEL memory cells are directly involved in this type of protection remains to be proven. However, rhesus macaques inoculated with rhesus cytomegalovirus vectors containing simian immunodeficiency virus (SIV)

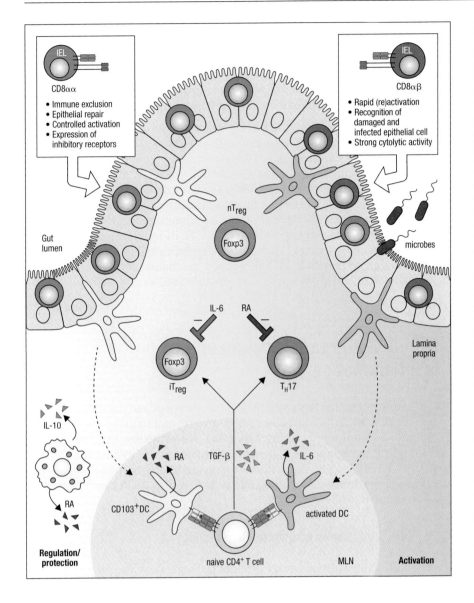

Figure 6.7 The functionally diverse T-cell populations of the intestine that are shaped by the gut environment are important players in sustaining the delicate immune balance between activation and regulation. The CD8αα⁺ intraepithelial cells (IEL) of the intestine play a crucial role in protection of the mucosal barrier. They are involved in maintaining and restoring barrier homeostasis by stimulating epithelial cell turnover. Upon pathogen entry rapid activation and high cytolytic activity of the CD8αβ⁺ IEL contribute to the prevention of pathogen spreading by killing infected epithelial cells. The activation of IEL is highly controlled through the expression of inhibitory receptors that may alter the threshold for activation. Under the influence of epithelial cells and IEL, lamina propria dendritic cells (DC) acquire the ability to produce retinoic acid (RA) thereby inducing gut-homing receptors during CD4 T-cell priming in the mesenteric lymph nodes (MLN). Additionally, gut-derived CD103⁺ DC and lamina propria macrophages favor the conversion of Foxp3⁺ induced regulatory T cells (iT$_{reg}$) in a RA, TGF-β, and IL-10 (in the case of lamina propria macrophages) mediated fashion, whereas activated DC promote the differentiation of IL-17-producing T$_H$17 cells via a combination of IL-6 and TGF-β. This pro- and anti-inflammatory immune deviation of iT$_{reg}$ and T$_H$17 is reciprocally controlled by RA and IL-6. Finally, the lamina propria is also home to agonist-selected Foxp3-expressing naturally occurring regulatory T cells (nT$_{reg}$). (Adapted from F. van Wijk and H. Cheroutre, *Seminars in Immunology* 21:130–138, 2009. With permission from Elsevier Ltd.)

epitopes are protected from progressive wild-type SIV infection after repeated limiting-dose intrarectal challenges. Moreover, the protected animals showed no evidence of systemic infection, suggesting that infection was contained locally, well before extensive viral replication or systemic spreading of the virus occurred.

6-19 TCRαβ CD4 iIELs are involved in protective immunity.

Though less prevalent, the loss of mucosal CD4 T cells in SIV/HIV-1 infections, which results in translocation of enteric bacteria and increased local and systemic infections, indicates important participation of CD4 iIELs for local protective immunity. Furthermore, the observation that most CD4 T cells display strong plasticity, together with the suppressive effects of iT$_{reg}$ cells (induced regulatory T cells), suggest that CD4 T cells contribute important regulatory roles as well.

CD4 iIELs contain the typical helper T cell (T$_H$) subsets, T$_H$1, T$_H$2, and T$_H$17, and the TGF-β-induced Foxp3-expressing T$_{reg}$ cells, the latter being especially enriched in the small intestine. At steady state, the gut environment is thought to favor CD4 T_{reg} conversion and mucosal CD103$^+$ DCs produce TGF-β and RA to promote Foxp3$^+$ T_{reg} generation rather than T_H17 cells. T_{reg} cells are potent regulators of intestinal inflammation and they suppress IELs, including TCRγδ nIELs and TCRαβCD4 iIELs. Nevertheless, under inflammatory conditions, CD4 T_H17 iIELs also participate in protective immunity, especially in host defense against extracellular bacterial and fungal pathogens.

There appears to be a TGF-β-dependent process *in vivo* that is characterized by the conversion of CD4 effector cells to CD8α$^+$CD4 CTL-like iIELs. The lineage-switch mechanism leads to the generation of protective cells but also represents a new form of immune regulation that prevents CD4 CTL from becoming pro-inflammatory T_H17 iIELs. The control of T_H effector CD4 iIELs while preserving their protective function makes reasonable sense at this mucosal interface where effective protective immunity has to coincide with the least inflammatory damage. These CD4 CTL iIELs are capable of dealing with viral infections in class II$^+$ enterocytes and they might also provide protection against viruses that have developed mechanisms to escape surveillance by the MHC class I-restricted CD8αβTCRαβ iIELs.

Potential aberrant function of IELs in inflammation.

The dynamic interaction between the gut environment and the mucosal adaptive immune system is normally geared toward maintaining a stable equilibrium and sustaining the barrier function. Nevertheless, their heightened activation status, together with their close proximity and intense interactions with the vulnerable epithelial cells, also provide IELs with the opportunity to become potent destructive pathogenic cells that may fuel chronic intestinal inflammation or promote cancers.

6-20 TCRγδ nIELs may be involved in pathology.

Even though TCRγδ nIELs may serve beneficial regulatory and protective roles, other evidence suggests a more pathogenic side of nIELs. There are increased numbers of TCRγδ cells in the epithelium of patients with active celiac disease and in this context, gliadin-mediated deregulation of IL-15 secretion may lead to overexpression of MICA/MICB on enterocytes and enhanced cytotoxicity of TCRγδ nIELs.

6-21 CD8ααTCRαβ nIELs may also damage the epithelium.

Autoreactive αβTCRs are preserved and enriched in the repertoire of the CD8ααTCRαβ nIELs. It thus remains possible that under inflammatory conditions or upon recognition of modified foreign antigens, these autoreactive T cells might react against self and induce autoimmune pathology. Furthermore, increased or uncontrolled inflammatory conditions, such as excessive production of IL-15, may trigger their heightened autoreactive cytotoxic effector function, directly jeopardizing the integrity of the mucosal barrier.

6-22 CD8αβTCRαβ iIELs may be involved in the pathogenesis of inflammatory bowel disease.

Cytotoxic CD8αβTCRαβ iIELs have been implicated in the progression or even the induction of inflammatory bowel disease (IBD). The destruction of epithelial cells by pathogenic CD8⁺ T cells may initiate the loss of epithelial barrier integrity, consequently leading to exposure to luminal antigens and bacteria, which may then attract other immune cells that further exacerbate the inflammatory pathology.

In support of this, in a hapten-induced mouse model of colitis, previously hapten-sensitized CD8⁺ CTL are the earliest inducers of uncontrolled inflammation. Furthermore, although epithelial-cell-specific CD8⁺ iIEL may not mediate persisting inflammation, some hapten-specific CD8⁺ effector cells survive long-term as effector memory T cells that initiate relapse disease upon secondary challenge. Although the pathogenic CTL may direct their responses toward epithelial-cell-expressed antigens, they are conventional selected cells, showing that a breakdown in self-tolerance in the periphery can lead to uncontrolled immune responses and self-destruction.

6-23 Aberrant functions of CD4 TCRαβ iIELs driving gut inflammation.

Aberrant CD4⁺ TCRαβ iIELs together with their lamina propria lymphocyte counterparts are probably the most significant contributors to immunopathology in the intestine and their roles in various inflammatory diseases such as celiac disease and IBD are extensively discussed in Chapters 28 and 31.

Overall, although all effector T cells, including the various IEL subsets, have the potential to initiate or propagate gut inflammation, aberrant CD4 effector cells are almost always involved and play a big part in the destructive immune pathology that targets and jeopardizes the barrier function of the mucosal epithelium.

Summary.

In many ways IELs are tightly regulated warriors and peacekeepers (Figure 6.7). On the one hand, the self-antigen-based functional differentiation of nIELs and the expression of inhibitory and activating NK receptors endows them with natural antigen-experienced ability and innate-like properties to sense stressed and transformed epithelial cells, and to respond in a protective yet controlled fashion even in the absence of preexisting antigen-induced immunity. On the other hand, the environmental antigen-induced effector memory iIELs provide rapid and effective protective immunity against repeated antigen challenges. This constant dilemma to prevent excessive and/or unwanted immune responses while at the same time providing optimal protection against invading pathogens, presents truly unique challenges that undoubtedly shape the complexity and sophistication of the IELs, and their multidirectional interactions with the surrounding epithelial cells. Because the intestinal mucosa is the major primary target for many pathogens, including HIV-1, and because a major branch of the immune system is formed by the IELs, it is critical that we elucidate and identify all the components and interactions involved in their networking, and that we understand the rules governing their migration, functional differentiation, activation, and regulation. This knowledge will aid in the eventual intelligent design of vaccines able to direct appropriate effector functions to the desired locations in order to prevent the entry and spreading of pathogens, and in generating strategies to regulate immunity and avoid detrimental inflammation or the development of cancers.

Further Reading.

Cheroutre, H.: **IELs: enforcing law and order in the court of the intestinal epithelium.** *Immunol. Rev.* 2005, **206**:114–131.

Cheroutre, H., and Lambolez, F.: **The thymus chapter in the life of gut-specific intra epithelial lymphocytes.** *Curr. Opin. Immunol.* 2008, **20**:185–191.

Cheroutre, H., Mucida, D., and Lambolez, F.: **The importance of being earnestly selfish.** *Nat. Immunol.* 2009, **10**:1047–1049.

Hayday, A.C.: **Gamma delta T cells and the lymphoid stress-surveillance response.** *Immunity* 2009, **31**:184–196.

Henderson, P., van Limbergen, J.E., Schwarze, J., *et al.*: **Function of the intestinal epithelium and its dysregulation in inflammatory bowel disease.** *Inflamm. Bowel Dis.* 2011, **17**:382–395.

Jabri, B., and Ebert, E.: **Human CD8+ intraepithelial lymphocytes: a unique model to study the regulation of effector cytotoxic T lymphocytes in tissue.** *Immunol. Rev.* 2007, **215**:202–214.

Jabri, B., and Sollid, L.M.: **Tissue-mediated control of immunopathology in coeliac disease.** *Nat. Rev. Immunol.* 2009, **9**:858–870.

Kronenberg, M., and Havran, W.L.: **Frontline T cells: gamma delta T cells and intraepithelial lymphocytes.** *Immunol. Rev.* 2007, **215**:5–7.

Kunisawa, J., Takahashi, I., and Kiyono, H.: **Intraepithelial lymphocytes: their shared and divergent immunological behaviors in the small and large intestine.** *Immunol. Rev.* 2007, **215**:136–153.

Laky, K., Lefrançois, L., Lingenheld, E.G., *et al.*: **Enterocyte expression of interleukin 7 induces development of gamma delta T cells and Peyer's patches.** *J. Exp. Med.* 2000, **191**:1569–1580.

Lefrançois, L., and Puddington, L.: **Extrathymic intestinal T-cell development: virtual reality?** *Immunol. Today* 1995, **16**:16–21.

Lefrançois, L., and Puddington, L.: **Intestinal and pulmonary mucosal T cells: local heroes fight to maintain the status quo.** *Annu. Rev. Immunol.* 2006, **24**:681–704.

Probert, C.S., Saubermann, L.J., Balk, S., *et al.*: **Repertoire of the alpha beta T-cell receptor in the intestine.** *Immunol. Rev.* 2007, **215**:215–225.

Shires, J., Theodoridis, E., and Hayday, A.C.: **Biological insights into TCRγδ+ and TCRαβ+ intraepithelial lymphocytes provided by serial analysis of gene expression (SAGE).** *Immunity* 2001, **15**:419–434.

van Wijk, F., and Cheroutre, H.: **Intestinal T cells: facing the mucosal immune dilemma with synergy and diversity.** *Semin. Immunol.* 2009, **21**:130–138.

Lymphocyte Populations Within the Lamina Propria

7

There are more T cells in the gut-associated lymphoid tissue (GALT) and intestinal lamina propria and epithelium than in the rest of the body. That they are absolutely essential for health is demonstrated when they are not present, as in children with severe combined immunodeficiency or adults with untreated human immunodeficiency virus-1 (HIV-1) infection: intestinal infections caused by cryptosporidia, isospora, cytomegalovirus, and other low-grade pathogens result in chronic diarrhea, wasting, and eventually death (Table 7.1). The advent of highly active antiretroviral therapy that helps maintain gut T cell numbers has seen gut infections and the enteropathy of HIV-1 infection diminish in clinical significance. Gut T cells are also at the forefront of protective responses to more aggressive pathogens such as *Salmonella* and *Shigella*. Identifying the pathways that allow gut T cells to react appropriately to commensal microbiota, yet respond to gut pathogens, remains one of the major challenges in biology. At the same time, over-reactivity of gut T cells to harmless foods, or the commensal microbiota, underpins food-sensitive enteropathies and inflammatory bowel disease (IBD).

In healthy individuals, T lymphocytes are a natural component of the cell types present at mucosal surfaces. The largest numbers are found in the small intestinal mucosa, primarily in the epithelium, the lamina propria cores of the

Table 7.1 Mucosal infections in T-cell-deficient patients

Mouth/esophagus
Candida albicans Human papillomavirus
Intestine
Cryptosporidium parvum *Isospora belli* *Giardia lamblia* Microsporidia *Enterocytozoon bieneusi* Cytomegalovirus Corona virus, rotavirus, adenovirus *Mycobacterium avium* complex
Airways
Pneumocystis carinii *Cryptococcus neoformans* Non-tuberculous Mycobacteria (e.g. *Mycobacterium avium* complex) Cytomegalovirus Bacterial pneumonia

villi, Peyer's patches (PPs), and isolated lymphoid follicles (Figure 7.1). They are somewhat less common in normal colon epithelium and lamina propria, but abundant in isolated colonic lymphoid follicles. There are also many T cells in the appendix. Normal gastric mucosa only contains a few T cells. At other mucosal surfaces, such as the upper airways and the upper respiratory tract, T cells are rather sparse in the connective tissue, but are abundant in nasopharynx-associated lymphoid tissue in rodents, and tonsils and adenoids in humans. It is important to emphasize that the organized mucosa-associated lymphoid tissues (MALT) are the inductive sites for mucosal immune responses, and the connective tissue between MALT is the site of expression of the mucosal immune response (Chapter 1). The majority of T cells in MALT are small non-activated T cells, largely part of the recirculating pool, whereas T cells in the lamina propria are activated cells, the progeny of T cells responding to antigen in the MALT.

The origin and phenotype of mucosal T cells.

In all mammalian species studied so far, CD4 T cells predominate in the lamina propria and CD8 cells in the gut epithelium. Intraepithelial lymphocytes are discussed in detail in Chapter 6; here we will focus on CD4 T cells. Embedded in the connective tissue matrix, lamina propria T cells receive signals from epithelial cells, stromal cells, and the matrix itself via integrin receptors, and are closely associated with dendritic cells and macrophages. These cells and signals likely control the lifespan and function of T cells in the gut, but there is also cross-talk between the T cells and other cell types; and, importantly, the presence of the microbiota, signaling via pathogen-associated molecular patterns and producing short-chain fatty acids in the distal bowel, produces an extremely complex series of interactions, which are only now being unraveled.

Figure 7.1 Immunostaining of human distal ileum for CD3, CD4, CD8, and CD20. The section contains three B-cell follicles (CD20). Note that surrounding the B-cell follicles are the T-cell zones, containing mostly CD4 cells.

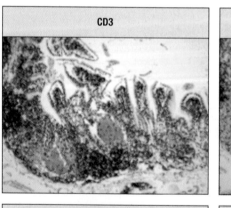

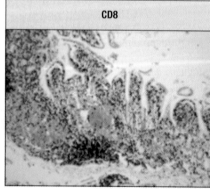

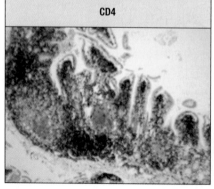

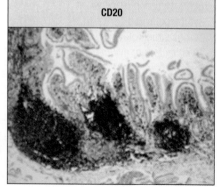

7-1   Mucosal T cells traffic from MALT to mucosal lamina propria.

It has been known for 40 years that the gut-homing patterns of small lymphocytes and immunoblasts differ. Small resting lymphocytes from mesenteric lymph nodes (MLNs) do not migrate to the lamina propria when transferred into normal recipients. However, in animals, T and B blasts from the mesenteric nodes or dividing cells in thoracic duct lymph readily migrate to the lamina propria. While this is more difficult to show in humans, analysis of T-cell receptor rearrangements in GALT and nearby lamina propria has shown the same clones of cells at each site, strongly supporting the notion that in humans, GALT T cells traffic via blood to lamina propria. The molecular basis for gut homing was initially elucidated 20 years ago when it was shown that gut homing of blasts was conferred by the expression of the $\alpha_4\beta_7$ integrin on T- and B-cell blasts and mucosal addressin cell adhesion molecule-1 (MAdCAM-1) on intestinal vascular endothelium. Later studies showed that chemokines were also involved. The small bowel epithelium secretes CCL25 and gut-homing cells express the CCL25 receptor CCR9. CCL25 is also tethered to vascular endothelium in the gut. In contrast, CCR9 and CCL25 interactions do not seem to be important in T-cell migration to the colon and in fact colonic epithelium does not express CCL25. Instead, colon homing may be controlled by CCL28 made by colon epithelium, $\alpha_4\beta_7$ integrin, and perhaps CCR10. It was, however, only relatively recently that the molecular basis for imprinting of gut homing by Peyer's patch T cells was elucidated, and it came from an unexpected angle, namely vitamin A.

Vitamin A is stored in the liver, so to make vitamin A deficient mice, pregnant mice are given a vitamin A deficient diet. There is enough vitamin A in the mother to allow the mouse pups to be born. However, maintaining the pups on a vitamin A deficient diet for 3 months has dramatic effects on mucosal T cells, in that these mice have marked depletion of small intestinal CD4 T cells and CD8$^+$ intraepithelial lymphocytes. Subsequent studies showed that in the Peyer's patches and mesenteric nodes, dendritic cells (DCs) express a retinal dehydrogenase (RALDH), which can convert all-*trans*-retinol to all-*trans*-retinoic acid. When added to activated T cells and B cells all-*trans*-retinoic acid is a potent inducer of both CCR9 and $\alpha_4\beta_7$. Thus the preferential expression of RALDH by antigen-presenting cells (APCs) in PPs and MLNs accounts for gut homing.

Interestingly, vitamin A deficiency had no effect on T cells in the airways, an observation of relevance to a notion which has gone out of fashion, namely the common mucosal immune system. This hypothesis suggested that immunization in the gut could protect other mucosal sites due to activated B cells and perhaps T cells migrating into all mucosal sites after enteric immunization. However, since respiratory tract vascular endothelium does not express MAdCAM, there is no longer a physiological basis for the hypothesis. A possible explanation for the homing of mesenteric lymph node cells to the airways may lie in the fact that the rules that govern homing become disrupted during inflammation. For example, gut-homing cells lodge in inflamed skin, but not normal skin. In the 1970s when the common mucosal immune system was suggested, virtually all mouse and rat colonies were infected with *Mycoplasma pulmonis*, a low-grade pathogen that induces lung inflammation.

7-2   Lamina propria T cells have the characteristics of activated lymphocytes.

Lamina propria mononuclear cells can be isolated relatively easily from human and mouse small bowel and colon. Many studies in humans have confirmed that the CD4 cells in normal gut lamina propria have the phenotype of activated effector T cells. Thus they are CD45RO$^+$, CD62^{low}, CD69^{high}, CD25$^+$, Fas$^+$, and FasL$^+$, and as expected from above, $\alpha_4\beta_7^+$ and CCR9$^+$.

It has also been established for many years that MALT are on the major route of lymphocyte recirculation. Small recirculating naive T cells (CD45RA⁺, L-selectin^high) can enter PPs because glycosylation of the mucin-like domains of the gut-specific addressin MAdCAM-1 allows it to bind L-selectin on naive T cells to allow them to cross high endothelial venules into the tissue. If these small naive T cells become activated by gut antigen in PPs, they lose L-selectin and express CD45RO. The presence of CD45RO⁺ cells co-expressing high levels of the $\alpha_4\beta_7$ integrin and L-selectin^low near microlymphatics in PPs suggests that these cells are leaving the PPs and are on their way to the lamina propria via the blood. The other site in PPs at which activated T cells are seen is next to M cells. These cells are also CD45RO⁺, L-selectin^low, CD69⁺ and are dividing.

All of these data are consistent with the idea that the gut-associated lymphoid tissues are the inductive sites of mucosal CD4 immune responses and that cells traffic via the thoracic duct and blood toward the lamina propria.

7-3   Cytokine production by mucosal T cells in healthy animals is T_H1/T_H17 dominated.

Germ-free mice have virtually no T cells in their gut lamina propria and the Peyer's patches in these mice only have primary follicles. Following colonization with bacteria, there is rapid immune activation and mucosal T cells rapidly reach the same levels as normal mice. Thus we can conclude that mucosal T cells are activated by gut microbial antigens and it would be expected that as lamina propria T cells are activated effector cells, they would be actively secreting cytokines. In humans, enzyme-linked immunospot (ELISPOT) analysis of freshly isolated PP and small intestinal lamina propria T cells shows a response dominated by interferon-γ (IFN-γ), with very few T cells secreting interleukin-4 (IL-4), IL-5, or IL-10. When human PP T cells are activated with mitogens or food antigens, the response is again IFN-γ dominated. These studies, however, pre-dated the discovery that IL-17A and IL-17F, made largely by a specialized helper T cell (T_H) subset (T_H17 cells), seem to be particularly highly expressed at mucosal surfaces. Thus CD4 T cells from healthy gut secrete equivalent amounts of IL-17A and IFN-γ when activated with anti-CD3/CD28. Mucosal T cells are therefore directed toward effector T_H1 and T_H17 responses; and in mice, in the absence of intestinal helminths, there are virtually no T_H2 cells in the small intestine. The ability of CD4 cells in the gut to make IFN-γ is probably very important in the control of gut infections such as cytomegalovirus and cryptosporidia, because, in the latter case, there is compelling evidence that immunity to this parasite depends on the ability of IFN-γ to render gut epithelial cells resistant to infection.

7-4   IL-17 and IL-22 may play an important role in protecting mucosal surfaces.

The strategic positioning of T_H17 cells at barrier surfaces reflects their importance in the neutralization of pathogens and commensal microbiota alike, both by coordination of neutrophilic inflammation and by maintenance or restitution of epithelial barrier integrity. Mucosal epithelia express receptors for IL-17 and/or IL-22, T_H17 cytokines that promote tight junction formation, mucus production, antimicrobial peptide production, and epithelial regeneration following injury. Elevation of IL-17 and IL-22 is characteristic of intestinal inflammation in murine colitis and IBD. Whereas IL-17 contributes to both neutrophilic inflammation and aspects of epithelial barrier function, IL-22 functions appear more limited to epithelial barrier function in the gut. Accordingly, IL-22-deficient mice have more severe intestinal inflammation and epithelial injury in mouse models of colitis, and are highly susceptible to colonic infection with the attaching–effacing natural mouse pathogen, *Citrobacter rodentium*.

Unlike the genes that encode IL-17A and IL-17F, expression of the gene encoding IL-22 is strictly dependent on the aryl hydrocarbon receptor, a ligand-activated transcription factor that binds a wide array of environmental and endogenous aromatic hydrocarbons, including tryptophan metabolites. Since tryptophan makes up 1–2% of the total protein in many foods, there is the intriguing possibility that dietary proteins may regulate immunity by the AroA pathway. In humans, CD4 T cells that express IL-22 in the absence of IL-17 cytokines have been identified, supporting the possibility that these cells are regulatory in nature and are dedicated to the maintenance and repair of epithelial integrity. However, their role in normal and inflamed gut has not been established.

7-5 Different types of gut microbiota appear to induce different cytokine responses in mucosal T cells.

In animals, where it is possible to experimentally control the gut flora, remarkable progress has been made on the texture of cytokine responses and their regulation by specific members of the microbiota (Figure 7.2). Germ-free mice monocolonized with the human gut bacterium, *Bacteroides fragilis*, show an expansion of T-cell numbers in the spleen and a shift in T-cell cytokine responses from a predominantly T_H2 response, to a T_H1 response. Remarkably, this effect is not seen if mice are mono-associated with *B. fragilis* lacking polysaccharide A, a highly unusual capsular polysaccharide which can be processed by APCs and induce CD4 T-cell activation. Polysaccharide A itself seems to induce IL-10-secreting regulatory T cells (T_{reg}) in the colon which can limit the extent of chemical colitis, both prophylactically and therapeutically.

Further support for the idea that specific types of bacteria can induce markedly different types of T-cell responses comes from the observations that mice from different suppliers have markedly different numbers of T_H17 cells in their small intestine, despite having identical numbers of T cells in their gut. It was shown that adult germ-free mice had virtually no T_H17 cells in their gut; however, on colonization with a microbial flora, T_H17 cells became abundant. If adult normal mice were treated with antibiotics to lower the flora, then T_H17 cell numbers were reduced by 50%. When mice from different vendors were examined, most had abundant T_H17 cells in their intestine, apart from mice from the Jackson Laboratory, where there were essentially none. Likewise, reconstitution of germ-free mice with the altered Schaedler flora, for many years considered to be the gold-standard specific pathogen free (SPF) flora, also did not result in mucosal T_H17 cells. Detailed analysis of the components of the microflora in mice from different vendors showed that the key organism that drives T_H17 responses was an unculturable segmented filamentous bacterium (SFB), related to clostridia. Mice harboring SFB are somewhat more resistant to *C. rodentium* infection than mice lacking SFB.

The final example of commensals which appear to induce regulatory responses are gut clostridial species. Colonization of mice with a flora restricted to clostridia induces somewhat more IL-10-secreting Foxp3+ inducible T_{reg} cells in the colonic lamina propria than in mice colonized with standard SPF flora. Increase in T_{reg} cells in the colon is associated with increased resistance to chemical colitis.

These studies are conceptually important because they indicate that different types of microbes may influence the generation of helper T cell lineages. They are also interesting in a historical context because the components of the Schaedler flora were chosen by Russell Schaedler in the 1960s on the basis of their lack of immunogenicity, and so a whole generation of immunologists who used animals colonized with the Schaedler flora were working on highly unusual and atypical mice. It will be intriguing to identify which bacterial components skew immune responses. At the same time, however, care has to be taken in extrapolating from a mouse monocolonized with a single bacterial strain to the complex microbiota (1000 species) of humans.

7-6   The developmental pathways of T$_H$17 cells are not well defined, especially at mucosal surfaces.

The developmental origins of T$_H$17 cells in mice and humans have stirred considerable controversy and a clear understanding of potential differences between the species is not resolved. Akin to the earlier debate over the presence of T$_H$1 and T$_H$2 subsets in humans after their discovery in mice, it may be that many of the apparent species differences in T$_H$17 biology reflect more the differences in the cell types investigated (blood T cells in humans, spleen T cells in mice), and experimental conditions, than intrinsic species differences, although this remains to be determined. In fact there is no a priori

Figure 7.2 Three examples of how specific microbes can influence T-cell subsets in the intestine. In panel a, colonization of mice with a specific pathogen free (SPF) flora containing a segmented filamentous bacteria (SFB) increases the numbers of T$_H$17 cells in the small intestine compared with mice colonized with the SPF flora alone. In panel b, mice colonized with an SPF flora containing a variety of clostridial species have many more IL-10-secreting T$_{reg}$ cells in the colon than mice colonized with an SPF flora alone. In panel c, mice colonized with *Bacteroides fragilis* expressing a capsular polysaccharide antigen (polysaccharide A) have many more IL-10-secreting T$_{reg}$ cells than mice colonized with a bacterium lacking polysaccharide A, because polysaccharide A appears to be able to convert Foxp3$^-$ CD4 T cells into Foxp3$^+$ regulatory cells in the gut.

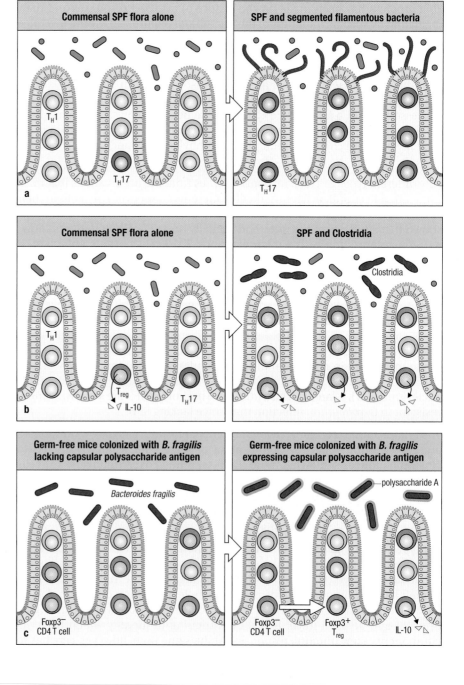

reason why different species should not adopt different strategies to generate effector T-cell responses appropriate for the lifestyle of the species, and the pathogens they face in their particular environments.

Whereas there is agreement over the requirement for pro-inflammatory cytokines in the differentiation of T_H17 cells, whether or not the same cytokines act similarly in both species is contentious. Even more contentious is the role of transforming growth factor-beta (TGF-β). In mice, where T_H17 cells were first described, IL-6 is indispensible for the induction of T_H17 cells and acts in concert with TGF-β to initiate T_H17 cell development from antigen-stimulated naive CD4 T cells. IL-23 acts downstream of IL-6 and TGF-β to reinforce T_H17 development in mice, but is not needed for the development of IL-17-producing T cells in mouse intestinal mucosa, despite the requirement for IL-23 in some mouse models of intestinal inflammation. IL-1β has been found to amplify T_H17 development in mouse T cells, and along with IL-23, elicits the production of T_H17 cytokines from mature T_H17 cells without a requirement for T-cell receptor stimulation, providing an antigen-independent mechanism for T_H17 cell responses.

In humans, initial studies that examined the generation of the T_H17 cells using naive (CD45RA$^+$) CD4 T cells derived from peripheral blood found that IL-1β and IL-23 induced production of T_H17 cells without a requirement for IL-6 or TGF-β, contrasting greatly with studies from mice. However, it was subsequently found, using naive CD4 T cells derived from human umbilical cord blood, that T_H17 development did require low levels of TGF-β as well as IL-1β and a STAT3-inducing cytokine—IL-6, IL-21, or IL-23—more closely resembling findings from mice. These studies emphasize that differences in the previous activation history of T-cell precursors in human and mouse studies may influence T_H17 cell development, although fundamental differences in the development of human and mouse T_H17 cells cannot be excluded. Despite these differences, there are major similarities in some aspects of T_H17 cells in both species, such as the expression of the transcription factor retinoic acid-related orphan receptor RORγt, IL-23 receptor, and CCR6.

In mice, T_H17 cells can be derived from naive precursors by the concerted actions of IL-6, IL-1β, and IL-23 in the absence of TGF-β, in agreement with some human studies. Importantly, the T_H17 cells generated under these conditions have a gene expression profile distinct from T_H17 cells generated in the presence of TGF-β, raising the possibility of further T_H17 subsets that might have different functions. While further studies will be needed, there appears to be different pathways involved in the generation of subtypes of T_H17 cells, the precise functions of which are yet to be defined. It is plausible that cells derived via the TGF-β-dependent pathway of T_H17 development may play a more important role in maintenance of barrier integrity at mucosal sites, particularly in the intestine where there is abundant active TGF-β made by epithelial cells and stromal cells and the greatest number of T_H17 cells are found in health; whereas a TGF-β-independent pathway plays a more prominent role under conditions of pathogen-induced or autoimmune inflammation.

7-7   The role of local differentiation or selective migration in T_H1 and T_H17 responses is unclear.

As for other CD4 T-cell subsets, the priming of T_H17 cell development is thought to occur in the GALT, where on exposure to antigen there is rapid downregulation in T cells of L-selectin and expression of CCR6 and CCR9, which favors trafficking to the intestinal lamina propria. As for T_H1 and T_H2 cells, the transfer of T_H17 cells into lymphopenic hosts has established that, at least in the setting of an 'empty' T-cell compartment, migration of T_H17 cells into the intestinal lamina propria can readily be seen. Whether this occurs similarly in a T cell-replete repertoire is not known. In fact, there is remarkably little known about the factors which control effector CD4 polarization in the MALT of experimental animals, and essentially no data for humans.

There is no doubt that the gut lamina propria is enriched for effector T_H1 cells, T_H17 cells, and T cells that make both IFN-γ and IL-17A. The critical question is whether the imprinting of gut homing in GALT following T-cell activation has another layer of complexity, with differences in homing of different CD4 cell lineages. Added to this is the issue of how plastic development in the lamina propria is; in other words, is lineage commitment settled in the lamina propria or is it pre-committed prior to extravasation, perhaps imprinted in GALT. At the heart of this debate is whether there is a defined T_H17 cell precursor, separate from virgin CD4 T cells that can become T_H1, T_H2, or T_{reg} cells. In humans it has been suggested that CD161 along with IL-23R, CCR6, and the transcription factor retinoic acid-related orphan receptor C isoform 2 are markers of T_H17 cells. CCR6 is of particular interest in the gut because it is the only receptor for CCL20. Normal colonic epithelial cells express CCL20, but expression is highest in inflamed epithelium and, constitutively, in the follicle-associated epithelium overlying Peyer's patches and cryptopatches. CCR6 is also expressed on some lymphoid tissue inducer (LTi) cells and mice lacking CCR6 have small Peyer's patches. Further experiments are therefore needed to determine if T_H17 and T_H1 cells differ in their ability to migrate into the gut. It is also unclear if the effects that different types of bacteria have on mucosal T_{reg}, T_H1, and T_H17 responses are controlled at inductive sites or the lamina propria.

On the other hand, when a T cell extravasates into the lamina propria, the local environment contains molecules known to affect T_H17 cell development. Epithelial cells produce active TGF-β and latent TGF-β is abundant in the matrix. Lamina propria DCs also secrete retinoic acid which may inhibit T_H17 generation and promote T_{reg} development. Mice lacking TGF-β have few T_H17 cells in their gut and spleen, suggesting that TGF-β has a global effect on T_H17 cell generation rather than a local effect in the gut lamina propria. A slightly different mouse which can make TGF-β but which cannot activate the cytokine in the matrix also has a reduced number of lamina propria T_H17 cells. Confounders in these experiments, however, are that mice lacking TGF-β have extremely strong T_H1 responses that inhibit T_H17 cell generation and there is the possibility that even at inductive sites, matrix bound TGF-β may be needed for T_H17 cell generation.

7-8 The T_H17 lineage displays considerable flexibility.

With the emergence of better technical approaches to track the fate of defined T-cell populations *in vivo*, it has become apparent that there is substantial flexibility, or plasticity, in the developmental programs of regulatory and effector T-cell subsets. CD4$^+$ T cells that co-express lineage markers of regulatory (Foxp3) and T_H17 (RORγt) T cells have been detected in the normal gut of both mice and humans, and the frequencies of these cells increase in the setting of inflammation. At present, it is unclear whether these 'dual-expressors' represent early precursors from which T_H17 or T_{reg} cells emerge, represent Foxp3$^+$ T_{reg} cells that are in transition to T_H17 cells in response to pro-inflammatory cytokines, are a functionally distinct T-cell subset, or a combination thereof. However, there are growing data from studies in murine models of infectious disease that T_{reg} cells can transition to progeny with effector properties, such as T_H17 or T_H1, under the influence of inflammatory cytokines. Reciprocal programming (i.e., effector T cells into T_{reg} cells) has not been well demonstrated, suggesting that the conversion of T_{reg} cells to T effectors might represent an evolutionary adaptation to override the dominance of homeostatic T_{reg} function at mucosal sites in the face of a pathogenic threat. It is currently unclear to what extent similar transitions might occur in the context of immune dysregulation during the evolution of IBD. However, the possibility that T_{reg} cells may have the potential to become effector T cells needs to be considered when thinking of adoptive T$_{reg}$ therapy for chronic inflammatory conditions, or even strategies to boost T$_{reg}$ numbers *in vivo*.

T-cell subset plasticity is not, however, limited to T$_{reg}$ cells. The transition of CD4$^+$ T cells of one effector subset to progeny with features of a distinct effector subset has been increasingly reported. This has been most clearly demonstrated for T$_H$17 cells, which can give rise to T$_H$1-like cells contingent upon the hierarchy of cytokine signals received following T$_H$17 cell differentiation. In particular, T$_H$17 cells exposed to the T$_H$1-polarizing cytokine IL-12 extinguish expression of genes that encode IL-17 cytokines in favor of IFN-γ. The transition of T$_H$17 cells into IFN-γ-producing T$_H$1 cells is dependent on STAT4 signaling and induction of the T$_H$1 lineage transcription factor, T-bet. Similarly to the transition of T$_{reg}$ cells to T effector cells, T$_H$17 to T$_H$1 transitions appear to be unidirectional. That is, T$_H$17 cells can transition to T$_H$1 cells, but T$_H$1 cells do not transition to T$_H$17 cells. In intestinal inflammation in mice, the transition of T$_H$17 cells into T$_H$1-type cells is associated with the development of colitis, suggesting that the mix of T$_H$17 and T$_H$1 cells typically found in the intestinal lamina propria, both at homeostasis and in IBD, might be due to diversion of some T$_H$17 precursors locally in the intestines (Figure 7.3).

The plasticity of T$_H$17 cells appears to be shared in humans. CD4$^+$ T-cell isolates from the intestine of patients with Crohn's disease contain a mix of distinct subsets of T$_H$17 and T$_H$1 cells, as well as IL-17$^+$ IFN-γ$^+$ T cells (Figure 7.4). Similar to findings in mice, T$_H$17 and 'T$_H$17/T$_H$1' cells cloned from intestinal T-cell isolates can be diverted to a T$_H$1 phenotype in response to IL-12. Interestingly, CD161, which was initially proposed as a marker for human T$_H$17 cells (mice do not express a CD161 homolog) is also expressed by a substantial fraction of IFN-γ$^+$ T cells in normal and inflamed intestines. Further,

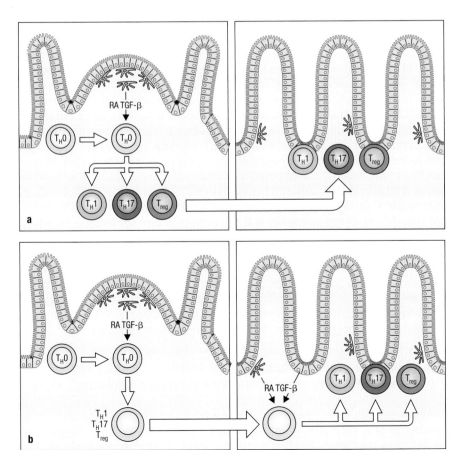

Figure 7.3 In the gut, the control of CD4 T cell differentiation into T$_H$1, T$_H$17, and T$_{reg}$ cells, and their evolution into cells with other cytokine profiles is still not clear. In one model (panel a), lineage commitment and imprinting of gut homing occurs in GALT and T$_H$1, T$_H$17, and T$_{reg}$ cells migrate as distinct subsets into the lamina propria. In an alternative model (panel b), terminal differentiation into distinct subsets occurs in the lamina propria under the influence of local cytokines made by dendritic cells and epithelial cells. In health, however, where the cytokine milieu is inhibitory for further differentiation and survival, T cells die and are replaced by cells from the blood. RA, retinoic acid; TGF-β, transforming growth factor beta.

Figure 7.4 Flow cytometric analysis of CD4 lymphocytes from normal human colon (HC), Crohn's disease colon (CD), and ulcerative colitis colon (UC). Cells making IFN-γ dominate, but in ulcerative colitis there are many cells making IL-17A only. In all groups, there are cells making both cytokines. Lamina propria mononuclear cells were stimulated with phorbol myristate acetate (PMA) and ionophore for 4 hours prior to analysis. (Adapted from L. Rovedatti et al., *Gut* 58:1629–1636, 2009. With permission from BMJ Publishing Group Ltd.)

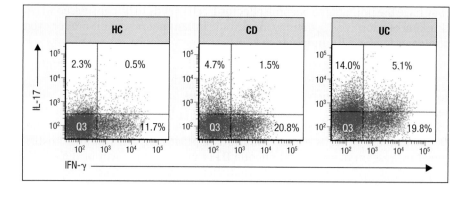

CD161+ human T cells give rise to a distinct subpopulation of IFN-γ+ cells under T_H17-polarizing conditions. Thus, as in studies in mice, human T_H17 cells may be phenotypically unstable and give rise to T_H1-like cells, whereas the converse has not been demonstrated. The implications of T_H17 plasticity for normal intestinal immune regulation and the development of IBD remain to be defined.

Regulatory T cells and the control of effector T-cell responses.

It is very well established that reconstitution of mice lacking T or B cells with small numbers of naive CD4+ T cells leads to colitis. If the recipients are germ-free, colitis does not develop. It is also well known that co-injection of natural CD4+ T_{reg} cells ameliorates the colitis induced by naive T cells. Thus it is often claimed that pathogenic effector T-cell responses against the gut microflora are controlled or dampened by T_{reg} cells, and a corollary of this statement, namely that IBD is somehow due to a failure of T-cell regulation, is often mooted. Suppression appears to be mediated by TGF-β, although other cytokines such as IL-10 and even IL-17A may be suppressive in the correct context. In reality, control of mucosal effector and regulatory T cells is likely to be multilayered and multifaceted, and there are almost certainly many different mechanisms that control immune responses in the gut, including the accessibility of antigen to the lamina propria. For example, patchy defects in the intestinal barrier leading to increased local intestinal permeability cause a patchy Crohn's-like lesion in mice with a normal immune system. The colonic epithelial cells in mice with a gut epithelial knockout of nuclear factor κB essential modulator (NEMO) undergo apoptosis and the barrier is broken, leading to colitis. Finally, in patients with neutrophil defects such as chronic granulomatous disease or glycogen storage disease Type 1b, the normal flora enters the lamina propria and persists and elicits a disease almost identical to Crohn's disease. In normal lamina propria, there is abundant active TGF-β made by epithelial cells and stromal cells and as relatively little gut antigen crosses the intact epithelial/mucus barrier, it is entirely possible that effector T-cell regulation in health is due to 'neglect,' rather than active suppression (Figure 7.5).

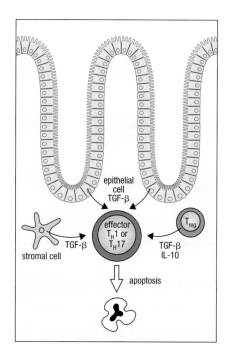

Figure 7.5 Different pathways probably downregulate effector CD4 T_H1 and T_H17 cell responses in the intestine in health and induce apoptosis of activated T cells. Epithelial cells are a major source of bioactive TGF-β. Stromal cells also make TGF-β which may be activated by the low levels of proteases in normal gut. CD103+ dendritic cells may also induce T_{reg} cells, which by making TGF-β and IL-10 dampen potentially damaging T_H1 and T_H17 responses. However, a major factor is probably the fact that very little antigen will cross the mucus layers and epithelium, and subepithelial macrophages (not shown) are highly efficient at killing and degrading bacterial antigens without evoking a pro-inflammatory response.

7-9   Human diseases due to single gene defects are informative in understanding gut inflammation and immune regulation.

Immunodysregulation, polyendocrinopathy, enteropathy, X-linked (IPEX) syndrome is a very rare X-linked disease where mutations in Foxp3 lead to defective regulatory T-cell activity. It is often used as an example of how regulatory T-cell defects cause colitis. Indeed, virtually all of these children have small bowel inflammation and some have colitis. The gut manifestations, however, go alongside chronic inflammation in the pancreas and other endocrine organs, kidneys, liver, and skin; there is often anemia and/or thrombocytopenia, autoantibodies, and a generalized lymphadenopathy. It is therefore possible that children with IPEX have a general autoreactivity to self-antigens, with the gut tissue as a target of autoimmune attack because the gut is also an endocrine organ. Indeed before IPEX and Foxp3 were identified, the disease was often referred to as autoimmune enteropathy. The scurfy mouse also has mutations in Foxp3, but does not appear to have colitis; instead, the mouse dies at about 3–4 weeks of age showing lymphohistiocytic infiltration of many tissues and lymphadenopathy. Both of these situations differ markedly from human IBD where, especially for Crohn's disease, pathological manifestations in tissues other than the gut are generally thought to be a consequence of the gut lesions. It is also worth noting that prior to the development of the transfer model of colitis in lymphopenic mice, there was an extensive literature in the rat where transfer of naive T cells into athymic animals caused generalized autoimmune disease, including colitis.

Almost 20 years ago it was shown that IL-10 null mice developed spontaneous enteritis, dependent on the microbiota. Mice with T-cell deficiency of IL-10 also develop a spontaneous colitis. Mice lacking IL-10 in the T_{reg} population also develop spontaneous colitis, but it is less severe than in mice with global loss of IL-10, suggesting that other sources of IL-10 may be important. Consistent with this, IL-10-producing Foxp3$^+$ cells are present in the intestinal lamina propria, and there are also data to suggest that T_H1 cells, as they differentiate, also begin to make IL-10. Extremely compelling evidence that IL-10 controls responses to the flora in humans comes from rare cases where there are defects in IL-10R. Very young children with IL-10R loss-of-function mutations develop a severe early onset inflammatory bowel disease with many characteristics of Crohn's disease. Successful therapy has been achieved using bone marrow transplantation.

7-10 Regulation of mucosal immune responses is complex and depends on whether the effector T cell can be regulated.

It is worth considering whether potentially pathogenic T_H1 or T_H17 responses in the gut are controlled by regulatory T cells. First, in Peyer's patches and mesenteric lymph nodes there is evidence that feeding protein antigens in intact mice leads to conversion of naive T cells into Foxp3$^+$ T_{reg} cells. However, the largest numbers of antigen-specific Foxp3$^+$ T_{reg} cells are found in the small bowel lamina propria. CD103$^+$ DCs from lamina propria can also convert antigen-specific T cells into T_{reg} cells and this process is dependent on retinoic acid production. In mice, where Foxp3 is a good marker for T_{reg} cells, these experiments suggest that there is generation of T_{reg} cells in response to gut protein antigens in GALT and cells then migrate to the lamina propria. In the small intestine there is evidence for heterogeneity in the lamina propria in that macrophages appear to be able to induce T_{reg} cells by secreting retinoic acid, IL-10, and TGF-β, whereas dendritic cells activate T_H17 cells. It is not known, however, whether the same type of response occurs to the antigens of the microbiota.

In humans there are difficulties in using markers such as Foxp3 and CD25 to examine T_{reg} cells in the gut. Both markers are expressed on activated effector T cells, so they are not particularly good tools to analyze T_{reg} function. In humans there are few data on GALT; however, there have been some functional studies on lamina propria mononuclear cells. CD4$^+$CD25^{high}cells from normal lamina propria are less proliferative and produce fewer cytokines than CD4$^+$CD25$^-$ lamina propria mononuclear cells. When the former population is added to mitogen-activated CD4$^+$CD25$^-$ blood T cells, there is a marked

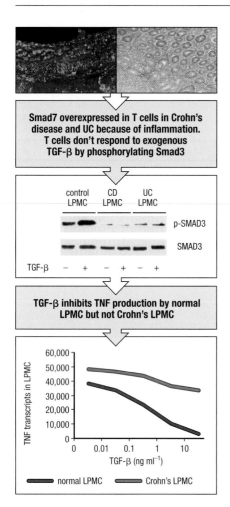

Figure 7.6 Smad7 overexpression in activated T cells in Crohn's disease (CD) and ulcerative colitis (UC) renders mucosal T cells resistant to TGF-β.
In humans, addition of exogenous TGF-β rapidly phosphorylates Smad3 (p-SMAD3) in lamina propria mononuclear cells (LPMC) from normal human gut, but has no effect on lamina propria mononuclear cells from IBD patients. Consistent with this, TGF-β can inhibit the production of tumor necrosis factor (TNF) transcripts by superantigen-activated normal lamina propria mononuclear cells, but has no effect on lamina propria mononuclear cells from Crohn's patients. Collectively these data show that expression of Smad7 in effector T cells regulates whether T cells are susceptible to regulation. (Adapted from G. Monteleone et al., *J. Clin. Invest.* 108:601–609, 2001. With permission from Elsevier Ltd.)

reduction in proliferation. This suggests that there are regulatory T cells in the lamina propria; however, regulatory activity is partially abolished by exogenous IL-2, which does suggest that at least some of the effect is due to cytokine deprivation, rather than active suppression. There does not appear to be any reduction in T_{reg} numbers in the gut in IBD.

An alternative approach to understanding CD4 T_H1 or T_H17 responses in the gut in idiopathic inflammation is to assume that the problem is an antipathogen response gone awry, directed at harmless microbes. In pathogen responses, it is important to prevent T_{reg} cells dampening effector cell function until the pathogen has been eliminated. Thus there is an advantage for effector CD4 cells in the gut to be resistant to suppression, regardless of the numbers of T_{reg} cells in the tissue. One way that effector T cells achieve this is by overexpressing Smad7, the intracellular inhibitor of TGF-β signaling. Smad7 functions in two main ways: by blocking the binding of Smad2/3 to ligand-activated TGF-β receptor; and by targeting the receptor itself for ubiquitination and degradation. As might be expected, T_{reg} cells and TGF-β protein are ineffective in dampening pathogenic T-cell responses from inflamed IBD tissues. However, knocking down Smad7 with an anti-sense oligonucleotide allows endogenous TGF-β to dampen T-cell responses. Overexpression of Smad7 in CD4 T cells makes animals more susceptible to experimental colitis because endogenous TGF-β and T_{reg} cells cannot dampen inflammation. Similarly, in the transfer model of colitis, T cells overexpressing Smad7 cannot be controlled by transferred T_{reg} cells. Paradoxically, however, in mouse models of colon cancer, overexpression of Smad7 in T cells is highly beneficial because the exaggerated T-cell responses translate into exaggerated anti-tumor responses, even though the mucosa remains highly inflamed (Figure 7.6).

One of the main difficulties in this area, especially when looking at the effects of the microflora, is the difficulty in generating T-cell responses to the flora. Although it was demonstrated many years ago that T cells from normal gut responded to microbial antigens from the flora of different individuals but were tolerant to their own flora, and that in IBD autologous gut T cells responded vigorously to an individual's own microflora, these results have not been followed up. In mice, T-cell lines reactive to gut bacterial antigens cause colitis when transferred into lymphopenic mice, and the colitogenic response can be inhibited *in vivo* by *in vitro*-generated T_{reg} cells with the same specificity.

T cell-mediated gut diseases.

Although IBD and celiac disease will be covered in other chapters, it is worth considering these conditions as situations where the 'balance' between CD4 T-cell activity in the gut and luminal antigens becomes skewed and chronic T-cell activation causes disease.

Celiac disease is of interest because there is no doubt that most, if not all, of the pathology is driven by CD4 T cells in the lamina propria responding to gliadin peptides presented in the context of HLA-DQ2 or DQ8. However, DQ2 is an extremely common haplotype in the general population and it is still not known what additional factors lead to only a few percent of DQ2+ individuals becoming gluten reactive. Environmental factors are clearly important. There are cases in the literature where identical twins have developed celiac disease years apart, and when the proband was diagnosed, the healthy sibling has been investigated thoroughly and shown to be normal. One notion is that viral infections in the gut not only break the barrier to allow antigen to enter the lamina propria, but can induce the local production of type I interferons, which are potent stimulators of T_H1 responses. Once sensitized, however, celiac disease is a lifelong condition, which implies T-cell memory. It is highly unlikely that long-lived gluten-specific memory T cells persist in the lamina propria; instead they are probably part of the recirculating memory T-cell pool, in which case they may encounter gluten in GALT. One feature of celiac disease, however, that has not been explained is the time to relapse when returning to a gluten-containing diet. This is highly variable between individuals and does imply that environmental factors may also be important at all stages of the disease.

Recently, genome-wide association studies have attempted to fill the gap between the high frequency of DQ2$^+$ individuals and the low numbers of these who develop disease. Virtually all of the polymorphisms identified are associated with genes that control the tone and texture of the T-cell responses, such as IL-2 and IL-21.

In Crohn's disease there is overwhelming evidence that the disease is mediated by activated CD4 T_H1 and T_H17 cells. Genome-wide association studies have identified many dozens of polymorphisms related to innate immune function as being important risk factors (Chapter 31). It needs to be remembered, however, that the penetrance of Crohn's disease is almost completely environmental. Thus Crohn's disease is very uncommon in India, but when families migrate to westernized countries with high incidences of Crohn's disease, the incidence in the children of immigrants is the same as that of the indigenous population.

The unifying factor in both celiac disease and Crohn's disease is the persistence of activated CD4 T cells in the intestinal lamina propria. In celiac disease, the T cells are antigen driven, as shown by a gluten-free diet restoring normality of gut structure. In Crohn's disease, however, the situation is much more complex. The contribution of antigen is not known, and there are very many cytokines overexpressed in Crohn's disease mucosa, such as IL-1, IL-2, IL-6, IL-7, IL-12, IL-15, IL-18, IL-21, and IL-23, which can signal to activated T cells, prevent apoptosis, and allow the cells to continue to produce IFN-γ, IL-17A, and tumor necrosis factor-α (TNF-α). Identifying which, if any, of these cytokines are critical, and thus worthy of the expense of therapeutic intervention, represents an extremely significant challenge. It may well be that so many factors are involved in driving CD4 T-cell survival in Crohn's disease that targeting cytokines may not be the best way to go, and that strategies to target the T cells themselves may be more productive. The caveat, however, remains that because CD4 T cells in the gut and airways are vital to maintain freedom from low-grade infections, any T-cell directed therapy has to be effectively monitored.

The role of CD4 T cells in ulcerative colitis remains opaque. There are some data which suggest overexpression of IL-13; however, in absolute terms IL-13, or IL-5 for that matter, is dominated by IL-17A and IFN-γ, which are present in at least tenfold excess compared with IL-13. Severe ulcerative colitis responds well to cyclosporin A, which clearly suggests a role for T cell-mediated damage in fulminant disease; however, trials of tolerizing anti-CD3 antibodies in ulcerative colitis have not been a success.

Innate lymphoid cells.

Although there is much focus on cytokines in the gut made by T cells, one of the most interesting recent observations in mucosal immunology has been on the pro-inflammatory and regulatory cytokines made by innate lymphoid cells (ILC). The prototypic ILC is the natural killer (NK) cell, originally identified through their ability to kill tumor cells, but which also recognize and kill cells expressing pathogen-associated and stress-associated ligands. NK cells are also hard-wired to secrete cytokines when activated, typically IFN-γ, TNF-α, and granulocyte–macrophage colony-stimulating factor (GM-CSF). NK cells display heterogeneity and can be subdivided into different functional subsets in humans by the expression of CD56 and CD16. Likewise, in mice different NK subsets based on, for example, CD27, CD127, and CD11b have different functional properties. Despite much investigation, however, a clear role for NK cells in mucosal health and disease, especially in the gut, has yet to be identified.

A second type of ILC is represented by lymphoid tissue inducer (LTi) cells. These cells were originally described as being key for lymphoid tissue organogenesis through their ability to activate stromal cells to express adhesion molecules and secrete chemokines needed to attract dendritic cells, T cells, and B cells into developing lymph nodes and Peyer's patches. Like T_H17 cells, LTi cells depend on RORγt, and produce IL-17A and IL-22. IL-22 is thought to play an important role in epithelial homeostasis and induces production of defensins. IL-22-deficient mice are highly susceptible to *Citrobacter rodentium* infection, and there is now evidence that IL-22 production by LTi cells may be more important in resistance to *Citrobacter* than T cell-derived IL-22.

However, matters are complicated by the fact that a third type of ILC has also been identified that may be an intermediate between NK cells and LTi cells. In mice they express NKp46, but in humans they express NKp44 and are CD56$^+$. These cells produce very little IFN-γ, and are not cytotoxic, but contain abundant IL-22 and have been termed ILC22 cells; they have also been implicated in resistance to *Citrobacter* infection. Other ILC have been identified in the gut, termed natural helper cells or nuocytes, which make T_H2-type cytokines. Unraveling the complexity of ILC will be a major challenge in the next few years.

Summary.

The mucosal surfaces are the battleground between the immune system, pathogens, and dietary and environmental antigens. Terms such as 'balance' and 'homeostasis' are used to describe the situation in health. These words, however, are descriptors and give no insight into mechanisms. The fact that many different transgenic and knockout mice, many with normal numbers of T cells and B cells, develop chronic gut inflammation, mediated in most cases by CD4 T cells, strongly suggests that control of mucosal T cells is complex. In health it would appear that activated mucosal T cells are generated in GALT (hence the large numbers of T_H1 and T_H17 cells in normal gut) in response to, and perhaps controlled by, the microbiota and perhaps even diet. When effector cells migrate to the lamina propria in health they are short lived and die by apoptosis. An unpredictable coalition of antigen(s) (perhaps due to a break in the epithelial barrier or a genetically determined inability to break down the cell walls of the microbiota) and perhaps infection changing the local cytokine milieu, may result in the rescue of CD4 T cells from the apoptotic path, and drive them toward survival. The inflammation generated by the activated CD4$^+$ T cells then tips the microenvironment into self-perpetuating inflammatory milieu, since the barrier is then broken by T cell-derived cytokines, and inflammatory cells producing survival cytokines move into the tissues. Nonspecific agents such as corticosteroids may dampen the inflammation, and anti-cytokine agents such as anti-TNF may also produce transient relief. However, disease persists at low levels or returns, presumably because of CD4 T-cell memory, about which we know very little in mucosae. Exciting recent developments concerning different subsets of ILC in the gut add a new dimension to the complexity of gut immunity and inflammation and it will be interesting to determine the role of ILC at mucosal surfaces in humans.

Further Reading.

Atarashi, K., Tanoue, T., Shima, T., *et al.*: **Induction of colonic regulatory T cells by indigenous *Clostridium* species.** *Science* 2011, **331**:337–341.

Balasubramani, A., Mukasa, R., Hatton, R.D., *et al.*: **Regulation of the *Ifng* locus in the context of T-lineage specification and plasticity.** *Immunol. Rev.* 2010, **238**:216–232.

Fantini, M.C., Rizzo, A., Fina, D., *et al.*: **Smad7 controls resistance of colitogenic T cells to regulatory T cell-mediated suppression.** *Gastroenterology* 2009, **136**:1308–1316.

Glocker, E.O., Kotlarz, D., Boztug, K., *et al.*: **Inflammatory bowel disease and mutations affecting the interleukin-10 receptor.** *N. Engl. J. Med.* 2009, **361**:2033–2045.

Hovhannisyan, Z., Treatman, J., Littman, D.R., *et al.*: **Characterization of interleukin-17-producing regulatory T cells in inflamed intestinal mucosa from patients with inflammatory bowel diseases.** *Gastroenterology* 2011, **140**:957–965.

Ivanov, I.I., and Littman, D.R.: **Segmented filamentous bacteria take the stage.** *Mucosal Immunol.* 2010, **3**:209–212.

Iwakura, Y., Ishigame, H., Saijo, S., *et al.*: **Functional specialization of interleukin-17 family members.** *Immunity* 2011, **34**:149–162.

Maynard, C.L., and Weaver, C.T.: **Intestinal effector T cells in health and disease.** *Immunity* 2009, **31**:389–400.

Murphy, K.M., and Stockinger, B.: **Effector T cell plasticity: flexibility in the face of changing circumstances.** *Nat. Immunol.* 2010, **11**:674–680.

Round, J.L., and Mazmanian, S.K.: **Inducible Foxp3+ regulatory T-cell development by a commensal bacterium of the intestinal microbiota.** *Proc. Natl Acad. Sci. USA* 2010, **107**:12204–12209.

Rubtsov, Y.P., Rasmussen, J.P., Chi, E.Y., *et al.*: **Regulatory T cell-derived interleukin-10 limits inflammation at environmental interfaces.** *Immunity* 2008, **28**:546–558.

Sakaguchi, S., Miyara, M., Costantino, C.M., *et al.*: **FOXP3+ regulatory T cells in the human immune system.** *Nat. Rev. Immunol.* 2010, **10**:490–500.

Sarra, M., Pallone, F., Macdonald, T.T., *et al.*: **IL-23/IL-17 axis in IBD.** *Inflamm. Bowel Dis.* 2010, **16**:1808–1813.

Spits, H., and Di Santo, J.P.: **The expanding family of innate lymphoid cells: regulators and effectors of immunity and tissue remodeling.** *Nat. Immunol.* 2011, **12**:21–27.

Zenewicz, L.A., Antov, A., and Flavell, R.A.: **CD4 T-cell differentiation and inflammatory bowel disease.** *Trends Mol. Med.* 2009, **15**:199–207.

Mucosal B Cells and Their Function

8

Mucosal epithelia separate the external environment from the tissues of the body. Research over the last 20 years has revealed that these surfaces have specific immune protection provided by antibodies in the external body fluids, that is, exocrine secretions such as tears, nasal secretions, saliva, intestinal juice, and breast milk. The protection of the approximately 400 m^2 of internal epithelial surface area in healthy adult humans involves antibody production by approximately 6×10^{10} mucosal plasma cells, comprising 80% of the total plasma cell population in the body. Thus the largest and most active compartment of the immune system is dedicated to defense of mucosal surfaces.

Mucosal barriers can be subdivided into those associated with either the inducer or the effector arms of the mucosal immune response. The immune inductive compartments consist of mucosa-associated lymphoid tissue (MALT; Chapter 1). These tissue compartments resemble lymph nodes in that they are highly dynamic in terms of lymphocyte traffic and lymphocyte proliferation. However, because MALT is constantly exposed to antigens coming directly from mucosal surfaces, these lymphoid structures are uniquely adapted to generate precursors of mucosal effector cells. In contrast, the immune effector compartment is diffusely located within a subepithelial stroma of mucosal connective tissue, termed the lamina propria of mucosae or the stroma of exocrine glands, that contains plasma cells and their immediate precursors, together with effector T cells, macrophages, dendritic cells, granulocytes, and mast cells. The most abundant effector molecule produced in the lamina propria is IgA. Indeed IgA is probably the best understood and most widely accepted mediator of mucosal immunity.

In this chapter the B cells and antibodies associated with mucosal immunity will be described, followed by a description of the progression of mucosal B-cell responses from a regulated inductive phase in MALT, through to the production of immunoglobulins by plasma cells.

Cells and proteins involved in humoral immunity at mucosal surfaces.

It is often not appreciated that there are fundamental differences between the structure of the lymphoid tissues in MALT and the other lymph nodes scattered throughout the body. Likewise, there are major differences in the control of their development and the molecules that drive their formation. In particular, peripheral lymphoid tissue develops in a largely antigen-free environment, whereas mucosal lymphoid tissues are exposed to the microbiota at birth and face constant antigenic stimulation throughout life.

8-1   MALT has major differences from conventional lymphoid tissue.

The uncapsulated organized lymphoid tissue in MALT is separated from the lumen by a specialized follicle-associated epithelium (FAE) containing micro-fold or membrane (M) cells. FAE is infiltrated by B cells that express high levels of co-stimulatory molecules CD80/CD86, CD4 T cells, and dendritic cells, and is therefore quite different from ordinary mucosal surface epithelium, for example on the villus surface, which contains mostly CD8⁺ T cells. Lymphoepithelial FAE is a portal of entry of antigens from the lumen and its presence is a defining characteristic of MALT.

It should be noted that despite the important role of mesenteric lymph nodes in intestinal immunity, these organs are not MALT structures, and should therefore not be designated gut-associated lymphoid tissue (GALT), according to the terminology recommended by the International Society for Mucosal Immunology and the International Union of Immunological Societies. In contrast to GALT, but like peripheral lymph nodes, the mesenteric lymph nodes are encapsulated, receive antigens via afferent lymphatics, and therefore do not sample antigens directly from the mucosal surface.

Exogenous stimuli from the gut lumen are transported into GALT via the M cells of the FAE, probably aided by dendritic cells, which may send processes into the lumen, penetrating the epithelium, and which, as well as taking up antigens, may be activated by bacterial components. When antigen reaches the intraepithelial and subepithelial compartments of GALT, it encounters a mixed leukocyte population of B and T cells in a microenvironment that is also rich in macrophages and dendritic cells expressing many innate receptors. This part of GALT is often termed the dome region. Macrophages and dendritic cells in this location form a critical first line of defense beneath the epithelium that can influence downstream immunological events and B-cell behavior.

The structure of GALT is similar throughout the gastrointestinal tract. It comprises prominent B-cell follicles with intervening T-cell zones. The B-cell zone in most GALT structures is composed of a germinal center of rapidly dividing B cells, surrounded by a mantle zone of IgD⁺ naive B cells. The broadest aspect of the often narrow, crescent-shaped mantle zone faces the direction of antigen approach from FAE. GALT contains a broad marginal zone that resembles the splenic marginal zone (Figure 8.1). Splenic and GALT marginal-zone B cells are similar morphologically, being medium-sized cells with cleaved nuclei.

Figure 8.1 Differences between human and rodent GALT. Panel a: human GALT stained with hematoxylin (nuclei) and eosin (cytoplasm). Panel b: human GALT double stained by immunohistochemistry with antibodies to IgD (brown) and CD20 (blue). IgD-negative B cells in blue are present in the germinal center and in the marginal zone surrounding the mantle. Panel c: rat GALT stained with hematoxylin and eosin for comparison with human GALT in panel a. Panel d: rat GALT stained with antibody to IgD (brown). This section has a blue nuclear counterstain. In contrast to the human GALT, the IgD-expressing zone of B cells in rats extends into the follicle-associated epithelium.

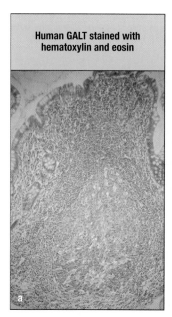

Human GALT stained with hematoxylin and eosin

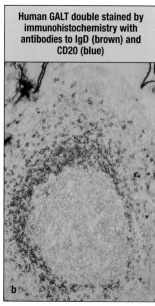

Human GALT double stained by immunohistochemistry with antibodies to IgD (brown) and CD20 (blue)

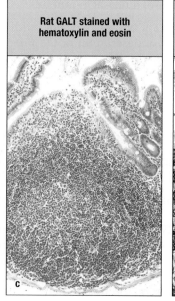

Rat GALT stained with hematoxylin and eosin

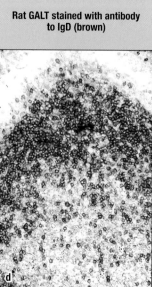

Rat GALT stained with antibody to IgD (brown)

They also do not express IgD. Splenic marginal-zone B cells have mutated IgHV genes (albeit not always heavily mutated) suggesting that they are post-germinal center memory B cells. In contrast, the GALT marginal zone contains both cells with no mutations in IgHV and cells with a large number of mutations. This property of heavily mutated IgHV genes of GALT marginal-zone B cells is shared with the adjacent plasma cell population in the lamina propria, and GALT germinal center cells, suggesting that they are both derived from germinal centers. The population of marginal-zone B cells in GALT with mutated IgHV genes is probably of memory cells.

MALT structures do not have a distinct boundary in the form of a fibrous capsule. Instead they diffuse gradually into the adjacent lamina propria, so it may be difficult to determine the boundary between the inductive site of GALT and the effector tissue of the lamina propria. A useful indicator of the margins of MALT in tissue sections is to stain with antibodies to mature B cells such as anti-CD20. Most B cells in MALT express CD20, but CD20$^+$ cells are very rare in the intestinal lamina propria (Figure 8.2). Researchers working on GALT and lamina propria cells should be aware of the microanatomy of the tissue. For example, studies claiming to characterize lamina propria B cells may in fact be analyzing the B cells in the periphery of GALT if the cells studied express mature B-cell antigens. The B cells on the periphery of GALT are also often large cells that may make connections with adjacent cells via cytoplasmic processes. There is also reduced expression of J chain by IgG-producing plasma cells at the GALT–lamina propria boundary region in comparison with the more distant lamina propria. This is consistent with the idea that the margins of GALT have distinctive properties, related to B-cell function or the stage of differentiation.

In the small intestine, the induction and regulation of the mucosal B-cell response mostly takes place in Peyer's patches, the clusters of organized lymphoid tissue concentrated in the terminal ileum in most mammalian species. In contrast, isolated lymphoid follicles are dispersed throughout the intestine. In mice, Peyer's patches and isolated lymphoid follicles develop sequentially. Whereas Peyer's patches are constitutive and induced to develop before birth by lymphoid tissue inducer (LTi) cells, isolated lymphoid follicles develop after birth in response to the microbiota that induces the accumulation of T and B cells around clusters of LTi-like cells, termed cryptopatches. Epithelial nucleotide-binding oligomerization domain 1 (NOD1) is crucially involved in the maturation of murine isolated lymphoid follicles through recognition of bacterial peptidoglycans as the gut becomes colonized with the microbiota.

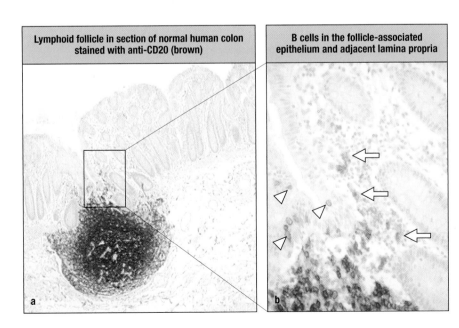

Lymphoid follicle in section of normal human colon stained with anti-CD20 (brown)	B cells in the follicle-associated epithelium and adjacent lamina propria

Figure 8.2 Paraffin section of normal human colon stained with anti-CD20 (brown). Panel a: an isolated lymphoid follicle is comprised mostly of B cells expressing CD20. In contrast, the lamina propria contains only very rare cells expressing CD20. Panel b: the area inside the rectangle in panel a illustrated at higher magnification to illustrate B cells in the FAE (arrowheads) and the infiltration of B cells into the adjacent lamina propria at the periphery of GALT (arrows).

The Peyer's patch component of human GALT is constitutive in humans, as in mice. In humans, constitutive GALT, where clusters of B cells express CD5, can be observed in fetal intestine from around 18 weeks of gestation. Although considered to identify the B-1 B-cell lineage in mice, CD5 can be an activation marker on human B cells (Figure 8.3). Consistent with this, fetal intestinal B cells are large activated cells, though there is no evidence of germinal center formation in the healthy fetal intestine. Like GALT in the postnatal intestine, B cells in fetal GALT infiltrate between the epithelial cells in the FAE, thus demonstrating that this property of the FAE is not a consequence of recognition of microbial antigens by any components of the GALT structure. The magnitude of this infiltration, however, depends on the impact of bacteria, as shown in originally germ-free rodents colonized with a conventional gut microbiota.

It is not known if isolated lymphoid follicles in human small bowel and colon are constitutive, or acquired, as in mice. However, it is known that GALT can be acquired in humans. For example, the normal stomach mucosa contains few lymphoid or plasma cells and no GALT structures. But in patients infected with the gastric bacterium *Helicobacter pylori*, GALT develops in the stomach, providing the necessary prerequisite background for the evolution of so-called MALT lymphomas. These B-cell malignancies are tumors of GALT marginal-zone B cells that in many cases are dependent on *H. pylori* for growth.

MALT is generally associated with B-2 (conventional) B-cell function. However, in mice, B-1 B cells also contribute significantly to the intestinal plasma cell population. The precise proportion is not clear, but it has been estimated that up to 50% of mouse gut plasma cells may be derived from B-1 B cells. B-1 B cells are a self-renewing and self-sustaining B-cell subset that reside predominantly in the peritoneal cavity and originate from B cells in the omentum, the fatty, blanket-like structure covering the intestine in the peritoneal cavity. B-1 B cells are associated with innate immunity through their production of polyspecific antibodies. In mice, it has been suggested that they may differentiate into plasma cells in the lamina propria rather than in GALT. This function may be related to the ability of dendritic cells to extend dendritic processes through the intestinal epithelium and sample antigens outside GALT. However, IgA responses in general are dependent on lymphoid-inducer cells that are founders and orchestrators of organized lymphoid tissues. It is therefore possible that even the B-1 B-cell axis of plasma cell production in mice may be dependent on GALT.

Figure 8.3 Frozen sections of human fetal intestine at 19 weeks of gestation. Panel a: section stained with antibody to IgD (brown) illustrating a cluster of B cells. Panel b: section stained with anti-CD3 to illustrate a surrounding T-cell zone in developing GALT. Panel c: section double stained with anti-CD3 (brown) and anti-CD5 (blue). Blue CD5 B cells are visible in the central B-cell zone of fetal GALT. The follicle-associated epithelium in this section is infiltrated by CD3-expressing T cells and CD5-expressing B cells.

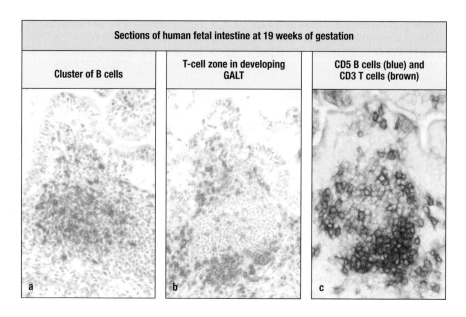

Sections of human fetal intestine at 19 weeks of gestation

Cluster of B cells	T-cell zone in developing GALT	CD5 B cells (blue) and CD3 T cells (brown)
a	b	c

There are no clear different B-cell lineages comprising B-1 cells and B-2 cells in humans, and there is no known contribution of peritoneal or omental B cells to human mucosal immunity. Despite this, there is some evidence that human IgA includes low-affinity cross-reactive antibodies, the production of which is associated with the B-1 B-cell lineage in mice. The origin of this IgA in humans is uncertain, but it may be induced by the continuously shifting antigen repertoire of the large commensal microbiota.

There are interesting similarities and differences between the structure of human GALT and GALT in other species. In rats and mice, Peyer's patches are bulbous structures clearly visible from the serosal surface of the intestine. In contrast, in humans, Peyer's patches are difficult to see, even from the mucosal surface in unfixed intestine. Histologically, in rodents and humans, the most prominent features of Peyer's patches are the secondary lymphoid follicles with germinal centers and the FAE. In rat and mouse Peyer's patches, the follicles often abut directly onto the muscularis mucosae, whereas in humans, where the wall of the intestine is thicker, the T-cell zone tends to mark the boundary on the serosal side (Figure 8.1). The FAE appears similar in structure in humans and rodents and is infiltrated by a mixed lymphoid population that includes medium-sized B cells that morphologically resemble marginal-zone B cells. However, in rats this lymphocyte subset expresses IgD, similarly to mantle-zone B cells, whereas GALT marginal-zone B cells in humans and other primates do not express IgD (Figure 8.1), implying a fundamental difference in the lineage of mucosal B cells between species.

The species differences in the Peyer's patch function are in fact more extreme in other species. In sheep for example, the large ileal Peyer's patch is a site of B-cell development with gene rearrangement in developing B cells.

8-2   IgA and IgM are the dominant mucosal immunoglobulins.

Mucosal epithelia are protected by secretory immunoglobulin A (SIgA) and secretory IgM (SIgM) antibodies made by plasma cells and/or plasmablasts in the lamina propria of secretory mucosae and the stroma of associated exocrine glands. These plasmablasts and plasma cells constitute the body's largest population of antibody-secreting cells. Although IgA is the major isotype produced at mucosal surfaces, there are striking regional variations and important local features. Mucosal plasma cells produce mainly dimeric IgA and some larger polymers of IgA (collectively called polymeric IgA, pIgA). In addition to the light and heavy chains, pIgA and pentameric IgM contain a 15-kDa polypeptide termed the 'joining' or J chain, which facilitates the spontaneous noncovalent interaction of these ligands with the pIg receptor (pIgR) expressed basolaterally on secretory epithelial cells (Chapter 9).

There are two IgA subclasses in humans, with IgA1 comprising 80–85% of total serum IgA. Only in tonsil and nasal mucosae do plasma cells predominantly secrete IgA1. In the gut, however, a relatively large proportion of plasma cells secrete IgA2 (29–64%) compared with a much lower fraction at other sites in peripheral lymph nodes, tonsils, and airway mucosae (7–25%). IgA2 is particularly abundant in the large bowel. SIgA2 may be particularly advantageous, because this isotype is resistant to several IgA1-specific bacterial proteases. The proportions of the two SIgA subclasses in various secretions correspond to the IgA1+:IgA2+ plasma cell ratios in the local secretory tissues, supporting the notion that both subclasses of pIgA are equally well exported by pIgR.

The molecular events underlying preferential IgA1 or IgA2 responses remain unclear. Secretory antibodies to lipopolysaccharide (LPS) are generally SIgA2, whereas protein antigens stimulate predominantly SIgA1 responses. The fact that jejunal IgA+ plasma cells are mainly of the IgA1 subclass (approximately 71%), in contrast to the IgA2 dominance in the colon (64%), may reflect the differences between a response to food antigens versus bacterial antigens.

Bacterial overgrowth in bypassed jejunal segments produces a change in the plasma cell isotype proportions, with an increase of IgA2 and a decrease of IgM, suggesting LPS-induced direct isotype switching from C_μ to $C_{\alpha2}$ or progressive sequential downstream switching of the IgC_H genes. This could be mediated by locally induced class switching in the effector microenvironment, or by differential induction of chemokines and chemokine receptors in MALT in different locations that guide cells to different effector sites.

IgM+ plasma cells constitute a substantial, but rather variable fraction of antibody-secreting cells in the adult human gut. The relatively high proportion (approximately 18%) of IgM plasma cells in the proximal small intestine may be related to the low levels of LPS (discussed above), and is in remarkable contrast to the much lower frequency in the upper airways and nasopharyngeal mucosa. This disparity is strikingly accentuated in IgA-deficient patients, who may have clinical problems caused by recurrent infections, which probably are due to a lack of a compensatory SIgM response in their airways.

IgG+ plasma cells form only 3–4% of the antibody-secreting cells in the normal human intestine, but there is a considerably larger proportion in gastric and nasal mucosae. This may be related to the underlying levels of low-grade inflammation at these last two sites, even in apparently healthy humans. Immunohistochemistry of human upper airways, as well as normal jejunal, ileal, and colonic mucosae, has demonstrated that IgG1 is the predominant mucosal IgG+ plasma cell isotype (56–69%), thus matching the dominance of this subclass in serum. However, IgG2+ plasma cells (20–35%) are generally more frequent than IgG3+ plasma cells (4–6%) in the distal gut, whereas the reverse is often true in upper-airway mucosa. Such IgG subclass disparity supports the idea that C_H gene switching may take different pathways in various mucosal regions as discussed later. Interestingly, the $C_{\gamma2}$ and $C_{\alpha2}$ genes are located on the same duplicated DNA segment, and many carbohydrate and bacterial antigens preferentially induce an IgG2 response in addition to IgA2, whereas proteins (which are usually T-cell dependent antigens) primarily generate IgG1 responses and IgA1.

In chronic inflammatory bowel disease the number of mucosal IgG+ plasma cells is dramatically increased. Moreover, although IgA+ plasma cells still remain predominant, the cells are unusual in that there are more IgA1-secreting cells and decreased J-chain expression, reflecting a more systemic type of plasma-cell pattern.

IgD+ plasma cells are rarely encountered in the human gut, but they make up a significant fraction (3–10%) in the upper airways. In IgA deficiency, this disparity is even more striking for IgD+ plasma cells than for IgM+ plasma cells. No active epithelial transport has been identified for IgD, although this isotype has been linked with antimicrobial activity through the recruitment of basophil polymorphonuclear cells.

IgE+ plasma cells are virtually absent from normal human mucosae. However, IgE plasma cells and especially IgE-bearing mast cells are commonly seen in the oropharyngeal mucosa in patients with allergic diseases.

Mucosal B-cell responses.

Interaction between dendritic cells and luminal contents sampled directly or transported into the MALT by M cells results in a change in gene expression that has several potential consequences. Dendritic cells activated by microbe-associated molecular patterns (MAMPs) directly affect B-cell function through modification of T-cell dependent and T-cell independent routes to the production of IgA and also the imprinting of the homing properties.

8-3   IgA responses in MALT germinal centers are T-cell dependent.

Dendritic cells in MALT are not only exposed to antigens for presentation to T cells and B cells, but their function may also be modified by bacterial products via pattern recognition receptors. T-cell activation induces the upregulation of CD278 (ICOS) and CD154 (CD40L), essential prerequisites for cognate B-cell–T-cell interactions. This T-cell dependent mechanism of B-cell activation initiates the formation of germinal centers. The importance of cognate interactions between B and T cells for germinal center formation is supported by the fact that no germinal centers are formed when the CD40–CD154/CD40L interaction is disabled in humans (CD154 mutation in X-linked hyper-IgM syndrome) or in CD40 knockout mice.

The importance of T-cell help for germinal center formation in mice has been documented by immunization with polysaccharide-based T-cell independent antigens. Although germinal centers are temporarily induced, the level of V-region mutation remains low and the germinal centers involute by massive B-cell apoptosis at the initiation of B cell centrocyte selection. Thus, germinal centers are clearly designed to generate conventional B-2 memory/effector cells for production of antibodies against T-cell dependent antigens; in this manner MALT structures can mount adaptive pathogen-directed responses of high affinity and specificity.

If B-cell activation occurs in a primary follicle where there is no existing germinal center, the proliferation of B cells activated with cognate T-cell help generates a cluster of large cells that form the precursors for formation of a mature germinal center. The cluster of dividing cells nudges the naive B-cell component of a primary follicle aside to form the mantle component of a secondary follicle. This has been demonstrated to occur within the first few days of life in human Peyer's patches, which invariably contain germinal centers throughout life. This, however, may not be the case with colonic isolated lymphoid follicles which frequently lack germinal centers. In germinal center responses in peripheral lymphoid organs and probably also in GALT, newly activated B cells can be recruited into existing germinal centers containing dividing B cells with other specificities providing that the cognate T-cell help has developed prior to their colonization to sustain specific germinal center B-cell growth. Germinal centers contain different functional compartments (Figure 8.4). B-cell stimulation in the dark zone produces exponential growth of blasts positive for the proliferation marker Ki-67, a process that is critically dependent on the transcriptional regulator Bcl-6. The dividing centroblasts then somatically hypermutate their IgV-region genes.

8-4   A high level of somatic hypermutation characterizes mucosal B-cell responses.

Germinal center responses support the modification of DNA sequences in B cells, and the formation of germinal centers involves the induction of activation-induced cytidine deaminase (AID), the enzyme that catalyzes the mutation of target DNA sequences (Figure 8.5). AID is upregulated in B cells following B-cell receptor engagement. AID is critically required both for somatic hypermutation to diversify and increase the antigen-binding potential of immunoglobulins, and for class-switch recombination to produce immunoglobulins with different heavy chains. Individuals with mutations in the functional domains of AID have autosomal recessive hyper-IgM syndrome, no germinal centers, and no IgA or other class-switched immunoglobulin isotypes.

AID deaminates cytidine in targeted segments of DNA in rapidly dividing centroblasts in the dark zones of germinal centers (Figure 8.5). The deamination of cytidine generates uridine that is a normal component of RNA, but not DNA. A subsequent balance in activity between cell division, removal of uridine by

Figure 8.4 Developmental events of B cells in the dark and light zones of germinal centers, leading to the generation of extrafollicular or distant plasma cells of various isotypes. The germinal center founder cell is activated in the extrafollicular compartment (see Figure 8.7) and then migrates to the dark zone where it proliferates and differentiates. The molecular cell–cell interactions and immune events are schematically depicted on the right, and further details are discussed in the text. GCDC, germinal center dendritic cell; IC, immune complex; T_FH, follicular B-cell helper T cells; Ki-67, proliferation marker; MHC-II, major histocompatibility complex class II molecules; TCR, T-cell receptor; FDC, follicular dendritic cell. (Adapted from P. Brandtzaeg, *Immunol. Invest.* 39:1–53, 2010. With permission from Taylor & Francis.)

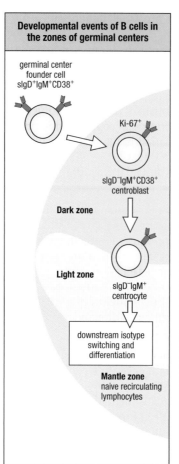

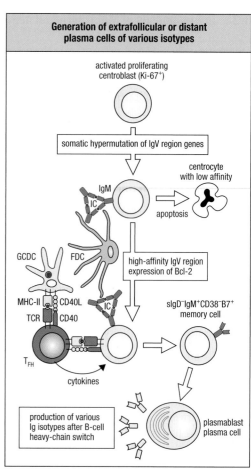

Figure 8.5 Activation-induced cytidine deaminase (AID) and cell division in human gut. Panel a: paraffin section of human colonic isolated lymphoid follicle stained to identify B cells (CD20 in pink) and AID (brown). The germinal center is identified by a dotted line. It is divided into the centroblast zone in which AID-mediated somatic hypermutation occurs (GC-CB) and the centrocyte zone where selection for immunoglobulin specificity takes place (GC-CC). Note the double staining for CD20 and AID in the centroblast zone. Large interfollicular B cells containing cytoplasmic AID are also apparent (black arrow in the insert). The very dark brown staining identified by a white arrow is a macrophage. This is a previously described artifact of the AID staining. Panel b: paraffin section of human ileum stained to identify B cells (CD20 in blue) and proliferation (Ki-67 proliferation antigen in brown). B-cell proliferation is observed in GALT inside and outside of the germinal center. B-cell proliferation is not observed in the adjacent lamina propria.

uracil glycosylase, and lesional repair by error-prone polymerases generates a highly diverse spectrum of mutational outcomes. Somatic hypermutation introduces mostly point mutations, but also some insertions and deletions, into the immunoglobulin heavy-(IgH) and light-chain (IgL) variable (V) region genes. Mutations are not introduced randomly, but tend to focus in hotspots in the complementarity determining regions (CDRs) that form the antigen-binding sites when the encoded proteins are folded. A very high frequency of somatic mutations is seen in the IgV genes in both murine and human Peyer's patches. Indeed, the very high mutation frequencies of Peyer's patch germinal

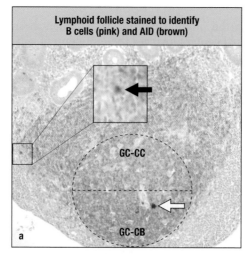

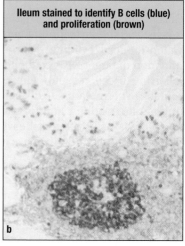

center cells in mice were helpful in elucidating the mechanisms of somatic hypermutation.

Somatic hypermutation changes on average over 5% of the variable region sequence of human mucosal immunoglobulins. Plasma cells in lamina propria throughout the gut, and other mucosal tissues such as the salivary gland, have more mutations in their immunoglobulin genes than plasma cells from the systemic compartment, including the splenic red pulp. This is not due to differences in isotype prevalence in the different tissues; IgHV segments from either IgM or IgA plasma cells individually still show higher levels of mutation in the intestine than in the spleen.

The number of mutations acquired during the mucosal germinal center response, whether observed in the plasma cell progeny or the germinal center cells themselves, provides a means of estimating the number of cell divisions such clones of cells undergo. The somatic hypermutation mechanism introduces approximately one mutation per two cell divisions into a recombined variable region sequence of between 250 and 300 nucleotides. A plasma cell with the average number of 15 mutations in IgHV is therefore likely to have undergone approximately 30 cycles of division. There is considerable B-cell death in the germinal center response. Nevertheless this number of doublings in B cells provides MALT with a tremendous capacity for generating plasma cell precursors. For example, 30 cycles of cell division would generate over 10^9 potential progeny from a single cell. It is not uncommon to observe intestinal plasma cells with 30 mutations that may be part of a clone that potentially could include up to 10^{18} cells, which is one million times more cells than in the human body. This therefore must mean that intestinal plasma cells are turning over rapidly, and also there must be extensive cell death. The gut lamina propria harbors a huge plasma cell population and there is no clear evidence that these cells, in contrast to the bone marrow plasma cells, are long lived. Indeed, there is a constant flow of lymph from the gut into the thoracic duct that contains immunoblast precursors of plasma cells derived from Peyer's patches. Nevertheless, when this lymph flow is interrupted in rats, following an initial decline, the intestinal IgA$^+$ plasma cell density is remarkably well maintained, perhaps implying sideways migration of precursor cells from GALT structures. Similar observations of highly localized B-cell responses to topically applied antigen have been made in a multiple-intestinal-loop model in lambs. Production of IgA antibodies in the gut lamina propria therefore probably depends on the potential of GALT to generate a continuous supply of plasma cell precursors, both laterally and by homing through lymph and peripheral blood. The cells at the periphery of GALT illustrated in Figure 8.2 may be involved in this process of lateral migration prior to differentiation.

The output of MALT germinal center responses therefore includes, almost by definition, the progeny of chronically dividing cells. If cell division occurred without somatic hypermutation, the large clonal output could be compromised in terms of the range of potential binding capacities, and the repertoire would potentially be bland and not matched to the variability of the luminal microbiota. The changes introduced into the antigen-binding sites of mucosal immunoglobulins by somatic hypermutation are likely to make an impact on the spectrum of diversity of the antigen binding potential of secretory IgA. In addition to somatic hypermutation, lambda light chain drifting probably also contributes to the diversity of the antigen binding potential of the gut antibodies; RAG-mediated recombination events that introduce new light chain are also likely to alter immunoglobulin specificity and contribute the diversity of IgA specificity.

The junctional sequences created in IgH genes by RAG-mediated recombination during B-cell development are unique to an individual B cell or B-cell clone. This unique 'clonal signature' can therefore be used to track the distribution and migratory potential of B cells and their plasma cell progeny. This methodology demonstrated that B cells in the germinal centers of human

Peyer's patches and lamina propria plasma cells can be developmentally linked, and that clones of plasma cells can be widely distributed along the gut, consistent with dissemination of GALT-derived plasma cell precursors via the blood. The fact that clonally related cells can be identified within the intestinal lamina propria at multiple sites is also consistent with the idea that GALT produces large numbers of progeny from a single founder B cell.

Somatic hypermutation of the heavy- and light-chain IgV genes in centroblasts has the potential to change the specificity of an immunoglobulin. Selection of the variants generated by somatic hypermutation with highest affinity for antigen results in the development of high-affinity secondary antibody responses. It is not known whether potential chronic exposure to an excess of antigen in the gut affects this process. However, there is no evidence of differences in germinal center structure that could be considered to reflect a difference in selection. Centroblasts give rise to surface (s)IgD⁻IgM⁺CD38⁺ centrocytes that then undergo apoptosis in the light zone unless selected by high-affinity binding to antigen-bearing follicular dendritic cells, via their sIgM/B-cell receptor (BCR). The centrocytes may also pick up and process antigen from follicular dendritic cell-derived iccosomes, and present foreign immunogenic peptides to cognate CD4⁺ T follicular helper (T$_{FH}$) cells (Figure 8.4).

8-5 Class-switch recombination generates IgA.

Somatic hypermutation in germinal centers may be followed by another AID-mediated DNA modification involving class-switch recombination (Figure 8.6). This DNA deletional mechanism results in switching from Cµ to downstream CH genes on chromosome 14, whilst maintaining the same variable region gene sequence. In most mammals, there is a single germline gene encoding the α constant heavy chain (CHα) domains at the 3' end of the Ig heavy chain locus. In humans and higher primates, however, there has been a duplication of the γ → α CH region, resulting in two functional CHα genes encoding the IgA1 and IgA2 subclasses whose plasma cell contributions are disparate throughout the mucosal and systemic immune systems.

In T-cell dependent IgA class-switch recombination, transforming growth factor-beta (TGF-β) is a potent IgA switch factor. A critical role of TGF-β is supported by the almost complete lack of IgA in mice deficient for the TGF-β receptor TGF-βRII. Mice deficient for the intracellular negative regulator of

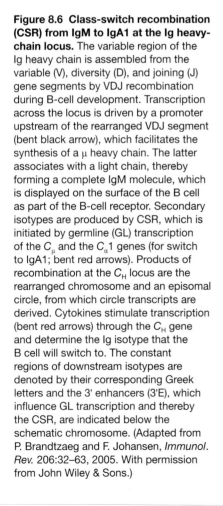

Figure 8.6 Class-switch recombination (CSR) from IgM to IgA1 at the Ig heavy-chain locus. The variable region of the Ig heavy chain is assembled from the variable (V), diversity (D), and joining (J) gene segments by VDJ recombination during B-cell development. Transcription across the locus is driven by a promoter upstream of the rearranged VDJ segment (bent black arrow), which facilitates the synthesis of a µ heavy chain. The latter associates with a light chain, thereby forming a complete IgM molecule, which is displayed on the surface of the B cell as part of the B-cell receptor. Secondary isotypes are produced by CSR, which is initiated by germline (GL) transcription of the $C_μ$ and the $C_α$1 genes (for switch to IgA1; bent red arrows). Products of recombination at the C_H locus are the rearranged chromosome and an episomal circle, from which circle transcripts are derived. Cytokines stimulate transcription (bent red arrows) through the C_H gene and determine the Ig isotype that the B cell will switch to. The constant regions of downstream isotypes are denoted by their corresponding Greek letters and the 3' enhancers (3'E), which influence GL transcription and thereby the CSR, are indicated below the schematic chromosome. (Adapted from P. Brandtzaeg and F. Johansen, *Immunol. Rev.* 206:32–63, 2005. With permission from John Wiley & Sons.)

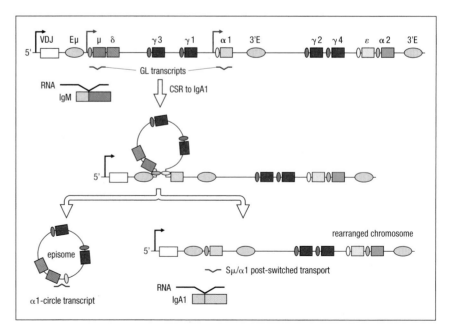

TGF-β signaling, SMAD7, have increased class-switch recombination to IgA, but reduced proliferation to LPS, whereas mice deficient in SMAD2, part of the TGF-β signaling pathway, lack IgA. Interleukin-4 (IL-4)-deficient mice show normal total IgA levels, but an impaired specific response to mucosal immunization with cholera toxin, while IL-6-deficient mice have defective IgA responses in some systems.

Dendritic cells enhance TGF-β receptor-mediated induction of IgA class switch by upregulating TGF-β receptor expression on B cells through the production of inducible nitric oxide synthase (iNOS). IL-4 as well as IL-2, IL-5, IL-6, and IL-10 have also been implicated as contributing factors supporting IgA class-switch recombination or differentiation to IgA plasma cells.

In particular, a proliferation inducing ligand (APRIL), but also B-cell activating factor belonging to the tumor necrosis factor family (BAFF), are potent IgA switch factors produced by GALT dendritic cells, and both favor class-switch recombination to both IgA1 and IgA2 in humans when combined with IL-4 and cell division. BAFF and APRIL share a high degree of homology and both are involved in B-cell maturation and survival. The BAFF/APRIL receptors, namely BAFF receptor (BAFF-R), B-cell maturation antigen, and transmembrane activator and calmodulin interactor, are expressed in GALT. Although, as will be discussed below, these receptor/ligand systems are associated with T-cell independent class-switch recombination, it should be noted that BAFF and APRIL and their receptors are all present in GALT and may contribute to IgA class switching (Figure 8.6). In all models, class-switch recombination is accompanied by cell division and signals promoting germinal center B-cell division, though CD40 is also likely to be an important component of the class-switch recombination process in germinal centers.

IgA switch factors initiate transcription through switch regions of C_α in cells expressing sIgM. The RNA produced is noncoding, thus this RNA may be referred to as a sterile transcript. Experimentally, the presence of sterile transcripts in a population of B cells reflects the potential for class-switch recombination. The function of sterile transcript production is to open up the DNA double helix in a transcription bubble, allowing the molecular mediators of class-switch recombination, including AID, access to single-stranded DNA substrate. The switch regions themselves are long DNA sequences comprised largely of repeated nucleotide motifs that include the DNA target substrate for AID. Deamination of cytidine on both DNA strands generates uracil that can be removed by uracil glycosylase. The abasic site can then be recognized by apurinic/apyrimidinic endonuclease that generates nicks in the DNA. The subsequent repair (by a nonhomologous end joining pathway related to that involved in repair of DNA strand breaks during RAG-mediated recombination) replaces the C_μ constant region with C_α, leaving a circular deleted episomal fragment. RNA transcripts of the circular DNA fragment (noncoding circle transcripts) are produced for a short time after the recombination event. Circle transcript production is therefore a feature of class-switch recombination in the recent past history of a B cell—a feature that can be exploited experimentally when investigating class-switch recombination events.

Activation of class-switch recombination is a negative signal for terminal B-cell differentiation within germinal centers. The decision whether activated B cells should continue down the memory pathway or differentiate along the effector pathway remains elusive, but interactions between CD38+CD27+ germinal center B cells and CD70 (CD27L)+ T cells may be a decisive event.

8-6 Chemokines and chemokine receptors are involved in the organization of MALT and germinal centers.

Several homeostatic chemokines have been identified as major orchestrators of lymphocyte trafficking and determinants of lymphocyte positioning in lymphoid tissue (Figure 8.7). The human B cell-attracting chemokine (BCA-1, CXCL13) attracts naive human B cells *in vitro* and is produced in human lymph node follicles. As discussed in Chapter 1, CXCL13 and its receptor CXCR5 are directly involved in the formation of organized lymphoid tissue. Thus, this chemokine upregulates $LT\alpha_1\beta_2$ on B cells, and establishes a positive-feedback loop to generate follicle formation and attract more B cells into the tissue. The follicular expression of murine CXCL13 is reportedly stronger in murine Peyer's patches than in peripheral lymph nodes. Other chemokines acting on B cells may also be important in human lymphoid tissue. For instance, stromal cell-derived factor 1 (SDF-1, CXCL12) is made by cells lining tonsillar germinal centers, and attracts naive and memory B cells expressing CXCR4 *in vitro*.

Expression of CXCR5 in follicular mantle zones at a relatively low level is consistent with the fact that CXCL13 attracts naive B cells *in vitro* with moderate effect; however, scattered T cells with strong CXCR5 expression occur within the follicles. These CD4+CXCR5+ cells are functionally defined as T_{FH} cells (Figure 8.7), and are distinct from T_H1 and T_H2 cells, but are particularly efficient in B-cell help. The T_{FH} cells also express PD-1 and CD57 and are required for the production of high-affinity antibodies (Figure 8.4).

The partial overlap produced by immunostaining for CXCL13 in human lymphoid tissue shows a similar pattern to staining for several traditional follicular dendritic cell markers, suggesting that this chemokine is deposited on peripheral extensions of follicular dendritic cells after secretion by another cell type (Figure 8.7). The main cellular source of CXCL13 does in fact appear to be the germinal center dendritic cells (Figure 8.4). Notably, however, both these germinal center dendritic cells and large CXCL13-producing cells in inflammatory bowel disease-associated B-cell aggregates and lymphoid neogenesis exhibit a phenotype compatible with macrophage derivation.

As discussed above, however, the germinal center reaction induced by the constant background challenge of the commensal gut microbiota may involve BCR-independent polyclonal stimuli via innate receptors to generate a response with a restricted IgA repertoire that exhibits broad reactivity, but with relatively low affinity. Such a dual model for antimicrobial IgA responses

Figure 8.7 Adhesion molecule- and chemokine-regulated steps in T- and B-cell migration to, and positioning within, gut-associated lymphoid tissue (GALT) compartments. Antigens are sampled from the gut lumen by M cells, and mesenteric lymph nodes receive antigens via draining lymph (either in soluble form or carried by dendritic cells; not shown). Naive T and B cells enter both GALT and mesenteric lymph nodes (left) via high endothelial venules (HEV) by interactions principally between CD62L/L-selectin (L-sel) and endothelial mucosal addressin cell adhesion molecule-1 (MAdCAM-1) or peripheral lymph node addressin (PNAd), distributed as indicated. Activated (memory/effector) T and B cells may to some extent reenter these sites by leukointegrin $\alpha_4\beta_7$-MAdCAM-1 interactions. The chemokines involved (right) at the level of HEVs are CCL21 (SLC) and CCL19 (ELC), provided by stromal cells and redistributed to the HEV endothelium, as indicated, to attract preferentially CCR7+ naive T cells and, less actively, B cells (broken arrow). SLC may also be involved in the exit of lymphoid cells from GALT via draining lymphatics. Naive B cells are CXCR5+ and extravasate mainly via modified HEVs presenting CXCL13 (also called BCA-1 in humans) juxtaposed to, or inside of, the lymphoid follicles; they are next attracted to the mantle zone where BCA-1 is deposited on dendritic elements such as the follicular dendritic cell tips. Also follicular B-cell helper T (T_{FH}) cells (CXCR5+CD4+CD57+) are attracted to the follicle by similar interactions. B cells are primed just outside of the lymphoid follicle by interaction with cognate T cells and antigen-presenting cells; they then reenter the follicle and end up as CCR7+ germinal center cells after interactions with follicular dendritic cells and T_{FH} cells (see Figure 8.4). The B cells may thereafter leave the follicle as memory or effector cells. (Adapted from P. Brandtzaeg and F. Johansen, *Immunol. Rev.* 206:32–63, 2005. With permission from John Wiley & Sons.)

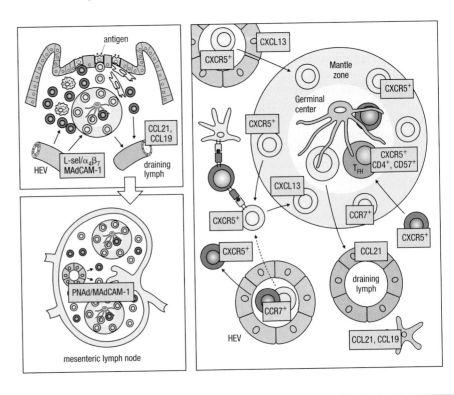

may be adequate for the host's coexistence with the indigenous bacteria, while at the same time providing protection against pathogens. Ongoing polyclonal activation of mucosal memory B cells in GALT could also be a means of maintaining SIgA antibody memory throughout a lifetime.

As mentioned previously, J-chain expression characterizes mucosal plasma cells. As a basis for secretory immunity, MALT must therefore favor the development and dispersion of B cells with J-chain expression, because this is a prerequisite for production of pIgA and pentameric IgM that can be exported by pIgR. The detailed mechanism for J-chain induction is not known, at least not in humans.

8-7   The exit of activated B cells from germinal centers is controlled by various factors.

The exit of the activated B cells from the germinal centers is most likely controlled by chemokines, and the actual cues appear to be extrafollicular, particularly ligands for CCR7 (Figure 8.7). Thus, germinal center B cells downregulate CXCR5 and upregulate CCR7, which in experiments with mice has profound consequences for cellular positioning. Most MALT-induced sIgD⁻IgM⁺CD38⁻ putative memory B cells continuously emigrate from germinal centers to extrafollicular sites, such as the tonsillar crypt epithelium or M-cell pockets in Peyer's patches, where they presumably present recall antigens to cognate memory T cells. Likewise, most plasma cell precursors (CD20⁻CD38^{hi}) become rapidly dispersed in extrafollicular compartments or migrate via lymph and blood to distant effector sites, where they undergo terminal differentiation to plasma cells (Figure 8.4), a process that depends on downregulation of the transcription factor Bcl-6 and upregulation of B lymphocyte-induced maturation protein-1 (Blimp-1). The egress of lymphoid cells to draining lymph is regulated by sphingosine-1-phosphate (S1P), but apparently not by a chemotactic process.

8-8   IgA responses can occur in the absence of cognate B cell–T cell interaction.

Antigens from the gut are rich in potential T-cell independent antigens, such as the polysaccharides abundant in the Gram-positive and Gram-negative bacteria. Mouse B cells can be activated by the type 1 T-cell independent bacterial antigen LPS that can induce polyclonal B-cell proliferation through ligation of toll-like recetor 4 (TLR4) in the absence of any form of T-cell help. Human B cells do not express TLR4, and there are no known TLR ligands that can induce human B-cell division to parallel the induction of murine B-cell division by LPS activation. However, there are a wealth of luminal bacterial antigens with repeating subunits; these typically induce type 2 T-cell-independent B-cell responses by binding such repeating epitopes via the BCR, thus cross-linking the BCR.

The germinal center response is dependent on signaling between CD154 (CD40L) on T cells and CD40 on B cells. Despite the absence of germinal centers in individuals with X-linked hyper-IgM syndrome who have mutations in CD154, these individuals do have IgA⁺ plasma cells in the intestinal lamina propria. CD40 knockout mice also have IgA plasma cells in the gut. In CD40 knockout mice, IgA class switch occurs only in MALT, but is independent of the germinal center response. Human and mouse GALT are known to express AID outside the context of the germinal center. In humans, AID expression has been observed in the cytoplasm of large interfollicular B cells in GALT, where it is considered to be in a storage compartment. Functional AID requires access to the nucleus to mediate the required DNA deamination events prior to class-switch recombination as observed in the germinal center. The location of cells expressing nuclear AID outside the germinal center remains unknown.

Dendritic cell-derived factors support T-cell independent IgA class switch. APRIL and BAFF are potent IgA switch factors produced by GALT dendritic cells and their receptors are expressed by B cells outside of the germinal center microenvironment. APRIL production by dendritic cells is enhanced following TLR-mediated dendritic cell activation via an iNOS. The neuropeptide vasoactive intestinal polypeptide (VIP) produced by epithelial cells also enhances IgA responses. GALT dendritic cells have RNA encoding the receptors for VIP, suggesting that dendritic cells may collaborate in VIP-mediated IgA production. Thus T-cell independent B-cell activators and IgA switch factors have been identified. What remains unclear is the nature of the B-cell proliferation stimulus that could underpin the recombination event.

The instructions involved in inducing GALT B cells to migrate to effector sites.

Migration to different anatomical destinations is a feature of the mucosal B-cell responses, allowing cells to home to appropriate effector sites. The ability of the GALT-derived B cells in blood to home via the lymphatics to the intestinal lamina propria is imprinted on B cells through binding of the vitamin A derivative retinoic acid to the B-cell retinoic acid receptor (RAR). Retinoic acid is generated by dendritic cells and macrophages exposed to pathogen-associated molecular patterns from the intestinal lumen, via the activity of induced retinal dehydrogenase 1 (RALDH1). RAR activation upregulates the expression of the integrin $\alpha_4\beta_7$ that mediates lymphocyte homing to the gut lamina propria by specific binding to its ligand mucosal addressin cell adhesion molecule-1 (MAdCAM-1), constitutively expressed by mucosal postcapillary endothelial cells. Binding of retinoic acid to B-cell RAR also results in the induction of chemokine receptor CCR9 which mediates preferential homing and retention in the small intestinal mucosa, because epithelial cells produce the CCR9 ligand, CCL25. As described in Chapter 1, migration of plasma cells to different destinations involves complex combinations of chemokines and their receptors to guide the plasmablasts to the effector sites. For example, migration to the colon is favored by the expression of CCR10 on B cells controlled by the production by colonic epithelium of CCL28, but is also dependent on high expression of the homing receptor $\alpha_4\beta_7$ to bind to colonic endothelial MAdCAM-1 to gain entry. This allows homing of plasmablasts that produce not only IgA and J chain but also small microbial peptides called defensins to the colonic microenvironment. As discussed below, it may allow the migration of plasmablasts biased toward the production of the IgA2 specifically into the colonic microenvironment.

B-lineage activity in the lamina propria— current controversies.

There are a number of highly contentious areas in the field of mucosal B-cell biology. Study of all aspects of mucosal biology is highly dependent on accurate assessment of the tissue microenvironment and, in all probability, the failure of researchers to agree on important issues in mucosal B-cell biology will be rooted in different ways of assessing microenvironments together with interspecies differences.

Analyzing the composition of the lamina propria lymphocyte population is particularly demanding. It is difficult to insure, for example, that cell suspensions of lamina propria cells are not contaminated by cells from isolated lymphoid follicles, because these lymphoid structures are not visible in fresh intestinal mucosa of any species. Analysis by histological methods may be distorted by the relatively small sample size compared with studies based on cell isolates, and also by the lack of knowledge of what might be in the next serial tissue section. The following fundamental questions in this area remain unanswered, probably as a consequence of these technical uncertainties:

- Can mucosal B cells be activated and mature to IgA-producing plasma cells outside of GALT?

- Can the switch event to human IgA1 or IgA2 production occur outside of MALT?

- Can B cells divide in the gut lamina propria before plasma cell differentiation, thereby giving the potential to reinforce specific antibody production locally?

8-9 The activation of mucosal B cells and their maturation to IgA-producing plasma cells outside of GALT is controversial.

The notion that the gut lamina propria could be a site of induction of B-cell responses gained support from three observations in mice. First, the observation that B-1 cells that are not associated with follicular B-cell structures contribute to the intestinal plasma cell population. However, it has become possible to generate mice that lack Peyer's patches and isolated lymphoid follicles. Animals that are compromised in this way do not have IgA+ plasma cells in the gut, implying that follicular structures are needed for all B-cell lineage routes to IgA production. Second, the observation that the lamina propria dendritic cells can collect antigen by passing their dendrites between the epithelial cells and sampling lumen contents directly without involving the FAE. Third, that isolated lamina propria cells include B cells that express AID (AID would be required for IgA class-switch recombination). The expression of AID in lamina propria is related to the next point and will be discussed in more detail below.

8-10 Evidence indicates that the switch event to human IgA1 or IgA2 production can occur outside of GALT.

As described earlier, there are marked differences along the human gastrointestinal tract in the percentage of plasma cells that produce IgA1 and IgA2. The proportion of IgA2-producing plasma cells is greatest in colon. This could be a consequence of either of two processes. First, B cells may class switch to IgA1 and IgA2 to different degrees in MALT in different locations and then, guided by relevant chemokines, home differentially to different mucosal sites. Consistent with this, different chemokines are known to be disparately produced by colonic versus small bowel epithelium. Chemokine production by epithelium may also be guided by developmental programming or by induction by factors in the microenvironment; it is possible that chemokine expression could be under the influence of local MAMPs. Second, switching to IgA2 may occur once the plasmablasts have extravasated into their effector sites. In support of this, it has been shown that signaling through TLR5 in epithelium induces the local production of IgA switch factor APRIL. This in turn could mediate sequential IgA class switch in lamina propria, resulting in a greater proportion of IgA2 in areas rich in TLR5 ligands, such as the colon. This model is elegant and has considerable support. However it has been disputed for two reasons. First, class-switch recombination requires cell division, which has rarely been convincingly shown in lamina propria cells. Second, this model requires local expression of AID. Immunofluorescence staining of human tissue sections has revealed abundant cytoplasmic AID in some lamina propria cells. Also, fluorescent AID reporter mice have demonstrated positive cells in the lamina propria. Critics of the reporter mice method suggest that such models can be flawed because AID would certainly have been expressed at some time in the history of plasma cells and that switching off of the fluorescence in such models can be difficult. Further to this, polymerase chain reaction (PCR)-based methods for the detection of AID in humans and mice, in systems where the tissue is checked to be free of lymphoid tissue, consistently fail to detect RNA transcripts of the AID gene in the intestinal lamina propria. This controversy therefore remains unresolved.

8-11 B cells in the gut lamina propria do not divide before plasma cell differentiation.

It is logical to propose that B-cell division occurs at effector sites prior to terminal plasma cell differentiation. Early animal studies suggested that this mechanism of local amplification of relevant B-cell responses occurred in response to local challenge, but it is possible that isolated lymphoid follicles were the source of clonal proliferation, since they would still be present after Peyer's patches had been dissected from mouse gut, thus falsely attributing local expansion to the gut lamina propria cells. Local clones of plasma cells can be detected in human lamina propria by PCR-based methods that identify clones of cells by sequence identity in the junctional regions of the immunoglobulin variable region genes generated by gene rearrangement. However, methods based on clonal amplification of DNA or RNA by PCR may be unreliable in this context because identical PCR products may be derived by PCR rather than from adjacent related dividing cells.

In addition, clonally related plasma cells in lamina propria may have disseminated laterally from local GALT structures, or they may be widely disseminated clonal progeny of distant GALT germinal center cells. However the main block to the notion that plasmablasts divide in lamina propria prior to terminal differentiation is the absence of Ki-67$^+$, IgA$^+$ cells in the lamina propria. These results are extremely powerful because Ki-67 is expressed by all dividing cells and, importantly, all the tissue sections contain positive cells in the crypts of Lieberkühn or colonic glands.

Summary.

The extensive mucosal plasma cell population is derived from B-cell responses in MALT initiated by antigens coming directly from the external environment. The antigens are acquired from the mucosal surface via the FAE and stimulate B-cell responses through T-cell dependent and T-cell independent mechanisms that initiate extensive cell division, diversification of antigen-binding capacity by somatic hypermutation, and commitment to IgA production by class-switch recombination. The subsequent dissemination and localization of the IgA responses to the mucosae is dependent on regional or compartmentalized migration regulated by lymphocyte homing receptors with affinity for local endothelial ligands and chemokines secreted by mucosal or glandular epithelia. The mucosal B-cell responses are orchestrated by antigen-activated dendritic cells in MALT that support T-cell dependent and T-cell independent routes to IgA class-switch recombination in mucosal B cells and induce molecular mediators of regional homing.

Further Reading.

Barone, F., Patel, P., Sanderson, J.D., *et al.*: **Gut-associated lymphoid tissue contains the molecular machinery to support T-cell-dependent and T-cell-independent class switch recombination.** *Mucosal Immunol.* 2009, **2**:495–503.

Baumgarth, N.: **The double life of a B-1 cell: self-reactivity selects for protective effector functions.** *Nat. Rev. Immunol.* 2011, **11**:34–46.

Brandtzaeg, P.: **Mucosal immunity: induction, dissemination, and effector functions.** *Scand. J. Immunol.* 2009, **70**:505–515.

Brandtzaeg, P.: **Function of mucosa-associated lymphoid tissue in antibody formation.** *Immunol. Invest.* 2010, **39**:1–53.

Brandtzaeg, P., and Johansen, F.E.: **Mucosal B cells: phenotypic characteristics, transcriptional regulation, and homing properties.** *Immunol. Rev.* 2005, **206**:32–63.

Brandtzaeg, P., Kiyono, H., Pabst, R., *et al.*: **Terminology: nomenclature of mucosa-associated lymphoid tissue.** *Mucosal Immunol.* 2008, **1**:31–37.

Brandtzaeg, P., and Pabst, R.: **Let's go mucosal: communication on slippery ground.** *Trends Immunol.* 2004, **25**:570–577.

Chorny, A., Puga, I., and Cerutti, A.: **Innate signaling networks in mucosal IgA class switching.** *Adv. Immunol.* 2010, **107**:31–69.

Hapfelmeier, S., Lawson, M.A., Slack, E., *et al.*: **Reversible microbial colonization of germ-free mice reveals the dynamics of IgA immune responses.** *Science* 2010, **328**:1705–1709.

Hooper, L.V., and Macpherson, A.J.: **Immune adaptations that maintain homeostasis with the intestinal microbiota.** *Nat. Rev. Immunol.* 2010, **10**:159–169.

Macpherson, A.J., McCoy, K.D., Johansen, F.E., *et al.*: **The immune geography of IgA induction and function.** *Mucosal Immunol.* 2008, **1**:11–22.

Maul, R.W., and Gearhart, P.J.: **AID and somatic hypermutation.** *Adv. Immunol.* 2010, **105**:159–191.

Spencer, J., Barone, F., and Dunn-Walters, D.: **Generation of immunoglobulin diversity in human gut-associated lymphoid tissue.** *Semin. Immunol.* 2009, **21**:139–146.

Suzuki, K., Kawamoto, S., Maruya, M., *et al.*: **GALT: organization and dynamics leading to IgA synthesis.** *Adv. Immunol.* 2010, **107**:153–185.

Yeramilli, V.A., and Knight, K.L.: **Requirement for BAFF and APRIL during B cell development in GALT.** *J. Immunol.* 2010, **184**:5527–5536.

Secretory Immunoglobulins and Their Transport

9

The early demonstration of the presence of antibodies in external secretions of the gastrointestinal tract and later in milk, saliva, tears, respiratory and genital tract secretions, and urine, stimulated extensive studies of their structure, function, and induction following immunization. In contrast to plasma, external secretions contain lower levels and a markedly different immunoglobulin (Ig) isotype distribution of antibodies with characteristic structural and functional properties. The dominant Ig isotype in tears, saliva, milk, and intestinal fluid is IgA, mainly in the form of secretory IgA (SIgA) (Table 9.1). By contrast, IgG predominates in secretions of the nose, lower respiratory tract, and genitourinary tracts of both females and males. Immunoglobulin levels vary between different external secretions and reflect the dominant source of Ig (local production or plasma). In addition, there can be pronounced variability in Ig levels within the same secretion, related to the method of collection, processing of the sample, expression of relevant receptors on epithelial cells involved in Ig transport, and distinct regulatory mechanisms that affect Ig class switching to different isotypes. The Ig content of mucosal secretions is also influenced by hormones (e.g., during the menstrual cycle in the female genital tract), nutritional status, age of the individual, damage of the mucosal barrier due to infection and inflammation, and other factors.

Features of secretory immunoglobulins.

All isotypes of Igs are present in external secretions, albeit at markedly different concentrations. Thus all immunoglobulin isotypes can be considered as secretory, because their entry into the lumen is largely determined by active transport mechanisms across epithelia. In health, paracellular movement of Igs is negligible because antibodies are too large to passively diffuse into the lumen. However, in inflammation when barrier function breaks down, macromolecules such as serum IgG and albumin are readily detectable in the lumen overlying all epithelia.

9-1 The relative distribution and molecular forms of Ig isotypes in external secretions display marked differences in comparison with plasma.

Immunoglobulins of all isotypes are present in external secretions; however, their relative distribution and molecular forms display marked differences in comparison with plasma. IgA occurs in plasma in a monomeric, four-chain form with two heavy (H) α chains and two light (L) κ or λ chains (Figure 9.1a–d). Only a small proportion (~1–5%) of IgA in plasma consists of polymeric (p) forms with two to four monomeric IgA molecules covalently linked by disulfide bridges between the α chains of monomeric IgA and an additional joining (J) chain (Figure 9.1e). In typical external secretions such as intestinal fluid, milk, tears, and saliva, the dominant form of IgA is SIgA which comprises two or four monomeric IgA molecules, one molecule of J chain, and an additional glycoprotein, secretory component, acquired during the receptor-mediated transport of J chain-containing pIgA through the epithelial cells (see below). The proportion of dimeric (two monomeric IgA molecules) to tetrameric (four monomeric IgA molecules) SIgA in saliva and milk is approximately 3:2. The structure of secretory IgA is shown in Figure 9.1f. A variable fraction of IgA in external secretions, particularly in bile, male and female genital tract secretions, and urine, occurs as monomeric IgA and pIgA devoid of secretory

Table 9.1 Levels of immunoglobulins in human external secretions

Fluid	IgM (μg ml^{-1})	IgG (μg ml^{-1})	IgA (μg ml^{-1})	IgA1 (%)	IgA2 (%)
Tear	0–18	Trace–16	80–400	63	37
Nasal secretions	0	8–304	70–846	71	29
Parotid saliva	0.4	0.4 2–5	15–39 120–319	ND	ND
Whole saliva	60–64	26–42	99–206	56	44
Bronchoalveolar fluid	0.1	13	3	67	33
Colostrum	610	100	12,340–53,800	40–77	23–59
Milk	50–340	40–168	470–1632	52–65	35–48
Duodenal fluid	207	104	313	ND	ND
Jejunal fluid	2	4–340	32–276	ND	ND
Colonic fluid	Trace–860	1	240–827	40	60
Intestinal fluid	8	4	166	70	30
Rectal fluid	30 49	0.9 379 297	143 3624* 2317**	30 34	70 66
Urine	ND	0.06–0.56	0.1–1.0	ND	ND
Pre-ejaculate	ND	0.6	1.7	ND	ND
Ejaculate	0–8	16–33	11–23	83	17
Uterine cervix	5–328	1–1200	3–330	50	50
Vaginal fluid	16	10–467	21–118	42	58

*Dry, **Wet. ND, not determined.

(From S. Jackson *et al.*, in J. Mestecky *et al.* (eds), *Mucosal Immunology*, San Diego, CA: Elsevier Inc, 2005: 1829–1839; Z. Moldoveanu *et al.*, *J. Immunol.* 175:4127–4136, 2005; S. Pakkanen *et al.*, *Clin. Vaccine Immunol.* 17:393–401, 2010.)

component. Studies of the structures of SIgA, pIgA, and monomeric IgA have revealed the dimensions and shapes of these molecules as shown in Figure 9.1.

Many species have a single IgA gene. However, humans and most hominoid primates have two IgA subclasses (IgA1 and IgA2) encoded by two distinct constant region genes. IgA1 and IgA2 display characteristic structural differences on their α1 and α2 H chains and exhibit distinct biological activities (Figures 9.1 and 9.2). In plasma ~84% of the molecules belong to the IgA1 and ~16% to the IgA2 subclasses. The proportions of IgA subclasses in external secretions display a characteristic distribution with, for example, nasal fluid dominated by IgA1 while large intestinal fluid contains more IgA2 than IgA1 (Figure 9.2).

In addition to IgA of various molecular forms, external secretions contain IgG, IgM, and trace amounts of IgD and IgE, depending on the fluid examined. IgM appears in secretions in its secretory form (SIgM) with secretory component acquired during the transepithelial transport mediated by the polymeric Ig receptor (pIgR), shared by both pIgA and IgM (see below). However, the level

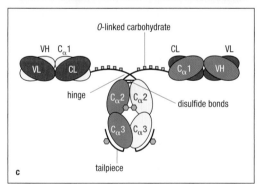

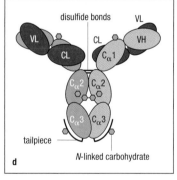

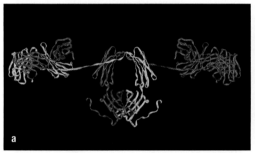

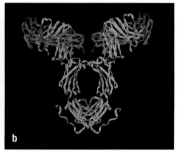

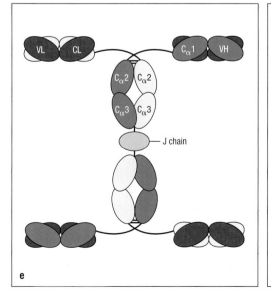

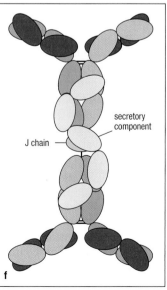

Figure 9.1 Structure of human IgA molecules. Panels a and b show ribbon diagrams of molecular models based on solution scattering studies showing the course of the backbones of the polypeptide chains. IgA1 is shown in panel a and IgA2 in panel b. Schematic representations of these structures are shown in panel c (IgA1) and panel d (IgA2), illustrating the four-chain composition and the separate domains comprising each chain. Domains in each polypeptide are similarly colored. N-linked and O-linked carbohydrates are shown in blue, as hexagons and small squares respectively. Panel e shows a schematic representation of dimeric IgA (in this case of the IgA1 subclass) where the IgA monomers are linked via their tailpieces to J chain. Panel f shows a schematic representation of secretory IgA (in this case of the IgA2 subclass). The five globular domains of secretory component, a cleavage product of the polymeric immunoglobulin receptor, form an integral part of this molecule.

Figure 9.2 Comparative distribution and properties of IgA-producing cells in systemic and mucosal compartments and metabolism of IgA. Panel a shows the relative distribution of cells producing IgA, IgG, IgD,or IgM in the systemic compartment, which is dominated by IgG, and the mucosal compartment, which is dominated by IgA. Note that the total number of Ig-producing cells is higher in the mucosal compartment (> 6×10^{10} cells) than in the systemic compartments (~ 2.5×10^{10} cells). Panel b shows the relative distribution of cells producing the IgA1 and IgA2 subclasses in systemic and mucosal tissues. Panel c illustrates the biosynthesis, properties, mucosal transport, and sites of catabolism of monomeric IgA (mIgA), polymeric IgA (pIgA), and IgA1 and IgA2 in humans.

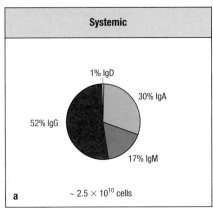

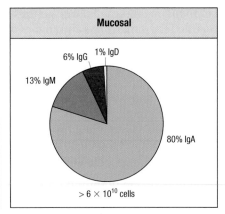

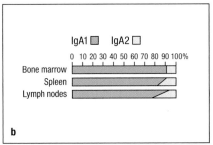

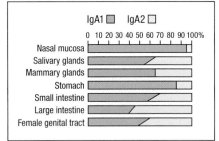

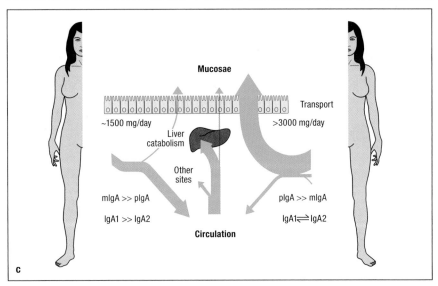

of SIgM is substantially lower than that of SIgA with the exception of individuals with selective IgA deficiency; in these individuals SIgM may replace and functionally compensate for deficient SIgA. The level and relative proportion of IgG to IgA varies among secretions with low levels of IgG in the distal intestinal fluid of the small intestine and colon, milk, tears, and saliva but dominance of IgG over IgA in the urine, semen, and cervico-vaginal secretions collected at various phases of the menstrual cycle. Monomeric IgG is actively transported into external secretions by the neonatal Fc receptor (FcRn) and can also diffuse paracellularly following damage to the epithelial barrier. IgE, which is transported across mucosal epithelial cells by FcεRII (CD23), has been detected in saliva and respiratory and intestinal secretions in extremely low concentrations. Nevertheless, it may participate in immune defense in parasitic infections and in local Type 1 hypersensitivity reactions in allergic individuals and in these circumstances can achieve very high levels. Low levels of IgD are found in human milk, but a specific transport mechanism has yet to be defined.

9-2   The distribution of mucosal immunoglobulins differs between species.

Significant differences in the isotypes, their levels, and their molecular forms exist among various vertebrate species which must be considered in the design of experiments, their interpretation, and their relevance to the mucosal immune system of humans. Common laboratory animals such as mice and rats have only a single IgA isotype that structurally resembles human IgA2 (in that it lacks a hinge region cleavable by bacterial IgA1 proteases—see below), and in their plasma almost all IgA occurs in the polymeric form. In contrast, the rabbit genome encodes 13 IgA subclasses but only one IgG subclass. Early milk (colostrum) and milk of species such as rodents, pigs, cows, horses, sheep, and goats contain IgG as the dominant isotype. Epithelial cells in the mammary glands of these species express FcRn, which mediates transport of IgG from the circulation into colostrum and milk. The presence of IgG in milk is of enormous functional significance for survival. Unlike humans and many other species, IgG is not transported through the placenta antenatally at significant levels in rodents, pigs, cattle, and horses and consequently the newborn of these latter species lack maternally acquired IgG in their circulation. Thus, without colostrum and milk they succumb within several days to infections with common environmental microorganisms. Ingestion of milk and selective, receptor-mediated transfer of IgG from the gut lumen into the newborns' circulation is essential for their survival. Such lumen-to-blood transport of IgG across the polarized epithelial cell is mediated by FcRn via a process called transcytosis. During the neonatal life of rodents (mice and rats), the lumen of the intestine is highly acidic and FcRn is expressed at extremely high levels by the intestinal epithelium. As discussed below, these acidic conditions are perfect for FcRn-mediated uptake of IgG and vectoral transport from an apical-to-basal direction across the simple, polarized epithelium. At the time of weaning (approximately 2 weeks of life), FcRn expression is downregulated nearly 1000-fold. Such developmentally regulated, functional expression of FcRn in rodents and other animals has led to naming this IgG receptor as neonatal. However, as will be discussed below, the expression of this receptor is not in the least restricted to neonatal life or to the epithelium of the mucosa, such that active low-level transport can be present throughout life as seen in humans, nonhuman primates, pigs, and likely other animals.

Igs that are structurally analogous to IgA and IgG appear in birds (e.g., chickens). External secretions of lower vertebrate species contain Ig analogous to IgM, as well as recently characterized alternative Ig isotypes that appear to have evolved independently in certain animal groups. Thus IgX is found in amphibian gut, and IgT has been recognized as a mucosally specialized Ig in teleost fishes.

9-3   The biosynthesis of mucosal IgA is distinct from circulating IgA.

In humans and several other species the production of IgA (~66 mg kg^{-1} body weight/day) far exceeds the combined synthesis of all other isotypes of Igs. This reflects both the distribution into mucosal secretions and markedly different mechanisms of catabolism. The lower serum concentration of IgA in comparison with IgG is the result of the distribution of IgA produced for the mucosal versus systemic pools: approximately two-thirds of IgA is produced mainly in its polymeric form in mucosal tissues, especially the gut, and is selectively transported into external secretions; approximately one-third is produced as monomeric IgA in the bone marrow, lymph nodes, and spleen and enters the circulation. Furthermore, the half-life of IgA in the circulation is considerably shorter than that of IgG (~5–6 days for IgA vs. ~20–25 days for IgG in humans) due to the much faster catabolism that takes place mainly in the liver, with the hepatocyte-expressed asialoglycoprotein receptor playing the dominant role in IgA binding. IgG is in fact the longest lived of the serum proteins due to its protection from catabolism by FcRn expressed in the endothelium and hematopoietic system (mostly monocytes, dendritic cells, B cells, and macrophages), which recycles IgG internalized by pinocytosis away from a degradative fate in lysosomes. The numbers of lymphoblast and plasma cells in mucosal tissues also exceed the number of cells in the bone marrow, spleen, and lymph nodes, with a characteristic distribution of Ig isotypes (see Chapter 1).

In the bone marrow, IgA is produced almost exclusively as monomeric IgA with a pronounced dominance of IgA1-producing cells. In the lymph nodes and spleen, cells producing a small proportion of pIgA are present. The α and L (κ or λ) chains are produced on separate sets of polyribosomes and assembled into monomeric IgA molecules through several possible pathways. The glycosylation of the α chains is initiated with the attachment of *N*-acetylglucosamine (GlcNAc) on the polyribosome, continues in the Golgi apparatus, and is terminated during the intracellular passage from the Golgi apparatus to the cell surface before secretion. In cells engaged in pIgA production, the J chain is complexed with the monomeric IgA molecules to generate IgA dimers or higher molecular forms. It is not certain whether the J chain initiates or terminates Ig polymerization; pIgA and IgM molecules may be assembled in the absence of J chain. It appears that the incorporation of J chain and generation of polymers occurs as a last step, shortly before secretion. Indeed the absolute majority of intracellular IgA in cells producing pIgA is present as monomeric IgA and IgA half-molecules. However, the incorporation of J chain into pIgA is essential in the subsequent binding of pIgA to the epithelial pIgR (see below). Thus, almost all pIgA-producing cells in mucosal tissues are positive for intracellular J chain, while in the bone marrow they are negative. Secretory component, the cleaved extracellular domain of pIgR, remains attached to the J chain-containing pIgA following transepithelial transport to form the SIgA complex (see below). Thus, SIgA is a terminal product of two structurally and functionally distinct cell types: plasma cells producing pIgA with J chain, and epithelial cells producing secretory component. Several findings provide evidence for a relative independence of the systemic and mucosal compartments of the IgA system. Levels of SIgA in external secretions reach adult levels within the first year of life while the plasma IgA levels reach adult levels at puberty. The distribution of cells producing IgA1 and IgA2 in mucosal tissue parallels the distribution of these two subclasses in corresponding external secretions. Mucosal, especially oral immunization with microbial antigens induces SIgA responses in external secretions but low levels of IgA responses in plasma. On the other hand, systemic immunization with antigens that induce initially dominant IgA responses in plasma (e.g., bacterial polysaccharide vaccines) does not induce parallel vigorous responses in external secretions. Finally, intravenous injection of radiolabeled pIgA or pIgA myeloma proteins present in high levels in plasma of patients with IgA myeloma does not result in an efficient appearance of IgA in external secretions. The latter point is of considerable importance because it indicates that pIgA specific for a desired antigen (e.g., human immunodeficiency virus-1 (HIV-1) or influenza virus) administered by a systemic route may be unlikely to lead to elevated levels of SIgA in external secretions.

9-4   IgA exists as two subclasses, IgA1 and IgA2.

As noted, structural and genetic studies have shown that IgA exists as two subclasses in humans and hominoid primates. Comparative structures of IgA from human and other species indicate that IgA2 represents the phylogenetically older form and that IgA1 was generated by the insertion of a gene segment encoding the IgA1 hinge region. Indeed, the structural difference between α1 and α2 H chains is minimal except for 13 amino acid residues in the IgA1 hinge region. In addition, IgA1 and IgA2 display differences in the number, composition and types of glycosidic bonds, and in the sites of attachment of their glycan side chains (see Figure 9.1). The proportion of IgA1 to IgA2 in various body fluids mirrors the distribution of IgA1- and IgA2-producing cells in corresponding tissues. Furthermore, studies relating antibody responses to different types of antigens to their IgA subclass have revealed marked differences. For example, antibodies specific for the influenza virus, HIV-1, or common protein food antigens are associated predominately with the IgA1 isotype while antibodies to bacterial polysaccharide, lipopolysaccharide (LPS), and lipoteichoic acid antigens are mainly present in the IgA2 subclass.

The longer hinge of IgA1 results in a greater distance between the antigen-binding sites formed by the pairing of VH and VL domains than in IgA2 (Figure 9.1a,b). Hence IgA1 is equipped for bivalent binding to antigens spaced at greater distances apart, which may offer advantages over IgA2 and other Ig isotypes in terms of high-avidity recognition of certain antigens. In turn, increased avidity is likely to result in longer residency of IgA1 on the antigenic surface, providing increased opportunities for triggering of effector functions that will eliminate the antigenic target being recognized. Other functional differences between IgA1 and IgA2 include their susceptibility to certain bacterial proteases (see below), inhibition of bacterial adherence mediated by IgA-associated glycans, and the ability to bind other proteins (e.g., fibronectin, lactoferrin, and a broad variety of enzymes) and interact with lectins. Structural aberrations in IgA1 molecules (altered glycan moiety or aggregation) may result in the conversion of IgA from a non-complement-activating Ig into a complement activator (lectin and alternative pathway) and in the formation of pathogenic immune complexes as demonstrated in the common glomerulonephritis known as IgA nephropathy (see Chapter 29).

9-5   IgA is heavily glycosylated with important functional consequences.

Analyses of IgA-associated glycans indicated that 6–10% of total molecular mass is contributed by *N*- and *O*-linked side chains that display a remarkable heterogeneity in their number and composition (Figure 9.1c,d). The most striking difference is the presence of three to five short *O*-linked glycans in the IgA1 hinge region composed of *N*-acetylgalactosamine (GalNAc), galactose (Gal), and sialic acid. The total glycan content of SIgA is even higher than that of serum monomeric IgA due to the presence of glycans on J chain and especially on secretory component, which is particularly rich in *N*-linked glycans (~22% of the total molecular mass). Glycans appear to play an important role in IgA catabolism, the ability to activate the lectin pathway of complement, and binding to bacteria with consequent inhibition of their adherence to epithelial cells (see below).

9-6   J chain is a unique protein that links together monomeric IgA to create polymeric IgA.

The human gene for J chain lies on chromosome 4. It encodes a 15–16 kDa polypeptide, unrelated to any other known protein. J chain is rich in acidic amino acids and contains eight Cys residues, six of which form conserved intrachain disulfide linkages. It is very highly conserved (~70%) across a wide range of species, from mammals through birds and reptiles to amphibians and fishes. While a detailed three-dimensional structure is still awaited, a number of models of J chain conformation have been proposed. These include a two-domain structure, with N-terminal β-sheets and C-terminal α-helical segments, a single domain β-barrel structure reminiscent of an Ig VL domain, and an alternative two-domain model with an N-terminal β-barrel domain and a C-terminal domain combining both α-helices and β-strands that takes into account the disulfide bond pairing arrangement.

Early electron micrographs revealed dimeric IgA to have a double-Y shape. The dimensions were consistent with an arrangement where the two IgA monomers were linked via the tips of their Fc regions, and sedimentation and viscosity experiments as well as more recent mutagenesis and molecular modeling studies support such an 'end-to-end' arrangement of the monomers with J chain interposed (Figure 9.1e). The arrangement is stabilized by disulfide bridges linking the so-called tailpiece regions of the two IgA monomers to J chain. Each tailpiece, a highly conserved stretch of 18 amino acids at the extreme C-terminus of the α H chain, carries a cysteine as its penultimate residue. One of these cysteines in each IgA monomer forms a disulfide bridge to one of the two Cys residues in J chain not involved in intrachain disulfide bonds (Cys14 and Cys68). N-linked sugars on both the tailpiece and J chain appear to influence dimer formation and the domains of IgA Fc also seem to play some role in determining the efficiency of dimerization. The fact that some J chain epitopes are not recognized in polymeric IgA suggests that parts of the Fc may partially obscure J chain. However, J chain is easily released from polymeric IgA by cleavage of interchain disulfide bonds, indicating that any noncovalent interactions between J chain and the Fc region are relatively weak.

Epithelial transcytosis of secretory immunoglobulins.

The single layer of columnar epithelial cells that form the linings of mucous membranes and exocrine glands are connected by tight junctions and other barrier structures that prevent paracellular transport of most solutes, especially large multisubunit protein complexes such as immunoglobulins (see Chapter 5). Locally synthesized polymeric IgA and IgM as well as monomeric IgG and IgE are therefore actively transported across mucosal epithelial cells into external secretions by specialized receptors. The receptor for polymeric IgA and IgM is cleaved at the luminal surface and its extracellular domain becomes a functional part of the SIgA or SIgM complex. The receptors for IgG and IgE mediate multiple rounds of Ig transport in both directions (so-called bidirectional transcytosis) across mucosa epithelia.

9-7   The polymeric immunoglobulin receptor delivers polymeric IgA and IgM into mucosal secretions.

Transport of polymeric immunoglobulins (pIgA and pIgM) across mucosal epithelial cells is mediated by a transmembrane glycoprotein called the polymeric immunoglobulin receptor (pIgR; Figure 9.3). The pIgR binds to the Fc region of polymeric antibodies, and can thus be classified as a type of Fc receptor. The requirement for pIgR for epithelial transcytosis of SIgA was demonstrated by the finding that pIgR-deficient mice have markedly reduced IgA in external secretions and elevated levels of serum IgA. The magnitude of pIgR-mediated antibody transport is greatest in the gut, resulting in delivery of approximately 3 g per day of SIgA into the intestinal fluids of the average adult. Lesser amounts of SIgM are transported into gut secretions due to the far greater abundance of IgA-producing plasma cells, although SIgM can partially compensate for loss of SIgA in IgA-deficient humans. The pIgR is synthesized as an integral membrane protein in the rough endoplasmic reticulum and is further modified in the Golgi apparatus, including extensive N-linked glycosylation. In the trans-Golgi network pIgR is sorted into vesicles that deliver it to the basolateral surface of the epithelial cell, where it binds pIgA and pIgM secreted by plasma cells in the lamina propria underlying the epithelium. With or without bound pIg, pIgR is endocytosed and delivered through a series of sorting vesicles to

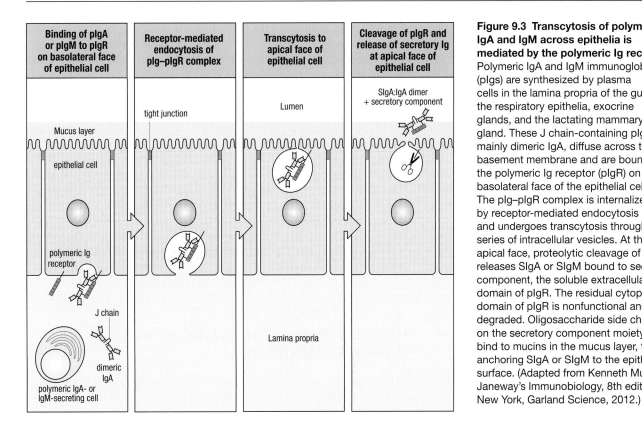

Binding of pIgA or pIgM to pIgR on basolateral face of epithelial cell	Receptor-mediated endocytosis of pIg–pIgR complex	Transcytosis to apical face of epithelial cell	Cleavage of pIgR and release of secretory Ig at apical face of epithelial cell

Figure 9.3 Transcytosis of polymeric IgA and IgM across epithelia is mediated by the polymeric Ig receptor. Polymeric IgA and IgM immunoglobulins (pIgs) are synthesized by plasma cells in the lamina propria of the gut, the respiratory epithelia, exocrine glands, and the lactating mammary gland. These J chain-containing pIgs, mainly dimeric IgA, diffuse across the basement membrane and are bound by the polymeric Ig receptor (pIgR) on the basolateral face of the epithelial cell. The pIg–pIgR complex is internalized by receptor-mediated endocytosis and undergoes transcytosis through a series of intracellular vesicles. At the apical face, proteolytic cleavage of pIgR releases SIgA or SIgM bound to secretory component, the soluble extracellular domain of pIgR. The residual cytoplasmic domain of pIgR is nonfunctional and is degraded. Oligosaccharide side chains on the secretory component moiety bind to mucins in the mucus layer, thus anchoring SIgA or SIgM to the epithelial surface. (Adapted from Kenneth Murphy, Janeway's Immunobiology, 8th edition, New York, Garland Science, 2012.)

the apical (luminal) membrane of the epithelial cell. During transcytosis a disulfide bond forms between pIgR and one of the two subunits of dimeric IgA, resulting in a permanent association between the receptor and antibody. Once the pIg–pIgR complex reaches the apical plasma membrane, the extracellular domain of pIgR is proteolytically cleaved to form soluble secretory component. Cleavage of unoccupied pIgR results in release of free secretory component, while cleavage of pIgR bound to pIgA or pIgM results in release of SIgA or SIgM. Many external secretions such as colostrum contain significant amounts of free secretory component, which contributes to mucosal homeostasis by inhibiting the binding of certain microbes and modulating the activity of pro-inflammatory factors. The seven *N*-glycan chains of human secretory component have a unique structure that is heavily fucosylated and sialylated, similar to those of the antibacterial milk protein lactoferrin. These *N*-glycans bind a variety of host-, pathogen-, and environment-derived substances with lectin-like activity. Carbohydrate-mediated binding of secretory component to intestinal mucins anchors free secretory component, SIgA, and SIgM to the mucus layer overlying the epithelium, thus creating an immunological barrier against infection. Secretory component also stabilizes SIgA by inhibiting the access of microbial proteases to the vulnerable IgA hinge region (see below).

In some animals, such as rodents and rabbits, SIgA can also be delivered into intestinal secretions via the hepatobiliary route. In these species, unlike humans, most of the circulating IgA is polymeric rather than monomeric. Much of this pIgA appears to originate from plasma cells in the gut lamina propria, where pIgA that is not transported across intestinal epithelial cells enters the blood circulation through the portal vein. In the liver, pIgR on the sinusoidal surface of hepatocytes binds pIgA and transports it to the bile canaliculus, analogous to the apical surface of mucosal epithelial cells. Proteolytic cleavage of pIgR causes release of SIgA into bile, which drains into the proximal small intestine. Because the pIgR in these species does not bind pIgM, antibodies of this isotype are not transported from blood to bile and remain in the circulation. Humans and other primates have evolved a distinct pathway for transport of secretory antibodies into bile. In these species, pIgR is expressed by epithelial cells lining the bile ducts, but not by parenchymal hepatocytes. Plasma cells in the lamina propria underlying the bile duct epithelium secrete pIgA (and to a lesser extent, IgM), which is then delivered into bile by pIgR-mediated epithelial transcytosis.

Several unique features of the pIg–pIgR transcytosis pathway contribute to the immune functions of secretory antibodies (see Figure 9.6). During transport, intracellular vesicles containing pIg–pIgR complexes can fuse with endocytotic vesicles internalized from the apical surface of epithelial cells. This intracellular co-localization allows antibody-mediated intracellular neutralization of endocytosed microbes and antigens. Another important feature of pIgR is its ability to transport pIgA or pIgM bound to antigens or whole microbes from the lamina propria to the luminal surface of mucosal epithelial cells. The ability of pIgR to bind and transport large pIgA immune complexes demonstrates that crosslinking of the Fab fragments with antigen does not appreciably alter binding of pIgR to the Fcα-J chain segments of pIgA. Locally produced pIgA antibodies within the mucosa can therefore trap and excrete antigens derived from the environment, diet, commensal microbiota, or infectious microorganisms.

9-8   Expression of pIgR by mucosal epithelial cells is regulated by microbial and host factors.

Because membrane-bound pIgR is proteolytically cleaved to soluble secretory component with each round of SIgA or SIgM transport, epithelial expression of pIgR must be maintained at high levels to support the demand for delivery of mucosal antibodies into external secretions. The single polypeptide chain of pIgR is encoded by the *PIGR* gene on chromosome 1 in humans and mice. Expression of pIgR in epithelial cells is regulated by a variety of microbial and host factors (Table 9.2). In the gut, cross-talk between the commensal microbiota and host cells is crucial for maintenance of pIgR expression. Expression of pIgR is eight- to tenfold higher in the colon than in the small intestine of humans and mice, consistent with the much greater bacterial content in the colon. Germ-free mice have been shown to have underdeveloped intestinal lymphoid structures (see Chapter 17), and also have lower than normal expression of pIgR in intestinal epithelial cells. Colonization of germ-free mice with *Bacteroides thetaiotaomicron*, a prominent commensal of the normal mouse and human gut, restores expression of pIgR. Expression of pIgR is regulated by direct effects of microbes and their products on intestinal epithelial cells, as well as by indirect effects of host factors produced in response to microbial stimulation. LPS shed by bacteria of the family Enterobacteriaceae, including both commensals and pathogens, upregulates transcription of the *PIGR* gene via signaling pathways initiated through an innate receptor for LPS termed Toll-like receptor 4 (TLR4) and activation of the transcription factor nuclear factor κB (NFκB). Mice deficient in the TLR adaptor protein MyD88 have reduced pIgR expression and IgA transport in the colon, demonstrating the importance of TLR signaling for regulation of pIgR. Expression of pIgR has

Table 9.2 Factors that regulate expression of the polymeric Ig receptor

Microbes/microbial factors	*Escherichia coli*, *Bacteroides thetaiotaomicron*, *Saccharomyces boulardii*, reovirus, LPS, butyrate, double-stranded RNA
Cytokines	IFN-γ, TNF-α, IL-1, IL-4, IL-17
Hormones	Estrogen, progesterone, androgens, glucocorticoids, prolactin

IFN-γ, interferon-γ; TNF-α, tumor necrosis factor-α; IL, interleukin.

also been shown to be enhanced by butyrate, a bacterial fermentation product that has anti-inflammatory properties. Introduction of double-stranded RNA into epithelial cells, as a result of infection with intestinal viruses such as reovirus, upregulates *PIGR* gene transcription via signaling through TLR3, another Toll-like receptor, and activation of NFκB as well as the interferon regulatory factor family of transcription factors. Thus the rate of transport of secretory antibodies into intestinal fluids can be regulated by constituents of the normal gut microbiota as well as pathogens, and epithelial pIgR serves as a link between innate and adaptive immunity.

Cytokines produced by host cells, in response to colonization with commensal microbes or infection with pathogenic microbes, regulate expression of pIgR in mucosal epithelial cells. The pro-inflammatory cytokines interferon-γ (IFN-γ) and tumor necrosis factor-α (TNF-α) upregulate *PIGR* gene transcription via JAK-STAT and NFκB signaling respectively. Interleukin-1 (IL-1), which activates a MyD88-dependent signaling pathway that shares many elements with the TLR pathway, synergizes with IFN-γ and TNF-α. Another pro-inflammatory cytokine, IL-17, has been shown to enhance pIgR expression in respiratory epithelial cells, and likely contributes to pIgR regulation in the gut. A unique aspect of pIgR regulation is cooperativity between the helper T cell T_H1-type cytokine IFN-γ and the T_H2-type cytokine IL-4, the effects of which are usually antagonistic. The ability of pIgR to be regulated by a wide range of host cytokines facilitates optimal transport of secretory antibodies in response to the broad spectrum of microbial and environmental stimuli encountered at mucosal surfaces.

Paradoxically, expression of pIgR in intestinal epithelial cells is downregulated during intestinal inflammation associated with inflammatory bowel disease (IBD) in humans and experimental colitis in mice. This loss of pIgR expression is associated with accumulation of IgA in the intestinal lamina propria and increased levels of circulating IgA, which may lead to increased exposure of the systemic immune system to intestine-derived antigens. Decreased expression of pIgR is also seen in dysplastic epithelium and carcinomas of the gastrointestinal and respiratory tracts. A common feature of these conditions is rapid proliferation of relatively undifferentiated epithelial cells, which may not respond properly to microbial and host factors that maintain pIgR expression under homeostatic conditions and enhance pIgR expression during infections.

Expression of pIgR is regulated by hormones in a cell-type specific manner. The effects of estrogen and progesterone are antagonistic in endometrial epithelial cells, and thus the expression of pIgR and transport of IgA varies during the estrous cycle. Androgens have been shown to upregulate pIgR expression in both male and female reproductive tissues. The lamina propria underlying the epithelium of the genitourinary tract is rich in IgA-secreting plasma cells, and SIgA in genital secretions has been shown to contribute to immunity against sexually transmitted diseases. The polypeptide hormone prolactin upregulates pIgR expression in mammary gland epithelial cells during pregnancy and lactation, thus enhancing the delivery of free secretory component and SIgA into colostrum and milk. The lamina propria of the mammary gland is populated by pIgA-secreting plasma cells that were originally stimulated by antigens in the gut-associated and nasopharynx-associated lymphoid tissues. Thus SIgA antibodies transported by pIgR across the mammary gland epithelium into breast milk provide protection against infectious agents and exogenous antigens in the shared environment of mother and child, and also shape the development of the intestinal microbiota in the breast-fed infant.

9-9   Bidirectional transport of IgG and IgE across mucosal epithelia is mediated by specialized Fc receptors.

IgG is transported across polarized mucosal epithelial cells by the neonatal FcRn, so named for its high level of expression by intestinal epithelial cells of

Figure 9.4 Structure and IgG binding of FcRn. Shown schematically in panel a, FcRn is an MHC class I-like molecule with a membrane-spanning α chain bound noncovalently to β₂-microglobulin. Panel b shows a ribbon diagram of a molecule of FcRn (blue and green) bound to one chain of the Fc region of an IgG molecule (red). The α_1 and α_2 domains of the FcRn α chain fold to form a surface that binds at the interface of the $C_\gamma 2$ and $C_\gamma 3$ domains of IgG. The dark blue structures attached to the $C_\gamma 3$ domain of IgG are oligosaccharide side chains. Although only one molecule of FcRn is shown, high-affinity binding of IgG requires dimerization of FcRn such that FcRn binds to IgG with a 2:1 stoichiometry. $\beta_2 m$, β_2-microglobulin. (Courtesy of P. Björkman.)

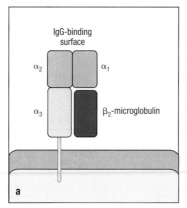
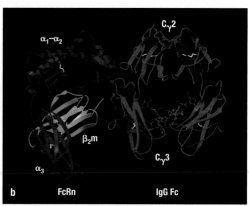

neonatal rats and mice. Initially characterized for its role in the metabolism of circulating IgG, FcRn (then known as the 'Brambell receptor' for its discoverer) was shown to protect internalized IgG from intracellular degradation in a variety of cell types. FcRn was subsequently shown to mediate transfer of IgG from the maternal to the fetal circulation in pregnant women, via transport across placental endothelial cells. The importance of this receptor for mucosal immunity was suggested by the discovery that FcRn could transport maternal milk-derived luminal IgG antibodies across the intestinal epithelium of neonatal rodents and into the systemic circulation. Whereas FcRn is expressed in intestinal epithelial cells of rats and mice only during the neonatal period, in humans it is expressed throughout life in enterocytes and a variety of other cell types, including epithelial cells of the respiratory tract, mammary gland, skin, kidneys, and eyes, as well as endothelial cells and hematopoietic cells (monocytes, macrophages, dendritic cells, neutrophils, and B cells). FcRn is a major histocompatibility complex (MHC) class I-like molecule with a transmembrane α heavy chain containing three extracellular domains that is noncovalently bound to soluble β₂-microglobulin (Figure 9.4a). The α heavy chain is encoded by the *FCGRT* gene outside of the MHC gene locus, on chromosome 19 in humans and chromosome 7 in mice. A major difference between FcRn and classical MHC class I proteins is that the α_1 and α_2 domains do not fold to form a peptide-binding groove, rather forming a platform that contacts the Fc region of IgG at the interface of the $C_\gamma 2$ and $C_\gamma 3$ domains (Figure 9.4b). Evidence from *in vitro* binding studies suggests that high-affinity binding of IgG requires dimerization of two FcRn molecules.

In polarized epithelial cells, FcRn directs IgG through a complex intracellular itinerary that can result in either apical-to-basolateral or basolateral-to-apical transcytosis (Figure 9.5). This pathway differs in three important ways from pIgR-mediated epithelial transcytosis of polymeric IgA and IgM. First, whereas pIgR-mediated transport is strongly biased in the basolateral-to-apical direction, FcRn-mediated transport can deliver IgG to both surfaces, with important functional consequences for mucosal immunity (see below). Second, FcRn is not cleaved at the plasma membrane like pIgR, and can mediate multiple rounds of IgG transport. Third, binding of FcRn is optimal at a slightly acidic pH of around 6.0, thus favoring association with IgG within acidic intracellular compartments and release of IgG at the neutral pH of the plasma membrane. An exception to this rule is the luminal plasma membrane of neonatal rodent enterocytes, where the naturally acidic environment favors binding of IgG to FcRn and uptake by receptor-mediated endocytosis. It is possible that this cell-surface FcRn–IgG binding may also occur at the surface of other epithelial cells within localized acidic microenvironments such as may occur during inflammation. However, for most cell types, IgG is primarily internalized via fluid-phase pinocytosis and delivered to acidic endosomal compartments where binding to FcRn occurs. Consistent with this, the majority of FcRn that

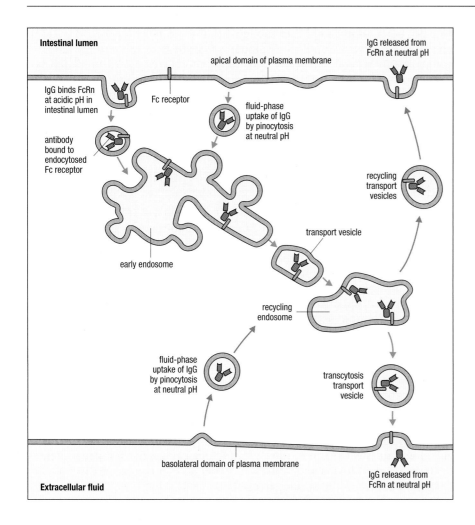

Figure 9.5 Bidirectional transcytosis of IgG across polarized epithelial cells by FcRn. Both apical-to-basolateral (green arrows) and basolateral-to-apical (blue arrows) transcytosis of IgG are mediated by FcRn. Most of the FcRn molecules are found in acidic intracellular endosomes and transport vesicles, where low pH promotes binding of FcRn to IgG. Some FcRn molecules reach the plasma membrane via recycling vesicles. In the acidic environment of the neonatal rodent intestinal lumen, IgG binds FcRn at the apical plasma membrane and is delivered to early endosomes via receptor-mediated endocytosis. In cells where fluids bathing the plasma membrane are closer to neutral pH, most of the IgG is internalized via fluid-phase pinocytosis. Fusion of pinocytic vesicles with acidic endosomes results in binding of soluble IgG to membrane-bound FcRn. FcRn–IgG complexes that are taken up by the basolateral and apical surfaces are delivered to the apical and basolateral surfaces of polarized epithelial cells, respectively, through a network of transport and recycling vesicles and endosomes in a process called transcytosis. At the neutral pH of the plasma membrane, IgG is released from FcRn into external secretions and interstitial fluids. Membrane-associated FcRn molecules then reenter the endocytic pathway, where they can mediate many rounds of IgG transport to the apical and basolateral cell surfaces. In humans, FcRn mediates transport of IgG across endothelial cells in the placenta, thus delivering IgG from the maternal to the fetal circulation. Expression of FcRn by a variety of cell types, including immune cells, prevents degradation of internalized monomeric IgG by directing these FcRn–IgG complexes away from the lysosome and into recycling vesicles. (Adapted from Bruce Alberts et al., Molecular Biology of the Cell, 5th edition, New York, Garland Science, 2008.)

is expressed within the epithelium is contained intracellularly within endosomes. Once bound, FcRn diverts monomeric IgG away from lysosomes and into a series of transport and recycling vesicles for delivery to the basolateral and apical surfaces. In nonepithelial cells such as endothelium and hematopoietic cells, recycling vesicles carry FcRn–IgG complexes back to the plasma membrane for release into interstitial fluids or the blood circulation. The ability of FcRn to protect internalized IgG from lysosomal degradation is the mechanism through which this receptor inhibits IgG catabolism and increases the half-life of circulating IgG. Interestingly, FcRn also binds serum albumin and appears to play a similar role in regulating albumin catabolism.

In contrast to the downregulation of pIgR expression and SIgA transcytosis observed in human and experimental IBD, the pro-inflammatory environment of the gut in IBD results in increased IgG production and absorption of luminal IgG-bound antigens by FcRn-mediated transcytosis. The combination of reduced SIgA-mediated immune exclusion and increased IgG-mediated antigen absorption may account for the increased levels of circulating IgG antibodies against self and bacterial antigens that are a common feature of human IBD and which may further promote the intestinal inflammation.

Compared with IgA, IgM, and IgG, the levels of IgE are normally very low in mucosal secretions. However, numbers of IgE-producing plasma cells in the lamina propria and levels of IgE in secretions of the respiratory and gastrointestinal tracts are elevated in individuals with allergic diseases such as asthma, allergic rhinitis, and food allergy, and during intestinal parasite infections. The high-affinity receptor for IgE, FcεRI, is expressed by a variety of immune cells and plays an important role in effector functions of IgE (see below). It has recently been appreciated that the low-affinity IgE receptor, FcεRII (CD23), mediates endocytosis and bidirectional transcytosis of IgE and IgE–antigen complexes in several cell types, including enterocytes. CD23-mediated IgE transcytosis may enhance transport of potential allergens across the intestinal epithelial barrier, with important implications for mucosal defense and allergy (see below).

Functions of mucosal immunoglobulins.

The characteristic distribution and biological properties of Igs of various isotypes in external secretions reflect functional requirements for optimal protection at diverse mucosal surfaces. SIgA and SIgM antibodies generally have a lower intrinsic affinity for antigens than do IgG antibodies, but due to the bonus effect of multivalency (e.g., four or eight antigen-binding sites for SIgA dimers and tetramers respectively) SIgA and SIgM have a higher avidity for and enhanced ability to cross-link antigens. Furthermore, the presence of bound secretory component protects SIgA and SIgM from proteolytic degradation, resulting in a longer half-life compared with IgG antibodies in protease-rich secretions such as those of the gastrointestinal tract. IgG antibodies may thus be more important for immune defense in secretions of the respiratory and genital tracts, where they are more stable. The direct protective role of mucosal antibodies was demonstrated *in vivo* in animal models of germ-free piglets and in mice with passively administered IgA specific for bacterial and viral pathogens or their products. Mucosal Igs effectively diminish absorption of antigens from mucosal surfaces, a process called immune exclusion. In the case of biologically active antigens, such as bacterial or plant toxins and viruses, mucosal Igs can neutralize their activity or block binding to the surface of epithelial cells (Figure 9.6). The antibody isotype profoundly influences the protective effect of mucosal Igs. SIgA, due to its inability to activate complement cascades, excludes antigens without generating local inflammation. In contrast, activation of complement by IgM– or IgG–antigen complexes may induce local inflammation, resulting in increased nonspecific uptake of antigens and pathogens due to the altered mucosal barrier. For example, antigen challenge of individuals with a selective IgA deficiency results in a higher level of circulating immune complexes than is seen in healthy individuals.

9-10 Secretory Igs protect mucosal surfaces against microbial invasion.

Mucosal Igs utilize a variety of mechanisms to enhance barrier function of mucosal surfaces (Figure 9.6). Coating of an enormously diverse range of mucosal microorganisms with SIgA via antigen-specific binding and glycan-mediated interactions results in inhibition of their attachment to receptors expressed on the surfaces of epithelial cells. *In vitro* experiments have indicated that glycans associated with the α heavy chains and the secretory component moiety of SIgA act as decoys, effectively preventing the attachment of Gram-positive and Gram-negative microorganisms to glycan-dependent receptors expressed on epithelial cells in the gastrointestinal and respiratory tracts. Such interactions confine Ig-coated bacteria to their natural biological niches and contribute to the formation of physiological 'biofilms' essential for their continued survival and establishment of a mutually beneficial mucosal microbiota. IgG antibodies also participate in immune exclusion, particularly in the lower respiratory tract and the genitourinary tract where they are more stable. These antibody activities may be further enhanced through synergy with innate factors of humoral immunity such as lactoferrin, lysozyme, and the peroxidase system.

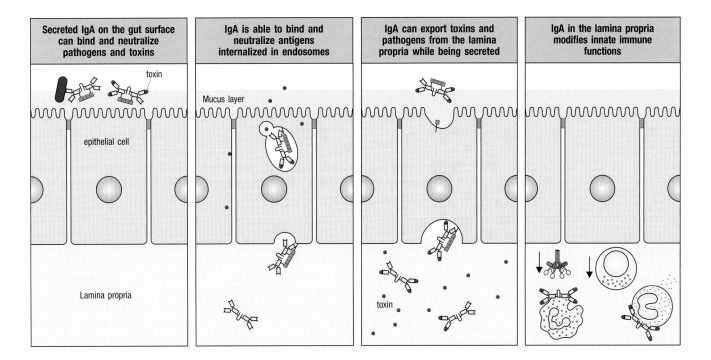

| Secreted IgA on the gut surface can bind and neutralize pathogens and toxins | IgA is able to bind and neutralize antigens internalized in endosomes | IgA can export toxins and pathogens from the lamina propria while being secreted | IgA in the lamina propria modifies innate immune functions |

9-11   Epithelial transcytosis of mucosal Igs enhances immune defense.

In addition to interactions with antigens on mucosal surfaces, SIgA antibodies can contribute to immune protection during the process of epithelial transport. Biologically active antigens such as toxins can be internalized from the apical surface of epithelial cells via receptor-mediated endocytosis and delivered to intracellular sorting vesicles. At the same time, transcytotic vesicles carrying pIgA–pIgR complexes from the basolateral surface fuse with the apical sorting vesicles, allowing the IgA antibodies to neutralize their specific antigens (Figure 9.6). For example, IgA antibodies have been shown to neutralize bacterial LPS within epithelial cells, thus dampening a potential pro-inflammatory response in the mucosa and limiting access of LPS to the systemic circulation. IgA antibodies have also been shown to interfere with the assembly of influenza virus and Sendai virus by binding to newly synthesized envelope glycoproteins within intracellular sorting vesicles. Mucosal IgA antibodies also contribute to clearance of antigens by pIgR-mediated transport of antigen–antibody immune complexes to the luminal surface of mucosal epithelial cells. In addition to immune complexes containing soluble protein antigens, pIgR has been shown to mediate transcytosis of whole viruses and bacteria complexed to pIgA. This excretory function of mucosal IgA serves to limit access of environmental, dietary, and microbial antigens to the systemic immune system, thus minimizing the generation of potentially pro-inflammatory IgG antibodies. Excretion of infectious microbes within epithelial cells or in the lamina propria can prevent their spread to other parts of the body.

FcRn-mediated transcytosis of IgG from the basolateral to the apical surface of mucosal epithelial cells delivers protective IgG antibodies into external secretions. In turn, transport of IgG–antigen complexes from the luminal to the abluminal surface of the epithelium associated with the intestines, lung, and genitourinary tract leads to absorption of intact antigens. Importantly, these IgG-antigen complexes can be phagocytosed by mucosal dendritic cells, which traffic to regional lymph nodes and present antigenic peptides to mucosal and systemic T cells. In experimental models, this FcRn-driven pathway of antigen absorption has been shown to induce tolerance to mucosal antigens in both the gut and the respiratory tract. This mechanism is likely to

Figure 9.6 Mucosal IgA has several functions in epithelial surfaces. First panel: IgA binds to the layer of mucus covering the epithelium, where it can neutralize pathogens and their toxins, preventing their access to tissues and inhibiting their functions. Second panel: antigen internalized by the epithelial cell can meet and be neutralized by IgA in endosomes. Third panel: toxins or pathogens that have reached the lamina propria encounter pathogen-specific IgA in the lamina propria, and the resulting complexes are reexported into the lumen across the epithelial cell as the dimeric IgA is secreted. Fourth panel: IgA in the lamina propria can inhibit (↓) complement activation, decrease (↓) natural killer cell activity, promote opsonization and phagocytosis of antigens by neutrophils, and trigger degranulations of eosinophils. (Adapted from Kenneth Murphy, Janeway's Immunobiology, 8th edition, New York, Garland Science, 2012.)

be particularly important in newborns, where IgG antibodies in maternal milk can promote absorption of intestinal antigens to which the mother is immune. Thus IgG antibodies, in addition to their roles in mucosal defense, may promote physiological tolerance to the commensal gut microbiota and regulate the development of allergic airway inflammation. Although it is yet to be formally demonstrated, it is possible that CD23-mediated absorption of IgE-bound antigens may contribute to mucosal tolerance and/or inflammation, particularly in allergic individuals with relatively high levels of mucosal IgE.

9-12 SIgA and secretory component possess innate immune functions through their carbohydrate modifications and regulate mucosal inflammation.

In addition to antigen-specific immune functions of IgA, SIgA and free secretory component also contribute to the regulation of innate, 'nonspecific' responses to pathogens. Many of these functions are mediated by binding of the unusual N-glycans of secretory component to bacterial and host factors. Free secretory component has been shown to limit infection or reduce morbidity by binding to bacterial components such as *Clostridium difficile* toxin A and fimbriae of enterotoxigenic *Escherichia coli*. During both innate and antigen-specific immune responses, secretory component and SIgA enhance mucosal homeostasis by limiting potentially pro-inflammatory immune responses. Because IgA, unlike IgG, is a poor activator of complement, IgA can neutralize antigens and excrete immune complexes without eliciting an inflammatory response that could cause collateral damage to mucous membranes and disrupt barrier function. Although artificially aggregated IgA or IgA with altered glycan moieties may under some circumstances activate the complement cascade by the alternative or lectin pathways, this mechanism does not appear to occur in mucosal tissues. Although external secretions contain complement components, they typically do so at levels that would not promote effective complement activation. Furthermore, the abundance of IgA relative to IgG antibodies in these secretions leads to competition for antigen binding and limitation of IgG-mediated activation of the classical complement pathway. Free secretory component can also prevent inflammation by forming a high-molecular-weight complex with the chemokine IL-8, limiting its activity as a neutrophil chemoattractant.

9-13 Mucosal Igs also interact with Fc receptors on immune cells.

By activating isotype-specific Fc receptors on phagocytes and other immune cells, mucosal Igs may facilitate protective immune responses against mucosal pathogens. Some of these immune cells, such as macrophages, are normal residents at the mucosal surface, while others tend to infiltrate mucosal sites only during times of infection or other insult. In humans, three classes of IgG-specific Fc receptors (FcγRI, FcγRII, and FcγRIII), an IgA-specific Fc receptor (FcαRI), and the high-affinity IgE Fc receptor (FcεRI) are expressed on various types of immune cells in lamina propria of mucous membranes. Ligation of these Fc receptors triggers elimination of Ig-coated targets through mechanisms such as phagocytosis, degranulation and release of antimicrobial factors, and release of activated oxygen species. Intriguingly, the mode of interaction of IgG with FcγR is very different from that of IgA with FcαRI. All of the FcγRs interact with a site on IgG lying at the N-terminal end of the Fc region in close proximity to the hinge region. In contrast, the interaction site for FcαRI lays close to the midpoint of the IgA Fc, at the interface between the $C_\alpha 2$ and $C_\alpha 3$ domains (Figure 9.1). Targeting of this receptor site on IgA by microbial proteins appears to offer an effective means for certain pathogens to evade elimination via FcαRI-mediated clearance mechanisms (see below). Intestinal macrophages, dendritic cells, and B cells also express FcRn. In these cell types, although FcRn mediates the recycling of monomeric IgG, it participates in directing polymeric IgG-containing antigen–antibody complexes into antigen processing compartments associated with MHC class II-restricted presentation which may be important in defending against mucosal pathogens or promoting intestinal inflammation as in inflammatory bowel disease. Binding of IgE–antigen complexes to FcεRI on mucosal antigen-presenting cells such as dendritic cells, macrophages, and basophils likely regulates IgE-mediated adaptive immunity. In the effector phase, cross-linking of cell-surface FcεRI by IgE–antigen complexes results in degranulation of mucosal mast cells, an important mechanism for elimination of parasitic infections.

Strategies employed by pathogens to evade IgA-mediated defense.

The relationship between IgA and the microbiota is complex. Commensal microbes, as well as pathogens, regulate the production and transport of IgA, and their colonization is, in turn, affected by IgA secretion. The importance of this relationship and the effectiveness of IgA as a form of immune defense are highlighted by the fact that a number of pathogens have targeted IgA as a means of evading the immune response.

9-14   Pathogens express IgA1 proteases that destroy IgA function.

Certain pathogenic bacteria, known to cause clinically important and life-threatening infections, secrete proteolytic enzymes that cleave specifically in the hinge region of IgA1. These so-called IgA1 proteases are produced by bacteria that invade mucosal sites to cause diseases such as acute meningitis (*Neisseria meningitidis*, *Haemophilus influenzae*, and *Streptococcus pneumoniae*), oral cavity disease (*Streptococcus sanguis*, *Streptococcus oralis*), and sexually transmitted infections (*Neiserria gonorrhoeae*). The proteases appear to be associated with virulence because closely related nonpathogenic strains do not produce them. Diversity in the structural and mechanistic features of IgA1 proteases suggests that they have arisen by convergent evolution. Presumably, their ability to disrupt IgA function affords the bacteria concerned with advantageous opportunities for colonization and invasion of mucosal surfaces.

IgA1 proteases cleave post-proline peptide bonds within the extended hinge region of IgA1, each protease cleaving at a particular proline–serine or proline–threonine site (Figure 9.7). IgA2 lacks the susceptible sequence and therefore resists cleavage. Proteolysis of the IgA1 hinge essentially disengages the antigen-recognition function of the Fab regions from the effector function capabilities of the Fc region, allowing any bacteria recognized by the free Fab fragments to escape the elimination processes normally triggered through the Fc region. Moreover, Fab binding blocks access of intact Igs, enabling the bacterium to evade the protective functions of mucosal Igs.

It has become clear that efficient recognition and cleavage by IgA1 proteases is governed both by the hinge amino acid sequence and the structural context of the hinge within the antibody as a whole. Thus mutagenesis experiments have shown that susceptible bonds must be positioned at a suitable distance away from the Fc region, and that motifs in the lower part of the Fc region ($C_\alpha 3$ domain) are required for substrate recognition by several IgA1 proteases. The recent solution of the first X-ray crystal structure of an IgA1 protease, the type 1 protease from *Haemophilus influenzae*, has suggested a binding mechanism in

Figure 9.7 Differences between the hinge regions of human IgA1 and IgA2. The amino acid sequences of the hinge regions are shown. Arrows above the IgA1 hinge sequence indicate the cleavage sites of particular IgA1 proteases. The *O*-linked oligosaccharides carried by the IgA1 hinge are indicated below the sequence.

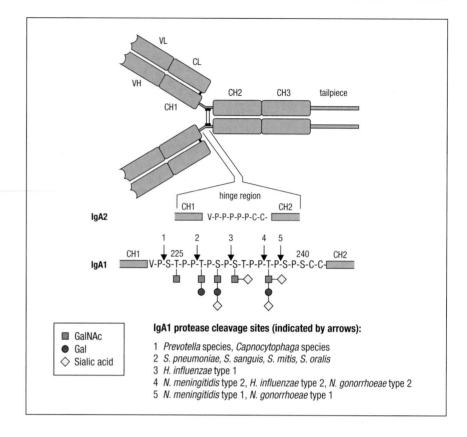

IgA1 protease cleavage sites (indicated by arrows):

1 *Prevotella* species, *Capnocytophaga* species
2 *S. pneumoniae, S. sanguis, S. mitis, S. oralis*
3 *H. influenzae* type 1
4 *N. meningitidis* type 2, *H. influenzae* type 2, *N. gonorrhoeae* type 2
5 *N. meningitidis* type 1, *N. gonorrhoeae* type 1

keeping with these findings. In this model, the binding of the Fc is postulated to stabilize the protease in an open conformation thereby allowing access of the hinge peptide to the active site so that cleavage may ensue.

9-15 Pathogenic bacteria express binding proteins specific for IgA and secretory component that are involved in virulence.

Another bacterial strategy to evade the mucosal IgA response employs surface proteins, termed IgA-binding proteins, that bind specifically to both serum and secretory forms of human IgA. Such proteins are expressed by strains of group A streptococcus (*Streptococcus pyogenes*), an important pathogen causing acute infections sometimes resulting in compromised heart and kidney function; group B streptococcus, a major cause of septicemia, meningitis, and pneumonia in newborn babies; and *Staphylococcus aureus*, responsible for skin infections, abscesses, pneumonia, bacteremia, and other illnesses, many of which are life-threatening. Remarkably, although the IgA-binding proteins produced by these organisms are structurally unrelated, they all interact with broadly the same region of IgA Fc. Their sites of interaction on the Fc region of IgA overlap with that bound by FcαRI and their binding has been shown to block binding of IgA to this receptor and inhibit triggering of FcαRI-mediated responses. This inhibitory capability suggests that these bacteria have evolved IgA-binding protein expression as a means to evade elimination mechanisms elicited by IgA through interaction with FcαRI.

Streptococcus pneumoniae, which causes diseases ranging from relatively mild otitis media to potentially fatal sepsis, pneumonia, and meningitis, expresses a surface protein called CbpA (also known as SpsA and PspC) that binds specifically to human pIgR and secretory component. *S. pneumoniae* uses this protein to enhance colonization and invasion of the nasopharynx by binding to pIgR on the surface of epithelial cells. Strains of *S. pneumoniae* that lack CbpA have reduced ability to colonize the nasopharynx, suggesting that its ability to bind pIgR may represent an important virulence factor. However, free secretory component and SIgA secreted by nasopharyngeal epithelial cells can bind to CbpA and inhibit its ability to bind to cell-surface pIgR. The balance between this bacterial virulence factor and host defense mechanism may contribute to variations in susceptibility among humans to nasopharyngeal carriage of *S. pneumoniae*.

Summary.

All isotypes of antibodies can function as secretory immunoglobulins, and as such play a critical role in mucosal defense. Secretory IgA is the chief antibody class at most mucosal sites, whereas IgG predominates in secretions of the nose, lower respiratory tract, and genitourinary tract. IgM is found at lower concentrations, but can compensate for IgA in selective IgA deficiency. Mucosal IgE can play a protective role in parasitic infections and a pathogenic role in food allergies and asthma, and IgD is commonly found in tonsillar secretions. Most of the mucosal IgA and IgM form polymers through covalent interactions with the J chain polypeptide. Polymeric IgA and IgM are transported across mucosal epithelial cells in the basolateral-to-apical direction by the polymeric immunoglobulin receptor, which is cleaved at the apical surface to form SIgA and SIgM. The extracellular domain of pIgR, a five-domain polypeptide also known as secretory component, remains bound to SIgA and SIgM and provides protection against proteolysis and enhances immune effector functions. SIgA is highly glycosylated, which increases its stability and some aspects of its protective function. IgG is transported across mucosal epithelial cells in both the apical-to-basolateral and basolateral-to-apical directions by the neonatal Fc receptor, and IgE is transported bidirectionally by FcεRII (CD23).

SIgA in mucosal secretions neutralizes biologically active antigens, prevents absorption of food antigens, and inhibits adherence of mucosal microbes to epithelial cells. SIgA also neutralizes pathogens and antigens during pIgR-mediated epithelial transcytosis, and clears antigen–antibody immune complexes from the lamina propria via transport to the luminal surface of epithelial cells. IgG antibodies facilitate uptake and presentation of mucosal antigens to link mucosal antigen exposures to systemic induction of immune responses and promote immune tolerance via FcRn-mediated transcytosis. IgG, IgA, and IgE antibodies engage specific Fc receptors on phagocytes and immune cells in mucosal tissues, triggering elimination of pathogens and antigens by mechanisms such as phagocytosis and release of antimicrobial substances. Some pathogenic bacteria inhibit SIgA defense mechanisms by secreting IgA-binding proteins or proteolytic enzymes that cleave IgA1.

Further Reading.

Baker, K., Qiao, S.W., Kuo, T., *et al.*: **Immune and non-immune functions of the (not so) neonatal Fc receptor, FcRn.** *Semin. Immunopathol.* 2009, **31**:223–36.

Conley, M.E., and Delacroix, D.L.: **Intravascular and mucosal immunoglobulin A: Two separate but related systems of immune defense?** *Ann. Int. Med.* 1987, **106**:892–899.

Corthésy, B.: **Roundtrip ticket for secretory IgA: role in mucosal homeostasis?** *J. Immunol.* 2007, **178**:27–32.

Gould, H.J., and Sutton, B.J.: **IgE in allergy and asthma today.** *Nat. Rev. Immunol.* 2008, **8**:205–217.

Johansen, F.-E., Yen, H.E., Dickinson, B., *et al.*: **Biology of gut immunoglobulins**, in Barrett, K.E., Ghishan, F.K., Merchant, J.L., *et al.* (eds): *Physiology of the Gastrointestinal Tract*, 4th ed. Burlington, MA, Elsevier Academic Press, 2006:1067–1090.

Kaetzel, C.S.: **The polymeric immunoglobulin receptor: bridging innate and adaptive immune responses at mucosal surfaces.** *Immunol. Rev.* 2005, **206**:83–99.

Kaetzel, C.S. (ed): *Mucosal Immune Defense: Immunoglobulin A.* New York, Springer, 2007.

Kaetzel, C.S., and Mostov, K.E.: **Immunoglobulin transport and the polymeric immunoglobulin receptor**, in Mestecky, J., Lamm, M.E., Strober, W., *et al.* (eds): *Mucosal Immunology*, 3rd ed. Burlington, MA, Elsevier Academic Press, 2005:211–250.

Kilian, M., and Russell, M.W.: **Microbial evasion of IgA functions**, in Mestecky, J., Lamm, M.E., Strober, W., *et al.* (eds): *Mucosal Immunology*, 3rd ed. Burlington, MA, Elsevier Academic Press, 2005:291–303.

Kuo, T.T., Baker, K., Yoshida, M., *et al.*: **Neonatal Fc receptor: from immunity to therapeutics.** *J. Clin. Immunol.* 2010, **30**:777–789.

Li, H., Nowark-Wegrzyn, A., Charlop-Powers, Z., *et al.*: **Transcytosis of IgE-antigen complexes by CD23a in human intestinal epithelial cells and its role in food allergy.** *Gastroenterology* 2006, **131**:47–58.

Mason, A.P., Thrall, R.S., Rafti, E., *et al.*: **IgG transmitted from allergic mothers decreases allergic sensitization in breastfed offspring.** *Clin. Mol. Allergy* 2010, **13**:8–9.

Mattu, T.S., Pleass, R.J., Willis, A.C., *et al.*: **The glycosylation and structure of human serum IgA1, Fab, and Fc regions and the role of *N*-glycosylation on Fcα receptor interactions.** *J. Biol. Chem.* 1998, **273**:2260–2272.

Mestecky, J., Moro, I., Kerr, M.A., *et al.*: **Mucosal immunoglobulins**, in Mestecky, J., Lamm, M.E., Strober, W., *et al.* (eds): *Mucosal Immunology*, 3rd ed. Burlington, MA, Elsevier Academic Press, 2005:153–181.

Peppard, J.V., Kaetzel, C.S., and Russell, M.W.: **Phylogeny and comparative physiology of IgA**, in Mestecky, J., Lamm, M.E., Strober, W., *et al.* (eds): *Mucosal Immunology*, 3rd ed. Burlington, MA, Elsevier Academic Press, 2005:195–210.

Royle, L., Roos, A., Harvey, D.J., *et al.*: **Secretory IgA *N*- and *O*-glycans provide a link between the innate and adaptive immune systems.** *J. Biol. Chem.* 2003, **278**:20140–20153.

Russell, M.W., and Kilian, M.: **Biological activities of IgA**, in Mestecky, J., Lamm, M. E., Strober, W., *et al.* (eds): *Mucosal Immunology*, 3rd ed. Burlington, MA, Elsevier Academic Press, 2005:267–289.

Woof, J.M., and Burton, D.R.: **Human antibody-Fc receptor interactions illuminated by crystal structures.** *Nat. Rev. Immunol.* 2004, **4**:89–99.

Woof, J.M., and Kerr, M.A.: **The function of immunoglobulin A in immunity.** *J. Pathol.* 2006, **208**:270–282.

Woof, J.M., and Mestecky, J.: **Mucosal immunoglobulins.** *Immunol. Rev.* 2005, **206**:64–82.

Woof, J.M., van Egmond, M., and Kerr, M.A.: **Fc receptors**, in Mestecky, J., Lamm, M.E., Strober, W., *et al.* (eds): *Mucosal Immunology*, 3rd ed. Burlington, MA, Elsevier Academic Press, 2005:251–265.

Ye, L., Zeng, R., Bai, Y., *et al.*: **Efficient mucosal vaccination mediated by the neonatal Fc receptor.** *Nat. Biotechnol.* 2011, **29**:158–163.

Yoshida, M., Claypool, S.M., Wagner, J.S., *et al.*: **Human neonatal Fc receptor mediates transport of IgG into luminal secretions for delivery of antigens to mucosal dendritic cells.** *Immunity* 2004, **20**:769–783.

Zhang, Y.A., Salinas, I., Li, J., *et al.*: **IgT, a primitive immunoglobulin class specialized in mucosal immunity.** *Nat. Immunol.* 2010, **11**:827–835.

Role of Dendritic Cells in Integrating Immune Responses to Luminal Antigens

10

Mucosal tissues constitute the major lymphoid compartments of the body. These tissues generate complex and unique immune responses that protect the body against invading microbes and prevent untoward responses to common mucosal antigens and symbiotic bacteria. This chapter focuses on the role of dendritic cells (DCs) in the induction and regulation of immune responses at mucosal surfaces.

DCs are divided into follicular and nonfollicular DCs. Follicular DCs develop from a non-bone-marrow-derived precursor and express high levels of receptors for immunoglobulin and complement. Antigen expressed on follicular DCs may be important for the germinal center reaction and for the maintenance of B-cell memory. Nonfollicular DCs comprise a family of mononuclear phagocytes that develop from bone marrow derived stem cells under the influence of FMS-like tyrosine kinase 3 (Flt3) ligand, granulocyte–macrophage colony-stimulating factor (GM-CSF), and other cytokines; these cells include conventional DCs (cDCs), monocyte-derived DCs (mDCs), peripheral CD103+ DCs, and plasmacytoid DCs (pDCs). Three immunologic compartments contain these cells: (1) nonlymphoid tissues, including the interstitium of internal organs, mucosal lamina propria, and skin dermis and epidermis (tissue DCs, Langerhans cells, and dermal DCs respectively); (2) lymphoid tissues such as the T-cell regions of draining lymph nodes (interdigitating DCs); and (3) lymph where they have extensive processes (veiled cells). Focusing on mucosal tissues, we discuss DC phenotype, localization, antigen uptake and trafficking, interaction with T and B cells, and role in the pathogenesis of inflammatory bowel disease (IBD).

Defining characteristics of dendritic cells.

In the early 1970s, Ralph Steinman and Zanvil Cohn discovered a population of 'tree-like' cells in cultures of splenic 'accessory cells' (cells required to generate antibody responses *in vitro*), which they called 'dendritic cells' derived from 'dendron,' the Greek word for tree. These cells were distinct in appearance and motility when compared with other members of the mononuclear phagocyte family, and were subsequently found to have unique functional properties. In particular, DCs were shown to act in the thymus to select T cells, and in systemic and mucosal tissues to capture antigens and induce and direct primary T- and B-cell responses.

10-1   Dendritic cells capture antigens and select and activate naive T-cell clones *in vivo*.

Immature DCs in the mucosa and skin are positioned to capture external antigens, whereas DCs in lymphoid tissues capture antigens in the blood or draining lymph. DCs in peripheral nonlymphoid tissues take up tissue-specific antigens. DCs express pattern recognition receptors, such as Toll-like receptors (TLRs) and C-type lectin-like receptors, and other receptors involved in antigen uptake, and have a distinct endocytic system that promotes efficient capture, processing, and the presentation of antigens via major histocompatibility complex (MHC) I and MHC II to CD8 and CD4 T cells respectively. Upon activation by pathogens, cytokines, or factors present in damaged or inflamed tissues, DCs undergo 'maturation,' which is characterized by dendrite elongation, increased processing of antigen, expression of MHC I– and MHC II–peptide complexes, and enhanced expression of co-stimulatory molecules. This maturation process coincides with DC migration from nonlymphoid tissues to T-cell zones of draining lymph nodes, where the DCs localize along connective tissue fibers and present the antigen to circulating naive T cells attracted by chemokines such as DC-derived CCL18. Mature DCs express adhesion molecules, including intercellular adhesion molecule-1(ICAM-1), ICAM-2, leukocyte function-associated antigen-1 (LFA-1), LFA-3, and intercellular adhesion molecule 3-grabbing nonintegrin (DC-SIGN, CD209). DC-SIGN binds transiently and with high avidity to ligands expressed by naive T cells, enhancing efficient T-cell sampling of DC-expressed MHC–peptide complexes. High levels of MHC complexes, adhesion molecules, co-stimulatory molecules, and secreted cytokines allow mature DCs to effectively drive naive T-cell activation and differentiation. This process of antigen sampling in peripheral tissues, migration, and presentation to naive T cells in lymph nodes results in a primary immune response and constitutes a major functional characteristic of DCs.

10-2   Dendritic cells determine the nature of lymphocyte responses.

During infection or inflammation, migrating or resident DCs drive the differentiation of effector CD4$^+$ T cells, including helper T (T$_H$) T$_H$1, T$_H$2, and T$_H$17 cells, and CD8$^+$ T cells. DC activation of T cells occurs through a combination of activating co-stimulatory molecules, including CD80, CD86, and OX40L, and cytokines, including interleukins (IL) IL-12p70, IL-23, IL-6, and IL-4, produced by the DCs or other cells in the infected or inflamed microenvironment.

In contrast to DCs in inflamed or infected tissue, immature DCs in healthy tissue constitutively take up self-antigens in the form of apoptotic bodies, soluble exogenous proteins such as food antigens, and commensal bacteria from the skin and mucosal tissues, and then migrate to the draining lymph nodes. DCs under noninflammatory conditions induce peripheral tolerance through a combination of antigen-specific T-cell deletion, the induction of regulatory T (T$_{reg}$) cell differentiation, and in the mucosa, the generation of IgA that prevents antigen and microbial uptake. These 'tolerogenic' DCs may have characteristics of 'mature' cells, reflected in their expression of high levels of MHC antigens, but express soluble mediators (e.g., IL-10, transforming growth factor-β (TGF-β), and retinoic acid, RA) and/or tolerogenic surface receptors (e.g., OX40L, CTLA-4 (CD152), programmed death ligand-1) that inhibit T-cell expansion, induce T-cell death, or drive the differentiation of T$_{reg}$ cells. Soluble protein antigens delivered into the circulation by intravenous injection or after oral administration of high doses of proteins, will become captured by DCs residing within lymphoid tissues. Resident DCs in nonmucosal sites that capture soluble protein antigens can induce T-cell anergy and deletion, likely by activating T cells in the absence of sufficient co-stimulation. Thus, depending on the antigenic stimulus, route of antigen administration, and local inflammatory conditions, DCs can induce either active immunity or systemic tolerance.

The ability to efficiently take up, process, and present antigens to naive lymphocytes, and to direct primary lymphocyte responses distinguishes DCs from other members of the mononuclear phagocyte family. In contrast to DCs, macrophages lack dendrites, do not migrate at high rates from tissues to draining lymph nodes, and are inefficient at activating naive T cells. The latter is likely due to the production of lysozymes and cathepsins, which rapidly degrade antigens, the absence of endocytic pathways that promote efficient presentation of MHC–antigen complexes on the cell surface, and the decreased expression of appropriate adhesion molecules. Macrophages are more specialized for scavenging dead and dying cells, repairing damaged tissues, and killing intracellular microbes. During infection and inflammation, newly recruited blood monocytes may contribute to local inflammation through the production of inflammatory cytokines such as tumor necrosis factor-α (TNF-α) and IL-1.

10-3   Dendritic cells are a family of cells with distinct subpopulations.

DC induction of T-cell responses is complex, as schematically depicted in Figure 10.1. This complexity is reflected in the subsets of DCs that arise from distinct precursors, the array of stimuli to which DCs are exposed during maturation and T-cell priming, and local microenvironmental factors that influence DC migration, maturation, survival, and the production of cytokines that influence T-cell differentiation.

Subpopulations of DCs have been defined based on unique surface marker expression, localization, gene expression, and function. Importantly, surface markers that define populations of DCs in mice and humans may differ, but similar DC populations have been identified in both species. The majority of tissue and resident DCs and are clearly distinguished from pDCs, which are not DCs *per se*, but plasmacytoid-shaped cells that are activated by viruses or nucleic acids to produce high amounts of type-I interferon and then differentiate into antigen-presenting DCs.

Surface markers commonly used to define DC populations in the mouse include CD11c, CD11b, CD8, CD4, CD 207 (langerin), CD205, 33D1, CD103, and CX3CR1; and in humans CD11c, CD209, CD1c (BDCA-1), CD141 (BDCA-3), CD14, CD207, CX3CR1, and CD103, in addition to expression of MHC II antigens and the absence of typical markers for lymphocytes and natural killer (NK) cells. F4/80 has traditionally been used to define macrophage populations in mice, but this antigen is shared by certain tissue DCs, such as Langerhans' cells and a subpopulation of mucosal mDCs. pDCs are distinguished by low levels of MHC II; the lack of CD19, T-cell, and NK-cell markers; moderate levels of

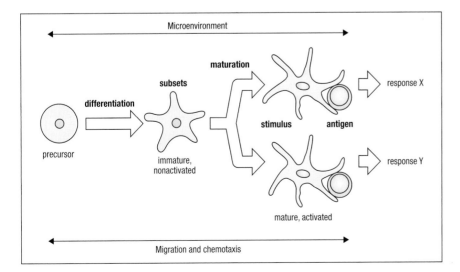

Figure 10.1 Factors affecting DC function. Bone marrow derived precursor cells differentiate into distinct subsets of DCs that mature and differentiate under the influence of microenvironmental factors. In steady-state conditions, the factors that stimulate maturation have not been defined, but in infected, inflamed, or injured mucosa, DCs mature after activation by microbial products, inflammatory cytokines, and/or products of damaged tissue. The signals that induce maturation also affect the functional phenotype of the DC population.

CD11c; and high levels of B220, Ly6C, and Siglec H in the mouse and CD123 (the α-chain of the IL-3 receptor) and CD303 (BDCA-2) in humans.

In mice, DC populations have been defined on the basis of DC ontogeny (Figure 10.2). According to this scheme, the three primary circulating DC precursors, namely monocytes, pre-cDCs, and pDCs, differentiate from hematopoietic stem cells in the bone marrow. Monocytes can be further distinguished by their expression of Ly6C, a glycosyl phosphatidylinositol-anchored surface glycoprotein. In the steady state, circulating Ly6C^{high} monocytes give rise to CD11c$^+$ CX3CR1$^+$ macrophages and mDCs in intestinal tissues and likely resident CX3CR1$^+$ cells in lymph nodes. During inflammation, Ly6C^{high} monocytes can enter tissues and develop into inflammatory mDCs or inflammatory (M1) macrophages, while Ly6C^{low} monocytes differentiate into either M1 or alternatively activated (M2) macrophages, which contribute to tissue repair and homeostasis, depending on the inflammatory conditions.

Pre-cDCs differentiate into cDC populations in lymphoid organs and CD103$^+$ DCs in tissues. These DCs include CD8$^+$CD11b$^-$, CD8$^-$CD11b$^+$, CD8$^-$CD11b^{low}, and CD8$^-$CD11b$^+$CD4$^+$ populations. CD103 is variably expressed by a fraction of cDC populations, depending on the lymphoid tissue. For example, CD103 is expressed by a small proportion of CD8$^+$ CD11b$^-$ mDCs in the spleen but by a high proportion in Peyer's patches (PPs). In the lamina propria of the intestine, CD103$^+$CD11b$^+$CD8$^-$ DCs are the primary cells that migrate to the mesenteric lymph nodes (MLNs). Consequently, CD103$^+$CD11b$^+$CD8$^-$ (or CD8^{low}) DCs in the MLN include those cells that have migrated from the intestinal lamina propria. CD103 and CX3CR1 expression on mononuclear phagocytes is mutually exclusive. Therefore, pre-cDCs give rise to CD103$^+$ and CD103$^-$ CX3CR1$^-$ DCs in lymphoid and nonlymphoid tissues, whereas monocytes give rise to CD103$^-$CX3CR1$^+$ tissue DCs, as well as tissue macrophages. The maturation of pre-cDCs and monocytes into their respective cell populations is differentially dependent on the growth factors Flt3 ligand, GM-CSF, and granulocyte colony-stimulating factor.

Figure 10.2 Mucosal mononuclear phagocyte populations. DC and macrophage populations arise from circulating bone marrow derived precursors. Gr1^{high} monocytes give rise to CX3CR1$^+$CD103$^-$ cells, whereas pre-cDCs give rise to CD103$^+$ DCs and likely CD103$^-$ mDCs. pDCs originate in the bone marrow. All cells depicted in the lamina propria and gut-associated lymphoid tissue (Peyer's patch (PP), mesenteric lymph nodes (MLN), and isolated lymphoid follicles (ILF)) express CD11c, with the exception of CD11c$^-$ CX3CR1$^+$ macrophages. CD103$^+$CD11b$^+$ DCs are the primary cells that migrate from the lamina propria to the MLNs.

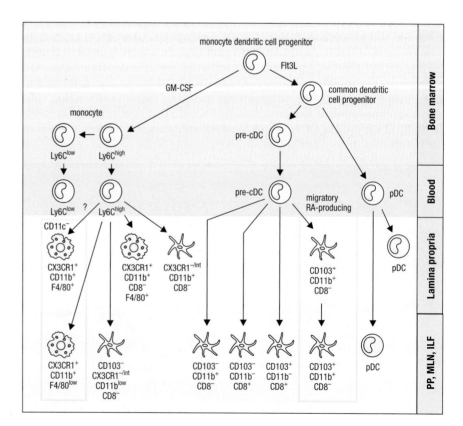

10-4   Dendritic cell function is influenced by activation signals.

The ability of DCs to drive a particular T-cell response is influenced by signals received during antigen uptake, activation, and T-cell priming. DC function also is influenced by the form of antigen (e.g., soluble, apoptotic-body-associated, CD205-directed, or pathogen-associated) and tissue factors (e.g., inflammatory cytokines or products of damaged cells), which in turn affects the type of T-cell response induced by the DC. For example, the microbial products cholera toxin, lipopolysaccharide (LPS) from *Porphyromonas gingivalis*, soluble egg antigens from *Schistosoma mansoni*, and hyphae from *Candida albicans*, and the mediators histamine and thymic stromal lymphopoietin (TSLP) appear to induce DCs, at least in part through suppressed IL-12 production, to drive $T_H 2$ responses. In contrast, TLR ligands more commonly induce high levels of IL-12 production by DCs to drive $T_H 1$ responses.

Further reflecting their remarkable plasticity, exposure of DCs to certain cytokines, including IL-10 or TGF-β, pathogens such as *Plasmodium falciparum*, or pathogen products such as *Bordetella pertussis* fimbrial hemagglutinin induces DCs that inhibit effector T cells and drive regulatory T (T_{reg}) cell differentiation. Finally, the T-cell phenotype (e.g., $T_H 1$ vs. $T_H 2$) and the level of homing molecule expression on naive T cells differentiated with DCs are influenced, at least *in vitro*, by the antigen dose and the ratio of DCs to T cells, with high antigen doses or DC/T cell ratios favoring $T_H 1$ responses and low antigen dose or DC/T cell ratios favoring $T_H 2$ responses.

Intestinal dendritic cell populations.

Most information regarding intestinal DC phenotype and function is derived from murine studies. Human intestinal DCs, particularly in healthy mucosa, have not been well characterized.

An additional issue with human studies is that the mucosa is studded with thousands of small isolated lymphoid follicles which cannot be seen, so when mononuclear cells are isolated from human gut for functional studies, they are likely to be a mixture of cells from the lamina propria and gut-associated lymphoid tissues (GALT).

10-5   Distinct populations of dendritic cells are present in mucosal inductive and effector sites and capture antigens by different mechanisms.

Primary sites for the induction of intestinal T- and B-cell responses are the GALT, which include PPs in the small intestine, isolated lymphoid follicles in the small and large intestines, and mesenteric lymph nodes (MLNs). In contrast, the diffuse lamina propria and the intraepithelial cell compartments are primarily effector sites. Distinct DC populations in these tissues capture antigens by different mechanisms, have different abilities to migrate to the MLNs, and have distinct functional capacities (Figure 10.3).

Luminal antigens, including macromolecules, bacteria, and viruses, gain access to PPs and isolated lymphoid follicles through specialized epithelial cells called M (microfold) cells present in the follicle-associated epithelium (FAE). M-cell transport of antigen is promiscuous and is initiated when the antigen binds to surface-expressed carbohydrates in regions with less overlying mucus, but can be enhanced by the presence of antigen-specific IgA. DCs are present in small numbers within the FAE and in large numbers in the subepithelial dome. At least two populations of DCs are present in the subepithelial

Figure 10.3 Localization, maturation, and migration of DCs. CX3CR1⁺ DCs and macrophages are present close to the epithelium in the lamina propria and PP, where the cells can extend dendrites into the intestinal lumen. CX3CR1⁺-TNF-α and inducible nitric oxide synthase (iNOS)-producing (Tip) DCs and CX3CR1⁺TLR5⁺ DCs are present in the lamina propria; and Tip DCs are also present in the PP, likely within the PP interfollicular region or B-cell follicle. Whether CX3CR1⁺ cells migrate to the interfollicular region in the PP is not known. CD103⁺ DCs are present in the small intestinal lamina propria, where they preferentially express CD11b, and in the isolated lymphoid follicles and PPs, where they primarily express CD8 and not CD11b. CD103⁺ DCs actively migrate from the lamina propria, and possibly from the isolated lymphoid follicles and PPs, to the MLNs. CCR6⁺ DCs in the PP subepithelial dome can migrate into the FAE or just beneath the FAE during infection with *Salmonella* Typhimurium and may migrate to the interfollicular region after a switch in chemokine receptor expression from CCR6 to CCR7. Whether CCR6⁺ DCs express CD103 is not yet clear. HEV, high endothelial venule.

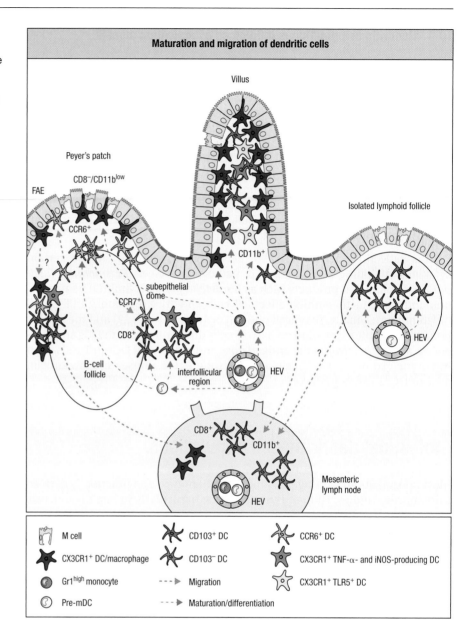

dome. CX3CR1⁺CD11b^{low/-}CD8⁻ DCs are located just beneath the FAE and can extend dendrites into the intestinal lumen to capture particulate antigens and bacteria. A proportion of these cells express high levels of lysozyme normally present in macrophages, suggesting a role in innate defense against incoming microbes. In addition, a population of CX3CR1⁻CD11b^{low}CD8⁻ DCs expresses CCR6 and is present in the subepithelial dome. Whether this population also expresses CD103 is not known. Following infection or activation with mucosal adjuvants such as cholera toxin, TLR2 agonists, or flagellin, subepithelial dome DCs can be recruited into the FAE and increase their extension of dendrites into the lumen. The recruitment of CCR6⁺ DCs into the subepithelial dome and FAE depends on the stimulated release of CCL20, the ligand for CCR6, from epithelial cells in the FAE. Recruited CCR6⁺ DCs may activate T cells in the subepithelial dome. These DCs form clusters with T cells in response to bacteria or may migrate to the isolated lymphoid follicles after a switch in chemokine receptor expression to CCR7 and migration to the CCR7 ligands CCL19 and CCL21, which are constitutively expressed in the T cell-rich interfollicular region.

In a murine model of *Salmonella* Typhimurium infection, CX3CR1$^+$CD11b$^{low/-}$ CD8$^-$ DCs containing the bacteria have been detected in interfollicular regions after oral inoculation. CX3CR1$^-$CD11b$^-$CD8$^+$ DCs, many of which co-express CD103, also are present in the interfollicular region and likely represent a stationary blood-derived population that enters this region through high endothelial venules. These DCs can process viral antigens for T-cell activation, but the mechanism by which interfollicular DCs acquire antigens has not been elucidated.

DCs are also present within isolated lymphoid follicles of the small intestine and colon. Isolated lymphoid follicles develop from rudimentary structures called cryptopatches, collections of lymphoid tissue inducer-like CD90$^+$ cells that are present at birth. Isolated lymphoid follicles and cryptopatches are collectively termed solitary isolated lymphoid tissue. Upon bacterial colonization of germ-free mice, DCs are recruited to cryptopatches and produce CXCL13 that attracts B cells to form an isolated lymphoid follicle. DCs present in isolated lymphoid follicles are both CD103$^+$ and CD103$^-$, and some express CD8. Intestinal bacteria gain entry into the isolated lymphoid follicles via M cells similar to PPs and induce the production of IgA.

A second site for antigen entry into the intestine is the nonfollicular absorptive epithelium, which overlays the lamina propria. Soluble antigens and bacteria gain access to DCs and macrophages in the lamina propria by trans- or paracellular epithelial transport, receptor-mediated trafficking such as through neonatal Fc receptors on epithelial cells, and DC extensions that reach between epithelial cells into the intestinal lumen. Antigens may also be sampled via DC uptake of epithelial cell exosomes and apoptotic bodies, from epithelial cells or after transport across M cells in epithelium not associated with GALT called 'villous M cells,' and through injured epithelium.

The intestinal lamina propria contains MHC class II^{high} macrophages and DCs. Lamina propria macrophages have a potent capacity to kill microorganisms, but do not present antigens to T cells or produce cytokines. In the mouse, F4/80$^+$ lamina propria macrophages express CX3CR1, can express high or low levels of CD11c, and constitutively produce high levels of IL-10. In contrast, lamina propria DCs express CD11c, CD11b, and CD103; and an additional population of mDCs express CD11c, CD11b, and intermediate levels of CX3CR1.

CD103$^+$ DCs and CX3CR1$^+$ cells (including mDCs and macrophages) are present in different locations in the subepithelial tissue. CX3CR1$^+$ lamina propria cells are associated with the epithelium and actively participate in antigen capture through transepithelial dendrites. The proximal intestine (duodenum and jejunum) displays a higher number of these dendrites than the terminal ileum under steady-state conditions. Transepithelial dendrites are dependent on the intestinal microbiota, as antibiotic treatment markedly reduces the number of these dendrites. In the terminal ileum, the number of transepithelial dendrites increases in response to *Salmonella*, suggesting that constitutive and inducible mechanisms contribute to dendrite extension. CX3CR1 is involved in dendrite extensions: CX3CR1$^{-/-}$ mice lack transepithelial cell protrusions and display defects in the internalization of noninvasive bacteria. Lamina propria transepithelial cell extensions in the terminal ileum are dependent on TLR engagement on epithelial cells; consequently, mice reared in germ-free conditions display reduced numbers of transepithelial dendrites. CX3CR1$^+$ macrophages, but not mDCs, have a relatively slow turnover rate and a low capacity to activate naive CD4 and CD8 naive T cells *in vitro*.

In contrast to epithelium-associated CX3CR1$^+$ macrophages and mDCs, CD103$^+$ DCs are present deeper in the lamina propria, and the ability of these DCs to extend dendrites into the intestinal lumen is unclear. CD103$^+$ lamina propria DCs have a high capacity to activate naive T cells *in vitro* and very high turnover rates, and actively migrate from the lamina propria to the MLN under steady-state conditions and after stimulation with TLR ligands *in vivo*. Lamina propria CD103$^+$ DCs, but not CD103$^-$ cells, capture soluble antigens administered orally and migrate in a CCR7-dependent manner to the MLN where they activate T cells. DCs that migrate from the lamina propria to the MLN carry fragments of intestinal epithelial cells, suggesting that luminal antigens taken up by, or bound to, epithelial cells may be carried to the MLN with epithelial cell fragments.

In addition to CD103$^+$CD11b$^+$ DCs that have migrated from the lamina propria, MLNs contain a large population of resident CD103$^+$CD11b$^-$ DCs and CD103$^-$CX3CR1$^+$CD11b$^+$ DCs. The CD103$^-$ DCs appear to capture antigens in the blood rather than the lamina propria and drive the differentiation of T$_H$1 cells *ex vivo*. In the steady state, CD103$^+$ lamina propria and MLN DCs drive the expression of the intestinal homing receptors CCR9 and $\alpha_4\beta_7$ on CD4 and CD8 T cells; they also induce the differentiation of Foxp3$^+$ T$_{reg}$ cells from naive CD4$^+$ T cells through production of TGF-β and retinoic acid. However, in the presence of intestinal inflammation the cells have decreased capacity to drive T$_{reg}$ differentiation and instead drive T$_H$1 cell differentiation.

Unique functions of intestinal dendritic cells.

Unique functions of intestinal DCs include the ability to drive the differentiation of T$_{reg}$ cells that are involved in tolerance to soluble oral antigens and commensal bacteria, the capacity to provide signals for the differentiation of IgA-producing B cells, and the ability to imprint lymphocytes with unique homing receptors that promote recirculation back to intestinal tissues.

These functions are likely to be altered in disease states and indeed dysregulated DC–T cell interactions may be critical in the development of gut inflammation. Reagents which block DC–T cell interactions may therefore have therapeutic potential.

10-6 Oral tolerance is mediated by intestinal CD103$^+$ dendritic cells and resident dendritic cells in the spleen and lymph nodes.

Oral administration of soluble protein antigens results in systemic tolerance to subsequent immunization and is referred to as oral tolerance (Figure 10.4). Oral tolerance is mediated by T-cell anergy, T-cell deletion, and differentiation of CD4$^+$ T$_{reg}$ cells. The mechanisms that mediate oral tolerance may be influenced by antigen dose and frequency with higher doses favoring anergy and deletion, and lower and repeated doses favoring the generation of T$_{reg}$ cells (Figure 10.4).

Resident DCs in lymph nodes and spleen drive T-cell anergy and deletion following oral antigen administration. In particular, CD11c$^+$CD8$^+$CD205$^+$ resident DCs are implicated in this process, as they have been shown to be important in tolerance to soluble protein antigens given systemically. In contrast, CD103$^+$ DCs in the lamina propria capture oral antigens and migrate to the MLN to induce Foxp3$^+$ T$_{reg}$ cells. CD103$^+$ DC induction of Foxp3$^+$ T$_{reg}$ cells is dependent on the production of TGF-β and the co-factor RA, a DC-generated metabolite of vitamin A that acts via RA receptors in the responding cell nucleus. DC expression of $\alpha_v\beta_8$ integrin, which functions to activate latent TGF-β and CCR7, is important for the *de novo* induction of Foxp3$^+$ T$_{reg}$ cells in MLNs. Foxp3$^+$ T$_{reg}$ cells are also induced in PPs; however, the details of how this occurs and the importance of this induction are not yet clear.

CD103$^+$ DCs also express indoleamine 2,3 dioxygenase (IDO), an enzyme required for their tolerogenic functions. IDO participates in tryptophan

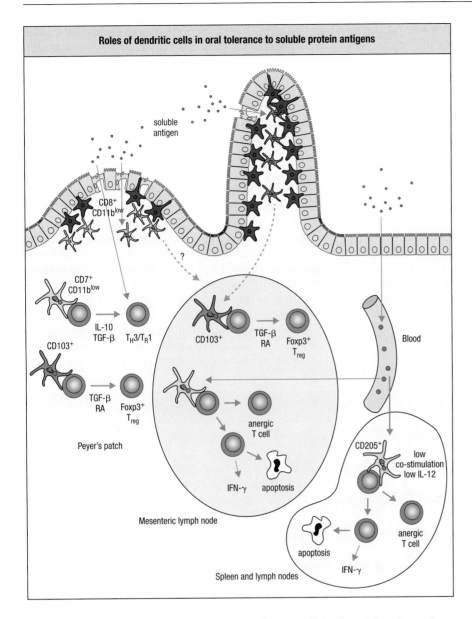

Figure 10.4 Role of DCs in oral tolerance to soluble protein antigens. Soluble protein antigens given orally can be taken up locally and presented to T cells in PPs. High doses of antigen that enters the circulation in the form of immunogenic peptides or protein derivatives are presented by cDCs in lymph nodes and spleen to T cells that are induced to undergo apoptosis or become anergic. Local CD103⁺ DCs in PPs or MLNs induce *de novo* generation of CD4⁺ Foxp3⁺ T_reg cells. CD11b⁺ DCs in PPs may also induce the differentiation of T cells that produce TGF-β or IL-10, originally called T_H3 cells. CD4⁺ Foxp3⁺ T_reg cells can secrete TGF-β and IL-10, and express TGF-β on their surface. RA, retinoic acid; IFN-γ, interferon-γ.

catabolism, and its immunosuppressive effects are linked to either the reduction of local tryptophan concentration or to the production of immunomodulatory tryptophan metabolites, such as kynurenine. IDO expression in CD103⁺ DCs inhibits T_H17 cell development and reduces T-cell proliferation, and likely allows for the upregulation of Foxp3 by TGF-β and RA.

Thus, high antigen doses favor anergy and deletion of specific naive T-cell clones at systemic sites, likely limiting the differentiation of T_{reg} cells in mucosal lymphoid tissues, whereas low (and repeated) antigen doses favor Foxp3⁺ T_{reg} induction in local mucosal tissues driven by specialized CD103⁺ DCs.

10-7 Dendritic cells regulate T-cell responses to intestinal bacteria.

The human intestine is colonized by approximately 10^{14} bacteria, roughly 100 times the number of cells in the body, and up to 1000 different bacterial species. These microbes are 'symbionts' in that they live together with the host in either mutualistic (beneficial to both host and microbe) or commensal (beneficial to either host or microbe) relationships. In contrast, pathogenic microbes have a parasitic relationship with their host (benefiting the microbe at the expense of the host). Normally mutualistic or commensal microbes can become pathogenic/parasitic when host factors such as the epithelial barrier or the innate or adaptive immune response are significantly compromised. This is particularly evident in patients with inflammatory bowel disease, in which it is now quite clear that dysregulated innate and adaptive immunity to normally symbiotic bacteria results in uncontrolled intestinal inflammation. How the intestinal immune system maintains control over symbiotic bacteria while promoting immunity to more pathogenic microbes, such as invasive bacteria and enteric viruses, is an area of intense investigation.

Intestinal DCs appear to play an important role in the regulation of T-cell responses to symbiotic bacteria and in the induction of bacteria-specific IgA (see below), both of which are critical for the maintenance of mucosal homeostasis (Figure 10.5). Under steady-state conditions, it is presumed that CD103$^+$ DCs capture bacteria and/or bacterial antigens in the intestinal lamina propria and migrate to the MLN where they are thought to induce the differentiation of Foxp3$^+$ T$_{reg}$ cells in response to TGF-β and retinoic acid produced by the CD103$^+$ DCs themselves or present within the intestinal microenvironment. Foxp3$^+$ T$_{reg}$ cells from the MLN migrate to the intestinal lamina propria to control effector T-cell responses and suppress pro-inflammatory cytokine production by innate immune cells. In addition, Foxp3$^+$ T cells are essential in controlling the expansion of pathogenic T$_H$1 and T$_H$17 effector cells within the MLNs themselves. Thus, in the absence of significant numbers of Foxp3$^+$ T cells, as occurs experimentally in the adoptive transfer of naive CD4$^+$ T cells

Figure 10.5 DCs in immune responses to intestinal microbiota. Symbiotic intestinal bacteria induce T$_H$1, T$_H$17, and Foxp3$^+$ T$_{reg}$ responses. Symbiotic bacteria can also drive the production of secretory IgA through mainly T-cell independent mechanisms that, in turn, limit the number of bacteria and bacterial contact with the intestinal epithelium. Symbiotic bacteria can also induce the release of TGF-β, B cell-activating factor (BAFF), and a proliferation inducing ligand (APRIL) by epithelial cells that are involved in IgA isotype switching. PP DCs release RA that participates in IgA class switching while TNF-α- and inducible nitric oxide synthase (iNOS)-producing DCs in the lamina propria enhance BAFF and APRIL production through the production of nitric oxide (NO). In the MLN and PP, T$_{reg}$ cells can differentiate into T follicular helper cells that help B cells in class switching. PP- and MLN-derived IgA$^+$ B cells reach the lamina propria via the circulation and differentiate into plasma cells that release dimeric and oligomeric IgA antibodies.

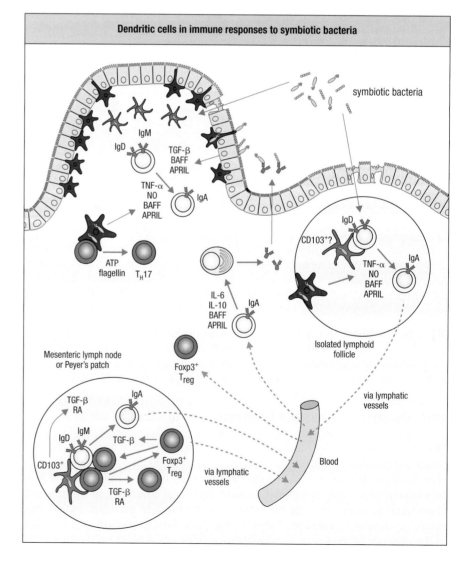

(lacking Foxp3⁺ T cells) to RAG$^{-/-}$ mice, CD103⁺ DCs drive the differentiation of pathogenic T cells in the MLN, resulting in colitis induction. How Foxp3⁺ T_{reg} cells control pathogenic T-cell expansion in the MLN and/or regulate local effector cell function and pro-inflammatory cytokine production in the lamina propria is not yet clear.

CX3CR1⁺ DCs may also be responsible for T_H1 and T_H17 induction. CD11c⁺ CX3CR1⁺ cells in the intestinal lamina propria express TLR5 and can induce the differentiation of T_H1 and T_H17 cells in response to flagellin or ATP produced by many symbiotic intestinal bacteria. T_H17 cells can produce IL-17 and IL-22 that can contribute to epithelial barrier function through the induction of innate defense factors from epithelial cells and induce epithelial cell proliferation.

It is also now becoming clear that specific species of symbiotic bacteria can preferentially induce certain helper T cell responses. For example, when given to germ-free mice, *Clostridium* species preferentially induce T_{reg} cells; *Bacteroides fragilis* induces T_H1 and Foxp3⁺ T_{reg} cells; and segmented filamentous bacteria (SFB) preferentially induce T_H17, with less T_H1 and Foxp3⁺ T_{reg} differentiation. The specific mechanisms responsible for these effects are not yet clear; however, intestinal DCs may be influenced by bacterial products and inflammatory signals to induce specific T-cell responses. SFB, for example, drive the production of serum amyloid A from epithelial cells. Serum amyloid can then stimulate IL-6 and IL-23 production by DCs, which in turn contributes to the differentiation and expansion of T_H1 and T_H17 cells. Similarly, DNA from some bacterial species can act through TLR9 to decrease Foxp3⁺ T_{reg} cell, and enhance T_H1, T_H2, and T_H17 cell differentiation in the intestine in mice by inducing IL-6 expression by CD103⁺ DCs.

In contrast to tolerogenic responses to symbiotic bacteria and innocuous antigens, mucosal pathogens induce rapid innate and acquired immune responses that eliminate the pathogen.

How the host discriminates between symbiotic bacteria and pathogenic microbes is poorly understood, in particular because many immune activating molecules (such as LPS and flagellin) are expressed by both symbionts and pathogens. Several possibilities have been proposed including differences in microbial recognition molecules (pathogen-associated molecular patterns, PAMPs) with more (pathogen) or less (symbiont) ability to activate host cells through pattern recognition receptors such as TLRs; unique PAMPs, produced by pathogens, such as those produced by *Nippostrongylus brasiliensis* that direct DCs to produce strong T_H2 responses; the production of factors by symbiotic microbes that actively induce tolerogenic signals; and the ability of pathogens to overcome innate epithelial defenses to invade and destroy tissues or induce 'stress' or 'danger' signals (such as the production of inflammatory cytokines and chemokines) that result in enhanced inflammation via the recruitment and activation of neutrophils, macrophages, and DCs to infected sites. Thus, the immune response to a pathogen is dictated by the subpopulations of DCs that are present or recruited, the pathogen response receptors engaged on the DCs, and the changes in the microenvironment, including the production of cytokines from innate cells that are encountered or recruited to the site of infection.

10-8 DCs are essential for both T-cell dependent and independent IgA B-cell responses in the intestine.

In addition to neutralizing toxins and providing short-term protection against infectious agents, secretory IgA in the intestinal lumen limits the growth and entry of symbiotic bacteria and is essential for maintaining intestinal immune homeostasis (Figure 10.5). Pathogenic microbes such as *Salmonella* Typhimurium and symbiotic bacteria such as nonpathogenic *Escherichia coli* can be taken up by mucosal DCs and transported to MLNs. In addition, bacteria, viruses, and particulates are taken up by M cells to enter PPs and (at least symbiotic bacteria) isolated lymphoid follicles, where naive B cells undergo IgA class switch and differentiation. IgA plasmablasts migrate to the intestinal lamina propria, where they differentiate into plasma cells and produce specific IgA that after transport across epithelial cells acts to neutralize toxins and viruses and limit bacterial colonization and entry.

DCs play several important roles in T-cell dependent IgA B-cell responses in PPs and MLNs (Figure 10.5). First, Foxp3+ T_{reg} cells provide T-cell help for IgA responses. Foxp3+ T_{reg} cells can become T follicular helper cells, where they produce TGF-β and possibly IL-21, and express CD40L, which signals class switching through CD40 expressed on B cells. Thus, CD103+ DC induction of Foxp3+ T_{reg} cells is important for IgA induction. Second, CD103+ DCs that produce RA and TGF-β can directly help CD40-activated B cells to differentiate into IgA-producing cells.

T-cell independent IgA B-cell switching occurs in isolated lymphoid follicles and the intestinal lamina propria through interaction with DCs. DCs in the isolated lymphoid follicles produce TNF-α and nitric oxide (NO) ('Tip' DCs) in response to TLR signaling. TNF-α induces proteinases that activate latent TGF-β produced by stromal cells, and NO drives the production of B cell-activating factor (BAFF) and a proliferation inducing ligand (APRIL) by DCs and stromal cells. BAFF and APRIL, together with activated TGF-β, drive IgA B-cell switching. In the intestinal lamina propria, BAFF and APRIL also are produced by the epithelium, and DCs release TGF-β and RA, in addition to NO, BAFF, and APRIL to drive IgA B-cell class-switch recombination. Furthermore, in mice, lamina propria DCs and macrophages produce IL-6 and IL-10 which contributes to IgA B-cell differentiation into plasma cells. Thus, DCs are important in T-cell independent IgA B-cell class-switch recombination and differentiation through an array of mediators that include TGF-β, BAFF, APRIL, TNF-α, NO, IL-6, and IL-10 produced in response to bacterial signals.

10-9 DCs drive intestinal homing receptors on T cells and B cells through their production of RA.

T and B cells are generally activated in lymph nodes, resulting in expression of tissue-specific homing receptors. Lymphocytes that home to the small intestinal mucosa express CCR9, the receptor for CCL25, which is highly expressed in the intestinal mucosa, and $\alpha_4\beta_7$, an integrin that binds mucosal addressin cell adhesion molecule-1 (MAdCAM-1) expressed on high endothelial venules in intestinal mucosa.

DCs isolated from PPs and MLNs have the capacity to imprint gut-homing properties to both CD4+ and CD8+ T cells, and IgA+ B cells; such imprinting is dependent on RA released by PP and MLN DCs. Thus, MLN and PP CD103+ DCs drive the expression of CCR9 and $\alpha_4\beta_7$ on lymphocytes.

Dendritic cell conditioning.

DCs and macrophages arise from distinct circulating precursor cells and differentiate into tissue-specific cells in the local microenvironment. Local environmental factors condition these populations (Figure 10.6). The source of these conditioning factors may be epithelium, fibroblasts, macrophages, and T cells and will undoubtedly be modulated by infection and inflammation.

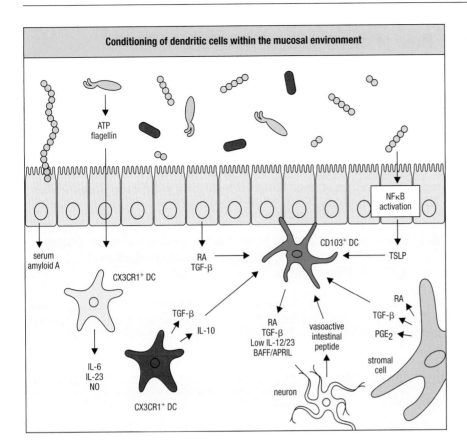

Conditioning of dendritic cells within the mucosal environment

ATP
flagellin

NFκB
activation

serum
amyloid A

CX3CR1⁺ DC

RA
TGF-β

CD103⁺ DC

TSLP

TGF-β

IL-10

RA
TGF-β
Low IL-12/23
BAFF/APRIL

vasoactive
intestinal
peptide

RA

TGF-β

PGE₂

IL-6
IL-23
NO

stromal
cell

CX3CR1⁺ DC

neuron

Figure 10.6 Conditioning of DCs within the intestinal microenvironment. Bacterial, epithelial cell, stromal cell, and neuron products condition DCs and/or CX3CR1⁺ DCs. Circulating pre-DCs can upregulate the expression of CD103 and acquire a tolerogenic phenotype in response to environmental factors. The role of the microenvironment in the differentiation of CX3CR1⁺ DCs and macrophages is unclear, although bacteria-derived products can influence their function. In inflamed mucosa, inflammatory cytokines activate or induce the differentiation of DCs that drive pro-inflammatory (pathogenic T_H1, T_H2, or T_H17) T-cell responses.

10-10 DCs are conditioned by the local microenvironment.

In the presence of symbiotic bacteria, nuclear factor κB (NFκB) signaling pathways are constitutively activated in epithelial cells, resulting in the release of factors that include TSLP, TGF-β, and RA. TSLP can inhibit DC production of IL-12 and induce DCs that drive the differentiation of T_H2 and T_{reg} cells and inhibit T_H1 cell differentiation. Epithelial cells also release TGF-β and RA, which drive the development of CD103⁻ cells into CD103⁺ DCs that are capable of inducing Foxp3⁺ T_{reg} cells. Experimental evidence indicates that the ability of epithelial cells to condition DCs is variably influenced by different species of bacteria. Thus, intestinal bacteria can impact DC function indirectly through their effects on epithelial cells.

In addition to epithelial cells, stromal cells play an important role in conditioning cells in the intestine. Stromal cells can produce TGF-β, RA, and prostaglandin E₂ (PGE₂). Stromal cell-derived TGF-β induces the differentiation of blood monocytes into intestinal macrophages, and stromal cell-derived RA is important in conditioning MLN DCs to then confer homing properties on T cells. In addition, stromal cells constitutively express cyclooxygenase 2 and produce PGE₂. PGE₂ inhibits IL-12 production by human DCs, impairs production of type I interferons by mouse pDCs in PPs, and promotes oral tolerance.

Intestinal macrophages also may influence DC function. Murine, but not human, intestinal macrophages produce high levels of IL-10, which can inhibit DC production of IL-12. When co-cultured with intestinal macrophages, lamina propria DCs may undergo a reduction in their capacity to induce T_H17 T cells independently of IL-10 production. Murine macrophages can also release GM-CSF, which may induce expression of *Aldh1a2*, the gene responsible for RA genesis from vitamin A, in DCs and thus promote Foxp3⁺ T_{reg} development.

Role of mucosal dendritic cells in IBD.

Crohn's disease and ulcerative colitis result from an inappropriate and exaggerated mucosal immune response to gut symbiotic bacteria in genetically susceptible persons. Observations in humans and mouse models suggest that DCs play a key role in these diseases (Figure 10.7). As mucosal DC function is regulated by the local microenvironment, including immune cells, non-immune cells, and luminal bacteria, dysregulation in these components or genetic changes intrinsic to DCs themselves may affect the function of DC populations, resulting in a shift from regulatory to effector T-cell induction (Figures 10.4 and 10.5).

10-11 Intestinal dendritic cells in mice with experimental colitis display impaired induction of T_{reg} cells and enhanced induction of T_H1 and T_H17 cell differentiation.

In the transfer model of colitis naive Foxp3⁻CD45RB^{high}CD4⁺ T cells are transferred into lymphocyte-deficient RAG$^{-/-}$ or SCID (severe combined immunodeficiency) mice, resulting in the accumulation of activated CD11c⁺ DCs in

Figure 10.7 Possible roles for DCs in the pathogenesis of IBD. In IBD, DCs likely contribute to the induction of T_{reg} or T_H17 cells through one or more of the following non–mutually exclusive mechanisms. (1) Commensal bacteria or higher concentrations of flagellated bacteria, possibly through increased production of ATP, induce more DCs that drive T_H17 cells than DCs that drive T_{reg} cells, leading to the local predominance of T_H17 cells. IL-17 released by the T_H17 cells mediates neutrophil recruitment and activation. (2) Impaired release of immunomodulatory factors by epithelial cells leads to failure to condition tolerogenic CD103⁺ DCs. (3) The differentiation of tolerogenic macrophages from recruited monocytes may be impaired and hence also the restimulation of T_{reg} cells. (4) Recruited monocytes differentiate into inflammatory E-cadherin⁺ (E-cad⁺) DCs that produce high amounts of IL-23, IL-6, IL-12, and TNF-α, driving T_H1 and T_H17 cell differentiation and tissue destruction through the release of matrix metalloproteinases by activated fibroblasts that contribute to epithelial damage and ulceration. T_H1 and DC-released TNF-α may also increase the endothelial expression of MAdCAM-1, favoring the recruitment of integrin $\alpha_4\beta_7^+$ T_H1 cells.

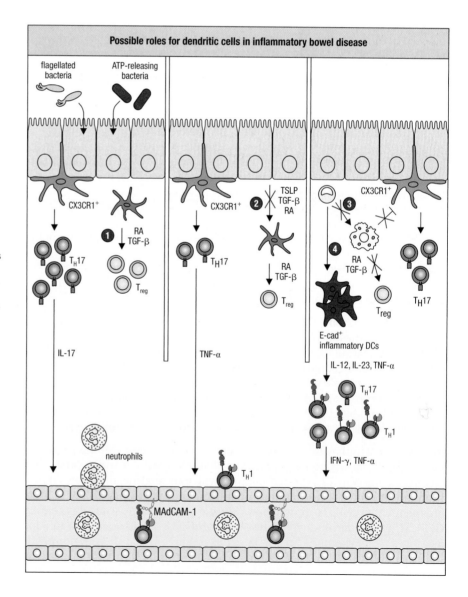

colonic mucosa and MLNs. CD103$^+$ DCs isolated from the MLNs of mice with induced colitis are deficient in their ability to induce Foxp3$^+$ T$_{reg}$ cells and instead induce T$_H$1 and T$_H$17 differentiation. Non-reconstituted RAG$^{-/-}$ mice have clusters of subepithelial CD11c$^+$ DCs in early isolated lymphoid follicles, where transferred naive CD4$^+$ T cells accumulate and possibly expand prior to the onset of colitis. Finally, a CD11c$^+$MHC II^{high}CD103$^-$ population of monocyte-derived cells that also expresses E-cadherin accumulates in the colonic mucosa and MLN and drives more severe inflammation when adoptively transferred into mice with early colitis. These cells broadly express TLRs and produce inflammatory cytokines and chemokines following anti-CD40 or LPS stimulation *in vitro*. Thus, activated DCs likely play a role in both the induction and perpetuation of inflammation in this model.

DCs also play a pathogenic role in T-cell independent models of colitis. For example, DC activation via CD40 causes gut inflammation in the absence of B and T cells through the release of IL-23, which induces production of IL-17 by innate lymphocytes. Oral administration of dextran sulfate sodium (DSS) induces an acute form of colitis due to the formation of pores in the epithelial barrier. Diphtheria-toxin-induced ablation of DCs in transgenic diphtheria toxin receptor mice during DSS administration ameliorates colitis. Mice reconstituted with monocytes that give rise to CX3CR1$^+$, but not CD103$^+$, DCs display exacerbated colitis.

Aberrant activation of resident DCs, recruitment of DCs that were not exposed to the local tolerogenic microenvironment, or an imbalance between tolerogenic and immunogenic DC populations have been associated with late-phase colitis in mouse models (Figures 10.4 and 10.5). Also, DCs that express higher levels of co-stimulatory molecules (CD40, CD80, and CD86) and increased amounts of IL-12p40 and IL-23p19 upon CD40 ligation are present in mouse colitis. Together, IL-12p40 and IL-23p19 form IL-23, which can enhance T$_H$1 and T$_H$17 differentiation by blocking the differentiation of Foxp3$^+$ T$_{reg}$ cells, and expanding the T$_H$17 cell population *in vivo*.

10-12 DCs are activated in the intestinal mucosa in IBD.

Limited studies in humans have revealed that activated DCs accumulate at sites of mucosal inflammation. The trafficking of DCs to these sites is thought to be due in part to the upregulated expression of chemokines such as CCL20 or addressins such as MAdCAM-1. DCs express CCR6 and integrin $\alpha_4\beta_7$ receptors for these gut-homing molecules.

In the lesions characteristic of Crohn's disease, DCs that express CD83, a glycoprotein associated with DC activation, are present in association with numerous CD83$^-$CD80$^+$DC-SIGN$^+$ DCs, producing IL-12 and IL-18. The expression of TLR2, TLR4, and CD40 is enhanced in DCs isolated from inflamed mucosa. Furthermore, mature DCs recruited to the lamina propria form clusters with proliferating T cells in the affected colonic tissue. M-DC8$^+$ cells, a subpopulation of DCs in human blood that expresses high levels of FcRIII (CD16) and secretes TNF-α, are increased in the inflamed mucosa, suggesting that DCs may be a source of TNF-α in Crohn's disease.

Nucleotide-binding oligomerization domain 2 (NOD2) mutations associated with an increased risk of Crohn's disease may decrease the response of monocyte-derived DCs to bacteria, suggesting defects in the intracellular recognition of bacterial muramyl dipeptide (MDP), which activates NOD2 to drive NFκB activation. In addition, DCs derived from NOD2-deficient Crohn's disease patients show an impaired capacity to induce IL-17 expression upon MDP triggering.

In Crohn's disease, activated DCs may migrate from the mucosa to MLNs. DC populations have been described in MLN from Crohn's disease patients that initiate potent T_H1 responses and have an enhanced ability to drive T_H17 differentiation. MLN DCs from patients release significantly higher amounts of IL-23 but lower amounts of IL-10 compared with MLN DCs from patients with ulcerative colitis or healthy control subjects. Thus, MLN DCs may be involved in Crohn's disease pathogenesis via the suppression of Foxp3$^+$ T_{reg} differentiation and the initiation of potent inflammatory T_H1 T cells. Finally, perturbation of the cross-talk between epithelial cells and DCs may disrupt intestinal immune homeostasis, thereby promoting gut inflammation. Notably, epithelial cells isolated from Crohn's disease patients are deficient in their production of TSLP, TGF-β, and ALDH1A1, the enzyme involved in the metabolism of vitamin A to RA. Consequently, epithelial cells fail to control DC production of IL-12 or to induce their capacity to drive the differentiation of Foxp3$^+$ T_{reg} cells.

The role of DCs in ulcerative colitis is less clear than their role in Crohn's disease. Numerous aggregates formed by lymphocytes and CD80$^+$ dendritiform cells resembling activated DCs, are present in the colonic mucosa of patients with ulcerative colitis. DCs derived from peripheral monocytes of patients may have an increased capacity to stimulate T cells in a mixed-lymphocyte reaction, suggesting an intrinsic defect in DC function in these patients. DCs are one of the sources of IL-27, an IL-12-related cytokine that has been implicated in the pathogenesis of ulcerative colitis. An increase in the expression of the IL-27 subunit EBI-3, which is produced by macrophages and DCs, has been detected in the lamina propria of patients with ulcerative colitis. Interestingly, EBI-3-deficient mice are resistant to oxazolone-induced colitis, a mouse model of natural killer T cell (NKT)-dependent ulcerative colitis. Together, these findings suggest that IL-27-secreting DCs may be involved in the pathogenesis of ulcerative colitis through the activation of NKT cells.

Summary.

DCs play a fundamental role in the maintenance of homeostasis of the intestinal mucosa under steady-state conditions and in the induction of immunity toward pathogenic microbes. These functions are associated with specialized DC populations that are either resident within lymphoid and peripheral tissues or that migrate from tissues to draining lymph nodes. DCs are conditioned by local environmental factors, many of which are induced by symbiotic bacteria, and in turn DCs contribute to the maintenance of a symbiotic relationship with the intestinal microbiota. Unraveling the biology of DC activation and response to the local microenvironment will further our understanding of intestinal immune disorders.

Further Reading.

Chorny, A., Puga, I., and Cerutti, A.: **Innate signaling networks in mucosal IgA class switching.** *Adv. Immunol.* 2010, **107**:31–69.

Coombes, J.L., and Powrie, F.: **Dendritic cells in intestinal immune regulation.** *Nat. Rev. Immunol.* 2008, **8**:435–446.

Geissmann, F., Manz, M.G., Jung, S., *et al.*: **Development of monocytes, macrophages, and dendritic cells.** *Science* 2010, **327**:656–661.

GeurtsvanKessel, C.H., and Lambrecht, B.N.: **Division of labor between dendritic cell subsets of the lung.** *Mucosal Immunol.* 2008, **1**:442–450.

Grainger, J.R., Hall, J.A., Bouladoux, N., *et al.*: **Microbe–dendritic cell dialog controls regulatory T-cell fate.** *Immunol. Rev.* 2010, **234**:305–316.

Hooper, L.V., and Macpherson, A.J.: **Immune adaptations that maintain homeostasis with the intestinal microbiota.** *Nat. Rev. Immunol.* 2010, **10**:159–169.

Iijima, N., Thompson, J.M., and Iwasaki, A.: **Dendritic cells and macrophages in the genitourinary tract.** *Mucosal Immunol.* 2008, **1**:451–459.

Kelsall, B.: **Recent progress in understanding the phenotype and function of intestinal dendritic cells and macrophages.** *Mucosal Immunol.* 2008, **1**:460–469.

Maldonado, R.A., and von Andrian, U.H.: **How tolerogenic dendritic cells induce regulatory T cells.** *Adv. Immunol.* 2010, **108**:111–165.

Matta, B.M., Castellaneta, A., and Thomson, A.W.: **Tolerogenic plasmacytoid DC.** *Eur. J. Immunol.* 2010, **40**:2667–2676.

Milling, S., Yrlid, U., Cerovic, V., *et al.*: **Subsets of migrating intestinal dendritic cells.** *Immunol. Rev.* 2010, **234**:259–267.

Rescigno, M., and Di Sabatino, A.: **Dendritic cells in intestinal homeostasis and disease.** *J. Clin. Invest.* 2009, **119**:2441–2450.

Sansonetti, P.J.: **To be or not to be a pathogen: that is the mucosally relevant question.** *Mucosal Immunol.* 2011, **4**:8–14.

Steinman, R.M., and Idoyaga, J.: **Features of the dendritic cell lineage.** *Immunol. Rev.* 2010, **234**:5–17.

Tezuka, H., and Ohteki, T.: **Regulation of intestinal homeostasis by dendritic cells.** *Immunol Rev.* 2010, **234**:247–258.

Intestinal Macrophages in Defense of the Mucosa

11

The gastrointestinal mucosa, the largest mucosal surface to interact with the external environment, contains the most abundant population of macrophages in the body (Figure 11.1). The three major functions of intestinal macrophages include protecting against pathogens and foreign substances that breach the epithelium, contributing to tolerance to commensal bacteria and food antigens, and scavenging apoptotic and dead cells in the lamina propria. Intestinal macrophages mediate these functions through powerful phagocytic and bactericidal capabilities, but, in contrast to macrophages in other tissues, are stringently restricted in pro-inflammatory capability. Specifically, intestinal macrophages display markedly downregulated pro-inflammatory cytokine production, decreased antigen presentation, and diminished oxygen radical production—a unique macrophage phenotype that serves to maintain low-level inflammation in normal intestinal mucosa. In this chapter, we discuss intestinal macrophage effector functions in the context of their evolutionary development and their differentiation in the intestinal mucosa from potentially pro-inflammatory mononuclear phagocytes in the circulation (monocytes) into uniquely inflammation-anergic resident intestinal macrophages.

Origin of lamina propria macrophages.

Lamina propria macrophages are part of the first-line defense mechanisms referred to collectively as the innate immune system, the system by which the host orchestrates initial responses to microorganisms and their products. These first-line mechanisms have evolved from an ancient defense system that arose perhaps a billion years ago in early multicellular organisms and are still present in protozoa, insects, plants, and animals. In animals, macrophages are most numerous in the gastrointestinal mucosa, where they reside in close proximity to the diverse multitude of microbial species that colonize the gut mucosa. Through complex cross-talk pathways involving the microbiota, epithelium, and stroma, intestinal macrophages contribute to the regulation of mucosal homeostasis. In contrast to the innate immune system, the acquired immune system, which is composed of second-line mechanisms of adaptive or acquired defense, arose approximately 400–500 million years ago, during the separation of vertebrates from invertebrates. Thus, the acquired immune system is present in fish, amphibians, reptiles, birds, and mammals, but not in invertebrate species.

The mucosal innate and acquired immune systems have two goals in common: the recognition and elimination of invading pathogens, and the recognition and tolerance of nonharmful (and often beneficial) commensal organisms in the lumen. Contributing to these goals at the level of the mucosa, intestinal and colonic epithelial cells form a structural barrier consisting of a single layer of cells that separates luminal microorganisms from the subepithelial lamina

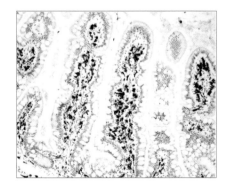

Figure 11.1 Human intestinal macrophages are distributed throughout the subepithelial lamina propria. Section of normal human small intestine stained for the macrophage marker CD68 (×20). (Photograph courtesy of T.T. MacDonald.)

propria, where intestinal macrophages reside. Goblet cells in the epithelium secrete a protective coat of mucus that contains antimicrobial peptides, including defensins and other proteins such as serine leukocyte protease inhibitor, that restrict microbial translocation. In the setting of mucosal injury or infection, microbes can translocate directly into the lamina propria, whereas in normal mucosa they enter via dendritic cell (DC) para-epithelial extensions, M cells, and epithelial cell transcytosis to initiate contact with subepithelial effector defense cells, especially lamina propria macrophages.

Effector cells of the innate immune system include macrophages, DCs, neutrophils, eosinophils, and natural killer cells. Effector T cells and B cells of the acquired immune system act in concert with the innate immune system. Innate responses are initiated within minutes and are directed toward conserved carbohydrate, lipid, and nucleic acid structures, referred to collectively as pathogen-associated molecular patterns (PAMPs), which are expressed by infectious microbes and are essential for microbe survival. Macrophages recognize PAMPs through predetermined repertoires of pattern recognition receptors (PRRs) that include the germline-encoded transmembrane Toll-like receptors (TLRs) and cytosolic nucleotide-binding oligomerization domain (NOD)-like receptors. The predetermined nature of PRRs facilitates rapid innate responses to microbial antigens but limits the diversity of ligands to which macrophages can respond. Indeed, the innate immune system cannot easily respond to novel molecules because memory for antigen recall is not a feature of macrophages or other innate immune cells. Adaptive responses to infectious agents by effector T and B cells are slower (hours to days) and use antigen-specific receptors, T-cell receptors, and immunoglobulins (Igs), respectively, to recognize antigenic epitopes on protein antigens. In contrast to the evolutionarily conserved genes for innate response receptors, the genes encoding antigen-specific receptors undergo somatic recombination, thereby enhancing ligand diversity and conferring on the acquired immune system the ability to identify and recall novel antigens. Cross-talk between the innate and acquired immune systems is accomplished through the presentation of foreign antigens by macrophages and DCs to cells of the acquired immune system. In this way, the innate and acquired immune systems complement and support one another to provide broad, rapid responses while developing specific, albeit delayed, responses to foreign cells and antigenic epitopes.

11-1   Mucosal macrophages are derived from pluripotent stem cells in the bone marrow.

All tissue macrophages, including intestinal macrophages, are derived from bone marrow stem cells through a complex and tightly regulated sequence of differentiation events. Beginning in the bone marrow, pluripotent stem cells are exposed sequentially to combinations of growth factors, hormones, and cytokines that regulate cell maturation and differentiation into common lymphoid progenitor cells and common myeloid progenitor cells. The cytokines interleukin (IL)-1, IL-3, and IL-6 together stimulate pluripotent stem cells to lose their capacity for self-renewal and become myeloid progenitor cells capable of differentiation into monocytic, granulocytic, megakaryocytic, or erythroid lineages. Myeloid progenitor cells become committed to a granulocyte/monocyte lineage and lose their ability to differentiate into the other myeloid lineages under the continued influence of IL-1 and IL-3. Further exposure to this combination of cytokines plus granulocyte–macrophage colony-stimulating factor (GM-CSF) induces the proliferation of both granulocyte and monocyte precursors. Exposure to IL-1, IL-3, GM-CSF, and then macrophage CSF (M-CSF) induces the sequential differentiation of monocyte precursors, monoblasts, promonocytes, and finally monocytes, which are released from the bone marrow and enter the circulation. Under homeostatic conditions, the same precursors can differentiate into monocytes, DCs, or polymorphonuclear neutrophils, but differentiation can be redirected in appropriate circumstances. In addition to the cytokine growth factors mentioned above, each stage of this differentiation pathway may be influenced by constitutive and inducible transcription factors, including PU.1 and AML1, which direct the expression of myeloid-specific genes involved in monocyte differentiation. PU.1 is particularly important because it regulates the expression of the receptor for M-CSF, which is critical for M-CSF-dependent differentiation. The transcription factors GATA-2, SCL, and c-Myb regulate myeloid cell survival. Additional transcription factors, including CCAAT-enhancer-binding protein β (C/EBPβ), HOXB7, and c-Myc, regulate the intermediate stages of myeloid differentiation, whereas C/EBPβ, EBR-1, IRF-1, NF-Y, and some Jun/Fos and STAT proteins regulate monocyte maturation.

11-2 Blood monocytes continually populate the lamina propria of the uninflamed healthy intestinal mucosa.

After leaving the bone marrow, monocytes circulate in the blood compartment for approximately 3–4 days, during which a proportion of cells migrate constitutively into tissues, where they differentiate into tissue macrophages. Recruitment, or homing, of monocytes into tissues, including the intestinal lamina propria, is directed by chemotactic cytokines (chemokines), which are divided into lymphoid and inflammatory chemokines. The lymphoid chemokines recruit cells into noninflamed tissue, whereas inflammatory chemokines recruit cells into inflamed tissue. In this connection, circulating monocytes express abundant chemokine receptors and adhesion molecules that facilitate their recruitment to, and retention in, the tissues. Reflecting the remarkable plasticity of mononuclear phagocytes, the differentiation of newly recruited monocytes into tissue macrophages rather than DCs is regulated by local cytokines, including IL-6 and M-CSF. During differentiation, monocytes enlarge as they increase their content of lysosomal and hydrolytic enzymes and their number and size of mitochondria to become terminally differentiated resident macrophages. Distinct functional roles have been reported for macrophages from different tissues, on the basis of macrophage cytokine and chemokine release and enzyme production. Thus, macrophages (particularly murine macrophages) have been characterized as M1 or M2, along the lines of the CD4$^+$ helper T cell T$_H$1 and T$_H$2 paradigm respectively. This polarization is not well defined for human macrophages, although intestinal macrophages seem to express some of the characteristics of M2 populations. Importantly, macrophages from different tissues may be distinguished phenotypically and functionally, suggesting that local tissue microenvironments influence macrophage differentiation and contribute to macrophage heterogeneity. For example, blood monocytes express high levels of CD14, the receptor for lipopolysaccharide (LPS) and LPS-binding protein, and CD89, the receptor for IgA, but promptly lose these receptors during differentiation in the mucosa.

In healthy intestinal mucosa, epithelial cells, mast cells, and stromal cells (fibroblasts, myofibroblasts, and pericytes) produce and release transforming growth factor-β (TGF-β), the most potent of all monocyte chemokines, and IL-8, which is chemotactic for monocytes as well as neutrophils. These chemokines bind to, and are released from, the lamina propria extracellular matrix (stroma) (Figure 11.2). *In vitro* studies on human intestinal tissues have shown that stromal TGF-β and IL-8 recruit blood monocytes, which express both TGF-β receptors I and II (TGF-β RI and RII) and IL-8 receptors (CXCR1,2). Once recruited to the lamina propria, monocytes take up residence in the extracellular matrix to become resident macrophages. The constitutive expression of these, and probably other, chemokines by mucosal cells promotes the continuous recruitment of blood monocytes to the mucosa, which, together with the long life span of resident macrophages, makes the gastrointestinal tract lamina propria the largest reservoir of macrophages in the body (see

Figure 11.2 Monocytes recruit to the lamina propria and differentiate into inflammation-anergic intestinal macrophages in the stroma. In the homeostatic conditions of normal mucosa, TGF-β and IL-8 produced by epithelial cells, mast cells, and stromal cells bind to and are released from the stromal extracellular matrix (first panel). Mucosal factors, including TGF-β and IL-8, promote the recruitment of resting but potentially pro-inflammatory CD14⁺ blood monocytes from the circulation into the mucosa (center panel). Newly recruited monocytes rapidly differentiate into inflammation-anergic intestinal macrophages as they embed in the lamina propria stroma (last panel). TNF-α, tumor necrosis factor-α.

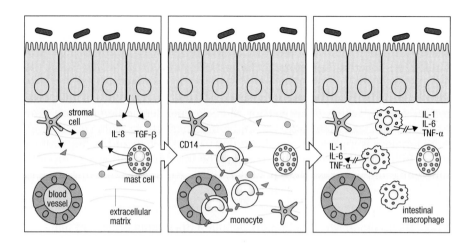

Figure 11.2). Importantly, human lamina propria macrophages do not proliferate, unlike Kupffer cells and alveolar macrophages, which have been shown to proliferate even under steady-state conditions. After terminal differentiation, the macrophages survive for weeks to months as resident macrophages in the lamina propria of the small intestine and colon, after which they undergo programmed cell death and are replaced by newly recruited blood monocytes.

11-3   Blood monocytes are the predominant source of macrophages in inflammatory lesions in intestinal mucosa.

In studies of macrophage accumulation in inflamed gastrointestinal mucosa, immunohistochemical analysis has shown that the endothelial cells lining small blood vessels in the mucosa of patients with Crohn's disease display high levels of CD34, a ligand that promotes the rolling of L-selectin⁺ monocytes in high endothelial venules. More recently, antibody blocking studies have revealed that P-selectin glycoprotein ligand-1, P-selectin, and vascular cell adhesion molecule-1 (VCAM-1) promote CD14⁺ monocyte rolling and adherence in the intestinal mucosa, particularly ileal mucosa, in a mouse model of spontaneous ileitis. Increased levels of intercellular adhesion molecule-1 (ICAM-1) and CD31, which facilitate the transendothelial migration of monocytes, are also present in Crohn's disease lesions. Thus, endothelial cells in mucosal vessels express molecules that promote the rolling, adherence, and subsequent transendothelial migration of circulating blood monocytes into inflamed gastrointestinal mucosa. In addition, recruitment factors, including CCL2 and CCL4, which selectively induce monocyte transendothelial migration and accumulation, may be released in inflamed and/or infected mucosa, enhancing the migration of monocytes into the mucosa. The interdiction of such recruitment is an attractive therapeutic strategy, but its localization to inflammatory sites will be difficult to achieve.

In the 1990s, macrophages in normal human colonic and intestinal mucosa were shown to lack CD14 (Figure 11.3). However, the expression of this receptor on a substantial proportion of macrophages in inflammatory lesions in the mucosa of patients with inflammatory bowel disease (IBD) suggested that the lesions were populated by newly recruited blood monocytes, which express CD14. In experiments designed to explore the origin of CD14⁺ macrophages in inflamed mucosa, investigators harvested CD14⁺ blood monocytes from subjects with IBD, labeled the cells with ⁹⁹ᵐtechnetium, inoculated the cells back into the subjects, and then examined the lesions: the labeled monocytes recruited to the inflamed mucosa, confirming that the CD14⁺ macrophages in the inflammatory lesions were newly recruited pro-inflammatory blood monocytes. The absence of respiratory burst activity in macrophages in normal gut mucosa, but the presence of this function in macrophages in the inflamed mucosa of patients with Crohn's disease, is consistent with blood monocyte recruitment to sites of mucosal inflammation. In addition, the presence of triggering receptor expressed on myeloid cells (TREM-1) on blood monocytes and mucosal macrophages in inflamed IBD mucosa, but not on mucosal macrophages in normal intestinal mucosa, is consistent with the notion that macrophages in inflammatory lesions are derived from blood monocytes. Furthermore, in contrast to monocytes, resident intestinal macrophages do not undergo chemotaxis, supporting the concept that inflammatory lesions are populated largely by newly recruited monocytes.

In addition to mucosal inflammatory diseases, the recruitment of monocytes to the gastrointestinal mucosa is accelerated during mucosal infections. For example, in comparison with the prevalence of lamina propria macrophages in normal gut mucosa, the number of mucosal macrophages, including those that contain internalized mycobacteria, in *Mycobacterium avium*-infected mucosa increases in response to bacterial invasion (Figure 11.4). Mucosal infection by cytomegalovirus, too, is accompanied by macrophage accumulation, even though the majority of such macrophages do not contain the virus. Whether the mucosal macrophages associated with or infected by opportunistic pathogens in immunosuppressed subjects exhibit a CD14+ phenotype has not been defined, but the cells probably represent monocyte-derived macrophages that have exited from the circulation at sites of infection in response to the elaboration of chemokines induced by the intracellular pathogen.

Role of lamina propria macrophages in intestinal homeostasis.

For millions of years, an astonishingly unhygienic external environment challenged the evolving intestine of early vertebrates, subsequently nonhuman primates, and eventually humans, with a spectrum of colonized microorganisms and infectious pathogens that were probably far greater in complexity and numbers than the 'typical' enteric microbiota of today. In such an inhospitable environment, a disrupted intestinal epithelium resulting from frequent infections was probably common. Because the exposure of myeloid effector cells to mucosal bacteria or their products can trigger a striking, and potentially life-threatening, pro-inflammatory transcriptional response, the gastrointestinal mucosa coevolved to downregulate inflammatory, but not host defense, responses to luminal microorganisms that breach the epithelium. During this coevolution, resident lamina propria macrophages emerged, at least in humans, that were unique for their capacity to phagocytose and digest microorganisms and innate material without an inflammatory response, a distinct selective advantage to the host. Extended to immune surveillance, unpublished findings indicate that intestinal macrophages also scavenge apoptotic mononuclear cells without the release of pro-inflammatory cytokines. Today, the unique phenotype and functional profile of gut macrophages described below probably contributes to the absence of mucosal inflammation in the small intestine, despite a 'cleaner' microbiota.

In the gastrointestinal mucosa, inflammation and immunity are tightly regulated to prevent inappropriate or exuberant responses to commensal bacteria and food antigens. Intestinal macrophages are intimately involved in this regulation and are strategically located adjacent to the epithelial basement membrane in close proximity to luminal bacteria, as well as throughout the lamina propria (see Figure 11.1). The number of luminal bacteria in the duodenum and jejunum is 10^3–10^4 bacteria ml^{-1} of intestinal contents, which is relatively low compared with that of the colon (10^{10}–10^{12}) but sufficient to activate

Phenotype	Blood monocyte	Intestinal macrophage
CD4	+	low
CD11a, b, c	+	−
CD13	+	+
CD14	+	−
CD16, 32, 64	+ (subset)	−
CD18/integrin β$_2$	+	−
CD25/IL-2Rα	−	−
CD33	+	+
CD36	+	+
CD40/TNFRSF5	+	−
CD69	− (transient)	−
CD80/B7.1	−	−
CD86/B7.2	+	−
CD88/C5aR	+	low
CD89/FcαR	+	−
CD123/IL-3Rα	+	−
HLA-DR	+	+
TGF-βRI	+	+
TGF-βRII	+	+
TREM-1	+	−
TLR1,2	low/+	low/−
TLR3–9	+/low	+/low
Function		
Phagocytosis	+	+
Killing	+	+
Chemotaxis	+	−
Respiratory burst	+	−
Antigen presentation	+	?
Cytokine production	+	−
Co-stimulation	+	−

Figure 11.3 Blood monocytes and intestinal macrophages are phenotypically and functionally distinct. (Adapted from P. D. Smith et al., *Mucosal Immunology*, 4:31–42, 2011. With permission from Nature Publishing Group.)

Figure 11.4 Monocytes migrate into the lamina propria in response to infection or inflammation. Panel a shows normal intestinal mucosa in which CD68⁺ macrophages are distributed throughout the lamina propria. Panel b shows infection of intestinal mucosa by *M. avium* (identified by red acid-fast stain), which results in striking increases in macrophage accumulation and phagocytosis of the bacteria in the lamina propria. Conditioned medium generated from inflamed (IBD) intestinal lamina propria contains increased levels of monocyte-targeted chemokines, including MCP-1/CCL2 (shown in panel c) and MIP-1β/CCL4 (shown in panel d), which contribute to the recruitment of monocytes into inflamed or infected intestinal mucosa. (Reprinted from P. D. Smith, *Mucosal Immunology* 4, 31–42, 2011.With permission from Macmillan Publishers Ltd.)

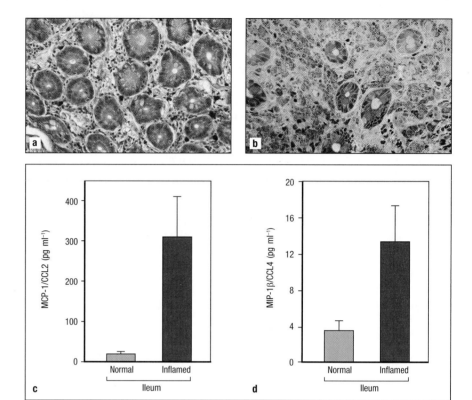

monocyte-derived macrophages *in vitro*. Nevertheless, normal intestinal tissue is characterized by strikingly low levels of inflammation, designated 'physiological inflammation,' suggesting that the large number of effector cells, including lamina propria macrophages, resident in the intestinal mucosa are actively downregulated by the mucosal microenvironment. We next discuss how the unique phenotype and profoundly downregulated pro-inflammatory function of intestinal macrophages, in concert with powerful defense mechanisms, contribute to the regulation of inflammation (homeostasis) in the normal intestinal mucosa.

11-4 Intestinal macrophages are uniquely inflammation anergic.

Circulating monocytes typically express surface molecules, including chemokine receptors, adhesion molecules, TLRs, the LPS receptor, and Fc receptors, which contribute to cell recruitment, retention, and PAMP recognition at sites of infection and/or inflammation. In addition to these key innate receptors, blood monocytes have powerful and inducible pro-inflammatory capabilities. Together, these features equip monocytes for rapid, pro-inflammatory defense after they take up residence in most organ tissues. In the gut lamina propria, however, macrophages display a unique innate receptor phenotype and potently downregulated pro-inflammatory capabilities. This phenotypic and functional profile reflects both the plasticity of mononuclear phagocytes and the unique microenvironment of the intestinal mucosa, and is termed inflammation anergy.

Intestinal macrophages display many, but not all, of the receptors expressed by blood monocytes. For example, intestinal macrophages and monocytes express high levels of the major histocompatibility complex (MHC) class II molecule human leukocyte antigen (HLA)-DR and the myeloid marker aminopeptidase N (CD13). Both populations also express surface receptors involved in the recognition of, and interaction with, potentially harmful microbes; these receptors include TLR1 and TLR3–9 (see Figures 11.3 and 11.5), as well as TGF-β RI and RII, which mediate recruitment (see above) and active Smad signaling (see below). However, in sharp contrast to blood monocytes, resident macrophages in healthy mucosa do not express the receptors for LPS (CD14), IgA (CD89), IgG (CD16, 32, and 64), CR3 (CD11b/CD18), CR4 (CD11c/CD18), growth factor receptors for IL-2 (CD25) and IL-3 (CD123), the integrin leukocyte function-associated antigen-1 (LFA-1) (CD11a), and TREM-1. Intestinal macrophages express very low levels of chemokine receptors CCR5 and CXCR4 and receptors for the chemotactic ligands f-Met-Leu-Phe and C5a. This unusual phenotype has profound functional implications. In particular, the absence of CD14 is consistent with the inability of intestinal macrophages to respond to LPS, a feature well suited to macrophages residing in a microenvironment potentially rich in immunostimulatory LPS. The absence of the receptors for IgA and IgG limits macrophage recognition and uptake of IgA- and IgG-opsonized cells or particles, thereby reducing macrophage integration of innate and adaptive immune responses. The striking downregulation of CCR5 and CXCR4 contributes to the nonpermissiveness of intestinal macrophages to CCR5-tropic (R5) and CXCR4-tropic (X4) human immunodeficiency virus 1 (HIV-1), the causative agent of the acquired immunodeficiency syndrome (AIDS). This mucosa-specific phenotype disables the interaction between intestinal macrophages and an array of inflammatory signals.

Intestinal macrophages also are functionally a unique population of mononuclear phagocytes. The cells are profoundly downregulated for the production of pro-inflammatory cytokines, first described for LPS-stimulated cytokine release and attributed to the absence of CD14. However, the inability of intestinal macrophages to respond to ligand stimulation is not solely dependent on absent or low-level expression of CD14, because intestinal macrophages express TLR3 and TLR5–9 (see Figure 11.5), yet the cells do not release detectable cytokines, including tumor necrosis factor-α (TNF-α), interferon (IFN)-α, IL-8, and IL-10 (Figure 11.6) or IL-1, IL-6, and IFN-β, in response to TLR-specific ligands. However, intestinal macrophages retain potent host defense function, reflected in their powerful phagocytic and bactericidal activities. Thus, despite profound inflammation anergy, intestinal macrophages retain antimicrobial activities through phagocytic and bactericidal function, enabling the cells to rapidly eliminate bacteria that penetrate the epithelium without promoting local inflammation.

11-5 Inflammation anergy of intestinal macrophages is due to inactive NFκB signaling.

A network of TGF-β-mediated mechanisms downregulates nuclear factor κB (NFκB) activation to cause inflammation anergy in intestinal macrophages. In the mucosa, latent TGF-β is produced by an array of cells, including epithelial cells, mast cells, stromal cells, regulatory T cells, and T cells undergoing apoptosis. Released into the lamina propria, TGF-β binds to the extracellular matrix, the network of secreted collagen, fibronectin, and laminin derived from stromal cells, establishing a 'TGF-β reservoir.' TGF-β released from mucosal cells or activated and released from the extracellular matrix engages the cognate receptors TGF-β RI and RII expressed on blood monocytes (and intestinal macrophages), promoting monocyte recruitment into the lamina propria and the cells' rapid differentiation into inflammation-anergic macrophages.

The inability of intestinal macrophages to release pro-inflammatory cytokines in response to TLR-specific ligands, despite the expression of TLRs, is consistent with defective downstream NFκB signaling. Indeed, molecular studies have shown that intestinal macrophages express markedly lower (or undetectable) levels of TRIF, MyD88, and TRAF6 proteins, leading to a failure to phosphorylate NFκB p65 (see Figure 11.6a,b). However, intestinal macrophages

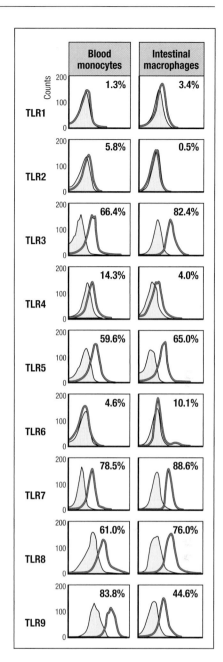

Figure 11.5 Intestinal macrophages express TLRs. Intestinal macrophages express TLR1 and TLR3–9, but not TLR2, as shown in the flow cytometric profiles. Open blue histograms indicate TLR expression on autologous intestinal macrophages and blood monocytes; yellow histograms indicate isotype control staining. (Adapted from L. Smythies et al., *J. Biol. Chem.* 285:19593–19604, 2010. With permission from The American Society for Biochemistry and Molecular Biology.)

Figure 11.6 MyD88-dependent and MyD88-independent signaling pathways are inhibited in intestinal macrophages. Inhibition of MyD88-dependent (panel a) and MyD88-independent (panel b) signaling pathways through blocked phosphorylation (and thus ubiquitination) of IκBα, reduced signaling through Bruton agammaglobulinemia tyrosine kinase (BTK) and Pelle-interacting protein (Pellino), subsequent reduced activation and nuclear translocation of nuclear factor κB (NFκB), and inhibition of interferon regulatory factor 3 (IRF3) signaling, results in blockade of pro-inflammatory cytokine release from intestinal macrophages. (TLR1, 2, 4–6 are expressed on the outer cell membrane; *TLR3, 7–9 are intracellular).

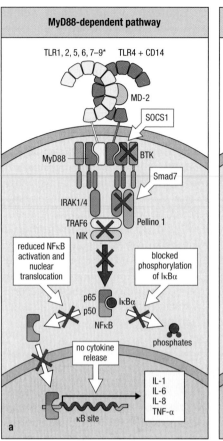
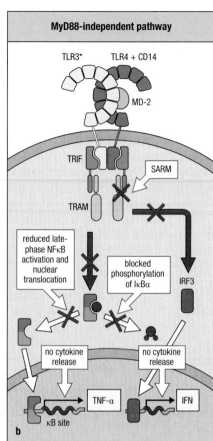

Signal molecule	Macrophage:monocyte fold difference in mRNA	
CD14	−194	
TLR4	−31.4	
MD-2	−64.9	
MyD88	−16.6	
TRAM	−17.6	
IRAK1	−15.3	
IRAK4	−3.9	
TRAF6	4	
NFκB p105	−9.1	
NFκB p65	−4.6	
IκBα	−5.9	
TOLLIP	2.1	
SOCS1	5	
SARM	9.1	
BTK	−4.1	
Pellino 1	−13.9	
CEPBα	−6.7	
CEPBβ	−28.8	
Sp1	−7.4	

Expression of intestinal macrophage mRNA relative to monocyte mRNA			
	−200	to	−101
	−100	to	−51
	−50	to	−6
	−5	to	+4
	+5	to	+49

express increased levels of mRNA for both suppressor of cytokine signaling (SOCS1), which promotes the degradation of MAL (MyD88 adaptor-like protein), and sterile and Armadillo motif-containing protein (SARM) (Figure 11.7), which inhibits TRIF signaling (see Figure 11.6b). MyD88 is a critical element in the NFκB activation pathway of all TLRs except TLR3, and TRIF mediates TLR3-induced RANTES and IFN-γ production, as well as TLR4-mediated MyD88-independent signaling. Thus, the inhibition of MyD88-dependent and MyD88-independent NFκB signaling powerfully decreases TLR-mediated pro-inflammatory responses in intestinal macrophages. Intestinal macrophages are similarly unable to activate NFκB through mitogen-activated protein kinase (MAPK) pathways involving phosphorylated(p) p38, p-ERK, or p-JNK, pathways that are dependent on TRAF6. These dysregulations lead to the inability of intestinal macrophages to activate NFκB and produce NFκB-dependent cytokines.

11-6 Active TGF-β/Smad signaling contributes to inflammation anergy of intestinal macrophages.

A second line of investigation has implicated active TGF-β signaling in the regulation of IκBα expression in intestinal macrophages, providing a mechanism

Figure 11.7 Intestinal macrophages and blood monocytes express markedly different levels of mRNA for signal proteins. mRNA for signal proteins is downregulated and mRNA for signal inhibitors is upregulated in intestinal macrophages compared with autologous blood monocytes.

for the inhibition of NFκB signaling in these cells. TGF-β mediates its functions through signaling via TGF-β receptors and the intracellular effectors of TGF-β signaling, the Smad proteins, which translocate into the nucleus and regulate the transcription of target molecules. In this respect, intestinal macrophages express TGF-β RI and RII and display constitutive active Smad signaling (Figure 11.8). Thus, intestinal macrophages express nuclear Smad4, a critical component of the TGF-β signal cascade that associates with the phosphorylated heterodimeric Smad2/3 complex and then translocates into the nucleus to initiate gene transcription. In addition, intestinal macrophages do not express cytoplasmic Smad7 (see Figure 11.8), the inhibitors of Smad2/3 phosphorylation, which prevent the nuclear translocation of the Smad2/3–Smad4 complex. Moreover, active Smad signaling in the macrophages promotes constitutive expression of IκBα, which sequesters NFκB in the cytoplasm and, together with the cells' inability to phosphorylate NFκB, blocks NFκB signal transduction, thereby inhibiting NFκB-mediated activities. Consistent with these findings, TGF-β also has been shown to block NFκB activation in response to TLR2, 4, and 5 signaling by facilitating the ubiquitination and proteasomal degradation of MyD88. Resting blood monocytes also express low levels of inhibitor Smad7 but, unlike intestinal macrophages, upregulate the expression of these inhibitors after exposure to LPS or other TLR-specific ligands, thereby permitting NFκB translocation into the nucleus and the subsequent production of pro-inflammatory cytokines. Thus, stromal extracellular matrix-derived TGF-β induces downregulation in NFκB signaling and NFκB-mediated function in blood monocytes, recapitulating the abrogated NFκB signaling in intestinal macrophages.

Taken together, ineffective NFκB signaling in intestinal macrophages and TGF-β-mediated Smad-induced IκBα upregulation provide a mechanism for the inflammation anergy of intestinal macrophages, promoting the absence of inflammation in intestinal mucosa despite the close proximity of lamina propria macrophages to luminal bacteria and food antigens. In this way, a breach in mucosal (epithelial) integrity would be met by potent host defense activity but not an inflammatory response in the normal intestinal mucosa.

11-7 Macrophages do not present antigen in normal intestinal mucosa.

As a consequence of their strategic location in the subepithelial lamina propria and their avid phagocytic activity, intestinal macrophages are assumed to interact with antigens translocated across the epithelium and to participate in accessory cell function. Consistent with this notion, intestinal macrophages express high levels of HLA-DR, whose major function is the presentation of antigenic peptides to T-cell receptors. However, human small intestinal macrophages from noninflamed mucosa lack constitutive and inducible expression of the co-stimulatory molecules CD40, CD80 (B7.1), and CD86 (B7.2), which have a key role in accessory cell stimulation (see Figure 11.3). The cells also lack the ability to produce the co-stimulatory cytokines IL-1, IL-10, IL-12, IL-21, IL-22, and IL-23, which participate in the priming and expansion of T-cell populations. Not surprisingly, therefore, intestinal macrophages pulsed with antigen (or mitogen) are markedly less efficient at stimulating T-cell proliferation and cytokine release than similarly pulsed autologous blood monocytes. Recent reports have suggested that colonic macrophages may have some antigen-presenting capability, particularly in inflamed colonic mucosa; however, because these cells are CD14+, they are probably recently recruited blood monocytes rather than resident macrophages. Together, these unique features of human intestinal macrophages may provide an indirect tolerogenic effect on the normal mucosal response to luminal antigens and the microbiota. An active tolerogenic role for intestinal macrophages has been identified in mouse macrophages, as discussed next.

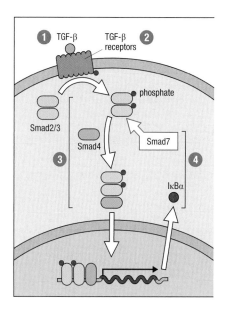

Figure 11.8 Active TGF-β/Smad signaling promotes inflammation anergy in intestinal macrophages. (1) Intestinal epithelial cells, mast cells, and stromal cells in normal mucosa produce TGF-β. (2) Intestinal macrophages express both TGF-β RI and TGF-β RII. (3) TGF-β signaling promotes the phosphorylation of Smad2/3, which complexes with Smad4, translocates into the nucleus, and drives the increased production of IκBα. (4) Absence of Smad7 inhibitor in resident intestinal macrophages permits active Smad signaling and increased constitutive expression of cytoplasmic IκBα, causing cytoplasmic sequestration of NFκB, resulting in inhibition of NFκB nuclear translocation and blockade of NFκB-mediated pro-inflammatory cytokine production (see Figure 11.6).

11-8 Murine intestinal macrophages have immunoregulatory function.

Murine intestinal macrophages are CD11b$^+$CD11c$^-$, in contrast to human intestinal macrophages, which are CD11b$^-$CD11c$^-$ (see Figure 11.3). Like human intestinal macrophages, murine intestinal macrophages are down-regulated for pro-inflammatory cytokine responses to TLR ligands. This downregulation is mediated, at least in part, by the constitutive secretion of IL-10, in sharp contrast to human intestinal macrophages, which do not produce IL-10. The IL-10 released by murine intestinal macrophages inhibits both T$_H$1 polarization through the blockade of IFN-γ production by CD4$^+$ T cells and also T$_H$17 polarization through the blockade of IL-17 production by co-cultured DC-T cells. Together, these findings suggest that intestinal macrophages in mice differ from human intestinal macrophages in phenotype and some functions, and raise the possibility of a tolerogenic role for murine intestinal macrophages.

Intestinal macrophages in host defense.

Consistent with their central role in host defense against microbes and noxious molecules that breach the epithelium, intestinal macrophages are avidly phagocytic for microorganisms and inert material. *In vitro* studies have shown that they phagocytose bacteria (*Salmonella enterica* serovar Typhimurium, *Escherichia coli*, and *Mycobacterium*), fungi (*Candida albicans*), apoptotic cells (Jurkat cells), and latex beads. The phagocytic and ensuing bactericidal activity is generally equivalent to or greater than that of autologous blood monocytes. An exception to the potent bactericidal activity of intestinal macrophages is their inability to kill *Tropheryma whipplei*, at least in patients with Whipple's disease. Indeed, some investigators have hypothesized that Whipple's disease is due to a genetic defect in macrophage processing. Importantly, macrophage phagocytosis and killing of engulfed bacteria do not induce inflammatory responses, whereas such activities in monocytes cause strong release of pro-inflammatory cytokines. Thus, in contrast to blood monocytes and macrophages in other tissues, resident intestinal macrophages perform host defense activities without inducing an inflammatory response, providing an evolutionary advantage that enables gut macrophages to protect against enteric microorganisms without contributing to inflammation in the mucosa. The mechanism by which intestinal macrophages kill bacteria is not known, but seems to be independent of reactive oxygen intermediates.

11-9 Intestinal macrophages scavenge apoptotic cells.

As discussed above, intestinal macrophages phagocytose microorganisms, inert material, and apoptotic Jurkat cells. Intestinal macrophages also express the scavenger receptor CD36, a receptor for lipoprotein and phospholipid, endowing intestinal macrophages with phagocytic activity for apoptotic cells. The ability of intestinal macrophages to phagocytose apoptotic mononuclear cells (and apoptotic cell debris) and clear this potential source of inflammatory material from the lamina propria is critical for the maintenance of mucosal homeostasis.

11-10 Intestinal macrophages participate in host defense against mucosal pathogens.

Macrophages contribute to host defense against a wide array of bacterial, viral, parasitic, and fungal pathogens. In the intestinal mucosa, however, the stringent downregulation of innate response receptors and pro-inflammatory signal transduction in lamina propria macrophages seems to influence the cells' response to some microorganisms such as HIV-1 and cytomegalovirus. For example, after HIV-1 is inoculated onto the gastrointestinal mucosa and subsequently translocated into the subepithelial lamina propria, the virus encounters the largest reservoir of macrophage and lymphocyte target cells in the body. However, the permissiveness of intestinal macrophages to HIV-1 is profoundly downregulated, in marked contrast to that of lamina propria lymphocytes, which are the predominant early target cell for infection and replication by R5 viruses, the transmitted species in the majority of acute HIV-1 infections. Insight into these confounding observations is provided by findings that intestinal macrophages express markedly reduced levels of CD4, the primary receptor, and CCR5, the co-receptor, for R5 virus entry and are incapable of NFκB activation, which is required for HIV-1 replication. Consequently, this unique phenotype and signal transduction profile restricts HIV-1 permissiveness in intestinal macrophages in primary HIV-1 infection.

In contrast to the direct mucosal entry of HIV-1, cytomegalovirus seems to enter the intestinal mucosa through the recruitment of cytomegalovirus-infected blood mononuclear cells, especially monocytes, to the lamina propria. The presence of mRNA and protein for macrophage-type cytokines in the cytomegalovirus-infected mucosa of immunosuppressed subjects suggests that cytomegalovirus reverses the characteristic noninflammatory profile of gut macrophages. Indeed, *in vitro* studies have shown that cytomegalovirus-infected blood monocytes become resistant to intestinal stromal downregulation of CD14 expression and NFκB activation during their differentiation into macrophages, in contrast to the potent stromal downregulation of CD14 expression and NFκB activation in uninfected monocytes. Thus, in immunosuppressed subjects with cytomegalovirus infection, bacterial LPS and flagellins that enter the intestinal lamina propria through a disrupted epithelium may stimulate resident cytomegalovirus-infected macrophages to release pro-inflammatory chemokines and cytokines, promoting the macrophage-driven inflammation characteristic of cytomegalovirus-infected mucosa.

Summary.

The gastrointestinal mucosa, the largest body surface to interface with the external environment, contains the major tissue reservoir of macrophages, recruited to the lamina propria stroma by extracellular matrix-associated TGF-β and IL-8 in normal mucosa and by inflammatory chemokines and microbial products in inflamed and/or infected mucosa. Reflecting a remarkable evolution and dynamic plasticity, the newly recruited monocytes undergo downregulation of some, but not all, innate response receptors and TGF-β-induced inactivation of NFκB and Smad-induced IκBα sequestration of NFκB. The resultant blockade of NFκB-mediated pro-inflammatory function promotes noninflammatory homeostasis in normal mucosa. In infected intestinal mucosa, however, recruited monocytes may retain pro-inflammatory function, promoting the containment of microbes and foreign molecules that have entered the lamina propria. In addition, the massive numbers of intestinal macrophages resident in the mucosa promptly phagocytose and digest invading microorganisms as well as apoptotic and dead host cells. Thus, the intestinal mucosa is host to continuous mononuclear phagocyte activity as huge numbers of blood monocytes recruit to the lamina propria, embed into the extracellular matrix, quickly undergo induction of inflammation anergy in normal mucosa but retain many pro-inflammatory activities in infected mucosa, and initiate a long-lived and vigorous phagocytic life of protection against harmful enteric microbes and potentially inflammatory apoptotic cells.

Further Reading.

Denning, T.L., Wang, Y.-C., Patel, S.R., *et al.*: **Lamina propria macrophages and dendritic cells differentially induce regulatory and interleukin 17-producing T cell responses.** *Nat. Immunol.* 2007, **8**:1086–1094.

Geissmann, F., Gordon, S., Hume, D.A., *et al.*: **Unravelling mononuclear phagocyte heterogeneity.** *Nat. Rev. Immunol.* 2010, **10**:453–460.

Gordon, S., and Martinez, F.O.: **Alternative activation of macrophages: mechanism and functions.** *Immunity* 2010, **32**:593–604.

Grimm, M.C., Pavli, P., and Doe, W.F.: **Evidence for a CD14+ population of monocytes in inflammatory bowel disease mucosa—implications for pathogenesis.** *Clin. Exp. Immunol.* 1995, **100**:291–297.

Hausmann, M., Kiessling, S., Mestermann, S., *et al.*: **Toll-like receptors 2 and 4 are upregulated during intestinal inflammation.** *Gastroenterology* 2002, **122**:1987–2000.

Kamada, N., Hisamatsu, T., Okamoto, S., *et al.*: **Unique CD14 intestinal macrophages contribute to the pathogenesis of Crohn disease via IL-23/IFN-γ axis.** *J. Clin. Invest.* 2008, **118**:2269–2280.

Rugtveit, J., Nilsen, E.M., Bakka, A., *et al.*: **Cytokine profiles differ in newly recruited and resident subsets of mucosal macrophages from inflammatory bowel disease.** *Gastroenterology* 1997, **112**:1493–1505.

Schenk, M., Bouchon, A., Seibold, F., *et al.*: **TREM-1-expressing intestinal macrophages crucially amplify chronic inflammation in experimental colitis and inflammatory bowel diseases.** *J. Clin. Invest.* 2007, **117**:3097–3106.

Shen, R., Meng, G., Ochsenbauer, C., *et al.*: **Stromal down-regulation of macrophage CD4/CCR5 expression and NF-κB activation mediates HIV-1 non-permissiveness in intestinal macrophages.** *PLoS Pathogens* 2011, **7**:e1002060.

Smith, P.D., Saini, S.S., Raffeld, M., *et al.*: **Cytomegalovirus induction of tumor necrosis factor-α by human monocytes and mucosal macrophages.** *J. Clin. Invest.* 1992, **90**:1642–1648.

Smith, P.D., Smythies, L.E., Shen, R., *et al.*: **Intestinal macrophages and response to microbial encroachment.** *Mucosal Immunol.* 2011, **4**:31–42.

Smythies, L.E., Maheshwari, A., Clements, R., *et al.*: **Mucosal IL-8 and TGF-β recruit blood monocytes: evidence for cross-talk between the lamina propria stroma and myeloid cells.** *J. Leukoc. Biol.* 2006, **80**:492–499.

Smythies, L.E., Sellers, M., Clements, R.H., *et al.*: **Human intestinal macrophages display profound inflammatory anergy despite avid phagocytic and bacteriocidal activity.** *J. Clin. Invest.* 2005, **115**:66–75.

Smythies, L.E., Shen, R., Bimczok, D., *et al.*: **Inflammation anergy in human intestinal macrophages is due to Smad-induced IκBα expression and NF-κB inactivation.** *J. Biol. Chem.* 2010, **285**:19593–19604.

Wahl, S.M., Hunt, D.A., Wakefield, L.M., *et al.*: **Transforming growth factor type beta induces monocyte chemotaxis and growth factor production.** *Proc. Natl Acad. Sci. USA* 1987, **84**:5788–5792.

Mucosal Basophils, Eosinophils, and Mast Cells

12

Mast cells, which are abundantly found in the skin and mucosal tissues, play an important role as regulators of multiple mucosal and mucosa-associated functions such as epithelial secretion, smooth muscle contraction, and local activation of nerves. Moreover, mast cells are involved in host defense against bacteria, viruses, and helminth parasites, and contribute to the progression of allergic and other types of inflammation at mucosal sites. In contrast, eosinophils and basophils are blood leukocytes, which can enter mucosal tissues under particular circumstances. In particular, eosinophils are found at mucosal sites, either under normal conditions (e.g., intestine) or under pathological conditions (e.g., esophagus and bronchial mucosa). Eosinophils, like mast cells, are found at elevated numbers during the course of allergic inflammation and helminth infection. They have multiple pro-inflammatory properties but are also involved in immunity to parasites and other infectious agents. Basophils, which are a relatively rare granulocyte population found primarily in the blood and spleen, were traditionally thought to have redundant roles to mast cells in immunity to parasites or the pathogenesis of allergic inflammation. However, recent studies have demonstrated that basophils express effector molecules distinct from mast cells and are found in the inflamed lung tissue of humans following exposure to allergens or lethal asthmatic attack. In addition to their effector functions, recent murine studies have demonstrated that basophils contribute to the development and maintenance of CD4$^+$ T-cell responses and allergic inflammation at multiple mucosal sites. Collectively these results suggest that mast cell, eosinophil, and basophil populations are critical regulators of mucosal immunity and inflammation. The focus of this chapter will be to discuss the molecules and pathways that control the development, activation, and effector functions of these three cell populations in mucosal tissues.

Biology of basophils.

Basophils are the least abundant granulocyte population and account for less than 1% of the leukocytes found in the blood, spleen, and bone marrow. Although originally discovered in 1879 by the German scientist Paul Ehrlich, almost a century passed before basophils were demonstrated to bind IgE and release histamine. Even after these findings, basophils were considered to be a redundant cell population with the same effector functions as mast cells. However, later studies that compared and contrasted the functions of basophils and mast cells discovered that these two populations differ in their gross phenotype, signal transduction pathways, and release of inflammatory mediators. More recently, studies have identified several previously unrecognized functions of basophils in multiple models of helper T cell (T$_H$)2 cytokine-dependent immunity and allergic inflammation. In addition to their known role as effector cells in inflamed tissues, findings now indicate that basophils express major

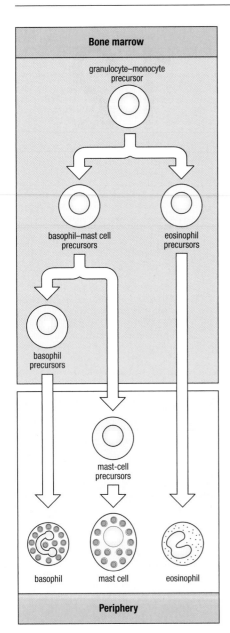

Bone marrow

granulocyte–monocyte
precursor

basophil–mast cell
precursors

eosinophil
precursors

basophil
precursors

mast-cell
precursors

basophil   mast cell   eosinophil

Periphery

Figure 12.1 Pathways of basophil, eosinophil, and mast-cell development.
The bone marrow resident granulocyte–monocyte precursor has the capacity to develop into eosinophils, mast cells, and basophils. Granulocyte–monocyte precursors can become eosinophil precursors and ultimately mature eosinophils that will enter the periphery. Granulocyte–monocyte precursors can also become basophil–mast cell precursors that have the capacity to become mast cells and basophils. Some basophil–mast cell precursors will develop into mast-cell precursors that will enter the periphery and ultimately develop into mature tissue-resident mast cells, while others will develop into basophil precursors and ultimately become mature basophils that will then enter the periphery.

histocompatibility complex (MHC) class II and co-stimulatory molecules, can migrate into draining lymph nodes, act cooperatively with dendritic cells to present antigen to naive CD4$^+$ T cells, and promote T_H2 cell differentiation. In this context, basophils have been shown to contribute to the optimal induction and propagation of T_H2 cytokine responses in murine models of helminth infection and allergic inflammation. In this section, we will discuss our current understanding of basophil biology in the context of immunity and inflammation and discuss the factors that regulate basophil development, activation, and effector functions and the role that they have during helminth infection, allergic inflammation, and some forms of autoimmunity.

12-1 Basophils arise from a common granulocyte–monocyte precursor in the bone marrow.

Basophils arise from a common granulocyte–monocyte precursor in the bone marrow that has the capacity to differentiate into eosinophils, basophil–mast cell precursors, mast-cell precursors, and basophil precursors *in vitro* (Figure 12.1). Both mast-cell precursors and basophil precursors generally remain in the bone marrow and give rise to mast cells and the majority of mature basophils in the bloodstream. Another source of basophils are basophil–mast cell precursors, which migrate to the spleen where they mature into basophils. Basophils exhibit a relatively short life span of 60–70 hours. It has thus been suggested that basophils are continually replenished by precursor populations in the bone marrow in order to maintain their presence in the periphery. Although relatively small in number and short in life span, basophils have the capacity to perform multiple effector mechanisms that will be introduced and discussed below.

12-2 Basophils provide effector functions by releasing preformed mediators as a consequence of IgE-dependent and IgE-independent stimuli.

Basophils are best known as effector cells that release preformed mediators in response to activation via surface-bound IgE. Circulating basophils bind IgE through the high-affinity IgE receptor FcεR1α and degranulate upon FcεR1α cross-linking. Basophils activated via surface-bound IgE produce histamines, leukotrienes, cytokines, and chemokines. However, basophils can also be activated by an array of stimuli in both antibody-dependent and antibody-independent fashions including via antibodies (IgE, IgG1, and IgD), cytokines (interleukin (IL)-3, IL-18, and IL-33), antigens, proteases, pathogen-associated molecular patterns (PAMPs), and complement components (Figure 12.2). Activated basophils are capable of secreting a variety of effector molecules including histamines, leukotrienes, IL-4, IL-6, IL-13, tumor necrosis factor (TNF)-α, and thymic stromal lymphopoietin (TSLP). In the following sections, we will discuss the stimuli that activate basophils and the bioactive molecules that are secreted as a consequence of this activation.

12-3 The effector functions of basophils can be activated by IgE, IgG1, and IgD antibody–antigen complexes.

As described above, basophils rapidly produce histamines, leukotriene C_4 (LTC$_4$), and cytokines in response to the cross-linking of FcεR1α via IgE. The rapid production of effector molecules by basophils in response to activation through surface-bound IgE supports their role as contributors to systemic anaphylaxis as a consequence of blood-borne antigen exposure in humans. Although similar studies have not identified a role for mouse basophils in IgE-mediated anaphylaxis, they are capable of producing platelet-activating factor

(PAF) as a consequence of IgG1 binding. The IgG1-mediated release of PAF by mouse basophils after sensitization to penicillin V leads to a pathway of alternative anaphylaxis that is dependent upon functional expression of FcγRII and FcγRIII. Thus, basophils can be activated by both IgE and IgG1 antibody–antigen complexes leading to release of preformed mediators that contribute to the development of systemic anaphylaxis in both mice and humans.

In addition to the antibody-mediated immediate activation of basophils described above, basophils can also be activated by IgE during certain types of chronic inflammation. Such IgE-mediated chronic allergic inflammation (IgE-CAI) is independent of mast cells and T cells, but dependent on basophil populations. Interestingly, while basophils account for only 1–2% of the cellular infiltrate at the site of such lesions in the skin, their depletion results in a dramatic reduction in inflammation. Pathologically, depletion of basophils results in decreased infiltration with eosinophils and neutrophils and a marked reduction in skin thickness. The loss of cellular infiltrates suggests that basophils produce chemokines and/or other factors that result in cellular recruitment in this model of chronic skin inflammation. Thus, basophils have a role in the initiation and maintenance of chronic IgE-mediated inflammatory responses.

In addition to IgE- and IgG1-containing immune complex activation of basophils via Fc receptor ligation, basophils can also be activated by IgD antibody–antigen complexes. In this case, human basophils can selectively bind IgD, a class of antibody produced early in B-cell development. Although the biological function of IgD remains enigmatic, it is interesting that IgD is highly expressed in the human upper respiratory tract where it can bind organisms such as *Haemophilus influenzae* and *Moraxella catarrhalis* and activate basophils to produce antimicrobial peptides that inhibit bacterial replication. IgD-mediated activation of basophils stimulates a distinct effector phenotype from that elicited by IgE-mediated activation. In comparison to cross-linking with IgE, IgD cross-linking of basophils results in enhanced expression of IL-4 and B-cell activating factor (BAFF), which facilitates B-cell class switching, and the production of antimicrobial peptides, which inhibits bacterial replication, but not the induction of histamine release.

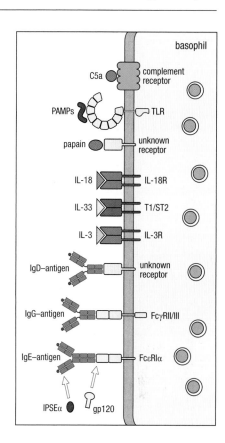

Figure 12.2 Mechanisms of basophil activation. Basophils can be activated by many signals including antigen–antibody complexes (IgE, IgG1, IgD), cytokines (IL-3, IL-18, IL-33), proteases, viral antigens, Toll-like receptor (TLR) agonists, and complement components. gp120, glycoprotein of human immunodeficiency virus; IPSEα, interleukin-4-inducing principle from *Schistosoma mansoni* eggs-α.

12-4 Basophils are activated by cytokines such as IL-3, IL-18, and IL-33.

Basophil development and effector functions can also be regulated by cytokines. For example, IL-3 is crucially important for the expansion, survival, and activation of basophils. IL-3 promotes the differentiation of basophils from bone marrow cells *in vitro* and administration of IL-3 *in vivo* induces their generation. Consistent with this, IL-3 is necessary for infection-induced basophilia in the bone marrow, blood, spleen, and liver following infection with organisms such as *Nippostrongylus brasiliensis* and *Strongyloides venezuelensis*. In addition, IL-3 can augment basophil functions induced by other factors such as IgE-dependent stimulation of IL-4 and IL-13 secretion.

IL-18 and IL-33, members of the IL-1 cytokine family, are also capable of activating basophils. IL-18 is produced by innate cells such as macrophages and Kupffer cells and is known to play a role in allergic disease and immunity to helminthic parasites. IL-33 is expressed by dermal fibroblasts, airway epithelial cells, and bronchial smooth muscle cells and is important in the induction of IL-4, IL-13, and IgE production *in vivo*. Both IL-18 and IL-33 are closely linked to T_H2 cytokine-mediated inflammation. Consistent with their ability to enhance T_H2 cytokine-associated immune responses, IL-18 and IL-33 are capable of activating basophils. Murine bone marrow derived basophils stimulated with IL-3 and IL-18 exhibit enhanced secretion of IL-4 and IL-13 compared with stimulation with IL-3 alone. Furthermore, administration of IL-18 *in vivo* increases the production of basophil-dependent IL-4 and histamine in mice. IL-18 also enhances the survival of murine basophils. Interestingly, although human blood basophils express the IL-18R, a role for IL-18 in the activation of human basophils remains to be established. Similarly to IL-18, IL-33 can stimulate the production of IL-4 and IL-13 from murine bone marrow derived basophils. IL-33 also induces the production of IL-4, IL-5, IL-6, and IL-13 from human blood-derived basophils.

12-5 Basophils can sense microbial products to provide innate immune resistance.

In addition to antibody- and cytokine-mediated activation, basophils are also activated by allergens, parasite antigens, and other pathogen-associated molecules. Specifically, secretions of the hookworm parasite *Necator americanus* and the house dust mite antigen Der p 1, which possess active protease activity, can induce the production of IL-4, IL-5, and IL-13 by human basophils. These activities are eliminated by protease inhibitors, suggesting that basophils may express protease-activated receptor-like factors that are capable of the proteolytic activation of such antigens (or organisms) and activate innate and adaptive immune cells.

In addition to proteases, basophils can be activated by a class of 'superantigens' that function independently of any known receptors. For example, the gp120 glycoprotein of the human immunodeficiency virus is able to interact with the VH3 region of IgE in a pathway that induces IL-4 and IL-13 production from human basophils. Similarly, murine basophils are activated by the schistosome-derived glycoprotein IPSE/alpha-1, which induces the production of IL-4 from basophils in a nonspecific manner.

12-6 Basophils are activated by Toll-like and complement receptors for innate immune function.

Basophils express Toll-like receptors (TLR) and complement receptors (CR). On human basophil populations these include TLR1, TLR2, TLR4, TLR6, TLR9, CR1, CR3, CR4, and CD88. This predicts that basophils may be capable of recognizing PAMPs and complement components. Consistent with this, there is evidence that TLR2 ligands can activate basophils and induce the production of IL-4 and IL-13 while the complement component C5a can induce histamine production. Despite these observations, this is likely an underestimate of the full innate immune capacities of basophils as will be discussed further below.

12-7 Basophils are recruited to inductive and effector sites throughout an immune response.

As we have seen, basophils are potent producers of type 2 cytokines such as IL-4 and IL-13 in response to a wide variety of stimuli. In addition, as we will see, they are able to do so throughout all phases of an immune response and consequently accumulate at sites of inflammation during an ongoing inflammatory response. Basophils were originally considered to be late-phase effector cells and thought to be excluded from lymph nodes and for the induction and early phases of inflammation. However, it is now appreciated that basophils may serve as liaisons between the innate and adaptive immune response and may even promote the latter. For example, MHC class II$^+$ basophils can migrate to draining lymph nodes following exposure to papain, *Schistosoma mansoni* eggs, or *N. brasiliensis* infection. Although basophils appear to be only transiently present in the lymph nodes, there is evidence that they are capable of directly interacting with lymph-node-resident CD4$^+$ T cells, B cells, and dendritic cell (DC) populations. Basophils may directly contribute to the induction of T$_H$2 cytokine-mediated inflammation. Consistent with this, depletion of basophils prior to papain challenge can prevent the subsequent development of an allergen-induced T_H2 cell response. Finally, it should be noted that the mechanisms by which basophils enter lymph nodes are poorly defined, although it appears to involve IL-3. For example, basophils are induced to accumulate in mediastinal lymph nodes during the first week of infection with *N. brasiliensis* and this recruitment is dependent upon IL-3 and IL-3 receptor signaling. Whether this relates to basophil recruitment in other circumstances is unknown but demonstrates that recruitment of basophils is doubtless directed.

12-8   Basophils possess antigen-presenting functions and cooperate with dendritic cells in the induction of T_H2 immunity along mucosal surfaces.

Recent studies have introduced the notion that basophils may be capable of providing the initial cellular and soluble stimuli for the development and induction of T_H2 cytokine-mediated immunity and inflammation independently of dendritic cells. For example, on the one hand, when MHC class II expression is restricted to DCs or when DCs are depleted, T_H2 cytokine-dependent immune responses may be impaired, but not eliminated, following exposure to mucosal helminth infection or mucosal exposure to allergens such as those associated with papain or allergic airway sensitization. This suggests that other cells may provide the adaptive signals necessary for induction of T_H2 cells. Consistent with this, there is some direct evidence that basophils may function as antigen-presenting cells in the context of mucosal exposure to helminths or allergens. In these studies, basophils have been shown to endocytose soluble antigens and IgE–allergen complexes, express MHC class II and co-stimulatory molecules, migrate to draining lymph nodes, and promote T_H2 cell differentiation in response to *S. mansoni* egg antigens. Depletion of basophils results in impaired protective immunity to the mucosal whipworm *Trichuris muris*, together with elimination of T_H2 cell development after mucosal challenge with allergens.

As a whole, multiple studies now indicate that both basophils and DCs are important for the optimal induction of T_H2 cytokine-mediated immunity and inflammation including those associated with mucosal tissues. The apparent identification of DC-dependent and DC-independent pathways of T_H2 cell differentiation suggests that there are multiple pathways by which T_H2 cell responses may develop, as summarized in Figure 12.3. For example, in some models of helminth infection or allergic inflammation, T_H2 cell differentiation may involve priming of naive T cells by DCs augmented with IL-4 or other soluble factors produced by basophils (Figure 12.3a). However, other models of antigen presentation by IL-4 and MHC class II-expressing basophils may be sufficient to promote T_H2 cell differentiation in the absence of DC populations (Figure 12.3b). Alternatively, basophils and DCs may present antigen to naive T cells cooperatively to initiate the propagation of optimal T_H2 cell responses (Figure 12.3c, d). Consistent with a cooperative model of antigen presentation, T_H2 cell development in response to immunization with papain is dependent on the presence of both DC and basophil populations. In this context, immunization with papain and soluble antigen results in DC populations releasing reactive oxygen species, which promote the production of oxidized lipids that are capable of triggering epithelial cells to produce TSLP, which suppresses the production of IL-12 by DCs. This cascade of events also results in the induction of DCs to produce CCL7, which mediates the recruitment of basophil populations to the lymph nodes, actions which together show how DCs and basophils cooperate in the induction of T_H2 cytokine-mediated responses (Figure 12.3d). It is likely that the nature of these responses will depend on the characteristics of the antigen being studied (e.g., helminth infection or allergens), the duration and site of antigen exposure, and whether the differentiation of T_H1, T_H17, or regulatory T (T_{reg}) cells is occurring simultaneously.

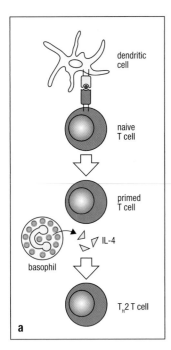
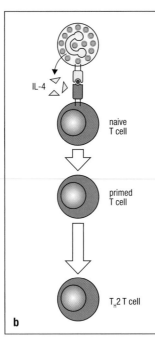
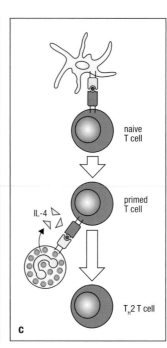
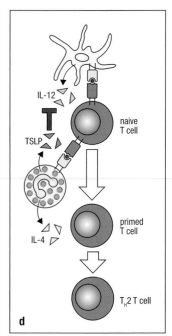

Figure 12.3 Basophils regulate T$_H$2 cell differentiation. There are several potential pathways by which basophils may interact with T cells to augment T$_H$2 cell differentiation. First, as shown in panel a, basophils may function as accessory cells that provide a source of IL-4, which helps to polarize T cells that were previously activated by DC-mediated antigen presentation. Second, in panel b, basophils may act independently of DCs and activate T cells by directly presenting antigen while simultaneously providing an innate source of IL-4. Third, basophils and DCs may act cooperatively to present antigen either in series (panel c), or in parallel (panel d). In the context of the model shown in panel d, basophils would augment DC-mediated antigen presentation and provide the innate IL-4 needed for T$_H$2 cell differentiation. In addition, basophils can produce thymic stromal lymphopoietin (TSLP) that is capable of inhibiting IL-12 production by DCs, thereby further promoting T$_H$2 cell differentiation.

Biology of mucosal eosinophils.

Eosinophil granulocytes are bone marrow derived leukocytes involved in inflammatory and immunoregulatory processes. This dual function of eosinophils is similar to that of basophils and mast cells. Obviously, inflammatory and immunoregulatory functions are frequently combined in the same cell type. Like basophils and mast cells, which will be discussed later in this chapter, eosinophils exert many biological functions by releasing humoral mediators that preferentially function locally in a paracrine manner. Both the factors that regulate eosinophil development and activation and the mediators that are released upon activation will be discussed. Consistent with the pleiotropic functions of eosinophils in health and disease, eosinophils are capable of releasing a wide range of granule-associated proteins and cytokines, upregulate a variety of cell-surface receptors and adhesion molecules, and potentially provide antigen-presentation functions for T cells. Eosinophils participate in parasitic and helminth infections, virus infections, asthma, gastrointestinal diseases, and neoplastic disorders of the leukocyte lineage and are thus an important component of mucosal immunity.

12-9   Development of eosinophils depends on lineage-specific transcription factors and cytokines.

Eosinophils are derived from pluripotent bone marrow cells, which first develop into a common precursor of basophils and eosinophils, and then into a separate eosinophil lineage. Eosinophil lineage specification is dictated by the interplay of at least three classes of transcription factors including GATA-1 (a zinc finger family member), PU.1 (an Ets family member), and c/EBP members (CCAAT/enhancer-binding protein family). The specificity of these factors for eosinophils is conserved across species. Of these transcription factors, GATA-1 is clearly the most important for eosinophil lineage specification as revealed by loss of the eosinophil lineage in mice harboring a targeted deletion of the high-affinity GATA-binding site in the GATA-1 promoter. Three cytokines, IL-3, IL-5, and granulocyte–macrophage colony-stimulating factor (GM-CSF,) are particularly important in regulating eosinophil development.

These eosinophilopoietins likely provide permissive proliferative and differentiation signals following the instructive signals specified by the transcription factors GATA-1, PU.1, and c/EBPs. Interestingly, these cytokines are encoded by closely linked genes on chromosome 5q31. They bind to receptors that share a common β-chain and possess unique α-chains. Of these three cytokines, IL-5 is the most specific to the eosinophil lineage and is responsible for selective differentiation of eosinophils. IL-5 also stimulates the release of eosinophils from the bone marrow into the peripheral circulation. The critical role of IL-5 in the production of eosinophils is best demonstrated by genetic manipulation of mice. Overproduction of IL-5 in transgenic mice results in profound eosinophilia and deletion of the IL-5 gene causes a marked reduction of eosinophils in the blood and lungs after allergen challenge. The overproduction of one or a combination of these three cytokines occurs in humans with eosinophilia, and diseases with selective eosinophilia are often accompanied by overproduction of IL-5. The critical role of IL-5 in regulating eosinophils in humans has been demonstrated by several clinical trials using a humanized anti-IL-5 antibody that lowers eosinophil levels in the blood and, to a lesser extent, in the inflamed lung.

12-10 Eosinophil survival and recruitment are largely determined by IL-5 and eotaxin.

The recruitment and survival of eosinophils are regulated by cytokines and chemokines. IL-3, IL-5, and GM-CSF support the survival of eosinophils; however, IL-5 has the most potent and most eosinophil-specific regulatory properties. IL-5 promotes the proliferation and maturation of eosinophils in the bone marrow and their release into the circulation. IL-5 also promotes the survival of eosinophils by diverting them from an apoptotic fate, primes them for responses to chemoattractant signals that are responsible for their mucosal recruitment, and promotes their degranulation in response to triggering agents.

The recruitment of eosinophils to mucosal sites is also critically dependent upon the local production of chemoattractants, especially chemokines. Eotaxin-1 (CCL11) is the most important and selective eosinophil chemoattractant. Eotaxin-1 is responsible for the physiologic recruitment of eosinophils to the intestines in healthy individuals, and is expressed constitutively in the intestinal lamina propria. In the absence of eotaxin-1, or its eosinophil receptor, CCR3, eosinophils fail to home to the gastrointestinal tract. Eosinophils also express the mucosal integrin $\alpha_4\beta_7$ on their cell surface, which is responsible for the selective recruitment of these leukocytes toward the mucosa of the intestines rather than other tissue compartments. Integrin $\alpha_4\beta_7$ binds to its specific ligand, MAdCAM-1 (mucosal vascular addressin cell adhesion molecule-1), which is preferentially expressed on the vascular endothelium of intestinal lamina propria venules.

12-11 Eosinophils are activated by a broad array of signals and function primarily through their ability to secrete a wide range of soluble mediators.

The granules of eosinophils contain major basic protein, eosinophil cationic protein, eosinophil-derived neurotoxin, eosinophil peroxidase, and other enzymes of uncertain significance, as well as proteins that include a broad range of preformed cytokines and chemokines (Figure 12.4). Major basic protein is cytotoxic for helminthic parasites and mammalian cells, activates the complement cascade, and leads to increased smooth muscle reactivity by causing dysfunction of vagal muscarinic M2 receptors. Moreover, major basic protein has been shown to stimulate substance P release from neonatal rat

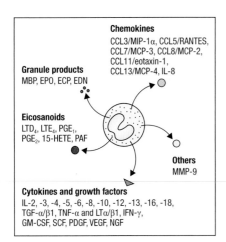

Chemokines
CCL3/MIP-1α, CCL5/RANTES,
CCL7/MCP-3, CCL8/MCP-2,
CCL11/eotaxin-1,
CCL13/MCP-4, IL-8

Granule products
MBP, EPO, ECP, EDN

Eicosanoids
LTD$_4$, LTE$_4$, PGE$_1$,
PGE$_2$, 15-HETE, PAF

Others
MMP-9

Cytokines and growth factors
IL-2, -3, -4, -5, -6, -8, -10, -12, -13, -16, -18,
TGF-α/β1, TNF-α and LTα/β1, IFN-γ,
GM-CSF, SCF, PDGF, VEGF, NGF

Figure 12.4 Eosinophil effector functions. Mediators of human eosinophils include cationic granule proteins, pro-inflammatory lipid mediators (leukotrienes D$_4$ and E$_4$, prostaglandins E$_1$ and E$_2$), cytokines (interleukins, IL; tumor necrosis factor, TNF-α; transforming growth factor, TGF; interferon, IFN; nerve growth factor, NGF; stem cell factor, SCF; platelet-derived growth factor, PDGF; vascular endothelial growth factor, VEGF), chemokines, and matrix metalloproteinases (MMP). MBP, major basic protein; EPO, eosinophil peroxidase; ECP, eosinophil cationic protein; EDN, eosinophil-derived neurotoxin; LT, lymphotoxin; MIP, macrophage inflammatory protein; MCP, monocyte chemotactic protein; 15-HETE, 15-hydroxyeicosatetraenoic acid.

dorsal root ganglia neurons. Eosinophil cationic protein, eosinophil peroxidase, and eosinophil-derived neurotoxin induce cytotoxic effects in helminthic parasites and mammalian cells by exerting ribonuclease activity or generating unstable oxygen radicals. Among the cytokines and chemokines produced by eosinophils are IL-1α, IL-2, IL-3, IL-4, IL-5, IL-6, IL-9, IL-10, IL-12, IL-16, interferon-γ (IFN-γ), GM-CSF, TNF-α, eotaxin (CCL11), IL-8 (CXCL8), macrophage inflammatory protein-1α (MIP-1α), RANTES (CCL5), nerve growth factor, stem cell factor (SCF), platelet-derived growth factor (PDGF), and transforming growth factor (TGF) α and β1, among others.

Most of the studies that have investigated the ability of human eosinophils to produce cytokines have been performed using cells isolated from peripheral blood, making it difficult to know whether these properties also relate to tissue-residing eosinophils. Eosinophils do indeed change when they migrate into mucosal tissues, as revealed by the induction of activation markers such as CD69, intercellular adhesion molecule-1 (ICAM-1), and CD25 on the cell surface when this occurs. In the few studies that have been performed, *in situ* expression of IL-3, IL-5, GM-CSF, IL-16, and TGF-β1 has been detected in eosinophils residing within the intestinal mucosa. In addition, human blood and intestinal eosinophils have been found to store vasoactive intestinal polypeptide, substance P, calcitonin-gene related peptide, and somatostatin. When activated, these preformed mediators are released and additional mediators generated, for example, reactive oxygen species and lipid mediators such as leukotrienes, platelet-activating factor, and prostaglandins.

Under homeostatic conditions eosinophils are present in a state of rest in order to avoid tissue damage from eosinophil-derived mediators. In contrast to the central role of IgE receptor cross-linking for activation of mast cells, as will be discussed below, eosinophil activation does not seem to depend upon a dominant mediator but seems to respond to a broad range of stimuli, most of which also mediate survival or chemotaxis. Activation of eosinophils involves initial priming and subsequent triggering of the cells for effector functions. Triggering is mediated by cytokines such as IL-5, IL-3, GM-CSF, or TNF-α; by chemokines such as RANTES, MIP-1α, monocyte chemotactic protein 3 (MCP-3), MCP-4, eotaxin, or IL-8; and by complement components, aggregated immunoglobulins, lipid mediators, or histamine. Interestingly, many of these agonists are also known as mast-cell secretory products suggesting that mast-cell activation might be involved in the rapid activation of eosinophils as observed during allergic reactions. Human eosinophils also have functions in phagocytosis and antigen presentation but these are poorly characterized. With regards to the latter function, studies in mouse models support a role for eosinophils in antigen presentation in view of their expression of MHC class II and accessory co-stimulatory molecules, and their ability to induce expansion of antigen-specific T cells. However, the biologic significance of this function remains to be firmly established.

12-12 Mucosal eosinophils collaborate with epithelial cells, T cells, mast cells, and nerve cells in mucosal immunophysiology.

A major function of eosinophils is through their contribution to the maintenance of mucosal barrier function and protection against invasion by mucosal pathogens. This is consistent with the general role of innate immunity in barrier function. For example, following direct exposure to bacterial products derived from the intestinal microbiota *ex vivo*, eosinophils release toxic mediators. Furthermore, in contrast to blood eosinophils, intestinal eosinophils exhibit an activation phenotype (expression of CD69, for example) in noninflamed mucosa in the absence of obvious degranulation suggesting that they are poised for immediate responses to commensal bacteria or pathogens in the event of loss of epithelial integrity.

The physiologic functions of eosinophils also depend on their capacity for coordinated interactions with other innate and adaptive immune cells. As noted, eosinophils likely receive stimulatory signals from mast cells. Eosinophils also interact with T cells in a bidirectional manner. At sites of mucosal allergic inflammation, mast cells and T_H2 cells secrete eosinophil-activating cytokines, including IL-5, which promote local survival and degranulation of eosinophils. Eosinophils are capable of presenting antigens to T cells through their expression of MHC class II and co-stimulatory molecules and as such stimulate or modulate T-cell function (Figure 12.5).

Eosinophils also interface directly and indirectly (through cross-talk with mast cells) with the enteric nervous system. Eosinophil-derived granule proteins, particularly major basic protein and other soluble mediators such as stem cell factor, induce human mast-cell activation, which leads to the release of tryptase, histamine, and prostaglandin D_2 (PGD_2). Eosinophils bind to nerve cells in a specific fashion, which results in nerve remodeling, changes in neurotransmitter activity, and the induction of genes involved in neurotransmitter metabolism.

Another physiological role of eosinophils could be in epithelial repair. Epithelial cell damage or necrosis is a potent signal for eosinophil chemotaxis, which prompts secretion of cytokines that possess important tissue repair and regulatory properties, such as TGF-β1 and fibroblast growth factors. Therefore, depending upon local signals, eosinophils may be incited to mediate either epithelial cell damage or repair.

Mucosal mast cells.

Unlike basophils and eosinophils, mature mast cells are only found within tissues, not in the blood. Although all three cell types (basophils, eosinophils, and mast cells) are derived from myeloid progenitors, mast cells leave the bone marrow in an immature state, with the tissue environment such as the

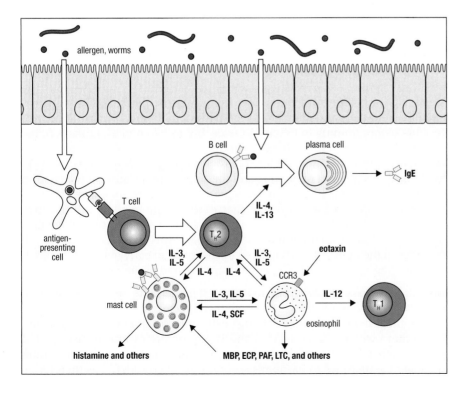

Figure 12.5 Interaction between eosinophils, mast cells, and T lymphocytes. The potential mechanisms that lead to immunological hypersensitivity and T_H2-type inflammation as a result of cross-talk between mast cells, eosinophils, and T cells in the intestinal mucosa are shown. LTC, leukotriene C. (Adapted from S. Bischoff and S. Crowe, *Gastroenterology* 128:1089–1113, 2005. With permission from Elsevier Ltd.)

mucosa necessary for full maturation. Mast cells have a general property of being strategically located at sites of the host–environment interface such as the skin and the mucosa of the respiratory, gastrointestinal, and urogenital tracts. This location suggests that a major function of mast cells is as immunological 'gate-keepers' or 'watch-dogs,' and indeed functional data support this idea as discussed below.

The normal human gastrointestinal tract contains numerous mast cells. The largest numbers are found in the lamina propria, where 2–3% of the cells are mast cells. There are two types of mast cells within mucosal tissues. The mast cells contained within the human lamina propria are considered to be the mucosal mast-cell subtype (MC_T) because they are tryptase positive and chymase negative and resemble a similar subtype in rodents. The second subtype of mast cells within human mucosal tissues is contained within the submucosa. Here, the mast-cell density is lower than in the lamina propria (about 1% of all cells) with the majority of cells being tryptase and chymase positive. This subtype of double-positive mast cells (MC_{TC}) is also observed in rodents.

Human mast cells are recognized as key effector cells in allergic inflammation, consistent with their strategic location within the skin and mucosal barriers and close relationship with the vascular system. Moreover, they bind IgE on their surface by virtue of their expression of the high-affinity IgE receptor that induces the release of histamine and other mediators upon cross-linking by surface-bound IgE in the context of an allergen. Not only are mast cells cellular mediators of allergy, they regulate many tissue functions such as blood flow and coagulation, smooth muscle contraction and peristalsis of the intestine, mucosal secretion, wound healing, and innate and adaptive immune responses, including that associated with immune tolerance. This explains why mast cells have been found to be involved in many different types of human disease in addition to allergic disorders. These include inflammatory diseases, neurological diseases, and functional diseases such as irritable bowel syndrome, functional dyspepsia, and fibromyalgia. Recent studies in rodents, and to some extent also in humans, have shown that mast cells also have a central role in host defense against bacteria and parasites, through the release of cytokines and other mediators that serve to recruit neutrophils, eosinophils, and T_H2 cells to the site of infection. These findings are consistent with the notion that mast cells are uniquely important in the defense of mucosal barriers.

12-13 Mast cells develop from immature bone marrow derived precursors in mucosal tissues.

Mast cells originate from immature, bone marrow derived CD34+ hematopoietic stem cells and circulate in the peripheral blood as committed progenitors before homing to tissues. In mice, homing to mucosal tissues of the intestinal lamina propria is dependent upon immature mast-cell expression of $\alpha_4\beta_7$ integrin and MAdCAM-1 expression on high endothelial venules. In humans, the regulation of homing and the stage of maturation at which mast cells migrate from the blood into the tissue remain largely unknown. Electron-microscopic studies have revealed that mast-cell progenitors are not only found in peripheral blood but also in tissues such as the intestine where they represent 5–15% of the total mast-cell numbers. This suggests that mast-cell densities in the intestine are regulated by the influx of early mast-cell progenitors and their growth-factor-dependent survival and proliferation.

Human mast cells develop from myeloid-cell progenitors under the influence of particular growth factors such as stem cell factor and IL-4, cytokines that also regulate the development of mast-cell subtypes. In addition to SCF and IL-4, other cytokines such as IL-6 and IL-9 promote mast-cell survival and proliferation. Consistent with a common progenitor, human mast cells exhibit significant similarities to basophils and have been considered as their tissue

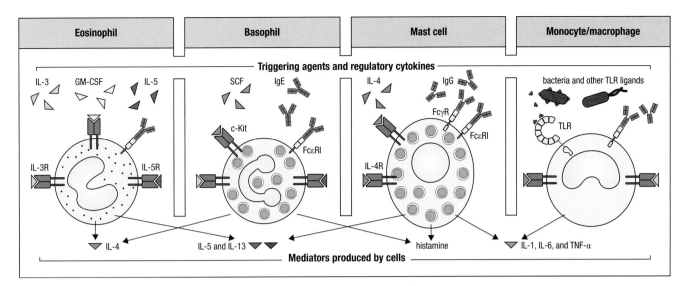

Figure 12.6 Comparison between human mast cells, eosinophils, and basophils. Mast cells exhibit similarities to, and differences from, eosinophil and basophil granulocytes and monocytes/macrophages. Mast cells share FcεRI expression and histamine release with basophils. Mast cells share responsiveness to cytokines and bacterial products as well as nuclear morphology with monocytes and macrophages. However, mast cells generally do not express CD14 like monocytes, or IL-3 or IL-5 receptors like basophils or eosinophils. Mast cells express almost exclusively the stem cell factor (SCF) receptor c-Kit. Triggering agents and regulatory cytokines of the four cell types are shown above each cell; mediators released from the cells after stimulation are shown below each cell. (Adapted from S. Bischoff, *Nat. Rev. Immunol.*, 7:93–104, 2007. With permission from Macmillan Publishers Ltd.)

equivalent. Both cells contain basophilic intracellular granules, release histamine, and express the high-affinity IgE receptor. Morphological and functional analyses, including gene expression studies, also support similarities between mast cells and monocytes or macrophages more than similarities with basophils (Figure 12.6). This is supported by the fact that although some murine mast-cell populations and human basophils respond well to IL-3, human mast cells either lack the IL-3 receptor or are poorly responsive. Thus, the developmental origin of human mast cells is incompletely known and may coincide with that of monocytes and/or basophils.

12-14 Mast cell effector function is initiated by IgE-dependent and IgE-independent stimuli.

The classical, and possibly most effective, mast-cell stimulus is cross-linking of cell-surface-bound IgE by allergen in sensitized individuals. This mechanism, which was first described shortly after the discovery of IgE in the late 1960s, is central to type I hypersensitivity reactions. Such responses can be readily modeled by stimulating mast cells with antibodies that cross-link IgE or the type 1 IgE receptor (FcεRI).

FcεRI-independent mast-cell regulators are typically growth factors or cytokines that either promote mast-cell development from progenitor states, or function as regulators of mediator release, or do both. For example, SCF acts not only as a mast-cell growth factor, but also as a regulator of mast-cell mediator release by either enhancing IgE-dependent mediator release or by directly inducing mediator release in mast cells maintained in a SCF-deprived milieu. The mechanisms of SCF effects in human mast cells are well understood (Figure 12.7). Binding of SCF induces autophosphorylation of its receptor CD117 (c-Kit) and subsequent activation of several signaling molecules including phosphatidylinositol-3-kinase (PI3K) and mitogen-activated protein kinase (MAPK). SCF-mediated signaling of mast cells derived from human CD34+ peripheral blood cells involves activation of a series of kinases, including the Src-related kinase Lyn and BTK, and phosphorylation of signal transducer of activation 5 (STAT5) and STAT6 that link c-Kit activation to specific transcriptional programs. c-Kit signaling also involves a transmembrane adaptor (NTAL), which is not only linked to c-Kit signaling but also to that associated with signaling by FcεRI. Both c-Kit and FcεRI signaling enhance IgE-dependent degranulation.

IL-4 is another important human mast-cell regulator. In contrast to SCF, IL-4 does not affect mast cells by itself, but acts synergistically with SCF on mast-cell

Figure 12.7 Stem cell factor (SCF) signaling pathways in mast cells. SCF binding to its receptor, c-Kit, results in its dimerization and autophosphorylation within the cytoplasmic tail. This results in recruitment of numerous adaptors and signaling modules to the phosphorylated cytoplasmic tail. Four general signaling pathways are activated including those associated with mitogen-activated protein kinases (MAPK), phosphatidylinositol-3-kinase (PI3K), Janus kinase-signal transducer and activator of transcription (JAK/STAT), and Src-family kinases that regulate mast-cell function, growth, differentiation, survival, mediator production, and chemotaxis. BTK, Bruton's tyrosine kinase; SHIP, SH2-domain-containing inositol-5-phosphatase; SHP2, SH2-domain-containing protein tyrosine phosphatase; SOS, son-of-sevenless homolog; TEC, tyrosine kinase expressed in hepatocellular carcinoma; SHC, Src homology 2 (SH2)-domain-containing transforming protein C; GRB2, growth factor-receptor-bound protein 2; PLCγ, phospholipase Cγ; MAPKK, MAPK kinase. (Adapted from A. M. Gilfillan and C. Tkaczyk, *Nat. Rev. Immunol.*, 6:218–230, 2006. With permission from Nature Publishing Group.)

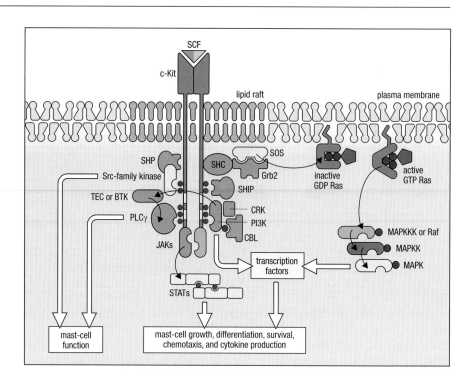

survival, proliferation, and IgE-dependent mediator release. Moreover, IL-4 alters the cytokine profile released by mast cells by reducing pro-inflammatory cytokines such as TNF-α and IL-6, and in turn enhancing secretion of type 2 cytokines such as IL-5 and IL-13. Considering that IL-4 also induces the development of T_H2 cells and IgE switching in B cells, this cytokine is a key mediator of allergic inflammation.

Little is known about other FcεRI-independent triggers of mediator release from human mast cells. Not only do the types of IgE-independent mast-cell agonists vary between human and rodent mast cells but also between human mast cells from different body sites. Some examples of such FcεRI-independent activators of human mast cells include the following. Human mast cells challenged with IFN-γ express FcγRI at sufficient quantities that allow them to become activated for mediator release upon FcγRI aggregation. This mechanism could be of relevance for the otherwise poorly understood IgE-independent allergic reactions as well as for nonallergic mast-cell activation during type III hypersensitivity reactions or infections. Mediators such as C3a, C5a, substance P and other neuropeptides, IL-8, nerve growth factor, SCF, bacterial products, and ultraviolet light have all been suggested to be mast-cell activators. However, most of these mediators only affect human skin mast cells and not mucosal mast cells and may require prior stimulation with SCF or IL-4. SCF and IL-4 may thus be considered as primary, co-stimulatory mediators since they not only enhance FcεRI-mediated signals but also induce the upregulation of secondary stimulatory receptors. One example is the receptor for substance P (NK-1) which is upregulated by SCF and IL-4. Such co-stimulation draws comparisons between mast-cell and T-cell signaling because both require the cooperation of two signals for optimal activation: an antigen-dependent signal such as the T-cell receptor or IgE bound to the cell surface, and a co-stimulatory molecule, for example, CD80 in T cells or SCF/IL-4 in mast cells.

Mast cells are also subject to regulation by inhibitory receptors which counter the activity of the aforementioned activating ligands and their receptors. These inhibitors include ligands of receptors that contain an immune tyrosine inhibitory motif such as FcγRIIB, gp49B1, SIRPα, and the human analogs LIR-5 and LILR B4. They also include ligands of the anti-inflammatory cytokines TGF-β1 and IL-10; CD200; intracellular signaling molecules such as NTAL or RabGEF1; and several other molecules including retinol, β2-adrenoceptor agonists, and extracellular matrix proteins binding to CD63.

12-15 Mast cells exert their biological functions almost exclusively by secretion of humoral factors.

Although there are a few reports describing mast-cell phagocytosis and other nonsecretory mast-cell functions, the biological activities of mast cells are mainly associated with the secretion of mediators. The array of mediators released by human mast cells is enormous and further explains how mast cells can be involved in so many different physiological and pathophysiological functions (Figure 12.8). The mast-cell mediators can be classified into small-molecule mediators (histamine, serotonin), protein mediators (cytokines, proteases), lipid mediators (leukotrienes, prostaglandins), and proteoglycans (heparin). Some of the mediators are stored in granules (histamine, proteases, proteoglycans, small amounts of TNF-α) and therefore can be released within seconds or minutes. Others are newly synthesized within minutes to hours upon stimulation of the cells (lipid mediators and most cytokines) and often require transcription.

12-16 Mucosal mast cells respond to extrinsic and intrinsic signals and thus regulate the function of the mucosal barrier and smooth muscle motor function.

The function of the mucosal barrier is under the influence of mast cells in response to extrinsic (e.g., infections and toxins) and intrinsic (e.g., acute or chronic stress) events. For example, in response to parasitic infection or stress, mast-cell proteases are released that increase epithelial paracellular permeability by inducing a redistribution of epithelial tight junction proteins. Psychological stress can also lead to enhanced release of mast-cell mediators such as histamine and tryptase in the proximal small intestine both in healthy individuals and to a larger extent in individuals that suffer from food allergies.

The basis for this likely resides in the fact that mast cells are in close contact with epithelial cells and nerve endings. Histamine, when released at low

Figure 12.8 Mast cell effector functions. Mediators of human mast cells include small molecules which primarily function by promoting inflammation (histamine, leukotriene B_4 and C_4, prostaglandin D_2), cytokines (interleukins, IL; tumor necrosis factor, TNF-α; transforming growth factor beta, TGF-β; basic fibroblast growth factor, FGF; monocyte chemotactic protein 1, MCP-1; macrophage inflammatory protein, MIP; vascular endothelial growth factor, VEGF), proteases (tryptases, chymase), and proteoglycans (heparin, chondroitin sulfate). The mediators displayed in the yellow section are mediators synthesized *de novo* upon stimulation of mast cells and those in the blue section exist within intracellular granules. The major biological effects of each mediator are indicated. ECM, extracellular matrix. (Adapted from S. Bischoff, *Semin. Immunopathol.*, 31:185–205, 2009. With permission from Springer.)

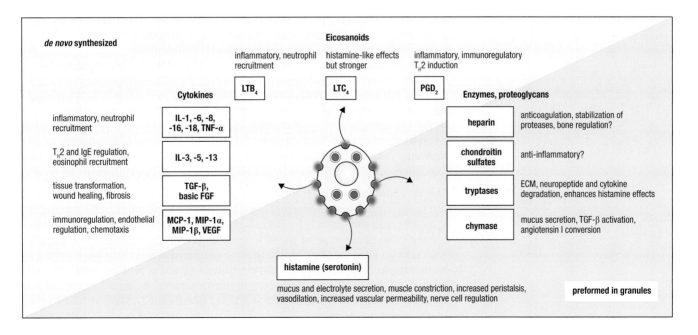

concentrations by mast cells, can bind histamine receptors on polarized epithelial cells resulting in increased transepithelial ion transport and mucus secretion. These effects on the epithelium are blocked by antihistamines and cyclooxygenase inhibitors, which in the latter case disrupt production of PGD_2 and LTC_4. This is clinically associated with diarrhea, which although abnormal in the context of food allergy would be beneficial in the case of exposure to microbes, toxins, and other harmful substances.

These physiologic activities of mast cells are further enhanced by the relationship between mast cells and nerve cells contained within all mucosal tissues. In intestines these nerve cells are highly organized and contained within the enteric nervous system. Information that is received by the enteric nervous system is derived from local sensory receptors, the central nervous system, and immune cells including mast cells. Specific IgE or IgG antibodies attach to the mast cells, enabling the mast cell to detect sensitizing antigens when they appear in the gut lumen, and are transcytosed by transporting immunoglobulin receptors (see Chapter 9). This allows for the mast cell to signal the presence of the antigen to the enteric nervous system. The enteric nervous system interprets the mast cell signal as a threat and releases neurotransmitters that drive the propulsive motor behavior of the gut tube which is in turn organized to rapidly and effectively eliminate the threat. Although operation of this alarm program protects the individual, it is at the expense of symptoms that include cramping abdominal pain, fecal urgency, and diarrhea in the intestines or bronchial constriction in the lungs. Mast cells use paracrine signaling, such as the release of histamine, proteases, and lipid mediators, to transfer the necessary chemical information to the neural networks. In addition to such humoral mediators, adhesion molecules such as ICAM-1 appear to be involved in mast cell–nerve interactions. It can be surmised that the regulation of epithelial secretion and smooth muscle motor function are due to the properties of MC_T and MC_{CT} respectively.

12-17 Mast cells provide host defense against microbes.

Human mast cells are capable of recognizing a large number of PAMPs and other bacterial products by TLRs and other pattern recognition receptors. Consistent with this, mast cells respond to a variety of microbially derived pathogenic factors. For example, *Escherichia coli* alpha hemolysin induces calcium influx in human intestinal mast cells leading to the release of histamine, sulfidoleukotrienes, and pro-inflammatory cytokines. Furthermore the type 1 fimbrial protein FimH, cholera toxins, and glucopeptides derived from parasites have been characterized as capable of activating mast cells. These factors are important in the host's response to infections with organisms such as *Trichinella spiralis* or *Nippostrongylus brasiliensis*, which are accompanied by substantial accumulation of mast cells in infected tissues. It is not clear whether this mast cell accumulation is triggered by the infectious agents or by the associated T_H2 response. Nonetheless, mast-cell proteases, such as monocyte chemotactic protein 1 (MCP-1), are critical mediators of the immune response to helminths. In the absence of MCP-1, for example, expulsion of *Trichinella* sp. and *Nippostrongylus* sp. is significantly delayed.

The mechanism by which mast cells sense parasitic infection likely involves not only pattern recognition receptors but also activation by FcεRI aggregation through cross-linking of IgE directed against parasitic antigens. This is consistent with the generation of parasite-specific IgE antibodies during a parasite infection and the fact that blocking mast-cell function leads to impairment in host defense against parasites despite the presence of anti-parasite IgE antibodies. Of the mast-cell mediators that are induced by FcεRI cross-linking, IL-5 (for eosinophil recruitment) and IL-13 (for B and T_H2 cell immunity) are of particular importance.

Mast cells also participate in host defense against bacterial pathogens. This has been demonstrated in mouse models of *Klebsiella pneumonia*-induced infections of the lung and in models of peritoneal sepsis. In response to bacterial infections, mast cells produce TNF-α and leukotrienes which serve to recruit neutrophils and as such result, along with other mechanisms, in increased bacterial clearance. Mast-cell activation during bacterial infection also induces dendritic cell migration, supporting a role for mast cells in linking innate to adaptive immune responses. In response to bacterial pathogens, mast cells are activated by both TLR and non-TLR (e.g., complement) dependent mechanisms.

Mast cells may also provide host defense against viruses. Stimulation of mast cells through TLR3 using polyinosinic/polycytidylic acid leads to the recruitment of CD8+ T cells. Human mast cells also express TLR3 and stimulation through this receptor leads to decreased mast-cell adhesion to extracellular matrix proteins and increased production of IFN-γ consistent with a role in host defense against viral infections.

12-18 Mast cells not only induce an immune response and inflammation but also regulate the immune response and the resultant processes of tissue repair.

Mast cells are also involved in effecting peripheral tolerance induced by CD4+CD25+Foxp3+ T_{reg} cells. Release of IL-9, a mast-cell growth and activation factor, by activated T_{reg} cells is associated with mast-cell recruitment and activation during the induction of tolerance. Finally, mast cells regulate wound healing, fibrosis, blood flow and coagulation, and protection against neoplasms (Figure 12.9). As such, mast cells are important in both the maintenance and reestablishment of tissue homeostasis.

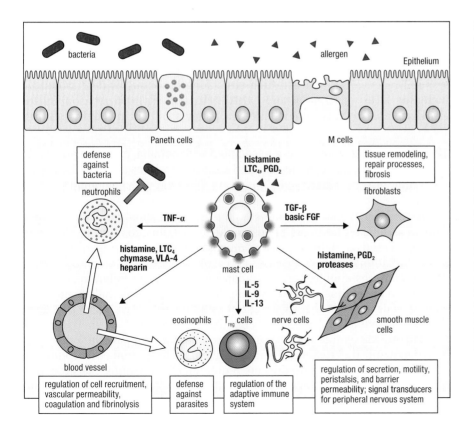

Figure 12.9 Physiological functions of mast cells at the mucosal barrier. Mucosal mast cells regulate tissue homeostasis (epithelial secretion and permeability, blood flow and vascular permeability, smooth muscle functions and peristalsis, wound healing and fibrosis), immune functions (recruitment and activation of neutrophils, eosinophils, and lymphocytes; induction of T_{reg} cells; and defense against microbes), neuronal functions (neuroimmune interactions, peristalsis, bronchoconstriction, and pain), and inflammation (allergy, asthma, inflammatory bowel disease, irritable bowel syndrome, infection). FGF, fibroblast growth factor; IL-5, interleukin 5; LTC_4, leukotriene C_4; PGD_2, prostaglandin D_2; TNF-α, tumor necrosis factor; TGF, transforming growth factor; T_{reg}, regulatory T cells; VLA-4, very late antigen 4. (Adapted from S. Bischoff, *Nat. Rev. Immunol.*, 7:93–104, 2007. With permission from Macmillan Publishers Ltd.)

Mucosal basophils, eosinophils, and mast cells in human disease.

Given their similarities in lineage development, sources of stimuli to which they respond, localization to mucosal tissues, and association with allergy and T_H2-dependent immune responses, it is not surprising that basophils, eosinophils, and mast cells exhibit overlapping but distinct relationships with mucosal diseases. The human diseases with which these cell types exhibit a unique relationship will be discussed below.

12-19 Mucosal basophils are involved in allergies of the skin and intestines, asthma, and autoimmunity.

The ability of basophils to be activated by multiple stimuli, coupled with their ability to migrate to draining lymph nodes with their many effector functions, suggests that they may be critical regulators of inflammation in multiple tissue sites. Accordingly, elevated basophil populations are correlated with the inflammation associated with allergies and asthma. For example, basophils are elevated in the airways of asthmatics, increase upon allergen challenge, and are observed at elevated numbers in the lungs during severe asthma. In addition, blood basophil counts are increased in asthma and elevated basophil populations directly correlate with airway hyperresponsiveness and decreased lung function.

Increased basophil populations are also associated with multiple allergic disorders of the skin. For example, elevated numbers of basophils are found in skin biopsies associated with atopic or allergic contact dermatitis and are reported to increase upon allergen challenge. Basophils are also implicated in chronic urticaria (hives), a common skin disease involving mast cells and basophil recruitment to the lesional sites. Although basophils are most commonly associated with allergic diseases, they likely play a role in autoimmune disorders as revealed by observations in systemic lupus erythematosus (SLE), an autoimmune disease characterized by autoantibody production that may affect mucosal tissues. More specifically, individuals suffering from SLE possess self-reactive IgE and activated circulating basophils as determined by increased levels of HLA-DR and CD62L. HLA-DR-expressing basophils are also detectable in the lymph nodes of SLE patients, but not in healthy controls.

12-20 Mucosal eosinophils are associated with asthma, parasitic and viral infections, mucosal allergy, and several hypereosinophilic disorders that are characterized by mucosal infiltration with eosinophils.

The respiratory mucosa, in contrast to the gastrointestinal mucosa, does not contain significant numbers of eosinophils in healthy conditions. However, eosinophil influx into the respiratory mucosa is observed during the course of several inflammatory lung diseases and especially asthma (see also Chapter 21). As such, eosinophils have been considered as major inflammatory cells involved in the pathogenesis of asthma. This paradigm was challenged when clinical trials revealed that treatment of asthma patients with anti-IL-5 antibodies substantially decreased the number of eosinophils in the lungs but failed to provide symptomatic benefit. More recent studies have, however, suggested that anti-IL-5 treatment benefits patients with sputum eosinophilia, as shown by a reduction in prednisone requirements or asthma exacerbations and improvement of asthma patients' quality of life. Therefore, the eosinophil is still viewed as a terminal effector cell in subgroups of allergic airway diseases. On the other hand, this multifunctional cell could be more involved in the initial stages of allergic disease development than was previously thought, particularly with regard to the ability of the eosinophil to modulate T-cell responses. Alternatively, eosinophils might play a role not only in initiating and perpetuating inflammation, but also in repair and remodeling processes. Indeed, there is a well-documented association of tissue eosinophilia and eosinophil degranulation with certain fibrotic syndromes and the eosinophil is the source of several fibrogenic and growth factors, including TGF-α, TGF-β1, fibroblast growth factor 2, vascular endothelial growth factor, matrix metalloproteinases, and selected cytokines.

Eosinophilia is commonly observed during helminth infection and typically associated with a strong T_H2 immune response. Eosinophils are able to effectively kill or damage larvae and adult worms. However, eosinophils are only partly effective in the control of helminth infection. This suggests that eosinophils have other functions during parasitic infection. Eosinophils, as described above, possess a range of immunomodulatory effects such as increased production of anti-inflammatory cytokines (IL-10 and TGF-β) and stimulation of regulatory T cells or alternatively activated macrophages. Therefore, as described for asthma, immunomodulation that allows parasite survival but reduces pathology may be a dominating function of eosinophils in association with parasitosis. Whether this is a primary function of the eosinophil or an immune-evasive function that is induced by the parasite is not known. However, it can be conjectured that such helminth (parasite)-induced immunomodulation may contribute to protection from allergic and autoimmune responses, as proposed by the 'hygiene hypothesis.'

Eosinophils are also important in viral infections. Human respiratory syncytial virus causes respiratory tract infection, primarily among infants and toddlers. The severity of infection can extend from mild upper respiratory symptoms to severe bronchiolitis and pneumonia, and may progress to acute respiratory distress syndrome and death. Human respiratory syncytial virus infections are associated with significant eosinophilia in the bronchoalveolar fluid. This is most likely because eosinophils are attracted to the site of inflammation by chemokines such as RANTES and MIP-1α. The role of eosinophils in this disease is uncertain, as there is no clear evidence from human studies as to whether they promote host defense or enhance pathology. Among these pathologies, there is a clear association between severe human respiratory syncytial virus infection, particularly among young infants, and the development of post-infection asthma.

Eosinophilic gastrointestinal diseases (EGIDs) comprise primary and secondary subtypes. Primary EGIDs are defined as gastrointestinal symptoms occurring in the context of increased numbers of mucosal eosinophils, and in the absence of other recognized causes of tissue eosinophilia, such as drugs, allergies, autoimmunity, malignancy, and parasitic infestation, which would be considered secondary EGIDs as discussed below. They include eosinophilic esophagitis, eosinophilic gastritis, eosinophilic gastroenteritis, eosinophilic enteritis, and eosinophilic colitis. The prevalence, or at least recognition of EGIDs, is increasing. The majority of patients (50–80%) with EGIDs are atopic. The close link with atopy, in combination with the increased likelihood of food-allergen-specific IgE in the serum or positive food allergen skin-prick tests, suggests that there is a pathologic role for an allergic inflammatory process in EGIDs. Allergic inflammation is characterized by the activation of T_H2 cells, which elaborate specific profiles of pro-allergic cytokines, including IL-4, IL-5, and IL-13, that promote the formation of IgE antibodies and support eosinophil and mast-cell recruitment, maturation, and activation. The potential importance of IL-5 has been highlighted by preliminary studies in which treatment of patients with eosinophilic esophagitis using anti-IL-5 antibody is associated with marked reductions in the numbers of eosinophils in the blood and esophageal mucosa.

Secondary EGIDs are inflammatory diseases that have recognized causes of gastrointestinal eosinophilia such as parasite infection, gastroesophageal reflux disease (GERD), inflammatory bowel disease (IBD), irritable bowel syndrome (IBS), allergic disease (see Section 12-21), drug reactions, and malignancy. The mild eosinophilia typically occurring in the distal esophagus in about 50% of patients with GERD has to be separated from the more pronounced and ubiquitous eosinophilia in patients with eosinophilic esophagitis.

Finally, hypereosinophilic syndrome comprises a heterogeneous group of disorders characterized by excessive numbers of circulating and tissue-infiltrating eosinophils resulting in organ dysfunction. Some hypereosinophilic syndrome subtypes are myeloproliferative disorders, whereas the etiology of other variants remains undefined. Involvement of the gastrointestinal tract is common and about 20% of patients experience diarrhea. Hypereosinophilic syndrome is a well-recognized cause of eosinophilic gastritis, gastroenteritis, and colitis, but in contrast to the other primary EGIDs, eosinophilic infiltration of other organs including the heart, skin, and nerves is characteristic. Consequently, hypereosinophilic syndrome might be considered to be a secondary EGID.

12-21 Mucosal mast cells are important mediators of the inflammation associated with allergy, asthma, celiac disease, inflammatory bowel disease, and diseases of unknown origin such as irritable bowel syndrome and systemic mastocytosis.

The inflammatory mediators produced by mast cells, basophils, and eosinophils are responsible for the clinical symptoms and the organ dysfunction that occur during IgE-mediated type I reactions. Type I allergic reactions, which are also taken as a basis for many cases of bronchial asthma, food allergy, and other atopic diseases, are divided into an immediate phase and a late phase occurring possibly some hours after the immediate phase (see Chapter 21). The immediate phase is characterized by the IgE-dependent activation of mast cells and basophils and the release of pro-inflammatory mediators—such as histamine, proteases, leukotrienes, and cytokines—from these cells. At the same time, the cells start to synthesize mediators leading to a more sustained release of mediators such as eicosanoids and cytokines. The late phase is characterized by the infiltration of the tissue with further inflammatory cells such as neutrophils, eosinophils, and lymphocytes. These cells are attracted by mediators such as TNF-α, IL-5, IL-4, and IL-3 released by mast cells and basophils upon IgE-dependent immediate-type activation. Most importantly, human mast cells induce the recruitment and local activation of eosinophils by expressing factors such as IL-5 upon IgE-dependent activation, and induce the recruitment of neutrophils by releasing IL-8 and TNF-α.

Human mast cells also likely participate in regulating lymphocyte functions during the course of allergic inflammation. Upon IgE cross-linking mast cells produce IL-13, a cytokine that supports the production of allergen-specific IgE by B cells. The release of IL-13 can be further increased by the presence of IL-4, which is known to shift the cytokine profile produced by human mast cells away from pro-inflammatory cytokines—such as TNF-α, IL-1β, and IL-6—to T_H2 cytokines, including IL-13. Human mast cells can also regulate T-cell functions through other mediators such as PGD_2, which almost exclusively derives from activated mast cells, is released during allergic reactions, and is particularly important at the onset, and in the perpetuation of, asthma in young adults. This lipid mediator evokes airway hypersensitivity and chemotaxis of T cells, basophils, and eosinophils through interaction with two receptors: the prostanoid DP receptor (PTGDR) on granulocytes and smooth muscle cells; and CRTH2 (chemoattractant receptor-homologous molecule expressed on T_H2 cells) on T_H2 cells. Furthermore, *PTGDR* has been identified as an asthma-susceptibility gene. Apart from PGD_2, other human mast-cell mediators, such as LTB_4, CCL3, and CCL4; OX40 ligand; and TNF-α, are involved in recruiting T cells and triggering T cell-mediated adaptive immune responses, including memory induction, that enhance and perpetuate allergic reactions.

Elevated levels of histamine, its metabolite methylhistamine, tryptase, eosinophil cationic protein, eosinophil-derived protein X (EPX), IL-5, and TNF-α are detectable in serum, urine, gut or bronchial lavage fluid, and stool from patients with allergy. This is further evidence that activation of mast cells, eosinophils, and to some extent basophils is involved in allergy. This is supported by histological studies showing degranulation and cytokine production by these cell types after allergen provocation tests. Mast cells stimulated by IgE cross-linking also trigger local nerve responses resulting in bronchial hypersecretion, pain, and diarrhea.

Mucosal mast cells also play an important role in specific mucosal diseases such as celiac disease (see Chapter 28) and IBD (see Chapter 31). In celiac disease, the histamine content and numbers of MC_T cells are increased. Moreover, gliadin challenge in celiac disease results in decreased mast cells consistent with mast-cell degranulation. Increased numbers of mast cells are also observed in the mucosa of the ileum and colon in IBD. Mast cells in IBD also exhibit significant increases in expression of histamine, tryptase, TNF-α, IL-16, and substance P. Such mast cells are found in close proximity to the basal lamina of the epithelial cell glands and in other layers of the gut wall. Studies with tryptase-deficient animal models suggest that mast cells function in the promotion of inflammation.

IBS is a disorder of unknown cause which is largely defined by clinical criteria and is thus a diagnosis of exclusion in the absence of any known gastrointestinal disease such as infection, IBD, celiac disease, food allergy, or malignancy. There is increasing evidence that IBS may represent a subclinical inflammatory disorder. Indeed, a low-grade mucosal inflammation can be detected in a subset of patients with IBS. This inflammatory infiltrate is mainly represented by increased numbers of T lymphocytes and mast cells that are located in the lamina propria. The close apposition of immunocytes to gut nerves supplying the mucosa provides a basis for neuroimmune cross-talk, which may explain the gut sensorimotor dysfunction and related symptoms in patients with IBS. Consistent with this, duodenal mast-cell hyperplasia is associated with IBS. However, the mechanisms that induce mast-cell hyperplasia and the presumed mast-cell mediator release in IBS still need to be defined.

The final disease to be discussed is systemic mastocytosis. Systemic mastocytosis is a disorder of unknown origin in which mast cells accumulate in multiple tissues, especially those in association with mucosal tissues. Consequently, gastrointestinal symptoms are common and consist of abdominal pain or diarrhea in the majority of patients. Systemic mastocytosis is also associated with gastric acid hypersecretion due to the hyperhistaminemia. This can result in ulcer disease that in turn can cause small intestinal mucosal damage due to the effects of the secreted acid and consequently disruption of digestion and absorption of nutrients. The malabsorption which occurs is similar to that observed in another disease which is associated with acid hypersecretion by the stomach—Zollinger–Ellison syndrome—which is a disorder due to tumors that secrete excessive gastrin, a hormone which stimulates acid secretion.

Systemic mastocytosis is characterized by the accumulation of neoplastic mast cells not only in the gastrointestinal tract but also in the bone marrow and other organs. Consequently, enlargement of the liver and the spleen, portal hypertension, and ascites occur frequently. Diagnosis of systemic mastocytosis is based upon histological and immunohistological examination of gastrointestinal tissue specimens using antibodies directed against tryptase, CD117, and CD25.

Summary.

Basophils, eosinophils, and mast cells are immunoregulatory cells with inflammatory potential that are derived from bone marrow myeloid progenitors. They are found in the blood and in tissues, especially mucosal tissues, under normal conditions but accumulate during the course of many diseases. These cells intercommunicate with each other and many other hematopoietic and parenchymal cells (e.g., epithelial cells and nerve cells) and contribute to protection from infections (e.g., parasitic infections) and pathologies associated with allergy and related conditions such as asthma. Their function during the course of these diseases is not always clear because they may either initiate and maintain the inflammatory process or be involved in tissue remodeling and healing. These cells have the common property of responding to IgE-complexed antigens and characteristically exert their biological functions by releasing inflammatory, cytotoxic, and immunomodulatory mediators which are tightly controlled by numerous agonists and antagonists.

Further Reading.

Abraham, S.N., and St John, A.L.: **Mast cell-orchestrated immunity to pathogens.** *Nat. Rev. Immunol.* 2010, **10**:440–452.

Bischoff, S.C.: **Role of mast cells in allergic and non-allergic immune responses: comparison of human and murine data.** *Nat. Rev. Immunol.* 2007, **7**:93–104.

Bischoff, S.C.: **Food allergy and eosinophilic gastroenteritis and colitis.** *Curr. Opin. Allergy Clin. Immunol.* 2010, **10**:238–245.

Bischoff, S.C., and Crowe, S.E.: **Gastrointestinal food allergy: new insights into pathophysiology and clinical perspectives.** *Gastroenterology* 2005, **128**:1089–1113.

Bochner, B.S., and Gleich, G.J.: **What targeting eosinophils has taught us about their role in diseases.** *J. Allergy Clin. Immunol.* 2010, **126**:16–25.

Brightling, C.E., and Bradding, P.: **The re-emergence of the mast cell as a pivotal cell in asthma pathogenesis.** *Curr. Allergy Asthma Rep.* 2005, **5**:130–135.

Hogan, S.P., Rosenberg, H.F., Moqbel, R., *et al.*: **Eosinophils: biological properties and role in health and disease.** *Clin. Exp. Allergy* 2008, **38**:709–750.

Iwasaki, H., and Akashi, K.: **Myeloid lineage commitment from the hematopoietic stem cell.** *Immunity* 2007, **26**:726–740.

Liu, C., Liu, Z., Li, Z., *et al.*: **Molecular regulation of mast cell development and maturation.** *Mol. Biol. Rep.* 2010, **37**:1993–2001.

Obata, K., Mukai, K., Tsujimura, Y., *et al.*: **Basophils are essential initiators of a novel type of chronic allergic inflammation.** *Blood* 2007, **110**:913–920.

Ohnmacht, C., and Voehringer, D.: **Basophil effector function and homeostasis during helminth infection.** *Blood* 2009, **113**:2816–2825.

Perrigoue, J.G., Saenz, S.A., Allenspach, E.G., *et al.*: **MHC class II-dependent basophil-CD4+ T cell interactions promote T(H)2 cytokine-dependent immunity.** *Nat. Immunol.* 2009, **10**:697–705.

Powell, N., Walker, M.M., and Talley, N.J.: **Gastrointestinal eosinophils in health, disease and functional disorders.** *Nat. Rev. Gastroenterol. Hepatol.* 2010, **7**:146–156.

Schroeder, J.T.: **Basophils beyond effector cells of allergic inflammation.** *Adv. Immunol.* 2009, **101**:123–161.

Stone, K.D., Prussin, C., and Metcalfe, D.D.: **IgE, mast cells, basophils, and eosinophils.** *J. Allergy Clin. Immunol.* 2010, **125**:S73–80.

M Cells and the Follicle-Associated Epithelium

13

The follicle-associated epithelium (FAE) overlying organized mucosal lymphoid tissues (Figure 13.1a) plays a major role in mucosal immunity. Representing a very small fraction of the total mucosal surface area of the gastrointestinal tract mucosa, the FAE contains a unique epithelial cell type, the M cell, whose primary function is to translocate luminal material across the epithelial barrier to dendritic cells (DCs) and lymphocytes within and below the epithelium. The FAE is separated from the underlying lymphoid follicle by a subepithelial 'dome' region filled with T and B cells, as well as DCs that efficiently capture materials transported by M cells (Figure 13.1b). Some DCs and lymphocytes migrate into intraepithelial pockets formed by the M cells, as described in more detail below. FAE and M cells are associated with the organized Peyer's patches in the small intestine, and with isolated follicles in the small and large intestine, appendix, rectum, crypts of the adenoids and tonsils, airways, and even the conjunctivae of the eyes.

Function of the FAE.

13-1 FAE promotes antigen sampling.

In contrast to the well-defended villus and surface epithelium of the intestine, the FAE lacks many defensive features and appears designed to allow macromolecules, particles, and microorganisms access to the apical surface. First, the secretions emanating from the FAE lack key protective molecules.

Figure 13.1 A specialized follicle-associated epithelium covers organized mucosal lymphoid follicles. Panel a shows follicle-associated epithelial cells and villus epithelial cells. The follicle-associated epithelium consists of special follicle-associated enterocytes and M cells. The villus epithelium is composed primarily of absorptive enterocytes and mucus-secreting goblet cells. Cells in the crypts located between a villus and a lymphoid follicle supply cells to both villus epithelium and FAE, differentiating along distinct phenotypic pathways. In panel b, a section of mouse Peyer's patch treated with labeled antibodies specific for the DC marker CD11c (brown) shows immature DCs congregated in the subepithelial dome region, where the cells capture incoming microbes and macromolecules transported by M cells. The DCs later migrate down the sides of the follicle to the interfollicular region to present antigen to naive T cells. (Panel a: Adapted from M. Neutra and P. Kozlowski, *Nat. Rev. Immunol.*, 6:148–158, 2006. With permission from Macmillan Publishers Ltd. Panel b: Adapted from S. Chabot et al., *Vaccine* 25:5348–5358, 2007. With permisssion from Elsevier Ltd.)

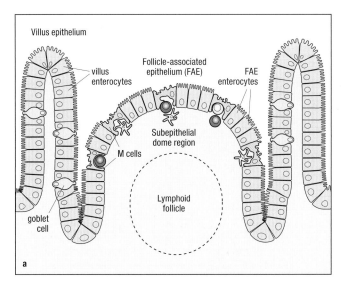
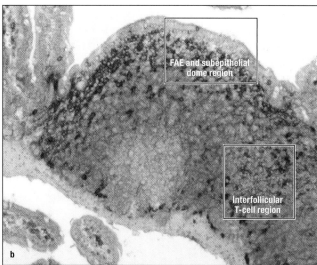

Whereas the villus epithelium is composed primarily of absorptive enterocytes and mucin-secreting goblet cells, the FAE contains few or no goblet cells and secretes little or no mucus (Figure 13.1a). Instead, it contains M cells along with enterocytes that display a FAE-specific phenotype that differs from villus enterocytes. Enterocytes of the FAE produce very low levels of digestive enzymes and alkaline phosphatase. The crypts surrounding the FAE lack defensin- and lysozyme-producing Paneth cells and do not secrete IgA into the lumen, as they lack polymeric immunoglobulin receptors.

FAE enterocytes have closely packed apical microvilli and a thick brush border glycocalyx that provides a barrier to large molecules and particles. Unlike M cells, FAE enterocytes do not participate in transport but appear to facilitate transport activities of neighboring M cells through distinct glycosylation patterns that allow recognition and adherence of certain microorganisms to the FAE. Gene expression studies indicate that the enterocytes of the FAE express specific extracellular matrix and matrix-interacting proteins. For example, the basal lamina of the FAE lacks laminin-2 subunits and contains large pores that presumably reflect the migration of cells into and out of the M-cell pockets in the epithelium.

13-2 FAE attracts specific populations of DCs and lymphocytes into organized mucosal lymphoid tissues.

FAE cells differ from villus cells in their ability to release certain chemokines that attract immune cells toward the FAE and thus to sites of organized lymphoid tissue. In the small intestine of mice and humans, for example, chemokine CCL20, also designated macrophage inflammatory protein-3α (MIP-3α), is constitutively expressed in the FAE but not in villus epithelium (Figure 13.2). CCL20 attracts subpopulations of DCs and lymphocytes that express the chemokine receptor CCR6. Mice that lack CCR6 have lower numbers of CD11c$^+$ DCs in the subepithelial dome regions of Peyer's patches and have an impaired humoral immune response to orally administered antigen and enteropathogenic viruses. Cells of the mouse FAE also express CCL9 (analogous to CCL23 in humans), which attracts CCR1-expressing myeloid DCs, and CXCL16, which attracts CXCR6-expressing B and T lymphocytes into Peyer's patches.

Gene expression in the FAE may be modulated by microorganisms that contact the FAE or are transported by M cells into the mucosa. In the small intestine, CCL20 gene expression is upregulated in the FAE by exposure to pathogenic bacteria or bacterial components. In the colon, where microbial populations

Figure 13.2 FAE mediates crosstalk between the luminal flora and the mucosal immune system. M cells conduct endocytosis and rapid transepithelial transport of intact antigens and microorganisms into intraepithelial pockets that contain B and T cells and DCs. The majority of FAE cells are enterocytes that do not transport antigens. FAE enterocytes may contribute to antigen sampling by sensing luminal pathogens and their products via Toll-like receptors (TLR) and then releasing cytokine and chemokine signals that attract and activate DCs. (Adapted from M. Neutra and P. Kozlowski, *Nat. Rev. Immunol.*, 6:148–158, 2006. With permission from Macmillan Publishers Ltd.)

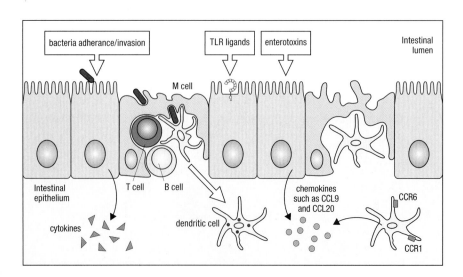

are more numerous, CCL20 is expressed on FAE and non-FAE epithelial cells. Like other intestinal epithelial cells, FAE cells express Toll-like receptors (TLRs), the microbial pattern recognition receptors that transmit intracellular signals and induce expression of pro-inflammatory and immunomodulatory cytokines and chemokines. In villus cells, TLRs are generally confined to basolateral cell membranes, allowing activation when microbes breach the epithelial barrier. In the FAE, however, transported microbes routinely contact both sides of the epithelium as well as subepithelial TLR-bearing DCs. Constitutive activation of TLRs in the FAE may contribute to the constitutive expression of chemokines.

M cells as gateways to the mucosal immune system.

The cardinal feature of the FAE is the presence of M cells. The major function of M cells is to deliver samples of macromolecules, particulate foreign material, and microorganisms by transepithelial transport from the lumen to organized mucosal lymphoid tissues. M cells, like all the epithelial cells of the gastrointestinal and respiratory tracts, are joined to their neighbors by tight junctions that seal the paracellular pathway. However, M cells provide functional openings in the epithelial barrier through transepithelial vesicular transport activity. M-cell differentiation is largely restricted to the FAE; consequently, transport of foreign material and microbes across the epithelial barrier is targeted to the organized, inductive sites of the mucosal immune system. M cells were first recognized microscopically by their unique morphology, especially their intraepithelial 'pocket' that provides a sequestered space for activated or memory B and T lymphocytes and DCs. The pocket shortens the transcytotic pathway and provides for rapid delivery of luminal samples to intraepithelial and subepithelial cells.

13-3 M cells are polarized epithelial cells with unique architecture.

M cells, like all epithelial cells lining the intestine, are highly polarized. The tight junctions that seal their apical poles prevent the passage of most molecules between cells and the lateral diffusion of integral membrane glycolipids and proteins between apical and basolateral domains of the plasma membrane. The basolateral domain is further subdivided into a lateral subdomain that adheres to the neighboring cell through homotypic interaction of adhesion molecules such as E-cadherin and a basal subdomain that adheres to the basal lamina through heterotypic interaction of integrins with extracellular matrix components. These interactions transmit signals internally, resulting in complex cytoskeletal organization and membrane polarization. The apical domain of typical intestinal epithelial cells is highly organized, with long, actin-supported microvilli, between which are small microdomains that include clathrin-coated pits and caveolae capable of limited endocytosis. Maintenance of the apical and basolateral domains involves sorting of newly synthesized membrane components in the trans-Golgi network and budding of vesicles destined for either the apical or basolateral domain.

As in other epithelial cells, M-cell tight junctions define the apical and basolateral plasma membrane domains. However, in M cells, the basolateral domain is expanded and modified to form a large, invaginated subdomain that lines a novel space, the intraepithelial pocket. Little is known about the molecular composition of the pocket domain, although it contains adhesion molecules that promote docking of DCs and intraepithelial lymphocytes as described below.

13-4 The M cell apical surface is designed for easy access and efficient endocytosis.

Importantly, M cell apical membranes lack the organized brush border and thick, protective glycocalyx of integral membrane mucins that blankets the microvillus tips on enterocytes (Figure 13.3a). M cells instead have widely spaced microvilli or microfolds, interspersed with large plasma membrane subdomains that are organized for endocytosis. Thus, microorganisms and particles can readily contact and adhere to the areas involved in endocytosis; this is important because adherence increases the likelihood of endocytosis. For example, polystyrene or latex beads measuring up to 1 μm in diameter can adhere nonspecifically to many cell surfaces, but they adhere avidly to M cells in Peyer's patches and are rapidly taken up into the mucosa.

Although M cells lack a thick glycocalyx, their apical membranes contain glycoproteins and glycolipids that can serve as binding sites. Many bacteria and viruses adhere to target host cells using specific adhesion molecules that recognize specific carbohydrate structures on host cell glycoproteins or glycolipids. Carbohydrate-based recognition mechanisms contribute to M cell-specific uptake into the mucosa in several ways. Most glycolipids and glycoprotein-associated carbohydrate structures are not unique to M cells but are ubiquitous, as they also are present in the mucin-like molecules of the brush border glycocalyx, thereby promoting microbe and particle adherence to intestinal epithelium. Particles that adhere to the enterocyte glycocalyx, however, do not interact with endocytic membrane domains and are eventually shed into the lumen. In contrast, particles that adhere to M-cell membranes are endocytosed and transported. Other glycoproteins and glycolipids present on membrane molecules of intestinal epithelial cells are masked by the brush border glycocalyx. Large macromolecules or microbes that cannot diffuse through the glycocalyx are unable to adhere to these enterocyte epitopes but can bind M cells. This was demonstrated using cholera toxin, whose pentameric B subunit binds to the ganglioside GM1, a membrane glycolipid present on all intestinal epithelial cells. Soluble cholera toxin diffuses through the glycocalyx to bind to the enterocyte membrane, triggering a cascade of events that leads to severe diarrhea. However, toxin immobilized on microparticles and administered into the intestine in particulate form adheres only to M cells (Figure 13.3b). Similarly, certain viruses are able to gain access to their receptors only on M cells, as described below.

Figure 13.3 The M cell apical surface is designed for easy access and efficient endocytosis. Panel a shows the apical poles of two FAE cells in a Peyer's patch that have been processed for electron micrographic visualization of membrane glycoproteins. The M cell (left) has irregular microvilli with a very thin surface coat. The microvilli of the enterocyte (right) are blanketed by a thick glycocalyx. A lymphocyte is present in the M-cell pocket. Cholera toxin B subunit (CTB) is a soluble protein that binds to a glycolipid receptor on all cell surfaces in the intestine. Panel b shows explanted rabbit Peyer's patch mucosa exposed to cholera toxin coupled to 15 nm colloidal gold particles. Electron microscopic analysis revealed that CTB-gold adhered to the apical surfaces of M cells but not to adjacent enterocyte microvilli. In M cells, gold particles were present in clathrin-coated pits and vesicles, indicating endocytosis and transport. (Adapted from A. Frey et al., *J. Exp. Med.*, 184:1045–1060, 1996. With permission from Rockefeller University Press.)

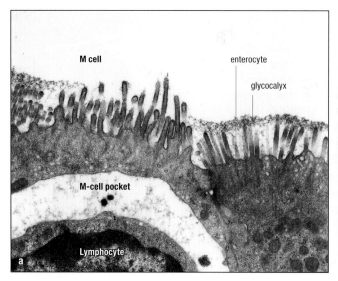

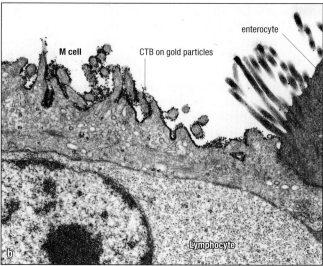

13-5 M cells display cell type-specific membrane molecules that may serve as receptors.

M cells show a unique 'face' to the lumen by displaying surface molecules that are not present on the apical surface of neighboring epithelial cells. For example, beta 1 integrins and intercellular adhesion molecule-1 (ICAM-1) are located on the apical surfaces of M cells but on the basolateral surfaces of neighboring epithelial cells, where they function in lymphocyte and matrix interactions. Structures expressed only on M cells include specific carbohydrates identified by lectin or antibody binding, reflecting M cell-specific glycosyltransferase expression. These M cell-specific carbohydrate epitopes vary among species, among different mucosal regions, and even within the same FAE. Also present on the apical surfaces of M cells are highly glycosylated, lipid-linked proteins that bind bacterial type 1 pili of Gram-negative bacteria, and proteins that bind proteoglycans of Gram-positive bacteria. Thus, M cell-specific oligosaccharides and glycoproteins likely serve as receptors for immune sampling of broad classes of microorganisms, but such receptors may be exploited by pathogens to invade the mucosa and cause disease, as discussed later.

The fine balance of host–microbe symbiosis in the intestine is crucial for normal mucosal immune function as well as maintenance of a healthy epithelium. Consequently, an important function of M cells in the absence of pathogens is uptake of commensal microorganisms that normally inhabit mucosal surfaces, especially in the intestine. Nonpathogenic bacteria are regularly delivered by M cells into Peyer's patches, and some may live within mucosal follicles without causing disease. In this regard, M cells express a putative IgA receptor that recognizes IgA and secretory IgA, potentially promoting the uptake of IgA-coated microbes and IgA–antigen complexes. Thus, through IgA-mediated uptake and selective microbial adherence, M cells participate in host interaction with normal luminal microflora, as well as with certain pathogens.

M cells display distinct proteins on their basolateral membrane to attract lymphocytes into the intraepithelial pocket. For example, M cells express on their basolateral membranes a membrane-bound form of the chemokine CXCL16, a molecule that interacts with CXCR6 on T and B cells. M cells also express CD137, an integrin family member that is recognized by certain B cells. Mice deficient in CD137 have Peyer's patches, FAE, and M cells with typical apical markers, but their M cells fail to form pockets and do not harbor intraepithelial lymphocytes.

13-6 Adherent macromolecules and particles are efficiently endocytosed and transported across M cells.

Epithelial cells move macromolecules and microbes from the apical cell surface to the basolateral surface by a process called transcytosis (Figure 13.4). Transcytosis is mediated by a complex series of events, including formation and fusion of endosomes and polarized recycling of membrane vesicles, with directional information provided by G proteins and adaptors and the highly polarized cytoskeleton. The first step in the transcytotic process is invagination of apical membrane microdomains to form intracellular compartments called early endosomes. The mechanisms by which M cells take up cargo are diverse. Adherent macromolecules, small particles, and viruses are taken up via clathrin-coated or noncoated pits and vesicles (Figure 13.4); soluble macromolecules are captured in the fluid phase. Large adherent particles are internalized in a process resembling phagocytosis that involves the assembly of organized submembrane actin networks. Electron-micrographic and immunocytochemical data suggest that certain proteases are delivered into M-cell

Figure 13.4 Reovirus adheres selectively to M cells in mouse Peyer's patch. This electron micrograph of mouse FAE shows reovirus bound to and taken up into the transepithelial transport pathway of an M cell. Some adherent virus particles have been taken up into endosomes. After transcytosis through the cell, endocytosed virus will be released at the modified basolateral membrane into the intraepithelial pocket that contains lymphocytes.

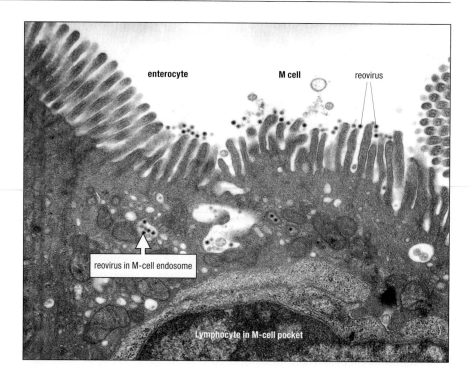

endosomes, and that the endosomal lumen is acidified to pH 6.0–6.2, perhaps allowing some ligands to be released from their receptors.

In model epithelial cells, membrane proteins and lipids are sorted in the early endosomes or in the trans-Golgi network for transport to multiple locations. Some are recycled to the cell surface and others transported along the degradative pathway via 'late endosomes' or 'multivesicular bodies' (that contain small internal vesicles) to lysosomes. Only a small amount of selected material enters vesicles directed to the basolateral membrane domain via transcytosis. In contrast, in M cells transcytosis traffics the majority of vesicles containing endocytosed material to a single destination: the specialized membrane domain lining the invaginated pocket. The short distance between the apical and basal cell surfaces, 1–2 μm, allows transcytosis to be completed in as little as 15 minutes. Fusion of vesicles with this membrane releases vesicle content into the intraepithelial space for uptake by DCs and lymphocytes. The extent to which the brief exposure to proteases in M-cell transport vesicles results in degradation or modification of content is not clear, but microorganisms arrive in the pocket intact and alive. Transcytosis by M cells is accelerated when the FAE is exposed to bacteria or TLR ligands, possibly due in part to release of macrophage inhibitory factor from subepithelial DCs.

13-7   Transported cargo is delivered to cells in the M-cell pocket and subepithelial dome region.

Antigens and pathogens released into the M-cell pocket immediately contact the intraepithelial DCs, B cells, and CD4+ T cells located in the pocket. The lymphocytes display markers typical of activated or memory cells. The B cells express major histocompatibility complex (MHC) class II, suggesting that they are capable of antigen presentation. DCs also migrate into the pocket from the subepithelial dome region and quickly take up transported material such as viruses and particles. DC migration toward and into the FAE is enhanced when the mucosa is exposed to TLR ligands or enterotoxins, likely due to release of chemokines such as CCL20 from the epithelium (Figure 13.2). Thus, the intraepithelial pocket containing subepithelial DCs and lymphocytes permits efficient antigen exposure, possibly without the influence of preexisting systemic antibodies, which appear incapable of diffusion into the pocket.

Mucosal lymphoid follicles consist primarily of a cluster of immature B cells, often including a germinal center and a few T cells, supported by a network of specialized follicular DCs. The follicles are flanked by interfollicular T-cell areas that contain a distinct population of antigen-presenting DCs, naive and antigen-sensitized T cells, and high endothelial venules that allow entry and exit of migrating cells (Figure 13.5). Between each mucosal lymphoid follicle and the FAE is a subepithelial dome region filled with lymphocytes and DCs. Antigens and pathogens transported across the FAE by M cells are efficiently

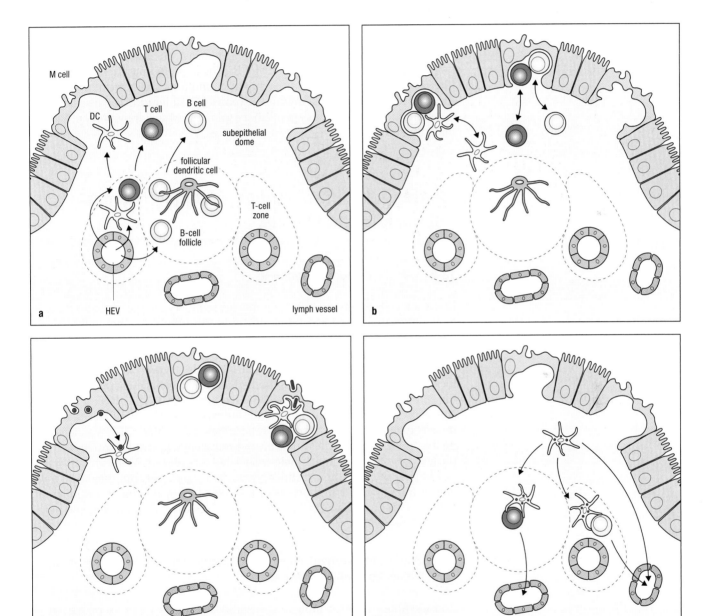

Figure 13.5 Cellular movements in organized mucosal lymphoid tissues are complex and only partly understood.
In panel a, naive and memory lymphocytes (B and T cells) and immature dendritic cells (DCs) enter the mucosa through high endothelial venules (HEV). Many of these cells are attracted to the subepithelial dome region by chemokines released from the follicle-associated epithelium. Panel b shows some B and T cells migrating into M-cell pockets where they express maturation or memory markers; the ultimate fate of these lymphocytes is not known. Most immature DCs remain in the subepithelial dome region but a few migrate into the FAE. In panel c, antigens and microorganisms transported by M cells are captured by DCs. In panel d, antigen capture along with other signals induces DC maturation and movement into interfollicular T-cell areas where they process and present antigen to naive T cells. DCs may also enter the B-cell follicle or migrate into draining lymphatics. (Adapted from M. Neutra and P. Kozlowski *Nat. Rev. Immunol.*, 6:148–158, 2006. With permission from Macmillan Publishers Ltd.)

captured by immature DCs in the pocket and subepithelial dome region. Thus, the site of initial antigen entry via M cells and capture by DCs occurs in close proximity to organized T- and B-cell zones.

Antigens and pathogens are endocytosed by immature DCs in the subepithelial dome region and translocated by DCs the short distance to the adjacent interfollicular T-cell zones (Figure 13.5), where DCs express maturation markers and antigen presentation occurs. Some DCs enter lymphatic vessels, migrating to the nearest draining lymph node, such as a mesenteric node, to present antigen in the context of information from broader mucosal areas and interface with the systemic immune system. DC migration is accelerated by uptake of live bacterial pathogens and enterotoxins, materials that are known to induce local cytokine and chemokine 'alarms,' but not by uptake of inert nonliving particles. DC migration and the location and nature of subsequent intercellular interactions are important determinants of the location and quality of the resulting mucosal immune response. DC migration can also facilitate dissemination of infectious agents that exploit the M-cell transport pathway to enter the mucosa.

Formation, differentiation, and maintenance of the FAE and M cells.

13-8 FAE is continuously renewed.

Epithelial cells emerging from multiple adjacent crypts migrate onto the surface of the underlying lymphoid follicle to form the FAE. The epithelial cell kinetics and sloughing at the crest of the dome resemble that of villus epithelial cells. Crypt cells destined for the FAE follow a distinct differentiation program, becoming FAE enterocytes and M cells (Figure 13.1a). Thus, a crypt located between a mucosal lymphoid follicle and a villus must provide cells to the villus on one side and the FAE on the other. In the crypt, cells on the villus side express the polymeric Ig receptor for export of IgA and differentiate into the cell types typical of the villus (enterocytes and goblet cells), whereas cells on the opposite wall of the same crypt do not secrete IgA and gradually acquire the features of M cells and follicle-associated enterocytes. M cells in the crypts can be identified by their distinct glycosylation patterns, but they do not yet have intraepithelial lymphocytes or pockets. As they emerge from the crypt, differentiating M cells form a pocket, acquire resident lymphocytes, and begin endocytic activity.

13-9 Signals from organized mucosal lymphoid tissues are required for induction of FAE and M cells.

Considerable evidence indicates that the distinctive patterns of FAE gene expression and M-cell differentiation are dependent on cells of the underlying follicle. Indeed, new sites of FAE and M cells appear *in vivo* where mucosal lymphoid follicles assemble, as observed when new isolated lymphoid follicles form in response to microbial challenge and experimentally by local injection of Peyer's patch cells into the mucosa. Factors produced by mucosal follicles appear to act very early in the differentiation pathway, inducing crypt cells to commit to FAE phenotypes. Evidence also indicates that during pathogen challenge from the lumen, some FAE enterocyte-like cells can be converted within hours to antigen-transporting M cells. Factors that signal crypt cells to follow the FAE differentiation program are discussed next.

The concept that lymphocytes play a role in differentiation of the FAE originated from indirect evidence. Immunodeficient SCID (severe combined immunodeficiency) mice, which lack both B and T lymphocytes, have no mucosal B-cell follicles, as well as no FAE or identifiable M cells. However, mucosal follicles with FAE appeared after adoptive transfer of Peyer's patch cells, especially when the transfer was enriched in B cells. B cell-deficient mice have T lymphocyte clusters in the mucosa and markedly reduced numbers of M cells. In contrast, T cell-deficient mice have small but recognizable Peyer's patches with follicles, FAE, and M cells. Some of the cytokine signals required for differentiation of FAE have been identified in mice with defined genetic disruptions. For example, lymphotoxin (LT) single-knockout and tumor necrosis factor (TNF)/LT double-knockout mice do not form Peyer's patches, indicating a role for LT and possibly TNF-α in Peyer's patch assembly. RANKL, a TNF family member produced by stromal cells, also is likely involved in M-cell differentiation.

The apparent requirement of lymphocytes and their products for the induction of FAE has been elucidated *in vitro* with the Caco-2 epithelial cell line. When grown on permeable filters, Caco-2 cells form polarized monolayers that resemble normal villus enterocytes with tight junctions, microvilli without a brush border glycocalyx, and variably expressed enterocyte enzymes and other molecules. When mouse Peyer's patch lymphocytes or cloned human B cells were added to the basolateral chamber of these cultures, some lymphocytes migrated through the holes in the filter and into the epithelium; and some of the Caco-2 cells showed M cell-like changes, including sparse microvilli, loss of the surface enzyme sucrase-isomaltase, and increased endocytic activity for the transport of latex particles and bacteria from the apical to the basolateral compartment. The addition of LT, TNF-α, and other factors to the culture medium also induced M cell-like changes in endocytic activity. Although the cells induced in this way do not resemble normal FAE M cells in all respects, the system has proven useful for elucidating the factors that differentiate the FAE and for identifying M cell-specific genes.

13-10 New organized mucosal lymphoid tissue and FAE form in response to antigenic challenge.

Solitary or isolated lymphoid follicles and associated T-cell clusters occur in many mucosal locations and are common throughout the gastrointestinal tract (Figure 13.6). The formation of these structures depends on exposure to antigens and microorganisms, as they appear only after birth and increase in number after microbial challenge. In the human digestive tract, isolated lymphoid follicles are most frequent where microbial populations are abundant such as the large intestine, cecum, and rectum. Isolated lymphoid follicles are common at mucocutaneous transitions such as the anal–rectal junction and near the ducts of secretory glands that empty onto mucosal surfaces. Isolated lymphoid follicles also form in the trachea and bronchi under conditions of antigenic challenge. Isolated lymphoid follicle formation involves a distinct sequence of events, including differentiation of local stromal cells, entry of migratory dendritic cells as well as distinct lymphocyte populations, and induction of a follicle-associated epithelium. The 'mature' isolated lymphoid follicle structure is functionally analogous to Peyer's patches for antigen sampling and initiation of immune responses.

Exploitation of M cells by microorganisms.

In addition to M-cell delivery of microorganisms into Peyer's patches for induction of IgA production and maintenance of a normal microflora in the

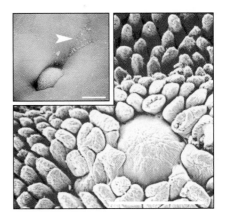

Figure 13.6 Isolated lymphoid follicles are covered by the FAE and M cells. This scanning electron micrograph of mouse intestine shows a single organized mucosal follicle surrounded by villi, as viewed from the lumen. (Scale bar: 500 μm.) The inset is a higher magnification view, showing the distinctive surface of an M cell (arrowhead). (Scale bar: 5 μm.) (Adapted from R. Lorenz and R. Newberry, *Ann. NY Acad. Sci.*, 1029:44–47, 2004. With permission from John Wiley & Sons.)

intestine, M-cell delivery of nonpathogenic bacteria into the rudimentary Peyer's patches of neonates contributes to maturation of the entire mucosal immune system. The potential harm of local infection is minimized by the translocation of microorganisms across the epithelial barrier via M cells at immune inductive sites, where phagocytes are abundant. Usually bacterial and viral pathogens transported by M cells are immediately killed by phagocytes and other innate components of the organized lymphoid tissues, or are cleared from the mucosa by the ensuing adaptive mucosal immune response. However, certain bacterial and viral pathogens successfully exploit this system to establish a mucosal infection and/or spread systemically and cause disease. In this section we discuss M cell–pathogen interactions involved in the initiation or control of disease.

13-11 M-cell transport of noninvasive, surface-colonizing bacterial pathogens can limit the duration of diarrheal disease.

Some pathogenic bacteria such as *Vibrio cholerae* and enterotoxigenic *Escherichia coli* cause severe diarrheal disease without invasion. *V. cholerae*, for example, adhere to the epithelial glycocalyx using adhesive toxin-co-regulated pili, form dense colonies, and then upregulate production of cholera toxin that diffuses through the glycocalyx to bind to a common glycolipid in enterocyte apical membranes. Endocytosis of the toxin initiates complex intracellular events that result in passage of the enzymatic toxin subunit into the cytoplasm, production of cyclic AMP, and apical secretion of chloride ions that drive massive loss of water into the intestinal lumen. Some vibrios, however, adhere to M cells and are rapidly endocytosed, transported into Peyer's patches, and processed by DCs (Figure 13.7). The resultant adaptive immune response leads to the secretion of IgA antibodies to *Vibrio* pili, lipopolysaccharide, and cholera toxin. These antibodies participate in the clearance of the adherent colonies and protection against subsequent challenge.

13-12 Invasive bacterial pathogens can exploit rapid M-cell transport to establish local and systemic infection.

M cells transcytose most particles that adhere to their surface. In the transcytotic vesicles, content may be acidified and undergo protease activity, but transport is rapid, and some microbes survive to infect the epithelium, underlying phagocytes, and other mucosal cells (Figure 13.7).

Salmonella species are able to infect many cell types, including epithelial cells and phagocytes. In humans, *Salmonella enterica* serovar Typhimurium causes a mucosa-limited diarrheal disease, but in mice the bacteria spread systemically and cause a lethal septicemia that closely mimics typhoid fever caused by *Salmonella enterica* serovar Typhi in humans. After adhering to the host-cell membrane, the bacteria use a type III secretion system to deliver proteins into the host cell, inducing massive actin reorganization and macropinocytosis, a process that disrupts normal cell architecture as bacteria are internalized. Because the FAE and M cells are relatively accessible, intestinal *Salmonella* preferentially adhere to M cells in both mice and humans, rapidly invade Peyer's patches, and are taken up by subepithelial DCs. Uptake of *Salmonella* has cytotoxic effects that can result in M cell death. In mice, some *Salmonella* may also be taken up from villus surfaces, captured by DCs that send extensions through tight junctions. *Salmonella enterica* serovar Typhimurium survive in special vacuoles in phagocytes and are carried by these migrating cells via the lymphatic vessels to the spleen, causing murine typhoid fever.

Shigella infections cause widespread mucosal damage to the colonic mucosa, but the bacterium does not spread systemically. Entry into the colonic mucosa

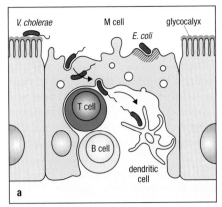

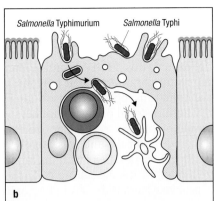

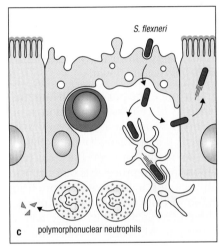

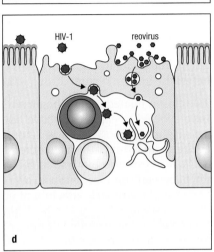

Figure 13.7 Microorganisms exploit M cells to cross the intestinal epithelial barrier. Panel a shows *Vibrio cholerae*, a noninvasive pathogen, adhering to the periphery of the glycocalyx of enterocytes (left) to form toxin-secreting colonies. In contrast, *V. cholerae* that adhere to M cells make close contact with M-cell apical membranes and are endocytosed and delivered to antigen-presenting cells. *Escherichia coli* RDEC-1 induces stable, actin-rich pedestals on M cells. In panel b, *Salmonella enterica* serovar Typhimurium and *Salmonella enterica* serovar Typhi adhere preferentially (but not exclusively) to M cells and initiate signal transduction events that alter the M cell apical cytoskeleton, causing active ruffling of the M-cell surface and uptake of bacteria by macropinocytosis. *Shigella flexneri*, in panel c, exploits the transport activity of M cells to enter the intestinal mucosa and invade the basolateral surfaces of epithelial cells. This invasion may trigger recruitment of inflammatory cells and local cytokine release, which in turn result in disruption of the epithelial barrier. In panel d, reovirus adheres selectively to M cells in mice, and transcytosis delivers the virus to the basolateral side where it can infect other epithelial target cells. Human immunodeficiency virus-1 (HIV-1) applied to mouse or rabbit mucosa adheres to epithelial cell surfaces. Virus adherent to M cells may be transcytosed into the intraepithelial pocket and subepithelial region, where it may infect target cells. (Adapted from M. Neutra et al., *Cell* 86:345–348, 1996. With permission from Elsevier Ltd.)

is highly efficient, as only a few *Shigella* organisms ingested orally are required to initiate infection. Studies in experimental animals showed that *Shigella* initially use M-cell transport to gain access to the basolateral side of the FAE and adjacent epithelium. Other studies on epithelial cells *in vitro* showed that *Shigella* are unable to infect enterocytes at the apical surface but readily attach to and are endocytosed at the basolateral membrane. The bacteria rapidly spread from cell to cell, usurping the host-cell cytoskeleton and signaling networks. Initial uptake into phagosomes occurs through a macropinocytic process that is induced by secretion of bacterial proteins through a type III secretory apparatus. *Shigella* then lyse the phagosome membrane and escape into the cytoplasm, where they initiate a complex actin-based motility process that propels the bacterium through the host-cell cytoplasm, form torpedo-like extensions on the lateral cell surface, and induce endocytosis by the neighboring cell. While spreading from cell to cell, *Shigella* induce activation of nuclear factor κB (NFκB), resulting in interleukin-8 production by epithelial cells and the rapid influx of polymorphonuclear leukocytes into the lamina propria. The resultant inflammatory process leads to disruption of the epithelial barrier, facilitating widespread *Shigella* invasion.

13-13 Endocytosis of viruses by M cells results in mucosal and/or systemic disease.

Viruses that adhere to cell surfaces may invade the host cell through endocytic pathways. Viruses adherent to M-cell surfaces are efficiently transcytosed through the cell. So far, only a few viruses are known to adhere specifically to M cells. Other viruses adhere to many epithelial cell surfaces but only M cells mediate their transcytosis. Still other enteric viruses exploit epithelial receptors that are located basolaterally, necessitating entry via epithelial breaks or capture by protruding intraepithelial DCs.

Indirect evidence suggests that M cells may be involved in human immuno-deficiency virus-1 (HIV-1) and simian immunodeficiency virus (SIV) entry. In the rectum and recto-anal junction, organized lymphoid tissues and M cells are numerous, possibly contributing to rectal transmission of HIV-1 and SIV. SIV administered orally to monkeys was detected first in organized lymphoid tissues of the tonsils. HIV-1 applied to mouse and rabbit Peyer's patch explants adhered to M cells and was rapidly endocytosed and transported across the epithelial barrier. Cells with morphological and functional features of M cells generated by the co-culture of Caco-2 cells and B cells transported HIV-1 in a galactosylceramide- and chemokine receptor-dependent manner. In rabbit Peyer's patch explants, some HIV-1 particles adhered to the glycocalyx on enterocytes but did not contact the cell membrane and were not endocytosed. Epithelial cells are not infected by these viruses but HIV-1 entry via M cells would provide rapid access to DCs and activated T cells in the M-cell pocket, and would facilitate DC delivery of virus to T cells. However, whether HIV-1 enters the human rectum through M cells *in vivo* is not known.

Two closely related viral pathogens, reovirus in mice and poliovirus in humans, have evolved to exploit selective adherence to M cells as an invasion strategy (Figure 13.7). Reovirus is a non-enveloped virus that adheres only to M cells and uses this pathway to enter the Peyer's patch mucosa of mice (Figure 13.4). Reovirus also binds to M cells in the colon and airways. The virus uses the pancreatic proteases trypsin and chymotrypsin to process its outer capsid and extend its trimeric attachment protein, sigma 1, a distance of 45 nm from the viral surface. Reovirus type 1 binding to M cells is mediated by interaction between the extended sigma 1 protein and a specific sialic acid-containing trisaccharide on the M-cell apical membrane. Although this determinant is present on epithelial cells, viral particles cannot pass through the overlying glycocalyx. On M cells, adherent reovirus is taken up by clathrin-coated vesicles and is released into the intraepithelial pocket and subepithelial dome region where it is endocytosed by DCs. Although reovirus is unable to adhere to the apical membranes of enterocytes in adult mice, it binds to a basolateral component near tight junctions called 'junction adhesion molecule-1' and proceeds to infect the entire epithelium from the basolateral side.

Poliovirus causes paralytic disease by infecting neurons, but it enters the body by the oral route. The ability of the virus to proliferate in Peyer's patches before spreading systemically suggested an M-cell entry strategy. Indeed, when explants of human Peyer's patch mucosa were exposed *in vitro* to wild poliovirus type 1, electron microscopy showed that the virus adhered selectively to M cells and then was endocytosed. The receptor for poliovirus on target cell membranes is CD155, a member of the immunoglobulin superfamily that mediates interactions between cells of the immune system; it is expressed on human intestinal epithelial cells, including M cells, primarily on basolateral but also on apical domains. That M-cell expression of CD155 is required for intestinal uptake of poliovirus is supported by studies in which transgenic mice that expressed CD155 on many cell types but not epithelial cells could be infected with injected but not orally administered poliovirus. The ability of poliovirus to exploit M-cell transport for uptake into Peyer's patches has lead to the testing of poliovirus-based vaccine vectors, including recombinant viral particles or pseudovirus particles, for mucosal delivery of foreign antigens.

13-14 M cells can be exploited for vaccine antigen delivery.

Given the importance of organized mucosal lymphoid tissues for the initiation of mucosal immune responses, there is now great interest in the delivery of vaccine antigens to these sites. However, vaccines administered orally or through other mucosal routes encounter host defenses, including mucus, proteases, and nucleases at the epithelial barrier, and dilution by gastrointestinal secretions. In addition, the marked infrequency of M cells reduces the likelihood of contact between an M cell and vaccine antigen. Consequently, much larger vaccine doses are usually required for oral immunization compared with injection. Also, endocytic and transcytotic activity in the FAE is efficient only for macromolecules and particles that adhere to M-cell apical membranes. In contrast, soluble, nonadherent antigens are taken up at low levels in the intestine and generally induce immune tolerance. Live or attenuated bacteria and attenuated viruses that adhere to mucosal surfaces, especially those that normally use M cells to invade organized mucosal lymphoid tissues, could theoretically make ideal mucosal vaccines or vaccine vectors. The biology of these vaccines and vectors, including the advantages and disadvantages of each, is discussed in Chapter 27.

Summary.

A specialized follicle-associated epithelium covers mucosal lymphoid follicles. The FAE contains a unique epithelial cell type, the M cell. M-cell differentiation is induced by cytokines from cells in organized mucosal lymphoid tissues. The major function of M cells is to transport samples of macromolecules, particles, and microorganisms across the epithelial barrier from the lumen to organized mucosal lymphoid tissues. A key feature of M cells is the invagination of the basal surface to form an intraepithelial 'pocket' that contains activated lymphocytes and DCs. M cells display apical surface features that promote adherence of microorganisms, and basolateral membrane proteins that promote lymphocyte docking in the intraepithelial pocket. After M-cell transport, antigens and pathogens are endocytosed by DCs that subsequently migrate to the adjacent interfollicular T-cell zones where antigen presentation occurs. The transepithelial transport activity of M cells is a 'double-edged sword,' playing a key role in initiating protective mucosal immune responses but also providing microbes with a rapid entry route into the mucosa. Thus, M cells are important in the pathogenesis of certain bacterial and viral diseases and could be exploited in mucosal vaccine delivery systems.

Further Reading.

Anderle, P., Rumbo, M., Sierro, F., *et al.*: **Novel markers of the human follicle-associated epithelium identified by genomic profiling and microdissection.** *Gastroenterology* 2005, **129**:321–327.

Debard, N., Sierro, F., Browning, J. *et al.*: **Effect of mature lymphocytes and lymphotoxin on the development of the follicle-associated epithelium and M cells in mouse Peyer's patches.** *Gastroenterology* 2001, **120**:1173–1182.

Frey, A., Lencer, W.I., Weltzin, R., *et al.*: **Role of the glycocalyx in regulating access of microparticles to apical plasma membranes of intestinal epithelial cells: implications for microbial attachment and oral vaccine targeting.** *J. Exp. Med.* 1996, **184**:1045–1060.

Fukuoka, S., Lowe, A.W., Itoh, K., *et al.*: **Uptake through glycoprotein 2 of FimH(+) bacteria by M cells initiates mucosal immune response.** *Nature* 2009, **462**:226–230.

Gebert, A., Fassbender, S., Werner, K., *et al.*: **The development of M cells in Peyer's patches is restricted to specialized dome-associated crypts.** *Am. J. Pathol.* 1999, **154**:1573–1582.

Helander, A., Silvey, K.J., Mantis, N.J., *et al.*: **The viral σ1 protein and glycoconjugates containing α2-3-linked sialic acid are involved in type 1 reovirus adherence to M cell apical surfaces.** *J. Virol.* 2003, **77**:7964–7977.

Herbrand, H., Bernhardt, G., Förster R., *et al.*: **Dynamics and function of solitary intestinal lymphoid tissue.** *Crit. Rev. Immunol.* 2008, **28**:1–13.

Kelsall, B.: **Recent progress in understanding the phenotype and function of intestinal dendritic cells and macrophages.** *Mucosal Immunol.* 2008, **1**:460–469.

Kraehenbuhl, J.P., and Neutra, M.R.: **Epithelial M cells: differentiation and function.** *Annu. Rev. Cell Dev. Biol.* 2000, **16**:301–332.

Lorenz, R.G., and Newberry, R.D.: **Isolated lymphoid follicles can function as sites for induction of mucosal immune responses.** *Ann. N. Y. Acad. Sci.* 2004, **1029**:44–57.

MacPherson, A.J., McKoy, K.D., Johansen F.E. *et al.*: **The immune geography of IgA induction and function.** *Mucosal Immunol.* 2008, **1**: 11–22.

Neutra, M.R., and Kozlowski, P.A.: **Mucosal vaccines: the promise and the challenge.** *Nat. Rev. Immunol.* 2006, **6**:148–158.

Neutra, M.R., and Kraehenbuhl, J.P.: **Immune defense at mucosal surfaces**, in Kaufmann, S.H.E., Sher, A., Ahmed, R. (eds): *Immunology of Infectious Diseases*. Washington D.C., American Society for Microbiology, 2010.

Neutra, M.R., Mantis, N.J., and Kraehenbuhl, J.P.: **Collaboration of epithelial cells with organized mucosal lymphoid tissues.** *Nat. Immunol.* 2001, **2**:1004–1009.

Neutra, M.R., Sansonetti, P., and Kraehenbuhl, J.P.: **Interactions of microbial pathogens with intestinal M cells**, in Blaser, M., Smith, P.D., Ravdin, J.I., *et al.* (eds): *Infections of the Gastrointestinal Tract*, 2nd ed. New York: Lippincott Raven, 2002: 141–156.

Terahara, K., Yoshida, M., Igarashi, O., *et al.*: **Comprehensive gene expression profiling of Peyer's patch M cells, villous M-like cells, and intestinal epithelial cells.** *J. Immunol.* 2008, **180**:7840–7846.

Lymphocyte Trafficking from Inductive Sites to Effector Sites

14

The initiation, maintenance, and resolution of innate and adaptive immune responses are critically dependent on immune cell migration, not only within tissues but often over long distances between organs. This process is highly dynamic and requires that immune cells interact with multiple vascular, lymphatic, and tissue environments in a tightly controlled and organized fashion.

Most infections will initially constitute a small number of pathogens and be localized to a small tissue area. If naive lymphocytes were tissue-resident cells dispersed at random throughout the body, the chances that a given antigen-specific lymphocyte would be at the right site—that is, at the site of pathogen entry—would be very slim. In a rough calculation, we might estimate that a T lymphocyte with a diameter of 10 μm and a volume of about 0.5×10^{-18} m^3 occupies less than 10^{-15}% of the body volume. Even if we suppose that several hundred T lymphocytes recognize a distinct antigen, the odds of an antigen-specific lymphocyte's being in the same location as a pathogen remain extremely low. Lymphocyte recirculation solves this problem by enabling the entire T lymphocyte population to scan antigen-presenting cells within lymphoid compartments.

Perhaps nowhere is this more obvious than during the induction of adaptive immune responses. Adaptive immunity is initiated when antigen-presenting cells, primarily dendritic cells (DCs), present antigen to lymphocytes in inductive immune compartments, such as lymph nodes and Peyer's patches. During mucosal immune responses, antigen-bearing DCs migrate from mucosal tissues through draining afferent lymph vessels to regional lymph nodes and into the lymph node T-cell zone. Conversely, to find a DC presenting relevant cognate antigen: major histocompatibility complex (MHC), naive lymphocytes continually traffic from the bloodstream into lymph nodes and back via efferent lymph to the venous blood. In combination, these migratory routes of DCs and lymphocytes allow frequent contact of both cell types and thus form the basis for the efficient induction of adaptive immune responses.

Besides the constitutive recirculation of naive lymphocytes, immune cells need to be directed to sites of inflammation. This holds true for cells of the innate immune system, including monocytes and granulocytes, which are rapidly recruited to the inflamed tissue during the initial phase of an immune reaction, and also for effector T cells and plasma cells, which are generated in the adaptive immune response.

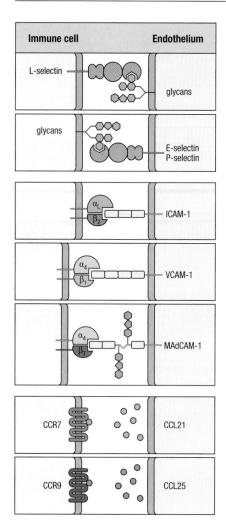

Figure 14.1 Adhesion molecules involved in immune-cell migration. Cell-adhesion molecules involved in immune-cell migration fall into three large groups: selectins, integrins, and members of the immunoglobulin superfamily of adhesion receptors. These adhesion molecules, together with chemokine/chemokine receptors, are expressed on immune cells and endothelial cells. Selectins bind sulfated glycans provided by a scaffold of glycoproteins. Integrins consist of an α and a β chain, forming heterodimers that bind to immunoglobulin superfamily members such as ICAM-1.

Basic concepts in immune-cell migration.

Cell adhesion molecules (CAMs) mediate intercellular interactions and cellular interactions with the extracellular environment. It is thus perhaps not surprising that CAMs also have essential roles in regulating immune cell function and, as described in detail below, form the basis for regulated immune-cell migration. One of the best-studied examples of such an interaction is that between immune cells and the endothelial cells that line blood vessels. We will therefore briefly describe the CAMs and chemoattractants stabilizing the interaction of immune cells with the microvascular endothelium. As we discuss later in this chapter, the dynamic regulation of these molecules permits the directed migration of immune cells and offers strategies to modulate the migration of immune cells therapeutically.

14-1 The adhesion molecules and chemoattractants involved in immune-cell migration.

CAMs involved in immune-cell migration fall into three main families: the selectin, integrin, and immunoglobulin gene superfamily of adhesion receptors. Although the precise molecules involved in immune cell adherence depend on the immune cell type, the tissue, and the inflammatory environment under study, generally each CAM needs to bind to its specific ligand expressed on the immune cell and endothelial cell respectively (Figure 14.1).

Selectins are glycoproteins of the C-type lectin group that bind to specific carbohydrate determinants on selectin ligands. Mammals express three types of selectins: L-selectin, P-selectin, and E-selectin. L-selectin is expressed by many types of immune cells and interacts with ligands expressed by vascular endothelial cells. Conversely, E- and P-selectin are expressed by endothelial cells, and the interacting carbohydrate selectin ligands can be expressed by immune cells. Selectin ligands comprise a heterogeneous group of molecules that frequently, but not exclusively, carry the sulfated tetrasaccharide sialyl-Lewis^x. Thus, the generation of functional selectin ligands depends on the post-translational modification of proteins by glycosyltransferases. Selectins bind their ligands with low affinity, facilitating the capture of immune cells freely flowing in the blood and resulting in their rolling along the vascular endothelium.

The subsequent strong adhesion of immune cells to vascular endothelial cells is mediated by integrins and their interaction with ligands of the immunoglobulin gene superfamily. Integrins control a wide array of cellular functions, including cell growth, migration, apoptosis, and differentiation. They are heterodimers comprising one α chain and one β chain. Both chains exist in multiple forms that, through dimerization, give rise to more than 20 distinct integrin-family members. All major integrins involved in immune-cell migration contain the β_1, β_2, or β_7 integrin chain and are expressed on immune cells, whereas their respective ligands are expressed on the vascular endothelium. Well-known integrin ligands include intercellular adhesion molecules-1 and 2 (ICAM-1, ICAM-2), vascular cell adhesion molecule-1 (VCAM-1), and mucosal addressin cell adhesion molecule-1 (MAdCAM-1) (see Figure 14.1).

Although ligation of selectins and integrins can activate intracellular signaling pathways, the chief molecules that mediate the critical signals driving immune-cell migration and arrest on vascular endothelial cells belong to the chemokine family. Chemokines are a large family of low-molecular-weight proteins (about 8–12 kDa) that are classified into groups depending on the positioning of cysteine residues at their amino terminus. The two major subgroups of chemokines are the CC and CXC chemokines, containing adjacent cysteine residues or cysteine residues separated by a single amino acid (X) respectively. Chemokines can be produced by a broad spectrum of cells and signal through seven-transmembrane domain G-protein-coupled receptors. Many of these receptors are promiscuous in that they can bind more than one chemokine. However, such promiscuity does not extend across CC or CXC chemokine-family members. Thus, chemokine receptors binding CC chemokine ligands (CCLs) are termed CCRs, and those binding CXCL chemokines are termed CXCRs. A major consequence of chemokine receptor activation in immune cells is alterations in integrin structure that occur within seconds of chemokine binding, through a process termed 'inside-out signaling,' and lead to an increased affinity of targeted integrins for their cellular and extracellular ligands. Together with chemokine-induced integrin clustering, these conformational changes result in a stable binding of the immune cell to the vascular endothelium. An additional important property of chemokines is their ability to bind proteoglycans, and chemokines are thought to be 'presented' in this form on the surface of vascular endothelium to circulating immune cells.

14-2 Immune-cell migration into tissues is a multi-step process.

A general model has been proposed to describe the entry of immune cells into tissue (Figure 14.2). This model is known as the 'multi-step adhesion cascade' and postulates the sequential involvement of several adhesion pathways mediated by selectins, chemokines, integrins, and their respective ligands. These molecular interactions enable the cells (1) to tether to the endothelial surface, resulting in a slow rolling of the cell on the endothelium; (2) to undergo an activation step enabling the cells to arrest stably on the endothelium; and (3) to traverse the endothelial layer in a process known as diapedesis. In what follows, we discuss first, as a paradigmatic example, the entry of naive T lymphocytes into lymph nodes, and subsequently the variations of this process as they occur in the mucosal immune system.

The entry of T cells from the blood into lymph nodes occurs across high endothelial venules (HEVs). HEVs are specialized postcapillary venules that constitutively express the CAMs and chemokines needed for the recruitment of naive lymphocytes, namely the molecules mediating the distinct steps of the multi-step adhesion cascade. During the first step, 'tethering and rolling,' L-selectin expressed on naive T cells engages oligosaccharides containing sialyl-Lewis[x] residues present on the luminal side of the HEV. This interaction is too weak to arrest the cells but it reduces their speed and manifests in their rolling along the HEV. The second step, 'arrest,' is initiated by chemokines

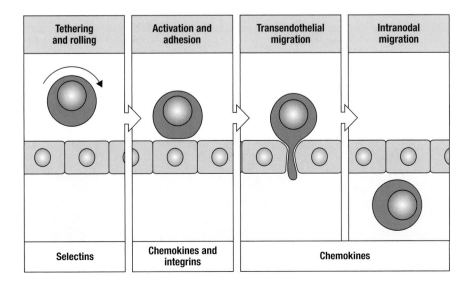

Figure 14.2 The multi-step adhesion cascade. Immune-cell entry into tissues is a multi-step process coordinated by tissue-homing molecules. The sequential interaction of cell adhesion molecules expressed on immune cells and specialized high endothelial venules allows the cells to tether and roll along the endothelium, to bind stably to the vascular endothelium, to transmigrate, and eventually to follow chemotactic gradients to localize to distinct subcompartments inside the lymph node.

presented on the HEV. In the context of T-cell homing into lymph nodes, the most important chemokine is CCL21, which binds to the chemokine receptor CCR7 expressed on the surface of all naive T cells. Signaling through the CCR7 receptor induces a high-affinity conformation of the $\alpha_L\beta_2$ integrin (also referred to as lymphocyte function-associated antigen-1 (LFA-1)) on the T cells. Activated $\alpha_L\beta_2$ integrin binds to ICAMs constitutively expressed on the HEV and mediates a stable arrest of the cells on the HEV that initiates the step of transendothelial migration. Thus, the constitutive expression of L-selectin, $\alpha_L\beta_2$ integrin, and CCR7 by naive T lymphocytes and their interaction with glycans, CCL21, and ICAMs constitutively present on HEVs permits their entry into lymph nodes (Figure 14.3). At the same time, naive lymphocytes are largely excluded from peripheral tissues such as the skin, lung, or gut, simply because in those tissues the essential molecular interaction partners required for the homing of naive lymphocytes are not expressed. We discuss later in this chapter the distinct patterns of CAMs and chemokine/chemokine receptors that direct immune cells into mucosal tissues and how changes in CAM expression come about.

Figure 14.3 Immune cells can use tissue-specific homing receptors to direct their entry into different tissues. Tissue-specific and immune-cell-specific expression of homing receptors enable rolling, activation, and adhesion. In peripheral lymph node high endothelial venules (HEVs), rolling is mediated by L-selectin interacting with sulfated glycans decorating CD34 and GlyCAM-1. In Peyer's patches and intestinal venules, rolling can be achieved by the interaction of nonactivated $\alpha_4\beta_7$ integrin with MAdCAM-1, whereas in mesenteric lymph nodes, $\alpha_4\beta_7$ integrin and L-selectin both contribute. Activation of integrins in HEVs is achieved by CCL21 signaling via CCR7, whereas in intestinal venules CCL25 signaling via CCR9 leads to integrin activation. Stable adhesion in HEVs is dominated by activated $\alpha_L\beta_2$ integrin interacting with ICAM, and $\alpha_4\beta_7$ integrin binding to MAdCAM-1 in intestinal venules.

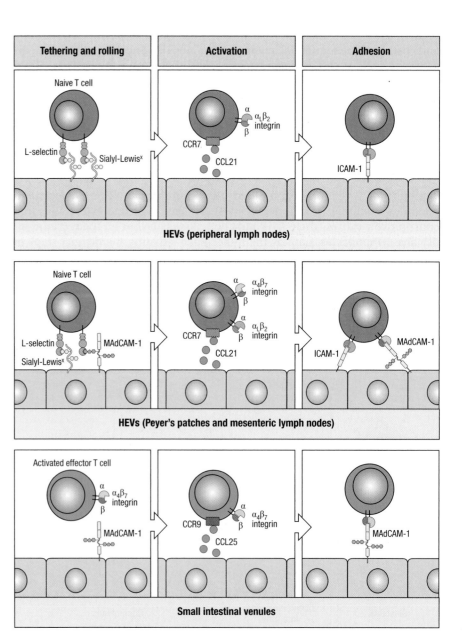

The CAMs and chemokines directing the immune response to distinct tissues have been studied in considerable detail. A particularly important technical approach in this regard had been the use of adoptive cell transfer, meaning the transfer of immune cells from donor animals to syngeneic or congenic recipients. Early experiments isolated immune cells from different compartments and studied their distribution after intravenous transfer into recipient animals. Results from such experiments revealed that lymphocytes derived from mucosal lymph nodes preferentially relocated to such lymph nodes in the recipient animals. Conversely, lymphocytes isolated from skin-draining lymph nodes retained a preference for the skin-draining lymph nodes of recipients. Similarly, plasma cells producing immunoglobulin A (IgA) but not IgG accumulated at mucosal sites such as the intestine and lung. These experiments demonstrated that the homing cues must differ between mucosal and other tissues and led to the concept of a distinct mucosal immune system, which was thought to differ from the systemic immune system principally in respect of cell homing properties, cellular composition, and function. In the light of the multistep adhesion cascade and with the forthcoming use of gene-deficient mice and selective CAM inhibitors, differential homing properties could be tracked down to the divergent expression of selectins, integrins, and chemokines by different immune cells and tissues. It is now apparent that immune cells are equipped with unique 'address codes.' Comparing mucosal and nonmucosal compartments, these codes show both differences and overlaps. Thus, the concept of a distinct mucosal immune system has been replaced with a refined definition of distinct homing codes, encoded chiefly by selectins, chemokines, and integrins. Still, we are far from understanding the mechanisms regulating the distinct homing properties of immune cells in all their detail, and it seems likely that in the future we will need to add further molecular players for an adequate description of the homing properties of immune cells.

14-3 Naive lymphocytes recirculate through MALT.

Mucosa-associated lymphoid tissue (MALT), such as the Peyer's patches and the appendix in humans, or follicular structures present in inflamed lungs, together with the mucosa-draining lymph nodes, serve as the inductive sites of mucosal adaptive immune responses. It is now clear that CAM expression on HEVs in certain MALT structures differs from that of other HEVs, for example those in skin-draining lymph nodes, and that selective CAM expression has an important impact on the mechanisms by which naive lymphocytes gain entry into these sites (Figure 14.4). The best example of this is the Ig superfamily member MAdCAM-1, which is constitutively expressed on HEVs in the gut-draining mesenteric lymph nodes and Peyer's patches but not on the HEVs of most extraintestinal lymph nodes. MAdCAM-1 can serve as a scaffold for sulfated glycans, thereby enabling the engagement of L-selectin, and can interact weakly with low-affinity $\alpha_4\beta_7$ integrin. Both interactions contribute to lymphocyte rolling on MAdCAM-1$^+$ venules. Subsequent to chemokine signaling, $\alpha_4\beta_7$ integrin binds MAdCAM-1 with high affinity, resulting in lymphocyte arrest on these HEVs. Interactions of $\alpha_4\beta_7$ integrin with MAdCAM-1 seem particularly important in mediating the entry of naive T lymphocytes into Peyer's patches, as demonstrated by the presence of hypocellular Peyer's patches in mice deficient for MAdCAM-1 or β_7 integrins, whereas mice with a deletion in both L-selectin and β_7 integrin display hypocellular gut-draining mesenteric lymph nodes. Thus, the constitutive expression of MAdCAM-1 and its binding to β_7 integrins are required for the efficient homing of naive lymphocytes into the mesenteric lymph nodes and gut-associated lymphoid tissue (GALT) (see Figure 14.3).

The homing of T and B cells to mucosal lymph nodes and Peyer's patches, as in peripheral lymph nodes, uses CCL21/CCR7 signaling (and can be helped by the chemokine receptor CXCR4). In addition to the CCL21-decorated HEVs

Figure 14.4 Tissue-specific expression of cell adhesion molecules and chemokines. Although immune cells can regulate the expression of cell adhesion molecules and chemokine receptors, their destination is determined by the divergent expression of the respective receptors and ligands in the target tissues. Constitutive tissue-specific expression of homing molecules is depicted in blue, and inflammation-triggered expression is depicted in red. The constitutive expression of ICAM and CCL21 in lymph nodes is critical in driving the homing of lymphocytes to lymph nodes, whereas the expression of MAdCAM-1 and CCL25 in the small intestine supports the entry of gut-homing immune cells. Under inflammatory conditions, VCAM-1, which serves as a ligand for $\alpha_4\beta_1$ integrin, becomes upregulated in numerous tissues including the lung and urogenital tract. In inflamed skin, upregulation of E- and P-selectin and VCAM on endothelium serves in the recruitment of skin-homing effector cells.

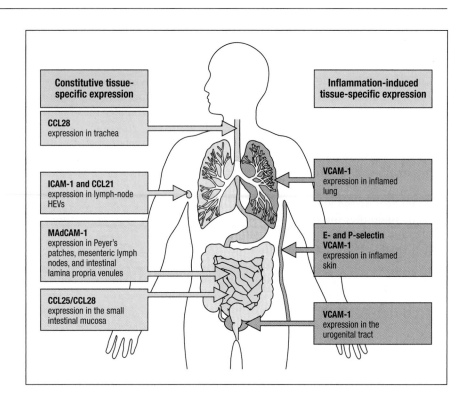

that are localized in the interfollicular T-cell zone of all lymph nodes, the B-cell follicles of Peyer's patches contain CXCL13-decorated HEVs. Naive B cells express the CXCL13 chemokine receptor CXCR5 and can use CXCL13/CXCR5 to gain direct entry into B-cell follicles in Peyer's patches. Such specialized HEVs are absent from lymph nodes and might have developed in response to the particular need of a B cell-rich compartment such as the Peyer's patches. Indeed, B cell-rich lymphoid tissues exist at various anatomical locations, including small follicular structures in the intestine, the rodent nasopharyngeal-associated lymphoid tissue, and the omentum (a unique streak of adipose tissue studded with numerous B cell-rich follicular structures known as milky spots), and CXCL13/CXCR5 contribute to B-cell entry at all these sites. These findings illustrate that, besides constitutive differences in the expression of integrin ligands on HEVs in intestinal inductive sites, the selective distribution of chemokines along HEVs can provide additional positional cues for migrating lymphocytes.

14-4 Lymphocytes traffic inside lymph nodes and leave to return to the blood.

On entry into the lymph node, lymphocytes are guided by chemokines into distinct anatomical localizations, most prominently the paracortical T-cell zone and the B-cell follicles (Figure 14.5). The backbone of the lymph node is made up of nonhematopoietic stromal cells, fibers, and extracellular matrix. Stromal cells constitute a heterogeneous array of nonhematopoietic cells that can be subdivided into endothelial cells (including cells forming the HEVs), follicular dendritic cells (FDCs) in the B-cell follicles, and fibroblastic reticular cells in the T-cell regions. Fibroblastic reticular cells are distributed throughout the T-cell zone, forming a three-dimensional network of cells that surround highly organized extracellular channels termed conduits. The conduits themselves are composed of a fibrous collagen core that is surrounded by a basement membrane, through which molecules of low molecular weight, including antigens, cytokines, and chemokines, can disperse. Of particular relevance to lymphocyte migration is the fact that conduits can channel chemokines to HEVs. Thus, HEVs can display chemokines produced at distant sites that are channeled to HEVs through the conduit system.

Pioneering technical advances in imaging and microscopy, and in particular the development of multiphoton microscopy, have allowed researchers to visualize the migratory paths of immune cells at depth within lymph nodes in living animals in real time. Such studies have provided fundamental insights into immune-cell migratory patterns and immune-cell interactions within lymph nodes. These techniques have demonstrated that naive T cells, after entering lymph nodes, are highly motile and actively scan antigen-presenting cells for their cognate antigens. Although early studies indicated that T-cell movement within the paracortical areas was random rather than directed, representing a 'random walk,' subsequent studies demonstrated that they migrate along the fibroblastic reticular cell network. Fibroblastic reticular cells express the CCR7 ligands CCL21 and CCL19, which seem to promote T-cell motility within the T-cell zone and thus the speed with which T cells interact with and scan the surface of resident DCs. In a similar fashion, naive B cells, although migrating on average more slowly than T cells, are thought to migrate along FDCs in the B-cell follicles in a process that is probably promoted through the CXCR5 ligand CXCL13.

In most cases, naive T cells will not find their cognate antigen:MHC on the surface of DCs and will leave the lymph nodes within a few hours via the efferent lymph. In contrast, if activated, T cells are initially retained within the lymph node, where they undergo rapid clonal expansion. A subset of these cells subsequently migrates via the efferent lymph back to the circulation, from where they can seed other organs. The underlying cellular and molecular mechanisms regulating the egress of lymphocytes from lymph nodes are only beginning to emerge. A key factor in the regulation of lymphocyte egress is the lysophospholipid sphingosine-1-phosphate (S1P). S1P concentrations are higher in blood and lymph than in lymph nodes, and these differential levels of S1P are needed to drive lymphocyte exit. S1P is produced by sphingosine kinases through the phosphorylation of sphingosine, and is degraded by sphingosine lyase. Thus, S1P concentrations in tissues, lymph, and blood are regulated by the differential activity of the respective S1P-producing and/or degrading enzymes as well as S1P release from intracellular stores. High S1P

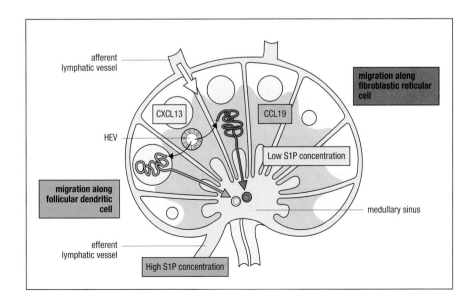

Figure 14.5 Intranodal migration of naive T and B cells. Naive lymphocytes enter lymph nodes via high endothelial venules. Within the lymph nodes, the chemokine CXCL13 guides B cells into the B-cell follicles, whereas T cells are guided by CCL19 into the T-cell zone. Within these subcompartments, lymphocytes migrate along stromal cells and scan resident antigen-presenting cells for their cognate antigens. In most cases, lymphocytes will not become activated and exit from the lymph nodes via sphingosine-1-phosphate (S1P) signaling.

levels in lymph are maintained by the expression of sphingosine kinase in lymphatic endothelial cells, whereas high S1P concentrations in blood rely on S1P release from red blood cells.

S1P binds to five G-protein-coupled receptors designated $S1P_1$ to $S1P_5$. Egress of naive T cells from lymph nodes depends on $S1P_1$, because $S1P_1$-deficient T cells fail to leave lymph nodes. $S1P_1$ is internalized on encountering high S1P concentration such as that found in blood and lymph, but is maintained on the cell surface in areas of low S1P concentration such as the lymph nodes. Thus, shuttling of $S1P_1$ between the cell surface and cytosol seems to regulate lymphocyte egress from lymph nodes, and the reappearance of $S1P_1$ on the cell surface during lymph node transit might override retention signals, possibly including CCL21–CCR7 interactions, and initiate lymphocyte egress. After their activation, T cells express the early T-cell activation marker CD69, a transmembrane protein of the C-type lectin family. Association between CD69 and $S1P_1$ induces $S1P_1$ internalization, leading to reduced responsiveness to S1P and inhibition of lymphocyte exit. Later, after extensive proliferation in the lymph node, T cells lose their expression of CD69 and reexpress $S1P_1$; as a consequence, they regain their ability to leave via efferent lymph.

Lymphocyte migration into mucosal tissues— generation of tissue-tropic lymphocyte subsets.

After recognizing their cognate antigen:MHC on the surface of antigen-presenting cells, activated T cells undergo extensive clonal expansion. In the context of appropriate co-stimulatory signals, T cells expand and differentiate into effector T cells, the functions of which seem to be determined, at least in part, by the environmental context in which antigen presentation and recognition take place. During this process, T cells alter their expression of CAMs and chemokine receptors, and thus their migratory potential.

A subset of CD4$^+$ T cells downregulate CCR7 and acquire expression of the chemokine receptor CXCR5, which allows these cells to migrate toward B-cell follicles and provide B cell help. Others, together with primed CD8$^+$ T cells, start to express CAMs and chemokine receptors involved in lymphocyte migration to extralymphoid tissues and sites of inflammation, allowing these cells to coordinate immune and inflammatory responses in peripheral sites, including mucosal tissues. Finally, a subset of T cells, termed central memory T cells, maintain their expression of CCR7 and CD62L, and thus a lymph node homing capability, and such cells probably contribute to the memory response in secondary lymphoid tissues after secondary infection with the same antigen.

14-5   Specific integrin–adhesion-molecule interactions and chemokines direct cell migration into mucosal tissues.

As with naive T lymphocyte entry into secondary lymphoid organs, the migration of effector/memory lymphocytes into peripheral tissues is coordinated through the interaction of CAMs and chemokine receptors on circulating lymphocytes with their respective ligands on vascular endothelial cells. Circulating effector/memory T cells are heterogeneous in their expression of CAMs and chemokine receptors. Thus, the subset(s) of circulating effector T cells that enter into any given peripheral tissue will depend on the array of receptor ligands expressed within that particular tissue. Because distinct mucosal tissues constitutively express different homing receptor ligands, it is currently believed that different mucosal sites preferentially attract different subsets of effector T cells from the circulation.

The intestinal mucosa is the best-studied example of a mucosal organ that preferentially recruits specific subsets of effector T cells from the circulation. Vascular endothelial cells that line the blood vessels within the intestinal mucosa constitutively express MAdCAM-1, and efficient migration of B and T lymphocytes into the small intestinal and colonic mucosa is dependent on interactions between MAdCAM-1 and $\alpha_4\beta_7$ integrin, which is expressed on a subset of circulating lymphocytes. Further specificity in effector T lymphocyte homing to the intestinal mucosa is provided by the chemokine CCL25. This chemokine is constitutively expressed at high levels by epithelial cells of the small intestine but not by epithelial cells in other peripheral or mucosal tissues, including the colon. The CCL25 receptor CCR9 is expressed on a subset of circulating T lymphocytes that co-express high levels of β_7 integrins, and on the majority of T cells in the small intestinal mucosa. T-cell adoptive transfer studies in mice have demonstrated an important and selective role for this chemokine/chemokine receptor pair in mediating the efficient recruitment of effector T cells into the small intestinal mucosa. Thus, expression of MAdCAM-1 and CCL25 in the small intestinal mucosa induces the selective recruitment of T cells co-expressing CCR9 and $\alpha_4\beta_7$ integrin from the circulating effector T lymphocyte pool.

The original concept of a common mucosal immune system postulated that all lymphocyte subsets primed in mucosal tissues are equally capable of migrating through all mucosal tissues. Although the example described argues against such a common system and emphasizes the known differences between mucosal tissues, we are in fact still lacking a detailed description of many of the factors regulating lymphocyte homing to mucosal tissues such as the lung, stomach, or urogenital tract. In keeping with the historic concept of a common mucosal immune system, specific 'address codes' for each of these organs may not exist or may at least be partly overlapping. The localization of IgA plasma cells to mucosal tissues is a good example of this. Epithelial cells at a range of mucosal sites including the intestinal tract, trachea, and salivary and mammary glands constitutively express the chemokine CCL28, and IgA-positive plasma cells at many of these sites have been shown to express the CCL28 receptor CCR10. Furthermore, CCR10 has a redundant role with CCR9 in mediating IgA plasmablast recruitment to the small intestine, has a non-redundant role in the recruitment of IgA plasmablasts into the lactating mammary gland and thus the passive transfer of IgA into the nursing neonate, and has been implicated in recruitment of these cells to the colon. Thus, the function of CCR10 in IgA plasma-cell recruitment seems to extend to a broad range of mucosal tissues. Nevertheless, according to the 'multi-step adhesion cascade' theory which proposes that lymphocyte trafficking from the blood into peripheral tissues requires the concerted action of several CAMs, IgA-positive plasmablast recruitment to distinct mucosal tissues probably involves the use of partly nonoverlapping CAM mechanisms.

14-6 Lymphocytes usually remain in tissues, but some reenter the circulating pool.

In contrast to naive lymphocytes, which as described above continually recirculate through secondary lymphoid organs, effector/memory cells that enter peripheral tissues in many cases never leave and may persist at these sites for long periods. The retention and survival of lymphocytes and their localization to distinct anatomical localizations within peripheral tissues probably require survival and retention signals provided by the local environment, including the involvement of CAMs and chemokine receptors. A striking example of divergent sublocalization of lymphocyte subsets in peripheral tissues is the differential localization of plasma cells and CD8+ lymphocytes to the lamina propria and intraepithelial compartments of the intestinal mucosa respectively. Although the underlying mechanisms regulating the differential localization of lymphocyte subsets in peripheral tissues are incompletely understood, T cells within epithelial tissues, unlike most of their lamina propria counterparts, express the E (for epithelial) β_7 integrin. The cellular ligand for $\alpha_E\beta_7$ integrin is not an Ig superfamily member like that of other integrins, but E-cadherin that is expressed on the lateral and basolateral surface of epithelial cells. Heterotypic interactions between E-cadherin on the epithelial cell and $\alpha_E\beta_7$-integrin-expressing intraepithelial lymphocytes (IELs) are believed to help maintain T cells within the epithelium and probably modulate IEL–epithelial cell cross-talk. $\alpha_E\beta_7$ integrin, at least on effector CD8 T cells, is not a classical 'tissue-homing' molecule, because it is induced on these cells once they have migrated into the epithelium in response to local signals. In addition to $\alpha_E\beta_7$, subsets of lymphocytes in mucosal tissues, including the lung and intestine, express the β_1 integrins $\alpha_1\beta_1$ and $\alpha_2\beta_1$, whose extracellular ligands include collagen IV and collagen I respectively, the former being a major epithelial basement membrane constituent. Although a role for $\alpha_2\beta_1$ integrin in lymphocyte localization is currently unclear, $\alpha_1\beta_1$ integrin has been shown to be important in CD8 effector/memory T-cell survival and/or retention within the lung mucosa.

Despite the findings described above, there is also clear evidence that subsets of effector/memory lymphocytes may leave peripheral tissues. Early studies in sheep demonstrated the presence of effector/memory cells in afferent lymph (lymph draining from peripheral tissues to draining lymph nodes). Thus, some effector/memory T cells are capable of exiting from peripheral tissues via the lymph and migrating back to the draining lymph nodes. Some effector/memory T cells within peripheral tissues have been shown to express the chemokine receptor CCR7, as do most T cells in sheep afferent lymph, and studies in mice have demonstrated that effector/memory T-cell migration from the lung and skin to draining lymph nodes is largely dependent on CCR7, whose ligand CCL21 is constitutively expressed by lymphatic vessels within peripheral tissues. It remains unclear whether these migratory cells are a subset of effector/memory cells that expressed CCR7 before their entry into peripheral tissues, or whether they are a subset of cells that selectively upregulated CCR7 after they received signals within the tissue environment. A second receptor–ligand system that seems to regulate T-cell exit from peripheral tissues is the $S1P_1$–S1P system, described above in regulating lymphocyte exit from secondary lymphoid organs, because $S1P_1$ stimulation inhibits T-cell entry into afferent lymph and promotes the accumulation of these cells in peripheral tissues. How these two pathways interact with one another to regulate T-cell exit from peripheral tissues is currently unclear.

14-7 Different homing molecules control lymphocyte migration to gut and skin.

Circulating effector/memory T-cell subsets are heterogeneous in their expression of CAMs and chemokine receptors, but the mechanism regulating the expression of these molecules on effector T-cell subsets and thus their homing potential to peripheral tissue has only recently become a focus of study. Certain chemokine receptors, including CXCR5, CXCR3, and CCR6, seem to be induced on T-cell subsets after their activation in all secondary lymphoid organs, whereas other tissue-homing receptors are induced on activated T cells in only a restricted set of lymph nodes. The best current examples of selective induction of tissue-homing receptors are found in the lymph nodes draining cutaneous tissues and the intestine-associated lymph nodes (mesenteric lymph nodes and Peyer's patches). By using T-lymphocyte adoptive transfer models it has been demonstrated that T cells primed in mesenteric lymph nodes are induced to express enhanced levels of $\alpha_4\beta_7$ integrin and CCR9, and thus acquire enhanced 'small intestinal homing' properties in comparison with T cells primed in nonintestinal lymph nodes. Conversely, T cells primed in skin-draining lymph nodes upregulate E-selectin and CCR4, two homing receptors involved in T-cell migration to cutaneous and extraintestinal sites of inflammation. It seems likely that this selective induction of gut- and skin-homing receptors in gut- and skin-draining lymph nodes, respectively, serves to enhance effector T-cell migration to the tissues in which infection initially takes place. However, it is currently unclear whether other mucosal-draining lymph nodes, such as the lung-draining mediastinal lymph nodes, generate effector T cells with enhanced tissue tropism for the tissues that they drain.

14-8 Dendritic cells are critical in the generation of tissue-tropic effector lymphocyte subsets.

All mucosal tissues contain large numbers of DCs. In the steady state, these cells continually sample their local environment taking up any self-antigen (from, for example, apoptotic epithelial cells) or innocuous antigen in their environment and transporting it to local draining lymph nodes for presentation to T cells. The steady-state migration of peripheral DCs to draining lymph nodes is dependent on the chemokine receptor CCR7, which is induced on DCs during their maturation, and the chemokine CCL21, which is constitutively expressed by lymphatic endothelial cells in peripheral tissues. In contrast to lymphocyte migration from the blood into lymph nodes, the migration of DCs into draining lymph nodes can occur in the absence of integrins. DCs migrating into draining lymph nodes in the steady state seem to have a central role in the induction of tolerance to peripheral self and innocuous foreign antigen. However, in the setting of mucosal inflammation and/or infection, there is a marked increase in CCR7-dependent DC migration into draining lymph nodes. Such increased migration of DCs can be mimicked in animal models by local or systemic administration of Toll-like receptor ligands, and under these conditions migratory DCs probably have a central role in the initiation of adaptive immune responses.

In addition to their role in initiating adaptive immune responses, DCs seem to contribute to the selective induction of tissue-homing receptors in secondary lymphoid organs. Again, most studies have focused on intestinal DCs, and it remains to be determined whether similar pathways are active at other mucosal sites. The major population of migratory DCs in the small intestinal lamina propria express the integrin heterodimer α_E(CD103)β_7. These cells, as well as their counterparts in the mesenteric lymph nodes, have been shown *in vitro* to efficiently induce expression of the 'small intestinal homing' receptors CCR9 and $\alpha_4\beta_7$ integrin on T lymphocytes compared with CD103$^-$ mesenteric lymph node DCs or DC subsets isolated from extraintestinal sites. Conversely, DCs isolated from the skin, or skin-draining lymph, induce enhanced levels of E-selectin ligand but not CCR9 or $\alpha_4\beta_7$ integrin. The expression of tissue-specific homing receptors seems to be reversible, because gut-homing T lymphocytes restimulated *in vitro* with skin DCs downregulate gut-homing receptors and adopt a skin-homing receptor profile, and vice versa. It seems likely that such plasticity in tissue-homing receptor expression serves to enhance protective immune responses to less tissue-specific pathogens on a second infection with the same pathogen. There is currently considerable interest in trying to understand the underlying molecular mechanisms regulating the tissue-homing receptor expression, because these pathways may provide interesting targets for modulating the tissue-homing potential of T cells for the treatment of inflammatory diseases and for vaccine development. Most progress in our understanding of tissue-homing receptor induction has been made with the gut-homing receptors CCR9 and $\alpha_4\beta_7$ integrin, and these are discussed in more detail below.

14-9   Vitamin A is required for the generation of gut-tropic effector T lymphocytes.

Vitamin A is a dietary vitamin that is taken up in the small intestine and transported to the liver, where it is stored as retinyl esters. Retinol is released into the circulation, primarily from the liver, in association with retinol-binding protein 4 and is taken up by tissue cells, where it is further metabolized. The major active metabolite of vitamin A is retinoic acid (RA). RA is generated by the consecutive oxidation of retinol to retinal and from retinal to RA. The oxidation of retinol to retinal is catalyzed by a family of alcohol dehydrogenases, which seem to be ubiquitously expressed in a wide range of tissues, and the oxidation of retinal to RA is catalyzed through an irreversible reaction by a family of retinal dehydrogenases, which seem to have more tissue- and cell-restricted expression patterns. Once generated, RA can be used by the cell generating it or released into the local environment to regulate the function of neighboring cells. RA signaling is mediated through retinoic acid receptor (RAR)-retinoid X receptor (RXR) heterodimers that function as ligand-induced transcription factors bound to DNA target sequences. The reason that there is so much discussion about vitamin A is because recent studies have shown that it has a major impact in regulating T-cell differentiation, and, perhaps more importantly in the context of this chapter, in regulating tissue-homing receptor expression by these cells.

The addition of low (subnanomolar) concentrations of RA to stimulated T lymphocytes *in vitro* is sufficient to induce expression of the gut-homing receptors CCR9 and $\alpha_4\beta_7$ integrin on responding T cells; more importantly, the ability of intestinal DCs to induce gut-homing receptors on responding T cells is blocked by the addition of antagonists of retinoic acid receptors. In this regard, small intestinal CD103+ DCs express higher levels of key enzymes involved in vitamin A metabolism and have been shown to induce enhanced RA signaling in responding T cells. Thus, one of the main reasons that small intestinal CD103+ DCs are so efficient at inducing CCR9 and $\alpha_4\beta_7$ integrin on T lymphocytes is because they have an enhanced ability to metabolize vitamin A (Figure 14.6). Numerous additional cells in the intestinal mucosa, including epithelial cells and stromal cells in the mesenteric lymph nodes, also seem to have the capacity to metabolize vitamin A and probably contribute to

Figure 14.6 CD103+ DCs induce the expression of gut-homing molecules. In the intestinal lamina propria under the influence of retinoic acid, CD103+ DCs are imprinted to express increased levels of retinal dehydrogenase (RALDH) enzymes. On migration into the gut-draining mesenteric lymph nodes, these DCs, in cooperation with lymph-node-resident fibroblastic reticular stromal cells, induce the expression of the gut-homing molecules CCR9 and $\alpha_4\beta_7$ integrin on activated T lymphocytes. This pathway is restricted to migratory DCs and does not extend to nonmigratory lamina propria cells expressing CX3CR1. TCR, T-cell receptor.

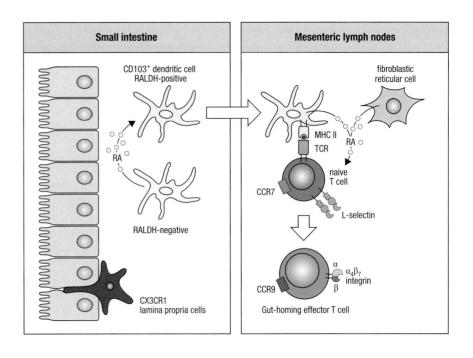

generating an RA-rich environment at these sites that helps drive the selective generation of gut-homing T cells in the mesenteric lymph node *in vivo*. The ability of RA to induce gut-homing receptors is also not restricted to T cells, because RA also induces the expression of gut-homing receptors on B cells. As one might expect from these findings, mice whose vitamin A stores have been depleted by being kept on a diet deficient in vitamin A for a long period have markedly reduced numbers of T cells and B cells in the intestinal mucosa. These findings probably explain some of the beneficial effects of dietary supplementation with vitamin A in reducing childhood mortality from persistent diarrhea induced by infectious disease in developing countries.

The ability of small intestinal CD103+ DCs to induce gut-homing receptors efficiently on responding T cells indicates that factors within the small intestinal mucosa are involved in selectively imprinting these cells with the enhanced ability to metabolize retinol. Although the nature of these factors currently remains to be fully elucidated, intestinal DCs from mice kept on a diet deficient in vitamin A express lower levels of retinal dehydrogenase (RALDH) enzymes and have lower aldehyde dehydrogenase activity, indicating that RA itself, either directly or indirectly, may be involved in small intestinal CD103+ DC imprinting *in vivo*.

Lymphocyte migration to sites of mucosal inflammation—therapeutic opportunities.

Although the recruitment of lymphocytes to mucosal tissues is a critical component of the immune response to mucosal pathogens, in certain susceptible individuals this process can contribute in a major way to the establishment and maintenance of chronic mucosal inflammation. Examples include mucosal exposure to allergens, such as gluten in celiac disease; airway allergens such as mite dust and pollens, causing asthma; and commensal bacteria, which when not properly tolerated can lead to inflammatory bowel disease (Crohn's disease and ulcerative colitis). Given the dual role of lymphocytes in protecting mucosal surfaces and in driving mucosal inflammation, there is considerable interest in determining the CAMs and chemokine receptors that regulate immune cell influx as potential therapeutic targets for enhancing or preventing immune-cell recruitment in these settings. During mucosal infection and inflammation, the expression of CAMs on vascular endothelial cells and the chemokine milieu of mucosal tissues are markedly altered, with enhanced expression as well as *de novo* expression of multiple chemokine family members. In this context, defining key molecular targets that regulate lymphocyte trafficking to sites of infection and inflammation has proven difficult. In particular, different sets of CAMs and chemokines seem to have sometimes distinct and sometimes overlapping roles depending on the infection/inflammatory model, lymphocyte subset, and mucosal tissue under study. Nevertheless there is evidence, at least in the intestinal mucosa, that some of the cellular adhesion receptors regulating lymphocyte recruitment in the steady state remain relevant in the setting of inflammation. Inflammatory bowel disease (IBD), as discussed in detail in several chapters of this book, comprises two major disease groups: Crohn's disease, which can affect all areas of the gastrointestinal tract, and ulcerative colitis, which is restricted to the colon and rectum. Importantly, vascular endothelial cells of the intestinal lamina propria in patients with IBD express enhanced levels of MAdCAM-1, and administration of neutralizing antibodies against $\alpha_4\beta_7$ integrin has been shown to alleviate inflammation in several animal models of chronic intestinal inflammation, suggesting that interactions between $\alpha_4\beta_7$ integrin and MAdCAM-1 remain operative in IBD. Similarly, CCL25 expression in the small intestinal mucosa is maintained in patients with ileal Crohn's disease, although the contribution of CCL25:CCR9 in animal models of intestinal inflammation is less clear. These findings have led to the intriguing possibility of targeting tissue-homing receptors or their endothelial ligands for the treatment of inflammation in a tissue-specific manner. As a consequence, several therapeutics have been developed that target $\alpha_4\beta_7$ integrin and CCR9, and these are currently being tested in human trials for the treatment of ulcerative colitis and Crohn's disease. If successful, such selective targeting may offer an important advantage over many of today's current treatments by alleviating inflammation in a tissue-selective manner without broadly inhibiting systemic immune function.

Summary.

Although it has been recognized for almost 50 years that small lymphocytes recirculate from blood into lymphoid tissues and reenter the blood via the lymphatic system, and that lymphoblasts migrate into tissues, especially the gut, it is only in the past decade or so that the molecular basis for specific homing has been elucidated. In addition to being of fundamental interest, this knowledge is being exploited for health benefits. The idiopathic inflammatory diseases of the modern world such as IBD, rheumatoid arthritis, and multiple sclerosis are all a consequence of the migration of immune cells into tissues and subsequent injury. Preventing lymphocytes from entering tissues is therefore a rational way of treating these serious conditions, and it is gratifying that antibodies against α_4 integrin are now routinely used in the clinic for the treatment of multiple sclerosis and that anti-$\alpha_4\beta_7$ antibodies are now being tested in humans.

Further Reading.

Agace, W.W.: **T-cell recruitment to the intestinal mucosa.** *Trends Immunol.* 2008, **29**:514–522.

Alvarez, D., Vollmann, E.H., and von Andrian, U.H.: **Mechanisms and consequences of dendritic cell migration.** *Immunity* 2008, **29**:325–342.

Ghosh, S., Goldin, E., Gordon, F.H., *et al.*: **Natalizumab for active Crohn's disease.** *N. Engl. J. Med.* 2003, **348**:24–32.

Gorfu, G., Rivera-Nieves, J., and Ley, K.: **Role of β_7 integrins in intestinal lymphocyte homing and retention.** *Curr. Mol. Med.* 2009, **9**:836–850.

Hart, A.L., Ng, S.C., Mann, E., *et al.*: **Homing of immune cells: role in homeostasis and intestinal inflammation.** *Inflamm. Bowel Dis.* 2010, **16**:1969–1977.

Kunkel, E.J., and Butcher, E.C.: **Plasma-cell homing.** *Nat. Rev. Immunol.* 2003, **3**:822–829.

Mora, J.R., and von Andrian, U.H.: **Role of retinoic acid in the imprinting of gut-homing IgA-secreting cells.** *Semin. Immunol.* 2009, **21**:28–35.

Mora, J.R., Iwata, M., and von Andrian, U.H.: **Vitamin effects on the immune system: vitamins A and D take centre stage.** *Nat. Rev. Immunol.* 2008, **8**:685–698.

Mora, J.R., Iwata, M., Eksteen, B., *et al.*: **Generation of gut-homing IgA-secreting B cells by intestinal dendritic cells.** *Science* 2006, **314**:1157–1160.

Polman, C.H., O'Connor, P.W., Havrdova, E., *et al.*: **A randomized, placebo-controlled trial of natalizumab for relapsing multiple sclerosis.** *N. Engl. J. Med.* 2006, **354**:899–910.

Sigmundsdottir, H., and Butcher, E.C.: **Environmental cues, dendritic cells and the programming of tissue-selective lymphocyte trafficking.** *Nat. Immunol.* 2008, **9**:81–87.

Stagg, A.J., Kamm, M.A., and Knight, S.C.: **Intestinal dendritic cells increase T cell expression of $\alpha_4\beta_7$ integrin.** *Eur. J. Immunol.* 2002, **32**:1445–1454.

Mucosal Tolerance

15

Immunologic tolerance has been a subject of fascination and study for as long as the immune system has been known. Immunologic tolerance is demonstrated by reduction or abrogation of an immune response by previous exposure to a given antigen. For decades this was largely accomplished by the intravenous injection of antigen such as soluble proteins. However, there has been a rebirth of a concept known since the beginning of the 20th century that tolerance can also be induced by feeding an immunogen prior to immunization, hence the term 'oral' or 'mucosal' tolerance, which is the topic of this chapter. The intestine and other mucosal surfaces have features and a microenvironment that are distinctly different from those of systemic lymphoid tissues. First, the mucosal surfaces are colonized with microbes, which are collectively called the microbiota. The microbiota has been found in recent years to have profound effects on the host, particularly on the host immune system. In addition, mucosal surfaces are exposed to exogenous antigens in the form of food in the intestine and aeroantigens and dust in the nasal and respiratory tract. The mucosal immune system has coevolved with these sources of exogenous antigen and thus has evolved mechanisms to deal with this continuous antigenic load. Considering the size of the antigen exposure, the dominant mucosal immune response mechanisms are those that dampen the immune and inflammatory responses to such antigens and thus limit inflammation that could injure the mucosal layer. Antigens transiently applied to a mucosal surface are necessarily encountered amongst a large number and variety of antigens associated with the microbiota that are continuously present. Multiple cell types contribute to this mucosal surface, as shown in Figure 15.1. There are many possible outcomes of antigen exposure at the mucosal surface, one of which is mucosal tolerance. This chapter will focus on the mechanisms of these processes, which have profound implications not only for understanding the biologic basis of many inflammatory disorders associated with the mucosal surfaces but also for the development of vaccines and other types of therapies.

General features of mucosal tolerance.

Mucosal tolerance is defined as a state of reduced immunologic responsiveness to an antigen induced by feeding or other mucosal exposure to that antigen. Most studies have used antigen feeding and thus the term 'oral tolerance' has been commonly applied. Other studies have used intranasal installation

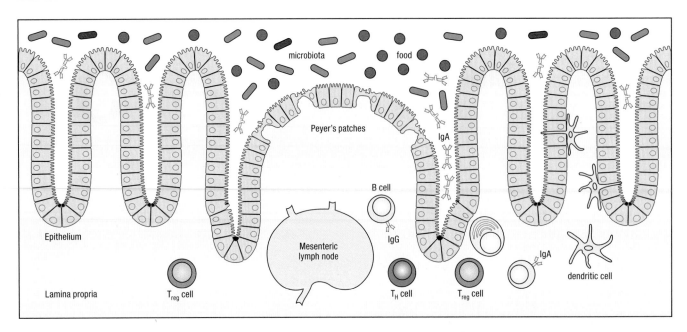

Figure 15.1 The mucosal microenvironment is complex and an antigen encounter at the mucosal surface has multiple possible outcomes. The mucosal surface is a very complex microenvironment. It is colonized by a large variety of commensal bacteria. The intestine also receives a variety of external food antigens (and digested proteins), whereas the respiratory and nasal mucosae are exposed to a variety of airborne foreign antigens. The intestine and other mucosal surfaces have the largest numbers of lymphoid cells in the body, most of which are likely responding to these environmental antigens. A large number of different cell types are present at the mucosal surface, including epithelial cells, dendritic cells, B cells, and T cells of the various subsets. Dendritic cells can extend dendrites between epithelial cells and into the lumen. An antigen encounter at the mucosal surface can result in a variety of different outcomes including a local mucosal IgA response, induction of helper T (T_H) and/or regulatory T (T_{reg}) cells, systemic immune hyporesponsiveness (tolerance), and an active systemic immune response. Various combinations of these outcomes are possible depending on the microenvironmental conditions, nature of the antigen, and dose. The microbiota that colonizes mucosal surfaces is now being recognized to have a profound impact on both the mucosal and systemic immune systems. An antigen applied to a mucosal surface is encountered in the context of this large and ongoing response to the microbiota. This may explain why mucosal tolerance seems to be impaired or absent in germ-free animals.

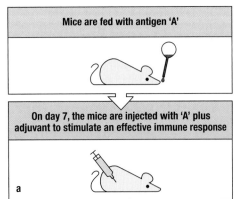

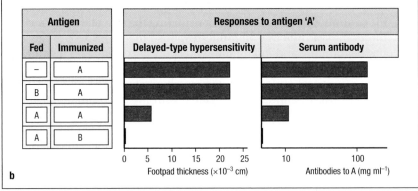

Figure 15.2 A generic oral tolerance experiment. In panel a, antigen 'A,' fed or applied to a mucosal surface prior to systemic priming immunization with antigen A plus adjuvant, which is generally followed by a booster immunization. Following the immunization the ensuing immune response to antigen A is measured. Two of the most common types of immune response are shown in panel b. Delayed-type hypersensitivity is measured by intradermal injection of small amounts of antigen into the footpad or skin dermis; the degree of induration, or thickening, 24–48 hours later is measured. Delayed-type hypersensitivity is mediated by T cells. Assays of antigen-specific T-cell proliferation or T-cell cytokine production are also utilized. The humoral or B-cell response to the antigen is measured by various antibody assays or measurement of antibody-producing cells using the ELISPOT (enzyme-linked immunospot) technique. In experimental disease models, measurements of clinical signs of the disease or of inflammation in the relevant tissue have been used. Tolerance is specific for the antigen that was fed or applied mucosally; that is, feeding a control antigen, denoted as 'B' in the figure, does not affect the systemic response to antigen A and vice versa. As shown, the antibody response usually is reduced by one log10 unit and the delayed-type hypersensitivity response is also reduced, but not abolished; that is, mucosal tolerance causes a reduction rather than abolition of the ensuing immune response.

of antigen in mice. In many intranasal studies the volume administered has been no more than a few microliters but has resulted in both nasal and respiratory mucosal exposure. A typical experiment demonstrating oral tolerance is shown in Figure 15.2. The tolerance induced by mucosal exposure is antigen specific, partial rather than complete, and abrogates an initial immune response better than reducing an established one. Many different types of antigens have been used to induce oral tolerance, including exogenous proteins, red blood cells, contact allergens, inactivated viruses, self-antigens, and alloantigens, to name a few.

15-1 Mucosal tolerance involves many mechanisms including anergy, deletion, and the induction of active regulatory pathways.

The major mechanisms underlying immunologic tolerance generally include deletion of antigen-reactive T cells, clonal anergy of antigen-reactive T cells, and induction of antigen-specific regulatory T (T_{reg}) cells. All of these different mechanisms have been demonstrated in mucosal tolerance. The dose administered at the mucosal surface is a crucial variable in this regard. Large doses of antigens, such as 20 mg or more per dose, given to an animal result in clonal deletion and/or anergy. Modest or 'low' doses, such as 1 mg per dose, particularly if given repeatedly, appear to induce T_{reg} cells as their dominant mechanism (Figure 15.3). Most of the mucosal administrations are of fairly short duration and it is unclear whether the same mechanisms would apply if

Figure 15.3 Induction of T_{reg} cells by antigen feeding. Panel a shows a typical experiment in which antigen A is fed to animal 1. Subsequently, mesenteric lymph node T cells from animal 1 are transferred to animal 2, which is immunized systemically with antigen A in adjuvant. Following the immunization, the immune response to the antigen is determined in animal 2, which had never been fed the antigen in question. The same types of assays as in Figure 15.2 are performed to measure delayed-type hypersensitivity responses to antigen A after its injection into the dermis or footpad (panel b). In addition, in vitro co-cultures of CD4+ T cells from animals fed antigen A with CD4+ T cells from animals parenterally immunized with antigen A have been done to test whether the former inhibit the latter, as measured by changes in proliferation or cytokine production (panel c). This is an in vitro assay for antigen-specific T_{reg} cells induced by feeding antigen. As shown in the figure, such T_{reg} cells are antigen-specific in that CD4+ T cells from animals fed control antigen B do not suppress CD4+ T cells of animals primed to antigen A. The dose of the antigen fed is a critical determinant of mucosal tolerance and largely determines the mechanism involved. In mice, very low doses (micrograms) of antigen do not induce oral tolerance, indicating that there is a threshold effect. Very high doses such as 20–25 mg, even given only once, induce mucosal tolerance due to anergy and/or deletion of antigen-specific T cells. Low doses such as 1 mg per dose given multiple times in mice appear to induce oral tolerance by inducing regulatory T cells specific for the fed antigen, although the specific dose needed varies with the antigen used. T_{reg} cells can be demonstrated by measurements of antigen-specific inhibition in in vitro assays, as shown, or by measurement of the production of inhibitory cytokines produced selectively by T_{reg} cells. T_{reg} cells induced by antigen feeding preferentially produce transforming growth factor-β (TGF-β); T_{reg} cells induced in the respiratory mucosa appear to be predominantly interleukin-10 (IL-10)-producing T_{reg} cells.

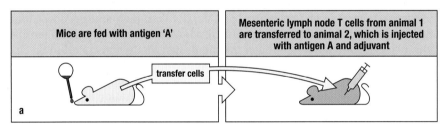

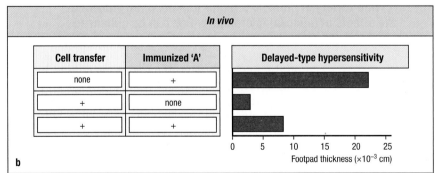

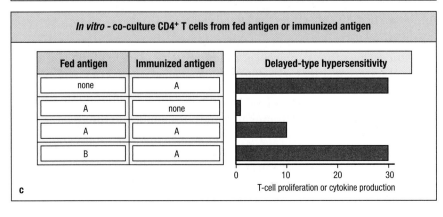

the antigen exposure were more chronic. Although most experimental studies have been performed with model antigens, such as ovalbumin and keyhole limpet hemocyanin, as will be discussed below similar mechanisms are also likely to apply to 'self' antigens.

15-2 Mucosal tolerance is likely induced in organized lymphoid structures associated with mucosal tissues and is disseminated widely throughout the MALT.

The site of induction of mucosal tolerance is also not entirely clear. The first candidates are the gut-associated lymphoid tissue, or Peyer's patches, and equivalent structures in the respiratory tract (nasopharynx-associated or bronchus-associated lymphoid tissue as described in Chapters 20 and 21). T_{reg} cells have been demonstrated in these structures shortly after mucosal exposure and subsequently in draining mesenteric lymph nodes. Mucosal epithelial cells have also been implicated in tolerance induction. Epithelial cells can express class I and II major histocompatibility complex (MHC) molecules as well as nonclassical MHC class I molecules (e.g., CD1d, HLA-E, and HLA-G); they can also process soluble antigen, produce cytokines, express co-stimulatory molecules, and appear to preferentially stimulate $CD8^+$ T cells, among which may be $CD8^+$ T_{reg} cells. A second possibility is the mesenteric lymph nodes, which contain dendritic cells (DCs) that induce tolerance. Mucosally applied antigen may cross the epithelium and reach these nodes via the lymph or 'tolerogenic' dendritic cells; for example, migrating $CD103^+$ DCs in the intestine may take up the antigen in the lamina propria and then migrate to the mesenteric lymph nodes to induce T_{reg} cells. A third potential location for induction of oral tolerance, particularly with high antigen doses that induce anergy, is the systemic lymphoid tissue. After large doses of antigen are administered, antigen and antigen fragments can be demonstrated to be circulating in the blood, and thus may reach systemic lymphoid sites. Consistent with this, after application of large doses of antigen, T-cell activation can be identified in the spleen soon after the mucosal administration of antigen.

15-3 A wide variety of immune effector functions are subject to the effects of mucosal tolerance, which can be enhanced by mucosal adjuvants.

Mucosal tolerance can inhibit a wide variety of immune responses, but the major commonality among these different types of immune responses is the T cell. There appears to be a gradient of sensitivity to mucosal tolerance with the T_H1 helper T cell being the most sensitive, T_H2 cell being of intermediate sensitivity, and B cells being relatively refractory. The reduction of antibody responses is largely due to inhibition of T_H cells rather than B cells directly. In fact, the feeding of T-cell independent B-cell antigens does not result in tolerance. The reduction of T-cell dependent antibody responses is largely due to inhibition of the T cell, rather than the B cell.

As an active form of the immune response, mucosal tolerance can be enhanced or reduced by various molecules coadministered with the antigen to the mucosal surface (Figure 15.4). One example of an agent that can enhance oral tolerance is cholera toxin B subunit (CTB). In order to amplify mucosal tolerance, CTB must be covalently coupled to the antigen and applied to the nasal or respiratory mucosa. CTB–antigen conjugates applied to the nasal mucosa result in mucosal tolerance at surprisingly small doses of antigen. Simple mixtures of CTB and antigen applied in a similar fashion do not have this effect. A variety of other substances have been shown to either enhance or reduce mucosal tolerance when coadministered with antigen and these are shown in Figure 15.4.

Following intranasal exposure the mice are immunized with antigen A and oral tolerance assessed

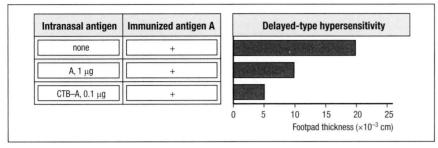

Intranasal antigen	Immunized antigen A
none	+
A, 1 µg	+
CTB–A, 0.1 µg	+

Delayed-type hypersensitivity

0   5   10   15   20   25
Footpad thickness (×10^{-3} cm)

Agents that can modulate mucosal tolerance	
Enhance tolerance	**Reduce tolerance**
IL-2, IL-4, IL-10	IFN-γ
Anti-IL-12	IL-12, IL-18
Cholera toxin B subunit (CTB)	Cholera toxin
Flt3 ligand	CD80 overexpression
TGF-β	Anti-T-cell receptor γδ
Anti-CD4	Cyclophosphamide
Lipopolysaccharide	Graft-versus-host reaction
Multiple emulsions	
Liposomes	

Figure 15.4 Mucosal tolerance can be enhanced or reduced by other compounds. A variety of compounds have been found to either enhance or reduce mucosal tolerance, demonstrating that mucosal tolerance is not the absence of an immune response, but an active immune response that itself can be enhanced or reduced. Enhancement of tolerance is shown by more profound inhibition of the subsequent immune response or by having equivalent tolerance at much lower doses of mucosally applied antigen. In this example, oral tolerance is enhanced by coupling cholera toxin B subunit (CTB) covalently to an antigen A and then applying it intranasally. Following intranasal exposure the mice are immunized with antigen A and oral tolerance is assessed in the same assays already described (top panel). In this figure, typical delayed-type hypersensitivity data are shown, demonstrating that the CTB–A conjugate induced greater tolerance than did intranasal antigen A alone, at one-tenth the dose. The lower panel shows agents that can either enhance or reduce mucosal tolerance. IL, interleukin; TGF-β, transforming growth factor-β; IFN-γ, interferon-γ.

Mucosal tolerance in experimental autoimmune and inflammatory disease.

Organ-specific autoimmune disease can be induced in rodents by parenteral immunization with an autoantigen in a strong adjuvant. For example, immunization with myelin basic protein, a constituent of brain and spinal cord, results in inflammation in the nervous system and subsequent neurologic symptoms and deficits, which are similar to those that occur in multiple sclerosis in humans. This model is called experimental allergic encephalomyelitis (EAE). Although EAE was initially induced in the Lewis strain of rat, it also occurs in various susceptible mouse strains, and can be induced by immunization with other nervous system proteins such as proteolipid protein. EAE has been used extensively to probe the mechanisms involved in this multiple sclerosis-like disease. Other organ-specific autoimmune diseases can be induced by immunization with various autoantigens such as collagen type II, which results in arthritis; retinal S-antigen, which results in uveitis; thyroglobulin, which results in thyroiditis; and the acetylcholine receptor, which results in a myasthenia gravis-like condition. There is also a spontaneous autoimmune disease that occurs in non-obese diabetic mice, which has been used extensively as a model of autoimmune diabetes in humans. Such models are useful for investigating the utility of mucosal (oral) tolerance in the treatment of autoimmune diseases.

Figure 15.5 Abrogation of experimental autoimmunity by mucosal tolerance. Mucosal tolerance has been used to abrogate experimental autoimmunity in multiple models. Most models involve feeding of the autoantigen prior to parenteral immunization with the tissue-specific autoantigen, but others involve spontaneous autoimmunity in a genetically susceptible host such as non-obese diabetic mice that develop type 1 diabetes. The sequence is similar to Figure 15.2 in which feeding precedes immunization with the tissue-specific antigen. Various measurements are made after the feeding and immunization, including a clinical score, a score of inflammation in the relevant tissue, or, in the case of non-obese diabetic mouse autoimmune diabetes, the blood glucose level. In the case of the non-obese diabetic mouse, there is no systemic immunization as the animal spontaneously develops the autoimmune diabetes. Feeding of a tissue-specific antigen after disease is already established is significantly less effective than feeding before the disease is established. Examples of the different models in which mucosal tolerance has been effective are provided in Table 15.1.

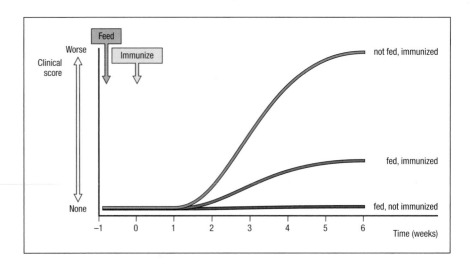

15-4 Mucosal tolerance can be elicited to autoantigens in experimental model systems.

Administration of the same autoantigen to the mucosal surface prior to the parenteral immunization reduces the frequency and severity of the subsequent autoimmune disease (Figure 15.5). This has been demonstrated with a diverse array of antigens and models (Table 15.1). The most common mucosal exposure has been by feeding the antigen, but the nasal/respiratory route has also been effective. The same mechanisms of mucosal tolerance have been manifest as for exogenous non-self antigens—namely clonal deletion, clonal anergy, and induction of T_{reg} cells—but most commonly antigen-specific T_{reg} cells appear to be the predominant mechanism. Mucosally applied antigen is most effective when delivered prior to the parenteral immunization. The delivery of autoantigen to mucosa after immunization has less or no effect on the disease, depending on the model.

Table 15.1 Oral tolerance as therapy of experimental autoimmune disease

Model	Antigen fed
EAE	MBP, PLP
Arthritis (CII)	Collagen type II
Uveitis	S-antigen, IRBP
Diabetes (non-obese diabetic mouse)	Insulin, GAD
Myasthenia gravis	Acetylcholine receptor
Thyroiditis	Thyroglobulin
Transplantation	Alloantigen (cells), MHC peptides

EAE, experimental allergic encephalomyelitis; MBP, myelin basic protein; PLP, proteolipid protein; CII, type II collagen-induced arthritis; IRBP, interphotoreceptor retinoid binding protein; GAD, glutamic acid decarboxylase

Decreased severity of disease
Delayed onset and decreased frequency of disease
Decreased frequency of MBP-specific T cells
Altered T-cell receptor usage of MBP-specific T cells
Decreased T-cell epitope spreading
Increased TGF-β-producing T$_{reg}$ cells
Decreased serum IgG anti-MBP
Increased salivary IgA anti-MBP

MBP, myelin basic protein

Table 15.2 Mucosal tolerance in experimental allergic encephalomyelitis: examples of immunological effects

Many experimental models have a clinical scoring system that allows repeated measurement of disease features that can be followed over time; for example, various neurologic deficits in EAE, the number and size of swollen joints in arthritis, and the blood glucose level in non-obese diabetic mice. Increasing clinical severity correlates with increasing infiltration of the relevant organ by immune and inflammatory cells, and with increasing T-cell and B-cell reactivity to the autoantigen. Mucosal tolerance can reduce the clinical score, organ inflammation, and diverse aspects of the antigen-specific immune response. Examples of the many effects of mucosal tolerance on EAE induced by immunization with myelin basic protein are given in Table 15.2; these are representative of effects observed in other models. In the affected organ, T$_H$ cell subsets and cytokines are generally altered, with a decrease in T$_H$1 cells and interferon-γ secretion, and an increase in T$_{reg}$ cells and in local production of interleukin-10 (IL-10) and transforming growth factor-β (TGF-β1).

Not only can mucosal tolerance be elicited against orally administered autoantigens in the treatment of experimental models of autoimmunity, but such tolerance can be elicited to other types of antigens associated with immune-mediated inflammation. For example, oral administration of trinitrobenzene sulfonic acid (TNBS) hapten-conjugated colonic proteins (including proteins associated with the commensal microbiota) results in the induction of mucosal tolerance which is associated with the prevention of TNBS-induced colitis. The mechanism involves induction of TGF-β-producing regulatory cells and indicates the potential broad range of clinically relevant antigens to which oral tolerance can be applied.

15-5 The mechanisms of mucosal tolerance that are induced in response to autoantigens are similar to those induced against model antigens.

Most of these models involve mucosal immunization of a known autoantigen followed by parenteral immunization with the same antigen. While these models have provided important insights into mechanisms of disease and its amelioration, tissues contain multiple autoantigens and during an autoimmune disease the immune system responds to many tissue autoantigens, not just one. This is illustrated by non-obese diabetic mice, which have T cells reacting to more than a dozen autoantigens of pancreatic beta cells, with insulin being only one of them. This would seem to limit the potential usefulness of mucosal tolerance for therapy of spontaneously occurring autoimmune or inflammatory disease. However, to the extent that T$_{reg}$ cells are generated by mucosal antigen exposure, this may not be the case. T$_{reg}$ cells bear T-cell receptors (TCR) specific for a single antigen, and must be activated via that antigen-specific TCR in order to express their inhibitory program, such as releasing the inhibitory cytokines such as TGF-β1, IL-10, and IL-35. Once released, these cytokines will inhibit T cells of other specificities in the local environment in an antigen-nonspecific manner. This phenomenon is named 'bystander suppression,' and it has been demonstrated in multiple experimental models (Figure 15.6) in which mucosal tolerance is induced by one organ-specific antigen and inhibits autoimmune disease induced by a different, 'unlinked,' autoantigen. In one of these models, insulin feeding was able to inhibit disease triggered by lymphocytic choriomeningitis virus (LCMV) infection in a mouse transgenically expressing a LCMV antigen in pancreatic islets. In these examples mucosal tolerance again preceded disease initiation, which is unlikely to apply to spontaneous autoimmune disease. However, if mucosal tolerance were induced early in the course of disease, in theory it may be able to interrupt epitope spreading and broadening of the immune responses, thus ameliorating disease, although this remains to be demonstrated.

Oral tolerance in the treatment of human immune-mediated diseases.

The effectiveness of oral tolerance in preventing experimental autoimmune disease raised interest in using oral tolerance to treat human autoimmune diseases. Oral tolerance is clearly quite complex and can be demonstrated only *in vivo* and not in cells tested *ex vivo*. Data on mucosal tolerance in humans are sparse. Experiments done in the 1960s with contact allergens exposed to the buccal mucosa of the mouth have demonstrated results that are compatible with the presence of mucosal tolerance to contact sensitizers in humans. Contact sensitizers, also known as contact allergens, are covalently reactive compounds that couple to host proteins to form a complex that provokes a cutaneous T cell-mediated inflammation.

15-6   Mucosal (oral) tolerance is a property of normal human mucosal tissues.

In order to determine whether oral tolerance to protein antigens exists in humans, keyhole limpet hemocyanin (KLH) has been fed to a group of human

Figure 15.6 Bystander suppression of experimental autoimmune disease. In experimental, and presumably in human, autoimmune disease more than one antigen is contributing to the pathogenic immune response. Moreover, in most cases, the actual antigen that might be initiating the disease is not known. Bystander suppression refers to a phenomenon in which activation of an organ-specific T$_{reg}$ cell can result in inhibition of the immune response to other, unlinked tissue-specific antigens. The typical protocol is shown in which animals are fed one organ-specific antigen (A), but are immunized with a different, unrelated antigen that is present in, and specific for, the same organ (B). The concept is that the mucosal exposure to antigen A induces antigen-specific T$_{reg}$ cells to antigen A. These T$_{reg}$ cells are then activated *in vivo* by encounter with antigen A in the tissue or organ. Once activated, the T$_{reg}$ cell specific for antigen A produces inhibitory cytokines that suppress T cells of other specificities, such as B, that are in the lesion or organ. Some examples in which bystander suppression has been demonstrated include (with the antigens used for feeding and immunization in parentheses) experimental allergic encephalomyelitis (myelin basic protein, proteolipid protein), arthritis (collagen type II, complete Freund's adjuvant), and autoimmune diabetes (insulin, lymphocytic choriomeningitis virus in an LCMV-islet cell transgenic mouse).

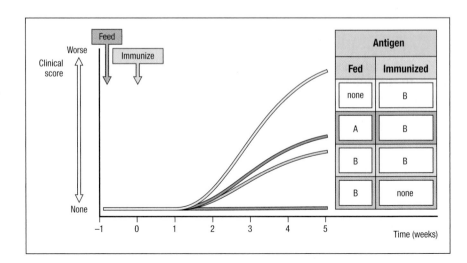

volunteers. KLH is a potent immunogen that has been used safely in humans to assess immunocompetence. The group fed KLH and a group not fed KLH were then parenterally immunized and the resulting systemic and mucosal immune responses were compared (Figure 15.7). KLH feeding resulted in significant reductions of the delayed-type hypersensitivity skin test response and of KLH-specific T-cell proliferation. Conversely, KLH feeding prior to immunization resulted in increased B-cell responses manifest by higher anti-KLH antibodies in serum and secretions compared with individuals not fed KLH. Such 'split tolerance' has been seen in some experimental models as well. Split tolerance may be biologically relevant as it can be imagined that secretory antibodies (e.g., secretory IgA) may further promote tolerance by excluding the uptake of 'allergic' or 'disease-producing' antigen.

15-7 Mucosal tolerance may be amenable to therapeutic manipulation in human immune-mediated disease.

Although the KLH feeding data confirmed that mucosal tolerance is achievable in humans, it is important to note that these studies, and the vast majority of animal models, involved the oral administration of exogenous antigen that has not previously been encountered, or, in the case of murine disease models, the application of self-antigen prior to the induction of disease. In contrast, in human autoimmune disease the immune response to the antigen has already been established. Despite this, the therapeutic potential demonstrated in numerous experimental animal models, as well as the noninvasive nature of the treatment and limited adverse effects, stimulated the development of clinical trials to address the use of mucosal tolerance as a therapy for human disease. Controlled clinical trials of oral tolerance have been done in rheumatoid arthritis, multiple sclerosis, autoimmune uveitis, and autoimmune diabetes (Table 15.3). Unfortunately, the results of these clinical trials have been ambiguous at best and have demonstrated only a limited efficacy in the treatment of these diseases. There are a number of possible explanations for the lack of efficacy in these trials. Because the antigens responsible for autoimmune and inflammatory disease in humans are unknown, the rationale for these studies was the induction of bystander suppression by T_{reg} cells specific for a tissue-specific antigen. The antigen in most of the trials was not of human origin, and the dose fed was based on extrapolations from animal studies rather than based on direct testing in humans. In all instances the antigen was administered without any agent that might enhance mucosal tolerance. The results would suggest that agents that can enhance mucosal tolerance would be necessary if oral tolerance is to succeed as a therapy for human immune-mediated inflammatory diseases.

There are human diseases in which failure of tolerance to food antigens or to the microbiota is the initiator of the pathologic process. Celiac sprue (or

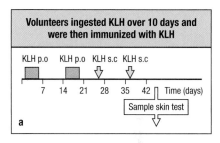

Volunteers ingested KLH over 10 days and were then immunized with KLH

a

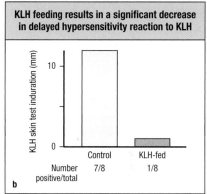

KLH feeding results in a significant decrease in delayed hypersensitivity reaction to KLH

b

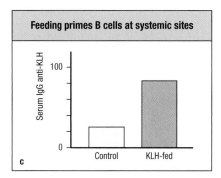

Feeding primes B cells at systemic sites

c

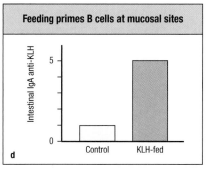

Feeding primes B cells at mucosal sites

d

Figure 15.7 Mucosal tolerance in humans. The existence of oral tolerance to protein antigens in humans was demonstrated using keyhole limpet hemocyanin (KLH), which is a strong immunogen, but not a food antigen. Panel a: human volunteers ingested 10 doses of 0.5 g KLH over 10 days and were then immunized with KLH by subcutaneous injection; controls received only the KLH immunization (p.o., *per os*, i.e. orally; s.c., subcutaneous). Panel b shows that KLH feeding resulted in a significant decrease in delayed hypersensitivity reaction to KLH, as well as reductions in antigen-specific T-cell proliferation by peripheral blood T cells (not shown). Although KLH feeding by itself did not induce significant antibody responses, feeding did prime B cells at both systemic (panel c) and mucosal (panel d) sites as evidenced by an enhanced antibody response at these sites upon parenteral immunization with KLH. This dissociation between T-cell and B-cell tolerance, so-called 'split tolerance,' has been observed after low-dose antigen feeding in animals. Subsequent studies showed that mucosal tolerance to KLH in humans was antigen-specific.

Table 15.3 Controlled clinical
trials of oral tolerance in humans

Disease	Antigen
Rheumatoid arthritis	Collagen type II, bovine/chicken
Multiple sclerosis	Myelin, bovine
Uveoretinitis	Retinal S-antigen, bovine
Diabetes mellitus type I	Insulin, human

gluten-sensitive enteropathy) and food allergies (peanut, egg, milk) are examples of the former; inflammatory bowel disease (IBD) is an example of the latter. Studies aimed at studying tolerance in IBD using KLH (as described above) found that not only do patients lack tolerance but also that family members share the same deficit. Thus while defects in tolerance do not necessarily result in disease, they contribute to susceptibility to disease.

15-8   Mucosal tolerance can be used in the treatment of human allergic disorders.

Although mucosal tolerance induction has been a relatively unsuccessful treatment for autoimmune disease, a voluminous body of literature has shown it to be a safe and effective therapy for many forms of allergy. Three primary routes of mucosal tolerance therapy have been utilized in the treatment of allergic disease: oral immunotherapy, in which antigen is swallowed; sublingual immunotherapy (SLIT), where antigen is held under the tongue for a short period of time before being either swallowed or spat out; and local nasal immunotherapy, in which the antigen is applied directly to the nasal mucosa typically as a dry powder. Oral immunotherapy is most commonly utilized for the treatment of food and contact allergies, while SLIT and local nasal immunotherapy are primarily used to treat inhaled antigens. This presumably has some rationale given the known preferences for mucosal homing to the site of antigen administration based upon the induction of specific homing molecules. Regardless of the route of administration, most clinical allergy trials are conducted in a similar manner: enrolled patients are subjected to a double-blind placebo-controlled challenge that serves to determine the minimum dose of allergen that elicits a response in a given individual. The treatment phase is then begun using some fraction of this 'eliciting' dose. Tolerance induction begins with a dose escalation phase where mucosal application of the antigen is carried out in increasing concentrations over a period of time. The escalation may occur over a period of weeks or months at the patient's home, or it may occur in a 'rush protocol' where the dose escalation is carried out over a period of hours or days at a medical institution. Dose escalation continues until the 'maintenance dose' is reached, which is then continued for the length of the protocol (typically months to years). Treatment efficacy is usually based upon clinical scores generated from patient self-evaluations and physician evaluations, rescue medication scores (scores assigned to medications taken to alleviate allergic symptoms), and allergen challenges conducted at the end of the trial.

While it is unclear why oral antigen would be a more effective therapy for allergy than autoimmune disease, there are a number of possible explanations. First, oral immunotherapy for allergy is directed against the same antigen that is responsible for causing the inappropriate immune response, while the antigens used for inducing tolerance in autoimmune disease may or may not be the antigen that caused the initial insult. Similarly, because tolerance is being induced to a single antigen in allergic disease, clonal deletion or anergy of antigen-specific cells will have a significant impact on the subsequent immune response; whereas therapy for autoimmune disease requires the generation of antigen-specific regulatory cells to suppress an immune response directed against multiple antigens and consequently requires bystander suppression, which is less potent. In addition, inflammation may impair tolerance induction or responsiveness of inflammatory cells to the tolerance-inducing factors (e.g., regulatory cytokines). Furthermore, allergens (and thus inflammation) can be avoided during the induction of tolerance, which is not possible in autoimmunity, where the antigen is always present, as are inflammatory cells. Moreover, oral tolerance protocols for allergy are carried out over significantly longer periods of time than those for autoimmune disease; it is possible that many tolerance protocols attempting to treat autoimmune disease are simply too short.

Summary.

Due to the antigenic assault (food and microbiota) that characterizes the mucosal immune system, avoidance of immune responses to innocuous antigens is necessary lest the immune system exhausts itself. Mucosal tolerance is the result of many different mechanisms (anergy, deletion, induction of T_{reg} cells), which specifically inhibit immune responses to a given antigen. One form of mucosal tolerance relates to the induction of T_{reg} cells that secrete TGF-β, IL-10, and IL-35 and is associated with the phenomenon of bystander suppression. This phenomenon has the potential to be used in the treatment of autoimmune and inflammatory diseases where the autoantigen is not known. However, initial successes in animal models have not been replicated in human diseases, despite the evidence that induction of mucosal tolerance can be elicited in humans. Whether these failures are due to the inability to shut down ongoing immune responses, especially in the context of ongoing inflammation, or to the dose and character of the antigen being incorrect, remains to be determined. Moreover, understanding mucosal tolerance has important implications for understanding diseases that are due to a loss of tolerance to environmental antigens such as those associated with food (e.g., gluten in celiac disease) or microbes (e.g., commensal microbiota in inflammatory bowel disease). In addition, the tendency of the mucosal surfaces to elicit suppression in response to mucosally applied antigens represents a major challenge for the development of mucosal vaccines and indicates the need for the development of mucosal adjuvants capable of bypassing the restrictive nature of mucosal tolerance (see Chapter 27).

Further Reading.

al-Sabbagh, A., Miller, A., Santor, L.M., *et al.*: **Antigen-driven tissue-specific suppression following oral tolerance: Orally administered myelin basic protein suppresses proteolipid protein-induced experimental autoimmune encephalomyelitis in the SJL mouse.** *Eur. J. Immunol.* 1994, **24**:2104–2109.

Burks, A.W., Laubach, S., and Jones, S.M.: **Oral tolerance, food allergy, and immunotherapy: Implications for future treatment.** *J. Allergy Clin. Immunol.* 2008, **121**:1344–1350.

du Pré, M.F., and Samsom, T.N.: **Adaptive T-cell responses regulating oral tolerance to protein antigen.** *Allergy* 2011, **66**:478–490.

Faria, A.M., and Weiner, H.L.: **Oral tolerance.** *Immunol. Rev.* 2005, **206**:232–259.

Hafler, D.A., Kent, S.C., Pietrusewicz, M.J., *et al.*: **Oral administration of myelin induces antigen-specific TGF-β1 secreting T cells in patients with multiple sclerosis.** *Ann. N. Y. Acad. Sci.* 1997, **835**:120–131.

Husby, S., Mestecky, J., Moldoveanu, Z., *et al.*: **Oral tolerance in humans. T cell but not B cell tolerance after antigen feeding.** *J. Immunol.* 1994, **152**:4663–4670.

Mayer, L., and Shao, L.: **Therapeutic potential of oral tolerance.** *Nat. Rev. Immunol.* 2004, **4**:407–419.

Nussenblatt, R.B., Gery, I., Weiner, H.L., *et al.*: **Treatment of uveitis by oral administration of retinal antigens: Results of a phase I/II randomized masked trial.** *Am. J. Ophthalmol.* 1997, **123**:583–592.

Pozzilli, P., Pitocco, D., Visalli, N., *et al.*: **No effect of oral insulin on residual beta-cell function in recent-onset type I diabetes (the IMDIAB VII).** IMDIAB group. *Diabetologia*, 2000, **43**:1000–1004.

Prakken, B.J., Samudal, R., Le, T.D., *et al.*: **Epitope-specific immunotherapy induces immune deviation of proinflammatory T cells in rheumatoid arthritis.** *Proc. Natl Acad. Sci. USA* 2004, **101**:4228–4233.

Rolinck-Werninghaus, C., Wolf, H., Liebke, C., *et al.*: **A prospective, randomized, double-blind, placebo-controlled multi-centre study on the efficacy and safety of sublingual immunotherapy (SLIT) in children with seasonal allergic rhinoconjunctivitis to grass pollen.** *Allergy* 2004, **59**:1285–1293.

Strober, W., Kelsall, B., and Marth, T.: **Oral tolerance.** *J. Clin. Immunol.* 1998, **18**:1–30.

Sun, J.B., Gerkinsky, C., Holmgren, J., *et al.*: **Mucosally induced immunological tolerance, regulatory T cells and the adjuvant effect by cholera toxin B subunit.** *Scand. J. Immunol.* 2010, **71**:1–11.

Vickery, B.P., and Burks, A.W.: **Immunotherapy in the treatment of food allergy: Focus on oral tolerance.** *Curr. Opin. Allergy Clin. Immunol.* 2009, **9**:364–370.

Weiner, H.L, Mayer, L., and Strober, W.: **Oral tolerance. New insights and prospects for clinical application.** *Ann. N. Y. Acad. Sci.* 2004, **1029**:1–421.

PART III | MICROBIAL COMMENSALISM

Recognition of Microbe-Associated Molecular Patterns by Pattern Recognition Receptors

16

The immune system has two major components: the innate or nonspecific immune system and the adaptive or specific immune system. After structural barriers, the innate immune system is the first line of host defense against microorganisms, whereas the adaptive immune system represents the second line. Both systems contribute to self and non-self discrimination, to protect the host from pathogens and to eliminate modified or host altered cells. Both systems also contain cellular and humoral components and interact with each other. Innate immunity is readily available at birth, whereas adaptive immunity arises later as a consequence of pathogen exposure. The two phases of immune host defense are pathogen recognition followed by pathogen removal.

In contrast to adaptive immunity, innate immunity uses defense mechanisms that promote the immediate detection and rapid destruction of microorganisms independently of the clonal expansion of effector cells or the generation of immunological memory. Consequently, the innate immune system is not antigen-specific and reacts to pathogen structures without previous exposure. Innate host defenses are limited and present in all multicellular organisms, in contrast to adaptive host defenses, which are diverse and present only in vertebrates. Tight control of innate immunity is critical to mucosal homeostasis in the intestine. After physical barriers (for example epithelia), chemical barriers (for example mucus), and biological barriers (for example competing microbiota) have been breached, innate immune responses are initiated.

Innate immunity relies on germline-encoded receptors that recognize structures common to pathogens and commensal microorganisms. These receptors are called pattern recognition receptors (PRRs) and are expressed by innate cells, including epithelial cells, macrophages, and dendritic cells. PRRs are grouped into Toll-like receptors (TLRs), NOD-like receptors (NLRs), and retinoic acid-inducible gene-I (RIG-I)-like receptors (RLRs) (Figure 16.1). In this chapter, we review the contribution of PRRs to innate immune defense of the intestinal mucosa, focusing on TLRs and NLRs. We also discuss the genetic imbalances in pattern recognition that may contribute to inflammatory bowel diseases.

Figure 16.1 Pattern recognition receptors (PRRs) and their cellular location. PRRs are grouped into Toll-like receptors (TLRs); NOD-like receptors (NLRs); and retinoic acid-inducible gene-I (RIG-I)-like receptors (RLRs). TLR1, 2, 4, 5, and 6 are cell-surface receptors, whereas TLR3, 7, 8, and 9 are located on intracellular endosomal membranes. The NLRs NOD1 and NOD2 are present exclusively in the cytosol, where they sense intracellular pathogens. The RLRs, including RIG-I, melanoma differentiation-associated gene 5 (MDA-5), and laboratory of genetics and physiology 2 (LGP2), are cytoplasmic helicases that recognize viral RNA.

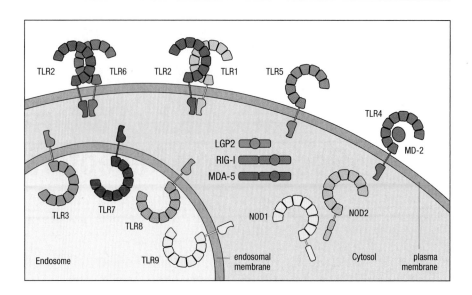

Principles of pattern recognition and signaling.

A key function of the mucosal innate immune system is to recognize invading pathogens and quickly mount defensive responses. Pathogen recognition is accomplished by PRRs that recognize highly conserved, pathogen-associated molecular patterns (PAMPs). PAMPs are expressed by microorganisms and not by mammalian host cells or tissues. The major features of innate cell PRRs are outlined in Table 16.1.

16-1   TLRs comprise a family of conserved receptors that recognize specific PAMPs.

TLRs comprise a major group of PRRs and have a key role in the innate cell recognition of PAMPs, leading to the induction of pro- and anti-inflammatory genes, phagocytosis, and control of adaptive immune responses. The *Toll* gene was originally discovered during the study of fruitfly (*Drosophila melanogaster*) embryogenesis. After the discovery of a role for the Toll pathway in controlling host defense through the induction of potent antifungal factors, a human homolog of the *Drosophila* Toll protein was identified. A constitutively active mutant of the Toll homolog transfected into human cell lines mediated the activation of the transcription factor NFκB and the expression of several pro-inflammatory genes. The first TLR, now designated TLR4, recognizes bacterial lipopolysaccharide (LPS). Since its discovery by Bruce Beutler, for which he received the 2011 Nobel Prize in Physiology or Medicine, 10 TLRs in humans and 13 in mice have been identified.

Table 16.1 Features of pattern recognition receptors in the innate immune system

Encoded within the genome (inherited)
Display limited diversity
Recognize broad, shared classes of conserved molecules on pathogens
Initiate responses within minutes
Expressed by all cells of a particular lineage (nonclonal)

Mammalian TLRs are type I transmembrane glycoproteins containing three common structural features: (1) a large extracellular domain with 19–25 consecutive leucine-rich repeats and one or two cysteine-rich regions; (2) a short transmembrane region; and (3) a highly conserved cytoplasmic domain (Figure 16.2). The extracellular domain is responsible for the recognition of PAMPs. The intracellular domain is highly homologous among the individual TLRs and contains a Toll/interleukin-1 (IL-1) receptor (TIR) domain that mediates homodimeric or heterodimeric interactions between TLRs and initiates downstream signaling cascades, as discussed below. The TIR domain also is present in members of the closely related IL-1, IL-18, and IL-33 receptor families.

Different classes of PAMPs activate different TLRs; that is, each TLR recognizes a limited number of specific molecular signatures in different classes of microorganisms (Table 16.2). Although the diversity of pattern recognition is limited, TLRs collectively recognize most pathogenic or commensal microorganisms that may invade the mucosa. For example, TLR2 signals the presence of bacterial lipopeptides and lipoteichoic acid, which are cell-wall constituents of Gram-positive bacteria. Select combinations of TLRs may act in concert to expand the repertoire of these PRRs; thus, TLR2 cooperates with TLR1 or TLR6. TLR4 is the major receptor for LPS signal activation but requires the presence of the accessory molecules MD-2 and CD14. Flagellins of flagellated bacteria are ligands for TLR5. Unmethylated CpG DNA present in prokaryotic genomes and DNA viruses is detected by TLR9. In addition, TLRs may recognize endogenous damage-associated molecular patterns such as heat-shock proteins, which are released by dying cells and function as alarm signals during inflammation.

In general, TLRs are expressed predominantly on cells that are the first to encounter the invading threat. Phagocytic cells, including macrophages and

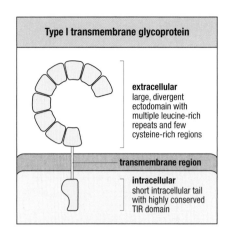

Type I transmembrane glycoprotein

extracellular
large, divergent ectodomain with multiple leucine-rich repeats and few cysteine-rich regions

transmembrane region

intracellular
short intracellular tail with highly conserved TIR domain

Figure 16.2 TLR structure. TLRs are type I transmembrane proteins with common structural features, including multiple leucine-rich repeats and one or two cysteine-rich regions in the large and divergent ligand-binding ectodomain, a short transmembrane region, and a conserved cytoplasmic domain that contains a Toll/interleukin-1 receptor (TIR) domain. TLRs mediate their receptor function as homodimers or heterodimers.

Table 16.2 Specific PRRs bind different molecular signatures or structures in microbes

Ligand	Receptor	Signaling adaptor
Peptidoglycan Triacylated lipoproteins Diacylated lipoproteins Glycosylphosphatidylinositol anchors	TLR1–TLR2 TLR6–TLR2	MyD88 MAL
Zymosan (yeast)	TLR2 (with Dectin-1)	
Double-stranded RNA	TLR3	TRIF
LPS Lipoteichoic acids	TLR4 (with MD-2, CD14, LBP)	MAL MyD88 TRIF TRAM
Flagellin	TLR5	MyD88 TRIF
Single-stranded RNA	TLR7	MyD88
G-rich oligonucleotides	TLR8	MyD88
Unmethylated CpG DNA	TLR9	MyD88
Profilin	TLR11 (mouse)	MyD88
iE-DAP	NOD1	
MDP	NOD2	

iE-DAP, γ-D glutamyl-meso-diaminopimelic acid; MDP, muramyl dipeptide.

dendritic cells, exhibit the broadest repertoire of TLRs, but TLR expression is not restricted to these cells. In the gastrointestinal tract mucosa, TLRs are constitutively or inducibly expressed by an array of cell types, including the four principal intestinal epithelial cell lineages, subepithelial myofibroblasts, as well as antigen-presenting cells in the lamina propria. TLR expression is influenced by the level of activation of the cell and/or surrounding tissue. In healthy intestinal mucosa, TLR expression is generally low, minimizing recognition of the omnipresent microflora. However, in the diseased intestinal mucosa, TLR expression may be increased, maximizing immune responsiveness to PAMPs.

TLRs are present in extracellular and intracellular forms. The TLRs that bind to surface bacterial antigens (TLR1, 2, 4, 5, and 6) are expressed on the outer membrane of the cell, whereas TLRs that bind to intracellular viral and bacterial structures (TLR3, 7, 8, and 9) are expressed on the endosomal membrane within the cell. Endosomes are phagolysosomes that degrade pathogens and their products. Another family of PRRs, the NLRs, exist exclusively in the cytosol and are discussed next.

16-2   NOD1 and NOD2 are NLR-family members that recognize peptidoglycan motifs.

NLRs comprise a second group of evolutionarily conserved host defense PRRs encoded in both animal and plant genomes. NLRs act as intracellular surveillance molecules, sensing microbial PAMPs and endogenous damage-associated molecular patterns that enter the cytoplasmic compartment. NLR orthologs were first discovered in plants and are called resistance (R) proteins because they protect against fungal, viral, parasitic, and insect pathogens.

Typically, NLRs contain three distinct functional domains: (1) a structurally variable amino-terminal effector domain that comprises caspase recruitment domains (CARDs), a pyrin domain (PYR), and a baculovirus-inhibitor-of-apoptosis repeat domain (BIR); (2) a central nucleotide-binding domain; and (3) a leucine-rich repeat ligand recognition domain at the carboxy-terminal end (Figure 16.3). The NLR family is large and encompasses more than 23 human and 34 murine cytosolic members. NLRs are divided into five subfamilies on the basis of the domain at the amino terminus.

Certain NLRs play complementary and nonredundant roles to TLRs in the innate immune detection of PAMPs. The participation of NLRs in host defense against pathogens that invade the cell, and in the regulation of NFκB and mitogen-activated protein kinase (MAPK) signaling and cell death, indicates that NLRs are important in the pathogenesis of a variety of inflammatory human diseases. Nucleotide-binding oligomerization domains 1 and 2 (NOD1 and NOD2) are the best-characterized NLR members in the intestinal mucosa.

Figure 16.3 NLR structure. NLRs are cytoplasmic proteins with conserved tripartite structures composed of an amino-terminal effector domain, a central nucleotide-binding domain, and a carboxy-terminal leucine-rich repeat (LRR) domain. The amino-terminal domain may contain caspase recruitment domains (CARDs), as in NOD1 and NOD2; a pyrin domain (PYR), as in NLRP1 and NLRP2; or a baculovirus-inhibitor-of-apoptosis repeat domain (BIR), as in NAIP4.

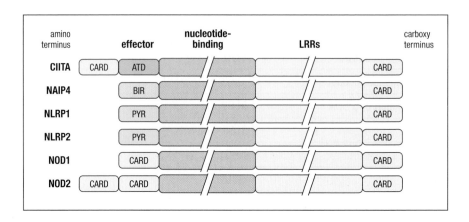

NOD1 and NOD2 were first discovered as mammalian members of the Ced4/Apaf-1 family of apoptosis regulators. Structurally, NOD2 shares significant homology with NOD1 but contains two, instead of one, CARDs at its amino terminus. The CARD recruits caspases, a family of intracellular proteases that induce apoptotic cascades. NOD1 is encoded by the *CARD4* gene and NOD2 by the *CARD15* gene. Both NOD proteins recognize peptidoglycan fragments released from the cell wall of bacteria: NOD1 senses a Gram-negative peptidoglycan derivative, γ-D-glutamyl-meso-diaminopimelic acid (iE-DAP), whereas NOD2 senses muramyl dipeptide (MDP), a minimal bioactive motif of peptidoglycan from both Gram-negative and Gram-positive bacteria. Thus, NOD2 acts as a general sensor of bacterial infection, whereas sensing via NOD1 is restricted to Gram-negative microorganisms. The ligands that activate NOD1 and NOD2 probably enter the cell by endocytosis through clathrin-coated pits. After enzymatic processing in endosomes, the peptide ligands enter the cytosol. NOD1 and NOD2 are implicated in the intracellular recognition of pathogens, including enteroinvasive *Escherichia coli*, *Shigella flexneri*, *Campylobacter jejuni*, *Helicobacter pylori*, and *Listeria monocytogenes*.

NOD2 is constitutively or inducibly expressed in monocytes, macrophages, T and B cells, dendritic cells, epithelial cells, and Paneth cells. In contrast, NOD1 is ubiquitously expressed in many tissues and cells. NOD2 protein has been detected throughout the cytoplasm of cells, but redistribution to the cell membrane is required for the induction of downstream effects after ligand recognition. NOD1 expression is regulated by interferon-γ (IFN-γ), and NOD2 expression by LPS, tumor necrosis factor-α (TNF-α), and IFN-γ. On activation, NOD1 and NOD2 induce the downstream production and release of chemokines and pro-inflammatory cytokines, such as TNF-α, IL-1β, and IL-6. NOD2 activation critically induces autophagy and mediates bacterial clearance. Both NOD1 and NOD2 interact with the inhibitor of apoptosis proteins cIAP1 and cIAP2.

16-3 TLR and NOD signaling pathways converge on downstream NFκB and MAPK.

The recognition of PAMPs by PRRs signals danger to host innate cells. Once PAMPs have been recognized, PRRs initiate intracellular signal transduction cascades, resulting in the activation of transcription factors (Figure 16.4). The transcription factor NFκB is a major common end result of TLR and NOD signaling pathways in innate immune cells.

Individual TLRs differentially activate distinct signaling events through adaptor proteins, resulting in specific immune responses. The 'classical' and first identified pathway involves the recruitment of the adaptor molecule MyD88 via the conserved TIR domain, and activation of the serine–threonine kinases of the IRAK family—leading ultimately to the degradation of IκB and the

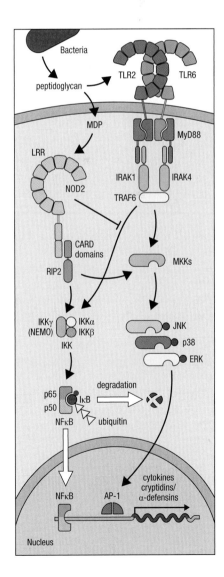

Figure 16.4 Schematic representation of NOD and TLR activation pathways. Bacterial peptidoglycan activates intracellular NOD2 via the interaction of muramyl dipeptide (MDP) with the leucine-rich repeat (LRR) in the carboxy-terminal domain. This interaction causes the oligomerization of NOD2 and recruitment of the kinase RIP2 through CARD–CARD interactions. RIP2 activates the IKK complex through IKKγ/NEMO, resulting in phosphorylation of IκB and its degradation through ubiquitination and release of the sequestered NFκB for translocation into the nucleus and initiation of gene transcription. Peptidoglycan also interacts with surface TLR2/6 to induce the MyD88 signal pathway, leading to activation of TRAF6, a key signal protein. TRAF6 can activate the IKK complex, leading to NFκB activation, and/or activate MAP kinases, including JNK, p38, and ERK, which translocate into the nucleus to induce AP-1 transcription factors. NFκB and AP-1 induce gene transcription for an array of pro-inflammatory cytokines and host defense proteins. MKKs, mitogen-activated protein kinase kinases. (Reprinted from B. Kelsall, *Nature Medicine* 11:383–384, 2005. With permission from Macmillan Publishers Ltd.)

translocation of NFκB into the nucleus. Importantly, all TLRs except TLR3 signal through the adaptor protein MyD88, and TLR4 uses both MyD88-dependent and MyD88-independent pathways. Subsequent transcriptional activation of TLR target genes that encode pro- and anti-inflammatory cytokines, chemokines, effector molecules, and type I interferons initiates the activation of antigen-specific and nonspecific adaptive immune responses. The resultant downstream effects promote mucosal homeostasis.

On ligand recognition, the NODs undergo conformational change and self-oligomerization. Both NOD1 and NOD2 enter into CARD–CARD interactions with the serine–threonine kinase RIP2, which leads to the activation of MAPKs (see Figure 16.4). NOD1 signals through TRAF2/5, whereas NOD2 signals through TRAF6. NOD1 and NOD2 signaling pathways converge downstream on the transcription factor NFκB to induce the expression of many important mediators of innate immune host defense, including cytokines, chemokines, and antimicrobial genes.

Function of pattern recognition molecules in healthy mucosa.

The normal gut flora has an essential role in promoting local immune balance, called homeostasis, in the intestinal mucosa. PRRs contribute to the maintenance of this balance by promoting the avoidance of excessive and deleterious activation of host cell inflammatory responses.

16-4 Negative regulation prevents prolonged and detrimental TLR/NOD signaling.

In healthy intestinal mucosa, tolerance is an essential defense mechanism for the maintenance of hyporesponsiveness to commensal microbiota and their products. However, when a pathogen disrupts or traverses the epithelial barrier and enters the underlying lamina propria, mechanisms that promote tolerance are switched off and regulators that promote PRR signaling to induce mucosal immune defenses are switched on (Figure 16.5). To this end, control mechanisms in the healthy intestinal mucosa regulate TLR expression, localization, and signaling cascades. In the healthy intestine, TLR2 and TLR4 are expressed at low levels on the apical surface of intestinal epithelial cells, limiting microbe recognition but maintaining a basal state of activation toward luminal commensals. The differentiation and polarity of epithelial cells influence TLR localization and ligand responsiveness. In differentiated intestinal epithelial cells, for example, TLR2 and TLR4 are present at the apical surface; however, in undifferentiated intestinal epithelial cells, both TLRs are expressed in cytoplasmic compartments. TLR5 is expressed exclusively on the basolateral surface of differentiated colonocytes, and is thus positioned to detect translocated flagellin when the epithelial barrier has been injured. TLR9 is expressed on both the apical and basolateral membranes of intestinal epithelial cells, and signaling cascades seem to vary in a pole-dependent manner.

Other mechanisms that contribute to the hyporesponsiveness of intestinal mucosa include intracellular molecules that block TLR-induced signaling cascades that activate NFκB and MAPKs. Examples of such molecules include Tollip, which inhibits IRAK activation, and A20, which downregulates pro-inflammatory NFκB. The main inhibitor molecules are depicted in Figure 16.5. In contrast to TLRs, expression of NOD1 or NOD2 is not downregulated in innate cells in healthy intestinal mucosa. Thus, normal cells remain responsive to invasive bacteria via NOD1 or NOD2. However, inhibitors of

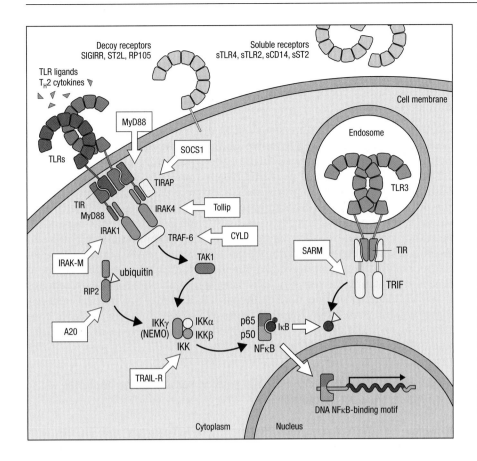

Figure 16.5 TLR signaling and inhibitors. TLR signaling is tightly regulated by signal proteins and inhibitors. TLR recognition of PAMPs on microorganisms initiates myeloid differentiating factor 88 (MyD88)-dependent or MyD88-independent cascades that culminate in the activation of NFκB and MAPK, transition factors which initiate pro-inflammatory and host defense response gene transcription. Regulation is achieved through inhibitors of the various signaling steps. The main inhibitors include Tollip, which inhibits IRAK4 and IRAK1; IRAK-M, a member of the IL-1 receptor-associated kinases in monocytes that blocks the formation of IRAK1:TRAF6 complexes; SOCS1, which inhibits TIRAP; CYLD, an enzyme that interferes with TLR2 signaling through inhibition of TRAF6 and TRAF7; A20, a cytoplasmic zinc-finger protein that blocks receptor interacting protein (RIP) activation; and SARM, which blocks TRIF to inhibit TLR3 and TLR4 signaling. T$_H$2, helper T cell type 2. (Adapted from O. Shibolet and D. Podolsky: *Am. J. Physiol. Gastrointest. Liver Physiol.* 2007, 292:G1469–G1473. With permission from The American Physiological Society.)

NOD1- and NOD2-induced NFκB activation that disrupt receptor oligomerization or block RIP2, regulating NOD-induced immune responses, have been identified. These NOD1 and NOD2 inhibitors include NOD2-S, an alternatively transcribed single-CARD-domain splice variant of NOD2; Centaurin β1, a GTPase-activating protein; and Erbin, a cell polarity protein.

TLRs and NODs may interact with each other through cross-tolerizing feedback loops. Thus, continuous exposure to TLR ligands can suppress exaggerated inflammatory responses after exposure to NOD ligands, and vice versa. IRAK has been identified as a central negative regulator of cross-tolerization between a variety of microbial components of both PRR families. In contrast, pro-inflammatory cytokines, including IFN-γ and TNF-α, may modulate PRR expression and downstream signaling to promote mucosal inflammation, as in inflammatory bowel diseases (IBDs).

16-5 TLR function is involved in the maintenance of mucosal barrier integrity.

The interaction between commensal microbes and the intestinal epithelium induces TLR signaling that promotes barrier preservation, cell survival, and restitution. Aberrant TLR signaling may alter the response to commensal microorganisms, facilitating injury and inflammatory responses. For example, mice in which the TLR2, 4, 5, or 9 or MyD88 genes have been deleted exhibit delayed or diminished tissue repair responses during acute dextran sulfate sodium (DSS)-induced colonic inflammation. Conversely, treatment with TLR2, 4, 5, and 9 ligands prevents or delays the onset of acute DSS-induced colitis. In this connection, treatment with the TLR2 ligand PCSK has been shown to protect tight-junction-associated barrier integrity and decrease intestinal permeability, thereby reducing DSS-induced colonic inflammation during the recovery phase. TLR2 also controls terminal goblet-cell differentiation by selectively regulating trefoil factor 3 (TFF3) expression in the colon, promoting anti-apoptotic protection of the intestinal mucosa. TLR-mediated epithelial cell proliferation in gut mucosa seems to be site-specific and developmentally regulated, possibly as a result of differences in the commensal microbiota in different regions of the gastrointestinal tract and during different stages of development.

Activation of innate immune signaling in the intestinal epithelium by commensal bacteria influences adaptive immunity through cross-talk between epithelial cells and immune cells in the underlying lamina propria. Resident or infiltrating dendritic cells, T cells, and B cells amplify antigen recognition and sorting, generate antibody production, and enhance host-protective inflammatory responses. In mice, bacteria-induced TLR signaling by epithelial cells drives subjacent dendritic cells to sample luminal bacteria. TLR-dependent secretion of mediators such as thymic stromal lymphopoietin (TSLP) and B-cell activating factor belonging to the TNF family (BAFF) may provide an additional link between intestinal epithelial cells and both dendritic cells and B cells, modulating the secretion of mucosa-protective IgA2 in the presence of IL-10. TLR ligation on intestinal epithelial cells also may promote IgG and IgA class switching via BAFF. Together, these TLR-dependent mechanisms contribute to bacterial clearance, limit deleterious pro-inflammatory responses, and thus promote mucosal homeostasis.

16-6 NOD2 modulates antimicrobial peptide secretion and bacterial clearance via autophagy.

Mucosal cell-derived antimicrobial peptides have a key role in the host response to luminal microbes. One important group of such peptides, the α-defensins, is produced by Paneth cells, which are located in the crypts of Lieberkühn in the small intestine. These positively charged, hydrophobic peptides are induced after PRR-mediated activation of NOD2 in Paneth cells. The released α-defensins interact with negatively charged phospholipids to kill bacterial cells efficiently. Activation of NOD2 in Paneth cells exerts antibacterial activity by limiting bacterial invasion and expansion through α-defensin production. In studies with intestinal epithelial cells expressing a functional NOD2 protein, the *in vitro* bacterial clearance of pathogenic *Salmonella* Typhimurium was strongly accelerated, whereas epithelial cells overexpressing a NOD2 mutant (L1007fsinsC) were unable to clear the pathogen. Mice that lack NOD2 or express a dysfunctional NOD2 show a decreased ability to clear intracellular bacteria such as *Listeria monocytogenes*. A NOD-mediated decrease in bacterial clearance may lead to persistent immune activation, dysregulated cytokine production, and subsequent T-cell activation in the lamina propria.

Emerging evidence indicates that autophagy is induced by innate immune pathways to defend against bacterial invasion and maintain cellular homeostasis. Autophagy involves complex catabolic signaling pathways to sequester cytosolic regions within double-membrane-enclosed compartments, called autophagosomes, which fuse with lysosomes for cargo degradation. Autophagy is important in the removal or recycling of aged macromolecules and damaged organelles under stress conditions and in the elimination of pathogens or virulence factors delivered into the cell through type III or IV secretion systems. NOD2 recruits the autophagic regulator ATG16L1 to the plasma membrane at the bacterial entry site. NOD2 then activates the autophagy pathway and directs bacterial trafficking to the autophagosome, which on fusion with the lysosome forms the autophagolysosome. The degraded cargo is loaded onto major histocompatibility complex (MHC) class II molecules and is able to stimulate CD4 T cells. Thus, the autophagic machinery links NOD2 to antigen presentation and subsequent adaptive immune responses.

Genetic alterations in pattern recognition.

Genetic variations in TLRs or NODs may influence ligand recognition, mucosal immune tolerance, and commensal microbe composition. Such variations may affect host susceptibility and disease progression in certain pathogenic processes (Table 16.3). For example, NOD2 mutations have been associated with increased susceptibility to ileal Crohn's disease. About 30% of patients of European ancestry have at least one of three NOD2 polymorphisms; in contrast, TLR variants are relatively rare. Although TLR polymorphisms may not predict overall disease risk, polymorphisms may influence phenotype severity in subgroups of patients with IBD.

16-7 *NOD2* is a major susceptibility gene for Crohn's disease.

In 2001, the first IBD susceptibility locus was found to involve NOD2. In some geographical regions, patients with Crohn's disease had a relatively high frequency of NOD2 variants. The major variants involving the leucine-rich repeat region include a frameshift mutation (L1007fsinsC) and two missense mutations (R702W and G908R), suggesting that a defect in peptidoglycan recognition may be associated with Crohn's disease. The most common mutation, L1007fsinsC, causes a truncated NOD2 protein that lacks the last 33 amino acids. Immune cells carrying such a mutant NOD2 have an impaired capacity to induce NFκB activation in response to ligand stimulation and fail to produce the anti-inflammatory cytokine IL-10. In addition, the presence of a

Table 16.3 Effects of genetic defects in PRRs

Defect	Association	Effect
Human genetic defects in NOD2		
NOD2-R702W NOD2-G908R NOD2-L1007fsinsC	Increased susceptibility to Crohn's disease (fibrostenosis/terminal ileal disease)	Loss of antibacterial activity as a result of impaired bacterial recognition and clearance, decreased defensin production and deregulated autophagy, TLR hypersensitivity, unbalanced T-cell activation with shift toward T_H1/T_H17 responses
Human genetic defects in TLR		
TLR2-R753Q	Ulcerative colitis pancolitis	Impairs intestinal epithelial cell restitution and communication
TLR4-D299G	Increased susceptibility to IBD	Interrupts LPS signaling, but phenotype in IBD so far unclear
TLR5-stop	Decreased susceptibility to IBD	Decreases adaptive immune responses to flagellin
Murine defects in IBD-associated genes, secondarily influencing TLR function		
IL-10$^{-/-}$	In mice. No association	TLR4 limits propagation of colitic effector CD4 T cells, thus ameliorating disease course
NOD2$^{-/-}$	In mice. No association	TLR4 limits propagation of colitic effector CD4 T cells, thus ameliorating disease course

Crohn's disease-associated NOD2 mutation may impair cross-tolerization by disturbing homeostatic TLR signaling and shift signal outcome toward pro-inflammatory hypersensitivity in antigen-presenting cells. Thus, impaired signaling via a variant of the *NOD2* gene could perpetuate disease by uncontrolled and excessive actions of TLR pathways in normally quiescent cells within the intestinal mucosa.

In Crohn's disease, expression of NOD2 is increased in Paneth cells and macrophages, possibly inducing pro-inflammatory cytokines. The production of cryptidins, homologs of α-defensins, is significantly diminished in mice that lack a *Nod2* gene, increasing susceptibility to infection with bacterial pathogens. Similarly, Crohn's disease-associated NOD2 mutants may represent loss-of-function phenotypes. In this connection, transfection studies in humans have shown that Crohn's disease-associated NOD2 mutants fail to constrain the survival and overgrowth of pathogenic bacteria. In patients with NOD2-mutant Crohn's disease, Paneth-cell-derived expression of the human α-defensins HD5 and HD6 is significantly diminished. Paneth-cell-mediated antimicrobial signaling may also be compromised by dysregulated autophagy during colonic inflammation. Crohn's disease-associated NOD2 mutants fail to recruit the critical autophagy protein ATG16L1 to the plasma membrane during bacterial invasion. Consequently, autophagy is defective on bacterial infection or stimulation with the peptidoglycan fragment muramyl dipeptide, resulting in decreased bacterial killing and defective antigen presentation.

Thus, aberrant bacterial handling and immune priming may act as combined triggers for initiating and perpetuating intestinal inflammation in patients with Crohn's disease with NOD2 mutations. Genetic defects in NOD2 may lead to changes in the composition of the microbiota as a result of defensin deficiency, which may alter homeostatic TLR signaling and promote inflammatory disease. However, NOD2 polymorphisms alone are not sufficient to cause Crohn's disease, supporting the concept that multiple factors contribute to the pathogenesis of this disease.

16-8   TLR polymorphisms may modulate IBD severity.

Cohort studies of the incidence and prevalence of TLR polymorphisms in selected populations with IBD have identified TLR1, 2, and 6 genes as being associated with distinct disease phenotypes. Patients with ulcerative colitis and the polymorphisms TLR1-R80T and TLR2-R753Q seem to have an increased risk for pancolitis. The TLR2-R753Q variant represents a loss-of-function mutation that mediates intestinal epithelial dysfunction *in vitro*. Goblet cells with TLR2-R753Q fail to induce TFF3, a goblet-cell-secreted protein important in mucosal defense, and enterocytes with TLR2-R753Q do not communicate properly with each other. Thus, TLR2-R753Q is associated with impaired epithelial cell restitution during wound healing and may contribute to more extensive disease in a subset of patients with IBD.

In active IBD, variant alleles in the *TLR4* gene could induce functional dysregulation of the LPS receptor. Two common mutations in the human *TLR4* gene, D299G and T399I, occur in up to 20% of Caucasians and have been associated with increased susceptibility to IBD. Although the D299G mutation has been shown to interrupt TLR4-mediated LPS signaling *in vitro*, the functional consequence of the mutation in the mucosa in IBD has not been determined.

Other TLR mutations have been identified in patients with IBD. A recently detected dominant-negative TLR5 polymorphism (TLR5-stop), which leads to a 75% loss of TLR5 function, decreases adaptive immune responses to flagellin and, in Jewish cohorts, protects against the development of Crohn's disease. A complete loss of TLR5 in mice results in the development of spontaneous colitis via aberrant TLR4 signaling in response to changes in the commensal composition.

Summary.

In the intestinal mucosa, cells of the innate immune system express an array of germline-encoded PRRs that detect highly conserved microbial structures. Composed of surface TLRs plus intracellular NLRs and some intracellular TLRs, PRRs participate in host defense and tissue repair responses. Together, TLRs and NLRs insure mucosal homeostasis in healthy mucosa, and rapid microbe elimination in infected mucosa. Conversely, defects in PRR genes lead to dysfunctional microbe recognition and may contribute to mucosal inflammation in subpopulations of patients with Crohn's disease.

Further Reading.

Abraham, C., and Cho, J.H.: **Inflammatory bowel disease.** *N. Engl. J. Med.* 2009, **361**:2066–2078.

Beutler, B.: **Microbe sensing, positive feedback loops, and the pathogenesis of inflammatory diseases.** *Immunol. Rev.* 2009, **227**:248–263.

Cario, E.: **Bacterial interactions with cells of the intestinal mucosa: Toll-like receptors and NOD2.** *Gut* 2005, **54**:1182–1193.

Cario, E.: **Toll-like receptors in inflammatory bowel diseases: a decade later.** *Inflamm. Bowel Dis.* 2010, **16**:1583–1597.

Franchi, L., Warner, N., Viani, K., *et al.*: **Function of Nod-like receptors in microbial recognition and host defense.** *Immunol. Rev.* 2009, **227**:106–128.

Fukata, M., Vamadevan, A.S., and Abreu, M.T.: **Toll-like receptors (TLRs) and Nod-like receptors (NLRs) in inflammatory disorders.** *Semin. Immunol.* 2009, **21**:242–253.

Kawai, T., and Akira, S.: **The role of pattern-recognition receptors in innate immunity: update on Toll-like receptors.** *Nat. Immunol.* 2010, **11**:373–384.

O'Neill, L.A.: **The interleukin-1 receptor/Toll-like receptor superfamily: 10 years of progress.** *Immunol. Rev.* 2008, **226**:10–18.

Palm, N.W., and Medzhitov, R.: **Pattern recognition receptors and control of adaptive immunity.** *Immunol. Rev.* 2009, **227**:221–233.

Philpott, D.J., and Girardin, S.E.: **Nod-like receptors: sentinels at host membranes.** *Curr. Opin. Immunol.* 2010, **22**:428–434.

Rosenstiel, P., and Schreiber, S.: **NOD-like receptors–pivotal guardians of the immunological integrity of barrier organs.** *Adv. Exp. Med. Biol.* 2009, **653**:35–47.

Shibolet, O., and Podolsky, D.K.: **TLRs in the gut. IV. Negative regulation of Toll-like receptors and intestinal homeostasis: addition by subtraction.** *Am. J. Physiol. Gastrointest. Liver Physiol.* 2007, **292**:G1469–G1473.

Ting, J.P.Y., Duncan, J.A., and Lei, Y.: **How the noninflammasome NLRs function in the innate immune system.** *Science* 2010, **327**:286–290.

Vijay-Kumar, M., and Gewirtz, A.T.: **Flagellin: key target of mucosal innate immunity.** *Mucosal Immunol.* 2009, **2**:197–205.

Xavier, R.J., and Podolsky, D.K.: **Unravelling the pathogenesis of inflammatory bowel disease.** *Nature* 2007, **448**:427–434.

The Commensal Microbiota and Its Relationship to Homeostasis and Disease

17

Vertebrates and bacteria have coevolved in intimate contact for more than 150 million years. This has created evolutionary pressures on both sides of the divide: on the one hand driving improvement and refinement of defense mechanisms employed by multicellular organisms to counter infection; and on the other hand prompting bacteria to develop more sophisticated evasive maneuvers to subvert the immune system. Selection pressures have also encouraged mutually beneficial relationships, and some bacterial species have adapted to coexist in harmony with the host, particularly at the epithelial barrier surfaces. Despite dwelling in such close proximity, these colonizing microbial communities, or microbiota, seldom cause disease in immunologically competent individuals and instead often confer survival advantage to the host. For instance, gut-residing bacteria liberate nutrients from otherwise indigestible dietary polysaccharides, synthesize essential vitamins, detoxify xenobiotics, and limit the growth of potential pathogens by competing for space and nutrients (Figure 17.1). For this symbiosis to flourish the host must curtail its potent immune responses that might otherwise be directed against the microbiota. However, the barrier surfaces are vast and include the skin and mouth, and

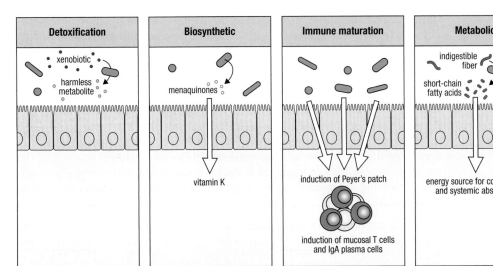
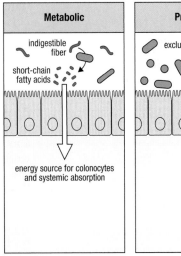
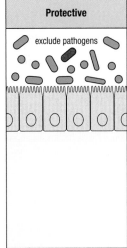

Figure 17.1 The intestinal microbiota confers many health benefits to the host. The intestinal microbiota performs a complex variety of functions that benefit the host. During detoxification, ingested xenobiotics are metabolized into inert byproducts that can pass harmlessly in the feces. Biosynthetic functions involve the synthesis of biomolecules that are absorbed and utilized by the host. For example microbial enzymes are important for the biosynthesis of vitamins such as vitamin K—a critical co-factor involved in hepatic biosynthesis of several key clotting factors. Immune maturation requires an intestinal microbiota: evidence from germ-free mice convincingly shows that in the absence of the intestinal microbiota there is impaired mucosal and systemic immune maturation (see Table 17.1). For example the microbiota facilitates the formation of Peyer's patches and isolated lymphoid follicles and increases the recruitment of lymphocytes to the lamina propria. Metabolic functions involve the fermentation of indigestible dietary fiber into energy-rich metabolites, such as short-chain fatty acids, which can be utilized by the host. These metabolites supply energy to colonocytes and can be utilized systemically. Protective functions are performed by excluding potential pathogens, both by competing for limiting nutrients and by physically excluding them from the epithelial barrier surface, thus protecting the underlying mucosa.

the respiratory, gastrointestinal, and genital tracts, providing an extensive target and potential portal of entry for pathogens. So at the same time as tolerating commensals at the barrier surfaces, the host immune system must also remain alert and poised to repel infections. The gravity of this task should not be underestimated, as worldwide the most common causes of human mortality are infections caused by organisms occupying or gaining entry to the body through the mucosal epithelia, including respiratory tract infections, intestinal infections, and human immunodeficiency virus-1 (HIV-1).

Mammals therefore provide a scaffold upon which numerous microbial communities are assembled, and subsequently provide residence to thousands of individual species. Almost every environmentally exposed surface of our bodies teems with microbes. These populations include bacteria, archaea, fungi, viruses, protozoans, and in some cases even multicellular helminthes. These polymicrobial communities play important roles in the architecture and function of the tissues they inhabit. The simple but profound concept is that the properties of our mucosal surfaces involve a true integration with their resident microbiota to create what has been called a 'supraorganism.'

Principles and definitions of the commensal microbiota.

As a starting point, the important definitions and principles of such microbial communities need to be explained. Three definitions and principles will be helpful. *Microbiome* refers to the members of a microbial community found in a particular anatomic habitat. The microbiome is defined by a combination of traditional microbiology that is dependent upon the ability to culture specific microorganisms under defined conditions and, more recently, the application of unbiased approaches based upon the genetic composition of the specific microbe. These unbiased approaches largely rely upon pyrosequencing (or next-generation sequencing) of conserved regions within the 16S rRNA subunit of microbes using region-specific primers and high-throughput DNA sequencing. Such approaches are changing the definitions of classical taxonomy such that an alternative definition of a species is as a phylotype as defined by 99% identity at the level of the conserved sequences within the 16S ribosomal subunit. Finally, it is important to note that the composition of a microbiome may change over time, even abruptly, due to physiologic, metabolic, or disease states, and be markedly different in different parts of the world. However, the composition of each anatomic microbiome is characteristic, reflecting the specialist microorganisms and their functions devoted to this habitat.

A second important definition is that of the *metagenome*, which refers to the aggregate of genes found in the microbiome. The metagenome is also defined by the application of high-throughput pyrosequencing. However, in this case, rather than focusing upon specific conserved regions of the 16S rRNA structure, next-generation sequencing is performed on random DNA fragments that are derived from the entire microbial community. Using bioinformatic approaches, a metagenome is assembled from the so-called shot-gunned sequenced DNA fragments that provides, in contrast to the high-throughput biased sequences of the microbiome that are focused upon 16S rRNA, an unbiased assessment of the genetic ensemble of the community which can then be organized into functional metabolic repertoires. Distinct sets of genes are found in each microorganism, such that the composite of genes in a microbiome is huge, complex, and overlapping. However, the concept of an 'aggregate' genome for the microbial community is important, because each microbiome operates in an integrated fashion, akin to the function of other multicellular

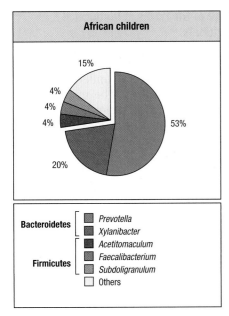

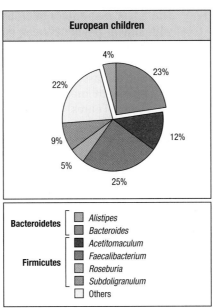

Figure 17.2 Culture-independent 16S ribosomal RNA gene sequencing is useful for mapping the composition of the intestinal microbiota. 16S ribosomal RNA was sequenced to disclose differences in the microbial composition in European children consuming a typical Western-style diet in comparison with children from rural villages in Burkina Faso who consume high-fiber diets resembling ancient human diets. The intestinal microbiome of children from Burkina Faso was enriched in Bacteroidetes and organisms from the genera *Prevotella* and *Xylanibacter* that are known to be a rich source of enzymes capable of degrading cellulose and xylose, consistent with the possibility that our intestinal microbiome has coevolved with our changing dietary patterns. (Adapted from C. De Filippo et al., *Proc. Natl Acad. Sci. USA* 107:14691–14696, 2010. With permission from National Academy of Sciences.)

organs of the body. A spectacular example of the use of this technique to assess the effect of environment comes from analysis of the microbiome in children in Italy and children in Africa (Figure 17.2). In this case Firmicutes make up 51% of the microbial community in Italy, but only 12% in Burkina Faso.

Inter-individual variation is another important feature of the microbiome (Figure 17.2). Because individuals are born germ-free, the composition is initially determined by microorganisms acquired in the local environment. As the intimate caretaker in most species and cultures, the mother's microbiome is the predominant microbial source for a newborn. Due to founder advantage, the composition of the maternally derived microbiota is preserved as a major fingerprint in the microbiome of her progeny even after they have reached adulthood. The genetics of an individual can also affect their microbiome. A substantial portion of the human genome is devoted to host functions that affect host–microbe interaction: host barrier properties, antimicrobial defense, metabolic stress, and inflammation. Genes that are involved in such 'gardening' of the microbiome are beginning to emerge. Indeed, global changes in the intestinal microbial community have been confirmed in mice for null mutations in molecules such as CD8 (T cell co-receptor), perforin (lymphocyte cytotoxicity), interleukin-10 (immunoregulation), CD1d (innate–adaptive immune interactions), and NOD2 (microbial sensing). This would suggest that each person's allelic mosaic would shape the detailed composition of that individual's various microbial communities. However, because composition to the genus level is nearly equally similar between monozygotic or dizygotic twins, the genetic effect may be less important than maternal and other childhood environmental factors, notably diet and antibiotics.

The microbial communities at mucosal surfaces.

Each mucosal surface has a unique physiological role, with different cell types and different environmental stimuli, which result in a different microbiota. These communities of bacteria evolve with the host and are highly specialized to occupy different niches. The microbiota of the gut is the best studied, but it is now also becoming clear that the microbiota of the airways and upper respiratory tract are also unique and diverse in health and disease.

17-1 The upper respiratory microbiome of the nares is distinctive with similar phyla between individuals.

The mucosa of the lower respiratory tract is minimally colonized by microbiota, although this assessment may change when it comes under study using culture-independent metagenomic analysis. In contrast, the microbiota of the nasal passage (nares) contains a rich, complex microbiome. The majority is derived from just two bacterial phyla, Actinobacteria and Firmicutes. A minor contribution is made by Proteobacteria, Bacteroidetes, and four other rare phyla. As expected, the nares microbiome is characteristic, and is more similar between individuals than it is to other anatomic sites within the same individual. However, inter-individual variation is also a feature, representing a relatively stable trait. Because aspects of the microbial fingerprint are shared among couples living together, the composition of this microbiome is likely to reflect a shared residential environment.

17-2 The oral microbiome is characterized by the formation of biofilms.

It is currently estimated that over 750 different species of bacteria reside in the human mouth. In contrast to the skin, the mouth is a mucosal tissue with various subanatomical locations, including epithelial cells of the cheeks, gums, tongue, and palate, as well as enamel on teeth. Therefore, the surfaces that can be colonized differ substantially from the dry squamous epithelium of the skin. The major groups of microbes that inhabit the oral cavity include *Streptococcus*, *Actinomyces*, *Veillonella*, *Fusobacterium*, *Prevotella*, and *Treponema*, among others. It is believed that diverse consortia of microbial classes reside in close association in the mouth, and assemble into true polymicrobial communities where different species of bacteria network cooperatively in interactions that sustain the integrity of the microbiota. Although the surfaces of the mouth are varied, communities appear to form in aggregates on all anatomical niches due to the formation of biofilms. Biofilms are higher-order structures composed of bacteria and their secreted products (mostly polysaccharides) and form intricate physical networks that sustain the microbial community. Within a biofilm, different microorganisms can exchange nutrients and metabolites that foster synergistic relationships that are required for the formation and maintenance of these microbial societies. Studies suggest that biofilm communities are highly structured, and that certain organisms (predominantly *Streptococcus*) are the primary colonizers that nucleate assembly and recruitment of other microbes into dental plaques. Within the biofilm, microbes appear to form organized partnerships and specific bacterial species have evolved adhesins not only for host tissues, but also for receptors on other bacteria. For example, *Fusobacterium nucleatum* can bind to epithelial cells as well as to *Streptococcus cristatus*, and this bacterial–bacterial interaction is required to suppress inflammatory responses against both microbes as a means to circumvent the host immune system during colonization. Specialized bacterial surface appendages have apparently evolved to mediate both bacterial and host associations in the oral cavity. Finally, the coordinated and regulated assembly of microbial communities suggests that perturbations in the 'normal' composition of these microbes (a process termed *dysbiosis*) may affect oral health such as in periodontitis (see below).

Commensal microbes within biofilms may be comprised of beneficial, symbiotic, or potentially pathogenic bacteria. In fact, inflammatory disease of the oral cavity is usually caused by members of the healthy microbiota. Periodontal diseases such as gingivitis and periodontitis result from immune responses to *Streptococcus*, *Actinobacteria*, and *Porphyromonas* species that are resident in the healthy mouth but which form abnormal configurations in relation to the mucosal tissues in these diseases. Furthermore, there have been strong associations reported between poor oral hygiene and non-oral diseases such as endocarditis and lung disease. Although the central cause-and-effect relationships between the oral microbiota and disease are yet to be established, these studies show that the bacterial composition of the oral mucosa may have a considerable impact on human health. Further understanding of the dynamic microbial interactions within oral biofilms may provide clues into how the microbiota affects immunologic outcomes during health and disease in the mucosal surfaces that comprise the gastrointestinal tract.

17-3   The gut microbiome is the most complex of the commensal ecosystems of the host.

Of all anatomical locations, the lower gastrointestinal tract of mammals has the highest density and diversity of commensal microorganisms, and is a habitat permanently colonized by members of five of the six kingdoms of life. Bacteria predominate, and reach astounding levels in the colon (~100 trillion cells; Figure 17.3). Parenthetically, this means that on a per-cell basis, more than 90% of a human being is actually composed of the colonic microbiota. The aggregate human microbiome contains between 1000 and 1150 bacterial species (among all individuals sampled), with a given person harboring approximately 160 bacterial species. However, the species are largely different between individuals and at the genus level the overlap is less than 30%, and indeed the overlap is only random. Thus, there does not yet appear to be a group of 'keystone species' of the gut microbiota. In addition, the composition of the mucosal-surface microbiota is distinct from that recovered in the feces. Therefore, it is likely that there are a number of microbial compartments in the gut, including different anatomic regions of the intestinal mucosa, and the maturing fecal stream.

At a metagenomic level, the intestinal microbiome contains an enormous number of genes, more than 150 times the number of unique (nonredundant) genes found in the human genome. This metagenome spans the entire range of enzymatic functions, and complements the human genome particularly in genes contributing to nutrient uptake and digestion. In contrast to the interindividual diversity of microbial composition, the representation of genes by function is highly conserved. Although less studied, the intestinal microbiome also includes diverse fungi, higher eukaryote organisms, and virus-like

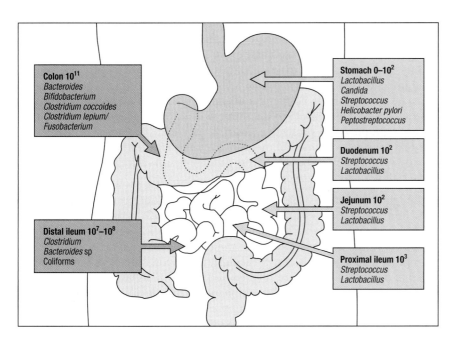

Figure 17.3 Microbial colonization of the gastrointestinal tract. The gut is home to a great density and diversity of commensal microbes. Although there is marked inter-individual variation, particular organisms predominate in particular anatomical locations. For instance, organisms such as *Lactobacillus* and *Streptococcus* dominate the stomach, whereas *Bacteroides* and *Clostridium* are commonly found in the distal gut.

organisms. Unlike the bacterial microbiota, the intestinal 'virome' seems to be a stable but highly individual trait, with minimal influence of either family or environment. It may also be of particular importance because its exceptional abundance and diversity may contribute an enormous genetic repertoire to intestinal function.

Host–microbe interactions.

The host must sense, react to, and discriminate between pathogenic and commensal organisms at the barrier surfaces. The epithelium itself plays a critical role in this process by providing a physical barrier between the external environment, where microbes are abundant, and the underlying tissues, to which their access is restricted.

17-4   The host perceives and responds to the microbiota.

Organisms occupying the gastrointestinal tract are sensed by the intestinal epithelium and cells of the innate immune system. These cells are equipped with a diverse repertoire of receptors to recognize different conserved microbial components termed pathogen-associated molecular patterns, or PAMPs (see Chapter 16). Typical agonists of the pattern recognition receptors are microbial molecules such as the bacterial cell wall components lipoteichoic acid, peptidoglycans, and lipopolysaccharide. The best known of these innate pattern recognition receptors are the Toll-like receptors (TLRs), with 10 family members in humans and an additional 3 in mice. In addition to the TLRs, there are several other pattern recognition receptor families, including cytosolic receptors such as the NOD-like family of receptors. Together the pattern recognition receptor families allow the host to respond to a diverse range of different microbes. By controlling the repertoire, cellular distribution, and tissue localization of these receptors the host can sense and respond to environmentally encountered microbes in different ways in different anatomical locations. For instance, TLRs may be differentially expressed on the apical surface of the mucosal epithelium where they interface with environmental bacteria, or alternatively on the basolateral aspect of the epithelium that in health is a 'privileged' tissue compartment inaccessible to gastrointestinal commensals. However, in the event of loss of containment, as might occur secondarily to epithelial damage, this 'hidden' pool of pattern recognition receptors becomes unmasked and exposed to bacteria richly endowed with PAMP signatures. A good example of this process is seen in acute gastrointestinal infection with invasive *Salmonella*. Infection with this flagellated pathogen results in colonic epithelial damage and *de novo* exposure of normally restricted basolateral TLR5 to the salmonella flagellin. Flagellin ligates TLR5, triggering a powerful protective host inflammatory response. Humans with homozygous loss-of-function mutations in the TLR5 gene are more susceptible to Legionnaire's disease, a severe lung infection caused by the flagellated bacteria *Legionella pneumophila*.

Although one of the key roles of the pattern recognition receptors is to initiate host immune responses following exposure to PAMPs, it remains to be fully understood how the host manages to prevent igniting its immune response pathways in the face of exposure to the enormous antigenic load present at the epithelial barrier surfaces, particularly in the gut. Indeed, it is often forgotten that mucosal surfaces are covered by a thick mucus layer, especially in the colon, so that the surface of the epithelium may in fact receive only minimal exposure to PAMPS derived from the microflora. Interestingly, defects in the mucus layer can lead to a spontaneous colitis in mice.

Despite their role in stimulating antipathogen immunity, it is also acknowledged that pattern recognition receptors interact with gut commensals to facilitate their relationship with the host. The critical importance of the cross-talk between the innate immune system and the gut microbiota for the maintenance of intestinal homeostasis is readily demonstrated by experimental perturbation of either of these components. Mice reared in germ-free conditions have developmental defects in the mucosal immune system and impaired mucosal immunity to common infections. Disrupting innate immune recognition of luminal organisms by the introduction of genetic deficiencies in particular pattern recognition receptors or their signaling pathways also results in disordered homeostasis and altered susceptibility to inflammatory insult.

17-5 Life without microbiota results in profound changes in the host.

Germ-free mice have proven to be a particularly useful experimental tool for studying the interaction between intestinal microbiota and the host immune system. The animals are housed in sterile chambers completely free from the presence of all environmental microbes. Air, food, and water are sterilized to prevent microbial contamination, and individual bacterial species or defined bacterial communities are selectively introduced experimentally. Mice with defined flora (or no flora) are termed gnotobiotic (from *gnotos*, Greek for 'to know,' and *bios*, Greek for 'life').

Since the 1960s it has been consistently noted that germ-free mice have numerous immunological, physiological, and anatomical differences from conventionally reared animals (Table 17.1). They require more calories because they lose the nutrients provided by microbial degradation of plant material in mouse chow (mice are coprophagic). The small intestine is longer and the

Table 17.1 Abnormalities apparent in germ-free mice

Anatomical/histological	Immunological	Functional
Enlarged cecum	Reduced secretory IgA production and low level of serum immunoglobulin	Increased susceptibility to infections: • *Salmonella* • *Listeria* • *Shigella* • *Leishmania* • *Bacillus anthracis*
Longer small intestine with taller villi in the duodenum and shorter villi in the ileum	Diminished numbers and activation of systemic T cells	
Poorly developed mesenteric lymph nodes	Reduced CD8+ T-cell cytotoxicity	
Poorly developed Peyer's patches	Impaired lymphocyte homing to inflammatory sites Reduced intestinal lamina propria lymphocytes and intraepithelial lymphocytes (IELs), especially αβ T-cell receptor IEL Impaired helper T cell T$_H$17 T-cell responses	Reduced susceptibility to autoimmune disease: • Experimental allergic encephalomyelitis • Arthritis • Some models of inflammatory bowel disease
Lower numbers of isolated lymphoid follicles		
Small spleen	Reduced ability of granulocytes to kill bacteria	Reduced ability to induce oral tolerance

villi in the duodenum are longer than in normal mice, but shorter in the ileum. The ceca of germ-free mice are grossly enlarged. There is poor development of gut-associated lymphoid tissue, including Peyer's patches which only contain primary follicles. Unlike conventionally colonized mice few T cells can be found infiltrating the intestinal lamina propria and there is an almost complete absence of secretory IgA. Intraepithelial lymphocyte numbers are very low, although there are more γδ T cells than αβ T cells. TLR and major histocompatibility class II expression on intestinal epithelial cells is also markedly reduced in germ-free animals, reflecting the reduced need for antigen detection and presentation.

Systemically germ-free mice have small spleens and very low levels of serum immunoglobulins. Granulocyte function in terms of their ability to kill bacteria is also reduced, and interestingly, is restored when mice are conventionalized. It is thought that systemic penetration of bacterial products into the bloodstream is responsible for this priming, and interestingly, priming is not seen in Nod1-deficient mice, implying that the effect is mediated via bacterial peptidoglycan.

Many studies have demonstrated that both the mucosal and systemic immune system returns to normal if germ-free mice are colonized with a complex flora, but in a fascinating observation made some 20 years ago, monoassociation with a single organism, namely an unculturable segmented filamentous bacterium, restored systemic and mucosal immune populations to normality.

17-6   T-cell responses appear to be modulated by the microbiota.

Early studies demonstrated defective T-cell function in germ-free mice. For example germ-free mice show an impaired ability to generate T cells that mediate delayed-type hypersensitivity. Germ-free mice also appear to have deficiencies in their ability to generate the regulatory T cells that dampen responses to orally administered antigens. Recent data also show that the differentiation of effector T-cell lineages is also modulated by the community composition of the intestinal microbiota. Since the initial reports of the importance of T_H17 responses in the pathogeneses of different autoimmune diseases, different groups have observed that genetically identical mice housed in different facilities did not appear to have the same susceptibility to experimentally induced autoimmunity. One group of researchers also noticed that C57/B6 mice supplied by Jackson Laboratories, the largest vendor of research mice in the United States, were unable to generate IL-17 responses, whereas C57/B6 mice acquired from an alternative supplier, Taconic Farms, readily produced IL-17. Intestinal T_H17 responses could be generated in Jackson mice if co-housed with mice from Taconic Farms. In germ-free mice, T_H17 cells were markedly reduced in the intestinal lamina propria. However, transplantation of the intestinal microbiota from Taconic mice effectively restored T_H17 cells, whereas transplantation of the intestinal microbiota from Jackson mice failed to induce T_H17 cells. Thus, the capacity of genetically identical hosts to initiate T_H17 responses was dependent on the community profile of the intestinal microbiota. For the first time, these studies confirmed that intestinal microbes are involved in the differentiation of helper T cell subsets, and that the microbial community interfacing with the host immune system may shape lineage commitment of helper T cells.

17-7   Segmented filamentous bacteria specifically drive T_H17 responses.

Sequencing of 16S ribosomal RNA genes confirmed differences in the intestinal microbiota community profile between Jackson and Taconic Farms mice. In particular, a bacterium from the genus *Arthromitus*, termed segmented filamentous bacteria (SFB), was present in Taconic Farms mice, but not in Jackson mice (Figure 17.4). This interesting *Clostridium*-related organism colonizes the terminal ileum where it tightly adheres to the ileal mucosa and in fact penetrates into the epithelial cells. Interestingly, this is the same organism that was shown many years ago to restore systemic and mucosal immunity when monocolonized into germ-free mice. SFB could be transmitted to Jackson mice if they were co-housed in the same cages as Taconic Farms mice, which also resulted in successful acquisition of mucosal T_H17 responses in Jackson mice. Critically, monoassociation of germ-free mice with SFB alone reinstated mucosal T_H17 cells. The T_H17-promoting property of SFB appears to be relatively selective because other T-cell lineages are less dramatically affected. SFB also reduced intestinal regulatory T cells, favoring a more inflammatory bias in gut-resident T cells.

The functional significance of SFB permissive T_H17 responses has been demonstrated in experiments with the natural mouse pathogen *Citrobacter rodentium*. This noninvasive organism colonizes the surface of epithelial cells causing distal colonic inflammation. In comparison with Taconic Farms mice, Jackson mice lacking SFB and T_H17 cells were highly susceptible to *Citrobacter* infection. However, colonization of Jackson mice with SFB reduced their susceptibility to *Citrobacter* infection and restored T_H17 responses.

The mechanisms responsible for T_H17 induction by SFB are incompletely understood, although the close contact formed between long filaments from SFB and intestinal epithelial cells in the ileum is likely to be a key feature of their capacity to modulate T-cell immunity. SFB appear to colonize the surface of Peyer's patch, where they may have a chance to orchestrate mucosal T-cell immunity at the inductive site. Studies analyzing the interaction between SFB and cells of the innate immune system, and especially with different intestinal dendritic cell populations, are being actively pursued.

Analysis of the gene expression profile of the ileum following *de novo* colonization of germ-free mice with SFB has been helpful in revealing which host genes are switched on or off in response to SFB colonization. SFB upregulates numerous immune response genes and genes involved in host defense, such as the antimicrobial peptide RegIIIγ and acute phase proteins, such as serum amyloid A. The latter has also been shown to promote T_H17 differentiation *in vitro*; however, the *in vivo* relevance of this potential mechanism awaits verification. It needs to be emphasized, however, that human gut does not contain bacteria which have the same intimate association with the epithelium as SFB has with mouse ileal epithelium.

Little is known about the SFB-expressed molecular determinants responsible for modulating T_H17 responses. However, there is an extremely interesting precedent where a single molecule from a commensal can have profound effects on host immunity, namely capsular polysaccharide A from the human commensal *Bacteroides fragilis*. This is a highly unusual molecule in that, despite being a polysaccharide, it can undergo processing by dendritic cells and be presented to T cells. In the transfer model of colitis used with lymphopenic mice and exacerbated by infection with *Helicobacter hepaticus*, co-colonization with *B. fragilis* ameliorates disease, but the effect is not seen with *B. fragilis* lacking polysaccharide A. Purified polysaccharide A given orally is also therapeutic in this model of colitis and in TNBS (trinitrobenzene sulfonic acid) colitis. It appears that polysaccharide A promotes the generation of IL-10-secreting CD4⁺ Foxp3⁺ regulatory T cells which inhibit pro-inflammatory T_H1 and T_H17 responses.

17-8 The intestinal microbiota influences autoimmunity.

Although CD4 helper T cell effector lineages are important in combating infections, inappropriate mobilization of these different T-cell subsets is also implicated in autoimmune and allergic disease. T_H17 cells, often in conjunction

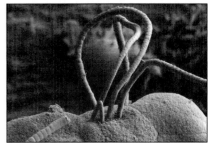
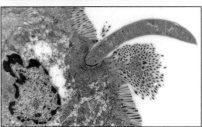

Figure 17.4 Segmented filamentous bacteria. Scanning and transmission electron micrographs of mouse terminal ileum reveal the presence of bacteria with long, segmental filaments in very close contact with the surface epithelium. These segmented filamentous bacteria (SFB) are observed in mice from Taconic Farms, mice that are good at generating mucosal T_H17 responses. SFB are not observed in mice from Jackson Laboratories that are poor inducers of T_H17 responses. Colonization of mice from Jackson Laboratories with SFB generates mucosal T cells that can make IL-17.

with $T_H 1$ cells, are expanded in inflammatory lesions of patients with rheumatoid arthritis, psoriasis, type 1 diabetes mellitus, multiple sclerosis, and inflammatory bowel disease. Similarly, misdirected $T_H 2$ cells, expressing IL-4, IL-5, and IL-13, support eosinophil and mast cell activation to drive allergic diseases such as asthma and rhinitis.

Although one might intuitively anticipate that intestinal microbes exert influence over helper T cell polarization in the gut, it is more surprising that gut microbes can modulate extraintestinal T-cell responses. Germ-free mice have impaired immunity and increased susceptibility to particular infections. However, they are also less sensitive to T cell-mediated autoimmune conditions. Experimental allergic encephalomyelitis (EAE) is a frequently studied animal model of autoimmunity that mirrors aspects of human multiple sclerosis. Disease is mediated by T cells responding to antigens present on the myelin sheath that coats nerves. EAE can be induced in rodents and primates by immunization with nerve tissue-derived proteins or peptides, such as myelin basic protein or myelin oligodendrocyte glycoprotein in adjuvant. About 2 weeks after immunization mice develop relapsing and remitting symptoms of weakness and paralysis. EAE is characterized by expansion of $T_H 17$ and $T_H 1$ cells in inflamed nerve tissue. It has been shown that germ-free mice are much less sensitive to EAE induction than mice colonized with a conventional intestinal microflora. Indeed, many germ-free mice failed to develop any neurological features. EAE resistance in germ-free mice is associated with an attenuated inflammatory response. These experiments strongly implied that gut microbes are necessary for at least some pathogenic T-cell responses outside the gut (Figure 17.5).

Although the germ-free state conferred protection from EAE, subsequent monoassociation with SFB reinstated the host's capacity to initiate $T_H 17$ responses and resulted in increased EAE development and severity. It is noteworthy that the recovery of $T_H 17$ cells and EAE susceptibility in SFB monoassociated mice did not quite reach the levels observed in mice colonized with a complete repertoire of intestinal bacterial commensals, implying that alternative non-SFB bacterial species also contribute to the maturation of effective T-cell responses and disease development.

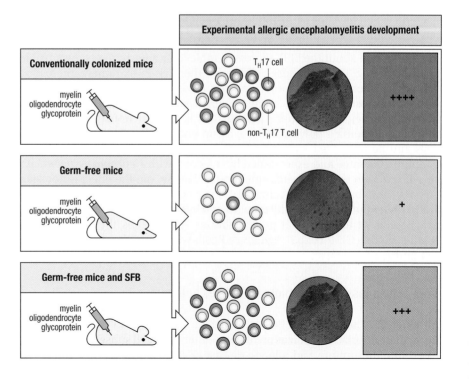

Figure 17.5 Segmented filamentous bacteria and autoimmunity. Immunization of conventionally colonized mice with myelin oligodendrocyte glycoprotein induces experimental allergic encephalomyelitis (EAE). Mice develop weakness and paralysis secondary to nerve damage characterized by leukocyte infiltration and erosion of the myelin sheath (H&E histology panels of inflamed spinal cords at the onset of EAE are shown in the right panels). This is associated with increased initiation of inflammatory T-cell responses, including $T_H 17$ cells (dark blue cells). Conversely, germ-free mice are largely resistant to EAE induction following myelin oligodendrocyte glycoprotein immunization with minor histological changes with fewer inflammatory $T_H 17$ cells. However, monoassociation of germ-free mice with SFB reinstates $T_H 17$ responses and susceptibility to EAE induction.

Very similar results have been observed in an experimental model of autoimmune arthritis. K/BxN mice overexpress a T-cell receptor that recognizes a self peptide derived from glucose-6-phosphate isomerase presented by the major histocompatibility complex class II molecule Ag7. Mice expressing the transgenic receptor spontaneously develop autoimmune joint disease from about 4 weeks of age. However, in the same way as germ-free mice are very resistant to EAE, germ-free K/BxN mice are also protected from the development of joint disease. It was also observed that disease-resistant germ-free mice had marked reduction in the expression of T_H17-related genes. It was possible to recover the typical autoimmune disease manifestations characteristically observed in K/BxN mice following reconstitution of germ-free mice with monocultures of SFB. The triggering of arthritis in SFB monoassociated germ-free mice is also associated with the emergence of lamina propria and splenic T_H17 responses.

17-9 The microbiota appears to be involved in inflammatory bowel disease, particularly Crohn's disease.

One of the most important examples of a disordered interaction between the host immune system and the intestinal microbiota is seen in inflammatory bowel disease (IBD). Crohn's disease and ulcerative colitis, the main forms, are characterized by intestinal inflammation and troublesome symptoms such as diarrhea, rectal bleeding, and abdominal pain, as well as potentially serious complications. Most investigators agree that IBD likely results from inappropriate immune activation in response to intestinal microbes. A number of lines of evidence have implicated the microbiota as key contributors to IBD pathogenesis (Table 17.2). In certain situations antibiotics are therapeutically

Table 17.2 Evidence implicating the microbiota in IBD

	Crohn's disease	Ulcerative colitis
Antibiotics useful?	Effective in management of fistulae Effective in penetrating disease with abscess formation Reduces postoperative recurrence	Effective in pouchitis
Probiotics useful?	Not helpful	Effective in pouchitis Effective in maintaining remission
Prebiotics useful?	Not helpful	Yet to be established, likely small/no benefit
Diversion of fecal stream?	Helpful	Not routinely performed for ulcerative colitis
Disease location	Most common location is ileo-cecal disease (areas of high microbe exposure)	Colon only (occasional 'backwash ileitis')
Adherent-invasive *E. coli*	Common	Uncommon
Faecalibacterium prausnitzii	Protective effect	Unknown
T-cell reactivity to microbes	Present	Present
Genetics	Innate immune molecules handling the microbiota (e.g. NOD2)	No innate components implicated to date

beneficial in patients with Crohn's disease, including patients with perianal disease or fistulae, and in the prevention of postoperative recurrence in patients with isolated ileo-cecal Crohn's disease. Antibiotics are also the first-line therapy for postoperative ulcerative colitis patients who develop pouchitis. Similarly, some probiotics can maintain remission in ulcerative colitis and can also be therapeutic in pouchitis. Probiotic bacteria appear to induce regulatory host immune responses favoring tolerance toward the intestinal microbiota. For instance, the probiotic mixture VSL3 induces production of IL-10 by dendritic cells and reduces pro-inflammatory cytokines such as IL-12. In severe IBD some patients can be effectively treated by surgically diverting the fecal stream away from inflamed intestinal segments, reducing the exposure of inflamed tissue to luminal microbes. It is also well recognized that many commonly studied mouse models of intestinal inflammation rely on the intestinal microbiota. For instance, the spontaneous colitis that develops in mice deficient in the regulatory cytokine IL-10 is not seen when the mice are housed under germ-free conditions. Indeed the penetrance of colitis and ileitis in these mice varies according to the cleanliness of the animal facility.

It is unlikely that a single member of the indigenous microbiota drives IBD. This is well demonstrated in animal models. *Bacteroides vulgatus* and *Bacteroides thetaiotaomicron*, but not *Bacteroides distasonis* or *Escherichia coli*, induce colitis in germ-free HLA-B27 rats. However, in germ-free IL-10-null mice, colonization with any of these species fails to induce colitis. In contrast, *Enterococcus faecalis* is highly colitogenic in IL-10-null mice. Viral infections may also be involved in colitis. Murine norovirus is endemic in laboratory mice. When mice expressing the Crohn's disease mutated *ATG16L1* autophagy gene are infected with murine norovirus, there is marked abnormality of Paneth cells and the animals are more susceptible to DSS (dextran sulfate sodium) colitis. Murine norovirus is related to several human norovirus species that cause acute gastroenteritis and have been associated with IBD.

An adherent and invasive *Escherichia coli*, which can invade intestinal epithelial cells and lamina propria macrophages, where they can survive and replicate, may be involved in Crohn's disease (Chapter 23). However, these organisms are not present in all patients and are often present in the intestine of patients without Crohn's disease. Conversely, another bacterial component of the intestinal microflora, *Faecalibacterium prausnitzii*, may be important in patients with Crohn's disease. *F. prausnitzii* is markedly reduced in the gut of such patients. Patients with high levels of *F. prausnitzii* are less prone to disease recurrence following surgical resection of an inflamed segment of intestine. There is intense interest in the gut microbiome in IBD, but it is not certain whether any changes are causal or secondary to inflammation.

Host responses to pathogen-associated intestinal inflammation are also influenced by the microbiota. For example, *Helicobacter hepaticus* infection induces colitis and liver inflammation in conventionally colonized animals. However, germ-free mice monoassociated with *H. hepaticus* do not develop disease, implying that components of the conventional microbiota are not necessarily innocent bystanders in the company of pathogens. Interestingly, patients with IBD have detectable circulating antibodies and T-cell specificities directed against normal constituents of their intestinal microbiota.

17-10 Defective immune system handling of intestinal microbes leads to intestinal inflammation.

Defects in the host immune system may result in disordered handling of the intestinal microbiota. The pattern recognition receptor families are crucial molecules involved in sensing and reacting to intestinal microbes and selective genetic deficiencies in particular TLRs, such as TLR5, result in spontaneous intestinal inflammation. Similarly, deletion of MyD88, an adaptor molecule that is responsible for facilitating effective signal transduction of almost all of the TLRs, also results in markedly enhanced susceptibility to DSS colitis.

Recently, a novel mouse model of IBD was described in mice lacking the key immune transcription factor T-bet in the innate immune system. In addition to selective deletion of T-bet, these mice also completely lack an adaptive immune system by additional knockout of the recombinase activating gene 2 (RAG2) that plays a critical role in the development of adaptive T-cell and B-cell receptors (Figure 17.6). These T-bet × RAG2 double-knockout mice develop spontaneous colitis that closely resembles many aspects of human ulcerative colitis, and they have been termed TRUC (T-bet × RAG2 ulcerative colitis) mice. The disease begins in the distal colon with relatively superficial inflammatory changes and in the long term, colitis is often complicated by colon cancer, similarly to the situation in uncontrolled human ulcerative colitis. The initial phase of the disease is characterized by overproduction of the inflammatory cytokine tumor necrosis factor (TNF)-α by CD11c-expressing lamina propria myeloid cells. Colitis can be effectively treated by blocking TNF-α with monoclonal antibodies, again recapitulating human IBD. Like other IBD models, intestinal inflammation in TRUC mice is dependent on the intestinal microbiota. TRUC colitis can be successfully treated with antibiotics and probiotics, and fails to develop if mice are reared under germ-free conditions. Furthermore, isogenic colonies of T-bet × RAG2 double-knockout mice housed at different laboratories are not equally sensitive to colitis development, suggesting that a specific microbial component present in some colonies, but not others, is responsible for driving intestinal inflammation. T-bet represses the transcription of the TNF-α gene in myeloid cells, and consequently in its absence there is a tendency for deregulated TNF-α expression that drives epithelial apoptosis and intestinal inflammation.

TRUC and other spontaneous IBD models have provided some important lessons regarding the mechanisms underlying human IBD. It is becoming increasingly clear that interactions between host factors, such as genotype, and environmental factors, such as the microbiota, act in concert rather than isolation to determine disease occurrence, progression, and outcome. In recent years considerable advances have been made in our understanding of the genetic basis of human IBD and genome-wide association studies have identified many single nucleotide polymorphisms that are associated with IBD (Chapter 31). In Crohn's disease genetic variations in genes encoding proteins involved in handling microbes have been identified, including NOD2, an intracellular pattern recognition receptor, and genes encoding autophagy pathway proteins, such as ATG16L1 and IRGM. Genetic variations in several components of the T_H17 pathway have also been singled out by genome-wide association studies, further underlining the critical importance of adaptive immunity in these complex diseases.

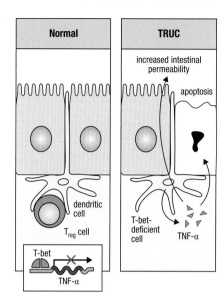

Figure 17.6 The TRUC (T-bet$^{-/-}$ × RAG2$^{-/-}$ ulcerative colitis) model of colitis. In the healthy gut, intestinal dendritic cells sample luminal contents but are prevented from initiating inflammatory responses by regulatory T cells (T_{reg}) and a tightly regulated transcriptional program. For example, the transcription factor T-bet represses the pro-inflammatory cytokine tumor necrosis factor (TNF)-α. However, in the absence of T-bet, de-repressed expression of TNF-α leads to excess production of TNF-α, epithelial apoptosis, and spontaneous colitis. (Adapted from G. Hecht, *N. Engl. J. Med.* 358:528–530, 2008. With permission from Massachusetts Medical Society.)

Influence of microbiota on host metabolism.

Western societies are currently experiencing an obesity pandemic that is resulting in considerable economic and health costs. Obesity is associated with metabolic disturbances, including hyperglycemia, hyperlipidemia, and insulin resistance, collectively termed 'the metabolic syndrome'. Patients with obesity and the metabolic syndrome have reduced life expectancy and are at high risk of developing diabetes, liver disease, hypertension, and cardiovascular disease. Although overeating and sedentary lifestyles play a central role in the development of obesity, recent insights indicate that colonic microbes also play a role in energy balance and contribute to the pathophysiology of obesity. Fat is also abundantly infiltrated with inflammatory macrophages secreting cytokines, thus the obese state is also a pro-inflammatory state with, for example, raised levels of IL-6. Although the pathways are only beginning to be established, diet can influence the microbiota, which can then influence the host, perhaps via PAMPS, in addition to the direct effect of excess nutrients on host metabolism.

Experiments with germ-free mice provided clues that the microbiota exerts influence on host metabolism. As stated above, germ-free mice eat more than genetically identical littermates with a conventional intestinal microflora because they cannot salvage nutrients from bacterial degradation of fiber in chow.

Colonization of germ-free mice with a conventional intestinal microbiota stimulates widespread changes in many processes involved in host metabolism. Intestinal glucose absorption is increased twofold and systemic changes in the host such as increased hepatic lipogenesis, increased adipose triglyceride synthesis, and increased leptin production are also observed following *de novo* intestinal colonization. Gene expression profiling of epithelial cells also revealed that after immune response genes, the second largest group of genes induced by conventional intestinal microbes are genes involved in metabolism. High-resolution analysis of the plasma of germ-free and conventionally colonized mice also reveals dramatic differences in the profile of hundreds of metabolites confirming the profound impact that the intestinal microbiota has on host metabolism.

In contrast to conventionally colonized mice that develop diet-induced obesity when they are fed a 'Western'-style diet rich in fat, germ-free mice are relatively resistant to these effects, implying that the influence of the microbiota on host metabolism can adversely affect health. Critical insights in to the role of the microbiota in obesity have come from studies evaluating the community structure of obese mice using 16S ribosomal RNA gene sequencing.

17-11 Alterations in the community structure of colonic microbes are linked to obesity and the metabolic syndrome.

Two distinct divisions of bacteria overwhelmingly dominate in the gut: Firmicutes and Bacteroidetes. However, the relative abundance of these two phyla is consistently altered in obesity. The *ob/ob* mouse has a spontaneous mutation in the leptin gene, a hormone responsible for regulating appetite. Although these mice have normal birth weights, excess food consumption invariably leads to obesity and its attendant complications, including the metabolic syndrome. Leptin is also involved in regulating immune responses. The relative abundances of Bacteroidetes and Firmicutes in the colon of *ob/ob* mice are very different to that of wild type mice, with a reduction in the abundance of Bacteroidetes and a reciprocal expansion of Firmicutes. Diet-induced obesity also prompted an increase in the ratio of Firmicutes to Bacteroidetes within the colonic microbiota, in keeping with observations in leptin-deficient *ob/ob* mice. Comparable shifts in the relative frequencies of Firmicutes and Bacteroidetes have also been observed in the colonic microflora of obese humans. Furthermore, obese volunteers who subsequently consumed a low-calorie diet and successfully lost weight restored the distorted ratio of the chief phyla back to normal.

Evidence directly implicating the colonic microbiota in a contributory role in obesity came from gnotobiotic studies. To directly demonstrate the role of the colonic microbiota in the development of obesity, an elegant series of experiments was conducted where the colonic microbiota was harvested from either obese or lean mice and then 'transplanted' to germ-free mice, so that these mice with no colonic commensals of their own subsequently became colonized with the colonic microbial communities that prevail in obese or

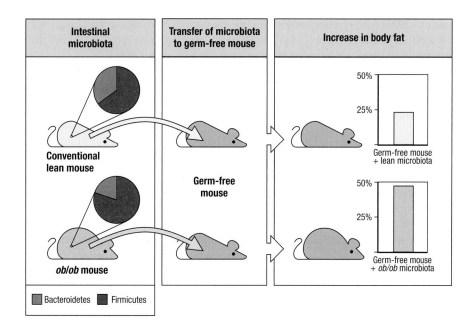

Figure 17.7 Obesity is associated with changes to the intestinal microbiota. It has been established that obese mice, and indeed humans, have an altered profile of the chief colonic bacterial phyla. Mice deficient in the appetite-controlling hormone leptin (*ob/ob* mice) develop marked obesity secondary to increased food consumption. In comparison with lean mice these obese mice have an increase in the relative frequency of Firmicutes (red) and a reciprocal contraction of Bacteroidetes (blue). Fascinatingly, if the intestinal microbiota is harvested from *ob/ob* mice and 'transplanted' to germ-free mice these mice put on more body fat than germ-free mice transplanted with the microbiota harvested from lean mice. These changes are associated with increased energy harvest from the feces by the microbiota derived from obese mice due to more pronounced bacterial fermentation.

lean mice (Figure 17.7). Crucially, germ-free mice colonized with the colonic microbiota from obese mice developed a 50% increase in fat deposition compared with mice colonized with the microbiota from lean mice. These experiments demonstrated that the colonic microbial community present in obese mice contributed to the increased fat deposition and metabolic complications observed in obese mice.

17-12 The host genome interacts with the microbiome to influence host phenotype.

Recent data have also highlighted how defects in host perception of the microbiota can also impact on host obesity. Pattern recognition receptors and especially the Toll-like receptors are the major sensing apparatus of the innate immune system. Fascinatingly, mice lacking TLR5 receptors are highly susceptible to obesity and its attendant metabolic complications. In this situation the structure of the major phyla appeared to be relatively preserved, hinting at the possibility that individual microbial species might also contribute to obesity and the metabolic syndrome.

Given the association of particular community microbiota profiles with disease states, it follows that correction or manipulation of the microbiota might be a credible therapeutic option to reverse the deleterious effects on health. Unsophisticated manipulation of the intestinal microbiota through administration of broad-spectrum antibiotics has been shown to improve diet-induced obesity and its metabolic complications in *ob/ob* mice. Oral administration of probiotic mixtures, or even a single organism such as *E. coli* Nissle, has also found therapeutic success in ulcerative colitis and antibiotic-associated diarrhea. It is conceptually attractive to contemplate ways of manipulating the microbiota with the implicit intention of influencing host health. For example, it is conceivable that an artificial microbiota constructed from polysaccharide A-expressing *Bacteroides fragilis* that favors regulatory T-cell expansion at the expense of inflammatory T_H17 cells could be harnessed to treat patients with IBD or other autoimmune diseases. In contrast, an artificial microbiota containing organisms that encourage inflammatory responses, such as SFB, might be employed to bolster protective immune responses against infectious agents. Alternatively, shaping the microbiota community profile by administering growth factors that encourage the growth of certain species at the expense of others might also allow us to shape inflammatory and regulatory immune responses *in vivo*.

Summary.

Prokaryotes are both our most enduring adversaries and our most constant allies. Present in phenomenal numbers in intimate contact with our barrier surfaces they rarely threaten health, instead performing a diverse range of activities that benefit the host. Experiments in germ-free mice have demonstrated that these organisms influence host immune responses in qualitative and quantitative ways, and in their absence the immune system fails to develop properly, rendering the host vulnerable to infections. In some ways the microbe-rich gastrointestinal tract might be considered a nursery or even *Ludus Magnus* for B cells and T cells, where they cut their teeth and subsequently mature into cells capable of mediating adaptive immunity. Although a number of these symbiotic activities can be carried out by many different intestinal commensals, it is now recognized that specific commensal species play a key role in shaping particular arms of adaptive immunity, including SFB for the initiation of T_H17 responses and polysaccharide A-expressing *Bacteroides fragilis* in promoting regulatory T cells. In the coming years it is likely that additional microbial factors, or unique species, will be identified that can also modulate specific aspects of host immunity. In recognition of the critical role played by the microbiota in influencing host health an ambitious 5-year research program called the 'Human Microbiome Project' was launched in 2008 by the National Institutes of Health costing in excess of $150 million. It is anticipated that these detailed analyses of the organisms that share our barrier surfaces will provide key insights in to the inner workings of our own immune system and may uncover novel therapeutic strategies for combating immune-mediated diseases.

Further Reading.

Cadwell, K., Patel, K.K., Maloney, N.S., *et al.:* **Virus-plus-susceptibility gene interaction determines Crohn's disease gene Atg16L1 phenotypes in intestine.** *Cell* 2010, **141**:1135–1145.

De Filippo, C., Cavalieri, D., Di Paola, M., *et al.:* **Impact of diet in shaping gut microbiota revealed by a comparative study in children from Europe and rural Africa.** *Proc. Natl Acad. Sci. USA* 2010, **107**:14691–14696. Epub 2010 Aug 2.

Garrett, W.S., Lord, G.M., Punit, S., *et al.:* **Communicable ulcerative colitis induced by T-bet deficiency in the innate immune system.** *Cell* 2007, **131**:33–45.

Goodman, A.L., and Gordon, J.I.: **Our unindicted coconspirators: human metabolism from a microbial perspective.** *Cell Metab.* 2010, **12**:111–116.

Hansen, J., Gulati, A., and Sartor, R.B.: **The role of mucosal immunity and host genetics in defining intestinal commensal bacteria.** *Curr. Opin. Gastroenterol.* 2010, **26**:564–571.

Ivanov, I.I., Atarashi, K., Manel, N., *et al.:* **Induction of intestinal Th17 cells by segmented filamentous bacteria.** *Cell* 2009, **139**:485–498.

Lee, Y.K., and Mazmanian, S.K.: **Has the microbiota played a critical role in the evolution of the adaptive immune system?** *Science* 2010, **330**:1768–1773.

Ley, R.E., Turnbaugh, P.J., Klein, S., *et al.:* **Microbial ecology: human gut microbes associated with obesity.** *Nature* 2006, **444**:1022–1023.

Powell, N., Canavan, J.B., MacDonald, T.T., *et al.:* **Transcriptional regulation of the mucosal immune system mediated by T-bet.** *Mucosal Immunol.* 2010, **3**:567–577.

Qin, J., Li, R., Raes, J., *et al.:* **A human gut microbial gene catalogue established by metagenomic sequencing.** Nature 2010*,* **464**:59–65.

Sokol, H., Pigneur, B., Watterlot, L., *et al.: Faecalibacterium prausnitzii* **is an anti-inflammatory commensal bacterium identified by gut microbiota analysis of Crohn disease patients.** *Proc. Natl Acad. Sci. USA* 2008, **105**:16731–16736. Epub 2008 Oct 20.

Thompson, G.R., and Trexler, P.C.: **Gastrointestinal structure and function in germ-free or gnotobiotic animals.** *Gut* 1971, **12**:230–235.

Turnbaugh, P.J., Ley, R.E., Mahowald, M.A., *et al.:* **An obesity-associated gut microbiome with increased capacity for energy harvest.** *Nature* 2006, **444**:1027–1031.

Vijay-Kumar, M., Aitken, J.D., Carvalho, F.A., *et al.:* **Metabolic syndrome and altered gut microbiota in mice lacking Toll-like receptor 5.** *Science* 2010, **328**:228–231.

Wu, H.J., Ivanov, I.I., Darce, J., *et al.:* **Gut-residing segmented filamentous bacteria drive autoimmune arthritis via T helper 17 cells.** *Immunity* 2010, **32**:815–827.

PART IV | GENITOURINARY TRACT

The Immune System of the Genitourinary Tract

18

The human genitourinary tract is characterized by a complex system of highly integrated tissues, finely tuned hormonal regulation, and an immune system that has some similarities to, but also distinct differences from, the gastrointestinal immune system. Innate and adaptive immune defenses in the genitourinary tract provide protection against sexually transmitted diseases. The important role of sex hormones in regulating immune protection is a feature that distinguishes reproductive mucosal immunology from mucosal immune responses at other sites.

Anatomy of the human male and female reproductive tract.

The human male reproductive system includes the testis, the epididymis, vas deferens, seminal vesicles, prostate, and the urethra (Figure 18.1, right). The testis of the male serves the dual function of producing both gametes (sperm) and sex hormones (testosterone). In the testis, androgen synthesis occurs in the Leydig cells, and spermatogenesis takes place in the seminiferous tubules. The two hormones (gonadotropins) produced by the anterior pituitary that regulate testicular function are follicle stimulating hormone and luteinizing hormone (LH), each controlled by hormones produced in the hypothalamus. Spermatogenesis and androgen synthesis in Leydig cells are regulated by a negative-feedback loop involving the hypothalamic–pituitary–gonadal axis. Sperm formed in the seminiferous tubules enter the vas deferens, an androgen-dependent organ that transports sperm into the pelvis. The vas deferens joins the seminal vesicles to form the ejaculatory ducts that enter the prostatic urethra. Just prior to ejaculation, the testes are brought close to the abdomen and fluid is rapidly transported through the vas deferens to the ejaculatory duct and subsequently into the prostatic urethra, the final pathway for both urine and semen.

The female reproductive tract has upper and lower components. The lower female reproductive tract structures (ectocervix and vagina) enable sperm to enter the body and protect the internal genital organs from potential pathogens. Internal reproductive structures include the endocervix, uterus, and fallopian tubes (Figure 18.1, left). The ovary produces both gametes (ova) and sex hormones (estradiol and progesterone). The fallopian tubes attached to the upper part of the uterus provide passage for the ovum from the ovary to the uterus. Following fertilization of an egg by a sperm in the fallopian tubes, the fertilized egg then moves to the uterus, where it implants into the lining of the uterine wall. Each site in the female reproductive tract functions to insure passage of sperm to the site of fertilization, as well as the route of passage of the baby at birth.

Figure 18.1 Internal and external organs of the female and male human reproductive tracts. Both the female and male reproductive systems are involved in the production of gametes and the secretion of hormones. In the female, the reproductive system produces ova or eggs and the hormones estradiol and progesterone. In the male, the reproductive system produces sperm as well as the hormone testosterone.

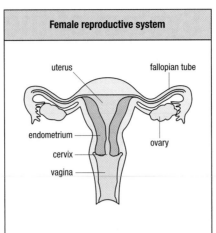

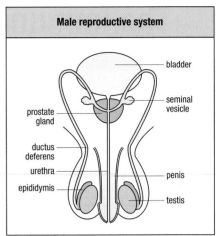

Hormonal regulation of female reproductive function.

The menstrual cycle can be divided into proliferative, midcycle, and secretory phases (Figure 18.2). Follicle stimulating hormone and luteinizing hormone made by the pituitary in response to hypothalamic signals are primarily responsible for the regulation of germ cell development and ovulation during the follicular phase. Follicle stimulating hormone stimulates follicle growth and maturation by its action on granulosa cells, but ovulation does not take place. For follicular growth to occur, follicle stimulating hormone acts in synergy with estradiol and LH. LH participates in the events leading to ovulation and acts on the preovulatory follicle to cause its rupture and the release of the ovum. Levels of these hormones in plasma are shown in Figure 18.2. The midcycle peak of LH is essential for ovulation, and occurs between 24 and 36 hours after estradiol levels peak in blood. Following ovulation, LH plays an important role in the formation and maintenance of the corpus luteum, the primary source of estradiol and progesterone during the secretory phase of the cycle. Human chorionic gonadotropin, with its intrinsic LH activity, rescues the corpus luteum during the cycle if fertilization occurs, and thereby maintains hormone production early in pregnancy. When fertilization does not occur, the corpus luteum involutes.

The mucosal immune system in the female reproductive tract contains an array of protective mechanisms that are under hormonal control. Resident epithelial cells and supportive stromal (lamina propria) cells, and the immune cells that migrate into the uterus, cervix, and vagina, confer immune protection to the female reproductive tract.

18-1 Cytokines, chemokines, and antimicrobial products contribute to protection of the female reproductive tract.

Soluble factors, including cytokines, chemokines, and antimicrobial products, are secreted constitutively in the female reproductive tract and in response to pathogenic challenge. Immune cells throughout the female reproductive tract express pathogen recognition receptors, including Toll-like receptors (TLRs), NOD-like receptors (NLRs), and RIG-like helicases (RIG-I and MDA-5). Antimicrobials in female reproductive tract secretions include α- and β-defensins, secretory leukocyte protease inhibitor, elafin, and LL37,

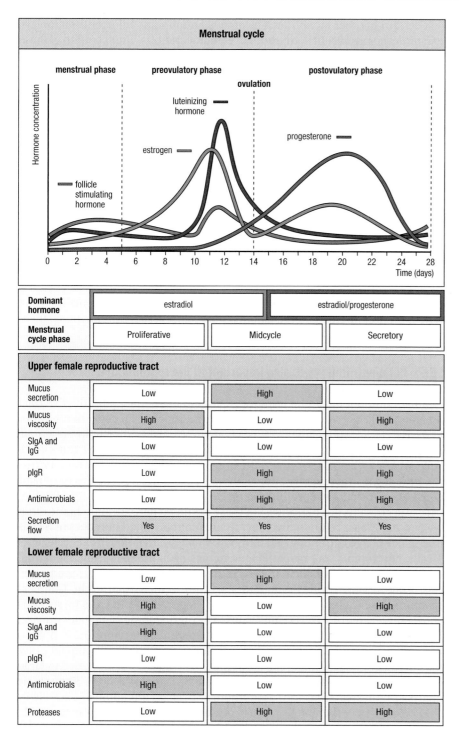

Figure 18.2 **Variable levels of sex hormones secreted during the menstrual cycle regulate the secretion of protective factors in the reproductive tract of women.** The menstrual cycle is divided into three phases: the proliferative phase during which estradiol is produced by the ovary; the midcycle phase when ovulation occurs; and the secretory phase during which estradiol and progesterone are the dominant hormones. These hormones in turn regulate the synthesis and secretion of mucus, antibodies, and antimicrobial products in the female reproductive tract. Soluble products optimize conditions for successful fertilization while protecting against potential pathogens. SIgA, secretory IgA; pIgR, polymeric immunoglobulin receptor.

all of which inhibit bacterial, fungal, and viral pathogens, including human immunodeficiency virus-1 (HIV-1). MIP-3α and elafin are dual mediators in that they have antimicrobial activity as well as a chemoattractant function for the recruitment of T cells and immature dendritic cells to sites of pathogen invasion. Secreted factors therefore contribute to initial mucosal protection through their antimicrobial activity and recruit and activate a second line of cellular immune protection. Among those cells pivotal in conferring immune protection are mucosal epithelial cells. In addition to secreting antimicrobial factors that are both bactericidal and viricidal, epithelial cells provide barrier protection and transport immunoglobulins (IgA and IgG) into female reproductive tract secretions.

A spectrum of other factors such as mucus, immunoglobulin, antimicrobials, and proteases is present in reproductive tract secretions. The secretion of these factors varies with the site (upper versus lower female reproductive tract) and the phase of the reproductive cycle (Figure 18.2). For example, the level and viscosity of secreted mucus varies with the cycle phase throughout the female reproductive tract. Cervical and vaginal mucus production is low with high viscosity during both the proliferative and secretory phases of the cycle. In contrast, at midcycle, estradiol leads to increased secretion of mucus with decreased viscosity to facilitate sperm movement from the vagina to the upper female reproductive tract. IgA and IgG levels are reduced in the upper tract throughout the cycle. In the lower tract, levels are low at midcycle and during the secretory phase. The polymeric immunoglobulin receptor (pIgR), which transports IgA from tissue to lumen, is elevated at midcycle in the uterus. Antimicrobials, including secretory leukocyte protease inhibitor, which has anti-protease activity, and the defensins, also vary with cycle stage in the female reproductive tract. In the upper tract, levels are elevated at midcycle and during the secretory phase. In contrast, they are lowest during this period in the lower female reproductive tract. Thus, sex hormones orchestrate a dynamic balance between humoral and innate immune components such that different components emerge as others are suppressed, optimizing chances for successful fertilization, while maintaining immune capacity.

18-2   Cellular immunity is also important in the reproductive tract.

Cellular components of the immune system in the female reproductive tract include dendritic cells, macrophages, natural killer (NK) cells, neutrophils, and CD4+, CD8+, and regulatory T cells (Figure 18.3). Leukocytes in the female reproductive tract make up approximately 12–40% of the total number of cells, with a higher proportion in the upper tract (fallopian tubes and uterus) than the lower tract (cervix and vagina). T cells and NK cells are the major constituents of female reproductive tract leukocytes in all tissues. The fallopian tube also contains granulocytes, but these are much less common in the other tissues. All female reproductive tract tissues contain small numbers of B cells and macrophages. Immune cell numbers vary with the menstrual cycle, with T cells, neutrophils, and macrophages increasing in number as the menstrual cycle progresses from follicular to secretory phase (Figure 18.3). Immune cell organization varies with tissue and stage of the menstrual cycle. Confocal scanning laser microscopy of vibratome-prepared uterine tissue sections shows aggregates of T cells exclusively in the uterus; these aggregates are oval in shape and located between glands in the functionalis region. The lymphoid aggregates contain a B cell central region surrounded by T cells and an outer halo of macrophages. The B-cell core (CD19+) is most often identified in large aggregates in the late proliferative and secretory stages of the cycle. Phenotypic analysis indicates that the surrounding T cells are almost exclusively CD3+CD8+. CD3+CD4+ T cells are also present, but usually are present outside the aggregates, mainly in the stroma. Macrophages (CD14+ cells) are present as a mantle around the T cells. Aggregates are smallest during the proliferative phase (300–400 cells per aggregate) and increase to 3000–4000 cells per aggregate during the secretory phase of the cycle. Similar menstrual cycle changes occur in the CD8+ cytotoxic T lymphocyte (CTL) population. Although absolute numbers do not change in the uterus, CTL activity is demonstrable during the follicular phase, but is suppressed during the secretory phase of the cycle, possibly to optimize conditions for successful implantation of the allogeneic fertilized egg. In contrast to CTL activity in the uterus, CTL activity in the cervix and vagina is sustained throughout the menstrual cycle, and is not affected by hormonal balance.

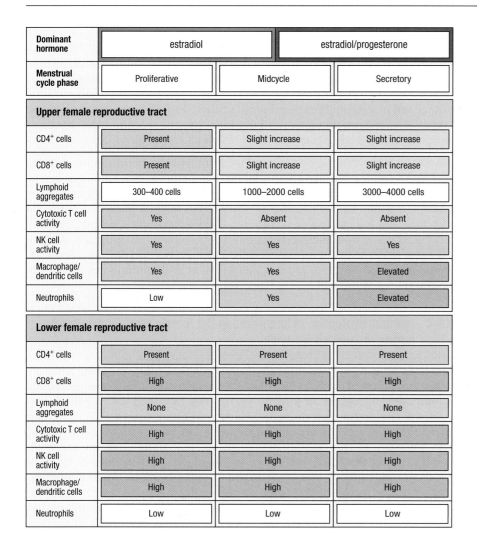

Dominant hormone	estradiol		estradiol/progesterone
Menstrual cycle phase	Proliferative	Midcycle	Secretory
Upper female reproductive tract			
CD4+ cells	Present	Slight increase	Slight increase
CD8+ cells	Present	Slight increase	Slight increase
Lymphoid aggregates	300–400 cells	1000–2000 cells	3000–4000 cells
Cytotoxic T cell activity	Yes	Absent	Absent
NK cell activity	Yes	Yes	Yes
Macrophage/dendritic cells	Yes	Yes	Elevated
Neutrophils	Low	Yes	Elevated
Lower female reproductive tract			
CD4+ cells	Present	Present	Present
CD8+ cells	High	High	High
Lymphoid aggregates	None	None	None
Cytotoxic T cell activity	High	High	High
NK cell activity	High	High	High
Macrophage/dendritic cells	High	High	High
Neutrophils	Low	Low	Low

Figure 18.3 Sex hormones released during the menstrual cycle regulate the recruitment and function of immune cells in the upper (fallopian tubes, uterus, and endocervix) and lower (ectocervix and vagina) female reproductive tissues. Estradiol and progesterone act directly and indirectly through growth factors, cytokines, and chemokines to regulate immune-cell trafficking and activation in a cyclic pattern of immune surveillance. In each tissue of the female reproductive tract, immune cells and their soluble products (Figure 18.2) play a key role protecting the mucosa from sexually transmitted pathogens.

Endocrine control of immune protection in the reproductive tract.

The reproductive tract in the female undergoes extensive remodeling during the menstrual cycle. This remodeling is regulated by an array of sex hormones. These hormones also regulate immune-cell function in tissues of the female reproductive tract as the immune system faces the unique challenge of protecting tissues which are changing shape, thickness, and vascularity.

18-3 Sex hormones directly and indirectly regulate immune-cell function in the reproductive tract.

Sex hormones (estradiol and progesterone) directly and indirectly regulate immune-cell function in the female reproductive tract. Estradiol receptors are expressed by innate and adaptive immune cells, such as epithelial cells, stromal cells, macrophages, dendritic cells, natural killer cells, and lymphocytes.

Estradiol effects on stromal cells, which then influence other cell types in the female reproductive tract, were initially suggested by experiments in which vaginal epithelia from neonatal mice was grown in the presence of uterine stroma. Under such conditions, the epithelia developed into a uterine-like structure. In other studies, neonatal mice were implanted with a mixture of epithelial cells lacking the nuclear estrogen receptor-α (estradiol receptor α) and stromal fibroblasts that expressed the receptor. Treatment with estradiol stimulated uterine epithelial-cell proliferation, leading to the conclusion that estradiol stimulates mitogenesis of uterine (estradiol receptor α positive) epithelial cells indirectly, via the underlying stroma.

Several growth factor–growth factor receptor systems, including epidermal growth factor, insulin-like growth factor-I, hepatocyte growth factor, keratinocyte growth factor, and transforming growth factor β (TGF-β), have been identified within the uterus and are involved in stromal fibroblast–epithelial cell–immune cell interactions. Within the local environment of the reproductive tract, interactions between estradiol, progesterone, and growth factors are complex and involve bidirectional cross-talk between ligands and receptors, influencing both the reproductive cell function and immune-cell migration, activation, and function.

18-4 Immune protection is integrated in the female reproductive tract during the menstrual cycle.

Sex hormones downregulate many immunological functions from midcycle and during most of the secretory phase of the menstrual cycle. Immunological suppression occurs as an integral part of the physiological processes that underlie successful reproduction. Selective immune suppression of certain components of the innate, humoral, and cell-mediated immune systems confers a window of vulnerability of approximately 7–10 days, raising the likelihood that viral, bacterial, and fungal pathogens can evade mucosal immune protection and infect the female reproductive tract. Immune suppression coincides with the recruitment of immune cells capable of hosting pathogens and the upregulation of receptors on target cells essential for pathogen infectivity.

18-5 Immune mechanisms contribute to protection of the male reproductive tract.

The male reproductive tract is less well studied than the female reproductive tract. There are, however, similarities in the restraints imposed on immune responses in both the male and female reproductive tracts. As the female reproductive tract must be regulated/suppressed to allow implantation and growth of a semi-allogeneic fetus, immunity to developing spermatozoa must be prevented in the male reproductive tract. Many genes expressed by developing male gametes are absent during late fetal and early neonatal life, the time when the developing immune system is 'educated' to learn self and non-self. Consequently, many sperm antigens escape tolerance, making the developing gamete susceptible to autoimmune attack at puberty. The male reproductive tract, particularly the testis, is an 'immune privileged' site. For example, survival of allografts in the testis is prolonged compared with survival in other tissues, and within the male reproductive tract multiple mechanisms prevent or suppress autoimmune responses to developing spermatozoa.

The male reproductive tract consists of the testis, the excurrent duct system (rete testis, efferent ducts, epididymis, and vas deferens), the accessory glands (seminal vesicles, prostate, and bulbo-urethral/Cowper's glands), the penile urethra, and the foreskin (Figure 18.4). The testis has a germinal compartment (the seminiferous tubules) containing Sertoli cells, immature germ cells and spermatozoa, and the interstitial tissues; the latter containing blood vessels, lymphatics, mast cells, and Leydig cells. To protect developing spermatozoa, the blood–testis barrier, formed by tight junctions between adjacent Sertoli

	Innate				Adaptive				
	Macrophage/ dendritic cells	Neutrophils	NK cells	Anti- microbials	CD4⁺ T cells	CD8⁺ T cells	IgA B cells	IgG B cells	pIgR
Seminiferous tubules	++	–	undetermined	++	++	++	+/–	+/–	+
Epididymis	++	+	undetermined	++	++	++	+/–	+/–	+
Prostate	++	+	++	+	+	+	++	+	++
Ductus deferens	++	+	undetermined	undetermined	++	++	+/–	+/–	+
Seminal vesicles	–	–	–	++	–	–	–	–	+
Penis (urethra)	+++	+	+/–	undetermined	+++	+++	+++	++	+++
Prepuce	++	++	+++	undetermined	+	+	–	–	–

Figure 18.4 Immune cells in the internal and external tissues of the male reproductive tract. Immune cells are present throughout the tissues of the male reproductive tract. Located at sites adjacent to the site of sperm production (seminiferous tubules), immune cells protect the male reproductive tract from pathogens that invade the tract to cause ascending infection.

cells, prevents entry of immunoglobulin, complement, and other macromolecules into the seminiferous tubules. Immune cells are rarely found in the germinal compartment, but macrophages and some T cells are common in the testicular interstitium. Cytokine secretion by interstitial macrophages probably contributes to both the suppression of immune responses in the testis, and the regulation of Leydig cell function essential for spermatogenesis. Multiple immune-suppressive mechanisms contribute to testicular immune suppression. FasL expression on Sertoli cells may play a role in deleting activated T cells by apoptosis. Sertoli cells may also contribute to maintenance of immune suppression through the secretion of FasL, the tryptophan-degrading enzyme indoleamine 2,3 dioxygenase, TGF-β 1 and 3, activin A, and inhibitors of complement activation and granzyme B. Maintenance of immune privilege may also depend upon the continuous release of germ-cell autoantigens, perhaps important in maintaining immune tolerance. Multiple mechanisms may limit adaptive immunity to developing spermatozoa, but inflammatory responses to infection occur in the testis, although infections in the testis appear to be less common than in other regions of the male reproductive tract. Important questions that remain to be answered include the degree to which immune privilege exists in other regions of the male reproductive tract, including the epididymis, where sperm mature and are stored prior to ejaculation, and whether these immune mechanisms could adversely affect vaccine-induced immunity to sexually transmitted organisms.

18-6 Innate immunity is an important component of the male reproductive tract immune system.

In the male reproductive tract, the suppression of adaptive immunity to maintain immune privilege complements enhanced innate immune protection against local infection. Seminal plasma has strong bacteriostatic and bactericidal activity, due to a variety of mediators of innate immunity, such as zinc, lysozyme, and transferrin, and also large numbers of antimicrobial peptides. The antimicrobial peptides include β-defensins, human epididymis peptide 2 (HE2/EP2/SPAG11), hCAP-18, proteolysis-inducing factor (PIF, a dermacidin), epididymal protease inhibitor (EPPIN), semenogelins, and surfactant protein D. HE2/EP2/SPAG11 has antimicrobial activity against *Neisseria gonorrhoeae* and *Enterococcus faecalis*, whereas surfactant protein D inhibits infection of prostate epithelial cells by *Chlamydia trachomatis*. Antimicrobial peptide expression has been documented in most regions of the male reproductive tract, including the testis (Sertoli cells), seminal vesicles, epididymis, and prostate. Similarly to estradiol regulation in the female reproductive tract, antimicrobial peptides may be regulated by androgens. NK cells have been identified in the testis and prostate (human and rat), but they are infrequently detected in the penile urethra.

18-7 Adaptive immunity in the male reproductive tract is mediated predominantly by CD8+ T cells.

Throughout the excurrent ducts, CD8+ T cells are present in the epithelium of the rete testis, the efferent ducts, and the epididymis, whereas CD4+ T cells are present in the lamina propria of these tissues. The majority of these T cells express the αβ T-cell receptor. Macrophages are abundant in most regions of the excurrent ducts, including the epithelium and lumen of the duct system. The pIgR, which transports polymeric IgA and IgM across mucosal epithelia, is expressed in the rete testis, efferent ducts, the epididymis, and the seminal vesicles. High levels of pIgR, as well as IgA and IgM plasma cells, have been identified in the prostate and the penile urethra. Most of these plasma cells contain J chain, suggesting that the secretory immune system is well developed at these sites. Large focal accumulations of lymphocytes, including CD8+ T cells, NK cells, and some B cells, are present in normal human prostate. Foxp3+CD4+ cells in the prostate may regulate responses to sperm or sexually transmitted pathogens. Antigen-presenting cells (dendritic cells and macrophages) and T cells are abundant in the human penile urethra. Most urethral intraepithelial T cells are CD8+ and many express the CD45RO marker characteristic of memory cells. A small population of CD4+ cells is present in the penile urethra. The presence of antigen-presenting cells and CD4+ cells in the penile urethra suggests that this tissue may be an inductive site, although this has not been demonstrated. The predominance of CD8+ cells at this site, however, is consistent with cell-mediated immunity to potential pathogens or, alternatively, immunosuppressive function. The function of T cells in the male reproductive tract has received little investigative attention, but adoptive transfer of immune CD4+ T cells from mice immunized intranasally with *Chlamydia* major outer membrane protein and cholera toxin into naive recipients provides partial protection against genital tract challenge with *Chlamydia*, raising the possibility that immune CD4+ cells, activated by intranasal immunization, can home to the male reproductive tract and mediate protective immunity.

IgG, not IgA, is the dominant isotype present in seminal plasma of healthy males. IgA in seminal plasma is predominantly IgA1, whereas IgA2 occurs in slight excess in female genital tract secretions. Seminal plasma IgA contains secretory IgA, polymeric IgA, and monomeric IgA, suggesting that local production and pIgR-mediated transport as well as serum contribute to the IgA present at this site. Both systemic and mucosal immunization can elicit antibody responses in seminal plasma. Systemic immunization of men with alum-adsorbed tetanus and diphtheria toxoids has been reported to elicit predominant IgG responses in seminal plasma (and serum) while oral immunization with a live attenuated *Salmonella enterica* serovar Typhi vaccine elicits secretory IgA (SIgA) in seminal plasma. In mouse models, intranasal immunization with chlamydial major outer membrane protein and cholera toxin as an adjuvant induces antigen-specific IgA plasma cells in the prostate and penile urethra, and *Chlamydia*-neutralizing IgA in prostatic secretions. Prostatic fluid from identically immunized pIgR knockout mice is unable to neutralize *in vitro* chlamydial infection, emphasizing the importance of SIgA in this model. Collectively, these findings indicate that both systemic and secretory antibodies contribute to humoral immunity in the male reproductive tract. IgG is predominantly serum-derived, whereas IgA in seminal plasma is derived from both local mucosal production and from serum, suggesting that both systemic and mucosal immunization approaches should be considered as strategies for the induction of humoral immunity to sexually transmitted infections.

Infection and immune protection against sexually transmitted diseases in the male and female reproductive tract.

Sex in mammals is invasive and offers a means by which pathogens can be transmitted into internal compartments of the body. Sexually transmitted diseases are a major and increasing health problem in humans, and are also important in many domesticated and semi-domesticated animals. The introduction of Gardasil™ and Cervarix™, systemic vaccines against human papillomavirus that protect against some human papillomavirus serotypes, genital warts, and hopefully cervical cancer, represents a major breakthrough.

18-8 *Chlamydia trachomatis* is the most common sexually transmitted disease worldwide.

Approximately 90 million new *Chlamydia trachomatis* infections occur annually, mostly in the young (15–29 years). The target cells are the columnar epithelial cells of the endocervix in women, and the epithelial cells of the penile urethra in men. In women, more than 75% of infections are asymptomatic. Important sequelae of infection include pelvic inflammatory disease, infertility, and ectopic pregnancy, all of which are due to ascending infection, and chronic inflammation that occurs over a period of months or years. In men, up to 50% of infections are asymptomatic, and infection can result in prostatitis, epididymitis, and infertility.

Chlamydia are obligate intracellular Gram-negative bacteria and have a unique biphasic life cycle. The infectious elementary body is metabolically inert and binds to unidentified receptors on target cells. Once internalized, the elementary body undergoes transformation to the replicative form, known as the reticulate body, which replicates by binary fission within a membrane-bound inclusion. After 24–72 hours, reticulate bodies convert back to elementary bodies and the newly formed elementary bodies are released either by lysis or extrusion. A third form, the nonreplicating, persistent, or aberrant body, can be induced by nutritional stress, certain antibiotics, or immune pressure by cytokines such as interferon-γ (IFN-γ). Upon removal of the stress, *Chlamydia* differentiates back to the normal replicating forms. Natural immunity following infection is serovar-specific and short lived. Local host defense against *Chlamydia* involves

Figure 18.5 Levels of immune protection against *Chlamydia trachomatis* in the female and male reproductive tract. Barrier protection and innate, humoral, and cell-mediated immunity provide integrated protection against genital pathogens such as *Chlamydia trachomatis*. Although *Chlamydia* has evolved to escape barrier protection, innate antimicrobial factors and adaptive responses restrict bacterial and viral infection of genital target cells. T cells also play a critical role in protecting against infection.

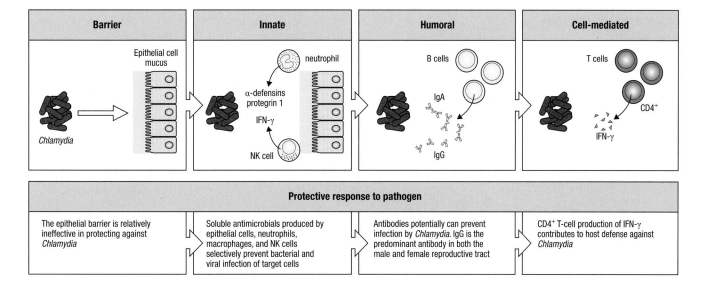

Figure 18.6 Integrated control of the mucosal immune system in the human female reproductive tract. The hormones estradiol and progesterone regulate immune cell and vascular cell receptor expression and immune cell function in tissues of the female reproductive tract. Communication between epithelial cells, local immune cells, and tissue cells, such as fibroblasts, is achieved through local release of cytokines, chemokines, and growth factors. NK cell and dendritic cell activity is downregulated, whereas CD8+ T-cell activity is maintained, except during the secretory phase of the cycle, in female genital tissue. Estradiol and progesterone also promote barrier protection, antimicrobial activity, local antibody secretion, and cell-mediated immunity. Acting in concert with cytokines, chemokines, and growth factors, sex hormones prepare the tract for fertilization, implantation, and pregnancy, while protecting the genital mucosa against bacterial and viral pathogens that may threaten reproductive health.

the mucosal barrier and innate, humoral, and cell-mediated responses (Figure 18.5). *Chlamydia* possesses multiple immune evasion mechanisms that regulate host innate and adaptive immune responses. Resolution of infection coincides with recruitment of CD4+ T cells into the infected tissues, and studies in both humans and animal models have identified IFN-γ-producing CD4+ T cells as essential for resolution of natural infection and prevention of experimental infections in animal models. Interferon-mediated protection in humans involves tryptophan depletion mediated by activation of indoleamine 2,3 dioxygenase, which starves the replicating *Chlamydia* of this essential amino acid. Adoptive transfer of both IgG and IgA monoclonal antibodies against chlamydial surface antigens can protect mice against chlamydial challenge, and studies in animal models have shown that specific antibody is essential for protection against reinfection. Activation of the immune system following chlamydial infection is a 'double-edged sword,' however, as both the innate and adaptive immune responses elicited by infection have been implicated in the development of inflammatory sequelae. Chlamydial infection of epithelial cells results in secretion of large amounts of pro-inflammatory chemokines and cytokines, including IL-8, GRO-α, GM-CSF, TNF-α, IL-1α, and IL-6. Unlike the short-lived innate immune response characteristic of most other bacterial infections, the secretion of inflammatory cytokines is prolonged in chlamydial infections, resulting in chemokine-mediated recruitment of inflammatory cells to the site of infection, and killing of fallopian tube ciliated epithelial cells by cytokines. Thus, the exaggerated innate immune response to chlamydial infection is a major contributor to the tissue scarring, fibrosis, and tubal occlusion associated with pelvic inflammatory disease, infertility, and ectopic pregnancy. Similarly, the adaptive immune response elicited by chlamydial infection, particularly antibody and T cell-mediated immune responses against chlamydial heat shock protein 60 (cHSP60), has been implicated in tissue damage. cHSP60 and human HSP60 share significant sequence identity, such that cHSP60-specific antibody and T cells can recognize and potentially damage host tissues that express human HSP60 (molecular mimicry). Currently no vaccines are available to prevent genital chlamydial infections in humans. Important issues that need to be resolved are the identification of critical chlamydial antigens to elicit protective immunity, without causing tissue pathology, and the best adjuvant(s) and routes of immunization for inducing protection in the female and male reproductive tracts.

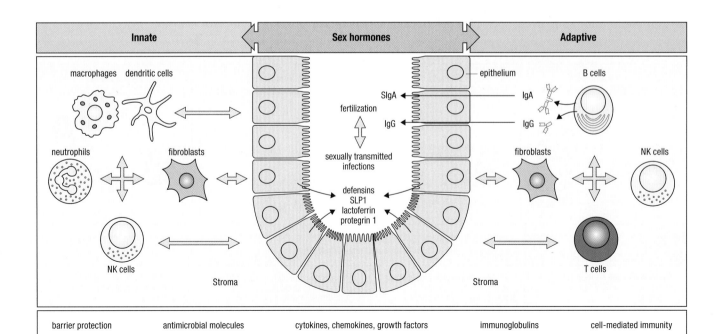

Summary.

The mucosal immune system in the male and female human reproductive tracts has evolved to meet the unique requirements of procreation and host defense against genital tract pathogens. Multiple levels of innate and adaptive immune protection in the female reproductive tract are regulated by sex hormones and optimize conditions for fertilization and pregnancy, while providing protection against potential pathogens (Figure 18.6). These ovary-derived hormones act directly and indirectly through soluble mediators on local immune cells, modulating local immune function in a manner compatible with successful reproduction, yet influencing the response to sexually transmitted pathogens such as *Chlamydia*. Clearly, our understanding of the immunobiology of reproduction and genital mucosal infections has advanced substantially in recent years. Further elucidation of the cellular and molecular mechanisms of host defense in the female and male reproductive tracts should inform new approaches for therapy and vaccine prevention of infection in the human genitourinary tract.

Further Reading.

Bulun, S.E., and Adashi, E.Y.: **The physiology and pathology of the female reproductive axis**, in Larsen, P.R., Foster, D.W., Kronenberg, H.M., *et al.* (eds): *Williams Textbook of Endocrinology*, 10th ed. Philadelphia, W. B. Saunders Company, 2003:587–664.

Com, E., Bourgeon, F., Evrard, B., *et al.*: **Expression of antimicrobial defensins in the male reproductive tract of rats, mice, and humans.** *Biol. Reprod.* 2003, **68**:95–104.

Ghosh, M., Fahey, J.V., Shen, Z., *et al.*: **Anti-HIV activity in cervical-vaginal secretions from HIV-positive and -negative women correlate with innate antimicrobial levels and IgG antibodies.** *PLoS One* 2010, **5**:e11366.

Griffin, J.E., and Wilson, J.D.: **Disorders of the testes and the male reproductive tract**, in Larsen, P.R., Foster, D.W., Kronenberg, H.M., *et al.* (eds): *Williams Textbook of Endocrinology*, 10th ed. Philadelphia, W. B. Saunders Company, 2003:709–770.

Hafner, L., Beagley, K., and Timms, P.: ***Chlamydia trachomatis* infection: host immune responses and potential vaccines.** *Mucosal Immunol.* 2008, **1**:116–130.

Hall, S.H., Hamil, K.G., and French, F.S.: **Host defense proteins of the male reproductive tract.** *J. Androl.* 2002, **23**:585–597.

Meinhardt, A., and Hedger, M.P.: **Immunological, paracrine and endocrine aspects of testicular immune privilege.** *Mol. Cell. Endocrinol.* 2011, **15**:60–68.

Moldoveanu, Z., Huang, W., Kulhavy, R., *et al.*: **Human male genital tract secretions: both mucosal and systemic immune compartments contribute to the humoral immunity.** *J. Immunol.* 2005, **175**:4127–4136.

Wira, C.R., and Fahey, J.V.: **A new strategy to understand how HIV infects women: identification of a window of vulnerability during the menstrual cycle.** *AIDS* 2008, **22**:1909–1917.

Wira, C.R., Fahey, J.V., Schaefer, T.S., *et al.*: **Innate and adaptive immunity in the human female reproductive tract: influence of the menstrual cycle and menopause on the mucosal immune system in the uterus**, in Aplin, J., Fazleabas, A.T., Glasser, S.R., *et al.* (eds): *The Endometrium: Molecular, Cellular and Clinical Perspectives.* London, Informa UK Ltd, 2008:493–523.

Wira, C.R., Sentman, C.L., Pioli, P.A., *et al.*: **Innate and adaptive immunity in male and female genital tract: cellular responses and interactions.** *Immunol. Rev.* 2005, **206**:306–335.

Mucosal Immune Responses to Microbes in the Genital Tract

19

Sexual reproduction provides an evolutionary advantage that generates a significant increase in the fitness of offspring. It is a fundamental activity for a species to procreate and survive in the ever-changing environment of its habitat. Not surprisingly, certain viruses, bacteria, fungi, and parasites have evolved to utilize sexual contact of the host organism for their own transmission. Sexually transmitted pathogens enter the human hosts through the male and female genital mucosae. The World Health Organization estimates that 1 million new cases of sexually transmitted diseases (STDs) occur daily. Yet, for the most part, vaccines that prevent STDs are not available.

Compared with the intestinal mucosa, female and male genital tracts are covered by distinct epithelial layer and mucus types, are inhabited by a unique set of microbial flora, and utilize distinct innate and adaptive effector mechanisms. In Chapter 18, we reviewed the anatomy and immunophysiology of the female and male genital mucosae. In this chapter, major human sexually transmitted pathogens will be described with respect to our current understanding of their epidemiology, mode of entry, innate detection, initiation of adaptive immune responses, and clearance.

Global prevalence of sexually transmitted pathogens.

Certain bacteria, parasites, and viruses use genital mucosa as a portal of entry into human hosts. Clinically relevant STDs (with the estimated number of people infected globally indicated in brackets) include chlamydia (89 million), gonorrhea (62 million), syphilis (12 million), trichomonas (170 million), human immunodeficiency virus type 1 (HIV-1; 34 million), herpes simplex virus type 2 (HSV-2; 315 million), human papillomavirus (HPV; 630 million), and hepatitis B virus (HBV; 350 million). The prevalence of STDs in general is much higher in the developing countries. Since its discovery in the early 1980s, acquired immunodeficiency syndrome (AIDS) has already claimed the lives of more than 25 million people, the majority of whom lived in sub-Saharan Africa. In endemic countries, death and a reduced fertility rate caused by HIV-positive women has dramatically affected population demographics. AIDS has generated 14 million orphans in sub-Saharan Africa alone.

Diseases caused by sexually transmitted pathogens.

Diseases caused by these sexually transmitted pathogens range from nuisance (warts) to lethal diseases (AIDS, cervical cancer). STDs caused by bacterial (*Chlamydia trachomatis*, *Neisseria gonorrhoeae*, *Treponema pallidum*) and parasitic (*Trichomonas vaginalis*) pathogens are curable by antibiotic treatment. In contrast, STDs caused by viral pathogens (HIV-1, HSV-2, HPV, HBV) are treated but not cured by antiviral therapy.

19-1   Chlamydia is a disease caused by the bacteria *Chlamydia trachomatis.*

Infection with *Chlamydia trachomatis* is the most common STD in the USA, with an estimated 2.3 million Americans aged 14–39 infected. Chlamydia can be cured with antibiotic treatment with tetracycline, azithromycin, or erythromycin. However, 80–90% of infected men and women are asymptomatic and often are unaware of their carrier status. In infected women, untreated infection may lead to pelvic inflammatory disease, which can cause scarring of the fallopian tubes and result in infertility, and also increases the likelihood of an ectopic pregnancy. In addition, during pregnancy the infection can cause premature labor and delivery. Moreover, the infant could develop chlamydial conjunctivitis and chlamydial pneumonia.

19-2   Gonorrhea is an STD that is caused by a bacterium, *Neisseria gonorrhoeae.*

Neisseria gonorrhoeae, a bacterium that replicates within the mucosae of the cervix, uterus, and fallopian tubes in women, and in the urethra in women and men, is the cause of gonorrhea. Gonorrhea can be cured by treatment with antibiotics including ceftriaxone and azithromycin. Untreated infection can lead to complications including pelvic inflammatory disease, fallopian tube damage leading to infertility, or increased chance of ectopic pregnancy. In men, gonorrhea can cause epididymitis that may lead to infertility. Gonococcal ophthalmia neonatorum, which could lead to blindness, is the most common manifestation in infants born to mothers with gonococcal genital tract infections.

19-3   Syphilis is caused by infection with the bacterium *Treponema pallidum.*

Syphilis is transmitted through direct contact with a sore, which can occur on the external genitals, vagina, anus, mouth, or in the rectum. Penicillin is used to treat syphilis. However, if untreated, syphilis can result in three stages of disease progression. The primary stage manifests with a single or multiple sores (chancres). The chancre lasts 3 to 6 weeks, and it heals without treatment. However, if untreated, the infection progresses to the secondary stage. Skin rash (often on palms of the hands and bottoms of the feet) and mucous membrane lesions appear which may be accompanied by fever, fatigue, weight loss, and swollen lymph nodes during secondary syphilis. The symptoms of secondary syphilis will resolve without treatment, but the infection will progress to the latent and possibly late stages of disease. The latent stage of syphilis can last for years. Upon reactivation, a tertiary stage of syphilis ensues, in which the disease may damage the internal organs, including the brain, nerves, eyes, heart, blood vessels, liver, bones, and joints. This damage can lead to difficulty coordinating muscle movements, paralysis, numbness, gradual blindness, dementia, and even death.

19-4 *Trichomonas vaginalis*, the cause of trichomoniasis, is a protozoal parasite that infects the vagina.

The vagina is the most common site of infection in women with *T. vaginalis*, and the urethra is the most common site of infection in men. While infected men rarely exhibit symptoms, infected women often experience vaginal discharge and discomfort during intercourse and urination. *Trichomonas* infection during pregnancy can cause premature birth or low birth weight. This infection can be cured by treatment with the antibiotics metronidazole or tinidazole.

19-5 Human immunodeficiency virus 1 (HIV-1), the cause of acquired immunodeficiency syndrome (AIDS), is an STD.

HIV-1 is a member of the lentivirus family of retroviruses. HIV-1 is transmitted through sexual contact (vaginal, penile, rectal, or oral), via blood, or by vertical transmission from infected mother to child. There is no cure or vaccine against HIV-1. HIV-1 infection is treated with highly active antiretroviral therapy in the developed world, but the treatment does not remove the latent viral pool and must be continued for the life of the infected individual. HIV-1 infection targets and kills CD4$^+$ T lymphocytes, whose function is essential in the adaptive immune system. Naive CD4 T cells differentiate into a variety of effector types and mediate protection against intracellular, extracellular, and fungal pathogens. CD4 T-cell help is also often required for generating primary and memory CD8$^+$ T-cell and B-cell responses. As a consequence of CD4 T-cell depletion, the later stage of HIV-1 infection will manifest in AIDS, which is characterized by immunodeficiency leading to life-threatening opportunistic infections, cancers, and death.

19-6 Genital herpes is caused by infection with herpes simplex virus-2 (HSV-2) and, less commonly, HSV-1.

HSV is an enveloped virus containing a double-stranded DNA (dsDNA) genome (~150 kb) within the capsid. HSV is transmitted through mucosal contact through vaginal, penile, and oral routes. Even though HSV-1 is traditionally associated with oral herpes, adults without immunity to HSV-1 who practice oral sex are at risk for genital HSV-1 infection. HSV-1 and HSV-2 enter the host through genital epithelia, and upon replication, enter the innervating ganglia where the virus establishes latent infection. Reactivation of the virus results in infection of the genital epithelia and external skin around the genitals and anus, leading to blisters followed by ulcerative lesions. However, most people infected with HSV are asymptomatic and are unaware of their carrier status. Vertical transmission of HSV-2 from an infected mother to newborn leads to neonatal herpes, which is often lethal. HSV infection can be treated with the antiviral drug acyclovir, but this does not remove the latent viral pool and does not cure disease.

19-7 Human papillomavirus (HPV) infects the stratified squamous epithelium.

HPV is a member of the *Papillomaviridae*, consisting of a capsid containing a circular dsDNA genome (~8 kb). Upon sexual contact, HPV establishes productive infections only in the stratified epithelium of the skin or mucous membranes. While the majority of the nearly 200 known types of HPV cause no symptoms in most people, 'high risk' types of HPV are the causal agents of over

90% of all cervical cancer, the second leading cause of death among women worldwide. High-risk HPV types, such as HPV-16 and HPV-18, also cause anal, penile, and vulvar cancer, and 'low risk' HPV types cause genital warts. Genital warts are treated with a topical ointment containing imiquimod (a Toll-like receptor 7 (TLR7) agonist) or an injection of type I interferon. HPV infection is also common, though usually asymptomatic, in men. HPV is the most common sexually transmitted disease in the world, causing 260,000 deaths annually, 80% of which occur in developing countries.

19-8 Hepatitis B virus (HBV) is a hepatotropic virus that is transmitted through sexual contact.

Hepatitis B is caused by hepatitis B virus (HBV), which is a member of the hepadnavirus family. It has a circular genome composed of partially double-stranded DNA. The acute illness causes liver inflammation, vomiting, jaundice, and, in rare cases, death. Most infected individuals clear acute HBV infection. A small percentage of infected individuals develop chronic hepatitis B, which may cause liver cirrhosis and liver cancer over many years. Chronic HBV infection can be treated with antiviral drugs and immune modulators such as interferon -α (IFN-α). These agents cannot clear the infection, but they can stop the virus replication, thus minimizing liver damage.

Invasion mechanisms employed by sexually transmitted pathogens.

All infectious pathogens must gain entry into the host by invasion of the skin or epithelium and sexually transmitted pathogens are no exception. Each of the aforementioned organisms invades specific types of the genitourinary epithelia (simple columnar or stratified squamous) by distinct mechanistic pathways. As will be seen below, although each of the strategies may vary, each microorganism seeks to either establish a niche within the epithelium or manage a means to traverse the epithelium in order to reach other desirable destinations, such as the bloodstream. In some cases, the organism has developed phasic shifts in its mechanisms of growth (e.g., *Chlamydia* infection) to achieve these objectives.

19-9 *Chlamydia* exhibits distinct infectious (elementary body) and replicative forms (reticulate body).

Chlamydia trachomatis is an obligate intracellular Gram-negative bacterium. *C. trachomatis* normally infects the single-cell columnar layer of the epithelium in the endocervix of women and the transitional epithelium of the urethra of men. Once inside the epithelial cell, *C. trachomatis* undergoes its entire life cycle within a vacuole, termed an inclusion. The chlamydial inclusion, which neither fuses with lysosomes nor acidifies, then trafficks to the Golgi region. The acute infection consists of infectious and noninfectious stages. The infectious elementary body is small and metabolically inactive. Within the inclusion, the elementary body transforms into a larger metabolically active reticulate body, which divides by binary fission, giving rise to over a thousand progeny per host cell. The reticulate bodies transform back into infective elementary bodies, which are subsequently released from the inclusion vacuole to infect neighboring cells. *C. trachomatis* encode a type III secretion system, which secretes proteins into the cytosol that bind to and transform the vacuole into a replication center.

19-10 Gonorrhea infects the apical surface of simple columnar epithelium and once invaded inhibits the function of the adaptive immune system.

Neisseria gonorrhoeae infects the cervix, uterus, and fallopian tubes in women, and the urethra in women and men. The bacterium can also be found in the mouth, throat, eyes, and anus. The initial attachment of the bacteria to the mucosal surface occurs via the hair-like pili, which recognize a putative receptor on mucosal epithelial cells. Subsequently, tight adherence to the apical surface of epithelial cells occurs and is mediated by opacity (Opa) outer membrane proteins. Adherence may also lead to the apoptosis and destruction of the epithelial layer, providing access to the basolateral compartment. Serum factors promote the infection of cells via the basolateral pole of the epithelium. Epithelial cell chemokines attract phagocytic cells, such as neutrophils and monocytes, which become readily activated and infected. The Opa proteins are also immunosuppressive by binding inhibitory proteins—such as carcinoembryonic antigen-related cell adhesion molecule 1 (CEACAM1), which contains immune tyrosine receptor-based inhibitory motifs as an immune-evasion mechanism—onto the cell surface of T cells. The release of inflammatory mediators, including reactive oxygen species, further increases inflammation and tissue destruction as well as upregulation of CEACAM1. *N. gonorrhoeae* uses its outer membrane porin (Por) molecules to bind the complement inhibitory proteins C4b-binding protein (C4BP) and factor H to evade killing by human complement. In addition, sialylation of gonococcal lipooligosaccharide also enables *N. gonorrhoeae* to bind factor H. Gonorrhea is a human-specific pathogen—rodent and primate species cannot be readily infected with this pathogen because gonorrhea does not bind to factor H or C4BP derived from these animals.

19-11 *Treponema pallidum* invades humans at mucosal epithelia where it causes infection and gains access to the blood and lymph systems.

Treponema pallidum is a pathogenic spirochete bacterium that has a helical-shaped body with flagella on both ends. Laboratory investigation of *T. pallidum* is hampered by the fact that the bacteria cannot be cultured on artificial media and must be grown in rabbit testicles. *T. pallidum* is able to pass through the endothelial cell monolayer through intercellular junctions without altering tight junctions. Primary infection typically results in a single chancre at the site of inoculation. Within hours of inoculation, and during the evolution of the primary stage, *T. pallidum* disseminates widely and organisms are deposited in a variety of tissues. Highly motile spirochetes may leave the circulation by invading the junctions between endothelial cells.

19-12 Trichomoniasis is initiated within either the female cervico-vaginal epithelium or the male urethral epithelium.

Trichomonas vaginalis is an anaerobic, flagellated protozoan. The only identified host for *T. vaginalis* is humans. This parasite encodes adherence factors that allow cervico-vaginal epithelium colonization in women. In men, *T. vaginalis* can infect the urethra but is rarely symptomatic. The adherence to vaginal epithelial cells is regulated by pH and temperature. When the healthy acidic pH of the vagina is altered to a basic pH, *T. vaginalis* can multiply through binary fission. The life cycle of this parasite does not have a cyst stage, and it cannot survive in the external environment for long periods of time.

19-13 HIV-1 gains access to the immune system by crossing the epithelial cell barrier.

HIV-1 uses the CD4 molecule as the viral receptor and CCR5 or CXCR4 as co-receptors to enter a target cell (CD4+ T lymphocytes). The viral genome is reverse transcribed into DNA which is integrated into the genome of the infected cells. Activation of CD4+ T cells induces the transcription of the provirus, leading to viral synthesis and release. Several modes of entry into the host from the initial inoculums have been suggested including transcytosis through M cells (only present over the mucosa-associated lymphoid tissue in type I mucosal epithelia), uptake by dendritic cells extending dendrites into the mucosal cavity, and entry through microabrasions in the epithelial layer. HIV-1 invasion is facilitated by physical abrasion, which occurs frequently even during consensual sex. Preexisting genital lesions caused by other STDs also contribute significantly to increasing risk of HIV-1 transmission. The majority of women acquire HIV-1 through heterosexual contact with infected men. Entry of the virus likely occurs near the transformation zone of the endocervix where the simple columnar epithelium of the endocervix transitions into the stratified squamous epithelium of the vagina. Here, the virus first replicates in local CD4+ T lymphocytes and sets up founder cells. From there, the virus can disseminate through the lymph to local lymph nodes and further to the systemic circulation. However, the vaginal route of transmission is estimated to be only successful in 1/200 to 1/2000 encounters, owing to the protective stratified squamous epithelial layer present in this tissue. Risk for HIV-1 acquisition from females to males is even less, particularly in circumcised males. In contrast, HIV-1 can enter through the rectal mucosa much more efficiently, with transmission rates estimated to be as high as 1/10. This difference in transmission rate likely reflects the fact that rectal mucosa is covered by only a single layer of columnar epithelia, and the fact that the lamina propria hosts many effector/memory CD4+ T cells that could serve as targets of HIV-1 infection.

19-14 Genital herpes infection initially targets the stratified squamous epithelium.

HSV-1 and HSV-2 enter the vaginal canal or the foreskin and first infect stratified squamous epithelial layers. HSV utilizes multiple cell-surface proteins for entry: heparan sulfate chains on cell-surface proteoglycans; a member of the tumor necrosis factor receptor family, herpesvirus entry mediator; and two members of the immunoglobulin superfamily related to the poliovirus receptor, nectin-1 and nectin-2. Primary replication in the keratinocytes results in the subsequent infection of sensory ganglia through nearby nerve endings. The de-enveloped virus migrates to the nerve cell body via retrograde axonal transport. Hiding in the ganglia, HSV establishes lifelong latent infection in the host. Immune suppression results in reactivation of the virus, causing lesions in the external genitalia. Notably, HSV virions are shed even in asymptomatic individuals, increasing the likelihood of transmission from carriers to the uninfected partners.

19-15 HPV infection requires access to the basement membrane for infection of basal keratinocytes of stratified epithelium.

HPV is thought to infect external genital skin, the vagina, or the cervix in women, and the penile shaft, glans, and foreskin in men. HPV cannot infect an intact epithelial layer but requires disruption of the epithelial integrity, which allows access to the basement membrane. HPV utilizes an unconventional mechanism for infection, in that it must first bind to the basement membrane for entry into the basal keratinocytes. Upon binding to the basement membrane, the virus undergoes conformational changes that result in cleavage of the minor capsid protein L2, followed by the exposure of an N-terminal L2 epitope and transfer of the capsid to the epithelial cell surface. Once within the basal keratinocyte stem cells, the virus uses the differentiation program of the epithelial cells to complete its life cycle. HPV virions are shed from the superficial layer of the stratified squamous epithelium of the mucosa and transmitted to sexual partners. Most subclinical infection with HPV regresses spontaneously. Only a very small percentage of HPV-infected women, if untreated, go on to develop cervical cancer over many years.

19-16 HBV is a mucosal pathogen in approximately one-third of infections in adults.

HBV is present in many mucosal secretions and in the blood of infected individuals and may reach 10^{10} copies per milliliter in serum. About 30% of HBV transmission occurs through sexual contact (the rest is related to perinatal, blood, or pericutaneous transmission). Due to the lack of infectable culture systems, the viral receptor responsible for entry is unknown. Once HBV enters the host circulation, the virus replicates in the hepatocytes. In 95% of the infected adults and in 10% of the infected neonates, HBV is cleared following acute infection. The remainder develop chronic viral infection. Hepatocellular carcinoma can develop in a small fraction of these patients over a period of 30 to 50 years.

The innate immune system of the genital mucosa and its relation to infections.

The genital mucosal surfaces are protected by a variety of mechanisms. Some of these mechanisms are described in Chapter 18. Here, we will focus on and summarize those host factors that are specifically relevant to protecting against mucosal infections of the genitourinary tract. A pathogen entering the genital tract must pass through several hurdles in order to reach the epithelial layer and subsequently the distal target organs (Figure 19.1). The first hurdle is provided by the mucus layer, which contains a variety of antimicrobial factors and has an inhospitable pH. A pathogen must then invade the multilayered epithelial cells of stratified squamous epithelium, reach the lamina propria,

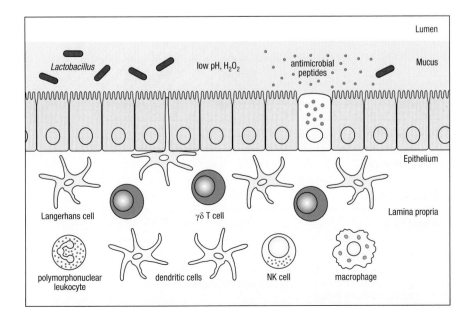

Figure 19.1 Innate immune system of the female genital mucosa.
A mucus layer containing a variety of antimicrobial factors and endogenous microbiota, *Lactobacillus* spp., provide an inhospitable environment for an invading pathogen. The epithelial layer and the lamina propria contain phagocytes and other innate cells that recognize and eliminate incoming microbes.

and find the target cells before being engulfed by phagocytes or attacked by circulating antimicrobial substances such as complement.

19-17 Mucus is the first line of epithelial defense.

The frontline of innate immune protection in the genital mucosa is provided by the physical barriers of the epithelium. In addition to the epithelial layer, mucus covers the internal surface of the vaginal tract, penis, and anal canal. Mucus is made up of mucins, which are complex high-molecular-weight glycoproteins that contain at least one, and sometimes multiple, protein domains that are sites of extensive *O*-glycan attachment. A major difference in the genital versus intestinal mucosae is the source of mucus-secreting cells. In the tissues that contain simple columnar epithelia (e.g., intestines, lung), goblet cells that are differentiated epithelial cells dedicated for mucus production and secretion are dispersed throughout the epithelial layer. Similarly, in simple columnar epithelium of the rectum and upper anal canal, goblet cells secrete mucus (Table 19.1). In the genital mucosa, which contains stratified squamous epithelium (male urethra, vagina, lower half of anal canal, foreskin), and in the lower anorectal canal, mucus is secreted from local mucus-secreting epithelial cells. During ovulation, crypts in the cervix secrete mucus that descends and covers the vaginal canal. In the male reproductive tract, the bulbo-urethral glands produce mucus, which is incorporated into semen. In the uncircumcised penis, the inner foreskin epithelium is kept moist by cells secreting mucin locally. In addition to mucins, mucus contains a variety of other defense molecules including immunoglobulins, complement, antimicrobial peptides, lysozyme, and lactoferrin (discussed further below).

19-18 Female and male genitourinary secretions contain antimicrobial factors.

Bacteria, fungi, parasites, and viruses within the mucus layer are confronted with a variety of antimicrobial factors in the genitourinary tract. Microbicidal molecules such as complement and antimicrobial peptides can directly bind and kill microbes before they can reach the host epithelial layer. The epithelial cells, cervical glands, and neutrophils contribute the majority of antimicrobial peptides to the vaginal fluid, including calprotectin, lysozyme, lactoferrin, secretory leukocyte protease inhibitor, human neutrophil peptides, and human β-defensins. The concentrations of these antimicrobial factors remain stable during different stages of the menstrual cycle. Calprotectin, highly abundant in vaginal and intestinal mucosae, is a calcium-binding protein that

Table 19-1 Distinct features of simple columnar and stratified epithelial mucosal surfaces

	Simple	Stratified
Epithelial	Polarized	Non-cornified
Presence of MALT	+	−
Presence of pIgR	+	−
Major Ig isotype	SIgA	IgG
Goblet cells	+	−
Langerhans cells	−	+

MALT, mucosa-associated lymphoid tissue; pIgR, polymeric immunoglobulin receptor; Ig, immunoglobulin; SIgA, secretory IgA.

inhibits bacterial and fungal growth, as well as bacterial adhesion. Lysozymes are a family of enzymes that catalyze hydrolysis of glycosidic bonds in peptidoglycans (present in the wall of Gram-positive bacteria). Lactoferrin, a member of the transferrin family of iron-binding glycoproteins, binds to ferric iron (Fe^{3+}), which deprives microorganisms of this essential factor and has direct antimicrobial activities against a large panel of bacteria, viruses, fungi, and parasites. Secretory leukocyte protease inhibitor is a double-edged sword—its C-terminus protects tissues by inhibiting proteases such as elastase, cathepsin G, and trypsin, and its N-terminus exerts antibacterial and antifungal capabilities. Human neutrophil peptides (α-defensins secreted by neutrophils) and human β-defensins (secreted by epithelial cells) are small cationic proteins with significant and broad antimicrobial activities.

19-19 The female and male genitourinary systems possess an endogenous (commensal) microbiota that provides colonization resistance against pathogenic infections.

Commensal bacteria are essential in shaping intestinal immune responses in both health and disease. Recent advances in 16S ribosomal RNA sequencing and deep metagenomic sequencing have enabled a richer taxonomic identification of human microbial communities. Of more than 50 bacterial phyla on Earth, human intestine-associated communities are dominated by only four phyla: Firmicutes, Bacteroidetes, Actinobacteria, and Proteobacteria. More than 1000 species of bacteria are estimated to live in the intestines of humans. In addition to providing key metabolic, trophic, and defense functions to the mammalian host, the microbiota shapes the intestinal immune system in a variety of ways. This topic is discussed in further detail in Chapter 17.

In comparison with the intestinal tract, a normal vaginal microbiota is predominately composed of Firmicutes and especially *Lactobacillus* species, which perform key functions for the female host. First, vaginal H_2O_2-producing lactobacilli prevent the outgrowth of harmful bacteria that can cause bacterial vaginosis. Second, *Lactobacillus* spp. maintain the vaginal fluid acidity (pH 3.8–4.0) through lactic acid production. The acidic vaginal pH protects against STDs, including *Haemophilus ducreyi*, HSV-2, *Trichomonas vaginalis*, and *Chlamydia trachomatis* infection. By virtue of causing vaginitis, all of these infectious agents lead to a significant increase in the transmission of HIV-1. In addition to *Lactobacillus* (phylum Firmicutes), phyla Proteobacteria and Actinobacteria (mainly the genera *Pseudomonas* and *Gardnerella* respectively) are also significant constituents of the commensal vaginal microbiota of healthy females. Within an individual, the vaginal microbiota is not homogeneous but differs significantly depending upon the anatomical location and other factors such as hormonal state of the host. The influence of the vaginal microbiota on the adaptive immune response to STDs is unknown.

19-20 Innate immune cells provide defense against invading pathogens within the female and male genitourinary tract.

Within the genital mucosa, a variety of innate immune cells provide defense against invading pathogens. At steady state, γδ T cells, macrophages, Langerhans cells, and dendritic cells (DCs) survey the stratified squamous epithelia of vaginal and anal canals. In the human vagina, the numbers of neutrophils, macrophages, and T cells remain relatively constant throughout the menstrual cycle. Upon infection, a number of cell types are mobilized into the vaginal tissue, including polymorphonuclear cells (PMNs), monocytes, plasmacytoid DCs (pDCs), and natural killer (NK) cells. Subsequently, antigen-specific T cells and B cells enter the tissues to provide pathogen-specific immune defense. A large number of PMNs are recruited to HSV-2-infected vaginal mucosa in response to chemokine signals, and are required for protection during primary and secondary challenge. NK cells sense infected cells by two different mechanisms: 'missing' self (lack of major histocompatibility complex (MHC) class I molecule expression) and 'altered' self (presence of NK activating factors on target cells such as 'stress' ligands including MHC class I chain-related gene A and B proteins in human). NK cells are rapidly recruited to the infected tissues where they are activated to lyse infected cells through perforin and Fas ligand-dependent mechanisms. NK cells are also a source of cytokines including interferon (IFN)-γ, which also has antiviral activity. Humans missing NK cells or having defective NK function often suffer from herpesvirus infections, indicating their role in defense against this family of viruses. The importance of NK cells in antiviral defense against HIV-1 is reflected by the evasion mechanism utilized by HIV-1 to specifically prevent NK recognition of infected cells. Viral nef protein downregulates HLA-A and B to avoid recognition by cytotoxic T lymphocytes (CTLs) but spares HLA-C and E, which are the primary ligands for inhibitory NK receptors. Intraepithelial γδ T cells are present within the simple columnar and stratified squamous epithelia, and although the nature of antigens they recognize is unclear, they are thought to sense altered self upon viral infection. These cells contribute to immune protection against vaginal challenge with HSV-2. Langerhans cells within the epithelium and DCs in the submucosa patrol the genital mucosa for invading pathogens. Of note, Langerhans cells in the mouse vagina originate from circulating bone marrow derived precursors, unlike the skin wherein Langerhans cells are derived from local noncirculating precursors. At steady state, these cells are highly phagocytic and express a variety of pattern recognition receptors (PRRs) that recognize a wide array of microbes (discussed further below). DCs and Langerhans cells depart the epithelium after capturing the relevant microbe and undergo a maturational program during migration to the draining lymph nodes to prime naive T cells and B cells.

Innate recognition of sexually transmitted pathogens.

Pathogens are detected by multiple families of PRRs expressed by innate immune cells and the infected target cells. Stimulation of cells through PRRs results in the activation of genes encoding immediate defense responses (antimicrobial factors, chemokines, type I IFNs) as well as those required to elicit adaptive immune responses (cytokines, co-stimulatory molecules, chemokine receptors). Innate recognition systems utilized by the human host to detect sexually transmitted pathogens are described in this section.

19-21 *Chlamydia trachomatis* is recognized by multiple PRRs.

While both TLR2 and TLR4 recognize *C. trachomatis*, TLR2 and MyD88 specifically co-localize with the intracellular chlamydial inclusion, suggesting that TLR2 is actively engaged in signaling from this intracellular location. TLR2 recognition of this organism is dependent on infection with live, replicating bacteria. In addition to TLRs, the NOD-like receptor, NOD1, is involved in sensing *Chlamydia* infection.

19-22 Host cells utilize immunoglobulin-related molecules such as CEACAM3 to initiate internalization and elimination of *Neisseria gonorrhoeae*.

Carcinoembryonic antigen-related cell adhesion molecule 3 (CEACAM3) mediates the opsonin-independent recognition and elimination of a restricted set of human-specific Gram-negative bacterial pathogens including *Neisseria gonorrhoeae*. CEACAM3 binds to the Opa protein expressed by *N. gonorrhoeae* and internalizes the bacteria for degradation. *N. gonorrhoeae* expresses lipoo-ligosaccharide, lacking the O antigen. This structure is recognized by TLR2, TLR4, and Triggering receptor expressed on myeloid cells-2A (TREM-2A). In addition, porins and Lip lipoproteins of *N. gonorrhoeae* are recognized through TLR2.

19-23 *Treponema pallidum* lacks lipopolysaccharide but contains internal lipoproteins which can stimulate Toll-like receptors.

The outer membrane of *T. pallidum* does not contain lipopolysaccharide and has relatively few surface-exposed transmembrane proteins. Therefore, the extracellular bacterium is not recognized efficiently by TLRs. However, upon degradation in phagocytes, the lipoproteins present under the outer membrane can become accessible and stimulate TLR2.

19-24 Innate immune recognition of HIV-1 occurs after cellular infection.

Innate recognition of HIV-1 or retroviruses in general remains unclear. pDCs can produce IFN-α in response to HIV-1 through a process that likely involves TLR7 (Figure 19.2). In infected cells, HIV-1 viral RNA is reverse transcribed into viral cDNA, which is recognized by a not-yet-characterized DNA sensor. The host factor TREX1, an exonuclease, inhibits innate immune detection of HIV DNA by metabolizing nonproductive reverse transcription products. A recent study showed that the innate response to HIV-1 in DCs depends on interaction between newly synthesized HIV-1 capsid and cellular cyclophilin A (Figure 19.3a). In addition, HIV-1 induces global disruption of the innate signaling pathway within infected cells by degrading IRF3, making directly infected cells incapable of IFN production. Studies in mice have shown that

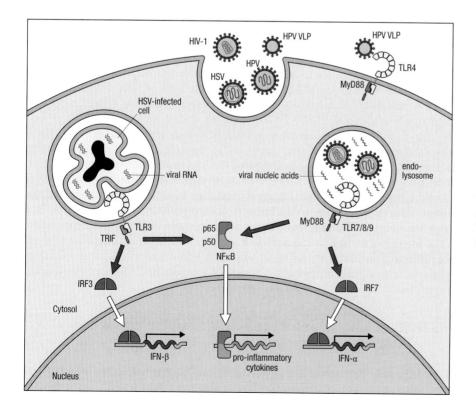

Figure 19.2 Cell-extrinsic recognition of sexually transmitted viruses by TLRs. Specialized viral sentinels such as the pDCs, B cells, and DCs recognize viral nucleic acids in the endolysosomes through TLR3, TLR7, TLR8, and TLR9. TLR3 recognizes higher order viral RNA from infected cells (HSV); TLR7 and TLR8 recognize viral RNA (HIV-1); and TLR9 recognizes viral DNA (HSV). HPV virus-like particle (VLP) also triggers MyD88-dependent signals. In B cells, HPV virus-like particles (VLP) stimulate TLR4/MyD88 to induce an antibody response. Signals transduced from these TLRs lead to the induction of IFN and cytokine genes.

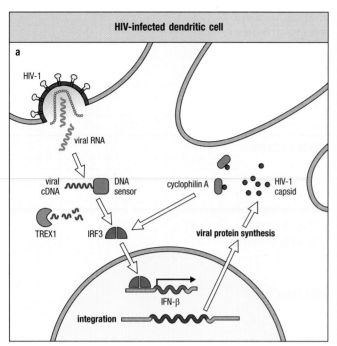

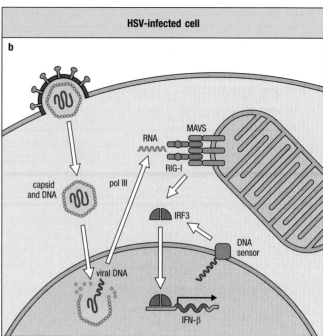

Figure 19.3 Cell-intrinsic recognition of sexually transmitted viruses. Virally infected cells recognize viral pathogen-associated molecular patterns in the cytosol. In panel a, HIV-1 infection is recognized through multiple mechanisms, which are efficiently counteracted by the virus. HIV-1 viral DNA synthesized by reverse transcription is recognized through an unidentified DNA sensor, activating IRF3. Exonuclease TREX1 degrades nonproductive reverse transcription products, inhibiting activation of IRF3. HIV-1 also degrades IRF3. In HIV-1-infected DCs, cyclophilin A binds to newly synthesized capsid proteins and induces IFN through the activation of IRF3. However, HIV-1 normally does not infect DCs, and thus avoids activating this pathway altogether. In addition, both HSV and HIV-1 trigger inflammasomes, leading to secretion of IL-1β and IL-18 (not shown). In panel b, HSV viral DNA is sensed by DNA sensors such as DAI and IFI16. HSV viral DNA is transcribed by RNA polymerase III (pol III), and the resulting RNA serves as a substrate for RIG-I. Signals transduced from these PRRs lead to the induction of IFN and cytokine genes. MAVS, mitochondrial antiviral signaling protein.

humoral immunity, but not cytotoxic T-cell responses, to a retrovirus (Friend murine leukemia virus) requires MyD88 and DCs.

19-25 Toll-related receptor and retinoic acid-inducible gene-related pattern recognition is involved in detecting genital herpes infection.

HSV infection is sensed by multiple PRRs. In pDCs, HSV is recognized by TLR9 in the endosome, while in other cell types HSV-derived RNA is recognized in the cytosol by a RIG-I-related pathway following transcription of HSV by DNA-dependent RNA polymerase III (Figure 19.3b). In the cytosol, viral DNA is sensed by a variety of DNA sensors, which may include DAI and IFI16 (Figure 19.3b). Certain strains of HSV also trigger TLR2 on DCs and macrophages. HSV-2 also activates the inflammasome, independently of AIM2. Even though caspase 1-deficient mice induce normal adaptive immunity to genital HSV-2 infection, IL-18 is required for innate protection against HSV-1 infection. In a mouse model of genital herpes, MyD88-dependent signaling in hematopoietic and nonhematopoietic cells is required for the induction of helper T cell type 1 (T_H1) immunity following genital HSV-2 infection. In addition, TLR9 responsiveness by radioresistant cells in the vaginal mucosa is required for the microbicidal effect of vaginally delivered CpG (cytosine-phosphate-guanosine oligodeoxynucleotides) in protection against HSV-2 challenge, highlighting the importance of direct pathogen recognition by nonhematopoietic cells in innate antiviral defense. In addition to these pathways, mutations in TLR3 and downstream signaling molecules are highly associated with uncontrolled HSV infection. In humans, a rare TLR3 P554S homozygous mutation and an L412F mutation, both resulting in impaired activity of TLR3, were found in patients with HSV-1 encephalitis and HSV-2-associated Mollaret meningitis respectively.

19-26 HPV infection is sensed by innate and adaptive immune cells through Toll-like receptors.

HPV virus-like particles stimulate DCs via MyD88, and adaptive immunity to virus-like particles is compromised in the absence of MyD88. B cells recognize the HPV virus-like particles through TLR4 in a MyD88-dependent manner to induce class switching and co-stimulatory molecule expression (Figure 19.2). The nature of the pathogen-associated molecular pattern that stimulates the PRRs awaits further clarification.

19-27 HBV impairs innate immune responses as a means of immune evasion.

The 3.2 kb genome of hepatitis B virus consists of partially double-stranded, circular DNA. HBV infection impairs innate immune responses triggered both by specialized cells (such as pDCs) and in infected hepatocytes by down-regulating functional expression of TLR. Remarkably, genetic analysis of HBV-infected cells showed no sign of innate gene activation. This 'invisible' nature of the HBV to the innate immune system reflects its replication strategy, which sequesters the transcriptional template in the nucleus, generating viral transcripts indistinguishable from normal cellular mRNA, and the fact that the replicating viral genome is sheltered within viral capsid particles in the cytoplasm. How HBV infection results in robust activation of T- and B-cell responses remains unclear.

Adaptive immune responses against sexually transmitted pathogens.

During the natural course of infection, sexually transmitted pathogens infect specific target cells and are taken up by local antigen-presenting cells. The type II mucosa is characterized by the absence of mucosa-associated lymphoid tissues. Priming occurs exclusively in the draining lymph nodes. Adaptive immunity that develops during the natural course of infection with sexually transmitted pathogens is required, but not always sufficient, to clear the primary infection. Effector mechanisms include antibody-mediated blockade of pathogen entry, antibody-dependent cellular cytotoxicity by NK cells, CD4+ T cell-mediated suppression of intracellular pathogens, and CD8+ T cell-mediated lysis of infected cells (Figure 19.4).

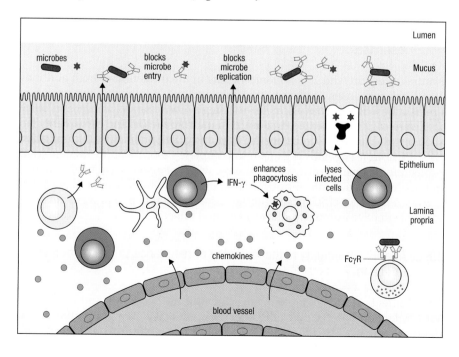

Figure 19.4 Adaptive immune system of the female genital mucosa. Once adaptive immune responses are initiated in the draining lymph nodes, B and T cells specific for the pathogen enter the infected tissue and eliminate the pathogen and the infected cells. B cells secrete antibodies to block the entry and further spread of pathogens. T_H1 CD4+ T cells secrete IFN-γ, which suppresses viral replication and enhances phagocyte clearance of bacterial and parasitic pathogens. CD8+ T cells lyse infected cells expressing MHC class I and limit the propagation of intracellular pathogens.

19-28 *Chlamydia* infection triggers the activation of local DCs to migrate to the draining lymph nodes and initiate T-cell activation.

While both CD4$^+$ and CD8$^+$ T cells can contribute to controlling *Chlamydia* infection, T$_H$1 responses against chlamydia play a dominant role for resolving genital infection in an IFN-γ-dependent manner. IFN-γ can upregulate the phagocytic potential of macrophages, thereby promoting the engulfment and destruction of extracellular elementary bodies. IFN-γ can also limit the growth of intracellular bacteria by upregulating nitric oxide synthase. In contrast, the role of B cells and antibody response in defense against chlamydia is less clear. High titers of *C. trachomatis*-specific antibodies do not correlate with resolution of infection in humans.

19-29 T$_H$1 cells mediate immunity against gonorrhea.

Protective immunity against *Neisseria gonorrhoeae* is mediated by T$_H$1-type responses, leading to the production of complement-fixing antibodies. Patients with defects in the complement pathways often suffer from recurrent and disseminated gonococcal infections. *Neisseria gonorrhoeae* expresses lipooligosaccharide lacking the O antigen, which activates TLR2, TLR4, and TREM-2A. Human DCs in genital tissues also use different C-type lectins, MGL and DC-SIGN, to detect the *N. gonorrhoeae* types and induce CD4$^+$ T-cell activation.

19-30 Protective immunity to syphilis is undefined but likely exists.

To date, no antigen has been definitively localized or identified in the outer membrane of *T. pallidum*, a factor contributing to the concept of the spirochete being a 'stealth pathogen.' This poses a challenge in the design of vaccines that can effectively bind to *T. pallidum* and block infection. However, patients who have been previously infected with *T. pallidum*, but have not been treated with antibiotics, appear to have at least some degree of immunity against repeated infection. In contrast, antibiotic-treated patients do not develop a similar degree of protective immunity, suggesting that either chronic infections or antigens exposed during the chronic stage of infections elicit a protective response against this pathogen.

19-31 Immunity to *Trichomonas* infection involves both B- and T-cell responses.

Natural infection with *Trichomonas* results in the induction of both antibody and T-cell responses. However, immunity generated following natural infection is at best only partially protective. Using a mouse model of infection, mice that were vaginally infected, treated with metronidazole, and then reinfected vaginally did not develop protective immunity. Subcutaneous immunization with a whole *T. vaginalis* organism in complete Freund's adjuvant conferred protection against intravaginal challenge with *T. vaginalis*, protection that is not achieved as a result of prior vaginal infection. These results suggest that *T. vaginalis* has an efficient immune-evasion mechanism to prevent development of effective immunity following natural infection.

19-32 Adaptive immunity is the mainstay of resistance to genital herpes infection.

HSV-2 infects stratified squamous epithelium of the vagina and ectocervix. Virus infection is detected by the infected epithelial cells and by submucosal dendritic cells in a MyD88-dependent manner. Recognition of infection by both hematopoietic and stromal compartments is required for successful induction of protective T_H1 immunity. However, directly infected cells are incapable of priming T cells because HSV-2 blocks MHC I and MHC II presentation and is highly lytic. Uninfected submucosal DCs pick up viral antigens and migrate to the draining lymph nodes, where they present antigenic peptides to cognate CD8+ and CD4+ T cells. Langerhans cells within the vaginal epithelium do not participate in T_H1 priming. Natural immunity that develops following genital HSV infection fails to clear the virus because the virus can invade and establish latent infection in the innervating ganglia prior to the onset of highly effective immunity. Latent virus cannot be cleared by T cells or antibody, although CD8+ T cells provide important immune surveillance of the infected neurons through a nonlytic mechanism. Natural immunity against HSV-1 or antibody responses generated by vaccines offer some level of cross-protection against acquisition of genital HSV-2 in women but not in men.

19-33 HPV engenders limited immune responses leading to persistent, latent infection.

Upon natural infection with HPV, only poor immune responses develop. This is owing to the ability of the virus to evade immunity by several mechanisms: (1) HPV E6 protein causes the depletion of Langerhans cells; (2) HPV E5 protein downregulates MHC I expression; and (3) HPV E7 protein blocks type I IFN signaling. Thus, activation of CD8+ T-cell immunity to HPV requires cross-presentation of infected cells by non-infected DCs. Although the nature of the DCs that are responsible for this cross-priming is unknown, because HPV infection occurs only within the stratified squamous epithelial layer, Langerhans cells within the epithelial layer, and possibly submucosal DCs that extend their dendrites, are most likely the ones participating in this process. HPV infection results in the induction of antibody and T-cell responses. Since the majority of HPV infections are cleared within two years, and because HIV infection results in an increased prevalence of HPV infections, it is likely that naturally induced immune responses are responsible for controlling HPV infection in most cases. It is noteworthy that neutralizing antibodies induced by HPV vaccines are remarkably efficient in preventing infection by HPV, as discussed below.

19-34 Cellular and humoral immunity is essential for immunity to HIV-1.

During HIV-1 infection, the type of DCs that prime CD4+ and CD8+ T-cell responses is unknown. Much of the focus has been on the role of DCs and Langerhans cells in enhancing HIV-1 infection. Langerhans cells can bind to HIV-1 gp120 through mannose C-type lectins. Using an *ex vivo* human organ culture system, Langerhans cells were shown to take up HIV-1 through endocytosis. As Langerhans cells exit the epithelium at the basal side, they transport intact virions, thereby enabling the infection to spread beyond the site of viral entry. All arms of the adaptive immune system are activated by HIV-1 infection. However, natural immunity to HIV-1 is not protective. HIV-1 superinfection can occur in individuals with a strong and broadly reactive virus-specific CD8+ T-cell response. The immunological correlates of protection against HIV-1 are currently unknown, making HIV/AIDS vaccine design difficult. However, recent studies have raised hope through the identification of broadly neutralizing IgG antibodies in long-term survivors. Some of these antibodies are able to bind to and neutralize more than 90% of infectious strains.

19-35 Adaptive immunity is critical to limiting and resolving HBV infection.

Natural infection with HBV results in the generation of T- and B-cell responses. Because HBV is a noncytopathic virus, disease caused by HBV is mediated by

T cells specific to HBV antigens. Once primed, HBV-specific CTLs migrate into the liver and recognize and kill HBV-infected hepatocytes. In addition, T cells secrete IFN-γ, which leads to the expression of chemokines that enable further leukocyte recruitment. An inefficient T-cell response that is unable to completely clear HBV from the liver leads to a sustained continuous cycle of low-level hepatocyte destruction. Over a long period of time, the chronic nature of these cycles may lead to recurrent immune-mediated liver damage, which contributes to the development of cirrhosis and hepatocellular carcinoma.

Challenges ahead.

Sexually transmitted pathogens continue to propagate and cause significant morbidity and mortality throughout the world. Infection by sexually transmitted pathogens represents a complex problem, not only with respect to host–pathogen interaction, but also with respect to reproductive and emotional health. In addition, unlike for other infectious diseases, vaccines and interventions for STDs are put through religious and socioeconomic considerations. Having said this, prophylactic vaccines are the only option available to combat the incurable viral STDs. In this regard, two recent vaccines against HPV, Gardasil™ (Merck) and Cervarix™ (GlaxoSmithKline), have demonstrated an overwhelming success, providing protection from infection in almost 100% of cases. Protection conferred by these HPV vaccines is mediated by neutralizing antibodies. The outcome of these trials highlights that successful vaccines against sexually transmitted viruses are possible. However, antibody-based vaccines are not effective against all STDs, and protective immunity to others such as HIV-1 and HSV-2 may require cellular immunity with the caveats described above. Future studies must focus on first identifying the protective immune mechanism tailored to each STD, and designing vaccines that elicit such immune responses at the relevant sites of exposure within the genital mucosae.

Summary.

STDs are caused by viral, bacterial, and parasitic pathogens. These pathogens have evolved to utilize the physiology of the female and male reproductive tracts for successful invasion, replication, and transmission. Currently, vaccines are not available for most of these pathogens and are desperately needed to combat the spread of incurable STDs. To this end, a clear understanding of the invasion mechanisms utilized by the pathogens, as well as innate and adaptive immune defense mechanisms for each sexually transmitted pathogen, would provide a rational basis for vaccine design. Innate recognition is well understood for some sexually transmitted pathogens (*C. trachomatis*, *N. gonorrhoeae*, HSV) but not for others (*T. pallidum*, HIV-1, HBV). Natural infections with sexually transmitted pathogens often do not result in protective immunity, likely due to the evasion mechanisms employed by the pathogens. Protective immune responses to some STDs can be induced by vaccines, which are mediated by neutralizing antibodies, cytotoxic CD8$^+$ T cells, and effector CD4$^+$ T cells. Future challenges to the development of vaccines include identifying immune correlates of protection for each sexually transmitted pathogen, and finding means of inducing such responses in humans.

Further Reading.

Akira, S., Uematsu, S., and Takeuchi, O.: **Pathogen recognition and innate immunity.** *Cell* 2006, **124**:783–801.

Altfeld, M., Allen, T.M., Yu, X.G., *et al.*: **HIV-1 superinfection despite broad CD8+ T-cell responses containing replication of the primary virus.** *Nature* 2002, **420**:434–439.

Dethlefsen, L., McFall-Ngai, M., and Relman, D.A.: **An ecological and evolutionary perspective on human-microbe mutualism and disease.** *Nature* 2007, **449**:811–818.

Frazer, I.H.: **Prevention of cervical cancer through papillomavirus vaccination.** *Nat. Rev. Immunol.* 2004, **4**:46–54.

Guidotti, L.G., and Chisari, F.V.: **Immunobiology and pathogenesis of viral hepatitis.** *Annu. Rev. Pathol.* 2006, **1**:23–61.

Haase, A.T.: **Targeting early infection to prevent HIV-1 mucosal transmission.** *Nature* 2010, **464**:217–223.

Hladik, F., and McElrath, M.J.: **Setting the stage: host invasion by HIV.** *Nat. Rev. Immunol.* 2008, **8**:447–457.

Iwasaki, A.: **Mucosal dendritic cells.** *Annu. Rev. Immunol.* 2007, **25**:381–418.

Iwasaki, A.: **Antiviral immune responses in the genital tract: clues for vaccines.** *Nat. Rev. Immunol.* 2010, **10**:699–711.

Medzhitov, R.: **Recognition of microorganisms and activation of the immune response.** *Nature* 2007, **449**:819.

Peeling, R.W., and Hook, E.W., 3rd.: **The pathogenesis of syphilis: the Great Mimicker, revisited.** *J. Pathol.* 2006, **208**:224–232.

Roan, N.R., and Starnbach., M.N.: **Immune-mediated control of *Chlamydia* infection.** *Cell. Microbiol.* 2008, **10**:9–19.

Rottenberg, M.E., Gigliotti Rothfuchs, A.C., Gigliotti, D., *et al.*: **Role of innate and adaptive immunity in the outcome of primary infection with *Chlamydia pneumoniae*, as analyzed in genetically modified mice.** *J. Immunol.* 1999, **162**: 2829–2836.

Stanberry, L.R.: **Clinical trials of prophylactic and therapeutic herpes simplex virus vaccines.** *Herpes* 2004, **11** Suppl. 3: 161A–169A.

PART V NOSE, AIRWAYS, ORAL CAVITY, AND EYES

The Nasopharyngeal and Oral Immune System

20

Mammals have evolved a sophisticated immune system consisting of an integrated network of tissues, lymphoid and mucous membrane-associated cells, and effector antibody molecules that serve to maintain homeostasis on exposed mucosal surfaces. It is noteworthy that this system is anatomically and functionally distinct from its blood-borne counterpart of the systemic (or peripheral) immune system and is strategically located at the portals by which most microorganisms enter the body. The development of this specific branch of the immune system alongside and separate from the peripheral immune system may have been necessitated by the size of the mucosal surfaces, which cover an area of ~400 m² in the adult human, as well as the large number of exogenous antigens to which these surfaces are exposed. Along with cytokines, chemokines, and their receptors, the effector antibody molecules, which are primarily of the IgA isotype, are key players in mucosal immunity and appear to function in synergy with innate host factors. Thus, in order to induce antigen-specific immune responses at these mucosal barriers, one must consider the mucosal immune system, which is composed of functionally distinct mucosal IgA inductive and effector tissues (Figure 20.1). Mucosal inductive sites include the nasopharynx-associated lymphoid tissues (NALT) and the Peyer's patches of the gut-associated lymphoid tissues (GALT) for initiation of mucosal secretory IgA (SIgA) antibody responses. The subsequent homing of memory/activated B and T lymphocytes from NALT to nasal passages and upper respiratory tract regions or from GALT to effector tissues, for example the lamina propria of the gastrointestinal tract, forms the basis for the mucosal immune system, which interconnects the inductive with the effector sites. This chapter will focus on the cellular and molecular roles of this nasal–oral immune system for the induction of SIgA antibody responses in the nasal and oral cavities, the portal of entry for the aerodigestive tract.

The nasopharyngeal–oral mucosal immune system.

In mammals, organized secondary lymphoid tissues have evolved in the upper respiratory and gastrointestinal tracts to facilitate antigen uptake, processing, and presentation; they are collectively termed mucosa-associated lymphoid tissues (MALT). Thus, uptake of foreign antigens can be blocked by the barrier function of the epithelium that covers these mucosal surfaces, or antigen can be selectively taken up into highly specialized inductive sites such as MALT for the initiation of an immune response (hence the term inductive site).

Figure 20.1 Nasopharyngeal–oral common mucosal immune system. Nasally administered antigens and vaccines are mainly taken up by the NALT as a mucosal inductive site. Antigen-presenting cells including dendritic cells process antigens and subsequently stimulate T cells. Further, B-cell follicles are formed and IgA class-switch recombination is induced. Antigen-stimulated T cells and IgA-committed (surface IgA positive (sIgA$^+$)) B cells dispatch into the systemic circulation via the cervical lymph nodes. These immune cells home to mucosal effector sites including nasal passages and exocrine glands such as submandibular glands. Antigen-specific T cell–B cell interactions induce sIgA$^+$ B-cell differentiation into antigen-specific IgA-producing plasma cells. The dimeric form of IgA antibody binds to polymeric immunoglobulin receptors on epithelial cells and ductal cells and the molecule is transported as secretory IgA (SIgA) onto the nasal and oral mucosa.

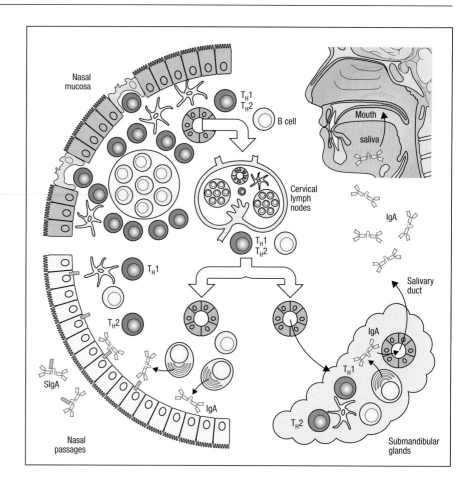

20-1 The MALT of the gut and upper airway share features of antigen uptake.

The Peyer's patches (PPs) and the NALT in humans and mice comprise a MALT network which, although integrated, is at best only partially characterized and understood. The MALT are covered by a lymphoepithelium containing micro-fold (M) cells. M cells are characterized by a pocket structure (M-cell pocket) at the basolateral side harboring a wide variety of immune cells such as dendritic cells (DCs), macrophages, and lymphocytes. A functionally unique feature of M cells is that, in spite of their capacities to take up luminal antigens via pino-cytosis and endocytosis, M cells contain very low numbers of lysosomes. In addition, IgA preferentially binds to the apical side of M cells. Thus, mouse IgA and human IgA2, but not IgA1, binds to mouse M cells via the Fc receptor for IgA, which differs from CD89 (FcαR expressed on human neutrophils and eosinophils). It has been considered that M cells contribute to the transport of luminal antigens to underlying antigen-presenting cells (APCs) without anti-gen processing. MALT contains well-organized regions, the subepithelium (dome), a B-cell zone with germinal centers, and adjacent T-cell areas with enriched APCs including DCs and high endothelial venules (HEVs). Naive, recirculating B and T lymphocytes enter MALT via the HEVs. Furthermore, antigen-primed, activated, and/or memory B- and T-cell populations then emigrate from the mucosal inductive environment via lymphatic drainage, circulate through the bloodstream, and then home to mucosal effector sites. In addition, it has been suggested that isolated lymphoid follicles in the mouse small intestine are a part of GALT for induction of mucosal immune responses. Thus, these isolated lymphoid follicles mainly contain B cells and DCs as well as M cells in the overlying epithelium. In addition to these follicle-associated epithelial (FAE) M cells in PPs and isolated lymphoid follicles, it was recently shown that M-cell clusters occur in non-FAE sites at the tips of the villi, hence called villous M cells, as alternative antigen-sampling sites.

20-2   NALT is a major IgA inductive site for the nasal and oral mucosa.

NALT is the major inductive tissue for nasal/inhaled antigens in humans, primates, mice, and rats (Figure 20.2). In rodents, NALT is found on both sides of the nasopharyngeal duct, dorsal to the cartilaginous soft palate. In humans, there are unpaired nasopharyngeal tonsils (adenoids) and the paired palatine tonsils, which play important roles for human airway immunity. The latter make up most of Waldeyer's ring. Furthermore, a NALT-like lymphocyte aggregate with follicles was identified in the human nasal mucosa, most notably in the middle concha in children less than two years old. In human tonsils, approximately one-half of the cells are B lymphocytes that mainly occur in follicle-containing germinal centers. Most human tonsillar B cells are surface IgG positive (sIgG+), but significant numbers of sIgM+ and sIgA+ B cells are also present (Figure 20.2). The human palatine tonsils also contain a distinct subepithelial B-cell population, similar to that seen underneath the FAE region of Peyer's patches, that differs from both germinal center and follicular mantle B cells (Figure 20.2). These subepithelial B cells located in NALT may play a crucial role in the immediate antibody production against antigen taken up through M cells. Indeed, both human tonsils and adenoids are equipped with M cells for inhaled antigen uptake.

Figure 20.2 Nasopharynx-associated lymphoid tissues (NALT) in humans and mice are mucosal inductive tissues. Both human palatine tonsils and mouse NALT are covered by an epithelium equipped with M cells. NALT consists of a follicle zone which contains mainly B cells and a parafollicular zone with T cells and DCs. Although sIgA+ B cells are not the major subset of B cells present, the immune features, including IgA class-switch recombination, are similar to the traditional mucosal inductive site, the Peyer's patches.

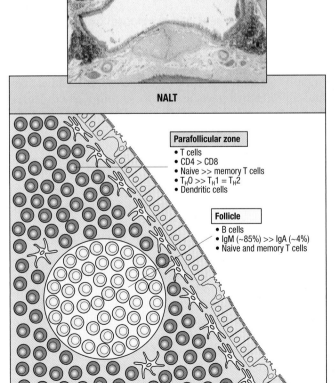

NALT

Parafollicular zone
• T cells
• CD4 > CD8
• Naïve >> memory T cells
• T_H0 >> T_H1 = T_H2
• Dendritic cells

Follicle
• B cells
• IgM (~85%) >> IgA (~4%)
• Naïve and memory T cells

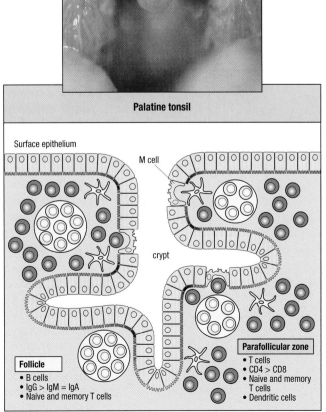

Palatine tonsil

Surface epithelium

M cell

crypt

Follicle
• B cells
• IgG > IgM = IgA
• Naïve and memory T cells

Parafollicular zone
• T cells
• CD4 > CD8
• Naïve and memory T cells
• Dendritic cells

Murine NALT consists of bilateral strips of non-encapsulated lymphoid tissue underlying the epithelium on the ventral aspect of the posterior nasal tract and exhibits a bell-like shape in frontal cross sections (Figure 20.2). Although dense aggregates of lymphocytes have been observed in the NALT of normal mice, germinal centers are absent but could be induced by nasal application of antigens. Mononuclear cells isolated from NALT have a high frequency (80–85%) of uncommitted B cells (sIgM$^+$) and only low frequencies of IgA (sIgA$^+$) and IgG (sIgG$^+$) committed B cells (3–4% and 0–1% respectively). In comparison with Peyer's patches, where a high frequency (10–15%) of sIgA$^+$ B cells occur, NALT contains fewer IgA committed B cells (Figure 20.2). Despite this, nasal immunization induces higher numbers of IgA$^+$ (than IgG$^+$) B cells among NALT B cells in the memory compartment, showing the propensity of NALT for the initiation of antigen-specific B-cell responses leading to the induction of mucosal SIgA antibody responses. Characterization of isolated NALT mononuclear cells revealed that approximately 30–40% of these cells are CD3$^+$ T cells with a CD4/CD8 ratio of ~3:1 (Figure 20.2). The majority of NALT CD3$^+$ T cells co-express CD45RB, suggestive of naive, resting T cells. Transcriptional single-cell analysis revealed the expression of mRNA for both helper T (T$_H$) cell T$_H$1 and T$_H$2 cytokines, the majority of CD4$^+$ T cells are considered T$_H$0 types (Figure 20.2). Furthermore, stimulation via the T-cell receptor–CD3 complex resulted in differentiation of both T$_H$1- and T$_H$2-type cells. These results support the notion that NALT exhibits characteristics of mucosal inductive sites.

NALT M cells are reported to be the site for entry of *Mycobacterium tuberculosis* into the host with subsequent uptake in draining lymph nodes. Furthermore, it has been shown that group A streptococci are transported across the NALT M cells. Thus, M cells act as a gateway to the outside environment, delivering antigen to the underlying immune cells for the subsequent induction of antigen-specific immune responses. Indeed, reoviruses initiate infection via this M cell, an ability that has been associated with the protein sigma one (pσ1). In this regard, DNA vaccines containing reovirus pσ1 target NALT M cells and produce high levels of antigen-specific IgA in the gut and serum IgG.

20-3 The nasal passage contains a new type of mucosal tissue.

After initial exposure to antigen in MALT, mucosal lymphocytes leave the inductive site and home to mucosal effector tissues. Effector sites for mucosal immune responses include the lamina propria regions of the gastrointestinal, upper respiratory, and reproductive tracts as well as secretory glandular tissues and intestinal epithelium. Nasal mucosa is also characterized by a diffuse collection of effector lymphoid cells. These include IgA-producing plasma cells, B and T lymphocytes, as well as APCs, such as DCs, and macrophages for the generation of antigen-specific acquired mucosal immunity. Nonclassical APCs as well as innate cells such as epithelial cells, eosinophils, basophils, natural killer cells, and mast cells occur and provide innate mucosal immunity. The interaction of these innate and adaptive effector cells with exogenous antigenic stimulation will typically result in the appropriate mucosal (SIgA) and systemic (serum IgG) antibody and/or T cell-mediated immune responses.

It has been shown that nasal washes from humans, nonhuman primates, rats, and mice contain significant levels of IgA antibodies. Nasal lamina propria contains a high frequency of IgA-producing plasma cells compared with those producing other isotypes of antibodies (Figure 20.3). Thus, the distribution of immunoglobulin isotypes is very similar to that seen in the intestinal lamina propria. At least two B-cell subsets, B-1 and B-2 B cells, are present in the mouse and human periphery. It is now well accepted that B-1 B cells, a minor subset comprising about 5% of the total B-cell population, arise during fetal development and have a restricted receptor repertoire. Furthermore, B-1 B cells differ from conventional B cells in cell-surface protein CD5 expression, anatomical localization, and functional characteristics. These B-1 B cells are also characterized by high levels of IgM, IgA, and IgG3 of low affinity and broad specificity for polysaccharides from Gram-positive bacteria and lipopolysaccharide (LPS) from Gram-negative bacteria. Although these responses can be enhanced/increased by T cells, they appear within 48 hours of exposure to antigen and so are not dependent upon T-cell help. Indeed, mononuclear cells in murine nasal passages contain approximately 20–25% of B cells consisting of both conventional B-2 and B-1 B cells (Figure 20.3). Although the precise immunobiological roles for B-1 and B-2 cells in humans remain to be elucidated, mouse B-1 B cells can be further classified into B-1a (IgM^{high}, IgD^{low}, B220^{low}, CD5^{+}) cells and B-1b (IgM^{high}, IgD^{low}, B220^{low}, CD5^{-}) cells based upon their expression of CD5. Thus, murine nasal passages contain both B-1b (~7%) and B-1a (~2%) cell subsets. Approximately 20% of nasal passage lymphocytes are CD3^{+} T cells. Half of these CD3^{+} T cells express the CD4 molecule and one-third are CD8 positive (Figure 20.3).

It has been shown that Langerhans-type (or DC-like) cells occur on the luminal side of the intestinal epithelium and could provide accessory functions including possible direct interactions with luminal antigens and epithelial cells. When confronted with microorganisms and even with soluble proteins which can traverse the tight junctions between epithelial cells, the lamina propria DCs in these effector sites may take up and process antigens, and in doing so induce antigen-specific B- and T-cell responses.

20-4   The salivary glands have features of both secretory and systemic immunity.

The oral cavity is protected by two arms of the immune system, namely mucosal SIgA and systemic (e.g., blood plasma IgG) immunity via saliva and gingival crevicular fluid respectively (Figure 20.4). Saliva has been extensively used as an easily accessible external fluid for the measurement of antigen-specific SIgA antibodies following mucosal immunization in both human and experimental animal models (Figure 20.4). When numbers of antibody-forming cells in submandibular glands were examined using the isotype-specific ELISPOT (enzyme-linked immunospot) assay, it was found that the dominant isotype of immunoglobulin-producing cells in submandibular glands was IgA, followed

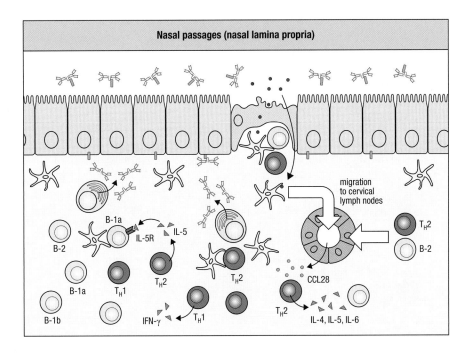

Figure 20.3 Nasal passages (nasal lamina propria) as a mucosal effector tissue. Nasal immunization induces antigen-specific T-cell and B-cell migration into the nasal passages. Antigen-specific CD4^{+} T cells produce T$_H$2 (IL-4, IL-5, and IL-6) type cytokines for antigen-specific B-cell activation, proliferation, and differentiation for the induction of IgA antibodies. Murine nasal passages contain both B-1a (~2%) and B-1b (~7%) cell subsets. B-1 B cells expressing IL-5R in nasal passages play a key role in the induction of innate IgA antibodies. The nasal epithelium is equipped with M cells for antigen uptake. It is possible that nasal DCs take up antigen and migrate into the cervical lymph nodes, providing a novel inductive site for the generation of antigen-specific IgA responses. IL, interleukin; IFN-γ, interferon-γ.

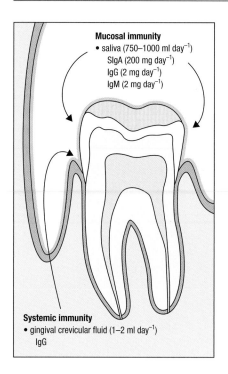

Figure 20.4 The uniqueness of oral mucosal immunology. The oral cavity is protected by two types of immune systems, namely mucosal and systemic immunity. Thus, saliva contains large amounts of SIgA antibody whereas the gingival crevicular fluid mainly contains IgG antibodies originally derived from the blood plasma.

by small numbers of IgM and IgG antibody-forming cells. Thus, comparison of their distribution with other IgA effector sites (e.g., the lamina propria of the small intestine and nasal passages) precisely followed the isotype pattern of immunoglobulin-producing cells taken from the lamina propria of the aerodigestive tract. Taken together, these findings suggest that the submandibular glands harbor all of the necessary components to be considered to be an IgA effector tissue, including the dominant occurrence of IgA-producing cells (Figure 20.5).

In addition to a predominance of IgA-producing cells, the submandibular glands also contain relatively high numbers of T cells (Figure 20.5). When mononuclear cells from these tissues were stained with immunofluorescence-labeled monoclonal antibody anti-CD3 and then analyzed by flow cytometry, approximately 50% of lymphocytes were CD3$^+$ T cells. When different subsets of CD3$^+$ T cells were assessed according to their expression of CD4 and CD8 molecules, three distinct subsets of T cells were observed in the submandibular glands: CD4$^+$CD8$^-$, CD4$^-$CD8$^+$, and CD4$^-$CD8$^-$. However, double-positive cells were essentially absent. Regarding CD4- and CD8-bearing T cells, approximately equal frequencies of CD4$^+$CD8$^-$ and CD4$^-$CD8$^+$ T cells were noted in this tissue. This finding was very similar to studies of rat salivary glands where the mononuclear cells were shown to contain approximately 60% T cells with a CD4 to CD8 ratio of 1.1:1.3. Furthermore, relatively high numbers of γδ T cells (approximately 15%) were found in the submandibular glands. The increased numbers of γδ T cells seems to be one of the unique characteristics of mucosa-associated tissues; other mucosal effector sites such as intestinal lamina propria and intestinal epithelium also contained a relatively high frequency of γδ T cells.

The T-cell cytokine profile in the submandibular glands is mixed, with many cells making interferon-γ (IFN-γ) or interleukin-5 (IL-5) or IL-6. When CD3$^+$ T cells were isolated from Con A or anti-CD3 stimulated submandibular gland lymphocyte cultures and then analyzed for cytokine synthesis, IL-2- and IL-4-producing cells were found in addition to elevated numbers of IFN-γ-, IL-5-, and IL-6-secreting cells (Figure 20.5). These findings demonstrated that although submandibular gland CD3$^+$ T cells are capable of secreting an array of T$_H$1 and T$_H$2 cytokines, these cells are programmed to produce selected T$_H$1 (IFN-γ) and T$_H$2 (IL-5 and IL-6) cytokines which provide a favorable molecular environment for IgA synthesis in mucosal effector tissues. Submandibular gland CD3$^+$ T cells help PP B cells to differentiate into IgM-, IgG-, and IgA-producing cells.

Figure 20.5 Submandibular glands as mucosal effector tissues for a functional oral immune system. Upon nasal immunization, antigen-stimulated CD4$^+$ T cells, B-2 B cells, and DCs in NALT migrate into the submandibular glands. T$_H$1 and T$_H$2 cytokines produced by CD4$^+$ T cells elicit conventional B-2 B-cell differentiation to antigen-specific IgA-producing plasma cells. Also shown are the potential roles for DCs in the induction of innate SIgA antibody responses in submandibular glands in the tissue in the lower lobe of the gland. Upon T-cell independent antigen stimulation, DCs in submandibular glands may produce APRIL for the induction of IgA class-switch recombination in B-1a B cells. IL-5-producing CD4$^+$ T cells induce sIgA$^+$ B-1a B cells to differentiate into IgA-producing plasma cells.

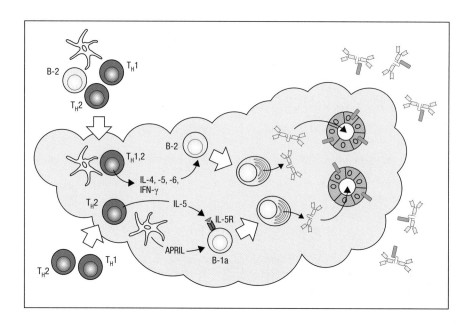

This is consistent with the notion that the submandibular gland CD3+ T cells seem to be programmed *in situ* to produce IgA-enhancing cytokines which can support B-cell responses including inducing sIgA+ B cells originally derived from an IgA inductive site (e.g., the NALT) to become IgA-producing plasma cells in mucosal effector tissues (e.g., the submandibular glands).

Another remarkable characteristic of the submandibular glands is that this tissue contains a significant number of B-1 B cells (Figure 20.5). In this regard, mucosal inductive lymphoid tissues, namely NALT and PPs, only contain conventional B-2 B cells, whereas mucosal effector lymphoid tissues, namely the intestinal lamina propria, nasal passages, and submandibular glands, are enriched in B-1 B cells. A high frequency of B-1a B cells is more typical of the submandibular glands, while a predominance of B-1b B cells is characteristic of the intestinal lamina propria and nasal passages of wild-type mice. Interestingly, up to 40% of IgA plasma cells in the mouse intestinal lamina propria arise from a pool of B-1 precursors derived from the peritoneal cavity. Thus, it is possible the B-1 B cells represent a distinct lineage from that of conventional B-2 B cells originating in mucosal inductive lymphoid tissues. In addition, up to 50% of human intestinal B cells express the CD5 molecule and a large number of these CD5+ B cells secrete IgA antibodies. Taken together, these findings suggest that B-1 B cells in submandibular glands could be an alternative but important source for IgA-producing cells in addition to conventional B cells.

Induction of acquired immune responses via the oral and nasal mucosal immune system.

Oral immunization has traditionally been the preferred route to generate protective mucosal immune responses. The upper airways, however, provide an alternative site because of ease of access and delivery through aerosols. Furthermore, the ease of directly targeting inductive lymphoid tissue in NALT, because of the relatively short distance between the external nares and the lymphoid tissue, offers distinct advantages over oral immunization, where enzymatic degradation and dilution play a role before antigen reaches the Peyer's patches and isolated lymphoid follicles.

20-5 Enterotoxin-based nasal adjuvants are the most effective method for the induction of antigen-specific immunity.

The induction of effective mucosal immune responses generally requires the use of mucosal adjuvants or of attenuated live viral or bacterial vectors, with adjuvants offering the advantage of eliciting mucosal as well as parenteral immune responses. Nasal delivery of purified antigen plus mucosal adjuvant has emerged as perhaps the most effective way to induce both peripheral and mucosal immunity, including salivary SIgA antibody responses (Figure 20.1). Most current protocols in mice instill vaccine into each nostril (usually 5–10 µl per nostril) and normal inhalation results in effective delivery of vaccine, presumably into NALT and nasal passages. Two bacterial enterotoxins—native cholera toxin (nCT) and heat-labile enterotoxin type 1—and their artificial nontoxic mutants—mutant cholera toxin (mCT) and mutant heat-labile enterotoxin type 1—are well-established mucosal adjuvants for the induction of both mucosal and systemic immunity to coadministered protein antigens. Mucosal administration of subunit (or purified antigen) vaccine together with nCT or nontoxic mCTs induces vaccine-antigen-specific CD4+ T_H2-type cells with characteristic serum antigen-specific IgG1, IgG2b, IgE, and IgA as well as mucosal SIgA antibody responses. Nasal immunization with the weak protein antigen ovalbumin plus mCT elicited ovalbumin-specific SIgA antibody responses in the submandibular glands and in other mucosal effector tissues. Furthermore, it has been shown that nasal vaccines containing tetanus toxoid and mCT induced protective immunity and the generation of tetanus-toxin-specific neutralizing antibodies. Mice immunized nasally with *Streptococcus pneumoniae* pneumococcal surface protein A (PspA) plus mCT revealed antigen-specific mucosal IgA responses associated with effective protection against capsular serotype 3 *S. pneumoniae* A66. These studies further showed that induction of an antigen-specific SIgA antibody response was associated with a polarized T_H2-type response when cholera toxin was used as an adjuvant in nasal vaccines. Nasal immunization with simian immunodeficiency virus p55gag plus cholera toxin resulted in selected T_H2- and T_H1-type cytokine profiles with antigen-specific IgA responses in mucosal secretions (e.g., cervico-vaginal secretions, rectal washes, and saliva) of nonhuman primates. Furthermore, high numbers of p55gag-specific IgA antibody-forming cells were detected in mucosal effector sites, namely uterine cervix, intestinal, and nasal lamina propria. It was also shown that nasal vaccine containing outer membrane proteins of *Haemophilus influenzae* and adjuvant cholera toxin induced antigen-specific SIgA that blocked attachment of the bacterium to epithelial cells. Similarly, when mice were immunized with fimbrial protein of *Porphyromonas gingivalis* and cholera toxin, fimbriae-specific IgA antibodies generated from the submandibular glands inhibited *P. gingivalis* attachment to, and reduced subsequent inflammatory cytokine production by, epithelial cells. These studies showed that nasal immunization effectively induced T_H1- and/or T_H2-type cytokine-producing antigen-specific CD4+ T cells in both mucosal effector tissues (nasal lamina propria and submandibular glands) and spleen for the generation of antigen-specific SIgA and plasma IgG responses respectively.

Studies in which inactivated respiratory syncytial virus (RSV) was given with cholera toxin nasally resulted in nasal IgA and serum IgG anti-RSV antibody responses, and analysis of IgG subclasses suggested that both IgG1 and IgG2a were induced. Interestingly, infection with RSV and subsequent T_H1-type responses are characterized by a favorable outcome, while T_H2-type responses were associated with significant pathology. An extensive series of studies using the influenza virus model demonstrated that the influenza vaccine provided more effective protective immunity when administered nasally compared with other routes of immunization. Nasal immunization with a trivalent influenza vaccine given with the B subunit of cholera toxin (CTB; a nontoxic portion of cholera toxin) containing a trace amount of nCT resulted in the induction of cross-protective SIgA antibodies against a broad range of viruses. It was also shown that nasal immunization with influenza vaccine together with the B subunit of heat-labile enterotoxin type 1 (containing a trace amount of heat-labile enterotoxin type 1) induced influenza-specific immune responses in humans. Taken together, these results suggest that the nasal mucosal immune system is an attractive site for the delivery of vaccine for the induction of protective immunity in both the mucosal and peripheral compartments.

20-6   Safe mucosal adjuvants have been developed as delivery systems for nasal vaccines.

The olfactory neuroepithelium in the nasopharynx constitutes approximately 50% of the nasal surface and has direct neuronal connections to the olfactory bulbs in the central nervous system. While diarrhea is the primary limiting factor for the use of oral enterotoxins as adjuvants in humans, major safety concerns with mucosal adjuvants and delivery systems for nasal vaccination are also important because they may enter and/or target olfactory neurons and, therefore, may bypass the blood–brain barrier and gain access to olfactory bulbs and deeper structures in the brain parenchyma. Studies with enterotoxin adjuvants and more recently with a recombinant adenovirus vector suggest that these adverse effects are in large part mediated by the ADP-ribosyl transferase activity and the nature of the cellular receptors targeted. Both nCT and heat-labile enterotoxin type 1 bind to GM1 on epithelial cells and require endocytosis followed by transport across the epithelial cell to reach the basolateral membrane. Heat-labile enterotoxin type 1 also binds to a lesser extent to GD1b and binds weakly to GM2 and asialo-GM1. GM1 gangliosides are also abundantly expressed by cells of the central nervous system and their concentration on neuronal and microglial cells varies during the development of various cell types and different regions of the brain. GM1 is thought to play a role in cell-to-cell and cell–matrix interactions and may act synergistically with several growth factors in development. GM1 ligation did not generate a calcium flux following exposure of microglial cells to CTB; however, binding of CTB to cerebellar granule cells *in vitro* improved survival and was associated with intracellular calcium increases through L-type voltage sensitive channels during the first 7 days of culture, with reduced calcium fluxes later in culture.

It has been shown that nCT or CTB, when given nasally to mice, enters the olfactory nerves/epithelium and olfactory bulbs by mechanisms that are selectively dependent upon GM1 ganglioside. These studies have also shown that nCT as adjuvant promotes the uptake of nasally coadministered, unrelated proteins into the olfactory nerves/epithelium. Nasal administration of mCT resulted in no sign of pathological changes in the central nervous system. Furthermore, no side effects were reported following nasal application of CTB to 12 IgA nephropathy patients. However, one has to be careful because a previous report showed that nasal vaccination of humans with a subunit-type influenza with heat-labile enterotoxin type 1 as mucosal adjuvant resulted in a high incidence of Bell's palsy (facial paresis) among volunteers. Thus, of the adverse events reported for approximately 3600 subjects participating in a clinical (safety) trial for a nonliving nasal influenza vaccine (Nasalflu™) in 2000, there were nine cases of Bell's palsy and one trigeminal neuralgia that developed into facial paresis. Furthermore, five cases of Bell's palsy were spontaneously reported in approximately 90,000 recipients of Nasalflu™ and the vaccine was thus withdrawn from the market. A relationship between Bell's palsy and the inactivated nasal influenza vaccine containing heat-labile enterotoxin type 1 was thus formally established.

To overcome these obstacles that prevent their clinical use, several groups have focused on developing nontoxic derivatives of cholera toxin or heat-labile enterotoxin. However, though these studies have been successful in producing nontoxic mutants, they did so at times by sacrificing and/or reducing the potent mucosal adjuvanticity associated with nCT. Efforts to develop safe but effective nontoxic forms of toxin-based mucosal adjuvants have proven mCT E112K (a mutated CT created by substituting a single amino acid in the ADP-ribosyl transferase active center; glutamate to lysine at position 112), which was constructed to be devoid of toxicity while retaining its adjuvant activity, to be a safe and stable adjuvant. Previous studies showed that nasal immunization with ovalbumin, PspA of *Streptococcus pneumoniae*, or diphtheria toxoid plus mCT E112K elicited antigen-specific IgA and IgG antibody responses in the mucosal and systemic immune compartments. Although these studies showed that mCT E112K did not elicit any neuronal damage based upon nerve growth factor-β1 production in the central nervous system as well as antigen redirection, these studies did not provide any direct information as to whether mCT E112K migrates into the central nervous system after nasal application. To further establish the safety of enterotoxin-based nasal adjuvants, double mutants of cholera toxin (dmCTs) were developed by introducing two potent mutations in the ADP-ribosylation activity center of the A1 subunit and the endoplasmic reticulum retention signal tetrapeptide KDEL in the A2 subunit. Thus, mutations in KDEL influence the movement of cholera toxin from the

Golgi apparatus to the endoplasmic reticulum. Confocal microscopic analysis showed that intracellular localization of dmCTs (E112K/KDEV or E112K/KDGL) differed from that of mCTs and the native form of cholera toxin (or nCT) in intestinal epithelial T84 cells. Furthermore, both dmCTs exhibited low toxicity in the Y1 cell assay and mouse ileal loop tests. When mucosal adjuvanticity was examined, both dmCTs induced enhanced ovalbumin-specific immune responses in both mucosal and systemic lymphoid tissues.

In addition to the approach which inhibits intracellular trafficking of the enterotoxin A subunit, others have developed a nCT-based safe and effective adjuvant that targets the immunoglobulin receptor on both naive and memory B cells in order to avoid GM1 ganglioside binding. Thus, the enzymatically active A1 subunit of nCT was combined with a dimer of an immunoglobulin binding element from *Staphylococcus aureus* protein A (CTA1-DD) and was coadministered with antigen. This CTA1-DD molecule directly bound to B cells of all isotypes but not to macrophages or DCs. Despite the lack of a mucosal binding element, the B cell-targeted CTA1-DD molecule was as strong an adjuvant as nCT. Thus, CTA1-DD adjuvant induced significant antibody and T-cell responses without central nervous system toxicity following nasal immunization. Notably, CTA1-DD promoted a balanced T_H1/T_H2 response with little effect on IgE antibody production. Furthermore, CTA1-DD did not induce inflammatory changes in the nasal mucosa, and most importantly did not bind to, or accumulate in, the olfactory bulbs or the central nervous system. Therefore, non-ganglioside-targeting adjuvants and delivery systems could represent new tools for the development of safe and effective nasal vaccines.

20-7   Nasal adjuvants activate the innate immune system.

It has been shown that nasal coadministration of non-ganglioside-binding regulatory cytokines or chemokines is effective for the augmentation of antigen-specific mucosal and systemic immune responses. For example, nasal immunization with tetanus toxoid or 2,4-dinitrophenol-ovalbumin and IL-12 enhanced antigen-specific antibody responses. In addition, nasal application of bicistronic DNA vaccines encoding IL-12 and antigens from *Yersinia pestis* induced protective antigen-specific antibody responses against pneumonic plague. Furthermore, both IL-1α and IL-1β are potent mucosal adjuvants. Finally, it has been shown that the human neutrophil peptides lymphotactin and RANTES possess nasal adjuvant activity for enhanced antigen-specific antibody responses.

Flt3 ligand has been used as a mucosal adjuvant in order to specifically target mucosal APCs, including DCs. The Flt3 ligand binds to the *fms*-like tyrosine kinase receptor Flt3/Flk2, a growth factor that dramatically increases the numbers of DCs *in vivo* without inducing their activation. Treatment of mice by systemic Flt3 ligand injection induced marked increases in the numbers of DCs in both systemic (i.e., spleen) and mucosal lymphoid tissues (i.e., intestinal lamina propria, Peyer's patches, and mesenteric lymph nodes). Other studies have now shown that Flt3 ligand treatment favors the induction of immune responses after mucosal, systemic, or cutaneous vaccine delivery. It was reported that plasmid DNA encoding Flt3 ligand (pFlt3-ligand) coadministered systemically with plasmids encoding protein antigens, or linked to the antigen itself, led to the induction of antigen-specific immune responses. Furthermore, nasal immunization with ovalbumin plus pFlt3-ligand or adenovirus expressing Flt3 ligand as nasal adjuvant specifically targets nasal DCs for the induction of antigen-specific antibody responses in both the mucosal and systemic compartments. These studies further showed that NALT DCs migrate into mucosal effector tissues including nasal passages and submandibular glands through the cervical lymph nodes to support the induction of antigen-specific immune responses. Importantly, coadministration of pFlt3-ligand as nasal adjuvant together with PspA elicited protective PspA-specific SIgA antibody responses against *Streptococcus pneumoniae* nasal colonization.

It has been shown that bacterial DNA and pathogen-associated molecular patterns contain a high frequency of unmethylated CpG motifs. The unmethylated CpG motifs are recognized by the innate immune system via Toll-like receptor 9 (TLR9), which is expressed by B cells and plasmacytoid dendritic cells (pDCs). Thus, CpG DNA induces the maturation and stimulation of professional pDCs as well as subsequent antigen-specific T_H1 and cytotoxic T lymphocyte (CTL) responses. In this regard, CpG oligodeoxynucleotides act as effective adjuvants for the induction of antigen-specific immunity. Indeed, CpG oligodeoxynucleotides enhanced both antibody and common mucosal immune system responses to ovalbumin in mice. In addition, when viral or toxoid vaccines were given with CpG oligodeoxynucleotides, significantly increased levels of antigen-specific antibody and CTL responses were seen. Mucosal delivery of CpG oligodeoxynucleotides plus formalin-inactivated influenza virus or hepatitis B virus surface antigen successfully induced antigen-specific antibody responses in both external secretions and plasma of mice. Furthermore, it was shown that mice given a nasal recombinant protective antigen of the anthrax lethal toxin plus CpG oligodeoxynucleotides exhibited high levels of protective antigen-specific IgG2a and IgA antibody responses in both plasma and external secretions. Importantly, these protective antigen-specific antibodies neutralized the lethal toxin *in vitro*.

Bacterial flagellin, which is the ligand for TLR5, has also been used as an adjuvant in vaccine delivery systems. BALB/c mice subcutaneously immunized with an enhanced green fluorescent protein fusion protein showed effective generation of green fluorescent protein-specific T-cell responses. In addition, *Salmonella* Typhimurium flagellin enhanced CD4$^+$ T-cell responses to co-intravenously administered ovalbumin peptides. Furthermore, nasal immunization with purified flagellin from *Salmonella* Enteritidis alone or conjugated to starch microparticles resulted in the induction of a mixed T_H1/T_H2-type response and high flagellin-specific SIgA antibody titers in fecal extracts. It has been shown that *Salmonella* Typhimurium flagellin increases the expression of co-stimulatory molecules (i.e., CD80) on DCs. Furthermore, *Vibrio vulnificus* major flagellin used as a nasal adjuvant, when coadministered with tetanus toxoid vaccine, induced significant tetanus toxoid-specific IgA antibody responses in both the mucosal and systemic compartments, and IgG antibody responses in the systemic compartment. The flagella filament structural protein of *Salmonella* Enteritidis was shown to stimulate human β-defensin-2 mRNA in Caco-2 cells by mechanisms involving mitogen-activated protein kinase. In this regard, nasal administration of human neutrophil peptide defensins enhances systemic IgG and fosters B- and T-cell interactions to link innate immunity with the adaptive immune system. A full molecular understanding of innate and acquired immune mechanisms initiated by microorganisms will most likely lead to the development of more appropriate, potent, and safe nasal adjuvants.

20-8 The mouth is a delivery site for the induction and modification of antigen-specific immune responses.

Current evidence suggests that the sublingual region of the oral cavity is an immunologically functional site for the initiation of antigen-specific immune responses. As discussed above, nasal immunization has been shown to be most effective in the induction of antigen-specific immunity in both the mucosal and systemic compartments; however, one must consider that some nasal immunization strategies risk antigen tracking into the olfactory tissues and the central nervous system. To obviate this potential concern, sublingual immunization may be an alternative mucosal antigen delivery system without the safety concerns associated with the central nervous system. Sublingual administration is also a noninvasive route which has the advantage of requiring lower doses of antigen than the oral route, due to the reduced exposure to proteolytic enzymes and lower pH of the stomach. Furthermore, vaccine uptake may be more efficient based upon the numbers of APCs present at the sublingual site.

Several recent studies have used the sublingual route for vaccine delivery against infectious agents. When plasmid DNA encoding hepatitis B surface antigen was sublingually administered to mice, the humoral and CD8+ CTL responses induced against hepatitis B virus were comparable to those elicited by intradermal injection. The sublingual delivery of 2,4-dinitrophenol-bovine serum albumin in starch microparticles, in combination with a penetration enhancer, resulted in good salivary IgA responses. It was also shown that sublingual immunization with influenza vaccine and mucosal non-toxin-based adjuvant mCT resulted in the generation of protective immunity. The sublingual delivery of lipopeptides induced elevated serum antibodies and T-cell responses in the spleen and inguinal lymph nodes of mice. Compared with subcutaneous administration of the same vaccine preparation, sublingual application preferentially induced IFN-γ-producing T cells and IgG2a antibody responses, whereas subcutaneous injection elicited IL-4 and IgG1 antibody responses. More recently, sublingual vaccination with the outer membrane protein of *Porphyromonas gingivalis* plus plasmid expressing Flt3 ligand cDNA elicited increased numbers of DCs in submandibular lymph nodes and protective immunity in the oral cavity. Furthermore, CCR7-expressing DCs in cervical lymph nodes also play an essential role in the induction of antigen-specific immunity initiated by sublingual immunization. These studies suggest that by using the appropriate form of antigen, mucosal adjuvant, and delivery system, the sublingual route is an attractive and alternative vaccination method for the induction of mucosal as well as systemic responses, without the safety issue of possible deposition of vaccine to the central nervous system.

20-9 The nasal immune system escapes mucosal aging.

It is well established that respiratory influenza virus and *Streptococcus pneumoniae* infections are sharply increased in the elderly and are often fatal. In sharp contrast to aging studies in the gastrointestinal tract, little work has assessed age-associated alterations in the upper respiratory tract. Although oral administration of ovalbumin along with nCT showed impairment in the induction of mucosal and systemic responses in one-year-old (aging) mice, the same aged mice did not exhibit similar immune depression when ovalbumin plus nCT was nasally administered. Thus, equivalent levels of ovalbumin-specific SIgA antibody responses were noted in the nasal cavity of both young adult and aging mice following nasal immunization. Furthermore, one-year-old mice given nasal tetanus toxoid and nCT vaccine were protected from the tetanus toxin challenge, as were nasally vaccinated young adult mice. It is often suggested that experimental mice should be at least two years old to be suitable and equivalent models for evaluating immunological aging effects that will provide useful information for the understanding of immunological changes in elderly humans. Systemic immunosenescence may also be overcome by nasal immunization. When two-year-old mice were immunized nasally with ovalbumin and nCT as adjuvant, they failed to undergo induction of antigen-specific SIgA antibody responses in secretions. However, the same mice showed similar systemic immune responses to young adult mice. For the control of infectious diseases in the elderly population, one must overcome this mucosal immunosenescence and seek to develop novel immune modulators that can maintain appropriate mucosal immunity in aged mice. In this regard, nasal application of CpG oligodeoxynucleotides together with pFlt3-ligand as a combined mucosal adjuvant successfully elicited antigen-specific SIgA antibody responses in aged mice. Importantly, nasal immunization of PspA plus a combination of CpG oligodeoxynucleotides and pFlt3-ligand resulted in the induction of mucosal SIgA antibody responses and prevention of *S. pneumoniae* nasal carriage.

20-10 T-cell independent mucosal IgA responses are induced in the nasal and oral cavities.

It was previously reported that two distinct lineages of sIgA⁺ B cells, developed from B-1 and B-2 B cells, are involved in the induction of SIgA antibodies for mucosal immunity. As we discussed above (Sections 20-3 and 20-4), the oral–nasopharyngeal mucosa, including nasal passages and submandibular glands, contains high frequencies of B-1 B cells. The nasal passages possess a higher proportion of B-1b B cells, which are most characteristic of the B-1 B cells in the intestinal lamina propria, whereas the submandibular glands show a more dominant B-1a B-cell repertoire. However, the majority of sIgA⁺ B cells in both submandibular glands and nasal passages were B-1a type, suggesting a similarity with the oral–nasopharyngeal mucosal effector tissues and a difference from the intestinal lamina propria. Murine B-1 B cells are capable of isotype switching as well as proliferation and differentiation in the presence of IL-5, even in the absence of CD40–CD40L signaling, which are distinctive characteristics compared with B-2 cells. Furthermore, the submandibular glands contain significant numbers of sIgM⁺ IgA⁻ B-1a B cells. Studies showing that DCs induced CD40-independent immunoglobulin class switching support their possible consideration for the T-cell independent induction of class-switch recombination in B-1 cells. Several studies showed that respiratory DCs play an important role in the induction of IgA antibody responses. DCs directly interact with B cells through the B-cell activation factor of the TNF family (BAFF), also called lymphocyte stimulator protein, and a proliferation-inducing ligand (APRIL) in order to induce sIgA⁺ B cells. It has been shown that B-1 B cells express receptors for these ligands, namely BAFF-R and transmembrane activator and calmodulin interactor respectively; it is thus possible that mucosal DCs in the nasal passages and submandibular glands play a key role in the induction of antigen-specific B-1 B cell IgA class switching in a T-cell independent manner, despite the fact that continuous IgA isotype class switching occurs effectively for the B-2 cells in the organized mucosa-associated tissues such as NALT, Peyer's patches, and isolated lymphoid follicles (Figure 20.5). In this regard, it was reported that nCT as a nasal adjuvant elicited increased levels of LPS-specific SIgA antibody responses through IL-5–IL-5-receptor interactions between CD4⁺ T cells and IgA⁺ B-1 B cells in murine submandibular glands and nasal passages. Thus, it is possible that DCs in the submandibular glands and nasal passages stimulated by T-cell independent antigen plus nCT play a central role in the induction of B-1 B cell IgA class-switch recombination for the enhancement of T-cell independent mucosal SIgA antibody responses. Taken together, these findings suggest that the B-1a B cells and DCs in submandibular glands and nasal passages play key roles in the induction of T-cell independent antigen-specific mucosal IgA antibody responses in the oral–nasopharyngeal mucosa in addition to the gastrointestinal tract.

Summary.

Nasopharyngeal and oral mucosae exhibit unique features in terms of their anatomical location and immune functions. Both nasal and oral cavities are closely connected with the open entrance of the aerodigestive tract, and are initially exposed to all antigens, pathogens, and allergens by inhalation and ingestion. Furthermore, nasal and oral mucosae are protected by two arms of the immune defense system including mucosal SIgA and systemic IgG antibodies in nasal washes and whole saliva. These features are totally distinct from those of the gastrointestinal and lower respiratory tracts despite the fact that they share some common features of the mucosal immune system. Indeed, their organogenesis and lymphocyte trafficking are distinctly regulated (Chapter 1). This supports the view that compartmentalization among the nasopharyngeal–oral cavity, gastrointestinal tract, and lower respiratory tract immune systems represents a framework for the mucosal immune system. The compartmentalization of immune systems is also evident because mucosal immunosenescence occurs earlier in the gastrointestinal than in the nasal immune system. Finally, compared with the peritoneal cavity and the intestinal lamina propria, nasopharyngeal–oral mucosa exhibits a different phenotype of B-1 B cells that express sIgA as innate immunoglobulin-producing cells. Understanding the common versus distinct characteristics of the nasopharyngeal–oral immune system from those present elsewhere will allow us to develop new strategies for mucosal vaccines to prevent mucosal infectious and inflammatory diseases.

Further Reading.

Brandtzaeg, P.: **Do salivary antibodies reliably reflect both mucosal and systemic immunity?** *Ann. N. Y. Acad. Sci.* 2007, **1098**:288–311.

Czerkinsky, C., and Holmgren, J.: **Topical immunization strategies.** *Mucosal Immunol.* 2010, **3**:545–555.

Fujihashi, K., and Kiyono, H.: **Mucosal immunosenescence: new developments and vaccines to control infectious diseases.** *Trends Immunol.* 2009, **30**:334–343.

Kiyono, H., and Fukuyama, S.: **NALT- versus Peyer's-patch-mediated mucosal immunity.** *Nat. Rev. Immunol.* 2004, **4**:699–710.

Kunisawa, J., Nochi, T., and Kiyono, H.: **Immunological commonalities and distinctions between airway and digestive immunity.** *Trends Immunol.* 2008, **29**:505–513.

Lycke, N., and Bemark, M.: **Mucosal adjuvants and long-term memory development with special focus on CTA1-DD and other ADP-ribosylating toxins.** *Mucosal Immunol.* 2010, **3**:556–566.

McGhee, J.R., Kunisawa, J., and Kiyono, H.: **Gut lymphocyte migration: we are halfway 'home'.** *Trends Immunol.* 2007, **28**:150–153.

Mora, J.R., Iwata, M., and von Andrian, U.H.: **Vitamin effects on the immune system: vitamins A and D take centre stage.** *Nat. Rev. Immunol.* 2008, **8**:685–698.

Moutsopoulos, N.M., Greenwell-Wild, T., and Wahl, S.M.: **Differential mucosal susceptibility in HIV-1 transmission and infection.** *Adv. Dent. Res.* 2006, **19**:52–56.

Pascual, D.W., Riccardi, C., and Csencsits-Smith, K.: **Distal IgA immunity can be sustained by $\alpha_E\beta_7^+$ B cells in L-selectin$^{-/-}$ mice following oral immunization.** *Mucosal Immunol.* 2008, **1**:68–77.

Schneider, P.: **The role of APRIL and BAFF in lymphocyte activation.** *Curr. Opin. Immunol.* 2005, **17**:282–289.

Takatsu, K., Kouro, T., and Nagai, Y.: **Interleukin 5 in the link between the innate and acquired immune response.** *Adv. Immunol.* 2009, **101**:191–236.

Taubman, M.A., and Nash, D.A.: **The scientific and public-health imperative for a vaccine against dental caries.** *Nat. Rev. Immunol.* 2006, **6**:555–563.

Williams, I.R.: **Chemokine receptors and leukocyte trafficking in the mucosal immune system.** *Immunol. Res.* 2004, **29**:283–292.

Yu, X., Tsibane, T., McGraw, P.A., *et al.*: **Neutralizing antibodies derived from the B cells of 1918 influenza pandemic survivors.** *Nature* 2008, **455**:532–536.

Bronchus-Associated Lymphoid Tissue and Immune-Mediated Respiratory Diseases

21

The respiratory tract can be divided into three compartments. The first is the upper airways, which includes the nasal passages, the sinuses, and the pharynx; these were discussed in Chapter 20. The other two immunologic compartments of the respiratory tract are the central airways, which include the trachea, bronchi, and bronchioles, and the lower airways, which include the respiratory bronchioles, alveolar ducts, alveolar sacs, and alveoli. The inductive sites within these two latter compartments are commonly termed the bronchus-associated lymphoid tissue (BALT), which is the focus of this chapter.

General anatomy and physiology of the central and lower airways.

In humans, BALT exists during childhood as isolated lymphoid follicles in close contact with the surface epithelium. However, such structures are not regularly identified in adults. Thus, in addition to BALT, mucosal-draining lymph nodes also have an important role in the induction of mucosal immunity in the airways. In this case, the superficially and deeply located cervical lymph nodes receive lymphatic vessels from both nasopharynx-associated lymphoid tissue (NALT) (see Chapter 20) and effector sites in the upper respiratory tree, whereas the airway mucosa distal to the pharynx, including the lung parenchyma, drains to the parabronchial, hilar, and paratracheal lymph nodes, where induction of immune responses can take place (Figure 21.1).

21-1 BALT is a potential inductive site.

Normal mice do not have a clearly identified BALT under homeostatic conditions, but a distinguishable BALT can be rapidly induced in the lungs in response to antigen challenge. After inducible BALT is established, lymphoid aggregates with germinal centers are maintained by the presence of dendritic cells (DCs), and efficiently induce primary immune responses to unrelated antigens. Importantly, mice with inducible BALT that lack peripheral lymphoid organs survive higher viral challenge doses than normal mice.

In comparison with the well-established role for NALT as an inductive site of mucosal immune responses (see Chapter 20), the role of BALT as an inductive site in humans is more controversial. The bronchial mucosa of infants and children, however, contains large numbers of isolated lymphoid follicles in close contact with the surface epithelium (Figure 21.2). These lymphoid aggregates contain many small B cells surrounded by naive T cells, regulatory T (T_{reg}) cells, and high endothelial venules, as well as many DCs. Under

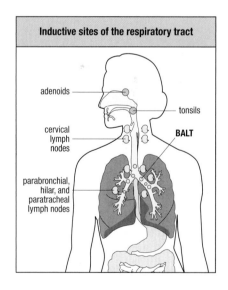

Figure 21.1 Inductive sites of the respiratory tract. Nasopharynx-associated lymphoid tissue (NALT), consisting mainly of the palatine tonsils and adenoids, is strategically located in the aerodigestive tract, whereas bronchus-associated lymphoid tissue (BALT) is located in the central airways. The cervical lymph nodes drain the upper airways, and the lymph nodes located adjacent to the bronchi drain the lower part of the airway mucosa as well as the lung parenchyma.

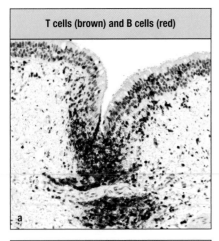

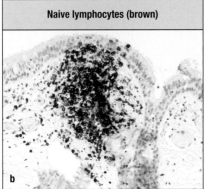

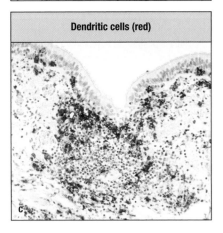

Figure 21.2 Micrographs of lymphoid tissue in the airway mucosa.
Immunostaining for CD3 indicating T cells (brown) and CD20 indicating B cells (red, panel a), CD45RA indicating naive lymphocytes (panel b), and CD11c indicating dendritic cells (panel c) that visualizes isolated lymphoid follicles in sections of bronchial mucosa of a 2-year-old child.

steady-state conditions only a minority of the aggregates show germinal centers. Thus, BALT may be an important inductive site for immune responses, especially during childhood.

In addition to NALT and BALT, regional lymph nodes that drain the airways are also important inductive sites for the respiratory tract. These lymph nodes receive afferent lymphatic vessels from both NALT and BALT, as well as effector sites, including the airway mucosa and the lung parenchyma (see Figure 21.1). T and B cells differentiated in NALT and BALT, and also immune cells within the effector sites of the airways, pass with free antigens through the lymphatics to the draining lymph nodes. Of particular interest for the induction of local immune responses is the dynamic population of DCs that are strategically positioned in the airway mucosa. These cells capture luminal antigens for transfer to the draining nodes. Such migrating DCs typically trigger helper T (T_H) cell T_H2-dominated responses under homeostatic conditions, which support B-cell class switching. However, through pathogen-associated molecular pattern (PAMP) recognition, cytokines, and other innate signals, DCs become activated and acquire the capacity to induce pathogen-specific T_H1 or T_H17 responses.

21-2   The airway mucosa contains effector sites that are similar to those in other mucosal tissues.

From the outer part of the nose to the respiratory bronchioles immediately distal to the trachea and bronchi, the airway mucosa is covered by a pseudo-stratified ciliated epithelium that is interspersed with goblet cells. Large numbers of exocrine glands are present in the underlying mucosa, which together with goblet cells produce mucus that covers the epithelial surface. The glands are surrounded by plasma cells that produce predominantly dimeric IgA, which is translocated from the basolateral to the luminal side of the epithelium by the polymeric immunoglobulin receptor. Secretory IgA, present in the mucus layer, participates in immune defense by neutralizing microbial toxins and pathogens and by preventing commensal bacteria from breaching the mucosal surface. The mucosal secretions of NALT and conducting systems of the upper airway also contain significant quantities of IgG that is transported by the neonatal Fc receptor (FcRn) expressed by the epithelium. The mucus and its content are constantly translocated upward by the ciliated cells, which provide mechanisms for mucociliary clearance of inhaled antigens, including antigens complexed with immunoglobulins.

The airway mucosa also contains DCs, macrophages, T cells, mast cells, and other minor populations of cells. In the human airway mucosa, both DCs and macrophages are strategically positioned within and beneath the surface epithelium. Airway mucosal DCs are sentinel cells that link the innate and the adaptive immune system by transporting antigenic material from the mucosal effector sites to the draining lymph nodes, where they interact with T and B cells. DCs located within the airway epithelium extend cellular projections into the airway lumen as they do within the gut. These extensions allow the DCs to sample antigens directly from the airway lumen through an intact epithelium, which may have important implications for the induction of immune responses under homeostatic situations and for the early detection of infectious threats. Subepithelial resident macrophages participate in the phagocytosis and elimination of incoming pathogens. Along the airway mucosa, large numbers of T cells are present within and beneath the epithelium. T cells in the epithelial compartment are mainly CD8, similar to intraepithelial lymphocytes in the intestinal mucosa, whereas most T cells in the lamina propria express CD4. Both T-cell subsets express CD45RO, consistent with a memory/effector phenotype, although their T-cell receptor repertoire is unknown. T cells with a regulatory phenotype (Foxp3+) are also present in the lamina propria of the airways.

Response of the lung to environmental challenges.

Both the lung and the gut interface with the environment, the lung to extract oxygen and eliminate carbon dioxide. This interface of the lung with the environment is enormous: about 160 m^2 in an adult, which is equivalent to the floor area of a medium-sized house. At this interface, the lung encounters the environment and interacts with pathogens, particulate matter including allergens, and inert material as well as oxidizing agents, related to the absorption of oxygen. Therefore, in a similar manner to the intestinal tract, the lung has developed several mechanisms to deal with these encounters, but these mechanisms sometimes result in inflammation and disease.

21-3 Two major chronic inflammatory diseases of the lung are asthma and chronic obstructive pulmonary disease (COPD).

Asthma and chronic obstructive pulmonary disease (COPD) are two major chronic inflammatory diseases of the lung. Asthma is a major public health problem, affecting 300 million persons worldwide, and has increased markedly in prevalence in westernized countries over the past three decades. Asthma is a complex trait caused by multiple environmental factors in combination with more than 100 major and minor susceptibility genes. Asthma is characterized by reversible obstruction of the airways, with symptoms of shortness of breath, coughing, and wheezing, and is a major cause of emergency room visits, hospitalization, and school absences. COPD is also a major public health problem; worldwide it is the fourth leading cause of death (shared with human immunodeficiency virus/acquired immunodeficiency syndrome). Both asthma and COPD are characterized by airway obstruction, which in asthma is variable and reversible, but in COPD is progressive and irreversible, and associated with symptoms of shortness of breath, which is progressive. Both chronic diseases are characterized by acute exacerbations, with a great increase in symptoms and inflammation; these exacerbations are frequently caused by infections or air pollution. In addition, the medications for these two diseases overlap, and patients with longstanding asthma may evolve into a picture that mimics COPD. Although many differences exist between asthma and COPD, the similarities between asthma and COPD suggest that asthma and COPD may be the ends of a single spectrum of inflammatory lung diseases modulated by environmental and genetic factors (the Dutch hypothesis). Alternatively, asthma and COPD may be fundamentally different diseases, with asthma caused by allergy, and COPD by cigarette smoking (the British hypothesis).

21-4 The inflammation in COPD is characterized as either chronic bronchitis or emphysema.

COPD, which is most commonly associated with a prior history of cigarette smoking, includes chronic obstructive bronchitis, emphysema, and small-airways disease, resulting in progressive, poorly reversible disease. Chronic bronchitis is diagnosed when there is chronic coughing, with sputum production due to inflammation in the large airways and increased airway mucus production, resulting in airway narrowing. Emphysema is characterized by enlargement and destruction of the air spaces (alveoli), with reduced elasticity of the lung tissue. The inflammation in COPD is primarily associated with smoking, other environmental irritants, oxidative stress, and aging, and is characterized by the presence of neutrophils, T_H1 cells, and CD8 T cells, with fibrosis around small airways and with alveolar disruption. Telomere length shortening is consistently found in COPD, as is dysregulation in the clearance of apoptotic epithelial cells, suggesting that cellular senescence is a characteristic of COPD.

21-5   Allergy is a central mechanism in the pathogenesis of asthma.

Classically, asthma has been considered a disease caused by adaptive immunity and allergy, associated with a predominantly allergic, eosinophilic inflammation in the airways. Allergy to aeroallergens is present in 70–80% of patients with asthma, and allergic asthma is the most common form of asthma. In addition, over the past 25 years, there has been increasing interest in the role of allergy in asthma. First, epidemiological studies demonstrated that the risk of developing asthma was directly related to the concentration of total serum IgE (Figure 21.3). Second, sensitization to aeroallergens has been shown to be a major risk factor for persistent wheezing, which is consistent with the idea that allergy is common in patients with asthma. Third, mast-cell infiltration of airway smooth muscle has been shown to be an important component of asthma. Fourth, allergen-specific T$_H$2 cells are thought to be present in the lungs of virtually all patients with asthma, particularly in the lungs of patients with allergic asthma. T$_H$2 cells produce an array of cytokines that greatly enhance the allergic inflammatory response (Figure 21.4). These cytokines include interleukin-4 (IL-4), which is essential for the differentiation of T$_H$2 cells; in the absence of IL-4, T$_H$2 cell differentiation is severely impaired. IL-4 is also required for the induction of IgE synthesis and for the upregulation of the low-affinity IgE receptor (CD23) on B cells and the high-affinity IgE receptor (FcεRI) on basophils. IL-4 also induces eotaxin expression by lung cells, and induces the production of mucus by airway epithelial cells.

IL-5 is another important T$_H$2 cytokine, coordinately produced with IL-4; together with IL-3, IL-5 promotes the differentiation and maturation of eosinophils, and the release of eosinophils from the bone marrow. Eosinophils are strongly associated with allergic responses and asthma, although their precise role is not fully understood. IL-13 is another important cytokine produced by T$_H$2 cells that has homology with IL-4 and shares one receptor chain with it—the IL-4R α chain—that signals through STAT6. Thus, IL-13 shares some functions with IL-4, including the induction of mucus secretion by epithelial cells

Figure 21.3 Sensitization to allergens and production of IgE in the airway. Allergens that enter the respiratory tract are taken up by dendritic cells at or below the epithelium. Disruption of the epithelial barrier enhances allergen uptake. Dendritic cells then mature and migrate to the draining lymph nodes, where they present peptides of the allergen to T cells. In the presence of interleukin-4 (IL-4); derived from basophils, mast cells, eosinophils, or natural killer T (NKT) cells), naive T cells differentiate into T$_H$2 cells producing IL-4 and IL-13, in a process that is enhanced by Jagged on dendritic cells. T$_H$2 cells then drive B cells to produce allergen-specific IgE either in the lymph nodes or in the airway mucosa. IgE then diffuses into the circulation and binds to the high-affinity receptor for IgE (FcεRI) on tissue-resident mast cells. (Adapted from S. Galli, M. Tsai, and A. Piliponsky, *Nature* 454:445, 2008. With permission from Macmillan Publishers Ltd.)

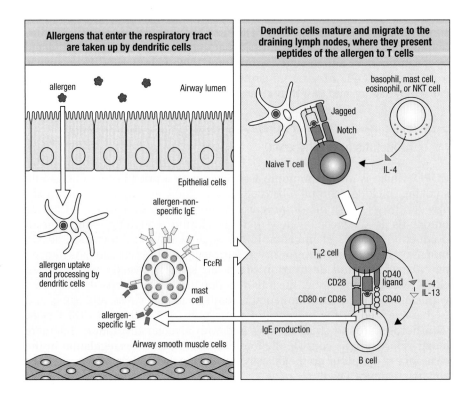

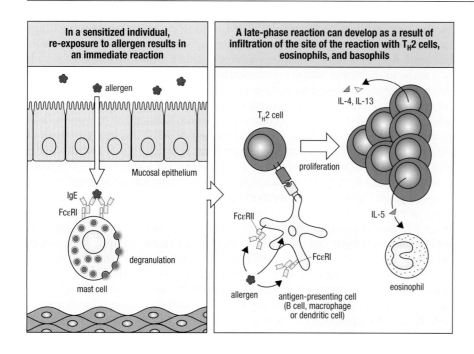

| In a sensitized individual, re-exposure to allergen results in an immediate reaction | A late-phase reaction can develop as a result of infiltration of the site of the reaction with T$_H$2 cells, eosinophils, and basophils |

Figure 21.4 Allergic reactions. In a sensitized individual, re-exposure to allergen results in an immediate reaction (early-phase reaction; left panel) that can occur within minutes of allergen exposure and is due to the cross-linking of allergen-specific IgE on the surface of a mast cell. This results in the release of preformed and newly synthesized mediators from mast cells, leading to localized reactions (acute asthma, allergic rhinitis, or gastrointestinal reactions in food allergies) or systemic reactions (anaphylaxis). Mast-cell mediators include histamine, proteases, prostaglandins, leukotrienes, platelet-activating factor, and cytokines, and cause vasodilation, edema, bronchoconstriction, mucus secretion, hives, and diarrhea. At 6–12 hours after the early-phase reaction, a late-phase reaction (right panel) can develop as a result of infiltration of the site of the allergic reaction with T$_H$2 cells, eosinophils, and basophils. This inflammatory reaction is characterized by swelling, itching, warmth, and erythema (redness), and in the lungs by airway narrowing, mucus hypersecretion, and the development of airway hyperreactivity. (Adapted from R. Valenta, *Nature Reviews Immunology* 2:446, 2002. With permission from Macmillan Publishers Ltd.)

in the lungs and intestines. However, the IL-13 receptor is absent on T and B cells, and IL-13 therefore has no effect on these cell types. In contrast, because the IL-13 receptor is expressed by airway epithelial cells, smooth muscle cells, and fibroblasts (and macrophages), and although IL-4 can bind to the IL-13 receptor, IL-13 but not IL-4 is thought to cause increased collagen deposition and airway fibrosis, as well as airway hyperreactivity (AHR), a cardinal feature of asthma (Figure 21.5). AHR is also commonly observed in COPD and is defined as an exaggerated bronchoconstrictor response to nonspecific stimuli and irritants, such as cold air, smoke, and odors. AHR correlates with asthma severity and can be induced directly in mice by the administration of IL-13, or the adoptive transfer of T$_H$2 cells or natural killer T (NKT) cells. The degree of AHR in a given patient can change over time, worsening with environmental exposures such as viral infection and allergic reactions. Airway remodeling, smooth muscle hyperplasia, collagen deposition, and fibrosis also contribute to the development of AHR. AHR can be quantitated by measuring decreases in pulmonary function in response to graded doses of bronchoconstrictors such as methacholine or histamine. These observations indicate that T$_H$2 cells and T$_H$2 cytokines have a major role in asthma.

Our understanding of the role of allergy and T$_H$2 cells in asthma has been extended by the study of mouse models of allergic asthma. In such models, AHR is abrogated by the elimination of CD4 T cells. Allergen-specific T$_H$2 cells

Figure 21.5 Depiction of airway hyperreactivity (AHR). AHR is a cardinal feature of asthma, because it is present in all forms of asthma and correlates with disease severity. AHR is characterized by hyperresponsiveness of the airways to nonspecific irritants, such as smoke, air pollutants, and cold air, and results in symptoms of asthma, such as wheezing and coughing. AHR can be measured by responsiveness to methacholine, a parasympathomimetic, which causes bronchospasm. A normal individual has only a small decrease in pulmonary function (measured as FEV$_1$, forced expiratory volume in 1 second) after challenge with increasing concentrations of methacholine, whereas an individual with asthma will have a significant decline in pulmonary function after challenge with even small concentrations of methacholine. An individual with mild asthma will have a smaller decrease in pulmonary function on challenge with methacholine. The level of AHR can change over time, worsening with viral infection or with chronic exposure to allergen.

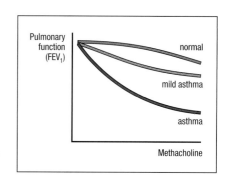

can be induced in mice by immunization, and when recruited into the lungs in various models, T_H2 cells can mediate the development of eosinophilic inflammation and AHR. Although allergen sensitization in humans, which is poorly understood, does not normally develop through immunization, the production by allergen-specific T_H2 cells of IL-4, IL-5, and IL-13 clearly orchestrates the inflammation in allergic asthma. T_H2 cells specific for allergens such as house-dust mite, cockroach extract, *Aspergillus fumigatus*, and ragweed pollen can mediate the development of AHR in mouse models of asthma, reflecting allergic pathways in patients with allergic asthma. In mice, T_H2 cells induced by immunization with allergen plus adjuvant can expand when exposed to allergen in the lungs, or they can be adoptively transferred into recipient mice, where they drive the development of many of the features of human asthma, such as airway eosinophilia, mucus secretion, goblet cell hyperplasia, AHR, and airway remodeling. T_H9 cells producing IL-9, which enhances mast-cell differentiation, may also have an important role in asthma. In contrast, allergen-specific T_H1 cells, when adoptively transferred, abolish airway eosinophilia and mucus production but contribute to airway inflammation, although not AHR. Taken together, these results strongly support the idea that T_H2 cells producing IL-4, IL-5, and IL-13 have a fundamental role in the development of asthma.

Characteristics of allergens.

Allergens are thought to be primarily proteins, which can be presented by DCs to T_H2 cells in the context of major histocompatibility complex (MHC) class II molecules. Allergens are thought to be small molecules that have high aqueous solubility and that can rapidly diffuse across mucosal tissue. However, many allergens have intrinsic protease activity or have associated endotoxin activity. This aspect of allergens is discussed in this section.

21-6   Allergens have intrinsic properties as defined by their ability to be directly recognized by specific receptors.

Allergens such as dust mite, fungi, and pollens have protease activity, which may be recognized by specific receptors on basophils; these cells respond to proteases by secreting cytokines such as thymic stromal lymphopoietin (TSLP). Endotoxin is associated with allergens such as cockroach and dust-mite fecal particles; in addition, dust-mite allergen, Der p 2, has structural homology with MD-2, the lipopolysaccharide-binding component of the Toll-like receptor 4 (TLR4) signaling complex. The endotoxin activity may be sensed by TLR4 not only on antigen-presenting cells (APCs) but also on epithelial cells in the airway, which, when activated, secrete large amounts of IL-25, IL-33, and TSLP (discussed further below). Finally, some allergens such as pollens cause oxidative stress, which can enhance mucosal immunity and sensitization.

Role of innate immune cell types in the induction of asthma.

A major advance in our understanding of the pathogenesis of asthma is the identification of many different effector cells that are involved in the initiation of the asthma phenotype. These cells include DCs, alveolar macrophages, mast cells, basophils, and eosinophils.

21-7   DCs and alveolar macrophages are the major APCs in asthma.

The induction of T_H2-biased adaptive immunity to inhaled allergens requires APCs such as DCs, which are the most potent type of APC (Figure 21.6). In the lung, several different subsets of DCs can be found throughout the conducting airways, interstitium, vasculature, and pleura, and in bronchial lymph nodes. Lung DCs also express numerous receptors such as TLRs, NOD-like receptors, and C-type lectin receptors, as well as CD11c, and can upregulate the expression of several co-stimulatory molecules (such as CD80 and CD86) and the production of chemokines (such as CCL17 and CCL22) and pro-inflammatory cytokines that attract T cells, eosinophils, and basophils into the lungs. DCs take up antigen that might accumulate in the lungs; they then migrate to the draining lymph nodes, where they encounter and activate naive antigen-specific T cells. In mouse models, depletion of DCs from the airways using CD11c-diphtheria toxin receptor transgenic mice or thymidine kinase-transgenic mice treated with the antiviral drug ganciclovir abolished the characteristic features of asthma, including eosinophilic inflammation and T_H2-cytokine production. Furthermore, in the absence of CD11c$^+$ DCs, T_H2 cells cannot produce IL-4, IL-5, and IL-13. In humans, monocyte-derived conventional DCs promote T_H2-cell responses by secreting pro-inflammatory cytokines and upregulating the expression of co-stimulatory molecules after stimulation with antigen. Taken together, these findings indicate that lung DCs are the primary APCs and are necessary for T_H2-cell stimulation during airway inflammation.

In the lung, another DC subset called plasmacytoid DCs (pDCs) is also present, although these pDCs are primarily found in the peripheral lung tissues rather than in the conducting airways. pDCs have a very important role in respiratory tolerance and in the induction of T_{reg} cells, which can suppress T_H2-biased adaptive immunity (discussed further below). Depletion of pDCs abolishes respiratory tolerance and exacerbates airway inflammation, whereas adoptive transfer of Feline McDonough Sarcoma-related tyrosine kinase 3 (Flt3L)-induced bone marrow derived pDCs, thought to be precursors of pDCs, enhances respiratory tolerance. Thus, lung DCs can not only initiate pulmonary inflammation but also downmodulate it.

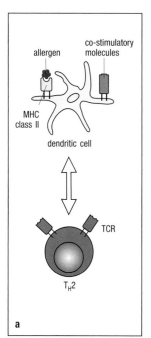
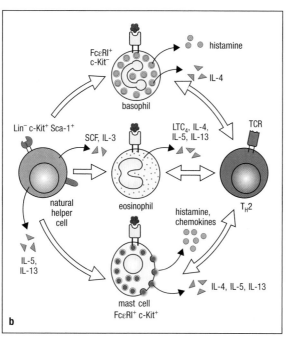

Figure 21.6 Dendritic cells (DCs) are key antigen-presenting cells (APCs) in the lung. After antigen challenge, lung DCs process antigen and induce antigen-specific T_H2 cell responses (panel a). Other cells can also function as APCs to activate T_H2 cells (panel b). Basophils, eosinophils, mast cells, and natural helper cells express MHC class II and co-stimulatory molecules. These cells of the innate immune system can therefore be potential APCs in the lung, and can also be the initial sources of T_H2 cytokines. TCR, T-cell receptor; SCF, stem cell factor; LTC$_4$, leukotriene C$_4$; Lin, lineage. (Adapted from H. Kim, R. DeKruyff, and D. Umetsu, *Nature Immunology* 11:577, 2010. With permission from Macmillan Publishers Ltd.)

Another important cell type in the lung that expresses CD11c is the alveolar macrophage, which must be distinguished from DCs. Alveolar macrophages are large, highly phagocytic cells that clear particulate matter or pathogens inhaled into the lungs. Alveolar macrophages, when activated by infection or interferon (IFN)-γ, produce large quantities of chemokines and cytokines, such as type I interferons (for example with influenza infection), and have an increased capacity to kill intracellular organisms (classical activation of macrophages). In contrast, alveolar macrophages activated by IL-4 and IL-13 express arginase 1 (Arg1), FIZZ1, CCL22, and IL-17RB (IL-25 receptor), and produce IL-13 (alternatively activated macrophages). For example, alternatively activated macrophages may be induced by infection with respiratory syncytial virus or with Sendai virus. However, the importance of alveolar macrophages in presenting antigen to T cells is not clear. Alveolar macrophages do not migrate to the draining lymph nodes, but because they express class II MHC antigen they may be able to activate memory antigen-specific T cells.

21-8 Mast cells, basophils, and eosinophils are the main innate immune cells that are recruited to the airways during asthma.

When the lungs are challenged with allergen, several inflammatory cell types, including basophils, eosinophils, and mast cells, are recruited to airways. Mast cells express FcεRI and c-Kit, and reside in tissues near mucosal surfaces and blood vessels. Classically, mast cells can be activated through cross-linking of antigen-specific IgE bound to FcεRI, by degranulating and releasing pre-formed mediators such as histamine and proteases, as well as newly formed mediators such as cysteinyl leukotrienes, cytokines (IL-1, IL-3, IL-4, IL-5, IL-6, IL-8, IL-10, IL-13, IL-16, tumor necrosis factor-α (TNF-α), and transforming growth factor-β (TGF-β)), and chemokines (for example IL-8, lymphotactin, TCA3, RANTES, MCP-1, and MIP-1α), and thus initiate and greatly amplify allergic reactions. In addition, mast cells can degranulate in response to innate immune signals, such as TLR agonists, or to cytokines such as IL-33 (discussed further below). In some but not all mouse models of allergic asthma, mast cells can enhance the development of asthma, and mast cell-deficient strains (for example W/W$_v$, FcεRI$^{-/-}$, and Kit$^{W-sh/W-sh}$) have decreased AHR, decreased airway inflammation, decreased goblet cell hyperplasia, and lower concentrations of IgE. Like basophils and eosinophils, mast cells can also act as APCs, but this function is poorly understood. IL-3 (which is essential for mast-cell growth), IL-4, and granulocyte–monocyte colony-stimulating factor (GM-CSF) increase the expression of MHC class II in mast cells, and allow mast cells to induce T-cell proliferation.

Basophils, which are circulating granulocytes that express FcεRI and CD49b (DX5) but not c-Kit, like mast cells, can also amplify immediate hypersensitivity responses by degranulating after IgE–FcεRI cross-linking, releasing pre-formed mediators (histamine-containing granules) and newly synthesized mediators (cysteinyl leukotrienes and cytokines, including large quantities of IL-4). In addition, several recent studies have highlighted a crucial new role for basophils as sensors of serine proteases (because many allergens have protease activity), and as APCs that specifically drive T$_H$2 responses through their expression of MHC class II and co-stimulatory molecules. Co-culture of naive T cells that possess a T-cell receptor specific for ovalbumin with bone marrow derived basophils and ovalbumin peptide in the absence of any other APC leads to MHC class II-dependent T$_H$2 cell differentiation, showing that basophils can provide both IL-4 for T$_H$2 differentiation and MHC class II for delivery of an antigen-specific signal. This is further shown by the fact that adoptive transfer of basophils into wild-type or *Ciita*$^{-/-}$ (which do not express MHC class II) mice followed by antigen challenge induces comparable IL-4 production from CD4 cells, whereas antibody depletion of basophils results in decreased IL-4 production. Although one recent study did not find that basophils were necessary for the development of T_H2 immunity, these results indicate that MHC class II-dependent interactions between basophils, which are prominent at sites of allergic inflammation, and CD4 T cells might be important in the induction of T_H2-mediated inflammation.

Another circulating granulocyte that is prominent at sites of allergic inflammation is the eosinophil. On stimulation, eosinophils have an important pro-inflammatory role by producing cysteinyl leukotrienes, as well as T_H1 (IFN-γ and IL-2) and T_H2 (IL-4, IL-5, IL-10, IL-13, and TNF-α) cytokines. In some but not all experimental models of asthma, eosinophils are required for AHR, possibly as a result of the activity of IL-13 and cysteinyl leukotrienes and owing to the toxicity of eosinophil granule proteins, such as major basic protein, which contribute to airway inflammation. In addition, eosinophils, like basophils, can also function as APCs. GM-CSF induces MHC class II and co-stimulatory molecule expression on eosinophils. Class II-expressing eosinophils, when loaded with antigen, promote the production of IL-4, IL-5, and IL-13 from antigen-specific CD4 T cells in a dose-dependent manner. Although this APC function of eosinophils has been contested, some of the discrepancies might be explained by the methods used for isolating eosinophils that can decrease antigen processing. Therefore, these studies together indicate that eosinophils have important effector cell functions, and might also modulate adaptive T_H2 immunity.

Late phases of asthma.

The T_H2-biased inflammatory response that underlies allergic asthma includes both an early-phase or immediate-phase reaction and a late-phase reaction (see Figure 21.4). The early-phase reaction occurs within minutes of allergen challenge, as a result of the binding of allergen to allergen-specific IgE bound to FcεRI on the surface of mast cells at mucosal sites, resulting in mast-cell degranulation and the release of preformed mediators, such as histamine, resulting in the symptoms that can include hives, sneezing, wheezing, vomiting, and anaphylaxis. The late-phase reaction occurs in some individuals 6–12 hours after the early-phase reaction and is caused by the infiltration of eosinophils, basophils, and lymphocytes into the tissues at sites of allergy. In the lung, the late-phase reaction is associated with significant bronchial obstruction that is often resistant to bronchodilator therapy, as well as with the development of AHR.

21-9 The late-phase response of asthma is characterized by the infiltration of the airways with inflammatory cells.

The initial mouse models of allergic asthma focused specifically on allergy, T_H2 cells, and adaptive immunity in mediating inflammation and symptoms in asthma. In this regard, the mouse model of allergic asthma is very similar to human allergen challenge models: both include airway eosinophilia, AHR, and dependence on T_H2 cells, DCs, and goblet-cell metaplasia. This model has strengthened the idea that T_H2 cells (which were first discovered in mouse models) and adaptive immunity can have very important roles in human asthma. It should be noted that in these models, mast cells are either dispensable or required, depending on how sensitization is performed, and eosinophils are also either dispensable (in BALB/c mice) or required (in C57BL/6 mice), suggesting that the function of some of the cell types can be enhanced or diminished by several factors, including genetic background and the strength of the signals provided by adaptive immunity. These conflicting results might reflect the heterogeneity seen in humans, in which treatments that specifically eliminate eosinophils have been effective in some, but not most, patients with asthma. In addition, in mice repeatedly challenged with allergen, a chronic form of experimental asthma develops, in which the relative roles of various factors, such as IL-4, eosinophils, and neutrophils, change.

21-10 T$_{reg}$ cells are also recruited or induced locally during asthma and provide restraint on the inflammatory response.

Cells that can downregulate the function of T$_H$2 cells and other airway effector cells are called T$_{reg}$ cells. They are thought to be very important in regulating allergic diseases including asthma, and atopic individuals are thought to have a relative deficiency in these cells. There are multiple subsets of T$_{reg}$ cells, including natural T$_{reg}$ cells, which develop in the thymus and are important in recognizing self-antigens and preventing autoimmunity. T$_{reg}$ cells that develop in the periphery in response to nominal antigens and allergens include adaptive T$_{reg}$ cells and T$_R$1 cells. The induction of adaptive T$_{reg}$ cells is related to the level of allergen exposure: lower doses of allergen induce greater numbers of allergen-specific adaptive T$_{reg}$ cells, and higher doses induce greater deletion of antigen-specific cells. The precise mechanism by which T$_{reg}$ cells control inflammation is not totally clear, but it includes the direct inhibition of T$_H$2 cells, or the inhibition of antigen presentation by DCs to T$_H$2 cells. TGF-β and/or the production of IL-10 by T$_{reg}$ cells may also inhibit the development of airway inflammation and AHR. Subsets of adaptive T$_{reg}$ cells may also develop under different conditions, involving inducible co-stimulatory molecule (ICOS)–ICOS ligand (ICOSL) pathways, IL-10, CTLA-4, and in the presence of IL-4 or IFN-γ, resulting in some T$_{reg}$ cells expressing T-bet and others expressing GATA-3. Allergen immunotherapy is thought to boost the number of allergen-specific T$_{reg}$ cells and decrease the number of allergen-specific T$_H$2 cells. Both natural and adaptive T$_{reg}$ cells express the transcription factor Forkhead box p3 (Foxp3), which is essential for the development and function of T$_{reg}$ cells. Mutations in Foxp3 result in multiorgan inflammatory disease, including autoimmune diseases, as well as food allergy with high levels of IgE and, in older children with IPEX (immune dysregulation, polyendocrinopathy, and enteropathy, X-linked), asthma. Allergen-specific T$_R$1 cells do not express Foxp3, develop on exposure to IL-10, and produce high levels of IL-10. Patients with steroid-resistant asthma are thought to have a deficiency of allergen-specific T$_R$1 cells, which in one study was improved by exposure to vitamin D$_3$.

Genetic basis of asthma.

21-11 Asthma is a complex genetic disease.

Asthma is a complex genetic trait, and more than 100 genes have been implicated. These genes function in a variety of ways and can be grouped into those that affect an atopic diathesis, versus those that affect processes in the lung, skin, or intestinal tract. For example, the genes *IL4*, *IL13*, *TNF*, *HLA-DR*, *FCER1B*, *TIM1* (*HAVCR1*), and *IL4RA* are expressed by immune cells and presumably predispose toward T$_H$2-biased immune responses and allergy, which underlie the most common form of asthma, allergic asthma (Table 21.1). In contrast, *ADAM33* and *GPRA* are expressed in the lung, whereas *FLG* is expressed primarily in the skin and predisposes to atopic dermatitis, which is a risk factor for asthma. The large number of genes involved in the predisposition to asthma suggests that the pathways to asthma are complex and may involve multiple routes, and that asthma itself may have multiple different forms and pathological pathways.

Table 21.1 Genes associated with asthma

Gene	Nature of polymorphism	Mechanism of association
IL4	Promoter variant	Enhanced production of IL-4
IL13	Promoter variant	Enhanced production of IL-13
IL4RA	Structural variant	Increased signaling in response to IL-4 and IL-13
TNF	Promoter variant	Increased production of TNF-α
HLA-DR	Structural variants	Enhanced presentation of allergens, enhanced sensitization
FCER1B	Structural variant	Enhanced activation of mast cells and dendritic cells
TIM1 (*HAVCR1*)	Structural variants	Altered regulation of T_H1/T_H2 balance
ADAM33	Structural variants	Altered response of airway smooth muscle
GPRA	Structural variants	Altered response of airway
β_2-Adrenergic receptor (*ADRB2*)	Structural variants	Reduced lung responsiveness to β-adrenergic medications
5-Lipoxygenase (*ALOX5*)	Promoter variant	Variation in leukotriene production and response to medications
FLG	Structural variants	Reduced barrier function in skin; enhanced sensitization

21-12 The marked increase in the prevalence of asthma supports the importance of environmental factors.

Over the past two or three decades, the prevalence of asthma and other atopic diseases has almost doubled in westernized countries. This increase in prevalence is not due to changes in the genetic composition of the population, but rather to environmental changes that have occurred in westernized countries. These environmental changes include increases in air pollution (sulfur dioxide, nitrogen dioxide, ozone, and particulate matter), increased exposure to indoor and outdoor allergens (due in part to increased time spent indoors, and to global warming, which increases pollen output), increased prevalence of obesity, increased exposure to acetaminophen, increase in vitamin D deficiency, and a decrease in infections. The variety of possible causes for this increase in prevalence of asthma is consistent with the idea that asthma is caused by multiple pathogenic pathways.

The notion that a decrease in infections, as a result of vaccinations, the use of antibiotics, and improved public health measures such as improved hygiene and indoor plumbing, is a possible cause for the increase in asthma prevalence is known as the hygiene hypothesis, first suggested by David P. Strachan in 1989. He noticed that atopic diseases (asthma, allergy, and atopic dermatitis) were less common in children from larger families, presumably because of greater exposure to infectious agents in larger families. Surprisingly, however, the specific infectious agents, and the mechanisms by which these agents might reduce atopy, are poorly defined. Data implicating infection specifically with hepatitis A virus, *Helicobacter pylori*, and *Toxoplasma gondii* are available; however, because these are not respiratory infections, the precise mechanisms by which asthma is reduced are not clear. Nevertheless, the idea that some infections can somehow limit asthma is supported by the observations that the treatment of children with antibiotics, and delivery by Caesarean section (decreasing exposure to vaginal bacteria), both increase the likelihood of developing asthma.

The complexity of asthma.

Although the T_H2 allergy paradigm can explain many features of asthma, particularly of allergic asthma, many clinical and experimental observations over the past 5 years suggest that asthma is much more heterogeneous and complex than suggested by the T_H2 paradigm. These observations include the finding that non-T_H2 factors such as IFN-γ, IL-17, and neutrophils are found in the lungs of patients with asthma, particularly in the lungs of patients with severe asthma or of patients with asthma resistant to corticosteroids. In addition, T_H2-targeted therapies, with anti-IL-4 monoclonal antibody (mAb), anti-IL-5 mAb, and IL-13 antagonists, have not been as effective as hoped in many clinical trials of asthma. Furthermore, as discussed above, nonallergic forms of asthma, triggered by environmental factors such as air pollutants (for example smoke, diesel particles, and ozone), viral infection, stress, and obesity, seem to cause asthma independently of T_H2 cells. Importantly, these 'nonallergic' forms of asthma respond poorly to treatment with corticosteroids, the most common therapy for asthma, which seems to primarily benefit patients with allergic asthma. These observations suggest that asthma is indeed heterogeneous, with distinctive phenotypes and with different pathogenic mechanisms, some dependent and some independent of T_H2 cells, and requiring different therapeutic approaches.

21-13 Consistent with its genetic heterogeneity, asthma involves other pathways beyond the T_H2 paradigm.

This heterogeneity in the clinical features of asthma has spurred the development of several additional distinct experimental mouse models of asthma that can isolate and focus on specific features and molecular pathways that initiate, worsen, or modulate the disease. Initially, clinicians, believing that human asthma was a single disease entity manifested by very diverse symptoms, criticized the initial mouse model of allergic asthma as not reflecting nonallergic aspects of human asthma. However, it has become apparent that allergic asthma, in which allergy and T_H2 cells are dominant, represents only one pathophysiological pathway to asthma. Thus, additional mouse models of asthma have been developed, demonstrating that other distinct pathways, independent of T_H2 cells and adaptive immunity, can indeed lead to the development of AHR. For example, viral respiratory infection, which precipitates symptoms of asthma in virtually all susceptible patients regardless of the presence of allergy, and which probably occurs independently of T_H2 cells, has recently been modeled in mice. In this model, infection with Sendai virus (a virus related to human parainfluenza virus) precipitates chronic lung disease associated with AHR independent of MHC class II-restricted adaptive immunity in mouse models. This pathway to AHR involves alternatively activated alveolar macrophages interacting with NKT cells. Evidence of such an NKT-cell–macrophage pathway is found in patients with severe asthma and in children with asthma undergoing viral infection.

Another pathway to asthma occurring independently of T_H2 cells and adaptive immunity that has been successfully modeled in mice is asthma associated with air pollution. Mice repeatedly exposed to ozone, a component of air pollution, developed severe AHR associated with airway neutrophils rather than eosinophils. The ozone-induced AHR occurred in MHC class II-deficient mice; importantly, this response required the presence of IL-17, a neutrophil chemotactic factor, because it did not occur in $Il17^{-/-}$ mice. These experiments demonstrate a specific pathway to experimental asthma that requires IL-17, which is not required for allergen-induced AHR; AHR is associated primarily with airway eosinophils rather than neutrophils.

Intrinsic AHR, a clinical feature of asthma present in some patients in whom asthma symptoms develop independently of any type of airway inflammation or allergic sensitization, has also been modeled in a particular strain of mice called A/J. These mice develop AHR spontaneously, without manipulation or treatment, independently of allergy, T_H2 cells, or adaptive immunity. In addition, this intrinsic AHR is associated with a chromosomal region that contains an asthma susceptibility gene, *Adam33*, that was identified in A/J mice before it was recognized in humans. In humans, *ADAM33* has been linked to asthma, in particular to AHR, and less so to asthma associated with elevated serum IgE or to asthma associated with elevations of specific IgE. Moreover, *ADAM33* is expressed in bronchial smooth muscle cells, suggesting that it has an important role in intrinsic asthma by directly affecting the responsiveness of airway smooth muscle cells. This characteristic may allow the expression of asthma symptoms in combination with any form of airway inflammation (such as that induced by allergy or infection), by converting the inflammatory effects into symptoms of asthma. Together, these additional mouse models of asthma have greatly helped to solidify the idea that asthma is heterogeneous with distinct pathways that might develop independently of T_H2 cells.

It should be noted that the use of linkage analysis first in mice and then in humans to identify *ADAM33*, and its use in a similar way to identify another asthma susceptibility gene, *TIM1*, are examples of how mouse models can be used effectively to study human asthma. The genetic and environmental diversity seen in humans can overwhelm the ability to study specific pathogenic pathways in human patients with asthma. However, modeling of a very specific pathway using reductionist methods in inbred and genetically manipulated mice can often overcome these problems, for example by allowing single environmental factors (allergy versus oxidative stress or viral infection) to induce experimental asthma, or by controlling for genetic background, to isolate and study specific pathophysiological effects. This is particularly advantageous when using genetically manipulated mouse strains (such as knockout or transgenic mice) to isolate the specific effects of single genes in the study of asthma. Of course, the experimental results from any given mouse model cannot be viewed in isolation, but must be interpreted in the context of the complexity of human disease.

21-14 Innate pathways and their associated cytokines can initiate asthma.

Although adaptive immunity and T_H2 cells are important in asthma, particularly in allergic asthma, it is becoming clear that innate immune mechanisms, involving a host of novel cytokines and cell types, can also induce AHR. The models of asthma described above that are independent of T_H2 cells and adaptive immunity are consistent with recent observations and studies of the innate immune system. Several components of the innate immune system, described below, seem to have a significant impact on the respiratory tract, and seem to profoundly affect the development of asthma (Figure 21.7).

Interestingly, some of the innate pathways associated with AHR and experimental asthma are associated with T_H2 cytokines, eosinophils, and basophils and seem to be allergic in nature, but these innate AHR pathways can occur in the complete absence of adaptive immunity and T_H2 cells. The availability of genetically manipulated mice to study airway disease has been key in understanding the role of several innate molecules, cytokines, and target cells. Many of these molecules and cells can induce AHR in mice, independently of

Figure 21.7 Innate immune mechanisms in the lung. Although adaptive immunity is critical for asthma pathogenesis, asthma also involves innate, cognate antigen-independent immune responses. IL-25 induces T_H2 cytokines such as IL-5 and IL-13 from natural helper cells in the absence of T_H2 cells and stimulates invariant NKT cells to produce IL-13, thereby promoting AHR and airway remodeling. IL-33 acts on multiple targets; it stimulates mast cells, eosinophils, basophils, natural helper cells, and NKT cells to elicit T_H2 cytokines. TSLP activates mast cells and NKT cells to secrete T_H2 cytokines. The finding of these cytokines, IL-25, IL-33, and TSLP, and cells of the innate immune system greatly extends the understanding of the pathogenesis of asthma. (Adapted from H. Kim , R. DeKruyff, and D. Umetsu, *Nature Immunology* 11:577, 2010. With permission from Macmillan Publishers Ltd.)

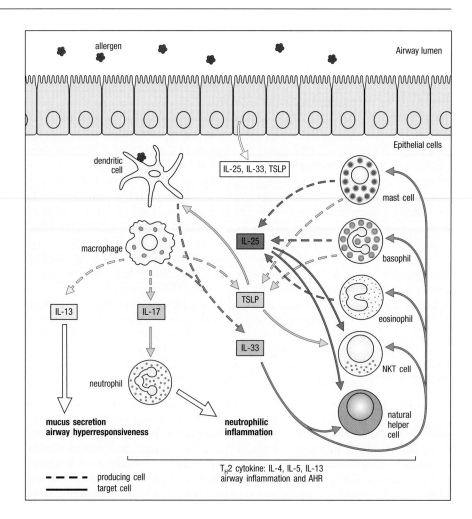

T_H2 cells or adaptive immunity, supporting the concept that multiple pathways to asthma exist, and that distinct asthma clinical phenotypes might be associated with specific molecular pathways. Although each of these distinct pathways can be induced and studied independently in mice, these pathways probably coexist in patients (for example, ozone- and virus-induced AHR can occur in patients with allergic asthma), which may explain why targeting a single pathway in patients is sometimes therapeutically ineffective. In contrast, in many patients a particular pathway can predominate (such as neutrophil-versus eosinophil-associated asthma). This gives rise to the concept of asthma phenotypes in which a specific form of asthma may predominate in a given patient, such as allergic, neutrophilic, or corticosteroid-resistant asthma that responds best to a specific therapy (Table 21.2).

Mediators involved in the development of asthma.

21-15 TSLP is a novel IL-17-like cytokine that promotes T_H2-related immune responses.

TSLP, which was originally cloned from a murine thymic stromal cell line, is released by human primary epithelial cells in response to certain microbial products such as peptidoglycans, lipoteichoic acid, and double-stranded RNA, or to physical injury or inflammatory cytokines (such as IL-1β and TNF-α), as well as to basophils activated by proteases. TSLP expression is detected in the airways of patients with asthma, and TSLP mRNA expression levels correlate with disease severity. In the lungs of allergen-challenged mice, TSLP mRNA expression is increased. Moreover, lung-specific expression of TSLP induces airway inflammation and AHR, although in a pathway that requires the presence of NKT cells. In contrast, TSLP receptor-deficient mice exhibit considerably decreased allergen-induced AHR. TSLP can activate DCs by increasing their OX40L expression and can therefore enhance T_H2 inflammatory responses via OX40 and OX40L interaction. These findings suggest that TSLP has an important role in the initiation of allergic and adaptive airway inflammation through innate pathways that initially activate airway epithelium or basophils.

21-16 IL-25 is an IL-17 cytokine family member that amplifies T_H2 responses.

IL-25, also known as IL-17E, is a member of the IL-17 cytokine family. Both mouse and human lung epithelial cell lines express IL-25 after exposure to allergens, particles, and helminth infection. IL-25 is also produced by activated eosinophils, by bone marrow derived mast cells, by basophils after FcεRI cross-linking, and c-Kit$^+$ cells after stimulation with stem cell factor (SCF) in the lung. Increased IL-25 has also been detected in eosinophil-infiltrated bronchial submucosa of patients with asthma, and IL-17RB (the receptor for IL-25) has been detected in human primary lung fibroblasts. In mouse models of asthma, several reports have shown that IL-25 amplifies T_H2 cytokine production and eosinophilia. In addition, sensitization and challenge with ovalbumin in wild-type BALB/c mice results in increased IL-25 mRNA expression in the lung. Moreover, treatment with soluble IL-25 receptor fusion protein or anti-IL-25 before sensitization and challenge with ovalbumin decreases AHR, airway inflammation, and levels of ovalbumin-specific serum IgE. In contrast,

Table 21.2 Asthma phenotypes

Asthma phenotype	Requirement for T_H2 cells	Associated genes	Clinical characteristics
Allergic	Yes	*IL4, IL13, IL4RA*	Eosinophils, allergic rhinitis, childhood onset, T_H2 cells, increased IgE, NKT cells, steroid responsive
Nonallergic			
Intrinsic	No		
Exercise, cold air	No	*ADAM33*	Heat transfer, change in mucosal osmolality, responds to β_2-agonists
Air pollution	No		Oxidative stress, small particulates, neutrophils, IL-17
Viral infection	No		Alveolar macrophages, neutrophils, NKT cells, IL-13
Bacterial infection	No	*TLR4, CD14*	Bronchitis, sinusitis, neutrophils
Aspirin	No		Nasal polyps, sinusitis, responsive to leukotriene antagonists
Obesity	No		Oxidative stress?
Steroid non-responsive	No		Poor response to corticosteroids, vitamin D deficiency, neutrophils, NKT cells?, non-eosinophilic, low IL-10
Eosinophilic	No		Eosinophils resistant to corticosteroid therapy

administration of IL-25 enhances ovalbumin-induced AHR, and IL-17RB⁺ NKT cells are required for the development of AHR in response to IL-25, suggesting that IL-25 functions, at least in part, through such CD1d-restricted T cells. Furthermore, administration of recombinant IL-25 induces the production of IL-4, IL-5, and IL-13 by an innate non-B/non-T c-Kit⁺ FcεRI⁻ cell population called natural helper cells that mediate the rapid expulsion of helminths in both wild-type and *Rag1⁻/⁻* mice (discussed further below). Taken together, these results show that IL-25 acts on both the innate and adaptive immune systems to amplify T_H2 and T_H2-like responses.

21-17 IL-33 is an IL-1 family member that enhances the activity of mast cells, basophils, and eosinophils.

IL-33 is a member of the IL-1 cytokine family, and IL-33 transcripts can be detected in many diverse cell types, including intestinal and lung epithelial cells. IL-33 synergizes with SCF and IgE receptor to activate human primary mast cells and basophils. IL-33 also enhances the survival of eosinophils and promotes eosinophil degranulation in humans. In mice, viral infection or exposure to pathogen-derived products, irritants, or allergen can enhance IL-33 production. In addition, ovalbumin challenge induces IL-33 protein production in the lung, and IL-33 expression correlates with maintenance of asthma. When IL-33 is administered with ovalbumin, IL-33 enhances airway inflammation in an IL-4-independent manner. Consistent with this, administration of neutralizing antibodies against IL-33 or against ST2, the receptor for IL-33, attenuates eosinophilic pulmonary inflammation and AHR. Furthermore, administration of IL-33 can induce AHR and even enhance airway inflammation in *Rag1*-deficient mice. Thus, IL-33 has an important role during the development of asthma in mouse models, but the specific mechanisms by which IL-33 functions in asthma are not fully understood.

21-18 IL-17 derived from T_H17 cells and innate immune cells enhances neutrophilic responses in asthma.

IL-17 (also known as IL-17A) is a member of the six-member IL-17 family of cytokines, IL-17A–F. The combination of IL-6 and TGF-β skews T-cell development toward T_H17 cell differentiation by suppressing Foxp3 and inducing the expression of RORγt, a lineage-specific transcription factor for T_H17 cells, which produce large amounts of IL-17A and IL-22. Besides T_H17 cells, γδ T cells, NKT cells, neutrophils, and macrophages produce IL-17, which is a potent neutrophil chemotactic agent. Importantly, in humans with asthma, the concentration of IL-17 in sputum correlates with the severity of AHR and with the presence of neutrophils, suggesting an important role for IL-17. In allergic asthma, IL-17 contributes to (but is not required for) the development of ovalbumin-induced AHR in some but not all models. In IL-17-dependent models, allergen sensitization through the airway, which can be enhanced with TLR4 stimulation, can prime strong T_H17 responses that promote airway neutrophilia and acute AHR. IL-17A-deficient or IL-17 receptor-deficient mice exhibit impaired T_H2-type allergic airway inflammation, suggesting that IL-17A can contribute to the development of some forms of experimental asthma. Because endotoxin induces IL-17 production, it is possible that in this model the induction of IL-17 enhances or complements the induction of T_H2 responses, resulting in greater AHR, although in other situations T_H2 responses alone are capable of inducing AHR. In IL-17-dependent models, IL-17 can be produced by T_H17 cells, or by macrophages, because depleting alveolar macrophages or neutralizing IL-17 decreases the number of inflammatory cells and the concentration of inflammatory factors in bronchoalveolar lavage fluid.

As mentioned earlier, IL-17 is required for AHR induced in a nonallergic asthma model triggered by the repeated exposure of mice to ozone. In this model, IL-17-secreting invariant NKT (iNKT) cells are required for the induction of AHR, which is associated with airway neutrophils. Although T_H17 cells are present together with IL-17-producing NKT cells in the lungs of ozone-exposed mice, the development of AHR can occur in MHC class II-deficient mice, indicating that this response is independent of adaptive immunity and T_H2 or T_H17 cells. These studies together suggest that IL-17A-secreting cells are important regulators of allergic and nonallergic asthma.

A unique aspect of IL-17 production in mice is that it seems to be entirely dependent on the presence of a specific commensal bacterial microbe, a *Clostridium*-related species called segmented filamentous bacterium (SFB). Although not yet identified in humans, the absence of SFB in germ-free mice results in a failure of T_H17 and other IL-17-producing cells to develop, and replenishment of this single phylotype fully restores the development of IL-17 production. This indicates that IL-17 production in the lung and intestine requires the presence of an infectious agent or commensal organism.

21-19 IL-22 derived from T_H2 cells promotes epithelial cell responses associated with the asthma phenotype.

IL-22, which is produced by T_H17 cells and NKT cells, and by some NK cells, $\gamma\delta$ T cells, and T_H1 cells, is a member of the IL-10 family of cytokines and has a very important role in mucosal immunology. Triggering of the IL-22 receptor, expressed by airway epithelial cells and also by intestinal epithelial cells, keratinocytes, and hepatocytes, activates STAT3, causing epithelial-cell proliferation and the production of antimicrobial peptides such as S100A7, RegIIIβ, β-defensins, and psoriasin. IL-22 therefore has an important role in host defense against extracellular bacteria such as *Klebsiella pneumoniae*. IL-22 also protects hepatocytes against apoptosis, but is thought to play a pathogenic role in psoriasis.

Innate effector cells in the development of asthma.

21-20 Lung epithelial cells may have a central role in the induction and maintenance of asthma.

Similar to intestinal epithelial cells, airway epithelial cells have an important role as a barrier that separates the lung from the environment and respond to environmental signals by producing a large array of cytokines and chemokines that regulate adaptive immunity. In general, this occurs by the ability of airway epithelial cells to respond to PAMPs associated with antigens in the airways. DCs in the airway not only respond to PAMPs but also play an important role in sampling the contents of the airway for antigens to present to T cells. For example, epithelial cells express TLRs, NOD (nucleotide-binding oligomerization domain)-like receptors, and C-type lectin receptors, which allow epithelial cells to respond to allergens and TLR agonists associated with cockroach, house-dust mites, and *Aspergillus* allergens. Signals generated by these receptors induce in lung epithelial cells the production of cytokines such as IL-25, IL-33, TSLP, and GM-CSF, as well as chemokines (CCL20 (MIP-3α; ligand of CCR6), CCL17 (TARC; ligand of CCR4), and CCL22 (MDC; ligand of CCR4)) and antimicrobial peptides such as defensins. Thus, epithelial cells actively participate in airway mucosal immunity.

21-21 Natural helper cells/nuocytes are lymphoid tissue inducer (LTi)-like cells that initiate T_H2-related inflammation.

Recently, a novel innate lymphocyte has been identified that is present in fat-associated lymphoid clusters and in mesenteric lymph nodes of helminth-infected mice, and which responds to both IL-25 and IL-33. This innate lymphoid cell type has been described in three independent studies, and has been referred to by several different names: 'natural helper cells,' 'nuocytes,' and 'multipotent progenitor cells.' Although their precise characteristics are still being defined, and although these three cell types do not have identical characteristics, they do have striking similarities. First, these cells, which for simplicity we will refer to as natural helper cells, do not express lineage (Lin) markers but do express c-Kit, Sca-1, IL-7R, IL-33R, and IL-17RB. After stimulation with IL-25 or IL-33, these natural helper cells are a major source of T_H2 cytokines in innate immunity. Furthermore, the cells have a potential to function as APCs because they express large amounts of MHC class II molecules and co-stimulatory molecules. In particular, the cells described as multipotent progenitor cells can differentiate *in vitro* in the presence of SCF and IL-3 into monocytes, macrophages, mast cells, eosinophils, and basophils. Previous studies have described a similar non-B/non-T c-Kit$^+$ FcεRI$^-$ cell population producing large amounts of T_H2 cytokines after IL-25 stimulation. Interestingly, innate-like non-B/non-T cells were identified after challenge with IL-25, and recent studies also show that these natural helper cells produce large amounts of T_H2 cytokines in response to IL-25 and IL-33. Both IL-25 and IL-33 induce AHR without other stimulation and in the absence of T_H2 cells, suggesting that natural helper cells might be one of several critical effector cells activated by IL-25 and IL-33, which mediate T_H2-like responses in the lung, even in the absence of T_H2 cells. These natural helper cells may be similar to another innate lymphoid cell type that drives IL-23-dependent innate intestinal pathology expressing Sca-1, RORγt, and the IL-23 receptor.

21-22 Alternatively activated macrophages promote T_H2 inflammation in asthma.

Macrophages are a major type of airway immunocyte and participate in lung immune homeostasis through phagocytosis and the release of mediators in response to antigens. However, their roles in asthma have not been well characterized. Originally, alveolar macrophages were reported as negative regulators, able to modulate immune responses by inhibiting antigen presentation by DCs, T-cell activation, and antibody production. In addition, alternatively activated macrophages or M2 cells, expressing arginase 1, chitinase-like molecules, and resistin-like molecule-α (also known as FIZZ1), can inhibit T_H2 cytokine production from CD4 cells. In contrast, there is increasing evidence that macrophages can actively participate in asthma, by producing both pro- and anti-inflammatory mediators, including T_H1, T_H2, and T_H17 cytokines and IL-33 under different conditions. Recent studies have shown that depletion of alveolar macrophages or deletion of IL-33 decreases IL-33-induced airway inflammation and ovalbumin-induced airway inflammation, suggesting that macrophages participate in T_H2 immunity by responding to IL-33, which enhances the development of M2 cells. Further, CD11b$^+$ F4/80$^+$ macrophages rather than T_H17 cells produce IL-17 in some models of ovalbumin-induced AHR. Moreover, depletion of pulmonary macrophages, but not CD4 T cells or CCR3$^+$ eosinophils, inhibits the development of prolonged AHR. These findings emphasize the importance of macrophages as innate immune cells that are involved in the development of asthma.

21-23 NKT cells have a central role in the initiation of the asthma phenotype in mouse models.

NKT cells comprise a unique subset of lymphocytes that express features of both classical T and NK cells. On stimulation, NKT cells produce large amounts of IL-4, IL-13, and IFN-γ, which have critical roles in the regulation of immune responses. Most NKT cells express an invariant T-cell receptor (TCR) α chain linked to a diverse set of TCR β chains that use a restricted range of TCRβ variable regions. These so-called invariant TCRs function as an innate pattern recognition-like receptor in recognizing both foreign and endogenous glycolipid antigens presented by the MHC class I-like molecule CD1d. These NKT cells, called iNKT cells, are required for the development of AHR in several mouse models of asthma, because NKT cell-deficient mice (*Cd1d*$^{-/-}$ mice, which lack NKT cells, or *Jα18*$^{-/-}$ mice, which lack the invariant TCR and lack iNKT cells) fail to develop AHR after challenge with allergen, exposure to ozone, or infection with Sendai virus. The glycolipid antigens that activate iNKT cells in these distinct models of asthma have not been defined, but could include endogenous glycolipids, which are expressed in the lung during inflammation or after oxidative stress, or exogenous glycolipids from bacteria or plants. The immunological pathways required for the development of AHR in these models of experimental asthma are quite distinct, with each requiring a phenotypically different iNKT cell subset for the development of AHR. For example, in allergen-induced AHR, CD4 iNKT cells producing IL-4 and IL-13 are required, whereas in ozone-induced AHR, CD4$^-$ iNKT cells producing IL-17 are required, and in Sendai virus-induced AHR, CD4$^-$ iNKT cells producing IL-13 are required. These observations suggest that several distinct pathways to asthma require the presence of iNKT cells. Moreover, innate cytokines, such as IL-25, TSLP, and IL-33, can stimulate NKT cells to enhance AHR. IL-17RB (IL-25 receptor)-expressing iNKT cells are essential for allergen-induced AHR, because they produce IL-13 and T$_H$2 chemokines on stimulation with IL-25. Wild-type mice depleted of IL-17RB$^+$ iNKT cells by IL-17RB-specific antibodies or iNKT cell-deficient *Jα18*$^{-/-}$ mice fail to develop AHR on stimulation with IL-25. In addition, adoptive transfer of IL-17RB$^+$ iNKT cells into *Jα18*$^{-/-}$ mice reconstitutes allergen-induced AHR. These studies suggest that IL-25 can exert at least some of its effects through iNKT cells. Other data suggest that TSLP and IL-33 might also mediate some of their effects on AHR through iNKT cells. Taken together, these results suggest that iNKT cells represent a common unifying element that is required for the development of multiple forms of asthma.

21-24 NKT cells may be central mediators of human asthma.

On the basis of the large number of animal studies, iNKT cells have been proposed to have an important role in human asthma. The number of iNKT cells in bronchial alveolar lavage fluid, endobronchial biopsies, and sputum samples from patients with asthma has been examined in several studies; most, but not all, have found that the number of iNKT cells is increased in the lungs of these patients. The differences in the number of pulmonary iNKT cells in patients with asthma in different studies, although generating some controversy, are probably accounted for by the heterogeneity in asthma. Thus, in one study, patients with severe, poorly controlled asthma consistently exhibited significant increases in the number of iNKT cells in bronchial alveolar lavage fluid, whereas patients with less severe disease were less likely to possess increased numbers of iNKT cells in bronchial alveolar lavage fluid. Clearly, functional studies in humans will be required for a fuller assessment of the role of iNKT cells in human asthma. Indeed, one recent study demonstrated that allergen challenge of patients with asthma resulted in a significant increase in the number of pulmonary iNKT cells associated with a significant increase in AHR, strongly suggesting that iNKT cells have an important role in at least some forms of asthma, including allergic asthma.

Summary.

Asthma and COPD are extremely important respiratory diseases, affecting very large numbers of individuals. Although it is clear that T_H2-biased immunity has a very important role in asthma, it has become increasingly apparent that asthma is a heterogeneous disease, and that many clinical forms or phenotypes exist, with distinct pathogenic mechanisms. These mechanisms include allergic (adaptive immunity) and nonallergic (innate) pathways, induced by allergen exposure, viral infection, and oxidative stress and involving a large number of cell types belonging to both the innate and adaptive immune systems, including eosinophils, basophils, T_H2 cells, DCs, neutrophils, NKT cells, natural helper cells, epithelial cells, T_H17 cells, and macrophages. In addition, asthma involves not only cytokines and chemokines produced by T_H2 cells, but also cytokines such as IL-17, IL-25, IL-33, and TSLP produced by innate cells including epithelial cells, $\gamma\delta$ cells, and NKT cells. Although asthma can develop through several independent pathways, these pathways can also coexist and interact. However, the recent recognition that asthma is heterogeneous with multiple molecular and cellular pathways and with diverse underlying disease mechanisms and a probable genetic basis provides us with a greater understanding of distinct asthma phenotypes. Such an understanding is likely to lead to new and more effective therapies for asthma, which must be individualized and personalized, taking into account the distinct pathogenic mechanisms that occur in each patient.

Further Reading.

Anderson, G.P.: **Endotyping asthma: new insights into key pathogenic mechanisms in a complex, heterogeneous disease.** *Lancet* 2008, **372**:1107–1119.

Barrett, N.A., and Austen, K.F.: **Innate cells and T helper 2 cell immunity in airway inflammation.** *Immunity* 2009, **31**:425–437.

Buonocore, S., Ahern, P.P., Uhlig, H.H., *et al.*: **Innate lymphoid cells drive interleukin-23-dependent innate intestinal pathology.** *Nature* 2010, **464**:1371–1375.

Gaboriau-Routhiau, V., Rakotobe, S., Lecuyer, E., *et al.*: **The key role of segmented filamentous bacteria in the coordinated maturation of gut helper T cell responses.** *Immunity* 2009, **31**:677–689.

Galli, S.J., Tsai, M., and Piliponsky, A.M.: **The development of allergic inflammation.** *Nature* 2008, **454**:445–454.

Gordon, S.: **Alternative activation of macrophages.** *Nat. Rev. Immunol.* 2003, **3**:23–35.

Holgate, S.T., and Polosa, R.: **Treatment strategies for allergy and asthma.** *Nat. Rev. Immunol.* 2008, **8**:218–230.

Ivanov, I.I., Atarashi, K., Manel, N., *et al.*: **Induction of intestinal Th17 cells by segmented filamentous bacteria.** *Cell* 2009, **139**:485–498.

Kim, E.Y., Battaile, J.T., Patel, A.C., *et al.*: **Persistent activation of an innate immune response translates respiratory viral infection into chronic lung disease.** *Nat. Med.* 2008, **14**:633–640.

Kim, H.Y., DeKruyff, R.H., and Umetsu, D.T.: **The many paths to asthma: phenotype shaped by innate and adaptive immunity.** *Nat. Immunol.* 2010, **11**:577–584.

Lambrecht, B.N., and Hammad, H.: **Biology of lung dendritic cells at the origin of asthma.** *Immunity* 2009, **31**:412–424.

Matangkasombut, P., Pichavant, M., DeKruyff, R.H., *et al.*: **Natural killer T cells and the regulation of asthma.** *Mucosal Immunol.* 2009, **2**:383–392.

Moro, K., Yamada, T., Tanabe, M., *et al.*: **Innate production of T$_H$2 cytokines by adipose tissue-associated c-Kit$^+$Sca-1$^+$ lymphoid cells.** *Nature* 2010, **463**:540–544.

Neill, D.R., Wong, S.H., Bellosi, A., *et al.*: **Nuocytes represent a new innate effector leukocyte that mediates type-2 immunity.** *Nature* 2010, **464**:1367–1370.

Perrigoue, J.G., Saenz, S.A., Siracusa, M.C., *et al.*: **MHC class II-dependent basophil-CD4$^+$ T cell interactions promote T$_H$2 cytokine-dependent immunity.** *Nat. Immunol.* 2009, **10**:697–705.

Saenz, S.A., Siracusa, M.C., Perrigoue, J.G., *et al.*: **IL25 elicits a multipotent progenitor cell population that promotes T$_H$2 cytokine responses.** *Nature* 2010, **464**:1362–1366.

Sokol, C.L., Chu, N.Q., Yu, S., *et al.*: **Basophils function as antigen-presenting cells for an allergen-induced T helper type 2 response.** *Nat. Immunol.* 2009, **10**:713–720.

von Mutius, E.: **Gene–environment interactions in asthma.** *J. Allergy Clin. Immunol.* 2009, **123**:3–11; quiz 12–13.

Xystrakis, E., Kusumakar, S., Boswell, S., *et al*.: **Reversing the defective induction of IL-10-secreting regulatory T cells in glucocorticoid-resistant asthma patients.** *J. Clin. Invest.* 2006, **116**:146–155.

Yoshimoto, T., Yasuda, K., Tanaka, H., *et al*.: **Basophils contribute to T$_H$2-IgE responses *in vivo* via IL-4 production and presentation of peptide–MHC class II complexes to CD4$^+$ T cells.** *Nat. Immunol.* 2009, **10**:706–712.

The Ocular Surface as a Mucosal Immune Site

22

Good vision is a very strong evolutionary selective pressure. Therefore, the eye and its associated structures (known as adnexa and including the eyelids, conjunctiva, and the tear producing and draining structures) have evolved not only anatomically, but also immunologically to protect vision. The process of inflammation, while important to eradicate infectious agents, is itself harmful to this tissue. Because even small perturbations of the light refracting, transmitting, and gathering ocular structures (Figure 22.1) can have very deleterious consequences to vision, the eye, perhaps more than any other tissue, depends on immune protection mechanisms that preserve tissue integrity and minimize collateral damage. The interior of the eye resists inflammatory processes and this phenomenon is known as immune privilege of the eye. The inside of the globe is devoid of immune cells and has no lymphatic drainage. The neural retina, where the light-sensitive photoreceptor cells are located, is protected by a tight blood–retinal barrier, which is largely impermeable to passage of cells and even of large molecules. As long as the eye is physically intact, this separation from the immune system is largely sufficient to protect the internal ocular structures from the environment. If immunocompetent cells enter the eye from a disrupted blood vessel, for example, they encounter a profoundly inhibitory microenvironment composed of cell-bound and soluble inhibitory molecules. If that proves insufficient to control inflammation, systemic eye-dependent regulatory processes are invoked to minimize its consequences.

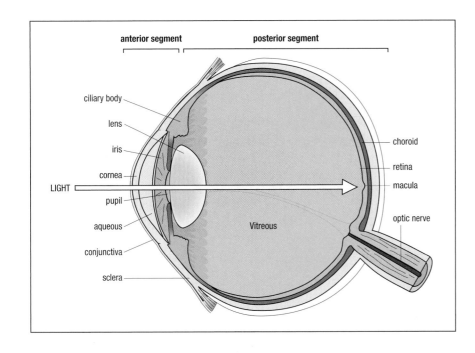

Figure 22.1 Scheme of an eye showing major anatomical structures and organization. Light passes through the ocular media and concentrates on the macula (area of retina responsible for sharp color vision), whereupon the photoreceptor cells sense the signal and transmit it to the brain via the optic nerve. Any damage to the ocular structures along the visual axis would likely result in a visual deficit.

On the other hand, the exterior of the eye is constantly exposed to the environment, and therefore contains a complex system of lymphoid structures whose primary role is to protect the health of the ocular surface and preserve the transparency of the cornea. Working in concert with the tear film that lubricates the ocular surface, efficient mucosal immune mechanisms resist microbial colonization of the warm and moist area while tolerating self-antigens and commensal microorganisms. Activation of immunity must be tightly controlled and requires precise cooperation of the innate and adaptive arms of the immune system. Although the mechanisms that are used to maintain homeostasis at the ocular surface and inside the globe are different, the purpose is the same: to prevent destructive inflammation that would compromise visual function. Thus, the notion of privilege should perhaps be extended to include also the ocular surface and its associated mucosal lymphoid tissue.

Organization of the eye-associated lymphoid tissue.

The eye-associated lymphoid tissue (EALT), which maintains homeostasis at the ocular surface, is a functionally important but relatively little-studied part of the mucosal immune system. It can be anatomically divided into conjunctiva-associated lymphoid tissue (CALT) and the tear (lacrimal)- duct-associated lymphoid tissue (variably known as TALT or LDALT), which lines the tear-draining system (Figure 22.2). Also included as part of EALT (but having no accepted acronym to its name) are the lymphoid cells found in the tear gland itself, which contain a rich population of B cells and plasma cells. Despite the nomenclature, which implies separateness, the lacrimal gland, CALT, and TALT systems are anatomically continuous with each other, and are, moreover, connected by the flow of tears. Thus, they necessarily all contribute to and share effector molecules, and together form a functional unit for antimicrobial defense of the ocular surface.

Figure 22.2 Eye-associated lymphoid tissue (EALT). EALT consists of, in order, lacrimal-gland-associated lymphoid tissue, CALT within the conjunctiva, and TALT (also known as LDALT) along the tear draining ducts. The arrows show the direction of movement of tears and of antigens encountered at the ocular surface. Diffuse lymphoid tissue found throughout the EALT contains numerous T lymphocytes and plasma cells. Additionally, the organized lymphoid tissue also contains lymphoid follicles with defined T- and B-cell zones. Cells primed within these follicles recirculate back to the tissue where they were primed, as well as to other mucosal lymphoid sites. (Adapted from E. Knop and N. Knop, *Chem. Immunol. Allergy* 92:36–49, 2007. With permission from Karger.)

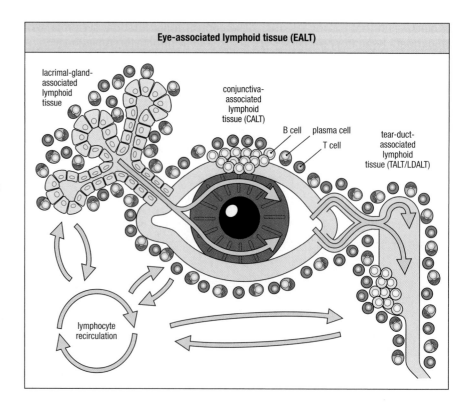

Eye-associated lymphoid tissue (EALT)

lacrimal-gland-associated lymphoid tissue

conjunctiva-associated lymphoid tissue (CALT)

B cell   plasma cell   T cell

tear-duct-associated lymphoid tissue (TALT/LDALT)

lymphocyte recirculation

22-1   The EALT contains organized and diffuse populations of lymphocytes.

As in other mucosa-associated lymphoid tissues (MALT), components of EALT reside within two tissue layers divided by a basement membrane: the epithelium and the underlying loose connective tissue of the lamina propria. The fine structure and cellular composition of these layers changes along the ocular tract, depending on location and degree of interaction with the external environment.

Again recapitulating MALT in other sites, EALT occurs in two forms: 'organized' and 'diffuse.' The diffuse EALT is mainly populated by differentiated effector cells capable of dealing directly with antigens and is therefore considered an efferent or effector site. This diffuse organization is typical of the tear gland. The organized EALT contains lymphoid follicles and serves as an afferent or inductive site where antigens from the ocular surface are sampled and acquired, immune responses initiated, and germinal centers formed. CALT and TALT, which come in direct contact with the environment, contain both diffuse and organized lymphoid tissue (Figure 22.3).

22-2   The lacrimal gland and the lacrimal gland-associated lymphoid tissue are anatomically located within the upper eyelid.

The lacrimal gland, located within the upper eyelid, contains acini whose secretory epithelial cells produce the tear fluid that continuously drains through 10–12 secretory ducts into the upper fold of the eyelid or fornix. Tear production is tightly controlled by neural input from the ocular surface tissues: the conjunctiva and cornea. In fact, the cornea is the most highly innervated surface tissue of the body and even very minor irritating stimuli activate copious production of tear fluid. A basement membrane separates the epithelium from the layer of loose connective tissue constituting the lamina propria, which is similar to and anatomically contiguous with the connective tissue of the conjunctiva, to which it connects through the secretory tear ducts. The connective tissue contains accumulations of lymphocytes as well as many plasma cells. The vast majority of lacrimal gland plasma cells in mice are dedicated to the secretion of IgA and are primarily of the B-1 lineage. The acinar epithelium of the lacrimal gland produces the polymeric immunoglobulin receptor (pIgR). As discussed in Chapter 9, pIgR binds to the J chain in the dimeric IgA

Figure 22.3 Organization of human EALT and associated structures (adnexa). As shown in panel a, mucosal immune tissue in the lacrimal gland and accessory lacrimal glands of Krause and Wolfring is continuous with the mucosal surface of the conjunctiva into the periacinar tissue of the lacrimal gland and the conjunctival accessory lacrimal gland. Organized accumulations of lymphocytes into lymphoid follicles occur in the conjunctiva. Panel b shows the components of the diffuse lymphoid tissue that occurs in the lamina propria and epithelium throughout the EALT. Intraepithelial lymphocytes (IELs) are mainly CD8 cells. In the lamina propria, there are CD4 and CD8 and plasma cells as well as monocyte–macrophages, mast cells, occasional neutrophilic granulocytes, and stromal fibroblastic cells. The epithelium is covered by the tear film and contains mucus-producing goblet cells. Panel c shows the organized lymphoid tissue in CALT and TALT. Antigens are sampled at the ocular surface and translocated into the underlying lymphoid follicles, composed of discrete T- and B-cell regions, where effector cells generated from lymphocytes enter lymphatic tissues through high endothelial venules (HEV) and leave via lymphatic vessels, from which they eventually reenter the circulation. FAE, follicle-associated epithelium.

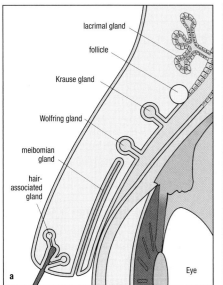
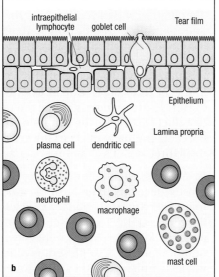
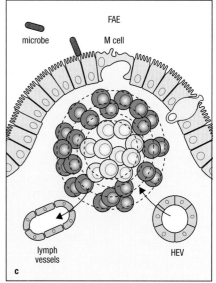

(and also pentameric IgM) and transports the antibody molecules which have been produced in the lamina propria through the epithelial cell layer into the lumen of the acini in covalent association with secretory component of pIgR. The secretory IgA mixes with the tear fluid where it provides protection to the eye surface. Lymphocytes other than plasma cells and B cells include intraepithelial lymphocytes (IELs) which are mostly CD8 cells and some CD4 cells as well as innate immune cells of various types. Of note are the unusually high numbers of natural killer T (NKT) cells and $\gamma\delta$ T cells in the lacrimal gland, which contribute to innate-like immune defense of the eye.

The lymphoid tissue in the lacrimal gland is diffuse and is not organized, as in the CALT and TALT, into well-defined follicular structures where priming takes place, which implies its effector nature. Lymphoid follicles are rarely observed and may not be physiologically relevant. Indeed, the tear gland comprises a normally sterile environment and studies have shown that there seems to be no retrograde transport of antigens from the ocular surface into the tear gland. Thus, the antibody-producing plasma cells most likely have reached the lacrimal gland by a process of recirculation, after having been primed and matured in lymphoid follicles elsewhere (Figure 22.2). While $\alpha_4\beta_7$ integrin is known to have a major role in directing homing to the gut, the homing molecules that direct the cells to the lacrimal gland are not well defined. That said, the lacrimal gland, CALT, and TALT express a complement of integrins, addressins, and chemokines that differs from those expressed in gut-associated lymphoid tissue (GALT), and that shows similarities to the nasopharynx-associated lymphoid tissue (NALT). The tear fluid leaves the ocular surface through the tear draining ducts at the inner corner of the lower and upper eyelids, which connect to the nasolacrimal duct leading into the nasal cavity (Figure 22.2). The NALT thus receives the antigens from the ocular surface and is the logical place, in addition to CALT and TALT, where priming of cells that end up in the tear gland could occur. Cells primed in the CALT, TALT, and NALT may then recirculate to the tear gland to secrete their effector molecules there. In addition to IgA, tears also contain many constitutively produced and inducible antimicrobial peptides such as lactoferrin and defensins; cytokines such as interleukin-1 (IL-1), IL-4, IL-6, and transforming growth factor-β (TGF-β); surfactant proteins; and mucins, including MUCs 4, 5AC, 5B, and 7 that are secreted into the tear film.

22-3   Conjunctiva-associated lymphoid tissue (CALT) covers the external surface of the eye with three anatomic regions (palpebral, bulbar, and fornix).

The conjunctiva is a continuous structure lining the inner surface of the eyelids and continuing over the front part of the eye to terminate around the cornea in a region known as the limbus. Its functions are protection, lubrication of the ocular surface by secretions from goblet cells and small accessory tear glands, and immunity. We recognize three principal conjunctival regions: palpebral, lining the inner surface of the eyelids; bulbar, covering the scleral surface; and the fornix, at the angle between the eyelids and the eyeball, where the secretory tear ducts discharge the tear fluid. The epithelium is thinnest in the palpebral part and gradually thickens toward the limbus of the eye. The lamina propria in all three regions contains organized as well as diffuse lymphoid tissue.

In humans, the conjunctiva has a well-developed lymphoid tissue, which mostly develops after birth, at around 3–4 months of age. Organized primary and typical secondary follicles with T cells, B cells, and high endothelial venules (HEV) are present in CALT, and the follicle-associated epithelium contains M cells which sample environmental antigens and transport them into the follicle. Also present is diffuse lymphoid tissue containing intraepithelial lymphocytes (IELs) and abundant plasma cells in the lamina propria).

Many of these plasma cells secrete IgA, and the conjunctival epithelial cells, as well as epithelium in the conjunctiva-associated crypts, express the pIgR that is responsible for the transcytosis of IgA. Thus, in keeping with its immunological context of direct contact with the environment, the CALT is both an inductive and an effector site.

In contrast to the IELs, which contain a large proportion of CD8$^+$ T cells as in other mucosal surfaces, the lymphocytes in the lamina propria of the conjunctiva are mostly CD4$^+$. About 20% of conjunctival leukocytes are plasma cells actively producing antibodies, primarily IgA. The conjunctiva contains about two-thirds as many plasma cells as the lacrimal gland. Therefore, the conjunctiva is able to contribute considerably to its own antimicrobial defense and does not appear to be solely dependent on a passive supply of IgA from the lacrimal gland.

Interestingly, in contrast to humans, the mouse and rat conjunctiva has a virtual absence of diffuse lymphoid cells including plasma cells, although some follicular structures and cells resembling M cells are present in the nictitating membrane of the eye. The nictitating membrane is part of the conjunctiva, is typically transparent, and slides across the eye like a third eyelid in some species that include rodents and other mammals as well as birds and reptiles. The CALT of larger animals, including rabbits, dogs, and cats, resembles the organization in humans more closely. It is likely that the tear fluid compensates for this lack of diffuse CALT and contains sufficient IgA and other effector molecules for the ecological situation of the rodent. The conjunctiva (including that of laboratory rodents) also contains many mast cells, whose mediators are released after binding allergen–IgE complexes to receptors expressed on their membrane, which are responsible for the unpleasant consequences of hay fever and allergy affecting the ocular surface.

22-4 Tear-duct-associated lymphoid tissue (TALT) is present in rodents and humans and is subject to a pathway of organogenesis that is distinct from GALT.

The lymphoid tissue associated with the tear draining apparatus, the lacrimal sac, and nasolacrimal duct is known as TALT or LDALT (tear-duct or lacrimal-duct-associated tissue). Unlike CALT, which is present in humans but absent in mice and rats, TALT is present and well developed in laboratory rodents. Therefore, its developmental requirements have been well studied at the molecular level. Interestingly, the developmental requirements of TALT are very different from those of other mucosal lymphoid tissues.

Organogenesis of secondary lymphoid tissues such as Peyer's patches (PP) and peripheral lymph nodes is driven by CD3$^-$CD4$^+$CD45$^+$ lymphoid tissue inducer (LTi) cells. The development of LTi cells themselves is dependent on inhibitor of DNA binding/differentiation 2 (Id2) and the retinoic acid-related orphan receptor (ROR) γt. Mice deficient in either one of these molecules lack LTi cells and do not develop lymph nodes and PP. LTi cells migrate to sites of lymphoid tissue genesis in response to chemokine signals including CXCL13, CCL19, and CCL21, and are triggered by IL-7 to produce lymphotoxin $\alpha_1\beta_2$ (LT$\alpha_1\beta_2$) that signals through the lymphotoxin β receptor (LTβR) to activate the mesenchyme in peripheral lymph nodes and PP anlagen. Production of CXCL13, CCL19, and CCL21 as well as adhesion molecules vascular cell adhesion molecule-1 (VCAM-1) and peripheral lymph node addressin (PNAd) then promotes migration and recruitment of leukocytes and permits formation of organized lymphoid tissue. Mice deficient in component(s) of these pathways largely or completely lack development of PP and peripheral lymph nodes, although they do develop nasopharynx-associated lymphoid tissue (NALT; Table 22.1).

Table 22.1 Presence of different mucosa-associated lymphoid tissues in mice deficient in genes controlling lymphoid organogenesis pathways

Genetic deficiency	Lymphoid tissue development						
	TALT	NALT	PP	CLN	MLN	ILF	Cryptopatch
Id2⁻/⁻	+++	−	−	−	−	ND	ND
Rorc⁻/⁻	+++	+++	−	−	−	−	+/−
Ltα⁻/⁻	+	+	−	−	+/−	−	+/−
aly/aly	+	+	−	−	−	−	++
Il7rα⁻/⁻	++	++	−	+/−	++	++	−
Cxcl13⁻/⁻	++	+	+/−	+/−	++	ND	ND
plt/plt	+++	++	++	++	++	ND	ND
Cxcl13⁻/⁻ plt/plt	+	+	+/−	−	++	ND	ND

CLN, cervical lymph node; MLN, mesenteric lymph node; ILF, isolated lymphoid follicle.

Note: *aly/aly* mice carry a null mutation of NFκB-inducing kinase (NIK), resulting in a failure to signal through LTβR and *plt/plt* mice have a null mutation of Ccl19 and Ccl21 genes. (Adapted from T. Nagatake *et al.*, *J. Exp. Med.* 206:2351–2364, 2009.)

Unlike PP and peripheral lymph nodes, which develop in embryonic life, TALT develops postnatally in mice. Despite its postnatal development, appearance of TALT does not require environmental microbial stimuli, as TALT develops normally even in germ-free mice. TALT genesis is also normal in both Id2-deficient and RORγt (*Rorc*)-deficient mice, which lack LTi cells. Furthermore, TALT formation is present, though reduced in size, in mice deficient in molecules downstream of LTβR signaling. This indicates that TALT formation utilizes different pathways than formation of classical secondary lymphoid tissues, and even than formation of other mucosal lymphoid tissues. Interestingly, LTi-like CD3⁻CD4⁺CD45⁺ cells are observed in TALT anlagen of the Id2- and RORγt-deficient mice, suggesting that the inducer cells which drive TALT formation may be a population that is distinct from the classical LTi which control organogenesis of secondary lymphoid tissues such as PP and peripheral lymph nodes.

Induction and expression of immunity at the ocular surface—the good and the bad.

Because the CALT and TALT contain lymphoid follicles into which antigens sampled by M cells and dendritic cells at the ocular surface are transported, processed, and presented, the ocular surface is a rich environment for inducing immune responses. Antigens from the ocular surface also drain with the tears through the lacrimal drainage system, that is, the nasolacrimal duct and its associated structures, where there is also an associated lymphoid tissue (see Figure 22.2). Thus, mucosal vaccination through the ocular surface inevitably also stimulates the nasal immune system. Furthermore, whether induction of immunity takes place locally in the mucosal lymphoid follicles, or occurs in the draining lymph nodes, the responses also become systemic such that antibodies to ocularly encountered antigens are present in the serum and immune cells can migrate not only back to the ocular surface, but also to other mucosal and lymphoid tissues. Conjunctival epithelial cells in culture can induce the expression of human mucosal lymphocyte antigen $\alpha_E\beta_7$ (CD103) on CD8$^+$ (and to a lesser extent also on CD4$^+$) lymphocytes. This may on the one hand allow the retention of CD8$^+$ and CD4$^+$ lymphocytes within the epithelial compartment of the conjunctiva and on the other hand play a part in homing of lymphocytes to other mucosal tissues. Immune responses induced in other mucosal tissues can have protective effects in the eye. Nasal vaccination of mice with herpes simplex virus type 1 (HSV-1) can ameliorate the ocular manifestations of HSV-1 and prevent herpetic stromal keratitis, for example.

That said, immune responses occurring at the ocular surface must strike a balance between protection from a pathogen and direct bystander damage to the ocular surface as a result of the inflammatory process leading to pathogen eradication. Ocular herpes is a prime example where an exuberant local response to eradicate infection causes serious harm to the corneal surface, known as herpes keratitis, which can culminate in opacification of the cornea and blindness.

22-5 Regulatory mechanisms exist in the EALT that restrict an inflammatory response.

To reduce inflammation-induced pathology, a number of mechanisms control the type and the intensity of the immune responses in the lacrimal gland and the ocular surface, thereby limiting inflammatory immunity in favor of less destructive effector mechanisms. Much of the ocular protective response is through secretory IgA (SIgA), which does not activate complement, but which is effective at neutralizing viruses and at promoting antibody-dependent cellular cytotoxicity by eosinophils bearing FcαR. The numerous CD8$^+$ T cells in the epithelial layer and lamina propria contain cytotoxic T lymphocytes, which kill in a targeted fashion through perforin and granzyme B rather than through effector cytokines such as tumor necrosis factor-α (TNF-α), interferon-γ (IFN-γ), and IL-17 that are harmful to many cells, promote recruitment of phagocytes, and result in considerable bystander tissue damage. Thus, the activity of helper T$_H$1 and T$_H$17 effector T cells which exert effector function through pro-inflammatory cytokines is fortunately dampened by 'natural' Foxp3$^+$CD4$^+$CD25$^+$ regulatory T (nT$_{reg}$) cells, which are found in conjunctival tissues and are believed to dampen or inhibit the inflammatory, autoimmune, and pathogen-directed immune responses on the ocular surface. Research in rabbits has revealed that nT$_{reg}$ cells present in the conjunctiva suppress virus-specific CD4$^+$ and CD8$^+$ effector T cells during experimental ocular HSV infection. Other T cells in the epithelium and lamina propria of the ocular surface that have been shown to be regulatory in other mucosal sites and control responses to commensal microbiota include CD8$^+$ T cells, $\gamma\delta$ T cells, and NKT cells; however, functional information about their relative importance in the eye is lacking.

The corneal epithelium contributes to immune homeostasis within the ocular surface in many ways. It secretes IL-1 receptor antagonist (IL-1RA) and anti-angiogenic factors. TGF-β, which can be secreted by many types of cells, both immune and nonimmune, is present at the ocular surface and together with IL-1RA promotes an anti-inflammatory environment by preventing dendritic cell (DC) maturation and inhibiting effector T-cell proliferation, differentiation, and survival. Antigen-presenting cells (APCs) from the ocular surface that acquire antigen in such an environment and transport it to regional lymph nodes would tend to induce antigen-specific T$_{reg}$ (iT$_{reg}$) cells that will further aid in controlling the inflammatory response. Vasoactive intestinal peptide (VIP), which is produced in many mucosal tissues, is secreted by sensory nerves in the cornea and promotes production of TGF-β and IL-10 while at the same time inhibiting production of IL-1, TNF-α, IFN-γ, and the monocyte-attracting chemokine CXCL2. Finally, the ocular surface endothelial cells express the programmed death ligand-1 (PDL-1), which can actively cause apoptosis of activated T cells. Acting in concert, these regulatory mechanisms permit induction of immunity, but dampen and control the intensity of inflammation at the ocular surface to maintain homeostasis and limit bystander damage.

22-6   Immunization through the ocular mucosal immune system can, however, induce an immune response.

It is possible to take advantage of the mucosal system of the eye for active induction of immunity by vaccination. Experiments in rabbits (which, unlike mice and rats, do have well-developed organized CALT) show that antigen-sampling M cells, present in the follicle-associated epithelium above the lymphoid follicles of the conjunctiva, bind and translocate SIgA from the tear film. Topical conjunctival immunization leads to generation of antibody-secreting plasma cells not only in the local but also in distant mucosae together with detectable antibody titers in serum. Natural antibodies, present in the tear film before immunization, may contribute to antigen uptake, potentiating immune responses induced through the eye.

Mucosal adjuvants such as cholera toxin (CT) can potentiate these responses, which they do at least in part by enhancing antigen uptake. CT has been well studied as an oral and nasal adjuvant. In mice, despite their paucity of organized CALT, eyedrop administration of a model antigen (ovalbumin) or microbial protein antigens plus CT induces immune responses in both mucosal and systemic tissues and affords effective protection against subsequent infection with the immunizing microbial agent. It is notable that the responses are induced even after occlusion of tear drainage from the eye to the nose and that they are CCR6 but not CCR7 dependent, thereby excluding the NALT and the draining regional lymph nodes (which require CCR7 for homing) respectively as necessary priming sites. A local priming mechanism is further supported by the finding that the eyedrops containing protein antigen plus CT induce organogenesis of CALT and increase the numbers of M-like cells on the nictitating membrane, which is part of the conjunctiva, though not in the conjunctiva proper. Moreover, efficient induction of CALT in the nictitating membrane also follows administration of a live microbial agent (*Chlamydia trachomatis*) without CT, supporting the notion that immunization through the ocular surface can be an effective route of mucosal vaccination.

Diseases associated with dysfunction of the ocular mucosal immune system.

Ocular surface inflammation can occur due to a wide range of causes: infection, chemical or mechanical injury, allergy, and autoimmune responses. This can manifest as conjunctivitis (i.e., inflammation of the conjunctiva), keratitis (inflammation of the cornea), or keratoconjunctivitis if both are involved. In many cases, dysfunction of the ocular mucosal immune system and ocular surface disease can be something of a vicious cycle, so that it becomes difficult to distinguish which came first.

22-7   Loss of tear fluid results in dry eye syndrome.

Dry eye is one of the most frequent ocular surface problems, whose primary manifestation is a dysfunction of the tear film, resulting in a gritty feeling, dryness, irritation, and inflammation of the ocular surface. Dry eye can stem from a number of causes: from neurogenic, due to dysfunction of parasympathetic nerves that control tear production, through autoimmune damage to the tear gland, to infections which can cause physical blockage of tear ducts or inhibition of tear production by inflammatory mediators. The classification of dry eye diseases, established in a 1995 National Eye Institute/Industry Workshop, provides a compartmentalized classification system for these syndromes (Figure 22.4), but in practice combined disorders are common and it is not always possible to determine a single cause for the disease. As an example, chronic irritation of the ocular surface due to evaporative dry eye can lead to secondary autoimmunization to autologous tissue components of the ocular surface and lacrimal gland components, shifting from 'simple' dry eye to Sjögren's syndrome.

22-8 Sjögren's syndrome is an autoimmune disease of the lacrimal and salivary glands.

Sjögren's syndrome, an inflammatory condition of the lacrimal and salivary glands that appears to be autoimmune, affects predominantly women (9 out of 10 Sjögren's syndrome patients are women) usually during or past middle age. There is extensive evidence that Sjögren's syndrome has an autoimmune basis. A major antigen has been reported to be α-fodrin but there are antibodies to other autoantigens such as Ro, La, antinuclear antibodies, and rheumatoid factor, resembling other systemic autoimmune diseases like systemic lupus erythematosus. Association of disease susceptibility with HLA-B8, HLA-Dw3, HLA-DR3, and with the DQA1*0501 allele supports the notion of autoimmunity, as these are antigen-presenting molecules. A recently developed mouse model, in which mechanical desiccating stress to the ocular surface results in induction of self-reactive lymphocytes (antigen currently unidentified) that can transfer the condition from a diseased host to healthy recipients, further supports the notion that autoimmune mechanisms can become induced by, and can perpetuate, dry eye disease. It has been found that sex and sex steroid hormones are critical factors in the regulation of ocular surface tissues, as well as in the pathogenesis of dry eye syndromes. Androgens reduce the inflammation in, and enhance the functional activity of, lacrimal glands in mouse

Figure 22.4 Classification of dry eye diseases. HIV, human immunodeficiency virus. (Adapted from P. Donshik, *CLAO J.* 21:214, 1995. With permission from Lippincott Williams & Wilkins.)

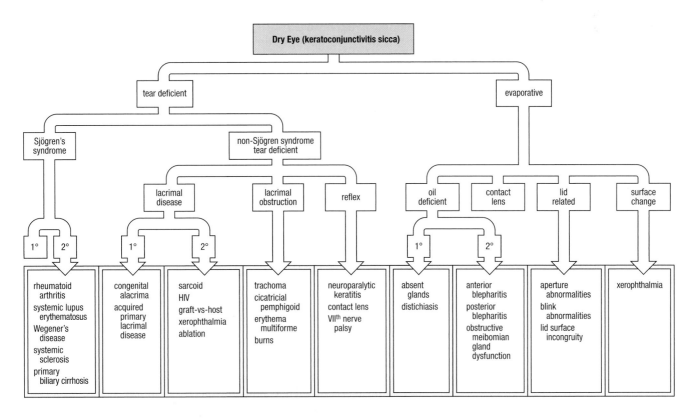

models of Sjögren's syndrome through unknown mechanisms that involve androgen binding to receptors in lacrimal gland epithelial cells, whereas estrogens have the opposite effect, providing a potential explanation for the predominance of the disease in women.

22-9   Ocular cicatricial pemphigoid is an autoimmune disease that is directed against components of the basement membrane.

Ocular cicatricial pemphigoid (OCP) is part of a group of systemic autoimmune diseases known as mucous membrane pemphigoid), characterized by circulating autoantibodies directed against a variety of epithelial basement membrane components, the major ones being bullous pemphigoid antigens 1 and 2 (BPAG1 and BPAG2), laminin 5 (epiligrin), and in some cases also laminin 6, collagen 7, and β_4 integrin subunit. As a result of antibodies binding to the basement membrane, the connection between the epidermis and dermis is weakened such that the two layers separate easily, forming blisters. The antibodies are mainly of the IgG4 subclass but also IgM (complement binding) and sometimes IgA. Anti-B cell treatment with rituximab (anti-CD20) has been reported to have a beneficial effect, supporting the role of anti-basement membrane autoantibodies in the pathogenesis. If blistering, a role for complement activation at the basement membrane has been suggested, but may not be necessary. A model of OCP has been developed by infusing anti-laminin-5 antibodies into neonatal mice. Although in humans complement C3 components at the basement membrane of conjunctival biopsies is strongly diagnostic of pemphigoid, mast cell- and complement-deficient neonatal mice still develop blistering, suggesting that the circulating autoantibodies can induce dermal–epidermal cleavage directly, without the need for complement activation. Demonstration of circulating autoantibodies in serum of OCP patients often requires sensitive detection methods, suggesting that most of them are bound to the tissue. Usually a patient will demonstrate only one antibody reactivity, suggesting that there are subsets of the disease, but clinical differences between patients with different reactivities have not been observed.

The morbidity in OCP is due to the progressive fibrosis and contraction of the conjunctiva that results from scarring when the blisters heal. This consequently results in shortening of the conjunctival fornices and blockage of the lacrimal ducts causing a secondary severe dry eye disease which leads to more keratinization and fibrosis, often ultimately causing blindness. Data from conjunctival biopsies reveal an increase in CD4+ T cells and DCs in the epithelium and activated IL-2R-bearing T cells (CD4+ and CD8+), macrophages, and B cells in the stroma. An increase in IL-17-producing cells has also been reported. T_H2 cells are present and may amplify the fibrotic process through production of cytokines, such as IL-13, that stimulate resident fibroblasts to produce pro-fibrotic cytokines, such as TGF-β, platelet-derived growth factor, basic fibroblast growth factor, and connective tissue growth factor, all of which are increased in the conjunctiva.

Several reports have estimated the incidence of OCP to be 0.87 cases per million of the population per year in Germany and 1.16 per million in France, with a 2:1 female predominance. Associations with several HLA antigens have also been reported, depending on the study and ethnic group, in keeping with the autoimmune nature of the disease.

22-10 Allergy is a common clinical manifestation in the eye due to its exposed surfaces.

The conjunctiva is a highly exposed environmental surface and contains many mast cells. In allergic individuals who have IgE antibodies directed against environmental allergens such as pollen, which bind to the FcɛR of mast cells and basophils, contact with the allergen through the ocular surface precipitates release of mast cell granules whose mediators cause enhancement of vascular permeability and conjunctival swelling (chemosis), itching, and tearing (see Chapter 12). While it is clear that allergy can be expressed when allergen encounters the ocular surface, evidence that allergy can be induced through the ocular surface is circumstantial, as it is difficult to exclude contact through additional mucosal surfaces, including the NALT and even the lung. In the classical animal models of ocular allergy, for example to ragweed pollen, induction is systemic by immunization with the allergen in alum, and instillation of eye drops containing allergen brings about allergic manifestations in the eye. However, as was stated above, it is possible to induce pathogen-related immune responses through the eye experimentally even when the ducts draining the tears into the nasal cavity are blocked, excluding significant participation of NALT. Furthermore, it has been suggested that local mucosal epithelial cells can help initiate ocular allergic inflammation by producing a novel proallergic cytokine, thymic stromal lymphopoietin (TSLP), which activates dendritic cells to prime T_H2 differentiation and allergic inflammation through the TSLP-TSLPR and OX40L-OX40 signaling pathway. NKT cells, probably through their production of IL-4, are needed for maximal expression of allergic conjunctivitis. Recent information also implicates IL-33, a cytokine within the IL-1 family that binds to ST2 together with IL-1 receptor accessory protein to form the IL-33 receptor complex on eosinophils and $CD4^+$ T cells and promotes T_H2 responses, in the induction and augmentation of allergic conjunctivitis.

Most types of allergic conjunctivitis are fairly benign, if unpleasant. However, there are exceptions. Vernal keratoconjunctivitis is a relatively rare but serious allergic response that can leave permanent scarring and visual deficit. It occurs most often in spring (vernal) in young boys around 8 years of age, and usually resolves spontaneously at puberty, indicating a hormonal influence. It is characterized by conjunctival giant papillae, hyperemia, and frequent involvement of the cornea. Although allergic mechanisms, including presence of typical T_H2-type immunity and IgE, are well documented in vernal keratoconjunctivitis, up to 50% of patients do not have a family or medical history of allergic diseases, and do not show IgE sensitization, suggesting that this disease is not solely IgE mediated. Recent studies have also pointed out the role of resident conjunctival nonlymphoid cells in the pathogenetic processes of vernal keratoconjunctivitis. Thus, the pathogenesis of the condition is likely to be multifactorial and complex, involving not only the ocular mucosal immune system but also, similarly to Sjögren's syndrome, the nervous and endocrine systems.

22-11 Herpetic stromal keratitis is a pathologic response against the cornea after herpes virus infection has been cleared.

Infection of the ocular surface with herpesvirus (HSV) causes keratitis which in about one-fifth of the cases induces an inflammatory reaction in the corneal stroma and permanent scarring of the cornea known as herpetic stromal keratitis (HSK). This is a classic example in which an exuberant and insufficiently controlled immune response to infection does as much harm as good, by causing serious bystander tissue damage. The pathology continues even after antiviral immunity has contained viral replication, suggesting a secondary autoimmune component. Mouse models of HSK have shown a strong participation of T_H1 cells and their mediators in the immune-mediated damage, as well as of innate immune cells recruited by them. The site of priming of these T cells may be local within the CALT as well as in the lymph nodes that drain the ocular surface, but the nature of the priming antigen(s) remains unknown. At least four different effector mechanisms of T-cell activation have been proposed. These include specific activation by viral antigens, secondary responses to cornea-derived self-antigens, nonspecific activation by CpG motifs in viral DNA, and bystander activation by inflammatory mediators or cytokines against corneal-derived autoantigens. These mechanisms are not mutually exclusive, and in fact may well occur together and synergize to cause immune-mediated tissue damage. Experimental results in the mouse model of HSK have shown that vascular endothelial growth factor and matrix metalloproteinase 9 (MMP9), which are produced by the immune cells invading the cornea, promote pathological formation of new blood vessels that encroach on the corneal surface (corneal neovascularization) and are an important part of HSK pathogenesis.

22-12 The interior of the eye is an immune privileged site due to powerful regulatory mechanisms that resist immune activation.

The internal cavity of the eye is not a mucosal system in the classical sense, nor does it contain lymphoid tissue resembling the mucosal lymphoid tissue. Nevertheless, there are some notable similarities in the biological requirement to control inflammatory immunity within the eye to protect the physical integrity and preserve the visual function of the neural retina, and in the molecular mediators that fulfill this requirement.

Ocular immune privilege is a term that has been coined to define the special immune environment within the eye. The healthy retina is protected by a tight blood–retinal barrier, composed of tight junctions between adjacent vascular endothelial cells, which prevents cells from freely entering the eye. The ocular fluids contain TGF-β, inhibitory neuropeptides (VIP, α-MSH, somatostatin), and a high concentration of retinoic acid; the ocular tissues constitutively express FasL, PDL-1, thrombospondin, galectins, and other inhibitory cell-surface molecules. These conditions help maintain resident ocular APCs in a non-activated state with low major histocompatibility complex class II expression and create an inhibitory environment against immune cells from the circulation that may enter the eye if the barrier breaks down due to physical trauma or inflammation. After injection of an immunogenic protein into the anterior chamber, tolerogenic APCs are also 'exported' from the eye into the spleen, where they induce CD4$^+$ and CD8$^+$ T$_{reg}$ cells; this phenomenon is known as anterior chamber associated immune deviation (ACAID) and may involve CD1d-restricted NKT cells. Another mechanism involves the direct conversion, within the ocular microenvironment, of conventional T cells specific to retinal antigens associated with the eye into Foxp3-expressing T$_{reg}$ cells. This latter process requires local antigen presentation in the presence of retinoic acid (which is abundant in the eye due to its function in the chemistry of the visual process) and TGF-β, and resembles T$_{reg}$ induction in mucosal tissues by presentation of antigen on retinoic acid-producing CD103$^+$ DCs.

Thus, the mucosal tissues and 'classical' immune privileged sites such as the interior of the eye share the need and the ability to actively control immune responses taking place in their 'territory' in order to maximize protection of the tissue. It therefore seems appropriate to extend the concept of immune privilege to the ocular surface and to its associated mucosal tissues, as well as to other mucosal sites.

Summary.

The ocular mucosal immune system has to fulfill the function of protection from microbial infection while also providing protection from the deleterious consequences of inflammation on vision. To do this, it has adapted to use immune mechanisms both at the humoral and the cellular level that are effective in antimicrobial defense, but yet minimize inflammation and the associated collateral damage to the surrounding tissue. Nevertheless, the system is not perfect, and an overactive effector response or defects in regulatory mechanisms can precipitate a vicious circle of immune, mechanical, and finally autoimmune or allergic processes that irritate and inflame the ocular surface. At best, they can be difficult to control and require chronic treatment, and at worst they bring about permanent ocular surface damage with disastrous consequences to vision.

Further Reading.

Knop, E., and Knop, N.: **The role of eye-associated lymphoid tissue in corneal immune protection.** *J. Anat.* 2005, **206**:271–285.

Knop, E., and Knop, N.: **Anatomy and immunology of the ocular surface.** *Chem. Immunol. Allergy* 2007, **92**:36–49.

Knop, N., and Knop, E.: **Regulation of the inflammatory component in chronic dry eye disease by the eye-associated lymphoid tissue (EALT).** *Dev. Ophthalmol.* 2010, **45**:23–39.

Mebius, R.E.: **Organogenesis of lymphoid tissues.** *Nat. Rev. Immunol.* 2003, **3**:292–303.

Nagatake, T., Fukuyama, S., Kim, D.Y., *et al.*: **Id2-, RORγt-, and LTβR-independent initiation of lymphoid organogenesis in ocular immunity.** *J. Exp. Med.* 2009, **206**:2351–2364.

Paulsen, F.: **Functional anatomy and immunological interactions of ocular surface and adnexa.** *Dev. Ophthalmol.* 2008, **41**:21–35.

Seo, K.Y., Han, S.J., Cha, H.R., *et al.*: **Eye mucosa: an efficient vaccine delivery route for inducing protective immunity.** *J. Immunol.* 2010, **185**:3610–3619.

Stern, M., Schaumburg, C., Dana, R., *et al.*: **Autoimmunity at the ocular surface: pathogenesis and regulation.** *Mucosal Immunol.* 2010, **3**:425–442.

Streilein, J.W.: **Ocular immune privilege: the eye takes a dim but practical view of immunity and inflammation.** *J. Leukoc. Biol.* 2003, **74**:179–185.

PART VI

INFECTIOUS DISEASES OF MUCOSAL SURFACES

Mucosal Interactions with Enteropathogenic Bacteria

23

The gastrointestinal epithelium is the main body interface with the microbial world. The density of bacteria increases along the gastrointestinal tract from 10^3 and 10^5 resident bacteria per milliliter luminal content in the stomach and duodenum respectively, to 10^8 in the ileum, and 10^{12} in the colon, one of the most densely populated microbial ecosystems that is known. In humans, the resident (or commensal) microbiota is composed of hundreds of species. The complexity is enormous and contains 10 divisions of bacteria as well as viruses, eukarya, and archaea. Despite this complexity, recent studies suggest that a surprising quantity of these bacteria can be cultured using specialized microbiologic approaches. Among the 10 dominant divisions (phyla), two represent more than 90% of the bacteria. These are the strictly anaerobic bacteria of the Firmicutes (Gram-positive bacteria) and the Bacteroidetes (Gram-negative bacteria) divisions. A few other phyla, such as Proteobacteria (which includes the Enterobacteriaceae order), Actinobacteria, and Verrucomicrobia, are minor components of the normal microbiota. In healthy adults, Proteobacteria represent less than 1% of the enteric microbiota. Yet, these Gram-negative aero-anaerobic (or facultatively anaerobic) microorganisms have a particular importance as they include a vast spectrum of pathogenic or opportunistic bacteria such as *Salmonella, Shigella, Helicobacter*, and *Escherichia coli* that are major causes of intestinal and extraintestinal diseases. Intestinal bacteria can thus be schematically divided into commensal and pathogenic bacteria even though the frontiers between these two categories have been blurred by genomic studies. Owing to their long coevolution, resident bacteria or symbionts (from the Greek for 'living with') and their hosts have developed mutualistic interactions that promote each other's fitness. Notably, resident bacteria stimulate a 'state-of-alert' within the innate and adaptive immune components within mucosal tissues of their hosts and form a competitive barrier that opposes mucosal colonization by pathogens (so-called colonization resistance). In contrast, pathogens behave as intruders that occasionally colonize and/or invade their eukaryotic hosts to promote their own life cycle. These bacteria possess virulence factors that encode a wide range of effectors that are able to subvert and control eukaryotic cellular functions in order to aid their adherence, replication, and/or dissemination and often eliminate resident bacteria. The continuous presence of resident bacteria, as well as the repeated intrusions of pathogens that are a major threat to their hosts, has compelled the mucosa-associated lymphoid tissues to undergo evolutionary adaptations to build a finely tuned and tightly regulated immune system that balances homeostasis with rapid responses to pathogens. The simple epithelium (see Chapter 5) of the gastrointestinal tract, which is the most well-studied mucosal interface, is the first line of defense against pathogenic intrusions and a central component of host innate immune defenses. This chapter compares interactions between resident bacteria and pathogens within the gastrointestinal epithelium, which is a model for understanding microbial interactions with other mucosal surfaces, and discusses the consequences for hosts and for bacteria (see Chapters 19, 20, 22, 24, 25, 26, and 31).

The gastrointestinal epithelium: a physical and physiological barrier against bacteria.

Besides its central function in digestion and absorption, the monolayer of polarized (simple columnar) epithelial cells that lines the luminal surface of the gastrointestinal tract forms a robust physico-chemical barrier that efficiently opposes the entry of luminal bacteria. Tight junctions, which maintain the cohesion between epithelial cells, are impermeable to bacteria. Hydrochloric acid produced by gastric parietal cells, together with bile salts released into the duodenal lumen, exert potent, nonimmunological bactericidal effects that reduce the density of bacteria in the upper part of the gastrointestinal tract; this density is further reduced by the very active small intestinal peristalsis. Mucus produced by gastric mucous cells and intestinal goblet cells, and microbicidal peptides produced by enterocytes, goblet cells, and ileal Paneth cells, play key and complementary roles to contain bacteria (and other microbes) within the lumen and reduce their contact with epithelial cells. Thus, in healthy individuals, most resident bacteria are contained within a central and relatively fluid mucus layer that is largely separated from the apical cell surface of the polarized epithelium. That said, the mucus layer varies significantly in depth along the length of the gastrointestinal tract, which emphasizes the importance of the redundant pathways of microbial containment that exist with mucosal tissues. In the colon, where the bacterial load increases massively, goblet cells are extremely dense and secrete an additional compact sterile layer of mucus that limits bacterial contact with the epithelial surface. In mice lacking Muc2, a major component of the colonic mucus, resident bacteria come into close contact with the epithelial surface and induce spontaneous colonic inflammation, attesting to the protective function of mucus. Production of microbicidal peptides by enterocytes and Paneth cells is enhanced upon stimulation of their pattern recognition receptors (PRRs) by bacteria-derived signals (see Chapter 16 and below).

23-1 The gastrointestinal epithelium acts as a sensor of luminal microbes and functions as an innate immune barrier.

Similarly to professional immune cells, epithelial cells express a vast array of PRRs including membrane-associated Toll-like receptors (TLRs) and intracellular nucleotide-binding oligomerization domain-containing protein (NOD)-like receptors (NLRs), and retinoic acid-inducible gene-related receptors (RIG) that recognize common structures derived from microorganisms (microbe-associated molecular patterns, or MAMPs). The six TLRs (TLR1–5 and TLR9) are able to recognize bacteria-derived motifs although their exact distribution along the intestine and within epithelial cells *in situ* is not fully delineated. Among the four intracellular PRRs known to interact with bacterial motifs, two of them, NOD1/CARD4 and NOD2/CARD15, have been described in epithelial cells. These two molecules act as cytosolic sensors for peptidoglycan-derived peptides derived from the bacterial cell walls of both Gram-positive and Gram-negative bacteria. NOD2 can also sense mycobacteria and viruses. NOD1 is constitutively expressed in most cell types including epithelial cells; NOD2 is primarily present in hematopoietic cells but is also expressed by Paneth cells and may be induced in enterocytes during inflammation.

Interactions of epithelial TLR and NLR with MAMPs trigger two intracellular cascades important for host defense against bacteria: the canonical NFκB and the mitogen-activated protein (MAP) kinase cascades. These synergize and activate several pathways that are useful to reinforce the epithelial antimicrobial barrier and maintain intestinal homeostasis. MAMP–PRR interactions in epithelial cells induce the synthesis of microbicidal peptides, notably of human β2- and β3-defensins by enterocytes and of RegIII proteins by Paneth cells. They also stimulate the production of chemokines and cytokines that recruit and/or activate immune cells of hematopoietic origin along the basolateral surface of the epithelium which, in turn, complement or sustain epithelial barrier functions. TLR activation of epithelial cells can also reinforce tight junctions (an effect described for TLR2) and promote humoral protection of epithelial cells by stimulating the synthesis of the polymeric immunoglobulin receptor (pIgR) and that of B-cell activation factor (BAFF) and a proliferation inducing ligand (APRIL), two cytokines which stimulate IgA production by B cells potentially independently of T cells (so-called T-cell independent B-cell switching). Finally, NLR and TLR stimulation can promote the production of reactive oxygen species. These properties underscore the role of epithelial PRRs in host protection against either resident or pathogenic bacteria. How resident bacteria and pathogens can respectively interact with epithelial PRRs and downstream intracellular cascades will be discussed later in this chapter.

23-2 The epithelium is a regulated gatekeeper for commensal bacteria.

Despite being an efficient and inducible physico-chemical barrier, the intestinal epithelium is not entirely impermeable even to commensal bacteria. Thus small numbers of resident bacteria can be detected in mesenteric lymph nodes (MLNs) of immune-competent mice. The best established mechanism of bacterial entry involves transcytosis across M cells in the specialized epithelium of Peyer's patches (PPs) or of isolated follicles, both structures which predominate in the lower part of the intestine and notably in the ileum (Figure 23.1). In the follicle-associated epithelium (FAE), reduced numbers of goblet cells and decreased expression of defensins and pIgR facilitate bacterial contact with the apical surface of the epithelium. Moreover, the lack of a brush border and the intense proteolytic activity of M cells allow for rapid transcytosis of intact bacteria and their delivery to underlying dendritic cells and macrophages (see Chapter 13). M cells may also express specific receptors. Thus, a microarray approach of M-cell specific molecules has recently identified glycoprotein 2 as a receptor for FimH, a component of type I pili expressed by many commensal and pathogenic *E. coli* strains (see below). Glycoprotein 2 binds FimH in a

Figure 23.1 Interactions of resident bacteria with host epithelial cells. Most commensal bacteria reside in the mucus and have little direct contact with epithelial cells. One exception is segmented filamentous bacteria, which can strongly adhere to epithelial cells in ileum and PPs. Small numbers of resident bacteria enter the host by transcytosis via PP M cells and initiate adaptive immune responses. A second potential pathway involves their capture by lamina propria CX3CR1+ phagocytes that extend dendrites in between epithelial cells. Bacteria-derived products released in the lumen, such as lipopolysaccharide and peptidoglycan, can interact with TLR on the epithelial cells. Some can be translocated across the epithelium and reach the lamina propria immune cells and bloodstream. In healthy individuals, this passage is very limited and its mechanism is not fully elucidated. It likely involves a transcellular pathway because tight junctions are impermeable to molecules with a molecular weight over 600 daltons. DC, dendritic cell. (Reprinted from N. Cerf-Bensussan and V. Gaboriau-Routhiau, *Nature Immunology* 10:735–744, 2010. With by permission from Macmillan Publishers Ltd.)

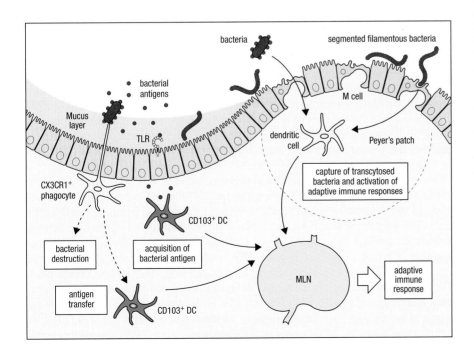

mannose-dependent fashion and seems necessary for the transcytosis of type I-piliated commensal *E. coli* and pathogenic *Salmonella* strains. Interestingly, expression of glycoprotein 2 by FAE and of FimH by bacteria seems necessary to observe efficient *Salmonella*-specific responses in PPs and MLNs. Thus, under physiologic conditions, M-cell sampling of commensal microbiota is important for the maintenance of tolerance pathways and induction of IgA secretion on the one hand and the initiation of anti-infective immune responses to invasive pathogens on the other hand.

On the basis of confocal and two-photon microscopy studies, a second mechanism of bacterial transepithelial passage has been described. This involves subsets of lamina propria dendritic cells (DCs) and potentially macrophages. Several studies, mainly performed in *in vivo* mouse models and *in vitro* cellular models with human epithelial cells, have shown that CX3CR1+ DCs can extend protrusions between the tight junctions of epithelial cells and capture commensal and pathogenic bacteria within the intestinal lumen. This process depends on the synthesis of CX3CR1 ligand by the epithelium and is dependent upon MyD88-associated signaling. In contrast to CD103+CD11c+ DCs, which mainly confer regulatory function on naive T cells within MLNs, CX3CR1+ DCs do not migrate to MLNs, are poor antigen-presenting cells, and might be closer to macrophages than to DCs. The role of CX3CR1+ cells and of their transepithelial dendrites may be to serve multiple purposes. They could perhaps pass captured bacteria to lamina propria CD103+ DCs, which subsequently migrate to MLNs and present bacteria-derived antigens to naive T and B cells and buttress regulatory pathways to commensal microbiota. Alternatively, CX3CR1+ cells may be to serve as a first line of defense by phagocytosing and killing bacteria adjacent to the epithelial surface or in the lamina propria. Accordingly, it has recently been demonstrated that CX3CR1+ cells can migrate into the lumen in response to a pathogenic *Salmonella* strain, but not a nonpathogenic strain, via a MyD88/TLR5-dependent mechanism, suggesting a role in cell-mediated immune exclusion.

In an immune-competent host, the entry of small quantities of resident bacteria typically has no deleterious consequences as such bacteria can be readily eliminated by phagocytes present in the intestinal mucosa and associated lymphoid tissues. Accordingly, such translocated commensal microbiota cannot be detected beyond the MLNs, and notably cannot be cultured from the spleen. Such steady-state, low-grade levels of bacterial translocation may be viewed, then, as a component of the pathways that generate tolerance to the commensal microbiota as well as the maintenance of commensalism. In fact, this physiological entry of resident bacteria is mandatory for the postnatal maturation of the gut-associated lymphoid system and stimulation of IgA-secreting plasma-cell and activated T-cell migration into the intestinal mucosa, where they reinforce the epithelial barrier and the composition of the intestinal microbiota through the pIgR-mediated transcytosis of IgA and thereby help maintain local homeostasis. One species of bacteria in mice, segmented filamentous bacterium (SFB), appears to play a special role in the postnatal maturation of innate and adaptive host responses. This unusual and as yet unculturable symbiont is related to Gram-positive clostridia. SFB has been identified in many species from arthropods to mammals (most notably rodents, but not humans) on the basis of its unusual morphology and/or by its 16S rRNA composition. In contrast to most resident species, which remain entrapped in mucus within the intestinal lumen, SFB can attach to the surface of intestinal epithelium within the ileum and of PPs via an as yet unknown receptor. This behavior mimics that of a pathogen (see below). SFB settles in the mouse intestine at the time of weaning and induces strong innate and adaptive immune responses, including the production of RegIII defensins, and the activation and migration of IgA plasma cells, CD8+ intraepithelial lymphocytes, and CD4+ lamina propria T cells that have been immune-mediated to either a T_H1, T_H17, T_H2, or regulatory function. SFB seems indispensable for the induction of intestinal

T_H17-secreting $CD4^+$ T cells. The mechanisms underlying its strong stimulatory effect are not yet elucidated but likely involve activation of an epithelial TLR and epithelial production of serum amyloid A protein, which can activate interleukin-23 (IL-23) production by dendritic cells. In immune-competent hosts, colonization by SFB reinforces the mucosal barrier and protects against enteropathogens and has been a very useful model for understanding host–commensal microbial interactions.

The epithelium: a gateway for enteropathogens.

In order for enteropathogens and other systemic pathogens to gain access to the host, they must first invade or traverse the intestinal (or mucosal) epithelium. For this purpose, they have evolved a wide variety of strategies based on modifications of their genome, notably through the loss and gain of mobile genetic elements and horizontal gene transfer. Bacterial pathogenicity thus largely depends on large clusters of virulence genes called pathogenicity islands that are either found on plasmids or integrated into the chromosome and are usually flanked by mobile genetic elements. Some virulence traits and particularly toxin-encoding genes are carried by prophages. Although each pathogen has developed distinctive ways to colonize the intestine, intestinal pathovars share many virulence strategies (Table 23.1). They use fimbriae and/or adhesins to harness host-membrane molecules as receptors, and promote their adherence to epithelial cells and/or their internalization. As for resident bacteria, the initial steps of adhesion and entrance preferentially take place in PPs across M cells. Many pathogens also use needle-shaped protein assemblages, called type III, IV, or VI secretion systems, to inject a wide spectrum of specialized proteins, called effectors, into the cytosol of host cells. These bacterial effectors can sequentially target multiple host cellular pathways in order to promote bacterial adhesion and/or entry into epithelial cells, evade phagocytic degradation and immune cell responses, eliminate resident microbiota from their intestinal niches, and ultimately facilitate their replication and spread.

Adherence to the epithelium by a pathogen is necessary to avoid mechanical clearance and is the first step of colonization by enteropathogens. Bacterial pathogens exhibit a large variety of cell-surface adhesins, including fimbriae (a term used to define pili with a function in adhesion, but not in conjugation) and afimbrial adhesins, some of which participate in a subsequent step of internalization in epithelial cells. We now will focus upon these various types of pathogenic mechanisms.

23-3 Some bacteria express fimbriae that allow for adhesion to the epithelium.

Fimbriae are pili with adhesive properties. Such pili are adhesive hairlike organelles that are exported by complex machineries to the bacterial surface. They comprise a scaffold-like rod anchored to the bacterial outer membrane and a bacterial adherence factor, or adhesin, located at the external tip. Type I pili expressed by many commensal and pathogenic *E. coli* strains possess FimH adhesins, the different variants of which bind to host mannose-containing glycoproteins. They notably bind to glycoprotein 2 expressed by M cells (see above). FimH from uropathogenic *E. coli* can also bind β_1 and α_3 integrins and thereby promote bacterial internalization, a process, however, only described in urinary epithelial cells. Type IV pili are another category of polymeric adhesive surface structures expressed by many Gram-negative bacteria including pathogens such as *Salmonella* spp., enteropathogenic *E. coli* (EPEC), enterohemorrhagic *E. coli* (EHEC), and *Vibrio cholerae*. In the latter species, type IV

Table 23.1 Main human enteropathogens

Enteropathogens	Site of colonization	Clinical expression	Virulence genes	Intestinal histology	Adhesion to epithelial cells		Replication and invasion	Toxins	Secretion system
					Pili/fimbriae	Bacterial adhesin/receptor			
Gram-negative Proteobacteria									
Enteropathogenic *E. coli* (EPEC)[1]	Small intestine	Diarrhea	35 kb LEE pathogenicity islands EAF plasmid	Attaching/effacing epithelial lesions forming pedestals	Bundle-forming pili	Intimin/bacterium encoded Tir	No replication at epithelial cell surface		T3SS
Enterohemorrhagic *E. coli* (EHEC) Common serotype: O157:H7	Ileum, colon	Bloody diarrhea Hemolytic uremic syndrome	92 kb LEE pathogenicity islands Shiga toxin encoding phage p0157 plasmid	Attaching/effacing epithelial lesions forming pedestals	Type IV pili?	Intimin/bacterium encoded Tir	No replication at epithelial cell surface	Shiga toxin	T3SS
Enterotoxigenic *E. coli* (ETEC)	Small intestine	Diarrhea			Multiple fimbriae, including type IV pili	Autotransporter TibA/epithelial cell receptor Tia adhesin/host heparan sulfate proteoglycan	Extracellular replication Invasion possible in epithelial cells	Heat-stable toxins (Sta, STb, AEST1) Heat-labile toxins (AB_5 toxin) Pore-forming cytotoxin	T3SS
Enteroinvasive *E. coli* (EIEC) *Shigella*	Colon	Bloody diarrhea (dysentery)	220 kb plasmid	Epithelial destruction/crypt abscesses	No dedicated structure	T3SS insertion into epithelial cell membrane IpaB effector/host CD44	M cells and PP macrophages Dissemination via laterobasal membranes of epithelial cells		T3SS
Enteroaggregative *E. coli* (EAEC)	Small intestine, colon	Diarrhea	100 kb plasmid	Aggregative adhesion/biofilms Exfoliating lesions (some isolates)	Dr-related aggregative fimbriae			Exfoliating Plasmid-encoded toxin	
Diffusely adhesive *E. coli*	Small intestine	Diarrhea	Undetermined	Diffuse adherence	Dr fimbriae Afa adhesins	Afa-Dr/host CD55 and CEACAM6	Extracellular replication Replication possible in epithelial cells	Secreted autotransporter toxin	
Adherent-invasive *E. coli* (AIEC) Prototype strain: LF82	Ileum	Associated with Crohn's disease (ileal lesions)	Undetermined		Type I pili	Type I pili (mannose residues)/CEACAM6 (induced by IFN-γ and TNF-α)	Epithelial cells Macrophages		
Uropathogenic *E. coli*	Bladder, kidney	Urinary infections	Multiple pathogenicity islands (depending on strains)	None	Type I pili	FimH/host β_1, α_3 integrins	Translocation into bloodstream		
Neonatal meningitis *E. coli* (NMEC)	Brain	Neonatal meningitis		None	Type I pili	FimH/host CD48	Translocation into bloodstream but extracellular replication	Cytotoxic necrotizing factor	

Intestinal *E. coli*

Extraintestinal *E. coli*

Enteropathogens	Site of colonization	Clinical expression	Virulence genes	Intestinal histology	Adhesion to epithelial cells		Replication and invasion	Toxins	Secretion system
					Pili/fimbriae	Bacterial adhesin/receptor			
Vibrio cholerae	Small intestine	Watery diarrhea	41.2 kb phage-derived pathogenicity islands Cholera toxin (CT)-encoding prophage		Type IV pili			Cholera toxin (AB$_5$)	
Yersinia enterocolitica *Yersinia pseudotuberculosis*	Ileum	Diarrhea Ileitis Mesenteric lymphadenitis Septicemia Autoimmunity	Chromosomal pathogenicity islands 70 kb virulence plasmid (pYV)	Peyer's patch abscess		Invasin/host β$_1$ integrins YadA/host extracellular matrix proteins	M cells and PP macrophages followed by extracellular replication in PP and MLN		T3SS
Salmonella Typhoid: *Salmonella* Typhi/Paratyphi Non-typhoid: *Salmonella* Enteritidis, *Salmonella* Typhimurium	Small intestine	Diarrhea, enteric fever Gastroenteritis	7–40 kb chromosomal pathogenicity islands (SPI-1 to 5, SPI-7) *Salmonella* Typhimurium: 90 kb virulence plasmid ± prophages	Variable epithelial destruction	Type IV pili	T3SS insertion into epithelial cell membrane	M cells, macrophages, epithelial cells		T3SS SPI-encoded for intestinal colonization T3SS SPI-encoded for survival in macrophages
Helicobacter pylori	Stomach	Gastritis Gastric carcinomas, lymphomas	Cag chromosomal pathogenicity islands	Chronic gastric inflammation		BabA/host Lewis[b] (MuC5AC) SabA/host Sialyl Lewis[x] (FUT4) CagL/host α$_5$β$_1$	Extracellular replication	*H. pylori* vacuolating cytotoxin	T4SS
Gram-positive bacteria									
Listeria monocytogenes	Brain/placenta	Abortion Meningoencephalitis in immunocompromised hosts	Chromosomal LIP1-1/ LIP1/2	None		Internalin A/host E-cadherin	Epithelial cells, macrophages	Listeriolysin 0	

LEE, locus of enterocyte effacement; EAF, EPEC adherence factor; MLN, mesenteric lymph nodes; PP, Peyer's patches; T3SS, type III secretion system; T4SS, type IV secretion system.

[1] Related strains: rabbit diarrheagenic *E. coli* (RDEC), murine pathogen *Citrobacter rodentium*, and recently identified human pathogen *Escherichia albertii* (formerly known as *Hafnia alvei*)

pili can aggregate laterally and form bundles. These bundles are required for the initial adhesion of EPEC and EHEC to epithelial brush borders as a prelude to the development of the typical 'attaching/effacing' lesions. A striking feature of type IV pili is their ability to retract through the bacterial wall while the pilus tip remains firmly attached to the surface of target cells. Another heterogeneous family of adherent structures is the Afa/Dr adhesin family identified in uropathogenic *E. coli* and in diffusely adherent *E. coli* strains, some of which were shown to assemble into fimbriae. Afa/Dr adhesion fimbriae bind brush border-associated complement decay-accelerating factor (also called CD55).

Still others can interact with carcinoembryonic antigen-related adhesion molecules (CEACAM), a family of receptors that may be preferentially associated with lipid rafts. These interactions trigger intracellular signals that can disrupt the brush border and/or promote bacterial internalization. An illustrative example is an *E. coli* variant that was originally isolated from human subjects with inflammatory bowel disease (IBD), so-called adherent-invasive *E. coli* (AIEC).

23-4   Adherent-invasive *E. coli* (AIEC) utilize fimbriae to adhere to CEACAM6 and invade epithelia.

It had long been known that patients with IBD possessed increased circulating IgG antibodies with specificities for *E. coli* and exhibited increased numbers of mucosa-associated *E. coli* with invasive properties or the presence of intramucosal *E. coli*. These pathogenic *E. coli*, compared with commensal *E. coli*, have acquired specific virulence factors that increase their ability to adapt to new niches and allow them to cause disease. *E. coli* strains associated with the intestinal mucosa from IBD, and in particular Crohn's disease, patients are highly adherent to intestinal epithelial cells and are also invasive. On the basis of the pathogenic traits of Crohn's disease-associated *E. coli*, a new potentially pathogenic group of *E. coli* was designated AIEC, for adherent-invasive *Escherichia coli*. The interaction between AIEC and intestinal epithelial cultured cells induces inflammatory responses such as upregulated expression of IL-8 and CCL20, leading to transmigration of polymorphonuclear leukocytes and dendritic cells in co-culture models. AIEC can also disrupt the integrity of the polarized epithelial cell monolayer, allowing bacteria to breach the intestinal barrier and to penetrate into the gut mucosa. AIEC are also able to survive and to replicate extensively within large vacuoles in macrophages. In contrast to many pathogens that escape from the normal endocytic pathway, Crohn's disease-associated invasive *E. coli* are taken up by macrophages within phagosomes, which mature without diverting from the classical endocytic pathway, and which share features with phagolysosomes. To survive and replicate in the harsh environment encountered inside these compartments, including acid pH and the proteolytic activity of cathepsin D, such bacteria have developed adaptation mechanisms in which acidity constitutes a key signal for activating the expression of virulence genes. Macrophages infected with AIEC release large amounts of tumor necrosis factor-α (TNF-α), and in an *in vitro* model of human granuloma formation, AIEC-infected macrophages aggregate, fuse to form multinucleated giant cells, and subsequently recruit lymphocytes. These characteristics of AIEC have many similarities to pathogenic *Shigella* spp. as summarized in Figure 23.2.

In clinical samples, AIEC can be isolated from ileal specimens in nearly 40% of Crohn's disease patients in comparison with less than 10% of controls. It is likely that the higher prevalence of AIEC bacteria in Crohn's disease patients might arise from a variety of factors. These include the abnormal expression of host factor(s) involved in the epithelial cell gut colonization, an inability of the intestinal mucosa to control AIEC infection, for example because of defects in Paneth cell function and subsequent decreased secretion of antimicrobial

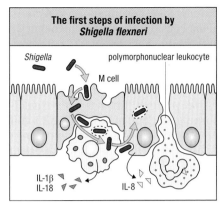

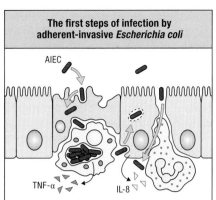

Figure 23.2 AIEC exhibit a similar mechanism of pathogenesis as *Shigella* infection. In the first panel, *S. flexneri* invades the epithelium from the intestinal lumen through M cells and is further phagocytosed by resident macrophages. *Shigella* escapes the phagosome by rapidly disrupting the phagosomal membrane, thereby avoiding phagosome–lysosome fusion and degradation. It induces macrophage cell death by triggering both caspase 1-dependent (pyroptosis) and caspase 1-independent cell death pathways. The dying macrophages release the pro-inflammatory cytokines IL-1β and IL-18, critical mediators of the acute and massive inflammatory responses. Together with IL-8 secreted from the invaded epithelial cells, they induce transmigration of polymorphonuclear leukocytes (PMNs) to the site of infection. Infiltrating PMNs destroy the integrity of the epithelial lining, thus enabling more luminal bacteria to reach the submucosa. In the second panel, similar to *Shigella*, AIEC have the ability to target M cells via binding to CEACAM6 to cross the intestinal barrier, invade intestinal epithelial cells through their basolateral cell surface, and induce the secretion of IL-8. The major difference between *Shigella flexneri* and AIEC is the ability of AIEC to replicate within macrophages without inducing cell death and to continuously activate infected macrophages to secrete large amounts of TNF-α.

peptides, and a loss of control of intracellular AIEC replication related to autophagy deficiencies (see Chapter 31).

Among the host factors that facilitate AIEC colonization is the abnormal ileal expression of CEACAM6, which has been observed in patients with ileal Crohn's disease. In contrast to CEACAM1, which is constitutively expressed on the apical cell surface of intestinal epithelial cells, CEACAM6 is an induced molecule. AIEC adhere to the brush border of primary ileal enterocytes isolated from Crohn's disease patients, but not those from individuals without IBD, due to the abnormal expression of CEACAM6. Most AIEC strains associated with Crohn's disease ileal mucosa express type I pili variants that increase the interaction between AIEC and ileal epithelial cells. Increased expression of CEACAM6 can result from interferon-γ (IFN-γ) or TNF-α stimulation and also from infection with AIEC bacteria, indicating that AIEC can promote their own colonization in Crohn's disease patients. The presence of AIEC bacteria and their ability to induce the secretion of pro-inflammatory cytokines by infected macrophages could lead to an amplification loop of colonization and inflammation. This has been confirmed *in vivo* with the LF82 strain of AIEC, but not nonpathogenic *E. coli* K-12, wherein LF82 persists in the gut of mice that transgenically express human CEACAMs and induces severe colitis.

Another host factor that participates in AIEC colonization of the ileal mucosa is the increased expression of the endoplasmic reticulum (ER) stress response protein Gp96 on the apical surface of ileal epithelial cells in patients with Crohn's disease. This glycoprotein is able to bind to *E. coli* outer membrane protein A present on the surface of AIEC outer membrane vesicles and serve as a means for the AIEC bacteria to deliver bacterial effectors into host cells and promote bacterial invasion. As such, AIEC bacteria can take advantage of the ER stress response occurring within the intestinal epithelium of patients with IBD. Because microbiota have an important role in the establishment of the ER stress response, AIEC, as opportunistic pathogens, might take advantage of such changes in the host innate immune response.

AIEC have another interesting effect upon the intestinal epithelium. Increased numbers of membranous, microfold cells (M cells) have been observed in Nod2-deficient mice. AIEC express long polar fimbriae that allow the bacteria to increase their interaction with mouse and human Peyer's patches and their translocation across M-cell monolayers. Thus, in the presence of NOD2 mutations as observed in Crohn's disease, higher numbers of AIEC bacteria are able to interact with Peyer's patches. This observation is particularly relevant because recurrent ileal Crohn's disease originates from small erosions in the FAE that lies over Peyer's patches.

Finally, the susceptibility of Crohn's disease patients to epithelial infection with AIEC is dependent upon a variety of genetic risk factors as discussed in Chapter 31. Moreover, AIEC bacteria can take advantage of defects in autophagy to replicate within host epithelial cells. Interestingly, the behavior of related *E. coli* strains, including nonpathogenic, environmental, commensal, or pathogenic enterotoxigenic *E. coli*, EPEC, diffusely adherent *E. coli*, and enteroinvasive *E. coli* bacteria with regard to autophagy, is different from that of AIEC bacteria. Defects in autophagy mainly affect the ability of the host to handle AIEC but not other pathogenic *E. coli* variants. This has been specifically shown for alterations due to hypomorphic function of both ATG16L1 and IRGM in epithelial cells, autophagy pathway proteins which are genetic risk factors for Crohn's disease. Hypomorphic function of these leads to increased numbers of AIEC LF82 intracellular bacteria, indicating that autophagy has a key role in controlling intracellular AIEC replication.

23-5 Bacteria have developed a large number of nonpolymeric structures that participate in afimbrial adhesion to and/or invasion of host cells.

Afimbrial adhesion occurs by many different mechanisms. So-called autotransporters, for example, possess C-terminal domains that form pores in the outer membrane, allowing the exposition of a passenger domain that can participate in adhesion to host cells. Thus, AIDA-A is expressed by diffusely adherent *E. coli* strains and is responsible for their diffusely adhesive property to epithelial cells. *Helicobacter pylori* BabA binds to the Lewis[b] blood group antigen (also called MuC5AC) in the gastric mucosa. *Yersinia* YadA mediates adhesion to collagen, laminin, and fibronectin.

A distinct original adhesion pathway has been developed by EPEC and EHEC strains that induce the characteristic 'attaching and effacing' epithelial lesions. After intimate attachment to intestinal epithelial cells via their adhesins and bundle-forming pili, these bacteria use their type III secretion system (T3SS) to rapidly translocate a protein called Tir (for translocated intimin receptor) into the cytosol of host cells (see below for T3SS description). Tir is displayed at the host-cell surface and acts as a receptor for the bacterial outer membrane protein intimin, resulting in Tir clustering and initiation of the cytoskeleton rearrangements that lead to attaching and effacing lesions (Figure 23.3).

Other pathogens have evolved molecules that are able to harness host transmembrane cell-adhesion receptors and trigger intracellular cascades that lead to epithelial cell invasion. This so-called 'zipper' strategy is used by *Yersinia* spp. and *Listeria* spp. to invade PPs (Figure 23.4). Depending on temperature, pH, and growth phase, two strains of enteropathogenic *Yersinia*, *Y. enterocolitica* and *Y. pseudotuberculosis*, express invasin (Inv), a 101 kDa outer membrane protein encoded by a 70 kb virulence plasmid called pYV. Bacterial Inv binds β_1 integrins (α_{-1}, α_{-4}, α_{-5}, or $\alpha v \beta_1$) on the luminal surface of M cells. Inv binding results in the clustering of β_1 integrins and subsequent activation of a signaling cascade that induces cytoskeleton rearrangements and bacterial internalization. Studies using *inv* mutants suggest that Inv is critical for dissemination from the intestinal lumen (even in animals lacking PPs) and for abscess formation in mesenteric lymph nodes but dispensable for overall virulence. Besides *Yersinia*, other bacteria may target host β_1 integrins to adhere to the mucosa. For example *H. pylori* uses its Cag type IV secretion system (T4SS) to target β_1 integrins and subsequently translocate its effector protein CagA, a protein with carcinogenic properties. Thus, an adhesin called CagL is present at the T4SS pilus surface. This adhesin was shown to bridge and activate the $\alpha_5\beta_1$ integrin on the basolateral membrane of gastric epithelial cells.

Listeria monocytogenes is a food-borne Gram-positive bacterium that makes use of two surface proteins, Internalin A and B, to engage in a species-specific manner host adhesion molecules E-cadherin and hepatocyte growth factor receptor Met respectively, and to induce its internalization. After entry within

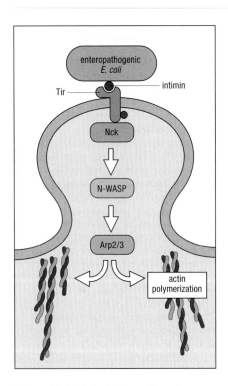

Figure 23.3 Tir/intimin-mediated interaction of enteropathogenic *E. coli* (EPEC) with epithelial cells. Adhesion to epithelial cells is initiated by EPEC bundle-forming pili which bind host *N*-acetyl lactosamine-containing host receptors. Intimate attachment is then triggered by injection of Tir protein into the epithelial cell via EPEC T3SS apparatus. Tir is displayed at the epithelial cell surface and acts as a receptor for bacterial intimin. Tir/intimin interaction induces clustering of Tir which is phosphorylated by various host kinases. Phosphorylated Tir recruits host adaptor Nck, which activates N-WASP and the actin nucleator Arp2/3 complex, resulting in actin polymerization and pedestal formation at the site of attachment. While stabilizing bacteria–host cell interactions, pedestal formation promotes T3SS-mediated injection into epithelial cells of additional effector proteins able to subvert host cell pathways. (Adapted from J. Pizarro-Cerda and P. Cossart, *Cell* 124:715–727, 2006. With permission from Nature Publishing Group.)

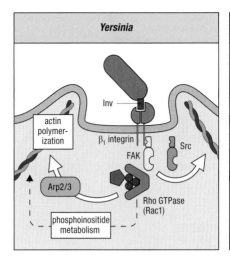
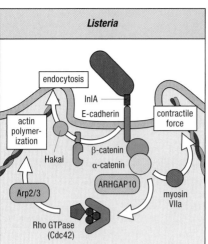

Figure 23.4 Invasion of epithelial cells by *Yersinia* and *Listeria* via hijacking of host adhesion receptors. In the left panel, *Yersinia* binds to β_1 integrin at the surface of PP M cells resulting in the activation of Rho GTPases, recruitment of the Arp2/3 complex, and actin polymerization. This process leads to the formation of a phagocytic vacuole which engulfs the bacterium. β_1 integrin-associated host kinases such as FAK or Src kinases participate in the process. In the right panel, *Listeria* Internalin A (InlA) interacts with host E-cadherin and activates the 'adherens junction machinery' leading to the successive recruitment of β-catenin, the Rho GTPase-activating protein ARHGAP10, α-catenin, and myosin VIIa. ARHGAP10 promotes the activation of Cdc42 and of the Arp2/3 complex, resulting in actin polymerization and cup formation around the bacterium. Myosin VIIa may generate the contractile forces required for internalization of engulfed bacteria. InlA-mediated *Listeria* internalization also activates the Src tyrosine kinase, inducing the recruitment of the ubiquitin ligase Hakai and the ubiquitination of E-cadherin. Clathrin-mediated endocytosis of ubiquitinylated E-cadherin further promotes bacterial internalization. (Adapted from T. Marlovits and C. Stebbins, *Current Opin. Microbiol.* 13:47–52, 2010. With permission from Elsevier Ltd.)

epithelial cells, *L. monocytogenes* can spread from cell to cell and disseminate to its target organs, the blood–brain and placental barriers in humans. Analysis of *inl* mutants indicates that only Internalin A is critical for crossing the intestinal epithelial barrier, while Internalin A and B both participate in bacterial crossing of the placenta. Internalin A is covalently anchored to the bacterial surface and possesses a leucine-rich repeat that interacts with the first ectodomain of human E-cadherin, a transmembrane glycoprotein that mediates homophilic interactions at adherent junctions of polarized epithelial cells. Thanks to Internalin A–E-cadherin interactions, *Listeria* can exploit the machinery involved in the formation of adherent junctions and induce bacterial entry into epithelial cells (Figure 23.4).

Beyond adhesion: bacterial injection systems.

Interestingly, some pathogens can bypass the first step of adhesion and interact directly with the cellular machinery by injecting effectors that regulate actin skeleton dynamics. *Shigella* and *Salmonella* use this 'trigger' mechanism. Contact between these bacteria and epithelial cells is mediated by T3SS, which inject effector proteins into host cells causing massive cytoskeletal changes that result in the formation of a macropinocytic pocket, loosely bound to the bacterial body. This step initiates the internalization of the bacteria.

23-6 At least six different (type I–VI) bacterial 'injection' systems have been defined.

Among the six secretion systems currently described in bacteria, three systems (type III, IV, and VI) allow penetration of host-cell membranes and intracytosol injection of effectors. The best characterized are type III systems (T3SS), which are evolutionarily related to the flagellar system and are well conserved across a variety of taxa (in contrast to the effector proteins which are generally specific to each pathogen and govern their adaptation to their host). T3SS are formed of 20–30 proteins, the assembly of which is tightly regulated. Following assembly of a basal body into the inner and outer bacterial membranes, early substrates including the needle subunit and the needle-length control protein are targeted across this apparatus and assemble to form the extracellular needle (Figure 23.5). Tip proteins are then secreted and assembled at the tip of the needle but further secretion remains blocked until host-cell contact. Upon host–bacterium contact, a second switch triggers the secretion of translocators that form a pore

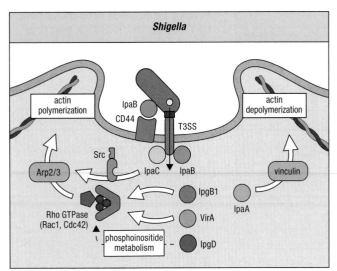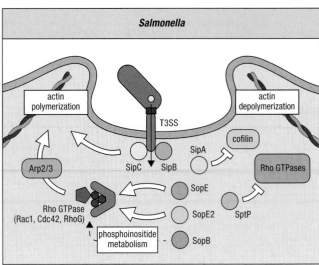

Figure 23.5 T3SS-mediated invasion of epithelial cells by *Shigella* and *Salmonella*. Upon contact with epithelial cells, *Shigella* and *Salmonella* T3SS are activated and deliver bacterial effectors via the IpaB/C and SipB/C pores respectively, thereby triggering internalization. In the left panel, the key effector IpaC activates Src kinases at the site of bacterial contact and ultimately recruits Arp2/3, causing actin polymerization and ruffle formation. This process is favored by effectors IpgB1 and VirA through the activation of Rac1. IpgD activates the phosphoinositide pathway and membrane ruffling while IpaA recruits vinculin and induces actin depolymerization, which allows closing of the phagocytic vacuole. IpaB (probably expressed at the tip of T3SS) can allow bacterial adhesion to the CD44 hyaluronic receptor at the basolateral membrane of the epithelial cell and further promote invasion. In the right panel, translocated effectors SopE, SopE2 (both acting as guanine nucleotide exchange factors), and inositol phosphatase SopB activate Rho GTPases Rac1, Cdc42, and RhoG, resulting in actin polymerization which is further promoted by SipC and SipA. SipC can nucleate actin while SipA antagonizes the function of cofilin, an actin depolymerizing factor. Subsequently, effectors such as SptP inactivate Rac1 and Cdc42, inducing actin depolymerization and closure of the phagocytic cup. (Adapted from J. Pizarro-Cerdá and P. Cossart, *Cell* 124:715–727, 2006. With permission from Elsevier Ltd.)

into the host-cell membrane. Bacterial effector proteins are then translocated into the host-cell cytosol. The mechanism activating this second switch has been deciphered in *Shigella*. In the anaerobic intestinal environment, *Shigella* is primed for invasion and expresses extended needles, which may help establish contact with the intestinal epithelium. These processes are tightly regulated. For example, a transcriptional regulator of anaerobic metabolism (called fumarate and nitrate reduction) represses effector secretion. This repressor is inactivated in the presence of oxygen. Because of the numerous capillaries underlying the epithelium, the luminal zone immediately adjacent to epithelial cells contains sufficient oxygen to inactivate fumarate and nitrate reduction. Thus when *Shigella* comes close to the epithelial surface, the anaerobic block of secretion is reverted, allowing secretion of Ipa effectors into epithelial cells and host invasion. Whether this mechanism can be generalized to other T3SS-expressing enteropathogens remains to be defined, but the presence of fumarate and nitrate reduction boxes upstream of genes required for T3SS functions in *Yersinia* spp. and serovars of *Salmonella* Enteritidis suggests that indeed this is likely to be the case.

Type IV secretion systems (T4SS) are large assemblies of at least 12 distinct proteins used by Gram-positive and Gram-negative bacteria for one-step transport not only of virulent proteins but also of DNA. Their needle-like structures exhibit pili of variable length. The T4SS encoded by the *cag* pathogenicity islands of *H. pylori* produces long and large pili that not only transport the CagA effector but also participate in adhesion to the gastric epithelium (see above).

Type VI secretion systems (T6SS) have been described only recently. They are expressed by many Gram-negative bacteria. In most cases, T6SS are not critical determinants of pathogenesis but rather promote transition toward the chronic phase of colonization and/or infection as in *Salmonella* Enterica where T6SS and T3SS are inversely regulated. In *H. hepaticus*, a bacterium often present in the murine microbiota, T6SS might even promote symbiotic relationships via the release of as yet unknown effectors downregulating host pro-inflammatory responses.

23-7 Bacteria-associated secretion systems inject specific bacterial effectors into the epithelium, which affects host membrane proteins and cytoskeleton.

The most commonly described cellular target of pathogens is the cytoskeleton. Each pathogen has its own repertoire of effectors and therefore of cellular targets. However, there are several common functions for such bacterial effector systems: to enable further firm adhesion, cell entry, and dissemination between cells. In this section we will discuss each of these.

Extracellular pathogens such as EPEC and EHEC induce the formation of actin-rich pedestal structures on the host epithelial surface where they can reside and grow. Via T3SS, they deliver the effector protein Tir, which embeds itself into the host membrane and binds the bacterial outer membrane protein intimin. Tir intracellular domain, either directly (EPEC) or indirectly via another bacterial effector (EHEC), stimulates an intracellular cascade that activates the host protein N-WASP (neural Wiskott–Aldrich syndrome protein). In turn, N-WASP recruits the seven-protein Arp2/3 complex that mediates actin polymerization and formation of the pedestal beneath the extracellular bacterium (see Figure 23.3).

Invasive pathogens such as *Salmonella* and *Shigella* use their T3SS to inject effectors which activate host Cdc42 and Rac1 GTPases albeit via different intracellular relays. Activation of these G proteins leads to the successive recruitment of N-WASP or related proteins, and of the complex Arp2/3 that stimulates actin polymerization and thereby the formation of a macropinocytic pocket. Another set of bacterial effectors, acting either directly or indirectly, can then mediate actin depolymerization, closing the macropinocytic pocket into an intracellular vacuole containing the bacterium (see Figure 23.5).

To invade host cells, *Yersinia* and *Listeria* use a different strategy. Following their binding to host adhesion receptors, they hijack downstream cellular machinery. Thus, binding of *Yersinia* invasin to β_1 integrin activates the focal adhesion tyrosine kinase and triggers a complex cascade implicating a Rac1–Arp2/3 pathway but also phosphoinositide-3-kinase (PI3K). These events lead to formation and then closure of a phagocytic cup (see Figure 23.4). In the case of *Listeria*, engagement of E-cadherin by Internalin A initiates activation of the adherens junction machinery and induces the recruitment of β-catenin, Rho GAP protein ARHGAP10, α-catenins, and unconventional myosin VIIa to the site of entry. Internalization is further mediated by Rac- and Arp2/3-dependent actin polymerization. Internalin A–E-cadherin interaction can also activate the kinase Src. This kinase promotes the recruitment of the ubiquitin ligase Hakai, allowing clustering and clathrin- or caveolin-dependent internalization of ubiquitinylated E-cadherin bound, or not, to *Listeria* (see Figure 23.4).

Finally, following their internalization, intracellular pathogens such as *Listeria* and *Shigella* can escape from the internalization vacuole, replicate in the cytosol and move within and between cells by recruiting and polymerizing actin. Actin polymerization provides the energy to propel bacteria across the cytosol. When they reach the plasma membrane, bacteria form protrusions into neighboring cells, which divide into two membrane vacuoles. These vacuoles can be lysed and the cycle repeated, allowing cell-to-cell spread and tissue dissemination. Thus *Listeria* uses a pore-forming toxin called listeriolysin O to escape from the vacuole; its surface protein ActA can then bind and activate Arp2/3. Similarly, the effector IpaB allows *Shigella* to escape from the intracellular vacuole. *Shigella* then expresses an outer membrane protein IcsA (also called VirG) which recruits N-WASP and Arp2/3.

Mechanisms of bacterially induced epithelial cell dysfunction.

Many enteric pathogens can inhibit intestinal absorption and/or stimulate secretion resulting in diarrhea, a mechanism thought to insure their dissemination outside their hosts. Bacteria accomplish this by the intracellular

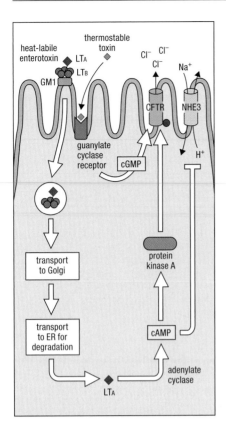

Figure 23.6 Alteration of epithelial transport by ETEC-derived toxins. Thermostable toxin binds to a guanylate cyclase receptor on the epithelial cell brush border, inducing production of cyclic GMP, which activates CFTR. Heat-labile AB_5 enterotoxin (LT) is a homolog of *Vibrio cholerae* toxin. Via its B subunits, heat-labile enterotoxin binds to epithelial cell surface GM1 ganglioside, inducing toxin internalization at lipid rafts. After retrograde shuttling through Golgi and endoplasmic reticulum (ER), the heat-labile enterotoxin A subunit is delivered into the cytosol where it activates adenylate cyclase and increases the level of cyclic AMP. cAMP activates protein kinase A, which in turn phosphorylates and activates CFTR, resulting in increased Cl⁻ secretion. cAMP can also inhibit Na⁺/H⁺ exchanger 3 (NHE3), resulting in decreased Na⁺ absorption.

injection of bacterial effectors as discussed above or via the intraluminal release of toxins that can bind epithelial cell-surface receptors. These various mechanisms are discussed below.

23-8   Bacteria secrete toxins that affect epithelial cell function into the lumen of the intestines.

Bacteria secrete many types of toxins that are capable of affecting epithelial cell function. The heat-stable toxin produced by some strains of enterotoxigenic *E. coli* (ETEC), for example, binds a brush border guanyl cyclase, stimulating the production of cyclic GMP that activates the cystic fibrosis transmembrane conductance regulator (CFTR) and Cl⁻ secretion (Figure 23.6). A more complex mechanism is used by *Vibrio cholerae* toxin and by the homolog heat-labile enterotoxin secreted by ETEC strains. Via their B subunits, these AB_5 toxins interact with monoganglioside GM1 at the surface of epithelial cells, allowing internalization of the A subunit via lipid rafts. After a complex intracellular traffic, the A subunit is released into the cytosol and ADP-ribosylates adenylate cyclase, resulting in increased cyclic AMP (cAMP) production. cAMP in turn stimulates CFTR phosphorylation and activation. cAMP can also decrease the activity of Na⁺/H⁺ exchangers and inhibit Na⁺ reabsorption. This effect combined with the activation of CFTR results in enhanced NaCl at epithelial cell surfaces.

23-9   Bacteria use their secretion machinery to directly inject toxic effector molecules that co-opt epithelial function for the pathogen's benefit.

A second mechanism to induce diarrhea through functional effects on the epithelium or effects on epithelial cell survival does not involve extracellular toxins, but bacterial effectors that are injected into or produced within host cells. Thus EPEC strains use their T3SS to inject two effectors (EspG, EspG2) that induce tubulin degradation and disruption of the host microtubule network, and subsequently the dramatic internalization of the apical Cl⁻/OH⁻ exchanger DRA (downregulated in adenoma).

In other cases, intestinal pathogens have derived many ways to manipulate host survival mechanisms. As an example, apoptosis induction is used by *Shigella* and *Salmonella* to promote their dissemination. Thus, following their translocation across M cells, both bacteria can be phagocytosed by underlying macrophages. Using their T3SS, the bacteria secrete the effector proteins IpaB and SipB respectively, which both activate the cysteine protease caspase 1 from the phagosome into the cytosol. This enzyme induces the apoptotic death of infected macrophages and the maturation of pro-inflammatory IL-1β and IL-18. Bacteria escaping dying macrophages can invade the basolateral side of enterocytes with the aid of their T3SS and subsequently propagate between epithelial cells, a process thought to be enhanced by the inflammatory reaction initiated by IL-1β and IL-18. In contrast, once free in the enterocyte, *Shigella* releases several bacterial effectors that prevent cell death and sloughing, providing a replication niche for *Shigella* to maintain infection. Thus, IpaB targets the anaphase inhibitor Mad2L2 and prevents intestinal epithelial cell renewal; OspE interacts with the integrin-linked kinase ILK and prevents epithelial cell detachment; and IpgD promotes cell survival via the activation of PI3K and Akt proteins. Beyond these examples, an ever-growing number of mechanisms have been unraveled through which pathogens and their effectors can manipulate host-cell survival.

Regulation of host epithelial innate immune responses by bacteria.

Bacteria have evolved multiple mechanisms to escape host innate and adaptive immune responses. This section focuses on their manipulation of innate epithelial responses. A complex cross-talk exists between bacteria and epithelial PRRs. For the host, PRR activation is necessary to regulate homeostasis through effects on epithelial barrier function on the one hand, and recruitment of inflammatory immune cells on the other hand, for the ultimate elimination of an invading pathogen. However, inflammation can lead to severe epithelial damage that emphasizes the need for tight control mechanisms in order to avoid excessive stimulation by resident bacteria. For bacteria, host inflammation is dangerous and must be dampened to avoid their own destruction. Yet several pathogens have been shown to use the host inflammatory response during the early phase of infection to eliminate competitors within the resident microbiota, colonize the emptied intestinal niches, and invade tissues. We will now consider the mechanisms by which pathogens stimulate PRRs to elicit these responses.

23-10 Pathogenicity, in comparison with commensalism, is often determined by the structure–function relationships between microbial MAMPs and host PRRs.

MAMPs derived from commensal and pathogenic bacteria are very similar. It has been suggested that MAMPs from commensals have undergone biochemical modifications that render them 'stealthy' to pathogenic PRR activation. This is notably the case for lipopolysaccharide (LPS) because its agonist effect on TLR4 depends on the number of acyl chains within its lipid A component. However, modification of lipid A acylation lessens or even abrogates LPS recognition by TLR4. Thus *Yersinia pestis* fully acylates its lipid A at 27°C but exhibits partial acylation at 37°C leading to significant loss of its agonistic effect on TLR4. A mutant strain expressing a fully acylated lipid A at 37°C becomes avirulent, indicating that avoiding TLR4 recognition facilitates invasion at body temperatures. Along the same line, *H. pylori* and *Campylobacter jejuni* produce flagellin molecules that are not recognized by TLR5 and this property is thought to promote chronic gastric colonization. In contrast, *Shigella* expresses an extra gene on its virulence plasmid that achieves full acylation of lipid A, thus significantly enhancing host release of IL-8 and recruitment of polymorphonuclear leukocytes in response to TLR4 stimulation, an event thought to promote *Shigella* dissemination within the epithelial layer, its main replication niche.

23-11 Commensal and pathogenic microbiota engage PRRs in a distinctive manner.

While further work is needed to evaluate the extent to which MAMPs from resident bacteria have evolved to avoid recognition by host PRRs, there is good evidence that hosts have evolved mechanisms that help discriminate signals from resident and pathogenic bacteria. There are many examples of mechanisms by which this may occur. One such mechanism involves the distribution and regulation of PRRs within the intestinal epithelium that may limit their activation by resident bacteria while preserving their activation by pathogens. At steady state, TLR4 is preferentially expressed by epithelial crypts and is localized in endosomes. Its co-receptor MD-2 is not, or only weakly, expressed and brush border alkaline phosphatase can dephosphorylate LPS lipid A, thereby reducing its stimulatory activity on TLR4. TLR5, a receptor for bacterial flagellin expressed by pathogens, but also by many commensal strains, is preferentially localized at the basolateral epithelial membrane at least in the colon away from its ligand. TLR9 ligands (CpG derived from bacterial DNA) can activate NFκB when applied at the epithelial basolateral membrane, but repress this cascade when applied at the apical membrane.

In contrast to the resident bacteria, which mainly remain within the intestinal lumen, enteropathogens can gain access to the basolateral membrane of epithelial cells where TLRs can be more readily activated. Furthermore pathogens can inject peptidoglycan moieties and thereby activate cytosolic NOD proteins and their downstream pro-inflammatory signaling cascades. Epithelial PRR activation by pathogens can thus increase the production of microbicidal products and reactive oxygen species, and stimulate the recruitment and activation of host immune cells, notably of phagocytes and DCs, which can in turn initiate adaptive immune responses; all of which are mechanisms that participate in clearing pathogens. As already alluded to above, some pathogens can exploit host inflammatory responses to their benefit. Thus, host inflammation induced by *Salmonella* Typhimurium infection results in the elimination of resident bacteria and promotes colonization. During *Shigella* infection, IL-8 induced via NFκB and MAP kinase (MAPK) activation of epithelial cells stimulates the recruitment and transmigration of polymorphonuclear leukocytes, which enhances access of *Shigella* to the basolateral membrane of the epithelium and thereby promotes bacterial dissemination.

In another set of examples, TLR activation by the microbiota can exert negative feedback on signaling pathways. One such pathway is that associated with NFκB (Figure 23.7). Thus, immediately after birth in mice, microbiota-derived LPS can downregulate expression of the interleukin-1 receptor-associated kinase-1 (IRAK1), the proximal activator of the NFκB cascade downstream of MyD88, resulting in desensitization of the TLR pathway. Via the induction of reactive oxygen species, commensal bacteria can also inhibit the common ubiquitin ligase E3-$^{SCF\beta\text{-TrCP}}$ and thereby prevent polyubiquitination and degradation of IκB, a key step in NFκB activation. Furthermore, TLR4 stimulation by LPS stimulates the expression of peroxisome proliferation-activated receptor-γ (PPARγ), which in turn can divert NFκB from the nucleus. Since PPARγ positively controls the expression of the colonic microbicidal peptide, human β-defensin-1, PPARγ can simultaneously maintain the gut barrier and prevent excessive activation of the NFκB cascade (Figure 23.7). Altogether these mechanisms may explain the overall beneficial effect of epithelial TLR activation and the corollary observation that selective inactivation of two key elements of the NFκB pathway (IKKβ or IKKγ) within the epithelium is highly detrimental. These observations further indicate that activation of epithelial PRRs by the microbiota limits epithelial damage and/or accelerates epithelial repair in several mouse models of colitis. In contrast, epithelial PRRs do not seem to participate in the induction of inflammation but rather the presence of PRRs in the hematopoietic compartment is necessary for the induction of intestinal inflammation within the colon. The beneficial effect of epithelial TLRs has been ascribed to the recruitment of immune cells producing cytokines that stimulate epithelial repair (e.g., IL-11, IL-22) or to the induction of antimicrobial peptides (e.g., defensins) that limit bacterial adhesion to the epithelium. Another beneficial effect of epithelial TLR stimulation by the microbiota may result from the induction of BAFF and APRIL, and of pIgR which respectively promote the production of secretory IgA (SIgA) and its translocation into the intestinal lumen, wherein SIgA can complex bacteria and promote their entrapment within the mucus. This mechanism is important to limit bacterial translocation and also to reduce epithelial cell production of reactive oxygen species in response to the microbiota.

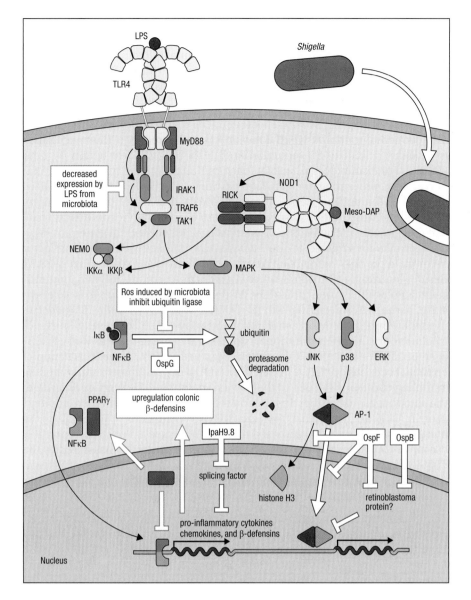

Figure 23.7 Modulation of epithelial cell pro-inflammatory signaling pathways by intestinal bacteria. NFκB and MAP kinases can be activated by bacteria-derived products via membrane-associated TLR and intracytosolic NOD1 leading to the production of microbicidal peptides and pro-inflammatory cytokines and chemokines. These pathways can be inhibited by commensal bacteria and enteropathogens. Steps inhibited by resident bacteria are indicated in blue. Steps inhibited by *Shigella*, a paradigm of an invasive enteropathogen, are shown in red.

23-12 Bacterial effectors from enteropathogens manipulate epithelial intracellular signaling cascades.

While host inflammatory responses can benefit some pathogens, at least at the early phase of infection, they are clearly a threat to the pathogen and many pathogens have evolved strategies to dampen host inflammatory responses in order to promote their persistence or dissemination. Thus many enteropathogens use their T3SS to inject effectors able to interfere with intracellular signaling cascades threatening their survival.

Many inhibitory effects of bacterial effectors have been described in hematopoietic cells and notably in macrophages. This is the case for the *Salmonella* secreted factor L, which impairs IκBα ubiquitination and NFκB activation in infected macrophages, and for *Yersinia* YopJ/P effectors, multifunctional proteins which inhibit NFκB, p38, ERK, and IRF3 pathways and thereby induce the apoptosis of infected macrophages.

Yet other effectors act on epithelial cells. A close homolog of YopJ, the *Salmonella* effector AvrA, is also a multifunctional protein with transacetylase and deubiquitinase activities. This protein is injected into epithelial cells during *Salmonella* Typhimurium infection and potently inhibits c-Jun N-terminal kinase and NFκB signaling pathways. *In vivo* analysis of mutants indicates that AvrA is a strong inhibitor of intestinal inflammation and epithelial apoptosis that promotes persistent infection. Two other effectors (SptP and SspH1) injected by *Salmonella* Typhimurium strains into epithelial cells can also inhibit the NFκB cascade.

In the case of *Shigella*, several effectors injected via T3SS into epithelial cells also cooperate to inhibit NFκB activation (see Figure 23.7). This mechanism is thought to limit IL-8 production and delay *Shigella* destruction by neutrophils. For example OspG impairs the ubiquitination and thus the degradation of IκBα, preventing nuclear translocation of NFκB. OspF is a phosphatase targeted to the nucleus where it impairs MAPK-dependent phosphorylation of histone H3. In addition, OspF can irreversibly dephosphorylate ERK. OspF can thus prevent the activation of numerous NFκB dependent genes. IpaH9.8, another effector targeted to the nucleus, can interact with a splicing factor involved in the maturation of inflammatory cytokines mRNA (see Figure 23.7). Besides interfering with NFκB, other effectors injected via T3SS into epithelial cells can block the production of the β3-defensin to which *Shigella* is very sensitive, perhaps by interfering with the JAK/STAT pathway. Finally, *Shigella* can inhibit autophagy, a cytoplasmic mechanism by which cells entrap internalized microorganisms into autophagosomes and degrade these as well as damaged organelles and misfolded protein aggregates. Upon *Shigella* infection of epithelial cells, autophagy is initiated via NOD1-dependent recruitment of the autophagy protein ATG16L1 to the plasma membrane. Delivery of *Shigella* to the autophagosome further depends on the interaction between the autophagy protein ATG5 and IscA, a bacterial effector that accumulates at one pole of the bacterium and induces actin-based motility. Wild type *S. flexneri* can partially escape autophagy using the IcsB effector, another protein translocated by the T3SS that competitively inhibits IscA binding with ATG5. Analysis of *Shigella icsB* mutant indicates that this camouflage role of IcsB is important for efficient intraepithelial dissemination of the bacterium.

These observations underscore the multiple mechanisms evolved by pathogens to manipulate host epithelial cells in order to promote their local survival and replication or their access to other host cells and tissues appropriate for their replication. For some enteropathogens, such as *Shigella* and many enteropathogenic *E. coli*, the infection remains limited to the intestine. If the host survives epithelial damage and diarrhea, the infection can be controlled by host innate immune cells, particularly by neutrophils, and by adaptive immune responses that include the production of specific SIgA and T-cell responses, notably helper T_H17 cells which recruit neutrophils. In contrast, other pathogens such as *Salmonella*, *Yersinia*, and *Listeria* have evolved multiple additional mechanisms to escape phagocytosis by macrophages, down-modulate host adaptive immune response, and disseminate toward other tissues, resulting in systemic infections.

Summary.

The gastrointestinal epithelium is the main body interface between the host and the microbial world. The epithelium in this location is exposed daily to 10^{14} resident bacteria and can be occasionally colonized by a wide spectrum of enteropathogens, most of which are Gram-negative Proteobacteria. The gastrointestinal epithelium has evolved as a central component of host innate immune defenses, and forms a potent and inducible physio-chemical and antibacterial barrier which can be finely tuned via the recognition of bacteria-derived motifs through pathogen-recognition receptors.

The gastrointestinal epithelium is, however, also a gateway for bacteria. A small fraction of resident bacteria can cross the epithelium, mainly via PP M cells. These bacteria are rapidly eliminated by host phagocytes and elicit adaptive immune responses that reinforce the epithelial barrier and promote their confinement into the intestinal lumen.

In contrast with resident bacteria, which mainly replicate and thrive within the mucus layer covering the intestinal epithelium, enteropathogens have the ability to adhere to and/or invade the gastrointestinal epithelium in order to promote their life cycles. For this purpose, they have evolved a wide variety of strategies depending on large clusters of virulence genes called pathogenicity islands.

In order to adhere to epithelium, enteropathogens express a large variety of cell-surface adhesins, including polymeric fimbriae and afimbrial adhesins. Some pathogens such as *Yersinia* spp. and *Listeria* spp. can harness host transmembrane cell-adhesion receptors and trigger intracellular cascades that lead to epithelial cell invasion. Other invasive bacteria, such as *Shigella* and *Salmonella*, bypass the first step of adhesion and use their T3SS to inject bacterial effectors and induce the formation of a macropinocytic pocket.

Following adhesion, enteropathogens use their T3SS to inject a wide spectrum of bacterial effectors which interfere with epithelial cell functions. Some effectors provoke massive cytoskeleton changes that reinforce adherence to the epithelial surface, or promote bacterial entry or cell-to-cell dissemination. Other T3SS-injected effectors influence host-cell survival or, together with toxins released in the intestinal lumen, alter epithelial transport functions, inducing diarrhea, which allows bacteria dissemination.

Finally bacteria, both resident and pathogenic, can regulate intracellular signaling cascades that control inflammation and autophagy. The distribution and regulation of PRRs in epithelial cells avoid their excessive activation by resident bacteria, so that their activation by the commensal bacteria results predominantly in the production of microbicidal peptides and/or of repair factors that preserve intestinal homeostasis. Enteropathogens are more easily recognized by epithelial PRRs and can induce strong inflammatory reactions which facilitate their elimination. Yet, they have evolved multiple mechanisms to downmodulate inflammatory cascades and/or autophagy, helping their escape from host mechanisms of defense.

Further Reading.

Abreu, M.T.: **Toll-like receptor signalling in the intestinal epithelium: how bacterial recognition shapes intestinal function.** *Nat. Rev. Immunol.* 2010, **10**:131–144.

Carvalho, F.A., Barnich, N., Sivignon, A., *et al.*: **Crohn's disease adherent-invasive *Escherichia coli* colonize and induce strong gut inflammation in transgenic mice expressing human CEACAM.** *J. Exp. Med.* 2009, **206**:2179–2189.

Cerf-Bensussan, N., and Gaboriau-Routhiau, V.: **The immune system and the gut microbiota: friends or foes.** *Nat. Rev. Immunol.* 2010, **10**:735–744.

Chassaing, B., and Darfeuille-Michaud, A.: **The commensal microbiota and enteropathogens in the pathogenesis of inflammatory bowel diseases.** *Gastroenterology.* 2011,**140**:1720–1728.

Cossart, P., and Sansonetti, P.J.: **Bacterial invasion: the paradigms of enteroinvasive pathogens.** *Science* 2004, **304**:242–248.

Croxen, M.A., and Finlay, B.B.: **Molecular mechanisms of *Escherichia coli* pathogenicity.** *Nat. Rev. Microbiol.* 2010, **8**:26–38.

Goodman, A.L., Kallstrom, G., Faith, J.J., *et al.*: **Extensive personal human gut microbiota culture collections characterized and manipulated in gnotobiotic mice.** *Proc. Natl Acad. Sci. USA* 2011, **108**:6252–6257.

Kim, M., Ashida, H., Ogawa, M., *et al.*: **Bacterial interactions with the host epithelium.** *Cell Host Microbe* 2010, **8**:20–35.

Pizarro-Cerda, J., and Cossart, P.: **Bacterial adhesion and entry into host cells.** *Cell* 2006, **124**:715–727.

Sansonetti, P.J., and Di Santo, J.P.: **Debugging how bacteria manipulate the immune response.** *Immunity* 2007, **26**:149–161.

Shames, S.R., Auweter, S.D., and Finlay, B.B.: **Co-evolution and exploitation of host cell signaling pathways by bacterial pathogens.** *Int. J. Biochem. Cell Biol.* 2009, **41**:380–389.

Sun, J.: **Pathogenic bacterial proteins and their anti-inflammatory effects in the eukaryotic host.** *Antiinflamm. Antiallergy Agents Med. Chem.* 2009, **8**:214–227.

Travassos, L.H., Carneiro, L.A., Girardin, S., *et al.*: **NOD proteins link bacterial sensing and autophagy.** *Autophagy* 2010, **6**:409–411.

Viswanathan, V.K., Hodges, K., and Hecht, G.: **Enteric infection meets intestinal function: how bacterial pathogens cause diarrhoea.** *Nat. Rev. Microbiol.* 2009, **7**:110–119.

Wells, J.M., Loonen, L.M., and Karczewski, J.M.: **The role of innate signaling in the homeostasis of tolerance and immunity in the intestine.** *Int. J. Med. Microbiol.* 2010, **300**:41–48.

Helicobacter pylori Infection

24

Helicobacter pylori is a spiral-shaped Gram-negative bacterial pathogen of the stomach that invariably induces gastric inflammation. A signature feature of the organism is its capacity to persist for decades in spite of strong innate and adaptive host responses. Sustained interaction between *H. pylori* and gastric mucosa causes chronic inflammation and increases the risk for peptic ulcer disease, atrophic gastritis, intestinal metaplasia, and distal gastric adenocarcinoma. Importantly, *H. pylori* colonization is the strongest identified risk factor for malignancies that arise within the gastric mucosa, leading to an estimated 700,000 cancer deaths per year worldwide. However, only 10–15% of colonized persons develop disease, as disease risk involves specific and well-choreographed interactions between pathogen and host. Gastric inflammation due to *H. pylori* infection can follow different patterns with specific disease outcomes. In subjects with inflammation of the (more distal) gastric antrum, overproduction of the gastric hormone gastrin (hypergastrinemia) stimulates acid production by parietal cells in the noninflamed gastric body, leading to enhanced acid production that increases the risk for duodenal ulceration. In subjects with inflammation of the gastric body (corpus-predominant gastritis) or with pan-gastritis, acid production is reduced despite hypergastrinemia, likely due to inflammatory destruction of parietal cells. Chronic inflammation, through direct and indirect effects of *H. pylori*, causes a fundamental remodeling of the gastric mucosa that involves loss of gastric glands (atrophy) and intestinal metaplasia with a further decrease in gastric acid production. Atrophy and metaplasia are pre-malignant conditions that can progress to intestinal-type gastric adenocarcinoma.

To establish persistence and disease, *H. pylori* has evolved multiple mechanisms to both evade and manipulate the immune response (Table 24.1). In this chapter, we discuss the mechanisms that permit *H. pylori* to persist and the host immune responses that attempt to limit *H. pylori* colonization and the associated inflammation.

Table 24.1 *H. pylori* adaptations that promote colonization

H. pylori adaptation	Host function evaded
Anergic LPS and flagella	Innate immune activation
Decreased production of heat-shock proteins	Innate immune activation
Residence in mucus layer	Host recognition
Membrane engraftment of cholesterol	Phagocytosis
VacA-induced T-cell suppression	Adaptive immunity

LPS, lipopolysaccharide.

H. pylori as commensal or pathogen.

Although the majority (80–85%) of *H. pylori* infections are asymptomatic, *H. pylori* colonization is an important risk factor for peptic ulcer disease and gastric cancer. Increased disease risk appears to be the consequence of a combination of bacterial virulence factor genotypes, host genetic factors such as polymorphisms for interleukin (IL)-1β and other pro-inflammatory genes, and environmental factors. The following characteristics of *H. pylori* are consistent with its classification as a commensal rather than as a pathogen.

24-1 *H. pylori* has coevolved with humans and persists for life.

Commensal bacteria coevolve with their hosts, which undergo changes in their habitats, nutrition, and behavior that, in turn, influence the microenvironment occupied by the commensals. Phylogeographic studies of human populations and their colonizing *H. pylori* strains have revealed that *H. pylori* has infected humans for at least 60,000 years and that the bacteria accompanied anatomically modern humans on their migration out of Africa. Interestingly, *H. pylori* also has a high short-term mutation rate, and new strains continue to develop through 'microevolution' within families and individuals even today.

Commensal bacteria typically colonize mucosal epithelial surfaces, are acquired early in life, and persist for the life of the host, all features of *H. pylori*. *H. pylori* is the dominant species that colonizes the human stomach and is endemic in certain populations, reflected in colonization rates of >80% in developing countries. In all other mammals, species-specific *Helicobacter* bacteria similarly colonize the gastric niche without causing disease. Only recently, with the development of hygienic 'Western' living conditions and the widespread use of antibiotics, has *H. pylori* become less common in humans.

24-2 Bacterial and host factors can prevent disease.

Both host and bacterium have developed strategies to minimize the gastric inflammatory response to *H. pylori*. Mechanisms used by *H. pylori* to escape the host immune response include the expression of lipopolysaccharide (LPS) O antigens that closely resemble human Lewis blood group antigens ('molecular mimicry'), an LPS lipid A that has reduced pro-inflammatory capacity, and flagella that are not recognized by Toll-like receptor 5 (TLR5). Also, *H. pylori* is usually noninvasive, residing predominantly in the gastric mucus layer, which limits contact with the epithelium. The host immune system responds to *H. pylori* with an influx of regulatory T (T_{reg}) cells that downmodulate inflammation and ulceration, particularly in infected children, in parallel with the induction of other adaptive responses. Additionally, *H. pylori* virulence factors may suppress certain immune cell functions.

Epidemiological data suggest that gastric colonization with *H. pylori* may have some beneficial effects for the host. For example, *H. pylori* colonization is associated with a decreased risk for childhood-onset asthma and possibly other allergies. Also, *H. pylori* colonization has been associated with partial protection from diarrheal diseases in children. In adults, chronic *H. pylori* infection may protect against Barrett's esophagus and other esophageal diseases. The mechanisms for these interactions, however, have not been elucidated.

H. pylori virulence factors.

A remarkably high level of genetic variability characterizes *H. pylori* isolates. Despite such variability, common virulence factors have evolved and include flagellin, urease, VacA, CagA, and outer membrane proteins. These bacterial components play a key role in *H. pylori* persistence, the host immune response to the bacteria, and the associated inflammatory disease in the gastric mucosa.

24-3   Flagellin mediates bacterial motility in gastric mucus.

H. pylori possesses several unipolar, sheathed flagella containing FlaA and FlaB proteins that provide *H. pylori* motility. This motility is necessary for bacterial colonization and persistence in the gastric mucus. Consequently, aflagellated mutant bacteria display a reduced ability to colonize the gastric mucosa of gnotobiotic piglets. Interestingly, in contrast to flagellin expressed by intestinal pathogens such as *Salmonella*, *H. pylori* flagellin does not bind to, or activate, TLR5, thereby promoting immune evasion.

24-4   Urease enables bacterial survival and adhesion.

Urease is a major *H. pylori* virulence factor expressed by all strains. Urease enables bacterial survival within the acidic environment of the gastric lumen by hydrolyzing urea into CO_2 and ammonia (NH_3), which neutralizes gastric acid. At the epithelial surface, the released ammonia disrupts tight junctions, thereby damaging the epithelial barrier. In addition to its enzymatic function, urease also acts as an adhesin for the bacteria and is essential to establish bacterial colonization. Also, urease can directly damage host cells, especially antigen-presenting and epithelial cells, by triggering apoptosis through a major histocompatibility complex (MHC) class II-dependent mechanism. As mammalian cells do not produce urease, the presence of this enzyme in the human stomach is highly correlated with active *H. pylori* infection in human patients, and a number of common diagnostic tests are based on the detection of urease.

24-5   *vacA* encodes a cytotoxin that mediates long-term persistence.

The *H. pylori* gene *vacA* encodes a secreted protein (VacA) that induces vacuolation in cultured epithelial cells, and this locus is linked to the induction of gastric malignancy. All *H. pylori* strains possess *vacA*, but *vacA* sequences vary markedly among strains. VacA is secreted and undergoes proteolysis to yield two fragments, p33 and p55, which represent VacA functional domains. The p33 domain contains a hydrophobic sequence involved in pore formation, whereas the p55 fragment contains cell-binding domains. VacA binds multiple epithelial cell-surface components, including the transmembrane protein tyrosine phosphatase PTPRZ1, fibronectin, the epidermal growth factor receptor, various lipids, and sphingomyelin, and CD18 (integrin β_2) on T cells. VacA also has stimulatory effects, including inducing mast cells to produce tumor necrosis factor-α (TNF-α) and IL-6 and inducing macrophages and neutrophils to express the pro-inflammatory enzyme cyclooxygenase-2.

In addition to inducing vacuolation and stimulatory effects, VacA exerts suppressive effects on the local immune response. Purified VacA inhibits processing and presentation of antigenic peptides to human $CD4^+$ T cells and can specifically block antigen-dependent T-cell proliferation by interfering with IL-2-mediated signaling. *H. pylori* co-opts CD18 β_2 integrin as a VacA receptor on human T lymphocytes. After entering the cell, VacA inhibits Ca^{2+} mobilization and downregulates Ca^{2+}-dependent phosphatase calcineurin activity, which in turn inhibits activation of the transcription factor called nuclear factor of activated T cells (NFAT) and thus expression of NFAT target genes such as IL-2 and IL-2 receptor. Additionally, VacA can suppress IL-2-induced cell cycle progression and proliferation of primary T cells in an NFAT-independent manner. Thus, VacA can inhibit the clonal expansion of T cells that have been activated by bacterial antigens, thereby promoting *H. pylori* evasion of the adaptive immune response.

24-6 The *H. pylori cag* pathogenicity island is a strain-specific locus that enhances the risk for disease.

The *H. pylori cag* pathogenicity island is an important determinant that influences disease. *cag*⁺ strains are more strongly associated with distal gastric cancer than *cag*⁻ strains. Genes within the *cag* island encode proteins that form a type IV bacterial secretion system (T4SS) that injects microbial proteins into host cells. After bacterial attachment, the protein product (CagA) of the terminal gene in the island is translocated via the T4SS into host epithelial cells, where CagA is phosphorylated.

Integrin receptors, particularly $\alpha_5\beta_1$ integrins, represent an entry point for the injection of CagA. CagL, a T4SS pilus-localized protein, bridges the T4SS to $\alpha_5\beta_1$ integrin on target cells and activates host cell focal adhesion kinase and Src to insure that CagA is phosphorylated directly at its site of injection. The *cag* proteins CagA, CagI, and CagY also bind β_1 integrin and induce conformational changes in integrin heterodimers, which permits CagA translocation.

After injection into epithelial cells, CagA undergoes targeted tyrosine phosphorylation by Src and Abl kinases at motifs that contain the amino acid sequence EPIYA (Figure 24.1a). Phospho-CagA activates a cellular phosphatase (SHP-2), leading to morphological aberrations such as cell scattering and elongation. Phospho-CagA can inhibit Src via recruitment of Csk, a negative regulator of Src, thus generating a negative-feedback loop that carefully controls the amount of intracellular phosphorylated CagA. However, non-phosphorylated CagA also exerts effects within the cell. Transgenic expression of CagA in mice leads to gastric epithelial cell proliferation and the development of carcinoma in the absence of inflammation, implicating CagA as a bacterial oncoprotein.

Figure 24.1 Epithelial morphogenic alterations induced by intracellular CagA. The *H. pylori cag* island in panel a encodes a type IV secretion system that protrudes from the bacterial surface and is induced upon contact with host cells. The *cag* pilus is covered on its surface by CagL, which binds to $\alpha_5\beta_1$ integrins on host cells to facilitate translocation of CagA. Tyrosine phosphorylation of EPIYA sites within the C-terminus of CagA leads to alterations in host epithelial cells, such as cell scattering and elongation. Variation in the number and sequence of these sites determines the degree of CagA phosphorylation and the intensity of cellular changes. CagA is phosphorylated by Src and Abl kinases, which activates SHP-2 and leads to changes in cell morphology. In panel b, non-phosphorylated CagA binds and inactivates Par1b, leading to dysregulated apical junctional complexes. CagA–Par1b complexes are tethered to the cell membrane by phosphatidylserine.

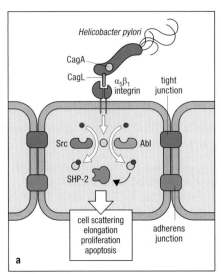
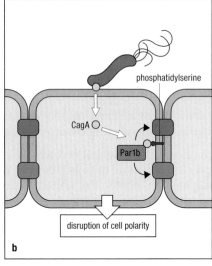

Unmodified CagA binds to the epithelial tight-junction scaffolding protein ZO-1 and the transmembrane protein junctional adhesion molecule-A to cause ineffective assembly of tight junctions at sites of bacterial attachment. CagA directly binds Par1b, a central regulator of cell polarity, and inhibits its kinase activity, an interaction that promotes loss of cell polarity (Figure 24.1b). Direct contact between *H. pylori* and epithelial cells induces externalization of the membrane phospholipid phosphatidylserine, which subsequently interacts with CagA via a conserved motif in the mid-region of the CagA protein. After translocation into the cell, CagA engages the cytoplasmic component of phosphatidylserine and binds Par1b, leading to junctional and polarity defects.

An important function of the *cag* secretion system is the delivery of *H. pylori* peptidoglycan into host cells. In the host cell cytoplasm peptidoglycans are sensed by NOD1, an intracytoplasmic pathogen-recognition molecule (Figure 24.2). NOD1 activation by *H. pylori* peptidoglycan stimulates NFκB, p38, and ERK, inducing expression of IL-8 and macrophage inflammatory protein-2 (MIP-2), chemokines that recruit neutrophils and monocytes to the gastric mucosa. Activation of NOD1 also leads to the production of type I interferon (IFN), which drives helper T (T_H) T_H 1-type inflammation. Thus, *cag*+ strains activate multiple signaling pathways involved in the regulation of inflammation and oncogenic cellular responses that may increase the risk for cell transformation during prolonged *H. pylori* colonization.

24-7 Additional *H. pylori* determinants influence disease pathogenesis.

The *H. pylori* genome contains an unusually high proportion of open reading frames that encode outer membrane proteins (OMPs). Consistent with sequence data, *H. pylori* strains express multiple paralogous OMPs, several of which bind to defined receptors on gastric epithelial cells but differ in expression and binding properties.

BabA, a highly conserved OMP encoded by the strain-specific gene *babA2*, is an adhesin that binds the Lewis histo-blood-group antigen Le^b on gastric epithelial cells (Figure 24.3). Infection by *babA2*+ strains is associated with an

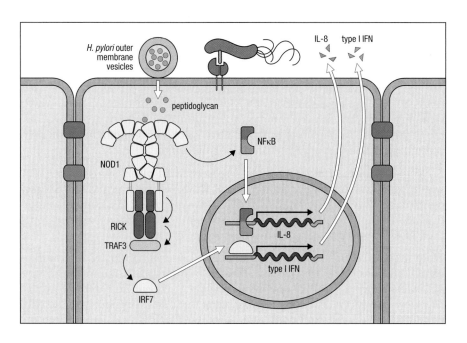

Figure 24.2 *H. pylori* delivery of peptidoglycan. After adherence, *H. pylori* can translocate components of peptidoglycan into the host cell. Peptidoglycan components can also be delivered into host cells via outer membrane vesicles. Peptidoglycan is sensed by the intracellular receptor NOD1, which activates nuclear factor-κB (NFκB) and IRF7, inducing release of pro-inflammatory cytokines.

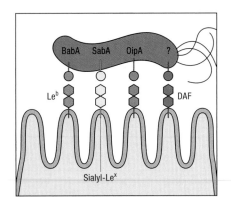

Figure 24.3 Interactions between pathogenic *H. pylori* and gastric epithelial cells. Several adhesins such as BabA, SabA, and OipA mediate binding of *H. pylori* to gastric epithelial cells. BabA and SabA bind to their cognate receptors Le^b and sialyl-Le^x respectively. The cellular receptor that binds OipA has not been defined. *H. pylori* can also adhere to decay-accelerating factor (DAF), but the bacterial adhesin required for DAF binding has not been identified.

increased risk for gastric cancer. SabA is an *H. pylori* adhesin that binds the sialyl-Lewis^x (Le^x) antigen, an established tumor antigen and marker of gastric dysplasia that is upregulated by chronic gastric inflammation. The O antigen of *H. pylori* lipopolysaccharide contains various human Lewis antigens, including Le^x, Le^y, Le^a, and Le^b. Inactivation of Le^x- and Le^y-encoding genes prevents *H. pylori* from colonizing the gastric mucosa of mice. *H. pylori* Lewis antigens can undergo phase variation *in vitro*, and their expression pattern is directly influenced *in vivo* by the Lewis expression pattern of the host. In Le^b-expressing transgenic or wild-type Le^b-negative control mice challenged with *H. pylori* that express Le^x and Le^y, only bacterial populations recovered from Le^b-positive mice expressed Le^b. These findings suggest that Lewis antigens allow *H. pylori* to escape host immune defenses by preventing immune responses to shared bacterial and host epitopes.

OipA is a differentially expressed OMP that also has been linked to disease outcome. OipA regulates expression of pro-inflammatory cytokines, including IL-8, IL-6, and RANTES, that may participate in disease pathogenesis. OipA also mediates adherence of *H. pylori* to gastric epithelial cells via an unidentified host receptor and triggers β-catenin activation. *H. pylori* binds decay-accelerating factor (DAF), a protein that protects epithelial cells from complement-mediated lysis. *H. pylori* transcriptionally upregulates DAF, and loss of DAF in murine models attenuates the development of inflammation in *H. pylori*-infected gastric mucosa. Thus, direct contact between *H. pylori* and gastric epithelial cells is pivotal for the induction of chronic inflammation and injury.

H. pylori and the innate immune system.

The innate immune system plays an important role in determining the host response to *H. pylori*. The initial innate immune cell encountered by *H. pylori* is the gastric epithelial cell. Contact between *H. pylori* and epithelial cells *in vitro* dysregulates signaling pathways, and similar *H. pylori* epithelial interactions occur *in vivo* in infected human gastric tissue. Since these pathways influence inflammation and oncogenesis, many studies have focused on aberrant epithelial responses induced by *H. pylori*.

24-8 *H. pylori* interacts with epithelial Toll-like receptors.

Toll-like receptors (TLRs) are pattern recognition receptors that play an essential role in initiating innate immune responses to pathogenic microbes. TLRs recognize distinct highly conserved pathogen-associated molecular patterns (PAMPs). Recognition of PAMPs by TLRs leads to activation of intracellular signaling pathways that culminate in the induction of host defense genes, including those that encode inflammatory cytokines, chemokines, antigen-presenting molecules, and co-stimulatory molecules.

Early studies of the innate immune response to *H. pylori* focused on TLR4, which is involved in the recognition of LPS. However, gastric epithelial cell recognition of *H. pylori* is independent of TLR4, as *H. pylori* LPS contains modifications in the LPS lipid A core and consequently is an ineffective activator of the TLR4/CD14/MD-2 receptor-induced signal pathway.

TLR2 also is expressed by gastric epithelial cells. Transfection studies using cell lines, including MKN45 gastric epithelial cells, indicate that live *H. pylori* activate NFκB in epithelial cells via TLR2 but not TLR4. However, *H. pylori* also interacts with epithelial cells through intracellular delivery of peptidoglycan, a cell-wall component, by the T4SS. After injection into the cell, the peptidoglycan activates NOD1, leading to pro-inflammatory signal transduction, as discussed above.

TLR5, a receptor for flagellin, the monomer component of bacterial flagella, is present on gastric epithelium. Like TLR4, TLR5 appears not to participate in the induction of the host response to *H. pylori*. *H. pylori* flagellin contains specific amino acid substitutions within the TLR5 recognition site that render it nonstimulatory. Compared with *Salmonella* Typhimurium flagellin, *H. pylori* flagellin is 1000-fold less potent in activating TLR5-mediated IL-8 secretion by gastric epithelial cells. Moreover, the ability of aflagellated mutants of *H. pylori* to induce a pro-inflammatory response in HEK293 cells and gastric epithelial cells further indicates that TLR5 likely is not involved in epithelial cell recognition of *H. pylori*.

24-9 Neutrophil accumulation is a characteristic feature of *H. pylori*-infected mucosa.

A signature feature of *H. pylori* infection is the migration of neutrophils into the gastric mucosa. Neutrophil-activating protein, a virulence factor secreted by *H. pylori*, promotes neutrophil recruitment and induces production of reactive oxygen species (ROS) by neutrophils. The mechanisms that allow the bacterium to survive in the presence of neutrophils are poorly understood but may involve the inhibition of opsonization by gastric acidity and mucins. Complement activation is inhibited by *H. pylori* urease and altered expression of membrane inhibitors of complement, such as DAF and homologous restriction factor (CD59).

H. pylori utilizes several mechanisms to reduce phagocytosis by neutrophils. Experimental disruption of *cag* pathogenicity island genes enhances bacterial engulfment, implicating *cag* gene products in limiting phagocytosis. Delayed phagocytosis may occur due to *H. pylori* induction of a novel host signaling cascade driven by atypical protein kinase Cζ (PKCζ). Finally, glycosylation of cholesterol in the outer membrane of *H. pylori* increases the ability of the bacterium to evade phagocytosis (Table 24.2).

Despite its ability to limit phagocytosis, once engulfment of *H. pylori* has occurred, survival within the phagocyte is facilitated by the ability of the bacterium to disrupt the nicotinamide adenine dinucleotide phosphate (NADPH) oxidase system that synthesizes ROS. Neutrophils containing *H. pylori* produce substantial amounts of ROS, but ROS do not accumulate inside the phagosomes, and NADPH oxidase assembly within the phagosomes is inefficient. Instead, the NADPH oxidase system assembles on the polymorphonuclear neutrophil (PMN) cell surface and releases ROS into the extracellular space,

Table 24.2 Evasion of neutrophil-mediated killing by *H. pylori*

Steps in neutrophil activation	Mechanisms of evasion by *H. pylori*
Opsonization	Utilization of low pH, mucins Urease production Upregulation of DAF, CD59
Phagocytosis	*cag* gene products Activation of PKCζ Altered cholesterol composition in bacterial membrane
Microbial killing within phagosomes	Inhibition of NADPH oxidase Release of reactive oxygen species extracellularly

increasing local inflammation. Thus, *H. pylori* actively recruits PMNs to the gastric mucosa but modulates phagocytosis and diverts ROS formation from the phagosome, thereby promoting bacterial survival.

24-10 Macrophages may contribute to inflammatory and carcinogenic responses to *H. pylori*.

Macrophages reside in the gastric lamina propria in close proximity to *H. pylori*-derived products that may cross the epithelium. One such product, *H. pylori* neutrophil-activating protein, stimulates neutrophils and monocytes to release IL-12 and IL-23, cytokines that drive T_H1 and T_H17 cells. In the setting of disrupted gastric epithelium, *H. pylori* may induce local pro-inflammatory cytokine production by newly recruited monocytes, as suggested by the ability of *H. pylori* to stimulate monocyte-derived macrophages to release IL-1, TNF-α, and IL-6. Similarly to its interaction with neutrophils, *H. pylori* avoids effective phagocytosis by macrophages. After phagocytosis and internalization into a phagosome, VacA toxin prevents the fusion of the phagosome with lysosomes required for bacterial killing.

H. pylori interaction with gastric macrophages, as reported by some investigators, may upregulate inducible nitric oxide synthase (iNOS), leading to the generation of nitric oxide (NO). Macrophages can kill *H. pylori*, even when the macrophages are separated from the bacteria, likely by an NO-dependent mechanism. However, *H. pylori* arginase, encoded by the gene *rocF*, can compete efficiently with macrophages for the iNOS substrate L-arginine, impairing host NO production and leading to enhanced survival of the bacterium. Bacterial arginase generates urea from L-arginine, and urea is hydrolyzed to ammonia, which neutralizes gastric acid, as discussed above. *H. pylori* also can escape the macrophage phagocytosis via glycosylation of cholesterol acquired from epithelial cells, as demonstrated by *H. pylori* mutant strains that do not process cholesterol having an increased susceptibility to phagocytosis by macrophages and failing to colonize mice.

Exposure of macrophages to *H. pylori* upregulates the enzyme arginase II, leading to at least three potentially pathogenic effects (Figure 24.4). First, arginase depletes substrate availability for iNOS, causing reduced NO production.

Figure 24.4 Regulation of iNOS synthesis and NO production in macrophages. Alterations in host responses due to the effects of ornithine decarboxylase (ODC)-derived spermine on L-arginine uptake are depicted. CAT2 is the primary transporter of L-arginine into macrophages activated by *H. pylori*. L-arginine transported into macrophages facilitates translation of iNOS, which induces NO to kill extracellular *H. pylori*. L-lysine can block L-arginine uptake. Induction of ODC by *H. pylori* results in the generation of the polyamines putrescine, spermidine, and spermine. Spermine can be back-converted to spermidine or block L-arginine uptake by CAT2. Once in the cell, L-arginine is converted to L-ornithine, the substrate for ODC generation of more polyamines. Inhibition of ODC with α-difluoromethylornithine (DFMO) blocks the overproduction of polyamines, and thus restores iNOS synthesis, NO production, and antimicrobial defenses against *H. pylori*.

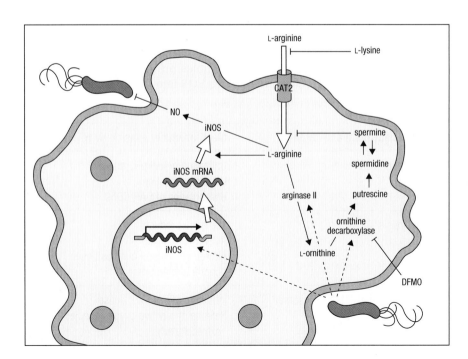

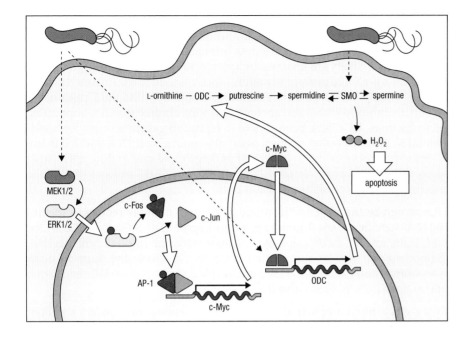

Figure 24.5 Model depicting mechanisms of apoptosis in *H. pylori*-stimulated macrophages. *H. pylori* activates ERK1/2 phosphorylation via MEK1/2, which leads to nuclear translocation of pERK, the induction of c-Fos and c-Jun, and the phosphorylation of c-Fos. pc-Fos and c-Jun form an AP-1 complex that binds to the *c-Myc* gene, leading to expression of c-Myc mRNA and protein. c-Myc in the cytoplasm translocates to the nucleus and enhances ODC transcription in concert with activation of a purported *H. pylori* response element. This induction of ODC leads to increased generation of the polyamines putrescine, spermidine, and spermine. In parallel, *H. pylori* upregulates the expression and activity of the enzyme spermine oxidase (SMO) that back-converts spermine to spermidine, a process that generates H_2O_2, which causes apoptosis through mitochondrial membrane depolarization.

Because NO is involved in the extracellular killing of *H. pylori*, reduced NO may contribute to enhanced colonization. Second, arginase II induces macrophage apoptosis through the metabolism of its product L-ornithine into polyamines such as spermine (Figure 24.5). Spermine then back-converts to spermidine by the enzyme spermine oxidase which is upregulated in *H. pylori*-treated macrophages, producing hydrogen peroxide (H_2O_2). In addition to promoting apoptosis, *H. pylori*-induced H_2O_2 metabolites, such as hydroxyl radicals, can cause DNA damage and thus contribute to the pathogenesis of gastric cancer in chronic *H. pylori* infection. Third, arginase II induction of ornithine results in increased substrate for the generation of polyamines by ornithine decarboxylase (ODC), which is also induced by *H. pylori*, resulting in inhibition of iNOS. Consequently, treatment of *H. pylori*-infected mice with an ODC inhibitor, α-difluoromethylornithine, results in marked increases in macrophage L-arginine uptake and NO production, and decreases in *H. pylori* colonization and the severity of gastritis. Further studies implicate the generation of polyamines in *H. pylori*-induced host-cell apoptosis (Figure 24.5). The production of putrescine by ODC results in the generation of spermidine and spermine by constitutive synthase enzymes.

H. pylori also upregulates the expression and nuclear translocation of c-Myc, and the binding of this transcription factor to the 5' untranslated region of the ODC promoter induces ODC transcription and associated apoptosis. This process is mediated by phosphorylation of ERK1/2, as nuclear translocation of pERK results in the formation of a unique activator protein-1 (AP-1) complex that consists of a heterodimer of phospho-c-Fos and c-Jun, which binds to the c-Myc promoter to activate c-Myc expression (Figure 24.5). In conjunction with activation of a putative *H. pylori* response element in the ODC promoter, c-Myc is a transcriptional enhancer for ODC. Importantly, inhibition of phosphorylation of c-Fos or activation of ERK results in prevention of *H. pylori*-induced macrophage apoptosis.

24-11 Dendritic cells initiate the immune response to *H. pylori*.

As antigen-presenting cells, dendritic cells (DCs) are innate immune cells that play an essential role in the induction of *H. pylori*-specific T-cell and B-cell responses, thereby bridging the innate and the adaptive response. Previously a matter of debate, human leukocyte antigen (HLA)-DR⁺ DCs have now been definitively identified in noninflamed human gastric mucosa, where they are distributed throughout the lamina propria, frequently adjacent to epithelial cells. Additionally, human gastric mucosa contains a subset of intraepithelial DCs, which likely are involved in luminal antigen sampling (Figure 24.6). Gastric DCs are more prevalent, mature, and activated in the inflamed lamina propria of *H. pylori*-infected subjects and also appear within the T-cell areas of tertiary lymphoid follicles that are formed during chronic gastritis. Supporting these findings, DCs have been shown to recruit to gastric mucosa in experimental *H. pylori* infection in mice. Isolated human gastric DCs exposed to live *H. pylori* phagocytose the bacteria and undergo rapid maturation, reflected in increased expression of surface CD11c, CD86, and CD83 and secretion of the pro-inflammatory cytokines IL-6 and IL-8.

DCs recognize *H. pylori* and its products through multiple receptors, including DC-specific ICAM-3-grabbing non-integrin (DC-SIGN), TLR2, TLR4, and TLR9, and *H. pylori* appears to induce signaling through both MyD88-dependent and MyD88-independent pathways. Additionally, *H. pylori* outer membrane proteins such as Omp18 and HpaA can stimulate DC maturation and antigen presentation capacity.

DCs are key initiator cells in the host immune response to *H. pylori*, and their cytokine secretion patterns determine the T-cell response. *H. pylori* infection is associated with a mixed T-cell response, with the T_H1 pathway dominant. Primary DCs isolated from human gastric mucosa induce T-cell secretion of interferon-γ (IFN-γ), but not IL-4 or IL-17 and only low levels of IL-10. T-cell IFN-γ secretion is generally driven by DC-derived IL-12. Although secretion of IL-12 by *H. pylori*-stimulated primary gastric DCs has not yet been shown,

Figure 24.6 Distribution of DCs in non-*H. pylori*-infected human gastric mucosa. Gastric mucosa contains intraepithelial and lamina propria HLA-DR⁺CD11c⁺ DCs, which are shown by schematic representation (a,b) and immunofluorescence (c,d) in longitudinal (a,c) and cross (b,d) sections of gastric mucosa. HLA-DR-expressing cells are green, CD11c-expressing cells red, and cell nuclei blue; HLA-DR/CD11c double-positive cells are yellow. (Panels c and d from D. Bimczok et al., *Mucosal Immunol.* 3:260–269, 2010. With permission from Macmillan Publishers Ltd.)

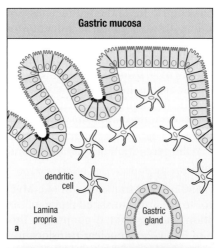
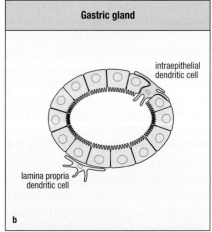
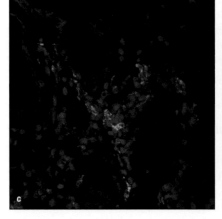
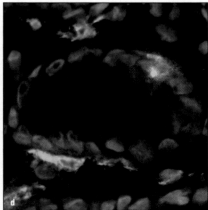

studies with human monocyte-derived DCs and murine bone marrow derived DCs indicate that *H. pylori* does induce DC IL-12 secretion. However, IL-12 secretion by DCs is tightly regulated and dependent on infection dose, time and length of stimulation, internalization of the bacteria, type of DCs analyzed, and bacterial factors. Moreover, different experimental conditions may generate T_H2, T_H17, or T_{reg} cells rather than T_H1 cells. Importantly, *H. pylori* expression of LPS Lewis antigen, which can be turned on or off in a process called phase variation, impacts the ensuing T-cell response. LPS Lewis-positive *H pylori* bind to DC-SIGN, which is expressed by 15–25% of human gastric DCs, and trigger DC IL-10 release, promoting a T_H2 response. In contrast, LPS Lewis-negative *H. pylori* induce strong DC IL-12 secretion and thus a T_H1 response.

Interesting murine studies have shown that the gastric response to *Helicobacter* can be modulated through interactions with DCs at extragastric sites. For example, intestinal colonization of mice with *H. bilis* (an intestinal *Helicobacter*) prior to gastric challenge with *H. pylori* results in attenuated severity of bacteria-associated gastritis, atrophy, metaplasia, and hyperplasia, and a dampened T_H1 response to *H. pylori* in co-infected animals. In addition, DCs in Peyer's patches phagocytose coccoid forms of *H. pylori*, leading to the priming of CD4$^+$ T cells that migrate to the stomach and initiate gastric inflammation.

Adaptive immune responses to *H. pylori*.

In addition to innate mucosal cell responses, adaptive immunity involving T and B cells characterizes the host response to *H. pylori*. The interdependent innate and adaptive immune responses to *H. pylori* are summarized in Figure 24.7. The T-cell response is a key determinant of the mucosal inflammatory response in the gastric mucosa of people infected with *H. pylori*.

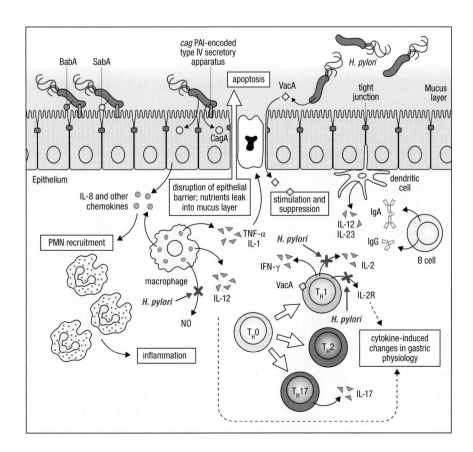

Figure 24.7 Gastric immune cell response to *H. pylori*. *H. pylori* binds to gastric epithelial cells through adhesins such as BabA and SabA. Utilizing a type IV secretory apparatus, the bacterium then inoculates CagA into the epithelial cell, leading to the release of the neutrophil chemokine IL-8 and other cytokines and alterations in epithelial cell integrity. The *H. pylori* product VacA induces apoptosis, local suppression of T-cell activation and proliferation, and together with other products may induce polymorphonuclear neutrophil (PMN) and macrophage pro-inflammatory responses, contributing to gastric inflammation and cytokine-induced changes in gastric physiology. PAI, pathogenicity island. (Adapted from D. Monack et al., *Nat. Rev. Microbiol.* 2:747–765, 2004. With permission from Macmillan Publishers Ltd.)

24-12 *H. pylori* induces a mixed T-cell response.

H. pylori induces a mixed CD4 T-cell response involving T_H1, T_H2, T_H17, and T_{reg} cells. Early studies in both mice and humans established that gastric lymphocyte populations in *H. pylori* infection contain increased numbers of IFN-γ-producing T cells, consistent with a T_H1 response. The number of IFN-γ-secreting cells in gastric mucosa correlates with the severity of gastritis. IFN-γ, which is also released by gastric CD8 T cells and natural killer (NK) cells, activates phagocytes, promoting inflammation. IFN-γ also induces gastrin, stimulating the secretion of gastric acid in early infection. However, chronic *H. pylori* gastritis causes gastric mucosal atrophy, a reduction in the parietal cell mass, and reduced acid secretion. Eventually, the T_H1-induced achlorhydria leads to reduced bacterial colonization.

H. pylori infection may induce increased numbers of T_H17 cells and higher gastric mucosal levels of IL-17 in some subjects. IL-17 secretion is triggered by *H. pylori*-induced DC release of IL-23, and increased levels of IL-23 have been reported in the gastric mucosa of some *H. pylori*-infected subjects. IL-17 recruits and activates neutrophils, which enhance inflammation but also may participate in bacteria clearance.

Although the gastric inflammation induced by *H. pylori* is predominantly a T_H1 response, the bacteria may also induce a strong T_H2 response. Such a response has been reported to ameliorate dyspetic symptoms, possibly through T_H2 secretion of anti-inflammatory IL-10.

During *H. pylori* infection, T_{reg} cells of the $CD4^+CD25^+Foxp3^+$ phenotype are more frequent in the gastric mucosa of *H. pylori*-infected subjects, particularly children. Both naturally occurring and inducible T_{reg} cells have been identified in the gastric mucosa in *H. pylori* infection. T_{reg} cells release transforming growth factor-β and IL-10, which have anti-inflammatory effects on macrophages and NK cells and inhibit proliferation of, and cytokine release by, other T cells. In mouse studies, depletion of T_{reg} cells increased gastritis and decreased *H. pylori* colonization and was associated with enhanced T_H1, T_H2, or T_H17 responses. In humans, an attenuated T_{reg} response is associated with an increased frequency of ulcer disease. In contrast, high numbers of T_{reg} cells correlate with increased gastric colonization with *H. pylori* and reduced pathology. Thus, the gastric T_{reg} response diminishes destructive inflammation but may promote *H. pylori* survival.

24-13 *H. pylori* infection induces a local and systemic antibody response that is not protective.

H. pylori infection initiates recruitment of IgA- and IgM-producing cells and induces strong local and systemic antibody responses. In chronic infection, lymphoid follicles that contain activated B cells and antibody-producing plasma cells form within the gastric mucosa. Remarkably, more than 300 antigenic *H. pylori* proteins have been identified, including urease, flagellin, CagA, HpaA, and various membrane proteins. However, the anti-*H. pylori* antibodies do not clear the bacteria.

Serum IgG anti-*H. pylori* antibodies can be detected within 21 days of initial infection. The IgG1 to IgG4 ratio of >1 is consistent with a T_H1 dominant response. In continuously infected hosts, *H. pylori* antibody responses remain stable. After eradication, however, the antibody titer typically declines by 20–50% within the first 9 months but may remain elevated for several years. Antibody titers do not differ between asymptomatic patients and those with duodenal ulcers. Importantly, *H. pylori* infection induces pathogenic autoantibodies in 20–30% of subjects, particularly to the gastric proton pump, which then contribute to achlorhydria or lead to autoimmune gastritis. Thus, antibodies to *H. pylori* do not offer protection and may actually be harmful.

24-14 *H. pylori* infection may lead to B-cell transformation.

B cells are important for the development of nonepithelial gastric malignancies associated with *H. pylori* infection. Individuals colonized with *H. pylori* have an increased risk of developing mucosa-associated lymphoid tissue (MALT) lymphoma and non-Hodgkin lymphoma of the stomach, lesions that are derived from transformed B cells. *H. pylori*-stimulated T-cell production of IL-2 supports the growth and proliferation of B lymphocytes in lymphomatous degeneration. Intriguingly, naive mouse splenocytes exposed to *H. pylori* are protected from spontaneous apoptosis and undergo proliferation in response to low, but not high, multiplicity of infection, and the responding cells are derived from the B-cell population. Furthermore, chronic gastric infection with *H. pylori* in mice protects splenic B cells from apoptosis, indicating a B-cell activation/survival phenotype that may have implications for MALT lymphoma. Although MALT lymphomas are rare in the United States, successful elimination of *H. pylori* leads to complete regression of these tumors in >80% of cases, indicating that removal of the bacterium can affect a clonal lesion.

Summary.

During thousands of years of coevolution with humans, *H. pylori* developed unique mechanisms for colonization and persistence in the human stomach. Once colonized, a combination of bacterial, host, and environmental factors promotes mucosal inflammation, the predisposing lesion for subsequent atrophic gastritis, leading in some subjects to intestinal metaplasia, dysplasia, and ultimately adenocarcinoma. The prevalence of *H. pylori* infection is declining in the United States and Europe, largely due to improved hygiene and frequent antibiotic usage. However, *H. pylori* infection is still the leading cause of gastroduodenal inflammatory disease in developed countries and the second leading cause of cancer-related mortality in many resource-poor nations. Further elucidation of the immunopathogenesis of *H. pylori* infection in the human gastric mucosa will provide new approaches to interdict the progression of infection-associated inflammation to gastroduodenal disease and gastric cancer.

Further Reading.

Amieva, M.R., Vogelmann, R., Covacci, A., *et al.*: **Disruption of the epithelial apical-junctional complex by *Helicobacter pylori* CagA.** *Science* 2003, **300**:1430–1434.

Asim, M., Chaturvedi, R., Hoge, S., *et al.*: ***Helicobacter pylori* induces ERK-dependent formation of a phospho-c-Fos c-Jun activator protein-1 complex that causes apoptosis in macrophages.** *J. Biol. Chem.* 2010, **285**:20343–20357.

Bimczok, D., Clements, R.H., Waites, K.B., *et al.*: **Human primary gastric dendritic cells induce a Th1 response to *H. pylori*.** *Mucosal Immunol.* 2010, **3**:260–269.

Chaturvedi, R., Asim, M., Hoge, S., *et al.*: **Polyamines impair immunity to *Helicobacter pylori* by inhibiting L-Arginine uptake required for nitric oxide production.** *Gastroenterology* 2010, **139**:1686–1698.

El-Omar, E.M., Carrington, M., Chow, W.H., *et al.*: **Interleukin-1 polymorphisms associated with increased risk of gastric cancer.** *Nature* 2000, **404**:398–402.

Gobert, A.P., McGee, D.J., Akhtar, M., *et al.*: **Helicobacter pylori arginase inhibits nitric oxide production by eukaryotic cells: a strategy for bacterial survival.** *Proc. Natl Acad. Sci. USA* 2001, **98**:13844–13849.

Kao, J.Y., Zhang, M., Miller, M.J., *et al.*: **Helicobacter pylori immune escape is mediated by dendritic cell-induced Treg skewing and Th17 suppression in mice.** *Gastroenterology* 2010, **138**:1046–1054.

Lewis, N.D., Asim, M., Barry, D.P., *et al.*: **Arginase II restricts host defense to Helicobacter pylori by attenuating inducible nitric oxide synthase translation in macrophages.** *J. Immunol.* 2010, **184**:2572–2582.

Necchi, V., Candusso, M.E., Tava, F., *et al.*: **Intracellular, intercellular, and stromal invasion of gastric mucosa, preneoplastic lesions, and cancer by Helicobacter pylori.** *Gastroenterology* 2007, **132**:1009–1023.

O'Brien, D.P., Israel, D.A., Krishna, U., *et al.*: **The role of decay-accelerating factor as a receptor for Helicobacter pylori and a mediator of gastric inflammation.** *J. Biol. Chem.* 2006, **281**:13317–13323.

Peek, R.M., Jr., Fiske, C., and Wilson, K.T.: **Role of innate immunity in Helicobacter pylori-induced gastric malignancy.** *Physiol. Rev.* 2010, **90**: 831–858.

Polk, D.B., and Peek, R.M., Jr.: **Helicobacter pylori: gastric cancer and beyond.** *Nat. Rev. Cancer* 2010, **10**:403–414.

Wilson, K.T., and Crabtree, J.E.: **Immunology of Helicobacter pylori: insights into the failure of the immune response and perspectives on vaccine studies.** *Gastroenterology* 2007, **133**:288–308.

Wroblewski, L.E., Peek, R.M., Jr., and Wilson, K.T.: **Helicobacter pylori and gastric cancer: factors that modulate disease risk.** *Clin. Microbiol. Rev.* 2010, **23**:713–739.

Xu, H., Chaturvedi, R., Cheng, Y., *et al.*: **Spermine oxidation induced by Helicobacter pylori results in apoptosis and DNA damage: implications for gastric carcinogenesis.** *Cancer Res.* 2004, **64**:8521–8525.

Viral Infections

25

Infection of mucosal surfaces by an array of viral pathogens is an important cause of morbidity, and occasionally mortality, in residents of both developed and developing countries. Among such viruses, certain RNA viruses have evolved unique properties by which they exploit mucosal surfaces to achieve or induce entry, survival, disease, and transmission to naive hosts. In this chapter, we discuss two RNA viruses: rotavirus, which infects infants, children, and adults, and human immunodeficiency virus-1 (HIV-1), which infects people of all ages but especially adults. We emphasize the unique bidirectional interactions between each virus and the mucosal immune system. Understanding the biology of rotavirus and interaction between the virus and the mucosal immune system has lead to the development of a highly effective vaccine, making rotavirus infection a vaccine-preventable disease. In contrast, the biology of HIV-1 and the interactions between HIV-1 and cells of the mucosal immune system are more complex and not fully elucidated, preventing the successful development of an effective vaccine.

Rotavirus infection.

Rotavirus infects nearly every child in the world and is the leading cause of life-threatening diarrheal disease among infants and young children in many countries (Table 25.1). Unlike many bacterial pathogens, rotavirus occurs with high frequency in both temperate and tropical climates and in both developed and less developed settings. Worldwide, severe rotavirus-induced gastroenteritis results in more than 2 million hospitalizations and 500,000 deaths annually, the latter primarily in developing countries, in children younger than 5 years of age. These grim statistics have stimulated efforts to develop rotavirus vaccines, some of which are licensed and being used now in young children.

Rotavirus causes an acute diarrheal disease, usually lasting about 5 days in immunocompetent individuals (Table 25.1). The clinical symptoms in young children include vomiting and diarrhea. The diarrheal stools do not contain blood, mucus, or inflammatory cells, hallmarks of inflammation that help distinguish viral from bacterial gastroenteritis. Rotavirus, which is environmentally highly stable, is shed in large numbers ($\sim10^{10}$ particles g^{-1}) in diarrheal stools. This explains why rotaviruses can be identified in the stools of ill children by electron microscopy, which requires approximately one million particles g^{-1} of stool for visualization, and why the virus is so infectious and easily spread by the fecal–oral route: as few as 10 particles can cause infection in a susceptible host. Asymptomatic infections frequently occur and multiple infections can lead to immunity. Treatment is supportive to alleviate the associated dehydration. Rotavirus disease has been significantly reduced by use of new, live attenuated vaccines.

Table 25.1 Clinical and epidemiologic features of rotavirus infection

Variable	Feature
Age predisposition	Primarily affects young children 6 months to 3 years old in developed countries; younger children affected in developing countries. Adults frequently infected, mostly asymptomatic
Seasonality	Seasonal infection in developed countries with epidemic peaks in cooler months of the year. No seasonality in tropical climates
Settings	Households, day care centers, hospitals, schools, favoring person-to-person spread
Asymptomatic infections	Common in adults and newborns. Can occur in all age groups
Incubation period	Generally 24–48 hours
Symptoms	Sudden onset of vomiting and diarrhea. Diarrheal stools do not contain blood, mucus, or leukocytes. Fever common
Severity of illness	Generally more severe than many other infectious diarrheal illnesses, leading to dehydration and hospitalization. Malnutrition increases disease severity
Duration of illness	Typically 3–5 days. Longer illness in immunocompromised
Virus shedding	Peaks 1–3 days after onset of illness. Shedding can be prolonged in immunocompromised individuals. Antigenemia and viremia detectable early after infection
Mode of transmission and vehicles	Fecal–oral; aerosol/vomitus; contact with fomites. Food, water, and environmental contamination may transmit infection
Immunity	Repeated infections with or without illness can occur with the same or different strains. Long-term immunity occurs with repeated exposure
Treatment	Supportive rehydration therapy to prevent dehydration. Live attenuated vaccines are available and highly effective in developed countries; vaccine effectiveness is less for children in developing countries
Reservoir	Humans and animal reservoirs

25-1   Rotavirus structure impacts pathogenicity in the intestinal mucosa.

Rotaviruses are members of the *Rotavirus* genus of the *Reoviridae* family of viruses, which contains viruses with segmented double-stranded RNA genomes. Rotavirus particles are large (1000 Å; Figure 25.1a) and complex with three concentric protein layers that surround the viral genome of 11 segments of double-stranded RNA (Figure 25.1b). The rotavirus genome segments code for six structural proteins that make up virus particles (viral protein or VP) and six nonstructural proteins (NSP; Figure 25.1b). The NSPs are synthesized in infected cells and function in some aspect of the viral replication cycle or interact with host proteins to influence pathogenesis and/or the immune response. Two rotavirus proteins, VP7, which makes up the outer capsid protein shell, and VP4, which forms spikes that emanate through the outer capsid shell, induce neutralizing antibody and are the basis of a binary classification system for viral serotypes (Figure 25.1c). Thus, VP7 (a glycoprotein or G-type antigen) and VP4 (a protease-sensitive protein or P-type antigen) are used to classify rotaviruses. To date, investigators have identified 19 G and 28 P types of rotavirus.

Rotaviruses exhibit unusual structural complexity and properties that are relevant to their success as gastrointestinal pathogens. Entry into, and replication in, intestinal epithelial cells is critical for this success (Figure 25.2). The stable outer proteins first bind to receptors on differentiated epithelial cells near the tips of the small intestinal villi. These mature intestinal epithelial cells express factors required for efficient infection and replication. The 60 spikes that protrude from the surface of the viral capsid act as initial attachment proteins to bind host receptors. The VP4 spike protein is susceptible to proteolytic cleavage, a common feature of attachment proteins of many viruses that infect mucosal surfaces. Proteolytic cleavage by trypsin at the mucosal surface induces a remarkable conformational change in the structure of the spike, exposing additional attachment sites on the spike and the surface glycoprotein of rotavirus for interaction with a series of host cell co-receptors. The multi-step attachment and entry process into intestinal epithelial cells results in virus delivery across the plasma membrane by endocytosis and into the cell where the outer capsid shell is removed and double-layered particles are delivered into the cytoplasm. These particles function as molecular machines and produce capped viral mRNAs that are extruded from transcribing particles into the cytoplasm to be translated to produce new proteins and replicated to produce new genomic RNA. The proteins in the core of the incoming particles possess all the enzymatic activities required to produce the viral transcripts from the viral genome dsRNA, because eukaryotic cells do not express RNA polymerases capable of transcribing mRNA from dsRNA templates.

25-2 Rotavirus replicates in intestinal epithelial cells.

Unique aspects of rotavirus replication include the incomplete uncoating of the virus, resulting in the absence of free dsRNA within the cell. Viral replication is restricted to the cell cytoplasm and occurs within specialized electron-dense structures called viroplasms localized adjacent to the cell nucleus and near the endoplasmic reticulum (Figure 25.2). Viroplasms are composed of nascent viral proteins, and their size and shape change during the replication cycle. Newly produced double-layered particles containing replicated dsRNA bud from viroplasms into the endoplasmic reticulum (ER) in an unusual process during which particles become transiently enveloped. The budding of particles into the ER is initiated by the binding of newly made double-layered particles to an intracellular viral receptor. This receptor consists of the cytoplasmic tail of a rotavirus nonstructural protein (NSP4) that is a transmembrane ER glycoprotein. The outer capsid proteins are incorporated into new particles during the budding process through protein rearrangements that occur as the transient envelope is lost. Mature viral particles are released from epithelial cells either by cell lysis or by delivery of particles to the apical plasma membrane of polarized cells by a nonclassical trafficking pathway. The epithelial cells die and are extruded from the villus surface, causing a potential, albeit transient, disruption in the epithelial barrier.

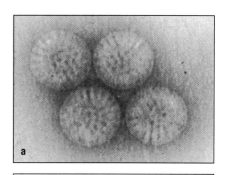

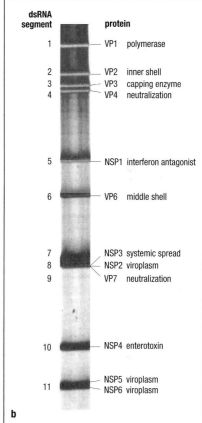

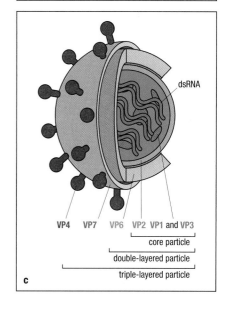

Figure 25.1 Rotavirus structure and proteins. Rotavirus particles are visualized in panel a by negative stain electron microscopy. Panel b shows the rotavirus genome of 11 segments of double-stranded RNA analyzed by polyacrylamide gel electrophoresis. Each gene encodes at least one protein with at least one major function. In panel c, a cartoon of a rotavirus particle shows the designated proteins that make up each concentric protein layer. (Adapted from H. Greenberg and M. Estes, *Gastroenterology* 136:1939–1951, 2009. With permission from Elsevier Ltd.)

Figure 25.2 Stages of the rotavirus life cycle. Interaction of rotavirus with epithelial cell receptors leads to rearrangement of the virus spike protein and virus entry and uncoating. Removal of the outer capsid proteins produces double-layered particles and activates the endogenous RNA-dependent RNA polymerase that produces mRNA transcripts from each dsRNA segment. The mRNAs are translated on ribosomes that are free or associated with the endoplasmic reticulum (ER). Some proteins associate to form viroplasms in which viral RNA is replicated and encapsidated into newly made particles. The nascent particles bind to the C-terminus of the NSP4, which acts as an intracellular receptor, and particles acquire the outer capsid proteins VP4 and VP7 while budding into the lumen of the ER. The transient membrane is lost and particles undergo maturation. Particles are released by cell lysis or from the apical surface of polarized cells by a Golgi-independent nonclassical vesicular transport mechanism.

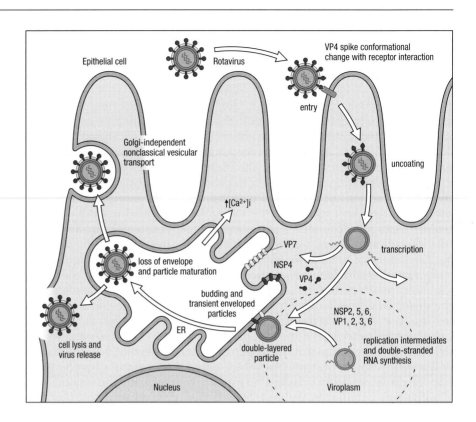

25-3   Rotavirus infection is calcium (Ca^{2+}) dependent.

Rotaviruses enhance and exploit Ca^{2+} signaling as a key cellular target to regulate replication, morphogenesis, and pathogenesis (Figure 25.3). Thus, rotavirus infection results in at least threefold increases in intracellular calcium [Ca^{2+}]i and up to tenfold increases in uptake of Ca^{2+} into cells. Ca^{2+} also plays an important role in virion assembly and disassembly processes. Ca^{2+} maintains the integrity of the rotavirus outer capsid layer; VP7 is a Ca^{2+} binding protein, and Ca^{2+} chelation is a mechanism that activates the endogenous RNA polymerase. NSP5 also is a Ca^{2+} binding protein, and viroplasm formation requires Ca^{2+}. Rotavirus morphogenesis is dependent on the presence of sufficient levels of [Ca^{2+}]i. In the absence of Ca^{2+}, virus morphogenesis is terminated at the double-layered particle step, and VP7 is excluded from heterooligomeric complexes made of NSP4 and VP4 that participate in the budding of double-layered particles into the ER. Furthermore, Ca^{2+} depletion of the ER by the sarco/endoplasmic reticulum Ca^{2+}-ATPase pump inhibitor thapsigargin inhibits VP7 and NSP4 glycosylation and virus maturation. NSP4 is the only rotavirus protein that mobilizes [Ca^{2+}]i in cells. Release of [Ca^{2+}]i from the ER alters plasma membrane permeability and compensatory entry of extracellular Ca^{2+} into cells. Changes in calcium homeostasis also alter other cellular functions that affect pathogenesis.

Rotavirus pathogenesis.

Rotavirus primarily infects small intestinal villus epithelial cells, sparing the less differentiated crypt cells. Infection typically resolves within 7 days. The outcome of infection is affected by both viral and host factors. The major host factor that affects the clinical outcome of infection is the age of the host. Thus, neonates may be infected with rotavirus, but they rarely have symptomatic

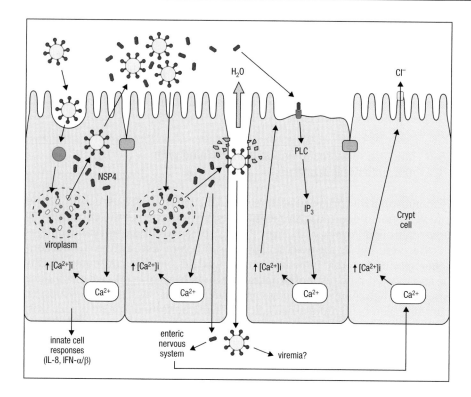

Figure 25.3 Mechanisms by which rotavirus causes diarrhea. Cellular and viral events that occur after rotavirus infection of enterocytes are shown in order from left to right. Rotavirus infection produces virus particles and viral proteins, such as NSP4, which are released by a nonclassical secretory pathway. Intracellular NSP4 also induces the release of Ca^{2+} from internal stores, primarily the endoplasmic reticulum, leading to increased levels of intracellular calcium, $[Ca^{2+}]i$. Innate intestinal epithelial cell responses to infection result in the release of pro-inflammatory cytokines. A cell is secondarily infected after rotavirus is released from the adjacent cell. Intracellular expression of NSP4 results in elevated levels of $[Ca^{2+}]i$ through a PLC-independent mechanism. The increase in $[Ca^{2+}]i$ also disrupts the microvillar cytoskeleton. Infected villus enterocytes may stimulate the enteric nervous system by the basolateral release of NSP4 or other effector molecules. NSP4 produced by the infection disrupts tight junctions, allowing paracellular flow of water and electrolytes (blue arrow). Viral particles may also gain access to the blood by this route. NSP4 released from previously infected cells binds to a specific receptor and triggers a signaling cascade through phospholipase C (PLC) and inositol 1,4,5-trisphosphate (IP_3) that results in release of Ca^{2+} and an increase in $[Ca^{2+}]i$. One receptor for NSP4 is $\alpha_2\beta_1$, and NSP4 binding to this receptor can elicit diarrhea in neonatal mice. A crypt cell can be acted on directly by NSP4 or NSP4 can stimulate the enteric nervous system, which in turn signals an increase in $[Ca^{2+}]i$ that induces Cl^- secretion. IL-8, interleukin-8; IFN, interferon. (Adapted from R. Ramig, *J. Virol.* 78:10213–10222, 2004. With permission from American Society for Microbiology.)

disease, likely due to the protection provided by transplacentally acquired maternal antibody. The waning of such antibody coincides with the age of maximum susceptibility of infants to severe rotavirus-induced disease (3 months to 2 years of age). As a consequence of acquired immunity, adults rarely experience severe symptomatic disease, although such illness might result from infections with an unusual virus strain or extremely high doses of virus.

25-4 Rotavirus virulence is mediated by five genes.

Rotavirus virulence is related to properties of the proteins encoded by a subset of the 11 viral genes. Virulence is multigenic and associated with the products of genes 3, 4, 5, 9, and 10 (Figure 25.1). The involvement of these genes in virulence is only partially understood. Gene 3 encodes the capping enzyme that affects levels of viral RNA replication, and genes 4 and 9 encode the outer capsid proteins required to initiate infection. Gene 5 encodes a protein (NSP1) that can function as an interferon antagonist and is discussed below (see rotavirus immunity). Gene 10 codes for the nonstructural protein NSP4, which regulates calcium homeostasis and virus replication and functions as an enterotoxin.

25-5 Diarrheal illness caused by rotavirus is multifactorial.

Diarrhea and vomiting are the main clinical manifestations of rotavirus infection in infants and young children. Disease pathogenesis is multifactorial, with malabsorption and diarrhea the consequences of virus-mediated destruction of absorptive enterocytes, downregulation of the expression of absorptive enzymes, and functional changes in tight junctions between enterocytes that lead to paracellular leakage (Figure 25.3). Our understanding of disease pathogenesis is based primarily on studies in animal models. A secretory component of rotavirus diarrhea is also recognized and is thought to be mediated by activation of the enteric nervous system and the effects of NSP4, the first identified virus-encoded enterotoxin. Studies of virus and NSP4 in cultured cells and animal models indicate that rotavirus-induced diarrhea results, in part, from activation of cellular Cl⁻ channels, inducing an increase in Cl⁻, and consequently water secretion (Figure 25.3). Cl⁻ secretion is induced by a mechanism that does not involve the cystic fibrosis transmembrane regulator, as rotavirus and NSP4 induce diarrhea in mouse pups lacking this channel and in children with cystic fibrosis. Villus ischemia and alterations in intestinal motility are recognized in some animal models, but their role in disease in children is poorly documented.

25-6 Mucosal rotavirus infection is associated with extraintestinal rotavirus.

Extraintestinal rotavirus can be detected in the blood for at least a short period after infection in immunocompetent hosts. Virus may enter the circulation by uptake through Peyer's patches or through breaches in the intestinal epithelium, accounting for the detection of virus in several tissues. The clinical consequences of systemic infection remain unclear, but one consequence may be enhanced innate responses, including the release of cytokines that induce fever. Although many case reports associate rotavirus with systemic illnesses, currently there is no proof of causation from extraintestinal spread of rotavirus.

Rotavirus immunity.

Innate and adaptive immune cells direct the host response to rotavirus in both primary infection and reinfection (Figure 25.4). Protection from reinfection and disease generally correlates positively with markers of mucosal immunity, including levels of anti-rotavirus intestinal IgA, enteric rotavirus-reactive antibody-secreting cells, and IgA memory cells.

Figure 25.4 Rotavirus interaction with intestinal cells. Rotavirus infects intestinal epithelial cells primarily at the tips of the villi. The virus enters the mucosa through several routes, including lamina propria dendritic cells, M cells located on Peyer's patches, and possibly breaches in the epithelium. The ensuing innate immune response helps control primary infection, facilitate rotavirus clearance, and regulate the adaptive immune response to the virus. During rotavirus infection, B cells are activated and proliferate. Protection from reinfection is due at least in part to rotavirus-specific antibody produced by IgA antibody-secreting cells that migrate back into the lamina propria. (Adapted from J. Niess and H. Reinecker, *Cell. Microbiol.* 8:558–564, 2006. With permission from John Wiley & Sons.)

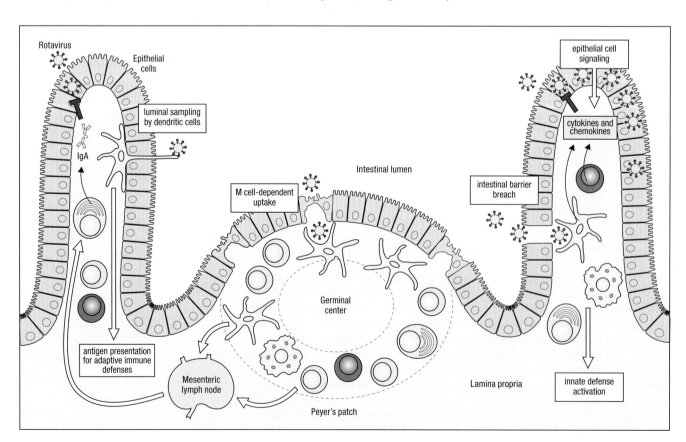

25-7   Multiple immune cells contribute to host responses to rotavirus.

CD8[+] T-cell responses contribute to rotavirus clearance and shortening the course of primary infection, whereas B-cell responses are the primary determinant of protection against rotavirus reinfection, as shown in mouse models. CD4[+] T cells provide help to CD8[+] T-cell and B-cell responses to rotavirus. After immunization with recombinant VP6, CD4[+] T cells appear to contribute to protection in an interferon-γ-dependent manner. Regulatory T cells appear not to affect the level of rotavirus replication and shedding or the severity of disease during primary infection in an animal model. Lymphocyte homing also plays a prominent role in the regulation of immune responses to rotavirus in the intestinal mucosa, and B-cell trafficking to the intestine appears critical for the induction of optimal protection in a mouse model of chronic infection.

Passive transfer studies of monoclonal antibodies in the mouse demonstrate that neutralizing antibody to either VP4 or VP7 can transfer either homotypic or heterotypic protection against rotavirus. Nonneutralizing IgA antibodies to the antigenically conserved VP6 protein contribute to protection, apparently via intracellular antiviral activity during viral transcytosis through intestinal epithelial cells. Rotavirus-induced diarrhea is also significantly reduced in mouse pups born to dams immunized with several forms of the enterotoxin NSP4. The role, if any, of these protective mechanisms in human immunity is unclear.

Innate immune responses may also participate in the control of rotavirus infection and the regulation of immune responses to the virus. Bacterial flagellin, which activates NFκB-mediated pro-inflammatory cytokine gene expression, provides temporary protection against rotavirus infection in mice. In a gnotobiotic porcine model, probiotic treatment with *Lactobacillus acidophilus* significantly enhances both B- and T-cell responses to attenuated live virus infection. However, the role of innate responses in the protection of the human small intestine remains to be determined.

25-8   Type I interferon responses contribute to rotavirus clearance.

The role of interferon-induced antiviral activity in rotavirus infection has been examined in *in vivo* and *in vitro* studies. Increasing evidence suggests that type I interferon (IFN) responses contribute to innate immune-mediated clearance of rotavirus. For example, elevated expression of a panel of genes related to type I IFN responses, as well as other pro-inflammatory genes, including interleukin (IL)-8, are detected in gene-profiling studies of rotavirus-infected intestinal epithelial cells. Pre-treatment of cultured cells with type I IFNs limits rotavirus infections. Levels of type I and II IFNs are elevated in rotavirus-infected children and animals, and administration of exogenous type I IFN reduces disease sequelae in cattle and pigs. Moreover, rotavirus NSP1 suppresses IFN signaling, and mutations in NSP1 that ablate rotavirus's ability to interfere with IFN-related signaling attenuate rotavirus's spread to uninfected cells, further supporting a role for IFN signaling as a potential hindrance to rotavirus infection. While mice deficient in type I and II IFN receptors are able to clear rotavirus, loss of the transcription factor STAT1, which mediates much of the gene expression induced by type I IFN, severely impairs the ability of the host to contain rotavirus.

Emerging evidence suggests a role for dendritic cells (DCs), particularly mucosal DCs, in the redundant host defense mechanisms against rotavirus. Human primary plasmacytoid DCs (pDCs), DCs in Peyer's patches in adult mice, as well as intestinal pDCs and conventional DCs from piglets are activated after exposure to rotavirus *in vitro* to produce both pro-inflammatory and in some cases anti-inflammatory cytokines. These responses are thought to play a critical role in controlling the infection and in limiting an excessive inflammatory response in the mucosa. DC responses to rotavirus likely regulate the subsequent adaptive immune response.

Rotavirus vaccines.

Remarkable progress has been made to reduce rotavirus-associated disease and deaths by the development of oral, live attenuated rotavirus vaccines. Two such vaccines are now licensed in many countries, and their use is increasing worldwide. These vaccines are the result of over 30 years of vaccine research and development that had initial success followed by unexpected complications.

25-9   The first available rotavirus vaccine was a live attenuated vaccine.

RotaShield® vaccine was the first live attenuated rotavirus vaccine to be licensed. Shortly after release, an increased relative risk (at least 25-fold) of intussusception within 10 days was reported to be associated with vaccination. The mechanism underlying this association is unknown but has been postulated to be specific to the rhesus rotavirus strain, because wild-type rotaviruses and other live attenuated vaccines have not been reproducibly associated with an increased rate of intussusception. Consequently, the RotaShield® vaccine was judged to be unsafe for routine use and withdrawn from commercial manufacture, resulting in high levels of rotavirus-induced morbidity and mortality for 7 years until two new rotavirus vaccine candidates became available. One of these vaccines is a pentavalent vaccine (Merck, RotaTeq®) that contains five separate viruses expressing human G1, 2, 3, or 4 VP7s and a human P[8] VP4 on a bovine WC3 backbone. The WC3 strain was initially studied as a stand-alone monovalent candidate (much like the earlier RIT 4237 vaccine) and was shown to be appropriately attenuated. However, clinical trials yielded varying efficacy rates, which led to the modification and inclusion of the various human G and P types. RotaTeq® is safe, not associated with intussusception, and is highly efficacious in developed countries. Efficacy rates do not appear to be affected by breast-feeding, and RotaTeq® administration does not interfere with immune responsiveness to other vaccines. RotaTeq® appears to be effective in preventing severe disease caused by a variety of rotavirus serotypes, including G9 strains, although a G9 component is not present in the actual vaccine. Additional evidence that serotype-specific immunity is not solely responsible for protection is the finding that type-specific neutralization rates after vaccination are much lower than the protection rates observed in clinical trials. Estimates from the Centers for Disease Control and Prevention indicate that vaccination is associated with a substantial delay in the annual onset of the rotavirus season and a greater than 50% decrease in the incidence of rotavirus infections. Thus, this vaccine may be able to reduce transmission and provide some level of herd as well as individual immunity.

25-10 Rotavirus is a vaccine-preventable enteric infection.

The second vaccine licensed in 2006 was a more traditionally prepared live attenuated vaccine. A virulent G1P[8] human rotavirus strain (89-12) was passaged multiple times in monkey kidney cell culture to acquire a suitable level of attenuation. This candidate was eventually licensed to GlaxoSmithKline and is now marketed under the trade name Rotarix®. The molecular basis for attenuation of this vaccine candidate is not known, although sequence analysis of the wild-type parent and the vaccine strain may eventually facilitate identification of the involved genes. Although a direct comparison between RotaTeq® and Rotarix® has not been performed, the human rotavirus vaccine candidate is apparently shed in substantially greater amounts than the bovine-derived vaccine, suggesting a higher likelihood of transmission from vaccinee to unvaccinated contacts. Better understanding of the genetic basis for attenuation and the degree of genetic stability of the vaccine following transmission will help resolve this issue. Despite the monovalent nature of the vaccine, it was 92% effective against homotypic G1 strains and 88% effective against heterotypic G3, 4, and 9 strains. At present, the efficacy of both vaccines is high with no reliable data to suggest substantial differences in efficacy between the two vaccines. Thus, rotavirus diarrhea is now a vaccine-preventable disease based on recent successful outcomes in vaccinated children in developed countries.

25-11 New rotavirus vaccines are under development.

Several third-generation rotavirus vaccines are currently being developed based on the following rationale. First, concern lingers about possible safety issues associated with the two licensed live virus vaccines. Consequently, several groups are pursuing inactivated or recombinant virus-like particle approaches. Parenteral immunization with inactive virus has proven effective in animal models, but no proof of principle for this approach exists for humans. Similarly, parenteral or intranasal immunization with recombinant, nonreplicating virus-like particle vaccines has been effective in all animal models tested, and these candidate vaccines are ready for phase I testing in humans. Second, rotavirus vaccines will never be fully affordable in the poorest countries until third-world manufacturers are able to compete with large pharmaceutical companies to produce an affordable vaccine. Thus, at least two third-generation vaccine candidates are currently being evaluated. A neonatal G9P[10] rotavirus isolated from a newborn nursery in New Delhi (strain 116E) will undergo pivotal phase III efficacy trials in India. This vaccine is produced by an Indian biotech company and was shown to be safe and highly immunogenic in initial phase I/II studies. A series of bovine reassortants with the UK bovine rotavirus strain as a backbone have been licensed by the National Institutes of Health to companies in China, India, and Brazil. Currently, this vaccine is being formulated as a tetravalent G1, 2, 3, 4 vaccine on the UK backbone. Earlier efficacy trials with this vaccine in Finland showed it to be non-reactogenic, highly immunogenic, and highly efficacious.

Several important basic and practical issues concerning rotavirus vaccine development have not yet been resolved, including whether the two currently licensed vaccines will be effective in very poor countries in Asia and Africa. Another important issue is whether the current restricted timing of administration of the first dose of these vaccines will limit their usefulness. Some children in the United States have not benefitted from rotavirus vaccination because the first dose needs to be administered by 2 months of age. Whether this timing restriction is suitable for infants in developing countries is unclear. Vaccine safety in children with immunodeficiencies also needs to be monitored, as chronic rotavirus infection has occurred in infants with severe combined immunodeficiency.

HIV-1 infection.

Mucosal surfaces of the genital and gastrointestinal tracts play fundamental roles in the pathogenesis of HIV-1 infection. First, these surfaces are the major route through which HIV-1 enters the body. Second, the intestinal mucosa is the site of the earliest and most intense viral replication, leading to profound local and systemic CD4$^+$ T-cell depletion. Third, as a consequence of compromised immunity, the mucosa is a site of disabling infections due to opportunistic viral, parasitic, bacterial, and fungal pathogens. Consequently, HIV-1 disease may be viewed as a mucosal infection with systemic manifestations. Accordingly, in this section of the chapter, we discuss the role of the mucosal surfaces of the genital and gastrointestinal tracts in HIV-1 pathogenesis.

25-12 Mucosal surfaces mediate HIV-1 entry.

Excluding infections acquired parenterally, virtually all HIV-1 infections are transmitted through a mucosal surface. In heterosexual transmission, HIV-1 enters the body through the genital tract mucosa, in male-to-male transmission through the rectal or upper gastrointestinal tract mucosa, and in mother-to-child (vertical) transmission through the upper gastrointestinal tract mucosa and placenta. Vertical transmission through the mucosa occurs when the fetus or infant ingests HIV-1-infected amniotic fluid *in utero*, infected blood or cervical secretions during delivery, or infected breast milk postpartum.

Worldwide, nearly 80% of the more than 30 million people infected with HIV-1 acquired the virus through heterosexual transmission. Transmission by this route occurs after cell-free and/or cell-associated HIV-1 in semen or secretions is inoculated onto recipient genital mucosa. Several factors contribute to HIV-1 transmission across genital, as well as gut, mucosal surfaces (Table 25.2). The donor is most infectious during primary HIV-1 infection and late-stage disease, when the level of virus in serum, and consequently semen and other body fluids, is highest. The level of virus in semen may be increased during mucosal inflammation or ulceration caused by sexually transmitted disease in the donor genital tract. Similarly, the recipient is more susceptible to HIV-1 when genital infection or inflammation is present, conditions that may increase the number and activation of local target mononuclear cells and their expression of HIV-1 receptors. Also, microtrauma to the vaginal mucosa may provide inoculated virus direct access to the microcirculation and lymphatic vessels of the subepithelial tissue. Microtrauma may be more injurious to single-cell columnar epithelium than multilayer squamous epithelium, and thus contribute to the higher HIV-1 transmission rate for rectal inoculation compared with genital inoculation. By extension, in simian immunodeficiency virus (SIV) infection of macaques, the clustering of SIV-infected cells in the endocervix (lined by columnar epithelium) and the transformation zone between the endocervix and the ectocervix (lined by squamous epithelium) may be due in part to the greater susceptibility of columnar epithelium to disruption.

Table 25.2 Factors that increase HIV-1 transmission across the mucosa

High viral load during primary and end-stage HIV-1 infection in the donor
Mucosal infection in the donor
Mucosal infection, trauma, inflammation, erosion, or ulceration in the recipient
Increased frequency of sexual contacts
Unprotected sexual contact
Receptive anal intercourse
Absence of circumcision in a male index partner

25-13 Cervico-vaginal mucosa provides both barrier function and pathways for HIV-1 entry.

Following inoculation onto the vaginal mucosa, HIV-1 first encounters cervico-vaginal mucus, which can trap both cell-associated and cell-free virus. Leukocytes that may carry HIV-1 are immobilized and then killed by normally acidic mucus; cell-free virus is trapped but not neutralized by acidic mucus, potentially limiting contact between inoculated virus and the underlying epithelium. In addition, cervico-vaginal secretions contain an array of antimicrobial factors with variable antiviral activity. Produced by local mucosal cells, these products include α- and β-defensins, lactoferrin, secretory leukocyte protease inhibitor, and macrophage inhibitory protein-3α. The next physical barrier encountered by HIV-1 is the non-keratinized squamous epithelium of the vagina and ectocervix in the lower female reproductive tract and the columnar epithelium of the uterus and endocervix in the upper tract.

The multiple layers of cells in the squamous epithelium of the vagina and ectocervix present a formidable barrier to HIV-1. However, the absence of tight junctions in squamous epithelium may permit diffusion of virions through gaps between cells in the loosely stratified upper epithelium. Vaginal and ectocervical epithelial cells express glycosphingolipids and heparan sulfate proteoglycans, adhesion molecules that can bind HIV-1, but, like columnar epithelial cells, squamous epithelial cells do not support viral replication. Importantly, in the absence of polarity, squamous epithelial cells also do not support transcytosis, the transcellular transport process characteristic of columnar epithelial cells. However, after diffusion through gaps between upper layer squamous cells, HIV-1 may encounter Langerhans cells and CD4+ lymphocytes, which support HIV-1 capture and productive infection respectively.

25-14 Vaginal Langerhans cells capture and internalize HIV-1.

The ability of Langerhans cells in the vaginal epithelium to capture and internalize HIV-1 has been conclusively demonstrated by electron microscopy. Langerhans cells are present in the upper layers of the epithelium and may extend processes between epithelial cells into the apical surface of the epithelium, positioning the cells to capture and internalize HIV-1 inoculated onto the mucosal surface (Figure 25.5). Langerhans cells express CD207 (langerin, a C-type lectin) and CD4 and CC-chemokine 5 (CCR5), the HIV-1 primary receptor and co-receptor, but blocking studies suggest that only CD4 and CCR5 are involved in HIV-1 endocytosis. Once captured, HIV-1-containing Langerhans cells can exit the upper epithelium in an *ex vivo* tissue system in which the underlying stroma has been removed. However, Langerhans cells do not migrate through explanted vaginal mucosa in which the subjacent stroma remains attached to the epithelium, and Langerhans cells have not been identified in the draining lymph nodes. Moreover, langerin, at least on skin Langerhans cells, mediates the binding and internalization of HIV-1

Figure 25.5 Pathways of HIV-1 entry in cervico-vaginal mucosa. HIV-1 can enter the ectocervix and vaginal tissue by multiple routes. In panel a, microtrauma provides cell-free (and cell-associated) virus direct access to target cells in the stratified squamous epithelium and, depending on the depth of the tear or disruption, access to the lamina propria and the microcirculation and lymphatic vessels. In panel b, the upper layer of loosely stratified epithelial cells allows cell-free virus to pass through gaps between the cells and penetrate into the epithelium, where cells such as DCs or Langerhans cells may take up the virus. In panel c, Langerhans cells in the epithelium bind and internalize HIV-1 via langerin, leading to virus degradation in Birbeck granules; dermal Langerhans cells migrate to lymph nodes but cervico-vaginal Langerhans cell migration is controversial. In panel d, DCs in the epithelium capture virus, migrate into the lamina propria, and *trans* infect target mononuclear cells, or continue to the draining lymph nodes, where the DCs also *trans* infect target cells. In panel e, HIV-1 infects epithelial CD4+ T cells in the epithelium through receptor-mediated fusion and the T cells become productively infected.

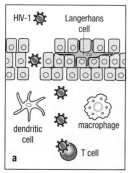

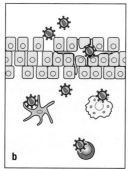

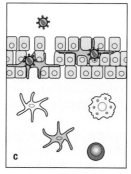

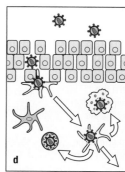

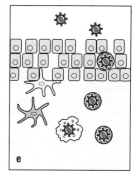

into Birbeck granules, which degrade the virus. Thus, dermal Langerhans cells inhibit T-cell infection by viral clearance through langerin-mediated internalization, but whether vaginal Langerhans cells also serve as a barrier to HIV-1 infection is unclear.

25-15 Vaginal CD4$^+$ T cells bind, take up, and support HIV-1 replication.

CD4$^+$ T cells in the vaginal epithelium also can take up HIV-1, likely by CD4/CCR5 receptor-mediated fusion, and support viral replication (Figure 25.5). In contrast, Langerhans cells do not support HIV-1 replication. Studies using epithelial sheets and fluorescence microscopy to detect green fluorescent protein-encoding pseudotyped HIV-1 indicate that HIV-1 can productively infect vaginal intraepithelial CD4$^+$ T cells independent of Langerhans cells. The HIV-1-infected T cells can emigrate from the outer layer of the epithelial sheets and likely can migrate and disseminate virus systemically. Thus, in the outer vaginal epithelium, CD4$^+$ T cells and Langerhans cells can take up or capture HIV-1, but of these two cell types, only the CD4$^+$ T cells support productive infection.

25-16 Cervico-vaginal dendritic cells (DCs) capture, disseminate, and *trans* infect mononuclear cells.

In contrast to Langerhans cells, few DCs are present in the outer layers of the epithelium (Figure 25.5). Rather, CD11c$^+$ DCs, which do not express langerin, are prominently distributed in the papillae, at the epithelial–stromal border and throughout the stroma. These myeloid DCs express high levels of human leukocyte antigen (HLA)-DR and moderate levels of C-type lectins, including intercellular adhesion molecule 3-grabbing non-integrin (DC-SIGN) and mannose receptor (CD206). The C-type lectins mediate DC binding to carbohydrates on immune cells and certain pathogens, critical functions by which DCs mediate HIV-1 *trans* infection of mononuclear cells. Vaginal DCs also express variable levels of CD4 and the co-receptors CCR5 and CXC-chemokine receptor 4 (CXCR4). In studies with explanted human cervix and vagina, DCs appear capable of transmitting HIV-1 to mononuclear cells by *trans* infection. Thus, in the vaginal mucosa, DCs in the deeper epithelial layers and lamina propria likely play a critical role in the uptake, dissemination, and possibly *trans* infection of HIV-1. Targets of DC *trans* infection in vaginal mucosa include resident lamina propria CD4$^+$ T cells and macrophages, in sharp contrast to the intestinal mucosa, where only lymphocytes support viral replication (see below). Nevertheless, much remains to be elucidated regarding the immunobiology of vaginal DCs in HIV-1 transmission, including whether the cells participate in the preferential selection of the R5 virus and genetically unique transmitted/founder virus that characterizes acute HIV-1 infection.

25-17 Cell-associated HIV-1 is transmitted across inner foreskin mucosa.

Female-to-male and male-to-male transmission of HIV-1 may occur after the inoculation of virus in cervico-vaginal and rectal secretions onto the male foreskin. Novel models of explanted foreskin epithelium and foreskin epithelium reconstructed from isolated primary epithelial cells have been used to elucidate the early events in this transmission pathway (Figure 25.6). Cell-associated, but not cell-free, virus is transmitted across the inner foreskin through the formation of viral synapses between the infected donor cell and apical keratinocytes. Synapse formation induces polarized budding of virus from the donor mononuclear cell and accumulation of Langerhans cells. After virus capture by the Langerhans cells, conjugates of Langerhans cells and T cells promote HIV-1 infection of the T cell. In contrast to the less keratinized inner foreskin, heavily keratinized outer foreskin traps virus within the keratin

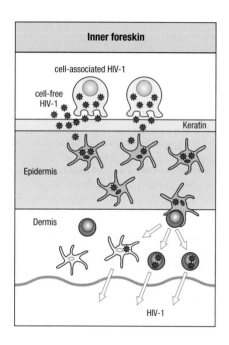

Figure 25.6 Cell-associated HIV-1 enters the inner, but not the outer, foreskin. The inner foreskin is the site of HIV-1 entry in female-to-male and some male-to-male transmissions. HIV-1-infected donor cells form viral synapses, resulting in viral budding and the concentration of virus in the synaptic cleft. Released virus can be captured, internalized, and degraded by Langerhans cells, or high levels of virus can induce Langerhans cell accumulation, HIV-1 capture, and migration to the epidermal-dermal junction, where the Langerhans cells can form conjugates with T cells and then transmit virus to the T cell. The conjugates may emigrate from the dermis or via productive T-cell infection transmit virus to DCs, which migrate to the draining lymph nodes. (Adapted from Y. Ganor et al., *Mucosal Immunol.* 3:506–521, 2010. With permission from Macmillan Publishers Ltd.)

layers, preventing transmission. These findings provide a potential explanation for the reduced transmission of HIV-1 associated with circumcision.

25-18 HIV-1 entry into gut mucosa is mediated by epithelial cells and DCs.

The small intestine, colon, and rectum are lined by columnar epithelium (Figure 25.7). The integrity of this single layer of cells may be altered in the small intestine by enteric pathogen infection or in the rectum by sexually transmitted disease or trauma associated with receptive anal intercourse. Tight junctions between epithelial cells promote polarity in which the cell is separated into apical and basal domains, the hallmark of columnar epithelium. Consequently, and in contrast to stratified squamous epithelium in the ectocervix and vagina, gut columnar epithelium is capable of transcytosis, the rapid, nondegradative process in which cargo endocytosed at the apical surface is transported in vesicles to the basolateral surface for release into the subepithelial space. In studies of cell line epithelial monolayers, the transcytosis of cell-associated HIV-1 is 100–1000-fold more efficient than the transcytosis of cell-free virus, likely due to the formation of viral synapses between the infected donor mononuclear cell and the epithelium, concentrating virus in the resultant pocket or cleft (Figure 25.7). The heparan sulfate proteoglycan agrin on epithelial cells participates in the formation of the synaptic cleft and also serves as an attachment molecule for a conserved region in the HIV-1 envelope glycoprotein gp41; attachment then promotes galactosylceramide-mediated endocytosis of the virus by the epithelial cells (see below). HIV-1 transcytosis across columnar epithelium in the intestine is relevant to vertical and genital–oral transmission and in the rectum to genital–rectal transmission.

Epithelial cell transcytosis of HIV-1 was first explored in epithelial cell lines. Subsequently, *in vitro* studies with primary human jejunal epithelial cells showed that intestinal epithelial cells express galactosylceramide, an alternative primary receptor for HIV-1, and CCR5, but not CXCR4; and that intestinal epithelial cells can efficiently transcytose infectious R5, but not X4, viruses for subsequent infection of CCR5+ cells. Importantly, the epithelial cells did not support productive HIV-1 infection. These findings suggest that human small intestinal epithelial cells may preferentially select R5 viruses for entry. In addition, *ex vivo* studies suggest that human rectal mucosa also transcytoses HIV-1 across the epithelium and that antibodies to the conserved membrane proximal external region of gp41 in the HIV-1 envelope can block transcytosis. Thus, antibodies to the virus, or perhaps the transcytosis machinery itself, could be developed to block this important entry step in HIV-1 transmission across gut mucosa.

In addition to the likely role of intestinal epithelial cells in HIV-1 entry, intestinal DCs have been shown in *ex vivo* studies to participate in HIV-1 entry (Figure 25.7). Intestinal CD11c+ DCs are distributed throughout the lamina

Figure 25.7 Pathways of HIV-1 entry through gut epithelium. Cell-associated and cell-free HIV-1 can cross gut columnar epithelium by four possible routes. In panel a, intestinal epithelial cells expressing galactosylceramide and CCR5 endocytose cell-associated or cell-free R5 HIV-1 and transcytose virus in vesicles to the basolateral surface, where fusion with the cell membrane leads to release of virus into the lamina propria. In panel b, lamina propria DCs with processes that extend between epithelial cells capture virus and migrate into the lamina propria or to the mesenteric lymph node where they *trans* infect mucosal or systemic CD4+ T cells respectively. In panel c, M cells translocate virus *in vitro* from the apical surface in vesicles by a nondegradative process, but this pathway has not been confirmed *in vivo*. In panel d, disrupted epithelium, in the small intestine due to enteric infection or in the rectum due to trauma or sexually transmitted disease, could provide HIV-1 direct access to lamina propria target cells and to the microcirculation and/or lymphatic vessels.

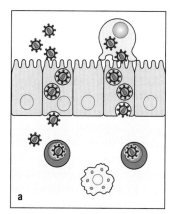

a

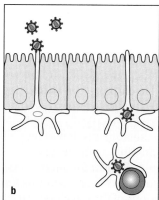

b

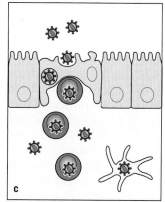

c

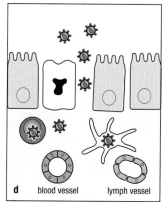

d   blood vessel   lymph vessel

propria and, as shown in immunohistochemical studies of normal human small intestine, extend processes between epithelial cells, a process that in mice is dependent on the activation of the myeloid differentiation primary-response gene 88 (MyD88) signal pathway. CD11c⁺ intestinal DCs, like vaginal DCs, express the C-type lectins DC-SIGN and moderate levels of CD206, potential attachment receptors for HIV-1. Importantly, approximately 30–50% of intestinal DCs are double positive for CD4/CCR5 and CD4/CXCR4, the receptors utilized by R5 and X4 viruses respectively for mononuclear cell entry. Monocyte-derived DCs express galactosylceramide, but whether primary intestinal DCs express the receptor is not known.

Primary human intestinal DCs rapidly take up HIV-1 in suspension, and, after inoculation onto explanted human jejunum, lamina propria DCs transport virus through the mucosa and *trans* infect intestinal and blood CD4⁺ T cells. However, whether DC processes capture the virus above the apical surface of the epithelium or capture virus transported by epithelial cells into the lamina propria has not been conclusively determined. Nevertheless, intestinal DCs are clearly capable of transporting HIV-1 through the intestinal mucosa and transmitting infectious virus to mucosal and blood target cells. In the intestinal mucosa, lamina propria CD4⁺ T cells are the exclusive target cell; lamina propria macrophages are nonpermissive to HIV-1 due to stromal downregulation of macrophage CD4/CCR5 expression and NFκB activation, requirements for cell entry and viral transcription respectively.

Another potential entry route for HIV-1 is through M (microfold) cells, which are present in the follicle-associated epithelium overlying Peyer's patches in the small intestine and isolated follicles in the rectum and tonsils (Figure 25.7). In contrast, genital mucosa does not contain inductive sites, and, consequently, M cells are not present in ectocervical or vaginal mucosa. M cells endocytose macromolecules and certain microorganisms at the apical membrane and then translocate the cargo in membrane-derived vesicles by a nondegradative, transcellular process to the basolateral cell surface for delivery to interdigitating DCs and lymphocytes (see Chapter 13). Rodent M cells can transport HIV-1, and model M cells created from Caco-2 cells and B cells transcytose HIV-1 in a rapid (<30 minutes), galactosylceramide/chemokine-dependent manner. Thus, M cell-mediated HIV-1 entry in human gut mucosa is theoretically possible but has not been established due to the inherent difficulty in tracking M cell transcytosis of HIV-1 in *in vivo* transmission studies.

25-19 Few HIV-1 virions are transmitted in acute HIV-1 infection.

Nearly all new HIV-1 infections are caused by CCR5-tropic (R5) viruses. In addition, approximately 80% of such infections are caused by a single virus (or virus-infected cell) and 20% by two to five viruses, which investigators have elegantly shown using single genome amplification of plasma virion RNA in combination with viral sequencing. Irrespective of the transmission route (male-to-male, male-to-female, female-to-male), transmitted/founder viruses productively infect CD4⁺ T cells but not monocyte-derived macrophages for as yet unclear reasons. The genotype- and R5 phenotype-restriction of *in vitro* studies show that transmitted/founder virus is strikingly different from that of genetically heterogeneous, CCR5-tropic and CXCR4-tropic virus populations that characterize chronically infected people. A mixture of such viruses is present in the body fluids of chronically infected people, and, consequently, a mixture of such viruses is typically inoculated onto the mucosal surface of a recipient. However, the transmission of only one or a few R5 viruses suggests that a 'bottleneck,' possibly at the level of the mucosae, restricts entry to the founder virus. The mechanism for such restriction has not been determined but may be linked to the level and composition of *N*-linked glycosylation on the V1–V3 variable loops of the envelope gp120.

T-cell depletion in early HIV-1 infection.

Within days of crossing the genital or gut epithelium, the single or few viruses that are transmitted infect target CD4+ T cells in the small intestinal lamina propria and viral replication is initiated (Figure 25.8). Because the gut mucosa contains the largest number of memory CD4+CCR5+ cells in the body, the intestinal lamina propria is the major site of viral replication and CD4+ T-cell depletion, regardless of whether the virus was acquired through genital or gut mucosa. The loss of this population of cells has profound immunological consequences for the host, as discussed next, and sets the stage for the subsequent acquisition of opportunistic infections, as discussed in conclusion.

25-20 HIV-1 causes rapid, profound, and prolonged CD4+ T-cell depletion in the intestinal mucosa.

HIV-1-induced depletion of gut CD4+ T cells is severe and persists through all stages of infection in the absence of antiretroviral therapy. The majority of activated memory CD4+CCR5+ T cells in the body are located in the intestinal lamina propria (<20% in blood and lymph nodes), and HIV-1 targets these cells, resulting in the profound depletion of CD4+ T cells. Gut mucosal T cells also express CXCR4, and CCR5-tropic HIV-1 is preferentially transmitted in acute infection. Thus, the gut mucosa plays a critical role in establishing HIV-1 infection and in the depletion of CD4+ T cells in humans and in SIV infection in rhesus macaques.

The rapid and severe CD4+ T-cell depletion in the gastrointestinal mucosa is due mainly to the HIV-1 cytopathic effect and partly to virus-induced bystander apoptosis. The loss of CD4+ T cells in gut mucosa coincides with a reciprocal increase in CD8+ T cells. Although major histocompatibility complex (MHC) class I-restricted cytotoxic T lymphocytes (CTLs) are present in the mucosa,

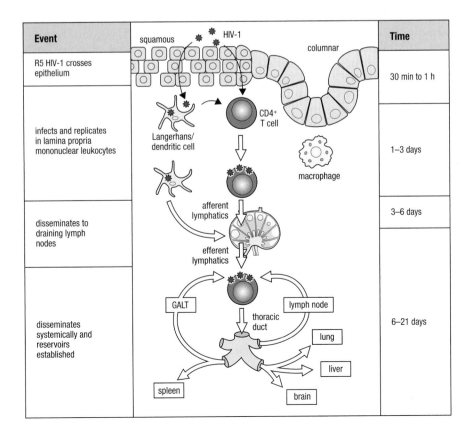

Figure 25.8 HIV-1 dissemination after mucosal entry. HIV-1 captured by DCs or Langerhans cells *trans* infects mucosal CD4+ T cells and possibly macrophages. The DCs also traffic via afferent lymphatic vessels to the draining lymph nodes to *trans* infect systemic target cells. Infected CD4+ T cells may support local viral replication or also enter the lymphatics and traffic to the lymph nodes, where massive HIV-1 replication occurs. Infected cells emigrate from the nodes, migrating through the efferent lymphatic vessels and thoracic duct to enter the circulation and disseminate HIV-1 to multiple tissues. GALT, gut-associated lymphoid tissue. (Adapted from M. Pope and A. Haase, *Nat. Med.* 9:847–852, 2003. With permission from Macmillan Publishers Ltd.)

they emerge too late to eradicate the HIV-1-infected cells and prevent the resultant massive CD4$^+$ T cell loss. Natural killer (NK) cells and $\gamma\delta$ T cells at mucosal surfaces may contribute to anti-HIV-1 defense, but with limited effect. Thus, intense viral replication and severe CD4$^+$ T cell loss progress relentlessly despite mucosal innate and subsequent adaptive antiviral responses.

The recently discovered helper T (T$_H$) T$_H$17 CD4$^+$ T-cell subset plays a critical role in host defense against bacterial and fungal pathogens at mucosal sites (see Chapter 7). T$_H$17 CD4$^+$ T cells are preferentially infected by HIV-1 (and SIV in macaques) and rapidly depleted due to the cytopathic effect of the virus. Linking gut HIV-1 infection with systemic manifestations of the disease, investigators have shown that the loss of mucosal T$_H$17 cells in SIV-infected macaques results in systemic dissemination of *Salmonella* Typhimurium from the gut in animals infected with this enteric pathogen. The T$_H$17 cells secrete IL-17, IL-6, tumor necrosis factor-α (TNF-α), IL-22, and granulocyte–macrophage colony-stimulating factor and promote the renewal of intestinal epithelial barrier by inducing ERK/MAPK-dependent synthesis of tight junction proteins and claudins. In addition, IL-17 in normal mucosa induces expression of antimicrobial defensins and chemokines in epithelial cells. Thus, loss of T$_H$17 cells likely contributes to an ineffective epithelial barrier and increased microbial translocation, enhancing systemic immune activation, a key feature of HIV-1 disease progression.

In contrast to the loss of memory CD4$^+$CCR5$^+$ and T$_H$17 T cells, the prevalence of regulatory T (T$_{reg}$) cells is increased in HIV-1 and SIV infections. T$_{reg}$ cells mediate suppressive function through tolerance to self and foreign antigens. Consequently, the increase in the number of T$_{reg}$ cells in gut mucosa may dampen adaptive immune responses and together with the decrease in the number of T$_H$17 cells contribute to local and systemic immune activation.

Highly active antiretroviral therapy (HAART) suppresses viral replication, resulting in restoration of the CD4$^+$ T-cell population in HIV-1-infected people. However, restoration of the CD4$^+$ T cells in gut mucosa is delayed compared with that of the peripheral blood compartment. This delay is likely due to incomplete suppression of viral replication in the mucosa despite suppressive activity of HAART. Residual HIV-1 in the mucosa limits local immune restoration and the eradication of viral reservoirs. In the SIV model, however, early initiation of antiretroviral therapy causes near complete restoration of gut mucosal CD4$^+$ T cells. Thus, early therapeutic intervention in HIV-1-infected subjects is necessary to restore effective immune function in the mucosal and systemic immune systems.

In contrast to the relentless progression of disease in the majority of subjects with HIV-1 infection, a small subset of HIV-1-infected people called 'long-term non-progressors' remain clinically asymptomatic for many years in the absence of therapy. Their plasma viral loads are usually undetectable or very low, and CD4$^+$ T cell numbers in peripheral blood are in the normal range. Long-term non-progressors retain normal levels of CD4$^+$ T cells and high levels of virus-specific cellular responses in the gastrointestinal mucosa, consistent with the critical role of mucosal immunity in limiting HIV-1 disease. In SIV-infected sooty mangabeys and African green monkeys, important nonhuman primate models of HIV-1 infection, maintenance of the integrity of the gut mucosal immune system correlates with nonpathogenic SIV infection. Thus, maintenance of the integrity of the gut mucosal immune system may become an important measure of the efficacy of HIV-1 vaccines and therapeutic agents.

25-21 HIV-1 alters intestinal epithelial permeability, promoting microbial translocation.

Early in the acquired immunodeficiency syndrome (AIDS) epidemic investigators showed that HIV-1 frequently causes low-grade atrophy of the intestinal mucosa with a maturational defect in enterocytes and reduced brush border enzyme activity. Studies of gut mucosa in SIV-infected macaques indicate that epithelial cell turnover is accompanied by increased expression of genes that regulate cell repair and regeneration. Epithelial cell apoptosis also occurs in the intestinal and colonic mucosa during early SIV infection. Together, villus atrophy, epithelial cell maturational defect, and epithelial cell apoptosis promote reduced absorptive capacity, manifesting as diarrhea and malabsorption, which can occur in HIV-1-infected subjects without an identifiable enteric pathogen. Termed AIDS enteropathy, manifestations of this syndrome improve or resolve with HAART-induced restoration of immune function, similar to the resolution of many opportunistic infections associated with AIDS during treatment with HAART.

People chronically infected with HIV-1 may have increased levels of circulating lipopolysaccharide (LPS), a measure of microbial translocation, which correlates with innate and adaptive immune activation. Translocation of gut bacteria and bacterial components into the circulation may help maintain a pool of activated immune cells as targets for HIV-1 infection. A possible mechanism for the microbial translocation is suggested by *in vitro* studies in which HIV-1 envelope glycoprotein (gp120) induced primary genital and intestinal epithelial cell lines to secrete TNF-α, which in turn dysregulated the tight junction proteins claudin, occludin, and ZO-1, causing a reduction in transepithelial resistance and enhanced permeability. The increased permeability permitted bacteria placed on the apical surface to leak into the subepithelial space. However, circulating levels of LPS did not increase during disease progression in a study of HIV-1-infected subjects in Africa. Nevertheless, the microbial translocation studies underscore the connection between compromised gut mucosa and systemic disease in HIV-1 infection.

Mucosal infections associated with HIV-1 infection.

The profound CD4+ T-cell depletion initiated in the intestinal lamina propria leads to severe local and systemic immunosuppression. HIV-1 also dysregulates the host defense activities of monocyte-derived macrophages, although the effect of virus on the defense capabilities of mucosal macrophages has not been reported. As a consequence of HIV-1-induced immunosuppression, the mucosa is a site of infection by an array of viral, parasitic, bacterial, and fungal pathogens.

Several principles govern the interaction between opportunistic pathogens and HIV-1 infection (Table 25.3). First, susceptibility to mucosal, as well as systemic, opportunistic infections is inversely related to the level of immunosuppression, and clearance of the infection usually occurs after restoration of immune function with antiretroviral therapy. For example, the presence of active cytomegalovirus infection is inversely proportional to the number of circulating CD4+ T cells, with mucosal cytomegalovirus-induced inflammation and ulceration usually occurring after the number of CD4+ T cells declines to less than 100 cells mm^{-3}, indicating severe immunosuppression; conversely, cytomegalovirus disease resolves with HAART-induced restoration of immune function. Second, replication of some opportunistic pathogens may enhance HIV-1 transcription and vice versa in dual-infected cells and in lymphoid tissue. For example, the bidirectional upregulation of *Mycobacterium avium* complex replication and HIV-1 transcription has been detected in dual-infected macrophages in lymph nodes. Bidirectional upregulation likely occurs in dual-infected immune cells in the mucosa. Third, the route of HIV-1 entry does not determine the site of mucosal infections. Thus, HIV-1 transmitted by the genital

Table 25.3 Principles of mucosal pathogen disease in people infected with HIV-1

Acquisition of mucosal opportunistic pathogens is directly related to the severity of the immunosuppression
Infection by certain opportunistic pathogens resolves with HAART-induced restoration of immune function
Opportunistic pathogen replication and HIV-1 transcription may be enhanced in a bidirectional manner
The route of HIV-1 entry does not determine the mucosal site of opportunistic infection
Acquisition of some mucosal infections reflects the pathogens endemic to a particular geographic area
Infection by certain mucosal pathogens reflects the sexual practices of the HIV-1-infected person

route can lead to opportunistic infections in the gastrointestinal mucosa and/or the genital mucosa. Fourth, the specific mucosal pathogen acquired reflects in part the pathogens endemic to a specific geographical area and the sexual practices of the HIV-1-infected host. For example, among HIV-1-infected people, *Isospora belli* is more common in Africa and the Caribbean region than in the United States, and *Histoplasma capsulatum* is more common in the Mississippi and Ohio River valleys. Regarding the influence of sexual practices on mucosal infection, *Chlamydia trachomatis* proctitis, for example, occurs only in men who practice receptive anal intercourse.

25-22 HIV-1-induced immunosuppression leads to opportunistic infection of gut mucosa by viral, parasitic, bacterial, and fungal pathogens.

The parasitic, viral, bacterial, and fungal microorganisms that cause enteric disease in HIV-1-infected people include opportunistic pathogens and common pathogens that acquire features of opportunistic pathogens in immunosuppressed HIV-1-infected people (Table 25.4). Features of mucosal opportunistic pathogens include (1) the ability to cause severe mucosal disease during immunosuppression but no disease in the presence of normal immune function, (2) resistance to common antimicrobial drugs, (3) prolonged infection, and (4) systemic manifestations. These features are illustrated by the enteric bacteria *Campylobacter jejuni*, *Shigella flexneri*,

Table 25.4 Mucosal pathogens in HIV-1 infection

Viral	Parasitic	Bacterial	Fungal
Cytomegalovirus	*Cryptosporidium* species	*Salmonella* species	*Candida albicans*
Herpes simplex virus	*Microsporidia* species	*Shigella flexneri*	*Histoplasma capsulatum*
Adenovirus	*Isospora belli*	*Campylobacter jejuni*	
Epstein–Barr virus	*Cyclospora*	*Mycobacterium avium* complex	
	Trichomonas vaginalis	*Mycobacterium tuberculosis*	
		Chlamydia trachomatis	

and *Salmonella* species, which normally cause easily treated diarrhea but in immunosuppressed HIV-1-infected subjects are associated with incapacitating and recurrent diarrheal illness, antimicrobial drug resistance, and bacteremia. Some opportunistic infections resolve after HAART restoration of immune function. Similarly, the AIDS enteropathy syndrome, which was common in the pre-HAART era of the AIDS epidemic, usually resolves after immune function has been restored by HAART.

Summary.

Nearly 50 years of basic and clinical research have provided fundamental insights into the molecular structure and biology of rotavirus and the host immune response to intestinal rotavirus infection. As a result, rotavirus diarrhea is now a vaccine-preventable disease, reflected in the successful outcomes of vaccinated children in developed countries. Despite this impressive progress, the mechanisms of virus-induced immunity and vaccine-induced protection, as well as the molecular basis for species restriction and cell tropism, have not been elucidated, but the development of a tractable reverse genetics system may help resolve these issues. In even less time, namely, since the first report 30 years ago of the syndrome eventually defined as AIDS, the pathogenesis, molecular virology, and immunobiology of HIV-1 have been substantially elucidated. Although the mucosa is recognized as playing critical roles in HIV-1 transmission, virus expression, CD4$^+$ T-cell depletion, and secondary opportunistic infections, a successful vaccine has eluded investigators. However, the accelerating pace of discoveries that continue to elucidate the biology of HIV-1 and the pathogenesis of mucosal HIV-1 infection should provide the scientific basis for a mucosal vaccine to halt the relentless HIV-1/AIDS epidemic.

Further Reading.

Rotavirus infection.

Angel, J., Franco, M.A., and Greenberg, H.B.: **Rotavirus vaccines: recent developments and future considerations.** *Nat. Rev. Microbiol.* 2007, **5**:529–539.

Aoki, S.T., Settembre, E.C., Trask, S.D., *et al.*: **Structure of rotavirus outer-layer protein VP7 bound with a neutralizing Fab.** *Science* 2009, **324**:1444–1447.

Blutt, S.E., Fenaux, M., Warfield, K.L., *et al.*: **Active viremia in rotavirus-infected mice.** *J. Virol.* 2006, **80**:6702–6705.

Blutt, S.E., Warfield, K.L., Estes, M.K., *et al.*: **Differential requirements for T cells in viruslike particle- and rotavirus-induced protective immunity.** *J. Virol.* 2008, **82**:3135–3138.

Deal, E.M., Jaimes, M.C., Crawford, S.E., *et al.*: **Rotavirus structural proteins and dsRNA are rewired for the human primary plasmacytoid dendritic cell IFNα response.** *PLoS Pathog.* 2010, **6**:e1000931.

Estes, M.K., and Kapikian, A.Z.: **Rotaviruses and their replication**, in Fields, B.N., Knipe, D.M., Howley, P.M., *et al.* (eds): *Fields Virology*, 5th ed. Philadelphia, PA, Lippincott Williams and Wilkins, 2007:1917–1974.

Greenberg, H.B., and Estes, M.K.: **Rotaviruses: from pathogenesis to vaccination.** *Gastroenterology* 2009, **136**:1939–1951.

Kim, B., Feng, N., Narváez, C.F., *et al.*: **The influence of CD4+ CD25+ Foxp3+ regulatory T cells on the immune response to rotavirus infection.** *Vaccine* 2008, **26**:5601–5611.

Li, Z., Baker, M.L., Jiang, W., *et al.*: **Rotavirus architecture at subnanometer resolution.** *J. Virol.* 2009, **83**:1754–1766.

Niess, J.H., and Reinecker, H.C.: **Dendritic cells in the recognition of intestinal microbiota.** *Cell. Microbiol.* 2006, **8**:558–564.

Patel, N.C., Hertel, P.M., Estes, M.K., *et al.*: **Vaccine-acquired rotavirus in infants with severe combined immunodeficiency.** *N. Engl. J. Med.* 2010, **362**:314–319.

Seo, N.S., Zeng, C.Q., Hyser, J.M., *et al.*: **Integrins α1β1 and α2β1 are receptors for the rotavirus enterotoxin.** *Proc. Natl Acad. Sci. USA* 2008, **105**:8811–8818.

Ward, R.L., Clark, H.F., and Offit, P.A.: **Influence of potential protective mechanisms on the development of live rotavirus vaccines.** *J. Infect. Dis.* 2010, **202**:S72–79.

HIV-1.

Alfsen, A., Yu, H., Magérus-Chatinet, A., *et al.*: **HIV-1-infected blood mononuclear cells form an integrin- and agrin-dependent viral synapse to induce efficient HIV-1 transcytosis across epithelial cell monolayer.** *Mol. Biol. Cell* 2005, **16**:4267–4279.

Brenchley, J.M., Price, D.A., Schacker, T.W., *et al.*: **Microbial translocation is a cause of systemic immune activation in chronic HIV infection.** *Nat. Med.* 2006, **12**:1365–1371.

Ganor, Y., Zhou, Z., Tudor, D., *et al.*: **Within 1 h, HIV-1 uses viral synapses to enter efficiently the inner, but not outer, foreskin mucosa and engages Langerhans-T cell conjugates.** *Mucosal Immunol.* 2010, **3**:506–521.

Ghosh, M., Fahey, J.V., Shen, Z., *et al.*: **Anti-HIV activity in cervical-vaginal secretions from HIV-positive and -negative women correlate with innate antimicrobial levels and IgG antibodies.** *PLoS One* 2010, **5**:e11366.

Hladik, F., and McElrath, M.J.: **Setting the stage: host invasion by HIV.** *Nat. Rev. Immunol.* 2008, **8**:447–457.

Hladik, F., Sakchalathorn, P., Ballweber, L., *et al.*: **Initial events in establishing vaginal entry and infection by human immunodeficiency virus type-1.** *Immunity* 2007, **26**:257–270.

Keele, B.F., Giorgi, E.E., Salazar-Gonzalez, J.F., *et al.*: **Identification and characterization of transmitted and early founder virus envelopes in primary HIV-1 infection.** *Proc. Natl Acad. Sci. USA* 2008, **105**:7552–7557.

Meng, G., Wei, X., Wu ,X., *et al.*: **Primary intestinal epithelial cells selectively transfer R5 HIV-1 to CCR5+ cells.** *Nat. Med.* 2002, **8**:150–156.

Nazli, A., Chan, O., Dobson-Belaire, W.N., *et al.*: **Exposure to HIV-1 directly impairs mucosal epithelial barrier integrity allowing microbial translocation.** *PLoS Pathog.* 2010, **6**:e1000852.

Pope, M., and Haase, A.T.: **Transmission, acute HIV-1 infection and the quest for strategies to prevent infection.** *Nat. Med.* 2003, **9**:847–852.

Raffatellu, M., Santos, R.L., Verhoeven, D.E., *et al.*: **Simian immunodeficiency virus-induced mucosal interleukin-17 deficiency promotes Salmonella dissemination from the gut.** *Nat. Med.* 2008, **14**:421–428.

Sankaran, S., Guadalupe, M., Reay, E., *et al.*: **Gut mucosal T cell responses and gene expression correlate with protection against disease in long-term HIV-1-infected nonprogressors.** *Proc. Natl Acad. Sci. USA* 2005, **102**:9860–9865.

Shattock, R.J., and Moore, J.P.: **Inhibiting sexual transmission of HIV-1 infection.** *Nat. Rev. Microbiol.* 2003, **1**:25–34.

Shen, R., Meng, G., Ochsenbauer, C., *et al.*: **Stromal down-regulation of macrophage CD4/CCR5 expression and NF-κB activation mediates HIV-1 non-permissiveness in intestinal macrophages.** *PLoS Pathog.* 2011, **7**:e1002060.

Shen, R., Richter, H.E., Clements, R.H., *et al.*: **Macrophages in vaginal but not intestinal mucosa are monocyte-like and permissive to human immunodeficiency virus type 1 infection.** *J. Virol.* 2009, **83**:3258–3267.

Infection-Driven Periodontal Disease

26

Periodontal diseases are defined as a series of inflammatory diseases of one or more of the periodontal tissues, including the alveolar bone, the periodontal ligament, the cementum on the tooth root surface, and the gingiva. There are many different diseases affecting the supporting structure of the teeth; however, bacterial biofilm (plaque)-induced inflammatory lesions comprise the vast majority of periodontal diseases. There are two major categories of plaque-induced periodontal diseases: gingivitis and periodontitis. The distinction between the two forms is based upon the reversibility of the lesion; tissue inflammation and connective tissue loss is reversible in gingivitis with removal of the offending organisms. In contrast, the tissue destruction associated with periodontitis is irreversible. Gingivitis may never progress to periodontitis; gingivitis always precedes periodontitis.

Periodontitis is a complex of multiple diseases that share common clinical manifestations. Within the plaque-induced periodontitis category, there is a chronic, slowly progressing form of disease that usually affects adults, termed chronic periodontitis; a rapidly progressive form that usually has early onset and exhibits an unusual distribution of affected teeth termed aggressive periodontitis; and necrotizing disease with marginal destruction of soft tissues.

The oral cavity and particularly the periodontium is the only organ system in the body where hard tissue traverses soft tissue into a contaminated external environment. The teeth are constantly bathed and colonized by microorganisms and their products. In health, there is a mild neutrophil infiltrate of the gingiva and the epithelium lining the gingival sulcus around the teeth. The *initial lesion* of periodontitis is a neutrophil infiltration in response to the bacteria colonizing the teeth. Loss of perivascular collagen is characteristic and an infiltration of macrophages and lymphocytes begins.

The *early lesion* of periodontitis is a more mature leukocyte infiltrate in which neutrophils no longer dominate. The absolute number of neutrophils does not decline, but the main infiltrating cell types are lymphocytes with increasing numbers of lymphoblasts, with a few peripheral plasma cells. Collagen loss may reach 80% and there is loss of fibroblasts and matrix. An exudate forms and flows through the gingival sulcus (gingival crevicular fluid).

The *established lesion* of periodontitis can also be termed the chronic lesion of gingivitis. It is characterized by plasma cells as the dominant inflammatory cell type, but there is no bone loss. There are extravascular immunoglobulins directed against plaque bacteria in the connective tissue and junctional epithelium.

The *advanced lesion* extends into the alveolar bone and periodontal ligament and loss of alveolar bone is seen. Loss of collagen continues and clinically there is a detachment of the gingiva from the tooth (pocket formation).

Etiology of periodontal disease.

In the mid-1960s classical studies in 'experimental gingivitis' showed that healthy volunteers who refrained from tooth brushing for 21 days developed gingivitis. The findings resulted in a major shift in defining the etiology of periodontal diseases. Gingivitis developed in all volunteers and reinstitution of daily dental plaque removal resulted in a return to gingival health. The accumulation of bacteria on the teeth caused gingivitis and removal of the bacteria returned the gingival condition to health.

Likewise in classic studies in the 1960s in dogs, dental plaque accumulation led to the development of periodontal disease. Plaque removal every other day for 18 months maintained clinically healthy periodontal tissues; however, teeth in the same mouth that were not cleaned developed gingival inflammation and attachment loss. Experimental gingivitis in most dogs progressed to periodontitis over a four-year period if dental plaque was not removed on a daily basis. Control animals with daily plaque removal during the same period did not develop gingivitis or periodontitis.

The infectious nature of periodontal disease was also established in the 1960s when it was shown that periodontitis was transmissible from diseased Syrian hamsters to healthy animals with *Actinomyces viscosus* isolated from the diseased animals.

Advances in microbiology resulted in the identification of clusters of indigenous bacteria in the subgingival microbiota that are associated with the presence and progression of periodontitis. Specific bacteria including *Porphyromonas gingivalis, Tannerella forsythia, Treponema denticola, Campylobacter rectus, Micromonas micros, Streptococcus intermedius, Eubacterium nodatum, Aggregatibacter* (formerly *Actinobacillus*) *actinomycetemcomitans*, and *Prevotella intermedia* are considered to be the causative agents of periodontitis.

The major conclusions of these early classic studies were that periodontal infections are polymicrobial; the causative agents are part of the indigenous (normal) microbiota; the amount of dental plaque is of etiologic importance; and some bacteria in dental plaque are more pathogenic than others.

26-1 Biofilms are important for the development of periodontitis.

Dental plaques are highly organized bacterial biofilms. Biofilms are complex polymicrobial communities that are resistant to externally applied antimicrobial agents and antibacterial host mechanisms. Other than mechanical removal, the disruption of biofilms remains a clinical problem.

Fifty percent or less of the oral microbiota is cultivable in the laboratory. It is highly likely that the other 50% comprises microorganisms that have etiologic importance with an important role in biofilm ecology. To further characterize the oral microflora, genetic techniques for bacterial detection are currently being used. The findings reveal that the oral flora in health and disease is extraordinarily diverse and that the level of variability appears to be the individual, not the disease, although there are marked differences in the composition of the flora in health and disease in the same individual.

Bacterial–host mucosal interactions in periodontal disease.

The current consensus is that periodontal disease is an inflammatory disease initiated by bacteria that causes the destruction of the supporting tissues of the teeth in a susceptible host. There are several bacteria associated with the disease, but which organisms actually initiate disease remains unknown. Organized biofilms enhance bacterial survival. Bacterial virulence factors further enhance survival. Molecules such as toxins, proteases, and glycosidases are classified as virulence factors that hide the bacteria from host detection and provide nutrients.

26-2 PAMPs are important in periodontal disease.

To combat the onslaught of microorganisms at mucosal surfaces and to maintain homeostasis in contaminated environments such as the oral cavity, the host has evolved mechanisms to detect bacteria: the recognition of structural components of the bacterial surface. Lipopolysaccharides (LPS), peptidoglycan (PGN), and other cell-surface components such as fimbriae are essential structural components of bacteria. Structural variation of these bacterial components between various species, or even between different strains of the same species, creates incredible structural diversity in prokaryotes. Despite structural heterogeneity, there are conserved motifs known as pathogen-associated molecular patterns (PAMPs). PAMPs and their receptors are considered in more detail in Chapter 16. Host cells have PAMP receptors termed pattern recognition receptors (PRRs). These innate immune receptors are highly conserved and presumably evolved to detect invading bacteria. Binding of PAMPs by PRRs initiates an inflammatory response. On mucosal surfaces colonized by commensal organisms, the balance between constant inflammatory stimulus and maintenance of homeostasis is often lost, resulting in local inflammation of colonized tissues.

The best characterized PAMPs include LPS, PGN, lipoteichoic acids, fimbriae, proteases, heat-shock proteins, formylated bacterial peptides, and toxins. The host's PRRs include Toll-like receptors (TLRs).

The oral cavity is the entrance to the gastrointestinal tract and is naturally colonized by a wide variety of bacteria. The colonization patterns and symbiotic function of the bacteria are physiologic. The tooth–gingival interface is the site of a variety of natural, innate host defenses. The regular shedding of epithelial cells, saliva, and the gingival crevicular fluid, and the continuous migration of neutrophils across the junctional epithelium into the gingival sulcus, maintains homeostasis: the sometimes delicate balance between the host tissues and the potentially pathogenic bacteria. Gingival inflammation results when equilibrium is lost and bacteria overgrow. Overt gingival inflammation may never proceed to destructive periodontitis, but the parameters that determine the transition remain unknown. In periodontitis, the emergence of pathogenic bacteria including *P. gingivalis*, *T. forsythia*, and *T. denticola* is common. Periodontal bacteria attach to epithelial cells via fimbriae; PRR recognition of PAMPs induces epithelial cell secretion of pro-inflammatory cytokines (tumor necrosis factor-α (TNF-α), interleukin (IL)-1β, IL-6) and the chemokine IL-8 in the connective tissue. The inflammation changes the growth environment of the biofilm leading to the overgrowth of pathogens. Several periodontopathogens (i.e., *P. gingivalis*, *A. actinomycetemcomitans*) can invade and transverse epithelial cells to the connective tissue, and PAMPs that are either shed or secreted further promote the development of the inflammatory lesion.

PAMP-mediated amplification of inflammation impacts leukocytes, fibroblasts, mast cells, endothelial cells, dendritic cells, and lymphocytes. Release of pro-inflammatory cytokines (TNF-α, IL-1β, IL-6, and IL-12), chemoattractants (CXCL8, CCL3, CXCL2, CCL2, and CCL12), and prostaglandin E$_2$ (PGE$_2$) by neutrophils, macrophages, fibroblasts, and mast cells in the connective tissue promotes collagen degradation. Mast-cell degranulation results in the secretion of histamine and leukotrienes.

Endothelial cells activated by the cytokines are active participants in the pathogenesis of periodontal inflammation. Secretion of endothelial cell chemokines (IL-8, monocyte chemotactic peptide-1 (MCP-1)) and expression of adhesion molecules (P- and E-selectin and intercellular adhesion molecule (ICAM)-1 and ICAM-2) on the surface of endothelial cells promote leukocyte extravasation. P- and E-selectins bind glycoproteins on leukocytes and promote rolling of cells along the vessel wall. Firm binding of leukocytes to endothelium is mediated by integrins (leukocyte function-associated antigen-1 (LFA-1)) and leukocytes attach firmly to ICAM-1 expressed on endothelial cells; TNF-α, PGE$_2$, and histamine increase vascular permeability and permit leukocyte diapedesis. Chemokines, such as IL-8, that are produced at the site of infection, along with bacterial chemoattractants (the tripeptide fMLP (N-formyl-methionine-leucine-phenylalanine)), form a concentration gradient for leukocytes to migrate to the focus of infection. Circulating pro-inflammatory cytokines from the site of inflammation activate hepatocytes in the liver to release acute phase proteins, including LPS-binding protein and sCD14 that are important for the recognition of LPS. Complement proteins and C-reactive protein are also released to opsonize bacteria for phagocytosis. Figure 26.1 illustrates the initiation sequence of inflammation at the gingiva.

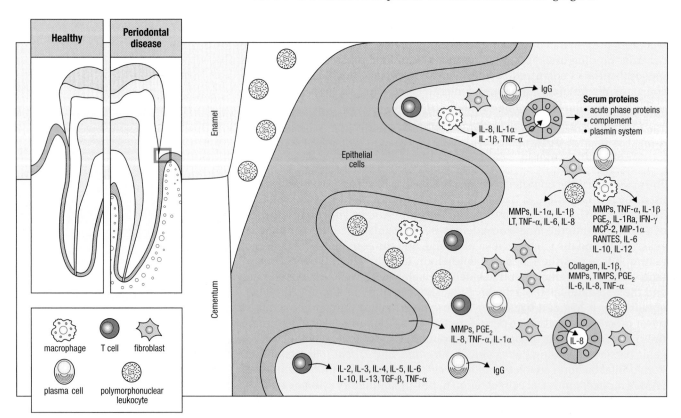

Figure 26.1 The inflammatory response to the bacterial biofilm. Healthy gingiva, by definition, does not contain inflammatory cells. In response to the buildup of a complex biofilm on the surfaces of teeth, an inflammatory lesion is established with the accumulation of inflammatory cells (T cells, plasma cells, polymorphonuclear leukocytes, and macrophages) resulting in the secretion of cytokines, chemokines, and matrix metalloproteinases (MMPs). The increased cytokine concentration in the environment causes fibroblasts to secrete collagen, MMPs, cytokines, and tissue inhibitors of metalloproteinases (TIMPs).

Acquired immunity and the chronic lesion.

Failure to clear the bacterial insult and persistence of the lesion leads to activation of the acquired immune response. Gingival dendritic cells take bacterial antigens to the draining lymph nodes in the neck. Antigenic peptide binding to a class II major histocompatibility complex (MHC) receptor is required for antigen presentation to antigen-specific effector T cells and antibody-secreting B cells are generated by clonal expansion and differentiation over the course of several days. Antigen-specific T cells are released into the blood and home to the site of infection. Likewise, macrophages that have engulfed bacteria present bacterial antigens to T cells. Antigen-specific T cells respond and activate macrophages. Secreted antibodies protect the host by (1) inhibiting toxicity or infectivity of pathogens (neutralization), (2) opsonizing pathogens and promoting phagocytosis, and (3) activating the complement system. Failure to clear the infection leads to further tissue damage. Activated macrophages produce oxygen radicals, nitric oxide, and proteases that are toxic to host cells. Moreover, recent work in mice reveals that the induction of an adaptive immune response to colonizing pathogens results in receptor activation of nuclear factor-κB (NFκB) ligand (RANKL)-dependent periodontal bone loss.

The trigger that causes the shift from tissue homeostasis to pathology remains unclear. Studies of the etiology of the periodontal lesion are cross-sectional and definitive cause/effect relationships have not been demonstrated. One recent longitudinal study of periodontal disease progression failed to implicate any single organism or group of organisms; longitudinal animal studies suggest that inflammation significantly impacts the composition of the biofilm. In clinical studies, inflammation is a stronger predictor of future periodontal attachment loss than the quality or quantity of the biofilm.

26-3   T cell and B cell immune responses are involved in periodontitis.

The induction of an innate immune response following the interaction between the periodontopathogens and host cell via PRR results in the production of cytokines and chemokines mediating an inflammatory response, as well as the induction of an acquired immune response. The TLRs that recognize ligands on periodontopathogens include TLR4 and TLR2 that heterodimerize with TLR1 or TLR6 to recognize a triacylated or diacylated form of lipoproteins respectively. Once the ligand interacts with its appropriate TLR, a sequence of cellular events follows including the recruitment of adaptor molecules, the phosphorylation of signaling molecules, and the activation and translocation of transcription factors, all of which results in the expression of gene products that are essential for activation of cells. In dendritic cells this process leads to the production of cytokines and chemokines, as well as an upregulation in the expression of co-stimulatory molecules, which are important for the induction of an adaptive immune response because they mediate the priming and differentiation of naive T cells that ultimately leads to the activation of B cells and antibody production. Furthermore, the nature of the cytokines produced by the activated dendritic cell influences the differentiation of the naive T cell into a helper T (T_H) T_H1, T_H2, or T_H17 cell, or regulatory T (T_{reg}) cell, depending on the cytokine profile they produce and their regulatory functions, which in turn affect the nature of the acquired immune response induced. In the case of chronic periodontitis, the early lesion is characterized by a cellular infiltrate of macrophages and T cells, then shifts to increased numbers of B cells and plasma cells in the more advanced lesion (*vide supra*). The histology of the early lesion is more reflective of a T_H1 response, whereas the advanced lesion is more reflective of a T_H2 response. Analyses of gingival tissues from periodontitis patients for cytokine mRNA have often revealed a predominance of T_H1 or T_H2 cytokines or a mixed T_H1 and T_H2 cytokine profile. This is not totally unexpected considering the complexity of the subgingival plaque microflora and their ability to activate an inflammatory response via TLRs. However, the interplay between the microflora and TLRs for the response has not been fully elucidated.

Systemic and local (gingival crevicular fluid) antibodies to periodontopathogens are present in periodontitis patients, and the nature of the antibody response, including the magnitude, specificity, and IgG subclass, has been extensively investigated in order to determine correlates with disease severity and local cytokine responses. Although some investigations have shown a relationship between the induction of antibodies to periodontopathogens, resulting from infection or from immunization, and a reduction in the microflora and/or progression of disease, other studies have revealed extremely variable results. The variability observed in these studies reflects the complexity of this disease, questions the functionality of the antibodies induced, and challenges future studies in this area.

26-4 Cytokines are involved in bone resorption in periodontitis.

A major consequence of the inflammatory response associated with periodontal disease is resorption of alveolar bone. Under normal conditions, bone is periodically resorbed while new bone forms. In the case of periodontal disease, there is a shift to bone resorption due to an increase in inflammatory cells and cytokines, and an increase in osteoclasts in the local tissue (osteoclasts are large, multinucleated cells derived from the monocyte–macrophage lineage, whose major function is to resorb bone). Differentiation of osteoclasts is regulated by macrophage colony-stimulating factor (M-CSF) and RANKL found on several cell types including T cells and osteoblasts. M-CSF promotes proliferation and survival of osteoclast progenitors, whereas RANKL promotes cells to differentiate along the osteoclast lineage and acts as an activating and survival factor for mature osteoclasts. RANKL binds to RANK present on osteoclast precursors and mature osteoclasts, resulting in the recruitment of TNF receptor associated factor (TRAF) family proteins, especially TRAF6, that leads to activation of various signaling pathways and transcription factors, and the secretion of bone-degrading factors.

Osteoclastogenesis is positively regulated by the RANK–RANKL signaling and negatively regulated by osteoprotegerin (OPG), a soluble decoy receptor. OPG is produced by various cells, including dendritic cells and osteoblasts, and binds to RANKL, thus preventing RANKL from interactions with its biologically active receptor RANK. A main mechanism regulating bone resorption and formation is the relative concentration of RANKL and OPG. When RANKL expression is high relative to OPG, RANKL can bind RANK on osteoclast precursors, which tips the balance to favor activation of osteoclast formation and bone resorption; whereas when OPG levels are higher than RANKL expression, OPG binds RANKL, inhibiting it from binding to RANK, which leads to a reduction in osteoclast formation and apoptosis of osteoclasts. *In vivo* studies have shown that overexpression of OPG in transgenic mice resulted in severe osteoporosis, while osteoporosis was seen in OPG-deficient mice due to an increase in the number of osteoclasts. Furthermore, analysis of RANKL and OPG levels in gingival tissue extracts or from gingival crevicular fluid has revealed a higher RANKL/OPG ratio from diseased sites than from healthy sites. During an inflammatory response, pro-inflammatory cytokines such as IL-1 or TNF-α can induce osteoclastogenesis by increasing the expression of RANKL, whereas the anti-inflammatory cytokine IL-10 inhibits RANKL but enhances OPG production.

Because osteoclast precursors express TLRs, it is likely that microbial components in dental plaque can affect tissue homeostasis toward a more destructive

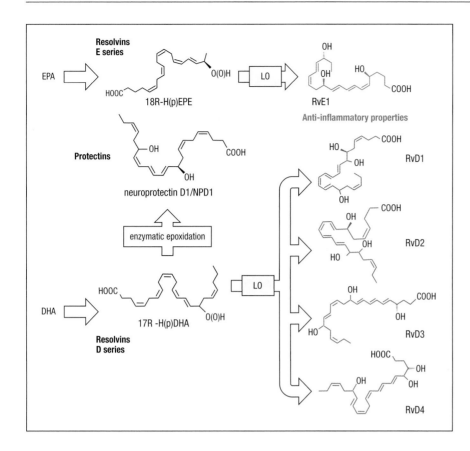

Figure 26.3 Metabolism of omega-3 PUFA. The same LO enzyme system that generates lipoxins from arachidonic acid metabolizes omega-3 PUFA into resolvins and protectins. The two major omega-3 PUFA are eicosapentaenoic acid (EPA), which is metabolized into resolvins of the E series, and docosahexaenoic acid (DHA), which is metabolized into resolvins of the D series and the protectins. The role of aspirin-modified COX-2 is also indicated. In the presence of aspirin, COX-2 is acetylated and becomes a new active enzyme, a 15R-lipoxygenase. Aspirin-modified COX-2 is responsible for the production of aspirin-triggered lipoxins, the 18R-resolvins of the E series, and 17R-resolvins of the D series.

In the context of periodontal disease, there is detectable LXA$_4$ in the gingival crevicular fluid of periodontitis patients, albeit at insufficient concentrations to exert clinical actions. These observations have suggested that periodontal diseases, and other inflammatory diseases, are characterized by a failure of resolution pathways in addition to, or instead of, overproduction of pro-inflammatory signals.

The main endogenous fatty acid substrate for the production of resolution agonists is arachidonic acid. However, dietary fatty acids are also substrates for the same metabolic pathways. The principal dietary fatty acids are omega-3 polyunsaturated fatty acids (omega-3 PUFA). Eicosapentaenoic acid, C20:5, and docosahexaenoic acid, C22:6, are the substrates for lipoxygenases generating a new genus of molecules termed the resolvins of the E and D series and docosatrienes. E- and D-series resolvins bind to distinct receptors on inflammatory cells to elicit anti-inflammatory and pro-resolution actions that are similar but not exactly the same as the lipoxins. For instance, the receptors for resolvin E1 (RvE1) are chemR23 on macrophages and BLT1 on neutrophils. It is also interesting to note that multiple ligands can bind to the same receptor: chemR23 is also a receptor for chemerin, a peptide that is chemotactic for macrophages. It is hypothesized that these omega-3 PUFA pathways contribute, in large part, to the systemic anti-inflammatory activity of omega-3 fatty acids providing protection from cardiovascular disease and other inflammatory conditions. The resolvins derived from eicosapentaenoic acid (RvE1–4), and the D-series (RvD1–4) and protectins derived from docosahexaenoic acid, are defined by their potent bioactivity and novel chemical structures. First identified in resolving exudates, the biologic activity of these compounds includes protective functions in neural tissues, retina, and corneal epithelial injury in the eye. The known actions of resolvins include reduction of neutrophil trafficking, regulation of cytokines and reactive oxygen species production, and a general dampening of the inflammatory response and enhancement of bacterial clearance. The protectins were so named because of their anti-inflammatory and protective actions in stroke, human Alzheimer's disease, and other neural systems.

Aspirin plays an interesting role in both the arachidonic acid and omega-3 PUFA pathways. Among the known actions of aspirin is inactivation of cyclooxygenase-2 (COX-2). Unlike other nonsteroidal drugs, aspirin modification of COX-2 through acetylation converts COX-2 into another active enzyme, a 15R-lipoxygenase. The stereochemical change in the resulting lipoxin and resolvin products yields resolving molecules that are more stable and longer acting. The potent anti-inflammatory properties of aspirin were recognized when it was discovered in the early 1900s; decades later, the inhibition of prostaglandin synthesis by aspirin was recognized as a mode of action. Discoveries further clarifying the mode of action of aspirin suggest that the action of aspirin is due in part to production of 'aspirin triggered' lipoxins and resolvins. The previously unappreciated aspects of the action of aspirin have interesting implications for drug design: the ligands are receptor agonists, not inhibitors or antagonist, precluding the requirement for maintaining an effective blood level, and the R-epimers of the natural molecules are resistant to degradation by specific endogenous enzymes.

26-8 The resolution phase of acute inflammation is as important as the onset phase.

The goal of the acute inflammatory response is the elimination of the injury or invader and return to homeostasis (Figure 26.4). The process is well known, involving leukocyte (neutrophils) recruitment, phagocytosis of invaders, recruitment and differentiation of mononuclear phagocytes into macrophages,

Figure 26.4 Failure of resolution pathways leads to chronic inflammation. The upper panel illustrates the normal successful resolution of an acute inflammatory lesion with the temporal emergence of resolution agonists that drive return to homeostasis. The lower panel illustrates a failure of resolution pathways leading to a chronic inflammatory lesion. 5-HT, 5-hydroxytryptamine.

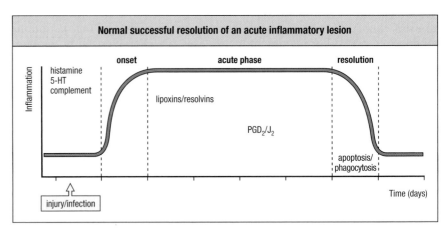

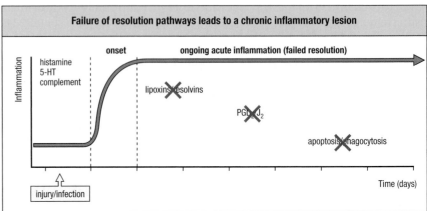

and clearance of the lesion. Return to homeostasis is not complete until all inflammatory cells, including neutrophils, are removed from the lesion. Failure to resolve because of persistent infection or destruction of the cellular matrix may result in scarring and fibrosis with incomplete healing or chronic inflammation. The early events in conditions such as periodontitis, for example, controlling neutrophil biology, can mediate conversion of acute gingivitis to chronic periodontitis. Scarring and fibrosis in periodontitis (repair rather than regeneration) prevents the return to homeostasis.

It is now known that resolution of inflammation is an active process and that anti-inflammation is not the same as resolution. Anti-inflammation is the downregulation, or pharmacologic blocking or inhibition, of pro-inflammatory mediators, such as PGE_2 and LTB_4. Resolution is the active recruitment of non-phlogistic macrophages to clear the lesion of apoptotic polymorphonuclear neutrophils and remaining microorganisms. The distinction is important because the current pharmacologic approach to the control of inflammation only addresses half of the process. Cyclooxygenase inhibitors such as ibuprofen or flurbiprofen block the production of PGE_2 and the cardinal signs of inflammation, but they do not promote resolution. In fact, this class of drug is resolution toxic, prolonging the time to resolve the lesion and extending tissue inflammation temporally.

Endogenous resolving molecules as therapeutic agents in periodontal disease.

A new hypothesis for the pathogenesis of periodontitis and other inflammatory conditions is that a failure of resolution contributes as much to inflammatory disease as overproduction of pro-inflammatory mediators. Proof-of-principle experiments for periodontitis were performed in a series of intervention studies in an animal model: *P. gingivalis*-induced periodontitis in the New Zealand White rabbit. Periodontitis was induced by tying 3/0 silk ligatures around the second premolars in the mandibular quadrants and 10^9 colony-forming units (cfu) *P. gingivalis* in a methyl cellulose slurry was applied three times per week for 6 weeks. Ligature alone is not sufficient to induce disease in this model and the disease progression induced by *P. gingivalis* was inhibitable by systemically administered metronidazole, confirming the bacterial etiology.

In separate experiments, prevention of disease and treatment of established disease with a resolution agonist as a monotherapy were assessed. In the prevention protocol, a parallel design compared application of resolving molecules three times weekly (applied at the same time as the *P. gingivalis*) with application of a placebo (ethanol vehicle) throughout the experimental period. In the treatment protocol, disease was induced for 6 weeks with topical *P. gingivalis*, which established a stable *P. gingivalis* infection; *P. gingivalis* was not applied during the treatment phase.

26-9 Resolvins can help prevent periodontitis.

The preventive regimen in the rabbit model was the application of 5 μl of a 1 mg ml^{-1} solution of RvE1 in ethanol topically three times per week; the control was ethanol alone (placebo). At the end of the 6 weeks, animals were sacrificed and periodontal disease progression was quantified morphologically and histologically. Placebo treatment had no impact on the progression of disease, which resulted in significant bone and attachment loss. RvE1 application prevented the onset and progression of periodontal destruction including all clinical signs of inflammation (Figure 26.5).

Figure 26.5 Prevention of periodontitis with exogenous topical application of a resolution agonist. The upper panel shows the experimental design of inducing rabbit periodontitis as described in the text. Two groups of animals were compared; with and without RvE1 application. The animals given the vehicle (lower left panels) show bone loss with pits and loss of soft tissue around the ligated tooth. In contrast, RvE1 treatment (lower right panels) prevents bone loss and the loss of soft tissue.

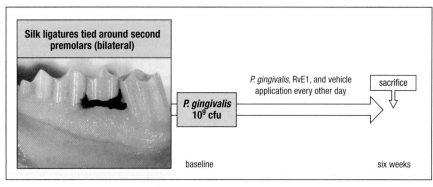

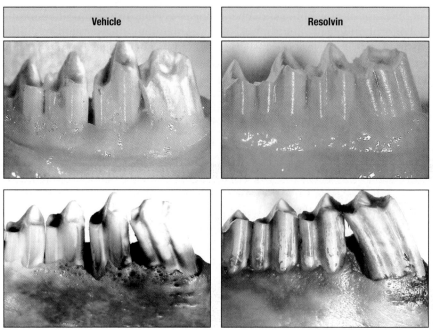

26-10 Resolvins can treat periodontitis.

Periodontal lesions induced by *P. gingivalis* were established over 6 weeks. After 6 weeks, application of *P. gingivalis* was discontinued. During a second 6-week interval, the established periodontal lesions were treated with topical RvE1 or placebo. At the end of the second 6 weeks, animals were sacrificed and periodontal destruction quantified. Periodontal disease characterized by soft tissue and bone loss was observed at 6 weeks and after an additional 6 weeks of placebo treatment. Progression of disease was significant with deepening of pocket depth and additional loss of crestal bone with infrabony pocketing. In the placebo group, disease progressed in the second 6 weeks although *P. gingivalis* application was discontinued; microbial analyses demonstrated the persistence of the *P. gingivalis* infection. In the RvE1-treated group, no clinical inflammation was observed; soft tissues returned to pre-ligature architecture. Bone regeneration to pre-disease architecture with restoration of crestal height, elimination of infrabony defects, and regeneration of new cementum and bone with an organized periodontal ligament was also observed (Figure 26.6).

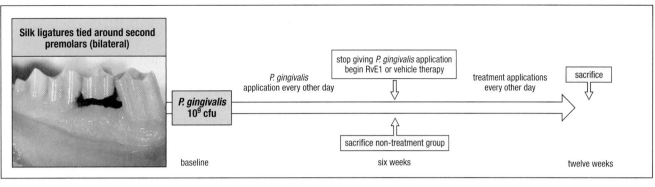

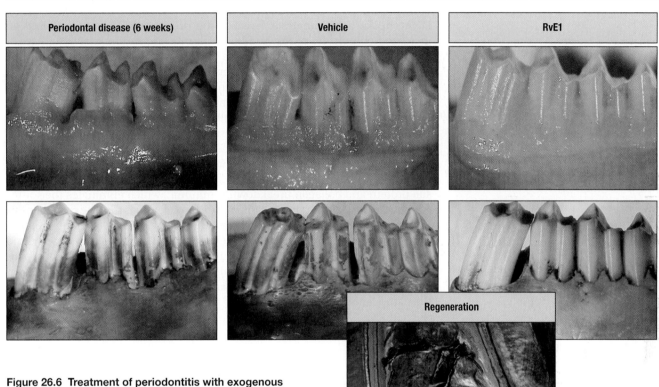

Figure 26.6 Treatment of periodontitis with exogenous topical application of a resolution agonist. The upper panel shows the experimental design for the induction and treatment of rabbit periodontitis as described in the text. After induction of periodontitis, one group of animals was sacrificed at 6 weeks to determine the extent of periodontal destruction. The other two groups were treated for an additional 6 weeks with RvE1 or placebo and sacrificed at 12 weeks. In the lower panels, the left images illustrate the soft and hard tissue periodontal destruction after 6 weeks of disease progression. The center images illustrate persistent soft tissue inflammation and bone destruction after 6 weeks of treatment with vehicle; the right images are after 6 weeks of RvE1 treatment. Note the appearance of healthy, pink soft tissue and the regrowth of bone after RvE1 treatment. The inset marked 'Regeneration' is a polarized light photomicrograph of an undecalcified section of regenerated tissues showing reestablishment of the periodontal ligament with new bone and new connective tissue fibers uniting the new bone and new cementum on the tooth root surface.

Biofilms and inflammatory periodontitis.

The relationship of the biofilm to the inflammatory response has been well documented as cause and effect based on intervention studies that clearly demonstrate that the biofilm induces inflammation. However, the initiating pathogens in the biofilm remain unknown. There is a large literature associating specific organisms in cross-sectional studies, but cause and effect have not been established. Based on the available data, a second explanation of the cross-sectional observations is that *P. gingivalis* is present in high numbers in periodontitis *because* of the deep pockets.

Longitudinal assessment of microbial changes with RvE1 treatment using DNA checkerboard analysis revealed the presence of *P. gingivalis* and an increased complexity of the biofilm in placebo-treated animals. RvE1 treatment eradicated *P. gingivalis* from the biofilm and the resident flora returned to its normal composition and numbers despite no antibacterial activity of the resolvins. The spontaneous elimination of the pathogen and the return of a 'normal' biofilm suggest that the composition of the biofilm is mediated by the inflammatory milieu providing nutrients to support the pathogenic organisms. For instance, *P. gingivalis* is an asaccharolytic organism that depends on essential amino acids as a food source. It is well appreciated that *P. gingivalis* has an armamentarium of proteolytic enzymes, the gingipains, that break down peptides into essential amino acids. In the absence of inflammation, tissue protein and collagen breakdown is minimal, which eliminates the source of peptides that support the organism. The implication is that the pathogenesis of periodontitis is not a linear progression of colonization by a pathogen followed by an inflammatory lesion. The dynamic interaction of the biofilm and the host response requires further study. Moreover, implications for therapeutic strategies that target etiology may also not be linear. Future therapeutics for periodontitis will likely target the inflammatory response to effect changes in the pathogenic biofilm.

Summary.

Periodontitis is an inflammatory disease initiated by specific bacteria present in the subgingival plaque that causes the destruction of the supporting tissues of the teeth in a susceptible host. The periodontal pathogens can bind and activate gingival epithelial cells via TLRs leading to the production of pro-inflammatory cytokines and chemokines. The inflammatory process is amplified following the activation of the underlying cells including macrophages, dendritic cells, lymphocytes, neutrophils, and fibroblasts via TLRs. This in turn also leads to the induction of an adaptive immune response to the colonizing pathogens. If left untreated, a major consequence of the host response to the periodontopathogens is resorption of alveolar bone, and a main mechanism regulating bone resorption and formation is the relative concentration of RANKL and OPG. When RANKL expression is high compared with OPG, RANKL binds RANK on osteoclast precursors, which results in osteoclast formation and bone resorption, whereas when OPG levels are higher than RANKL expression, OPG acts as a decoy receptor for RANKL leading to a reduction in osteoclast formation and apoptosis of preexisting osteoclasts. The lipoxins, lipid mediators derived from arachidonic acid, as well as other fatty acid products termed resolvins, have now been shown to play an important role in the resolution of inflammation by binding to specific receptors on inflammatory cells and elicit anti-inflammatory activity, such as a reduction of neutrophil trafficking, regulation of cytokine production, and a general dampening of the inflammatory response and enhancement of bacterial clearance. This active process differs from that seen with pharmacologic anti-inflammatory agents, such as ibuprofen, which block the production of PGE$_2$, but do not promote resolution. Our growing knowledge of the relationship between the etiologic agents of periodontitis and the inflammatory pathogenesis will have a major impact on treatment strategies and pharmacotherapeutics for the treatment of inflammatory diseases.

Further Reading.

Colombo, A.P., Boches, S.K., Cotton, S.L., *et al.*: **Comparisons of subgingival microbial profiles of refractory periodontitis, severe periodontitis, and periodontal health using the human oral microbe identification microarray.** *J. Periodontol.* 2009, **80**:1421–1432.

Darveau, R.P. **Periodontitis; a polymicrobial disruption of host homeostasis.** *Nat. Rev. Microbiol.* 2010, **8**:481–490.

Dewhirst, F.E., Chen, T., Izard, J., *et al.*: **The human oral microbiome.** *J. Bacteriol.* 2010, **192**:5002–5017.

El-Sharkawy, H., Aboelsaad, N., Eliwa, M., *et al.*: **Adjunctive treatment of chronic periodontitis with daily dietary supplementation with omega-3 fatty acids and low-dose aspirin.** *J. Periodontol.* 2010, **81**:1635–1643.

Kolenbrander, P.E., Palmer, R.J., Jr., Periasamy, S., *et al.*: **Oral multispecies biofilm development and the key role of cell-cell distance.** *Nat. Rev. Microbiol.* 2010, **8**:471–480.

Lorenzo, J., Horowitz, M., and Choi, Y.: **Osteoimmunology: interactions of the bone and immune system.** *Endocrine Rev.* 2008, **29**:403–440.

Manicassamy, S., and Pulendran, B.: **Modulation of adaptive immunity with Toll-like receptors.** *Semin. Immunol.* 2009, **21**:185–193.

Marsh, P.D., and Devine, D.A.: **How is the development of dental biofilms influenced by the host?** *J. Clin. Periodontol.* 2011, **38** Suppl. 11:28–35.

Schenkein, H.A.: **Host responses in maintaining periodontal health and determining periodontal disease.** *Periodontology 2000* 2006, **40**:77–93.

Serhan, C.N., Chiang, N., and Van Dyke, T.E.: **Resolving inflammation: dual anti-inflammatory and pro-resolving lipid mediators.** *Nat. Rev. Immunol.* 2008, **8**:349–361.

Takeuchi, O., and Akira, S.: **Pattern recognition receptors and inflammation.** *Cell* 2010, **140**:805–820.

Van Dyke, T.E.: **Proresolving lipid mediators: potential for prevention and treatment of periodontitis.** *J. Clin. Periodontol.* 2011, **38** Suppl. 11:119–125.

Principles of Mucosal Vaccine Strategies

27

Mucosal immunization was introduced in the 1950s with the Sabin live attenuated oral polio vaccine. The success of this vaccine, attributable not only to its needle-free character, but foremost to its ability to stimulate protective immunity, has not been followed by the series of efficient mucosal vaccines one might have anticipated. Rather, only a handful of commercially available mucosal vaccines exist today (Table 27.1). In fact, these vaccines, directed against cholera, typhoid fever, rotavirus, poliovirus, and influenza virus, have little in common, except that a majority are live attenuated vaccines, which could in part explain their efficacy. We lack a coherent understanding of the requirements for effective mucosal vaccines. Mucosal vaccine candidates that in animal models show strong protection often fail in clinical trials. For example, trials in humans with *H. pylori* vaccine have failed despite animal studies showing great promise. We believe there are several explanations as to why this is the case, the most important being the lack of safe and effective clinically acceptable vaccine adjuvants for mucosal administration. At the same time it should be emphasized that highly protective vaccines against a number of mucosal infections are administered by a parenteral rather than a mucosal route, and still convey strong protection, especially those that stimulate strong serum IgG antibodies that cross the mucosal epithelium by neonatal Fc receptor (FcRn)-mediated transport and perhaps transudation (see Chapters 5 and 9 and Table 27.2).

Principles of mucosal vaccination.

Today we have very precise information about signals necessary for the stimulation of adaptive immune responses. This involves the stepwise stimulation of naive T cells through recognition by the T-cell receptor of major histocompatibility complex (MHC) class I or II together with peptide, and the more complicated second signals that are provided by cytokines and co-stimulatory molecules on the membranes of antigen-presenting cells (APC). Co-stimulatory molecules such as CD40, CD80, CD86, and OX40L have been found critical for driving T-cell dependent immune responses. The linking of innate responses to the induction of adaptive immunity is the key to understanding how to construct effective mucosal vaccines. It is noteworthy that while we have a wealth of data relevant to the induction and preservation of an immune response (both systemic and mucosal) in animals, there is a significant lack of complementary mechanistic data as to how this information pertains to the licensed vaccines, irrespective of whether they are mucosal or parenteral vaccines, proven to provide significant protection against pathogens of mucosal surfaces (Tables 27.1 and 27.2).

Table 27.1 Licensed mucosally administered vaccines for diseases that initiate at mucosal sites

Target	Primary disease/spread	Composition of vaccine	Target of protection	Dosage/formulation	Mechanism of protection	Efficacy	Comments
Cholera	Severe diarrhea Fecal–oral spread	Whole killed bacteria ± cholera toxin B subunit (multiple strains used)	Cholera toxin B (CTB) subunit and lipopolysaccharide (LPS) (and perhaps bacterial wall proteins)	Liquid given with buffer as 2–3 doses orally	Intestinal IgA antibacterial, mainly anti-LPS and antitoxin antibodies. Serum vibriocidal antibody correlates with protection but unlikely effector mechanism	With CTB 85% short term, without 60–70%; protection after 1st year is 60–70% for vaccine with and without CTB, lasts ~3 years but falls off quicker in children <5 years	CTB component also provides short-term protection against enterotoxigenic *Escherichia coli* diarrhea. Evidence of herd protection may increase overall effectiveness beyond the estimated efficacy
Cholera	Severe diarrhea Fecal–oral spread	Live recombinant cholera with toxin A subunit deleted	Cholera B subunit and bacterial cell wall LPS and other antigens	Given as a single oral dose of reconstituted liquid plus buffer	Intestinal IgA antitoxin and antibacterial, mainly anti-LPS antibodies. Serum vibriocidal antibody correlates with protection but unlikely effector mechanism	~70% against El Tor cholera in volunteer setting; not effective in field trial	Not on the market since 2004
Type A (H1 and H3) and B influenza viruses	Influenza Respiratory spread	Live viral reassortant with current hemagglutinin (HA) + neuraminidase (NA) genes on an attenuated donor strain genetic backbone	Unclear, presumably HA and NA antigens on locally replicating virus	In adults, single dose of live virus administered intranasally as a trivalent blend of H1, H3, and B strains; in young children, 2 doses	Unclear, presumably local anti-HA and anti-NA response. Poor correlation of protection with serum antibody response	>85% in children, variable in adults	Current data indicate that live intranasal (IN) vaccine is more efficient in stimulating heterotypic immunity than trivalent inactivated influenza vaccine (TIV) and that the live vaccine is more effective than TIV in young children but possibly less effective than TIV in adults
Oral poliovirus vaccine	Polio, flaccid paralysis Fecal–oral and respiratory spread	Live attenuated polioviruses type 1, 2, 3 as a trivalent vaccine. Monovalent type 1 and 3 vaccines and a type 1 + 3 bivalent vaccine were recently licensed for use in developing countries	Surface neutralizing epitopes on polio types 1, 2, 3	3 doses administered orally to young children without buffer	Local and systemic neutralizing antibody to discontinuous regions mainly on VP1 and VP3 inhibit replication and dissemination from gastrointestinal tract	>90% in most of the world, lower in some poor countries	Vaccine has very low but detectable frequency of reversion to virulence that led to discontinuation of use in US. Also type 1 and 3 vaccine strains are less effective in very poor countries for unknown reasons. Bivalent type 1 + 3 is adequately effective
Rotavirus	Diarrhea Fecal–oral spread	Monovalent live attenuated human rotavirus and multivalent animal/human reassortant rotavirus	Unclear, likely both homo- and heterotypic neutralizing antigens on VP4 and VP7. Other targets possible	Administered orally as 3 doses with pH buffer to young children	Generation of mucosal and perhaps systemic neutralizing immunoglobulins (Igs) and perhaps other Igs and cellular responses	In developed countries >90% against severe disease; in less developed countries 70% or less	First-generation rotavirus vaccine was associated with increased risk of intussusception. The mechanism of decreased protection in poor countries is unclear, as it is with oral poliovirus vaccine
Salmonella Typhi	Typhoid fever Fecal–oral spread	Live attenuated *Salmonella* Typhi (Ty21A) (note: a purified polysaccharide Vi antigen administered parenterally)	Target of protection in live vaccine unknown but the effector mechanism appears to be CD8 cytotoxic lymphocytes. Vi antigen is target of the parenteral polysaccharide vaccine	Ty21A given orally with buffer as 3 or 4 doses in a liquid formulation	Both systemic and mucosal responses of both cellular and humoral nature. For Ty21A, mucosal IgA and systemic cell-mediated immunity to O and H antigens predominate	Vi efficacy variable but generally >50%. Ty21A variable, depending on dose, enteric coating or liquid, and length of follow-up but optimally >80%	Vi conjugate parenterally administered vaccine not currently conveniently available but experimental vaccine highly effective

Table 27.2 Licensed parenterally administered vaccines for diseases that initiate at mucosal sites

Vaccine target	Primary disease/spread	Composition of vaccine	Target of protection	Dosage and formulation	Mechanism of protection	Efficacy	Comments
Anthrax	Pneumonia, diarrhea, cutaneous. Respiratory, oral, and skin spread	Sterile filtrate of anthrax culture supernatant	Several bacterial toxins, primarily protective antigen (PA)	3 intramuscular (IM) doses with alum adjuvant	Presumed serum immunoglobulin (Ig) anti-PA acting in lung	Not well studied, likely >85%	Vaccine effective against cutaneous and inhalation anthrax, efficacy against gastrointestinal disease not well studied
Diphtheria	Respiratory and systemic illness. Respiratory spread	Denatured diphtheria exotoxin	Diphtheria exotoxin	Formulated as a combination with pertussis, tetanus, and other vaccines such as hepatitis B virus (HBV) plus alum administered as 3 IM doses + boosters; boosters of older children and adults	Presumed serum Ig neutralization of exotoxin acting in respiratory tree and systemically	Likely >80% but not studied well in modern era	
Haemophilus influenzae type b (Hib)	Meningitis, sepsis, and pneumonia. Respiratory spread	Capsular polysaccharide (PS) alone or conjugated to carrier protein	Capsular polysaccharide	2–3 IM doses + boosters, sometimes formulated with alum adjuvant	Serum Ig anti-PS promotes bacterial phagocytosis and death at mucosal surfaces and systemically	>90% vs. invasive Hib disease	Realization that conjugating a PS antigen to a protein would enhance immune response (especially in young children) and elicit immunologic memory was a critical advance in vaccine development in the 20th century. Serum Ig appears to lower mucosal colonization
Hepatitis A	Acute hepatitis. Fecal–oral spread	Formalin-inactivated and purified whole virus grown in cell culture	Viral surface proteins, primarily VP1 and VP3	2 IM doses with alum adjuvant	Serum neutralizing Ig, primarily acting systemically	>95% for clinical hepatitis	Vaccine may have some efficacy when administered post exposure where it can only act systemically. Role of immunization in blocking initial replication in gastrointestinal tract not well known
Human papillomavirus (HPV)	Cervical, vulvar, vaginal, and anal cancer, genital warts. Sexual transmission	Recombinant virus-like particle	Structural L1 protein of several HPV genotypes including types 16 and 18	3 IM doses administered with alum or ASO4 adjuvant	Presumed serum Ig-mediated neutralization of HPV	>95% efficacy in preventing cervical intraepithelial neoplasia grade 2 or 3	Some evidence for modest cross protection against non-vaccine oncogenic HPV types. Protection at level of cervix from systemic antibody, presumably via transudation
Type A (H1 and H3) and B influenza viruses	Influenza. Respiratory spread	Inactivated virus, partially purified virus envelope, or more purified viral envelope proteins hemagglutinin (HA) and neuraminidase (NA)	Primarily to the viral HA, some protection of antibodies to the NA	Given annually as 1–2 IM doses against projected circulating viruses in strains H1/H3/B. No adjuvant for seasonal vaccine	Complex but felt to be serum neutralizing Ig	Highly variable depending on strain match and age of vaccine. Generally >60%.	Protection following inactivated vaccine correlates well with serum neutralizing response. However, after live viral vaccine, this is not the case. Hence the correlates of immunity can vary
Measles virus	Measles, respiratory disease followed by an exanthem. Respiratory spread	Live attenuated virus	Primarily the hemagglutinin protein but the F protein may play a minor role	Generally combined with live attenuated mumps and rubella. 2 subcutaneous (SQ) doses, no adjuvant	Presumed neutralizing serum Ig plays an important role but cell-mediated immunity may also be involved	>90%	A parenterally administered killed vaccine induced short-lived immunity and was associated with later development of atypical (very severe) measles that was felt to be immunologically mediated
Meningococcus (several types including A, C, Y, and W135)	Meningitis, sepsis, and pneumonia. Respiratory spread	Capsular PS alone or conjugated to carrier protein	Capsular polysaccharide of several types	Both as multivalent PS vaccines and mono- and multivalent conjugate vaccines, often with alum adjuvant	Serum Ig to capsule PS promotes complement-mediated lysis and phagocytosis of bacteria	>85% in first year after vaccination then decreases	Question of why generally benign nasal carriage of bacteria can, in rare cases, result in systemic infection remains unclear but presence of serum anti-PS limits this occurrence. No vaccine yet against group B PS
Pertussis	Severe and prolonged paroxysmal cough. Respiratory spread	Originally whole killed bacteria, now replaced with purified bacterial products	Several bacterial products associated with pathogenesis. Most important are pertussis toxin (PT), filamentous hemagglutinin (FHA), and fimbriae	Several combinations of PT with FHA ± fimbriae and pertactin with alum adjuvant	Presumed serum Ig inhibition of several bacterial products	>70% but highly dependent on case definition of pertussis	Frequency in adults is increasing likely due to waning immunity after childhood vaccination. Different manufacturers make a number of different vaccines with different components complicating analysis. There is currently a resurgence of epidemic pertussis in various locations
Pneumococcus	Bacteremia, meningitis, pneumonia. Respiratory spread	23 purified capsular PS or 13 PS protein conjugates	Capsular PS of multiple types	Multivalent mixtures of common capsular PS alone or conjugated to proteins with alum adjuvant. 3 IM doses + boosters	Likely presence of serum Ig anti-capsular Ig at mucosal as well as systemic sites	Varies depending on disease endpoint, >90% against invasive disease caused by vaccine serotypes	Conjugate pneumococcal vaccine (like the Hib vaccine) reduces mucosal carrier state. In addition, vaccine has substantial herd immunity effects by inhibiting transmission from vaccinated children to unvaccinated adults
Inactivated poliovirus	Fecal/oral or respiratory spread. Polio (flaccid paralysis)	Inactivated and purified poliovirus	Surface proteins of polio types 1, 2, and 3	3 doses administered either IM or SQ	Serum neutralizing Ig	>80% against paralytic polio	Another example of mucosal immunity induced by parenteral immunization inactivated polio vaccine (IPV) eliminated enteric infection in Netherlands), presumably by transudation of serum Ig. IPV also creates herd immunity
Tuberculosis. Respiratory spread	Primarily pulmonary disease but all organs can be infected. Respiratory spread	Live attenuated bacteria derived from multiple strains of *M. bovis* Bacillus Calmette-Guérin (BCG)	Targets of protective immunity unclear but likely to be multiple proteins and other antigens	Wide range of schedules; World Health Organization recommends single intradermal dose at birth	Cell-mediated immunity critical	50–100% efficacy against meningitis and disseminated disease. Much less vs. pulmonary disease.	One of few vaccines where protection not transferred by serum. Vaccine very widely used around world but functions poorly, not routinely used in US

The mucosa-associated lymphoid tissues (MALT) are comprised of anatomically defined lymphoid microcompartments, which serve as the principal mucosal inductive sites where immune responses are initiated. The Peyer's patches (PPs) in the small intestine, and the tonsils and adenoids at the entrance of the aerodigestive tract, are examples of such inductive sites. Also viewed primarily as inductive sites are the colon patches, appendix, and abundant small lymphoid follicles in the upper respiratory tract, small and large intestines, and the mesenteric lymph nodes (MLNs), although their importance for IgA B-cell immunity is still incompletely understood. By contrast, immune effector sites within the MALT are localized to the lamina propria and intraepithelial compartments of the mucosal membranes. Although there is an ongoing controversy as to whether B cells can undergo class-switch recombination from IgM to IgA and, perhaps, somatic hypermutations that increase antibody affinity *in situ* in the lamina propria, most evidence favors the inductive sites, especially PPs, as most important for IgA class-switch recombination in the gut. There is general agreement that all T-cell dependent responses require organized lymphoid tissues, primarily the PPs in gut-associated lymphoid tissue (GALT).

IgA antibody production is a hallmark of mucosal immune responses and important for the barrier function of the mucosal membrane. In particular, the neutralizing function of secretory IgA (SIgA) for antitoxin protection has been well documented, while protection against bacterial colonization has been less well established. There is good correlation between IgA levels and resistance against infection, as shown for rotavirus and cholera. SIgA also controls translocation of the microbiota to the draining mesenteric lymph nodes.

Of critical importance to the induction of mucosal immunity is the presence of a specialized epithelium over the organized lymphoid follicles. This follicle-associated epithelium (FAE) in the gut, for example, can sample antigen from the lumen and allow APCs in the PPs to activate specific CD4 or CD8 T cells, under strict MHC I and II control. As more extensively discussed in Chapter 13, antigens may either penetrate or be taken up in mucosal inductive sites through a variety of mechanisms. The so-called microfold (M) cells are most prominent over the PPs and have special properties for engulfment and transcellular transport of antigens across the epithelial barrier. In the subepithelial dome of the PPs, APCs can take up antigen and activate T cells present in the interfollicular region. An alternative mechanism has been proposed, namely that antigens can be taken up by dendritic cells (DCs), which protrude antigen-sampling dendrites across the tight junctions of the intestinal epithelium, but there is little direct evidence for an important function of such dendrites *in vivo*. What role this mechanism could have, if any, for IgA B-cell responses at mucosal membranes is also unclear, as DCs transport antigen to the draining MLNs. The inductive function of the MLNs for IgA responses in normal adult individuals is still not firmly established. In mice, IgA class-switch recombination is much less in the MLNs than in the PPs, and the activation-induced cytidine deaminase enzyme, required for class-switch recombination, is poorly expressed. However, the function of the MLN as a host for IgA memory B cells is an emerging field of substantial interest to vaccine development. Thus, while antigens may either penetrate or be taken up at mucosal inductive sites in several ways, most observations favor uptake via specialized FAE as the most important for IgA B-cell responses. An exception to this general rule appears to be the genital tract mucosa, which seems to lack FAE, and may rely on local DCs in the lamina propria for antigen sampling, perhaps similar to that reported for DCs in the intestinal lamina propria.

27-1 The problem with mucosal vaccines is tolerance.

In most animal model systems, the vast majority of foreign antigens encountered at mucosal membranes generate unresponsiveness, that is, tolerance

(Chapter 15). Classically, oral administration of soluble proteins results in unresponsiveness to a subsequent systemic challenge with the same protein. Oral tolerance is not only seen in the gastrointestinal mucosa, but, in fact, most mucosal membranes have tolerance as their default response pathway. Although oral administration of antigen has been reported to concomitantly induce both mucosal IgA and systemic tolerance, in most studies a strong specific IgA response requires breaking the tolerogenic pathway. This is achieved when pathogens colonize or invade the mucosal membrane because they elicit 'danger signals,' which are recognized by pathogen recognition receptors, leading to IgA B-cell responses and CD4$^+$ T-cell help, while avoiding tolerance. It can be assumed that live attenuated vaccines enhance their potency by expressing a selection of pathogen-associated molecular patterns (PAMPs), a property probably shared by the inactivated whole-cell vaccines that induce mucosal immunity (Tables 27.1 and 27.2). Therefore, one strategy is to incorporate one or more PAMPs into nonliving mucosal vaccines that mimic 'danger signals' observed during natural infection, and in this way overcome tolerance and increase immunogenicity.

The dominant mechanism behind mucosal tolerance is the induction of regulatory T (T_{reg}) cells following mucosal antigen exposure. These cells have a dampening effect on the effector phase of an immune response following systemic, and probably also local, mucosal immunization. For example, T_{reg} cells can reduce priming of T_H1 or T_H17 effector CD4$^+$ T cells and limit interferon-γ and interleukin (IL)-17 production, thereby preventing tissue damage. Inducible T_{reg} cells express the transcription factor Foxp3, and produce transforming growth factor-β or IL-10, which inhibit T-cell effector functions. The T_{reg} populations that exert dampening effects after mucosal antigen exposure have been incompletely studied. Nevertheless, modulation of the function of T_{reg} cells may prove to be a promising avenue in the search for future vaccine adjuvant strategies. For example, blocking Foxp3 T-cell development could be an effective strategy to enhance T-cell priming and antibody production, as has been initially explored for anti-cancer vaccines. But, at present, few adjuvant strategies have included modulation of T_{reg} cells as a means to achieve systemic immunity, let alone mucosal immunization.

27-2 There are unique compartmentalization and cell-migration pathways in mucosal immune responses.

In the early days of mucosal immunology, it was assumed that immune responses initiated at one mucosal site would be disseminated widely to multiple mucosal tissues. Had this common mucosal immune system existed, it could have meant that immunization by any mucosal route, such as oral immunization, could be used for inducing effective immune responses, not only in the gastrointestinal tract, but also in the airways and the urogenital tract. However, further work showed that mucosal immune responses were generally compartmentalized, not only between separate mucosal organs, but also between regions from the same mucosal organ, such as the gut proximal duodenum and rectum. Irrespective of sampling mechanism, antigens taken up at the mucosal surface are transported to draining lymph nodes by DCs or are directly captured by FAE and delivered to professional APCs in the PPs and presented to CD4$^+$ and CD8$^+$ $\alpha\beta$ T cells. Also, it appears that certain antigens may be processed and presented directly by epithelial cells to T cells located close to mucosal membranes, although the significance of this pathway is poorly understood, especially in man.

Antigen-triggered B and T cells in the inductive site, for example the nasopharynx-associated lymphoid tissue (NALT) or the PPs, leave the site of initial antigen encounter, transit through the lymph, enter the blood circulation, and then seed the non-organized mucosal membrane at selected sites (Figure 27.1). Hence, there is a distinct compartmentalization for antigen-activated

B and T cells, which home preferentially to the mucosa of origin where they differentiate into effector cells. The anatomic affinity of such cells is determined through site-specific integrins—'homing receptors' ($\alpha_4\beta_7$) and chemokine receptors (CCR9, CCR10) on mucosal lymphoid cells—and complementary tissue-specific endothelial cell adhesion molecules ('addressins') and chemokines that are expressed differentially in the various mucosal tissues. For instance, gut-homing PP IgA B cells and plasmablasts, as well as mucosal

Figure 27.1 The organization of the MALT relevant to vaccination. The inductive sites at the PP and NALT take up antigen through the FAE, which is a specialized type of epithelium with M cells channeling antigen to underlying DCs in the subepithelial dome. These cells process and present antigen to naive CD4+ T cells in the interfollicular spaces. The activated CD4+ T cells can differentiate into different types of effector cells: T_H1, T_H2, and some will become follicular helper cells (T_{FH}), which enter the germinal center to support and direct the IgA class-switch recombination and somatic hypermutations of antigen-activated B cells in the follicle. Following differentiation in the germinal center, the antigen-triggered B cells and T cells migrate to the mesenteric or cervical lymph nodes and then enter the circulation via the lymphatics. From the circulation, these cells can distribute widely to mucosal effector sites and home back to the lamina propria, preferentially to the tissue from where the cells originated, but also to other mucosal sites. Some activated B cells also traffic to the spleen, where memory B cells may reside after oral immunizations.

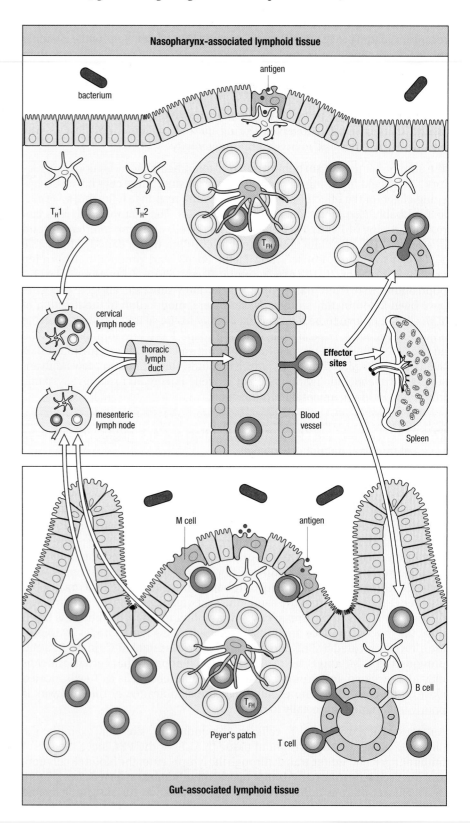

CD4+ and CD8+ T cells, express $\alpha_4\beta_7$ integrin that can specifically attach to vascular mucosal addressin cell adhesion molecule-1, a tissue-specific addressin expressed selectively on the high endothelial venules in the gut intestinal mucosa. As aforementioned, the homing properties are imprinted on the antigen-activated lymphocytes by mucosal DCs acting at the inductive sites, such as the MLNs. Interestingly, this property is the function of CD103+ DCs and is directly dependent on their ability to metabolize vitamin A because the concentration of retinoic acid appears to be responsible for the induction of *Aldh1a2* expression and aldehyde dehydrogenase activity, which have been linked to the imprinting properties of these DCs. During this process DCs can also be influenced by epithelial cells that produce co-factors in response to pathogens, which may then indirectly affect homing and differentiation of specific B and T cells. However, it is still unclear as to whether epithelial cells provide the retinoic acid to the DCs in the lamina propria. Moreover, chemokines produced by epithelial cells promote chemotaxis of immune cells with cognate chemokine receptors. For instance, colon CCL28, also known as mucosa-associated epithelial chemokine, selectively attracts IgA B cells and plasmablasts expressing the chemokine receptor CCR10; whereas CCL25 (TECK), which is abundantly produced in the small intestine, can selectively attract B and T cells expressing the CCR9 receptor to preferentially disseminate from the blood into the small intestinal mucosa. Thus, homing and chemokine receptors selectively expressed on antigen-activated lymphocytes (e.g., $\alpha_4\beta_7$, CCR9, and CCR10) and addressins and chemokines produced in the mucosal membrane help explain both the segregation of secretory immune responses from systemic immune responses, and the preferential dissemination of mucosal responses to mucosal sites.

27-3   There is preferential dissemination of mucosal immune responses after different routes of vaccination.

Compartmentalization within the mucosal immune system places constraints on the choice of vaccination route for inducing effective immune responses at the desired sites. Administration of antigens by rectal, vaginal, and more recently sublingual routes has been explored, but only for experimental purposes, and in humans mainly for studying SIgA antibody responses. In general the strongest immune response is usually obtained at the site of initial vaccine exposure and in anatomically adjacent mucosal sites. However, a few notable exceptions have been found, such as intranasal immunizations for genital tract antibody responses, which allows for more practical vaccine administrations than would otherwise be required. This has obvious practical implications for the development and deployment of mucosal vaccines against sexually transmitted diseases.

Traditional routes of mucosal immunization in humans include the oral and nasal routes. If appropriate antigens with inherent immunogenicity are used, either alone or coadministered with an effective adjuvant, oral immunization may induce a substantial antibody response in the proximal part of the small intestine, the ascending colon, the stomach, and in the mammary and salivary glands (Figure 27.2). Oral immunization, however, is relatively inefficient at evoking an IgA antibody response in the distal segments of the large intestine, tonsils, lower airway mucosa, or reproductive tract mucosa. Conversely, rectal immunization evokes a strong local antibody response in the rectum and sigmoid colon, a weaker response in the descending colon, and little, if any, response in the proximal colon or small intestine. In contrast, nasal or tonsillar immunizations in humans result in antibody responses in the upper airway mucosa and regional secretions (saliva, nasal secretions) without evoking an immune response in the gut. Nasal immunization, however, has been found to give rise to substantial IgA and IgG antibody responses in the human cervico-vaginal mucosae. The magnitude of the response achieved in the genital mucosa of women after intranasal immunization appears to be fully

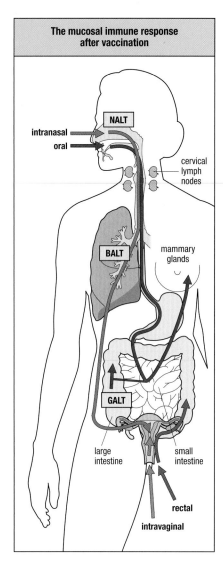

The mucosal immune response after vaccination

Figure 27.2 The compartmentalization of the mucosal immune response following oral and intranasal vaccination. This figure depicts the different routes used for mucosal immunizations. In humans, the most common routes are the oral and the nasal routes. A clear difference in the distribution of specific B-cell responses is elicited by different routes, such that oral immunization (red line) leads to preferential expression of IgA plasma cells in the proximal part of the small intestine, in the ascending colon, the stomach, and in the mammary glands. Intranasal vaccination (blue line) results in IgA responses in the upper airway mucosa, regional secretions (saliva, nasal secretions), and the cervico-vaginal mucosa without evoking an immune response in the gut. Rectal vaccination (purple line) gives strong IgA immunity in the rectum and sigmoid colon. Intravaginal vaccination (orange line) effectively stimulates B- and T-cell responses in the genital tract.

comparable to that seen when the vaccine is given by topical vaginal application. Whether this is also the case for priming of CD4+ and CD8+ T cells in the genital tract has been incompletely investigated, but recent studies in mice support vaginal rather than intranasal immunization for strong genital tract T-cell immunity.

Apart from the anatomical differences in the dissemination of SIgA antibody responses induced by orogastric and intranasal immunization respectively, the kinetics of the responses also appear to be markedly different. Several studies have shown that the intestinal immune response after oral immunization is rapid and relatively short lasting, but it is associated with long-lasting immunological memory. Data from field trials have shown that after oral cholera vaccination, protection from the acute intestinal IgA response appears to vanish after 6–9 months, while overall protection lasts for several years. It has been proposed that the longer duration of protection reflects the ability of IgA memory B cells to elicit a rapid recall response to a renewed exposure to the pathogen and subsequently the capacity to shut off an infection before it causes disease. In both mice and humans, the kinetics of the antibody-secreting cell recall response to a mucosal challenge is rapid, within days, as documented after oral booster vaccinations in previously cholera-vaccinated subjects, who exhibited strong duodenal-specific IgA antibody-secreting cell responses, peaking after 1 week and decreasing over a 5-month period. This finding is consistent with the notion that immunological memory following oral mucosal vaccination in humans can last for at least 5 years.

Because of the relative limitations of primarily oral vaccination for specific IgA immunity elicited at distant nonintestinal locations, other alternative routes have been explored. In this context, the potential of the sublingual route for administering vaccines is gaining interest. This is based on recent studies indicating that the sublingual route may promote induction of broad and exceptionally disseminated mucosal and systemic immune responses. Several studies have shown that sublingual administration of a variety of soluble and particulate antigens, including live and killed bacteria and viruses, can evoke a wide spectrum of immune responses in mucosal and extramucosal tissues, ranging from SIgA and systemic IgG antibody production to systemic cytotoxic T lymphocyte (CTL) responses. Although mainly studied in murine models, sublingual administration appears to require ten- to fiftyfold lower antigen doses than oral immunizations. Sublingual immunization induces strong immune responses and protection in the lungs of mice, as seen with both killed and live attenuated influenza vaccines. Similarly to nasal immunization, sublingual vaccination appears to effectively stimulate specific IgA and IgG immune responses and CTL in the genital tract. Also, it was recently reported that sublingual immunization with virus-like particles from human papillomavirus provided protection against a genital human papillomavirus challenge infection. Moreover, sublingual immunization with an experimental *Helicobacter pylori* vaccine or the enterotoxigenic *Escherichia coli* (ETEC) and *Vibric cholerae* whole-cell vaccines given together with cholera toxin (CT) adjuvant effectively induced B- and T-cell responses in the stomach and intestine respectively, resulting in significant protection against challenge infections.

Mucosal vaccine formulations and delivery systems.

The underlying strategies behind the still relatively few licensed mucosal vaccines for human use are largely empirical and draw attention to the need for novel approaches to mucosal vaccine development that are currently being actively pursued by researchers across the globe. Significant problems have been encountered in translating successful vaccines in animal models into clinical practice. It remains a challenge to carry out clinical studies on new vaccines in target populations, usually in developing countries with poor infrastructures.

27-4 The ability to divide, present native antigens, and stimulate innate immunity are key features of live attenuated vaccines.

Live attenuated bacteria and viruses have several attractive features for use as mucosal vaccines. Being based on the target pathogen, they usually have properties allowing them to colonize and multiply at the appropriate mucosal site. Their ability to self-replicate also provides sufficient antigen concentrations to overcome the requirement for relatively large amounts of antigen to stimulate mucosal IgA immunity. These amounts are more readily reached using a self-replicating live vaccine, which, as an extra advantage, continuously produces antigen, thereby prolonging stimulation of the mucosal immune system in comparison with inactivated or subunit vaccines. In addition, live vaccines, by definition, express all or most of the antigenic targets of protective immunity present in the parental pathogen. Finally, an important advantage compared with mucosally administered subunit vaccines in particular, is that the live attenuated vaccines are generally 'self-adjuvanting' by providing an appropriate selection of PAMPs. Hence one strategy to explore in nonliving mucosal vaccines would be to mimic the 'self-adjuvanting' potency of the few live attenuated mucosal vaccines that are listed in Table 27.1. Nevertheless, the preparation of an attenuated live mucosal vaccine is generally a multi-step process that requires obtaining sufficient attenuation to render the vaccine safe and genetically stable while retaining sufficient replicative capacity to permit immunization.

The development of the licensed rotavirus vaccines available on the market today serves as an example where considerations were brought into place to launch a successful live attenuated mucosal vaccine. It was already known that a natural rotavirus infection conveyed strong protective immunity against subsequent severe disease, irrespective of serotype. Hence, a live attenuated rotavirus vaccine was predicted to do the same. Therefore, attenuation by passage of the candidate bovine or human virus strains through culture with host cells in multiple steps was undertaken. The final result was a stable attenuation, although the molecular basis for the attenuation is currently unknown. A more refined way to achieve attenuation is the introduction of selected gene deletions as used in the *Salmonella* Typhimurium-based experimental live vectors for expression of recombinant vaccine antigens. It should be noted, though, that even when trying to mimic already licensed attenuated vaccines, such as the oral Ty21A typhoid vaccine, selected gene deletions can fail to generate a safer vaccine. Hence, as with all live attenuated vaccines, including the licensed rotavirus vaccines, there are always concerns about stability and safety. Another difficult problem to address, especially for live attenuated vaccines, is the extent of attenuation without loss of immunogenicity. This problem has been especially prominent for live attenuated bacterial vaccines against enteric infections intended for use in both industrialized and developing countries: marked differences have been noted in adverse reactions and poor vaccine 'take,' relating to nutritional status, gut microflora, and preexisting natural or maternal antibodies. To date these problems have prevented development of effective safe live vaccines against a number of enteric bacterial pathogens, for example *Shigella*.

27-5 Inactivated whole-cell bacterial oral vaccines are effective.

There is only one type of inactivated mucosal vaccine licensed for human use, namely the oral whole-cell cholera vaccine with or without added cholera toxin B subunit (CTB); it is therefore difficult to define with certainty which particular features of this vaccine explain its protective properties. Thus, at the moment we cannot define which properties of this vaccine could be further explored to successfully launch other oral mucosal vaccines. The following facts are, however, well established. (1) This vaccine induces substantial IgA antibacterial immunity: mainly anti-lipopolysaccharide (LPS) antibodies in the intestine and, when CTB is included, an IgA antitoxin together with long-lasting memory IgA responses to both these antigens. (2) As shown in experimental challenge models, and supported by observations in breast-fed children in Bangladesh, both anti-LPS and anti-CTB antibodies can independently protect against cholera disease by inhibiting bacterial colonization and toxin binding respectively. These antibody specificities have been found to act synergistically, strongly improving protection. (3) As observed in a large field trial, CTB-specific immunity added significantly to protection stimulated by the whole-cell-only vaccine for the first 9 months of follow-up, which closely corresponded to the IgA antibody response. By contrast, protection at later times—namely 50–60 % protection for the next 2–3 years in adults—was similar, irrespective of the CTB component, arguing for a predominant antibacterial IgA long-term memory function, perhaps reinforced by repeated bacterial exposure. Finally, (4) the antitoxin response induced by the CTB component also conferred significant short-term protection against ETEC. Based on these observations an analogous mucosal vaccine against ETEC diarrhea has been developed, which is composed of a cocktail of inactivated *E. coli* bacteria engineered to express large amounts of four different colonization antigens together with a heat-labile enterotoxin B-like B subunit. This vaccine, when given orally to mice, induces strong intestinal IgA antibody responses against each of the different components in the vaccine, especially when given together with an enterotoxin-based adjuvant. The vaccine is now being produced under good manufacturing practice conditions for further studies of safety and immunogenicity in humans. Other oral mucosal vaccines that might be developed along similar principles include vaccines against, for example, *Campylobacter*, *Salmonella*, and *Shigella* infections.

27-6 Plant-based 'edible' vaccines may represent a strategy for mucosal vaccination.

For more than a decade there has been work exploring plant expression of relevant antigens for use in edible vaccines. However, despite considerable efforts and substantial investments such vaccines are still at an early stage for clinical use. Successful production of protective vaccine antigens has been achieved in many types of plants including tobacco, tomatoes, lettuce, potatoes, and rice. An attractive aspect of recombinant protein expression in plants is that large amounts of protein can be produced at very low cost. Clinical trials initially reported very convincing immunogenicity data, but optimism has waned because of difficulties with regulatory authorities and the fact that most successful trials targeted norovirus or enterotoxin-induced diarrheal diseases, while trials of other vaccine candidates were not successful. A fundamental problem has been dosage of vaccine and quality control of the vaccine production. Therefore, the early ideas that inexpensive and effective vaccination could be instituted simply by eating, for example, a vaccine-containing banana have long since been abandoned. A difficult technical problem has also been that glycosylation sites differ between proteins expressed in mammalian and plant cells, resulting in immune responses directed against 'incorrect' polysaccharides rather than against the target pathogen protein. Nevertheless, clinical trials are underway against several pathogens, but the issue of tolerance remains controversial in the edible vaccine field. The experimental MucoRice-CTB edible vaccine against cholera and ETEC holds promise, and it appears to provide a stable and cold-chain-free alternative for mucosal vaccination.

27-7 Targeted mucosal vaccines can use lectins and microparticles.

Because uptake of orally administered antigen into PPs is important for inducing immune responses in the gastrointestinal tract, an attractive strategy to achieve gut intestinal immunizations would appear to be to facilitate vaccine targeting to M cells that overlay the PPs. Three principal strategies have been used. The first has explored lectins that bind to M cells, such as the *Ulex europaeus* agglutinin-1, which binds to the apical surface of the M cell. Experimental studies have shown that lectin-coated beads can be effectively taken up by PPs and be strongly immunogenic. Unfortunately, *Ulex europaeus* agglutinin-1 binds to human enterocytes, not selectively to M cells, and no specific lectin has so far been found suitable for targeting of human M cells. Perhaps more promising is the second strategy based on the generation of monoclonal antibodies that specifically bind to M cells through recognition of surface molecules such as $\alpha(1,2)$-fucose moieties. However, no clinical testing of this strategy has yet been done. A third alternative is the delivery of the vaccine in particles that target the inductive site. This was considered a very promising technology when it was first demonstrated in the 1990s that synthetic biodegradable microparticles were taken up by M cells, a function which was shown to be dependent on size, surface charge, and hydrophobicity. Despite technical difficulties and other problems, interest still remains in synthetic micro/nanoparticle delivery of oral vaccines. However, realistically this approach will likely entail the combination of particles and lectins targeting antibodies or substances, such as CTB or invasins, that specifically bind to M cells. Thus, we are still far from the potentially exciting approach of a mucosal vaccine based on M-cell targeting.

Limitations of mucosal vaccines due to unique conditions in the tropics and the age of the vaccinees.

Although initially noted for oral polio vaccine, several other live oral vaccines, including oral vaccines against rotavirus and cholera, have underperformed in developing countries compared with industrialized countries. This feature, sometimes referred to as the 'tropical barrier' to oral vaccines, is attributed mainly to chronic environmental enteropathy, also called tropical enteropathy. This is characterized by malabsorption, which is associated with mucosal inflammation and blunting of small intestinal villi. Factors that may contribute to chronic environmental enteropathy include poor sanitation and intestinal flora overgrowth, as seen in children living under conditions of extreme poverty. Other factors that might reduce the efficacy of oral vaccines in developing countries include deficiencies in micronutrients such as vitamin A (retinoic acid) and zinc, which can influence the response to oral adjuvants and vaccines by affecting discrete subpopulations of intestinal DC and T cells. Also, persistent activation of the gut immune system by infectious agents, such as helminths, or concomitant immunosuppressive viral and bacterial infections could simply exhaust the immune system. Likewise, in breast-fed infants and children, in developing countries, antibacterial antibodies in the breast milk could interfere with vaccine immunogenicity or 'take.' Possible strategies for overcoming the reduced vaccine efficacy in developing countries could include, for example, giving the vaccines together with agents that improve gut integrity, such as zinc, vitamin A, and possibly probiotics. Of course, withdrawing breast milk shortly before oral vaccination and providing antihelminth treatment prior to oral immunization could be other means of improving mucosal vaccination. Finally, it may be that immunization routes such as transcutaneous or sublingual are not affected by tropical and age barriers and may therefore be more suitable routes for vaccination against mucosal disease in developing countries.

Mucosal adjuvants and their function.

Over the last decade our knowledge about different substances that exert adjuvant effects has grown immensely. This has, however, not helped in the design of more effective mucosal vaccine adjuvants. CT and ETEC heat-labile enterotoxin holotoxins are among the most effective mucosal adjuvants, and their adjuvant function was first described in the 1980s. Unfortunately, these holotoxins are too toxic to be used in the clinic. Despite this, it is believed that the mechanism for the adjuvanticity of CT and heat-labile enterotoxin can be modified and exploited so that better and safer nontoxic constructs can be produced.

Because of the requirement for adjuvants in most nonliving vaccines, there is a growing interest in adjuvant development. However, we must conclude that little information is available as to how adjuvants work *in vivo*. All types of immune responses can be augmented: antibody formation as well as cell-mediated immunity, including CTL activity. Historically, adjuvants are mostly of microbial origin and frequently have induced various degrees of inflammation. In fact, the adjuvant efficacy could in many ways be linked directly to the level of local inflammation. Thus, in this regard adjuvants can also be responsible for tissue destruction, pain, and distress, which are key components to consider in any vaccine development. Hence, the selection of adjuvant is often as critical as which antigen or combination of antigens to include in a vaccine. Adjuvants profoundly affect the vaccine properties and function, such as quality and longevity of the vaccine response. Recent experimental and human data have documented that the choice of adjuvant can dramatically affect the long-term protective effect of a vaccine. For example, the formation of germinal centers appears to directly relate to the size of the long-lived plasma cell pool in the bone marrow as well as to the presence of memory B cells following vaccination. Because of the recent progress in identifying surface markers for human (CD27) and mouse memory B cells (CD73, CD80, and PD-L2) one can predict that many new studies will identify strategies for generating

Figure 27.3 An ideal system for mucosal adjuvant function. The system outlined schematically in the figure is the hypothetical ideal construct for an effective optimally adjuvanted mucosal vaccine based on the cholera toxin molecule. It capitalizes on the specific targeting of the antigen and adjuvant linked together in a complex that can bind to and be taken up by antigen-presenting cells (APCs) in the mucosal immune system. The complex provides sufficient shielding to protect the antigen and adjuvant against degradation as well as facilitating the specific targeting to the APC across the mucosal barrier. The CTA1 adjuvant component may be exchanged for a TLR agonist such as CpG or mutated to reduce its ADP-ribosylating function. On the other hand the targeting element could be either the CTB moiety of the holotoxin or an alternative targeting unit, such as the DD dimer derived from *Staphylococcus aureus* protein A, which by virtue of its ability to bind to surface immunoglobulin, targets vaccines to B cells and follicular dendritic cells.

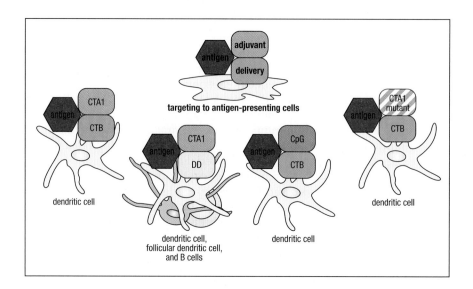

functionally effective memory B cells and principles will be defined for adjuvant selections that convey long-term protective immunity.

In principle, adjuvants may exploit three types of modulating effects on the innate immune system that will impact on the adaptive immune response and promote improved immunogenicity (Figure 27.3). The first type of effect occurs with adjuvants that contain Toll-like receptor (TLR) or NOD-like receptor (NLR) agonists. These are dramatically dominated by the former, such as MPL, a chemically modified derivative of LPS and a TLR4 agonist, or CpG-oligodeoxynucleotide (TLR9 agonist), which have been found to be the most effective. However, relative to the bacterial holotoxins CT and heat-labile enterotoxin, they are poor mucosal adjuvants, especially for mucosal IgA responses (Figure 27.3). The holotoxins, on the other hand, do not use TLRs or NLRs to activate the innate immune system. A third category of mucosal adjuvants are functionally less distinct. These are oil-in-water emulsions, immune stimulating complexes containing cytokines, such as IL-1, or mucoadhesive substances, such as chitosan. This third category exploits many different mechanisms; eventually it may be shown that some can act through TLR binding, but in most cases their function is unknown and, hence, we cannot explain the mechanism of action of most mucosal adjuvants.

27-8 The ADP-ribosylating toxins and their derivatives are mucosal adjuvants.

The bacterial enterotoxins or derivatives thereof, typified by CT and *E. coli* heat-labile enterotoxin, are AB_5 complexes and carry an A1 subunit that is an ADP-ribosylating enzyme and five B subunits that bind distinct ganglioside receptors present on most nucleated mammalian cells. Because gangliosides reside in the cell membrane of all nucleated cells the binding is promiscuous and hence the enzyme can affect virtually all cells in the human body. Following binding to the receptor the holotoxins ADP-ribosylate the Gsα membrane protein and stimulate adenylate cyclase, leading to an increase in intracellular cyclic adenosine monophosphate. This property makes them highly toxic, but also contributes to their excellent adjuvant function *in vivo*. Thus coupling antigens to CT and repeated oral immunization is a highly effective way to generate mucosal antibody responses to soluble antigens and haptens (Figure 27.4). The B subunit of CT and heat-labile enterotoxin is responsible for binding to the ganglioside receptors on the cell membrane. Following mucosal administration, CT and heat-labile enterotoxin have been reported to host variable effects on priming of T_H1, T_H2, T_H17, and T_{reg} cells. Whereas CT has mostly been associated with T_H2 responses and somewhat weaker T_H1 immunity, heat-labile enterotoxin has been found to have a more balanced effect on priming of T_H1

Figure 27.4 Specific gut IgA antibody responses require multiple oral immunizations. Oral mucosal immunizations require multiple administrations of the vaccine in order to stimulate strong antigen-specific IgA plasma cell responses in the gut lamina propria. The first chart in panel a depicts the specific IgA response against the hapten NP (4-hydroxy-3-nitrophenyl) conjugated to CT. The IgA anti-NP antibody-secreting cell numbers (SFC, spot forming cells) in the small intestinal lamina propria increase dramatically after a second or third immunization. CT on its own elicits no response. NP-specific IgA B cells in Peyer's patches undergo significantly more hypermutation after a third immunization (3×), as shown in the second chart. Using the 107 G to T mutation as a marker of high affinity maturation, the third chart shows that repeated immunization increases the affinity of the antigen-specific IgA response to NP. The image in panel b shows the presence of NP-specific IgA plasma cells in the lamina propria following three oral immunizations with NP-CT. A PP labeled with green fluorescent antibody to GL7, a sialic acid glycan moiety on the surface of germinal center cells, is clearly visible at the bottom of the image. In the lamina propria, roughly 10% of all IgA plasma cells are NP-specific (red). These cells exhibit strong clonal relationships with a relatively low number of dominant clones.

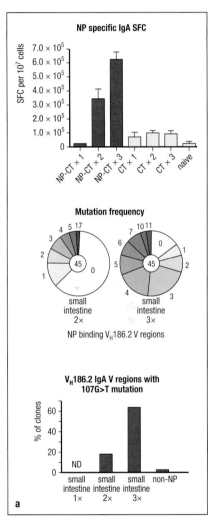

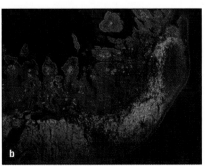

and T_H2 types of immunity. Moreover, CT has been found to depend on IL-17 for some adjuvant functions, but whether this depends on the antigen or is a general requirement for T_H17 priming is too early to tell. Also, a few studies have pointed to the conclusion that T_{reg} cells may be selectively targeted by CT, but this needs to be confirmed. Other studies have indicated that CD8$^+$ T cells and DCs are selectively lost following CT administration.

A major limitation of the holotoxins is the promiscuous binding to the ganglioside receptors present on all nucleated cells, including epithelial cells and nerve cells. This binding has rendered the holotoxins unattractive for clinical use because when used for intranasal immunizations it was found that an intranasal flu vaccine with heat-labile enterotoxin as an adjuvant caused facial nerve paralysis (Bell's palsy) in some patients; moreover, when used orally, some cases of diarrhea were observed in vaccinated subjects. The problem with toxicity has prompted a search to find alternative strategies to lower toxicity but retain adjuvant function. A strategy to introduce single-amino-acid point mutations, most frequently in the A1 subunit, dramatically lowers toxicity but still retains adjuvant function. Unfortunately, mutants, such as the LTK63 adjuvant, are also associated with cases of Bell's palsy in clinical trials after intranasal administration. More promising, though, are holotoxin derivatives that are mutated in the A1 binding region that connects to the A2 region, leaving the nicotinamide adenine dinucleotide-binding and enzymatically active site intact. In clinical trials the LTR192G mutant and, recently, the LTR192G/L211A double mutant were effective mucosal adjuvants. Studies are underway to dissect their adjuvant function in humans and preliminary findings convey optimism.

A second alternative strategy has capitalized on the fact that the CTA1 enzyme is the most critical component for the adjuvant function. Therefore, a novel molecule was constructed by combining the enzymatic activity of CTA1 with a synthetic dimer of the D fragment of *Staphylococcus aureus* protein A in a gene fusion protein. The resulting CTA1-DD adjuvant was completely devoid of the ganglioside binding because it lacked the CTB element of CT, while it had retained the full enzymatic activity of the CTA1 enzyme. CTA1-DD was found to be nontoxic and safe in mice and monkeys, and importantly was not found to interact with the nervous system following intranasal administration. No inflammatory response was seen at the site of administration and, hence, CTA1-DD represented a truly noninflammatory vaccine adjuvant. Numerous studies have testified to its potency as a mucosal adjuvant, comparable to that of CT and heat-labile enterotoxin, but completely nontoxic and safe. Thus, CTA1-DD represents an exciting new generation of targeted vaccine adjuvants, which holds promise for further exploitation in human vaccines. Following intranasal, rectal, or parenteral immunizations in mice it effectively augmented a wide range of immune responses: from IgA antibodies at mucosal sites to IgG antibodies in the serum, or CD4$^+$ T cell and CTL immunity in different tissues. The adjuvant effect of CTA1-DD has been shown to depend on an intact CTA1 enzyme. In fact, the disruption of this activity through single point mutations in the CTA1 completely blocked the adjuvant function. Unexpectedly, though, this inactivation facilitated the induction of tolerance against incorporated peptides when such constructs were given intranasally. Thus, the CTA1R7K-OVA-DD, hosting a peptide from ovalbumin, was shown to specifically stimulate tolerance mediated by IL-10-producing T_{reg} cells, and in this regard the mutant molecule represented a vector for antigen-specific tolerance induction, potentially useful therapeutically against autoimmune disease.

27-9 Autoimmune disease can be treated by mucosal vaccination.

Theoretically, targeting of DC subsets via, for example, the CD103 surface molecule may be a means to stimulate mucosal immunity by parenteral vaccination, because DCs will imprint mucosal homing of primed T cells. Immature

DCs residing in mucosal tissues are known to take up antigen and, if maturation occurs, migrate to regional lymph nodes. In the secondary lymphoid tissues, DC immigrants expressing co-stimulatory molecules are highly effective at priming T cells. However, whether poorly activated DC immigrants or resident DCs that take up antigen via the blood promote tolerance in the draining lymph nodes by stimulating T_{reg} differentiation has been little investigated. It appears that CD103+ DCs that travel from mucosal tissues under steady-state conditions support T_{reg} development and tolerance. Hence, exploiting this DC function for treatment of autoimmune conditions may be interesting.

Mucosal administration, mostly oral or intranasal, has been found effective at ameliorating or preventing autoimmune disease in a large number of animal, primarily murine, models. For example, oral feeding of collagen or insulin can prevent collagen-induced arthritis, a model for rheumatoid arthritis, and type 1 diabetes respectively in mouse models. Using mucosal tolerance for treatment of various autoimmune diseases has attracted much attention. The last decade has seen many clinical trials and, in particular, rheumatoid arthritis patient trials have proven that this interest may be justified. However, it has consistently been difficult or impossible to achieve the same results in the human trials as in the rodent models of rheumatoid arthritis or multiple sclerosis. Notwithstanding these negative findings, the strategies to vaccinate against autoimmune diseases in humans using mucosal pathways are still being pursued. Except for administering large doses of antigen, few tolerogenic vectors that require small amounts of antigen have been described. Type II collagen, for example, has been the antigen of choice in arthritis oral clinical trials principally because it is the structural component of cartilage against which both B- and T-cell immunity is probably directed in rheumatoid arthritis patients. Therefore, one strategy to improve the efficacy of treatment would be to more specifically target the antigen to the tolerogenic pathway and in this way enhance unresponsiveness and curb effector T-cell activity and tissue destruction. This could be achieved by exploiting certain targeting and immunomodulating proteins. Two such strategies have been described. The first is using CTB or heat-labile enterotoxin B molecules as carriers of whole protein to stimulate mucosal tolerance. A vast literature has documented the efficacy of this strategy not only in mice, but also in a human trial, where CTB linked to a Behçet's disease-associated peptide was given orally to patients to treat uveitis. The second strategy explores the enzymatically inactive mutant from CTA1-DD, the CTA1R7K-DD molecule, as aforementioned. Incorporating known peptides from disease-associated endogenous proteins, such as collagen, was found to greatly reduce autoimmunity in a mouse model of rheumatoid arthritis. Hence, a vaccine based on the enzymatically inactive CTA1R7K-COL-DD molecule could induce antigen-specific tolerance. The CTA1R7K-X-DD platform is now being explored in several models of autoimmune diseases such as multiple sclerosis and type 1 diabetes. Collectively, CTB/heat-labile enterotoxin B and CTA1R7K-X-DD could be termed tolerance-inducing vectors. It may be speculated that such molecules used in tolerance-inducing vaccines would introduce a shift of paradigm in the treatment of autoimmune diseases.

Summary.

Safe and effective mucosal vaccines currently in the market are few (Table 27.1), but many more are being developed and some are ready for clinical trials. However, it is fair to say that despite early success with the live attenuated oral polio vaccine only limited success has been seen in the nearly 50 years that have followed. This is partly due to a lack of safe mucosal adjuvants that would make nonliving vaccines more effective and the fact that many live attenuated vaccines are either genetically unstable or insufficiently or overly attenuated.

Therefore, intense research is ongoing to identify better and more effective mucosal adjuvants and studies in animals and in humans are dissecting routes of administration, antigen dose requirements, and formulations. However, in recent years live attenuated mucosal vaccines against influenza virus, rotavirus, and typhoid fever have been launched and a parenteral vaccine against human papillomavirus has also been highly successful. An inactivated oral cholera vaccine has provided proof-of-concept that mucosal vaccines based on nonliving components can be quite effective. But why this vaccine is effective while other nonliving vaccines are not is far from clear. Presently, we also suffer from a knowledge gap as to why so many candidate vaccines in experimental small animal models have been highly effective, while so few have proven to confer protection against infection in human trials. Notwithstanding this, the overall prospects for mucosal vaccine administration now look brighter, as the mechanisms responsible for priming of mucosal immune responses are unraveled. Based on the newly acquired knowledge on adjuvant activation of innate immunity at mucosal membranes and a better understanding of the buildup of IgA B-cell responses and long-term memory, we should expect many new and effective mucosal vaccines in the future.

Further Reading.

Castellino, F., Galli, G., Del Giudice, G. et al.: **Generating memory with vaccination.** *Eur. J. Immunol.* 2009, **39**:2100–2105.

Czerkinsky, C., and Holmgren, J.: **Enteric vaccines for the developing world: a challenge for mucosal immunology.** *Mucosal Immunol.* 2009, **2**:284–287.

Greenberg, H.B., and Estes, M.K.: **Rotaviruses: from pathogenesis to vaccination.** *Gastroenterology* 2009, **136**:1939–1951.

Holmgren, J., and Czerkinsky, C.: **Mucosal immunity and vaccines.** *Nat. Med.* 2005, **11**:S45–53.

Holmgren, J., and Kaper, J.: **Oral cholera vaccines,** in Levine, M.M. (ed): *New Generation Vaccines.* New York, Marcel Dekker, 2009:499–509.

Kuolee, R., and Chen, W.: **M cell-targeted delivery of vaccines and therapeutics.** *Expert Opin. Drug Deliv.* 2008, **5**:693–702.

Lambrecht, B.N., Kool, M., Willart, M.A., et al.: **Mechanism of action of clinically approved adjuvants.** *Curr. Opin. Immunol.* 2009, **21**:23–29.

Li, Z., Palaniyandi, S., Zeng, R., et al.: **Transfer of IgG in the female genital tract by MHC class I-related neonatal Fc receptor (FcRn) confers protective immunity to vaginal infection.** *Proc. Natl Acad. Sci. USA* 2011, **108**:4388–4393.

Lycke, N.: **Mechanisms of adjuvant action,** in Singh, M. (ed): *Vaccine Adjuvants and Delivery Systems.* Hoboken, NJ, John Wiley & Sons, 2007:53–79.

Lycke, N., and Bemark, M.: **Mucosal adjuvants and long-term memory development with special focus on CTA1-DD and other ADP-ribosylating toxins.** *Mucosal Immunol.* 2010, **3**:556–566.

Pasetti, M.F., Simon, J.K., Sztein, M.B., et al.: **Immunology of gut mucosal vaccines.** *Immunol. Rev.* 2011, **239**:125–148.

Plotkin, S.A.: **Correlates of protection induced by vaccination.** *Clin. Vaccine Immunol.* 2010, **17**:1055–1065.

Pulendran, B.: **Learning immunology from the yellow fever vaccine: innate immunity to systems vaccinology.** *Nat. Rev. Immunol.* 2009, **9**:741–747.

Pulendran, B., Li, S., and Nakaya, H.I.: **Systems vaccinology.** *Immunity* 2010, **33**:516–529.

Sette, A., and Rappuoli, R.: **Reverse vaccinology: developing vaccines in the era of genomics.** *Immunity* 2010, **33**:530–541.

Ye, L., Zeng, R., Bai, Y., et al.: **Efficient mucosal vaccination mediated by the neonatal Fc receptor.** *Nat. Biotechol.* 2011, **29**:158–165.

Yoshida, T., Mei, H., Dorner, T., et al.: **Memory B and memory plasma cells.** *Immunol. Rev.* 2010, **237**:117–139.

PART VII

SPECIFIC IMMUNE-MEDIATED DISEASES OF MUCOSAL SURFACES

Celiac Disease

28

Celiac disease is a chronic, T cell-mediated, inflammatory disorder of the small intestine resulting from an inappropriate immune response to gluten. Gluten is a complex of proteins present in wheat, barley, and rye. Wheat gluten proteins consist of α-, γ-, and ω-gliadins as well as high- and low-molecular-weight glutenin subcomponents. Both gliadin and glutenin proteins can induce the disease, which is treated with a lifelong gluten exclusion diet. The mucosal pathology associated with celiac disease is localized to the proximal small intestine and is characterized by villus atrophy and prominent infiltration of leukocytes into the epithelium and lamina propria. The mucosal absorptive surface is reduced as a consequence of villus atrophy, leading to malabsorption that may result in anemia and steatorrhea. Additional symptoms, including fatigue, infertility, neurological manifestations, and enamel defects, may not indicate an intestinal disease, and some patients have few or even no symptoms. The eclectic nature of the symptoms is a challenge to clinicians, and consequently many patients are not diagnosed or experience long delays before diagnosis.

The celiac lesion.

Even though celiac disease is an intestinal inflammatory disorder induced by a food antigen, the disease and its associated histology is more similar to organ-specific autoimmune disorders than to food allergy or intestinal inflammatory bowel disease. Celiac disease, like type 1 diabetes, is mediated by CD4$^+$ and CD8$^+$ T cells, and is associated with the presence of autoantibodies that can precede the disease and are not thought to be directly pathogenic. The multifaceted aspect of celiac disease makes the disorder particularly fascinating and highly relevant to mucosal immunology in particular. We begin with a discussion of the inflammatory lesion associated with celiac disease.

28-1 Villous blunting and crypt cell hyperplasia are characteristic histological features.

The intestinal lesion in celiac disease is characterized by loss of the finger-like villous structures that normally increase the absorptive surface of the gut and by increased cell division of epithelial cells in the crypts (crypt cell hyperplasia). The development from the normal state to an overt lesion follows a uniform pattern, often described as Marsh stages (Figure 28.1). Marsh 0 is the normal state. The Marsh 1 stage is characterized by an increased number of intraepithelial lymphocytes but without crypt cell hyperplasia or villous

Normal	Marsh 1 - Increased numbers of IELs	Marsh 2 - Crypt cell hyperplasia	Marsh 3c - Absence of villi, crypt cell hyperplasia
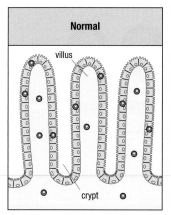	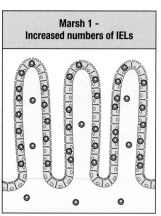	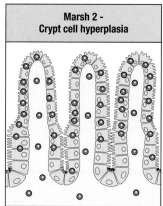	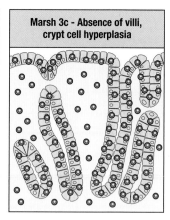

Figure 28.1 Schematic depiction of histological alterations of the small intestinal mucosa in celiac disease. The normal mucosa with slender intestinal villi and few leukocytes is shown on the left, and the Marsh stages of development of the celiac lesion are shown on the right. Marsh 1 is characterized by increased numbers of intraepithelial lymphocytes (IELs) but no crypt cell hyperplasia; Marsh 2 by crypt cell hyperplasia but no villous blunting; and Marsh 3c by absence of villi, crypt cell hyperplasia, and massive infiltration of leukocytes into the epithelium and lamina propria. The degree of villous blunting increases from Marsh 3a to Marsh 3c, with total loss of villi in Marsh 3c.

blunting. In Marsh 2, crypt cell hyperplasia occurs without villous blunting. Marsh 3 is often divided into Marsh 3a, Marsh 3b, and Marsh 3c. The degree of villous shortening increases from Marsh 3a to Marsh 3c, with total loss of villi in Marsh 3c. Different biopsies from the same patient may reveal varying degrees of mucosal alteration because the celiac lesion is not uniformly expressed throughout the intestine.

28-2   Intraepithelial lymphocytes are prominent.

The histological lesion in active celiac disease is strikingly different from that of other T cell-mediated enteropathies, such as graft-versus-host disease and autoimmune enteropathy. In celiac disease, the crypts are intact and regenerative, but the surface epithelium is profoundly altered. Furthermore, a major infiltration of lymphocytes in the epithelium can develop into enteropathy-associated T-cell lymphoma in rare cases. In contrast, crypt destruction is common in other enteropathies, and intraepithelial lymphocytosis is minor or appears late in the disease. Enteropathy-associated T-cell lymphoma has not been reported in intestinal inflammatory disorders other than celiac disease. These histological differences suggest that the surface epithelium is preferentially targeted and intraepithelial lymphocytes are abnormally activated in celiac disease.

In the normal human small intestine, three major lineages of intraepithelial lymphocytes (IELs) participate in the defense of the mucosal surface. The most prominent IEL is the cytolytic CD8αβ$^+$CD4$^-$ T-cell receptor (TCR)αβ population; CD8$^-$CD4$^+$ TCRαβ$^+$ and CD8$^-$CD4$^-$ TCRγδ$^+$ populations are also present (Table 28.1). The CD8$^+$ TCRαβ$^+$ IELs in humans simultaneously encompass characteristics of conventional adaptive T cells (expression of heterodimeric CD8αβ co-receptor and antigen-driven variable T-cell receptors) and innate T cells (expression of natural killer receptors recognizing inducible nonclassical major histocompatibility complex (MHC) class I molecules). In the mouse, the adaptive and innate properties are carried by two distinct subsets: a conventional TCRαβ CD8αβ$^+$CD4$^-$ IEL subset restricted by classical MHC class I molecules and the absence of natural killer receptors; and an innate CD8αα$^+$CD4$^-$ TCRαβ$^+$ IEL subset restricted by nonclassical MHC class I molecules and the presence of natural killer receptors. Interestingly, unlike in the peripheral blood and the liver, human IELs selectively express natural killer (NK) receptors of the C-type lectin family. The most represented natural killer receptors are NKRP1-A, receptors of the CD94/NKG2 family, and the NKG2D receptor. Under physiological conditions, IELs express high levels of inhibitory CD94/NKG2A receptors and low levels of activating NKG2D receptors. As IELs are effector cells, they can exert TCR-mediated cytolysis *ex vivo*. However, the level of cytolysis is low in the absence of additional stimuli. Furthermore, the IELs cannot kill NK cell targets. Another important difference between mouse

Table 28.1 Characteristics of human and mouse intraepithelial lymphocytes (IEL) in celiac disease

	Human	Mouse
Conventional TCRαβ⁺ CD4⁺	<15%	<15%
Conventional TCRαβ⁺ CD8αβ⁺	70–80%	20–30%
Innate TCRγδ	5–10%	50–70%
Innate TCRαβ CD8αα⁺ CD8αβ⁻	Undetermined	30–50%
Expression of NK receptors	On innate TCRγδ as well as on conventional TCRαβ CD8αβ cells	Only on innate-like TCRαβ and TCRγδ cells

and human IELs is that TCRγδ IELs represent a small proportion of human IELs (less than 20%), whereas they represent more than half of mouse IELs.

The numbers of CD8⁺CD4⁻ TCRαβ⁺ and CD8⁻CD4⁻ TCRγδ⁺ cells are increased in celiac disease. In contrast, the number of CD8⁻CD4⁺ TCRαβ⁺ cells is unchanged. In addition to increased prevalence, CD8⁺CD4⁻ TCRαβ IELs acquire strong cytolytic properties (see below). When patients adopt a gluten-free diet, the number of CD8⁺ TCRαβ⁺ IELs returns to normal within weeks or months, whereas the density of TCRγδ⁺ IELs remains elevated for a prolonged period. The TCRγδ cells in celiac disease are biased toward usage of the Vδ1 gene segment, suggesting recognition of a particular, yet unknown, ligand. The identification of this ligand may increase understanding of the elusive role of TCRγδ cells in celiac disease.

28-3 Immune cells infiltrate the lamina propria.

An array of immunocompetent cells is present in the small intestinal mucosa in celiac disease. In the celiac lesion, large numbers of leukocytes infiltrate into the epithelium and lamina propria. The increase of plasma cells in the lamina propria is particularly striking, but the number of T cells also increases (Figure 28.2). The number of plasma cells in the lesion more than doubles, and, as in normal mucosa, IgA is the dominant isotype. Some of the lamina propria

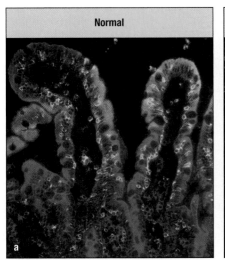
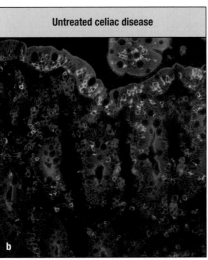

Figure 28.2 Distribution of T cells and plasma cells in the duodenal mucosa. The numbers of CD3-positive T cells (green) and CD138-positive plasma cells (red) in healthy mucosa (panel a) increase in active celiac disease (panel b). The epithelial cells stained in blue also express some CD138. In celiac disease, the number of plasma cells increases substantially in the lamina propria. (Photographs courtesy of A.-C. R. Beitnes.)

CD4$^+$ T cells are specific for gluten antigens and, as described later, are master regulators of the immune response that initiate formation of the celiac lesion. Few memory B cells are present in the lamina propria of celiac patients or normal subjects. The celiac lesion contains dendritic cells and macrophages that express high levels of human leukocyte antigen (HLA)-DQ molecules and thus should be able to present gluten epitopes to CD4$^+$ T cells, in contrast to epithelial cells, which do not express HLA-DQ molecules.

Celiac disease genetics.

Like most chronic inflammatory diseases, celiac disease has a multifactorial etiology, as both genes and environmental factors contribute to disease pathogenesis. In this regard, exclusion of gluten from the diet results in resolution of the disease, and the absence of the HLA predisposing genes is associated with reduced likelihood of developing celiac disease (99% of patients have the HLA-DQ2 or HLA-DQ8 haplotypes, see below). A high degree of familial clustering (10% of first-degree relatives affected compared to a population prevalence of about 1%), and a large difference in the concordance rate between monozygotic and dizygotic twins (75% versus 10%), suggest a genetic contribution. Both HLA genes and non-HLA genes are implicated in the disease. Most people who express HLA risk genes do not develop the disease, and the concordance rate among HLA-identical dizygotic twins that share their HLA genes in addition to half of their non-HLA genes is much lower than that of monozygotic twins.

28-4   HLA genes are associated with celiac disease.

Celiac disease has one of the strongest HLA associations. Although several genes within the HLA gene complex on chromosome 6 may contribute, the primary association is with the DQA1*05 and DQB1*02 alleles. This pair of genes is part of the autoimmune-associated HLA haplotype DR3-DQ2 and encodes the heterodimeric HLA-DQ2 molecule often termed DQ2.5 (Figure 28.3).

Figure 28.3 HLA association in celiac disease. The majority of celiac disease patients express the HLA-DQ2.5 heterodimer encoded by the DQA1*05 and DQB1*02 alleles. These two alleles are carried on either the DR3-DQ2.5 haplotype or on opposite chromosomes in individuals who are DR5-DQ7 and DR7-DQ2.2 heterozygous. Most DQ2.5-negative patients express HLA-DQ8 encoded by the DR4-DQ8 haplotype.

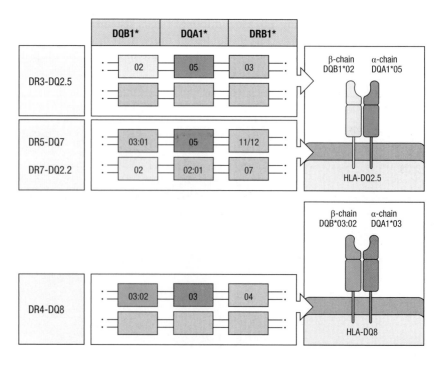

The DQ2 variant DQ2.2, encoded by the DQA1*02:01 and DQB1*02 alleles of the DR7-DQ2 haplotype, is barely associated with celiac disease on its own. However, people heterozygous for DR7-DQ2.2 and DR5-DQ7 are at high risk for disease, and in some populations a third of celiac disease patients express this genotype. The DR5-DQ7 haplotype carries the DQA1*05 allele, and the DR7-DQ2.2 haplotype carries the DQB1*02 allele, indicating that these individuals also encode the DQ2.5 molecule, but by genes located on opposite chromosomes. The few patients lacking this HLA-DQ2.5 heterodimer are usually DQ8, encoded by the DQA1*03 and DQB1*03:02 alleles. Interestingly, expression of HLA-DQ2.5 and HLA-DQ8 molecules predisposes to both celiac disease and type 1 diabetes; HLA-DQ2.5 molecules are predominant in celiac disease, whereas HLA-DQ8 molecules are predominant in type 1 diabetes.

HLA-DQ2.5 and HLA-DQ8 molecules both lack an aspartic acid at position beta 57. Consequently, the P9 binding pockets of both molecules are positively charged. This polymorphism is required for the development of type 1 diabetes and plays a critical role in the amplification of the HLA-DQ8-restricted anti-gluten T-cell response in the humanized HLA-DQ8 mouse model. It remains to be determined whether this polymorphism is also required for amplification of the HLA-DQ2.5-restricted anti-gluten immune response. Almost all celiac disease patients, as well as many healthy subjects, express HLA-DQ2.5 or HLA-DQ8 molecules. Therefore, certain HLA genes are necessary but not sufficient factors for disease development (Figure 28.4, top panel).

28-5 Non-HLA genes also contribute to celiac disease predisposition.

Whereas HLA is the single most important locus that predisposes to celiac disease, the combined effect of the non-HLA genes is larger than that of the HLA locus. Thus far, 39 chromosomal regions carrying non-HLA risk genes have been identified by genome-wide association studies. Notably, these regions usually contain several genes in strong linkage disequilibrium with each other, making it difficult to pinpoint the genes involved in disease pathogenesis. The size of the effect of each of the identified non-HLA loci is small: the 39 established non-HLA risk loci altogether account for about 15% of the genetic variance (Figure 28.4, bottom panel). In contrast, HLA contributes about 40% of the genetic variance, suggesting that many risk factors have not yet been discovered. A large number of common non-HLA gene variants with even lower risks than those already identified, along with highly penetrant rare variants at the established loci and elsewhere in the genome, are likely to contribute to the remaining genetic variance.

Most of the established regions harbor candidate genes with immune functions. These genes can be classified into different pathways (Table 28.2). Many of the genes belong to pathways of T and B cell co-stimulation. Such genes include *CTLA4*, *CD80*, *SH2B3*, *PTPN2*, *TAGAP*, *ICOSLG*, and *CD247*. Also frequently represented are cytokine and cytokine receptor genes, including *IL2/IL21*, *IL12A*, and *IL18RAP*; genes involved in migration of immune cells such as the chemokine receptors *CCR3/CCR5/CCR1*, and *CCR4*; and the integrin gene *ITGA4*. Another interesting pathway involves molecules important for T-cell development in the thymus such as THEMIS, which plays a role in both positive and negative T-cell selection during late thymocyte development, and RUNX3, which is involved in CD8⁺ T-cell differentiation. One network includes genes involved in NFκB signaling, such as *REL*, which encodes a component of the NFκB complex, and *TNFAIP3*, which encodes a molecule that inhibits NFκB activity. Finally, a pathway involving molecules implicated in innate immune detection such as Toll-like receptor 7 (TLR7), TLR8, and IRF4 has been identified. Both TLR7 and TLR8 recognize viral RNA, whereas IRF4 is a transcriptional activator that is part of the TLR7 pathway. This latter finding suggests involvement of viruses in the pathogenesis of celiac disease, likely

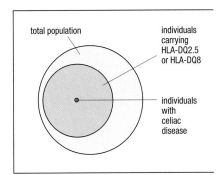

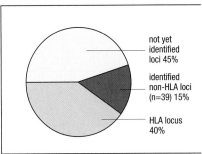

Figure 28.4 Key features of celiac disease genetics. The top panel illustrates that HLA is a necessary but not sufficient factor for the development of celiac disease. Almost all people with celiac disease express HLA-DQ2.5 or HLA-DQ8, but only a fraction of subjects who express HLA-DQ2.5 or HLA-DQ8 develop the disease. People who do not express HLA-DQ2.5 or HLA-DQ8 rarely develop celiac disease. The bottom panel shows diagrammatically the relative contributions to the total genetic variance of celiac disease by loci. HLA is the single most important locus contributing about 40% of the genetic variance. A total of 39 non-HLA loci that harbor celiac disease genes have been identified and collectively explain about 15% of the genetic variance. Most of the genetic variance has not been explained, suggesting the presence of additional non-HLA celiac disease genes.

Table 28.2 Pathways implicated in the pathogenesis of celiac disease on the basis of genetic association studies

Pathway	Candidate genes	Function
Antigen presentation	HLA-DQA1, HLA-DQB1	Presentation of antigenic peptides
T cell and B cell co-stimulation	CTLA4	Receptor for CD80 and CD86 and a negative regulator of T-cell activation
	CD80	Ligand for CD28 and CTLA4
	SH2B3	Involved in regulation of T-cell receptor signaling
	PTPN2	T-cell protein tyrosine phosphatase
	TAGAP	Involved in regulation of cytoskeletal changes in activated T cells
	ICOSLG	Ligand for the ICOS receptor that is involved in T-cell activation
	CD247	Zeta chain of the T-cell receptor CD3 complex
Cytokine	IL2/IL21	T-cell cytokines affecting T cells and B cells and other immune cells
	IL12A	Subunit (p35) of the IL-12 cytokine
	IL18RAP	Subunit of the IL-18 receptor
Cell migration	CCR3/CCR5/CCR1	Chemokine receptors involved in immune cell migration to inflamed sites
	CCR4	Chemokine receptor involved in immune cell migration to inflamed sites
	ITGA4	α_4 integrin subunit involved in cell anchoring
T-cell differentiation	THEMIS	Involved in positive and negative T-cell selection
	RUNX3	Involved in CD8+ T-cell differentiation
NFκB signaling	REL	Component of the NFκB complex
	TNFAIP3	Inhibits NFκB activity
Innate immune detection	TLR7/TLR8	Toll-like receptors binding viral RNA
	IRF4	Transcriptional activator of the TLR7 pathway

Candidate genes of regions that demonstrate genetic association with celiac disease are listed; the list of genes is not complete. (Data mainly from publications by David van Heel and co-workers.)

initial events important for inducing the anti-gluten CD4+ T-cell response. To date, few of the associated single nucleotide polymorphisms that have been discovered affect protein sequence, but many are associated with differences in the expression of nearby genes, suggesting that they influence immune gene expression.

Interestingly, many of the regions identified in the genome-wide association studies of celiac disease have also been identified in the genome-wide association studies of autoimmune diseases and other chronic inflammatory traits. The shared risk factors are particularly striking for type 1 diabetes, suggesting that these two diseases have common pathogenic pathways.

Adaptive immune response to gluten in celiac disease.

The adaptive immune response plays a fundamental role in celiac disease pathogenesis. Central to this response are gluten-reactive CD4+ T cells, which are present in the intestinal mucosa. These T cells recognize gluten epitopes presented by HLA-DQ2.5 or HLA-DQ8, but not other HLA molecules, suggesting that the mechanism underlying the HLA association in celiac disease is preferential antigen presentation.

28-6   Gluten-reactive CD4+ T cells in the intestinal mucosa recognize specific gluten residues.

Consistent with the complexity of the gluten proteins are the many similar, but distinct, T-cell epitopes (Figure 28.5). Strikingly, most gluten-reactive T cells preferentially recognize epitopes that have become post-translationally modified by the enzyme tissue transglutaminase (tTG). This enzyme selects specific glutamine residues in polypeptides and either cross-links the glutamine to a primary amine, such as lysine, in another polypeptide or converts it to glutamate via a process called deamidation (Figure 28.6). Deamidation generates gluten peptides with negatively charged glutamate residues that bind much better to the positively charged binding pockets of HLA-DQ2.5 and HLA-DQ8 than their native (non-deamidated) counterparts. Even though HLA-DQ2.5 and HLA-DQ8 molecules share physico-chemical properties, for example, both lack an aspartate at position beta 57 and show a preference for negatively charged peptides, a smaller repertoire of gluten peptides appear to be presented by HLA-DQ8 than by HLA-DQ2.5. Furthermore, the HLA-DQ2.5 and HLA-DQ8 restricted epitopes are distinct, as the preference for negatively charged anchor residues is different between the two HLA molecules; HLA-DQ2.5 prefers a negative charge in positions P4, P6, or P7, whereas HLA-DQ8 prefers a negative charge in P1 or P9 (Figure 28.5). Rye and barley contain related epitopes that are recognized by CD4+ T cells in patients with celiac disease.

Notably, HLA gene dosage influences risk for celiac disease. Homozygosity for both HLA-DQ2.5 and HLA-DQ8 is associated with higher risk. HLA-DQ2.5-expressing individuals are at a higher risk of developing disease even when the second copy of the DQB1*02 allele derives from the DR7-DQ2 haplotype (i.e., DR3-DQ2.5/DR7-DQ2.2 individuals). For HLA-DQ2, the gene dose effect is related to the magnitude and breadth of gluten-specific T-cell responses. On the basis of these observations, threshold effects have been suggested for disease development, with HLA-DQ expression and the available number of T-cell stimulatory gluten peptides being critical limiting factors.

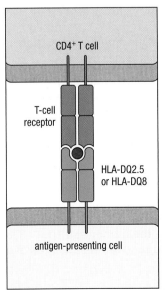

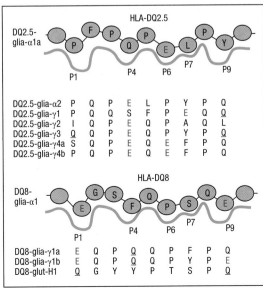

Figure 28.5 T-cell recognition of deamidated gluten epitopes in the context of the HLA-DQ2.5 and HLA-DQ8 molecules. The sequence of the DQ2.5-glia-α1a and DQ8-glia-α1 epitopes, as well as some other epitopes, that bind to HLA-DQ2.5 or HLA-DQ8 are shown. The HLA molecules have pockets at the relative positions P1, P4, P6, P7, and P9 that accommodate amino acid side chains of the peptides. Glutamate residues (E) induced by tissue transglutaminase (tTG) deamidation of glutamine residues are important for peptide binding and/or recognition by T cells and are marked in red. Glutamine residues targeted by tTG but not important for T-cell recognition are underlined. Note the many proline residues (P) in the epitopes. In the DQ2.5 restricted epitopes, the critical glutamate residues are at the P4, P6, or P7 positions, whereas in the DQ8 restricted epitopes, the critical glutamate residues are at the P1 or P9 positions.

Figure 28.6 Transamidation and deamidation reactions catalyzed by the enzyme tissue transglutaminase (tTG). tTG targets certain glutamine residues of polypeptides and either cross-links them to primary amines (transamidation) or converts them to glutamate by hydrolysis (deamidation). If the primary amine of the transamidation reaction is the lysine residue of another polypeptide, a covalent linkage of polypeptides is formed.

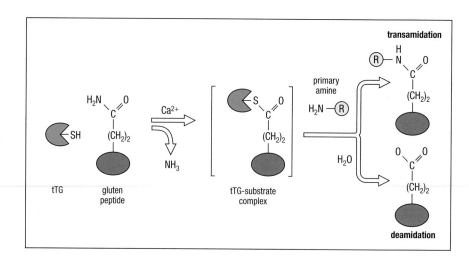

Threshold effects also may be implicated in the differential risk for celiac disease associated with DQ2.5 compared with DQ2.2. Despite the huge difference in genetic risk, the two molecules have very similar peptide-binding motifs and both present gluten epitopes to T cells derived from DQ2.5 patients. The functional difference between the two molecules has been mapped to residue 22 of the DQα chain. DQ2.5 carries a tyrosine at this position, which, in contrast to the phenylalanine of DQ2.2, forms hydrogen bonds with the peptide main chain. This leads to the differing ability of DQ2.5 and DQ2.2 to keep their loaded peptide cargo, and DQ2.5 has a stronger ability for sustained antigen presentation compared to DQ2.2. There are known threshold effects for priming of naive T cells by the amount of peptide:MHC complexes. DQ2.5 is likely better equipped to reach this threshold in the priming situation in mesenteric lymph nodes or in the Peyers' patches, which would then translate into a higher risk.

28-7 Tissue transglutaminase is involved in celiac disease pathogenesis.

The discoveries that tTG was the target of autoantibodies in celiac disease and increased the binding affinity of gluten peptides to HLA-DQ2.5 and HLA-DQ8 molecules placed the enzyme centrally in celiac disease pathogenesis. Gluten peptides are excellent substrates for tTG, and T-cell epitopes are among the preferred substrates among the peptides in a digest of gluten, suggesting that tTG plays a key role in T-cell epitope selection. The finding that gluten is a good substrate for tTG also provides interesting insights into the generation of gluten-dependent anti-tTG antibodies in celiac disease (see below).

Despite the recognition that tTG-mediated deamidation plays an important role in celiac disease pathogenesis, whether it plays a role in the initiation and/ or the amplification of the anti-gluten immune response is unclear. Indeed, observations in children and the humanized HLA-DQ8 mouse model suggest that deamidation is not required for the initiation of the immune response, but may be involved in the amplification of the immune response (see above). Furthermore, tTG is inactive in the intestinal environment, posing the question of how its activation occurs and whether it precedes or follows the initiation of the anti-gluten T-cell response. In one scenario, the inflammatory anti-gluten immune response itself is responsible for the initial activation of tTG. This scenario is compatible with a study in the humanized HLA-DQ8 mouse, which showed that deamidation is not required for the initiation of a

HLA-DQ8-restricted anti-gluten CD4+ T-cell response and that HLA-DQ8 can amplify the anti-gluten immune response by allowing the recruitment of cross-reactive TCR that recognizes native and deamidated peptides. In the other scenario, an environmental trigger, such as an intestinal viral infection, activates tTG and initiates the adaptive anti-gluten immune response. This scenario is compatible with reports suggesting that repeated rotavirus infections are associated with an increased incidence of celiac disease and that double-stranded RNA (as during rotavirus infection) can trigger tTG activation in mice either directly or indirectly by causing tissue damage. It is important to point out that these different scenarios are *not* mutually exclusive, and that one scenario can predominate, depending on the genetic background.

The role of post-translational modifications in generating immunogenic peptides is also reported for other autoimmune diseases. For example, in the case of rheumatoid arthritis, the post-translational modification and the MHC molecules associated with the disease are citrullination and DR4 respectively. The overall underlying theme involves a favorable combination of MHC molecule, tissue enzyme, and antigen that optimizes binding of the causative antigen to the disease-associated MHC molecule and promotes T-cell responses of sufficient amplitude to induce a pathogenic immune response associated with tissue damage. Because post-translational modifications of antigen can promote loss of immune tolerance to that antigen, enzymes promoting post-translational modifications must be under tight control. Understanding the regulation of tTG and its role in celiac disease pathogenesis will also enhance more general understanding of the role of post-translational modifications by tissue enzymes in autoimmunity.

28-8   Autoantibodies are present in celiac disease.

Serum IgA and IgG antibodies to tTG, which are dependent on gluten exposure, are a hallmark of celiac disease. Detection of these antibodies is increasingly important for the diagnosis of celiac disease due to the sensitivity and specificity of the antibodies. The formation of anti-tTG antibodies is also dependent on the presence of HLA-DQ2.5 and HLA-DQ8, as people who do not carry either of these HLA molecules do not make the antibodies. The strict HLA and gluten dependence suggest that gluten-reactive CD4+ T cells are involved in the generation of the anti-tTG antibodies. This could be explained by a mechanism in which complexes of tTG and gluten, which readily form when incubated together, act as hapten–carrier complexes so that gluten-reactive T cells can provide help to tTG-reactive B cells (Figure 28.7).

28-9   Oral antigen induces an inflammatory response.

Resistance of gluten to proteolytic digestion and post-translational modifications lead to the generation of peptides of higher affinity for HLA-DQ2.5 and HLA-DQ8, which then are presented to T cells. However, why inflammatory anti-gluten CD4+ T-cell responses occur in an environment where oral antigens normally lead to tolerogenic T-cell responses is less clear. To explain this phenomenon, epidemiological studies have shown the presence of high levels of interferon-α expression in the intestinal mucosa of celiac disease patients, and genome-wide association studies point to genes involved in viral responses, suggesting that viral infections are involved in the initiation of pathogenic anti-gluten T-cell immunity; however, the precise mechanisms leading to loss of tolerance have not been determined. Alternatively, gluten may have intrinsic innate properties that lead to activation of dendritic cells and production of pro-inflammatory mediators.

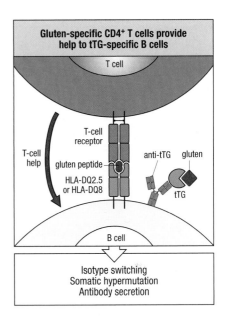

Figure 28.7 Model depicting how gluten-specific CD4+ T cells can provide help to tTG-specific B cells. Production of antibodies to tTG is dependent on ingestion of gluten and only occurs in subjects who carry HLA-DQ2.5 or HLA-DQ8. The formation of the tTG antibodies is likely dependent on T-cell help, probably provided by gluten-specific T cells. Gluten peptides and tTG readily form complexes, which when bound by antibodies on the surface of a tTG-specific B cell, are internalized, processed, and the peptide presented in the context of HLA-DQ2.5 or HLA-DQ8 to a gluten-specific CD4+ T cell. B cells expressing low-affinity anti-tTG immunoglobulins can receive T-cell help and undergo maturation to plasma cells, producing high-affinity IgA and IgG anti-tTG antibodies.

Innate immunity and the stress response to gluten.

Whether gluten induces an innate immune response has been debated. Gluten-mediated innate effects could explain some elements of celiac disease pathogenesis. However, to date no reproducible gluten-induced innate signaling pathway has been reported. Numerous gluten-mediated innate effects have been reported, but these effects are eclectic and difficult to integrate in a single biological pathway. Nevertheless, numerous studies point toward a role for innate immunity in the cytolytic intraepithelial T-cell response that is likely involved in epithelial cell destruction.

28-10 Innate immune cell responses may be induced by gluten.

Interestingly, TCRγδ IELs increase in number after rectal gluten challenge in HLA-DQ2.5⁻ and HLA-DQ8⁻ siblings of celiac disease patients, suggesting that celiac disease can be induced in the absence of HLA-DQ2.5 or HLA-DQ8 molecules. Another striking finding is the upregulation of mucosal interferon-α and IL-15 in subjects with celiac disease while on a gluten-containing diet. Organ culture studies have suggested that IL-15 can be induced in response to the α-gliadin 31-43 peptide (LGQQQPFPPQQPY), which does not induce an adaptive anti-gluten T-cell response. Interestingly, this peptide also was implicated in mitogen-activated protein kinase activation, inhibition of epidermal growth factor receptor degradation, and upregulation of the nonclassical stress-inducible MHC class I molecule MIC. However, structure–function studies and identification of a receptor are still lacking. In addition to the α-gliadin 31-43 peptide, other peptides with the ability to induce maturation of antigen-presenting cells, chemokine receptor activation, and upregulation of the nonclassical MHC class I molecule HLA-E have been described. If innate effects of gluten are confirmed, additional studies will be required to explain why gluten mediates these effects only in some individuals. Indeed, the level of tolerance to the innate effects of gluten may be genetically determined.

28-11 Intraepithelial cytotoxic T lymphocytes with NK cell receptors, epithelial cells with nonclassical MHC class I molecules, and IL-15 contribute to disease pathogenesis.

The induction of an inflammatory anti-gluten immune response may not be sufficient to induce epithelial cell destruction, tissue remodeling, and ultimately villous atrophy. In this connection, another histological hallmark of celiac disease is the dramatic increase in cytolytic IELs (CD8αβ⁺TCRαβ⁺) that reverts under a gluten-free diet and correlates with recovery of villous structures. Indeed, the role of IELs was disregarded before gluten-specific IELs could be identified. In contrast, gluten-specific CD8αβ⁺TCRαβ⁺ T cells were identified in the lamina propria. Strong evidence now indicates that IELs do not mediate destruction by direct recognition of gluten peptides presented by MHC class I molecules but by killing epithelial cells in a noncognate manner through NK receptors (Figure 28.8). This noncognate destruction of epithelial cells could be mediated in a TCR-dependent and TCR-independent manner. Regarding the former, NK receptors would lower the TCR activation threshold in such a way that IELs would recognize low-affinity self or microbial peptides. Regarding the latter, killing would be mediated only via NK receptors that recognize their ligands on epithelial cells. Two NK receptor–ligand pairs are involved in celiac disease pathogenesis, but other yet-to-be-identified NK receptors also may be involved.

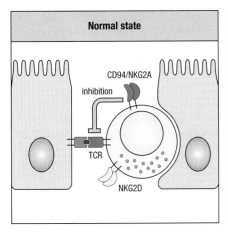
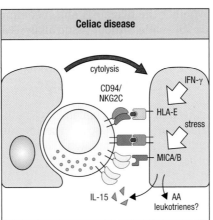

Figure 28.8 Natural killer receptors and their ligands in intestinal mucosa. Under physiological conditions, epithelial cells lack expression of nonclassical MHC class I molecules and release low levels of IL-15. IELs express inhibitory CD94/NKG2A receptors and low levels of activating NKG2D receptors. In active celiac disease, IELs lose expression of the inhibitory CD94/NKG2A receptors, acquire activating CD94/NKG2C receptors, and upregulate NKG2D receptors. These natural killer receptors recognize the nonclassical MHC class I molecules HLA-E and MIC induced on epithelial cells, inflammatory and stress signals respectively. Furthermore, epithelial cells express high levels of IL-15, which license intraepithelial lymphocytes to kill by co-stimulating the NKG2D cytolytic pathway and decreasing the activation threshold of the T-cell receptor. IFN-γ, interferon-γ; AA, arachidonic acid.

NKG2D is a C-type lectin receptor that associates selectively with the adaptor molecule DAP10 in humans. This adaptor molecule has a phosphoinositol-3-kinase binding motif but lacks an immunoreceptor tyrosine-based activation motif. However, NKG2D can mediate direct cytolysis in T cells via a cytolytic signaling pathway that involves cytosolic phospholipase A2 and arachidonic acid. In order for NKG2D to mediate direct cytolysis, IELs must be activated by IL-15 expressed on intestinal epithelial cells. IL-15 acts as a co-stimulatory molecule for the NKG2D cytolytic pathway. In addition, IL-15 upregulates NKG2D and DAP10 expression.

The second well-described NK receptor involved in celiac disease pathogenesis is CD94/NKG2C. Interestingly, normal IELs predominantly express the inhibitory CD94/NKG2A receptor that can block TCR and NKG2D signaling. In IELs of celiac disease patients, CD94/NKG2A receptors are conspicuously decreased and activating CD94/NKG2C receptors are expressed. Importantly, CD94/NKG2C receptors are associated with the adaptor molecule DAP12 that contains an immunoreceptor tyrosine-based activation motif and consequently can induce cytokine secretion and T-cell proliferation in addition to cytolysis. Furthermore, analysis of IELs expressing NKG2C showed that IELs had actually undergone a global natural killer reprogramming, resulting in the expression of a panel of natural killer receptors and molecules normally only found in natural killer cells. The ligand for CD94/NKG2C receptors is the nonclassical MHC class I molecule HLA-E that is induced in the presence of interferon-γ. Celiac disease epithelial cells upregulate HLA-E, MIC, and IL-15, indicating that they have received inflammatory and stress signals. Together, these observations suggest that ingested gluten interacts with epithelial cells, leading to the expression of IL-15 and nonclassical MHC class I molecules in genetically susceptible individuals. IL-15 and nonclassical MHC class I molecules then activate NK receptors that induce IEL activation and epithelial cell destruction. The role of tissue cells in licensing cytolytic T cells to mediate tissue damage may constitute a 'checkpoint' to limit tissue destruction in the presence of harmless antigens.

Immunological tests in the diagnostic workup of celiac disease.

The diagnosis of celiac disease is based on the histopathological assessment of small intestinal biopsies, supplemented by other newly available tests. These tests include serum antibodies to tTG and gluten and HLA genotyping.

Whether the diagnosis and recommendation to adhere to a gluten-free diet can be made without tissue analysis is controversial.

28-12 Antibodies to tissue transglutaminase and gluten are used to diagnose celiac disease.

Detection of serum IgA and IgG antibodies to tTG and gluten is widely used in the diagnosis of celiac disease. Measurements of IgA antibodies to tTG and IgA antibodies to deamidated gliadin peptides are the most sensitive and specific tests, depending on the study population. In selected populations with high risk for the disease, the specificity and sensitivity are usually higher than 99%, but these values become lower when testing general populations. Patients with IgA deficiency are over-represented among patients with celiac disease, and for such patients IgG antibodies to tTG and deamidated gliadin represent reliable markers for disease. A point-of-care test based on the detection of IgA serum antibodies to tTG has also been developed.

Plasma cells in the mucosa produce anti-tTG antibodies, and serum IgA-tTG antibodies likely represent a 'spill-over' from the gut. The antibodies produced in the mucosa react with the tTG-antigen expressed locally, and thus IgA anti-tTG can be detected in the mucosa without elevated levels of tTG antibodies in serum. The presence of IgA in a typical subepithelial distribution is typical of celiac disease and is an early marker of disease activity.

28-13 Other tests have a complementary role in the diagnosis of celiac disease.

HLA is a necessary but not sufficient factor for celiac disease development. HLA typing can help exclude the diagnosis of celiac disease. Specifically, the absence of DQB1*02 is a strong negative predictor of the disease. Importantly, detection of either DQ2.5 or DQ8 has no diagnostic value, as both DQ2.5 and DQ8 are frequent in the normal population.

An elevated level of $\gamma\delta$ T cells in the epithelium is a typical feature of celiac disease but may accompany other inflammatory conditions of the upper small bowel. As this is not a very specific test, its use in clinical practice is diminishing.

The diagnosis of celiac disease in a person on a gluten-free diet is problematic, because in the absence of gluten ingestion the clinical signs, serology, and morphology return to normal. Thus, several months of gluten ingestion is required for a proper diagnosis. Many patients are reluctant to engage in a prolonged gluten challenge, but such a challenge may not be necessary in the future. After oral gluten consumption for 3 days, most celiac disease patients have T cells specific to deamidated gluten peptides in their peripheral blood. Such T cells can be detected by ELISPOT (enzyme-linked immunospot assay) or by staining with HLA-DQ2–gliadin peptide tetramers. These tests have not yet been performed in clinical settings, and whether blood tests can replace biopsies is controversial. Anti-tTG antibodies can be detected in patients with extraintestinal diseases, such as dermatitis herpetiformis, but biopsy is required when the levels of anti-tTG antibodies are low.

28-14 Immune-based therapy may have a future in celiac disease.

Because a gluten-free diet is difficult to follow and more than half of patients on such a diet have incomplete recovery, three immune-based therapeutic strategies have been considered (Figure 28.9). The first strategy seeks to prevent recognition of gluten peptides by CD4+ T cells by increasing gluten digestion using bacterial enzymes that can cleave at proline and glutamine

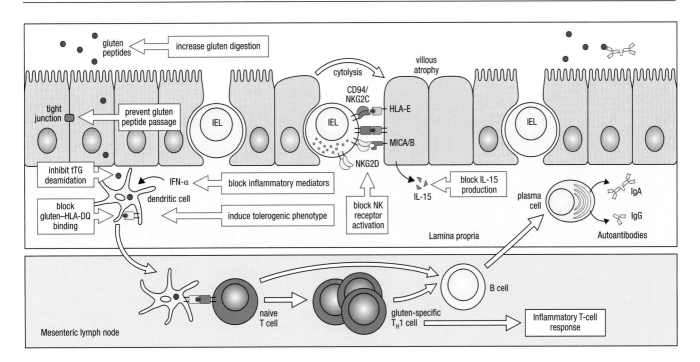

residues, by decreasing intestinal permeability with pharmacological reagents that act on tight junctions, by inhibiting tTG activity, or by blocking binding of gluten peptides to HLA-DQ2.5 or HLA-DQ8 molecules using high-affinity competitive ligands. The second strategy aims to delete anti-gluten CD4⁺ T cells and/or divert inflammatory, anti-gluten immune responses into regulatory, anti-gluten immune responses in which T cells produce anti-inflammatory cytokines such as IL-10 or transforming growth factor-β. This strategy employs agents that induce a tolerogenic phenotype in dendritic cells or repeatedly present low concentrations of immunodominant gluten peptides. The third strategy blocks the activation of intraepithelial lymphocytes by inhibiting IL-15 or NK receptor activation, such as NKG2D, with either blocking antibodies or molecules that interfere with their signaling pathway. Although still in their infancy, immune intervention therapies offer hope to patients with celiac disease that therapy other than gluten restriction may someday be possible.

Figure 28.9 Pathogenesis of celiac disease with targets for therapy. Three main therapeutic strategies are currently envisioned. In the first strategy, decreased gluten peptide presentation can be achieved by increasing digestion of gluten via introduction of bacterial enzymes, preventing passage of gluten peptides into the lamina propria by targeting tight junctions, inhibiting deamidation by tTG, or blocking the binding of gluten peptides to HLA-DQ2.5 or HLA-DQ8 molecules using competitive HLA ligands. The second strategy aims to prevent the differentiation of pathogenic inflammatory anti-gluten T cells by blocking inflammatory mediators such as interferon-α, educating dendritic cells to acquire a tolerogenic phenotype, or injecting low concentrations of immunodominant peptides multiple times in the absence of adjuvant. A third strategy seeks to prevent lysis of epithelial cells by blocking IL-15, which licenses intraepithelial lymphocytes to kill, or by using antibodies or molecules that block activating natural killer receptors.

Summary.

Celiac disease is caused by hypersensitivity to the food protein gluten and resembles features of T cell-mediated autoimmune disease. Manifestation of the disease is multifactorial, involving environmental factors such as gluten and possibly viral infections, as well as genetic factors. HLA is the single most important genetic factor, reflected in a strong association between the disease and the presence of HLA-DQ2.5 and HLA-DQ8. The total number of non-HLA genes is estimated to represent 60% of the total genetic risk. To date, only a fraction of these genes, which mainly encode molecules with important immune functions, has been identified. HLA-DQ2.5 and HLA-DQ8 molecules present gluten epitopes to CD4⁺ T cells in the lamina propria. The epitopes are harbored in proline- and glutamine-rich gluten peptides that are resistant to intestinal proteolysis and that have become post-translationally modified by conversion of certain residues from glutamine to glutamate. This process, termed deamidation, is mediated by the enzyme tTG, the presence of which is dependent on carriage of HLA-DQ2.5 or HLA-DQ8, and dietary exposure to gluten. The tTG enzyme is also the target of disease-specific IgA and IgG autoantibodies. Gluten-specific CD4⁺ T cells are necessary but not sufficient to induce the celiac lesion, which is characterized by villous blunting, crypt cell hyperplasia, and leukocyte infiltration in both epithelium and the lamina propria. Several effector mechanisms appear to be involved in the lesion formation, but CD8+ T cells with NK cell receptors infiltrating in the epithelium appear particularly critical. These cells lose inhibitory and acquire activating NK cell receptors under the influence of IL-15, licensing the cells to kill enterocytes expressing stress-induced ligands of the NK cell receptors. Immunological assays play an important role in the diagnostic workup of celiac disease, and novel therapies to interdict the key immune processes are under development.

Further Reading.

Anderson, R.P., Degano, P., Godkin, A.J., *et al.*: **In vivo antigen challenge in celiac disease identifies a single transglutaminase-modified peptide as the dominant A-gliadin T-cell epitope.** *Nat. Med.* 2000, **6**:337–342.

Arentz-Hansen, H., Körner, R., Molberg, O., *et al.*: **The intestinal T cell response to alpha-gliadin in adult celiac disease is focused on a single deamidated glutamine targeted by tissue transglutaminase.** *J. Exp. Med.* 2000, **191**:603–612.

Dieterich, W., Ehnis, T., Bauer, M., *et al.*: **Identification of tissue transglutaminase as the autoantigen of celiac disease.** *Nat. Med.* 1997, **3**:797–801.

Fallang, L.E., Bergseng, E., Hotta, K., *et al.*: **Differences in the risk of celiac disease associated with HLA-DQ2.5 or HLA-DQ2.2 are related to sustained gluten antigen presentation.** *Nat. Immunol.* 2009, **10**:1096–1101.

Hovhannisyan, Z., Weiss, A., Martin, A., *et al.*: **The role of HLA-DQ8 β57 polymorphism on the anti-gluten T cell response in celiac disease.** *Nature* 2008, **456**:534–538.

Hüe, S., Mention, J.J., Monteiro, R.C., *et al.*: **A direct role for NKG2D/MICA interaction in villous atrophy during celiac disease.** *Immunity* 2004, **21**:367–377.

Jabri, B., and Sollid, L.M.: **Tissue-mediated control of immunopathology in celiac disease.** *Nat. Rev. Immunol.* 2009, **9**:858–870.

Kagnoff, M.F.: **Celiac disease: pathogenesis of a model immunogenetic disease.** *J. Clin. Invest.* 2007, **117**:41–49.

Maiuri, L., Ciacci, C., Ricciardelli, I., *et al.*: **Association between innate response to gliadin and activation of pathogenic T cells in coeliac disease.** *Lancet* 2003, **362**:30–37.

Meresse, B., Chen, Z., Ciszewski, C., *et al.*; **Coordinated induction by IL-15 of a TCR-independent, NKG2D signaling pathway converts CTLs into natural killer-like, lymphokine activated killer (LAK) cells in celiac disease.** *Immunity* 2004, **21**:357–366.

Meresse, B., Curran, S.A., Ciszewski, C., *et al.*: **Reprogramming of CTLs into natural killer–like cells in celiac disease.** *J. Exp. Med.* 2006, **203**:1345–1355.

Molberg, O., Mcadam, S.N., Körner, R., *et al.*: **Tissue transglutaminase selectively modifies gliadin peptides that are recognized by gut-derived T cells in celiac disease.** *Nat. Med.* 1998, **4**:713–717.

Siegel, M., Strnad, P., Watts, R.E., *et al.*: **Extracellular transglutaminase 2 is catalytically inactive, but is transiently activated upon tissue injury.** *PLoS One* 2008, **3**:e1861.

Trynka, G., Hunt, K.A., Bockett, N.A., *et al.*: **Dense genotyping identifies and localizes multiple common and rare variant association signals in celiac disease.** *Nat. Genet.* 2011, **43**:1193–1201.

Vader, L.W., de Ru, A., van der Wal, Y., *et al.*: **Specificity of tissue transglutaminase explains cereal toxicity in celiac disease.** *J. Exp. Med.* 2002, **195**:643–649.

IgA Nephropathy

29

IgA nephropathy is the commonest form of primary glomerulonephritis and an important cause of renal failure. IgA nephropathy is a mesangioproliferative glomerulonephritis defined by mesangial deposits of IgA1-containing immune complexes. It was initially described in 1968 by Berger and Hinglais on the basis of its unique renal immunohistologic features as an IgA:IgG immune-complex renal disease.

Clinical presentation of IgA nephropathy.

IgA nephropathy is a relatively common disease and accounts for 5–20% of renal diseases diagnosed by native-kidney biopsy in Europe and the United States, and up to 30–45% in China and Japan. In contrast, IgA nephropathy is a rare renal disease in central Africa. Some individuals probably have subclinical disease for years. A study in Japan demonstrated mesangial IgA deposits in 82 of 510 renal allografts (16%) at implantation, of which 19 showed mesangioproliferative glomerulonephritis by light microscopy. The next section describes the typical clinical features of this entity.

29-1 IgA nephropathy is characterized by sporadic and familial forms.

Clinical features initially are manifested most frequently in adolescents and young adults, with a usual 2:1 predominance of males over females. Asymptomatic proteinuria and hematuria are typical clinical presentations. Painless macroscopic hematuria is common in children and often coincides with mucosal infections, including infections of the upper respiratory tract and digestive system.

It has been recognized that there are two forms of IgA nephropathy: sporadic and familial. The familial form is defined by the presence of at least two biopsy-proven cases in a family. Clinically, the two forms are indistinguishable. Regardless of whether IgA nephropathy is sporadic or familial, 20–40% of patients with IgA nephropathy progress to end-stage renal disease within 20 years of diagnosis, and such patients require functional renal replacement; that is, dialysis or transplantation. However, the disease recurs in 50–60% of patients with IgA nephropathy within 5 years after transplantation.

Secondary forms of IgA nephropathy occur in association with diseases of other organ systems, for example hepatobiliary diseases (such as cirrhosis due to alcoholic liver disease or viral hepatitis), gastrointestinal diseases (such as ulcerative colitis), or infectious diseases (such as infection with human immunodeficiency virus).

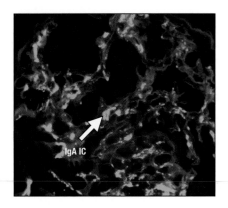

Figure 29.1 Immunofluorescence confocal microscopy image of a glomerulus from a patient with IgA nephropathy stained for IgA. Example of a granular IgA-containing immunodeposit (IgA IC, arrow). (Photograph courtesy of L. Novak.)

There is no disease-specific treatment of IgA nephropathy. Current therapy focuses on minimizing proteinuria and maintaining normal blood pressure, with emphasis on the use of agents to suppress the effects of angiotensin II pharmacologically. In this chapter, we focus on primary IgA nephropathy.

29-2   Definitive diagnosis of IgA nephropathy requires a renal biopsy.

Evaluation of a renal biopsy specimen is required for a definitive diagnosis of IgA nephropathy as well as for the assessment of severity of the tissue injury. Typically, samples of the kidney cortex obtained by percutaneous needle biopsy are divided into three portions and processed for immunofluorescence, light, and electron microscopy.

By immunofluorescence, predominant or codominant IgA deposits are observed in the mesangium (Figure 29.1), usually with complement component C3, and IgG and/or IgM co-deposits. The IgA is exclusively of the IgA1 subclass. Complement component C1q is absent or minimally present in most cases. Analysis of immunofluorescence staining for IgA and C3 by confocal microscopy has revealed co-localization of these components. About 50% of renal biopsy specimens have IgG co-deposits, by routine immunofluorescence or immunohistochemistry. In a minority of patients, IgA is the sole immunoglobulin detected.

29-3   IgA nephropathy is characterized by mesangial deposits of IgA.

By light microscopy, mesangial proliferation and expansion of the extracellular matrix are characteristic for IgA nephropathy (Figure 29.2). Glomerular sclerosis and interstitial fibrosis are associated with progressive disease that leads to renal insufficiency. Recently, a new 'Oxford classification' for IgA nephropathy was developed by an international group of physicians and pathologists on the basis of the identification of specific light microscopy features that more accurately predict the risk of progression of renal disease in adults and children. Four pathological features (mesangial hypercellularity, segmental glomerulosclerosis, endocapillary hypercellularity, and tubular atrophy/interstitial fibrosis) are each independently associated with clinical outcome. This classification system is currently being tested in additional cohorts; if validated, it will probably become a required part of a pathology report for IgA nephropathy.

Electron microscopy typically shows electron-dense deposits in the mesangial or paramesangial areas of glomeruli, confirming the presence of immune complexes detected by immunofluorescence (Figure 29.3). Electron microscopy may also reveal ultrastructural changes, such as the effacement of podocytes in association with proteinuria.

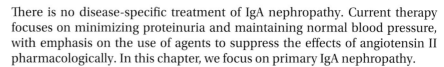

The immunopathogenesis of IgA nephropathy.

There is considerable evidence that the mesangial deposits originate from circulating IgA1-containing immune complexes, including: (1) the frequent recurrence of the disease in patients who receive a renal allograft; (2) in the few cases in which a kidney was transplanted from a donor with subclinical IgA nephropathy into a patient with non-IgA nephropathy renal disease, the immune deposits cleared from the engrafted kidney within weeks; (3) patients with IgA nephropathy have elevated levels of IgA and IgA1 immune complexes in the circulation; (4) idiotypic determinants are shared between the circulating complexes and the mesangial deposits, although a disease-specific

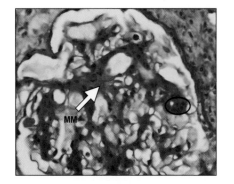

Figure 29.2 Light microscopy image of a glomerulus from a patient with IgA nephropathy. MM, expanded mesangial matrix; oval marks a cluster of proliferating mesangial cells. (Photograph courtesy of L. Novak.)

idiotype has not been identified; and (5) IgA1 in the circulating immune complexes and in the renal deposits shows an unusual chemical composition, with galactose (Gal) deficiency in the *O*-glycans of the hinge region unique to IgA1.

29-4 IgA, the target of IgA immune complexes, is the major immunoglobulin isotype produced in the body.

The daily production of IgA (about 66 mg kg^{-1} body weight per day) far exceeds the combined biosynthesis of immunoglobulins of all other isotypes. About two-thirds of IgA is produced in a polymeric form in mucosal tissues, particularly in the intestinal tract, and then selectively transported by a receptor-mediated pathway into external secretions (see Chapters 8 and 9). The IgA subclass distribution of IgA1- and IgA2-producing cells and corresponding IgA subclasses in secretions displays a characteristic pattern: IgA1 is more abundant than IgA2 in tears, saliva, respiratory tract secretion, and the fluid in the small intestine. In contrast, secretions of the large intestine and of the female genital tract contain a slight excess of IgA2. Most IgA in the circulation is represented by monomeric (m) IgA with a dominance of IgA1 (about 85% of total IgA). In contrast to IgG, circulatory IgA has a short half-life—about 5 days—and only trace amounts of mIgA1 from the circulatory pool enter external secretions. In humans, IgA-producing cells in the bone marrow and, to a lesser degree, in lymph nodes and spleen are the most important sources of the circulatory IgA pool. Experiments performed in mice and in primates clearly indicate that the liver is the major organ involved in the catabolism of IgA in both subclasses as well as of other glycoproteins. The asialoglycoprotein receptor (ASGP-R) expressed on cell surfaces of hepatocytes binds glycoproteins through their terminal Gal- and *N*-acetylgalactosamine (GalNAc)-containing saccharides. The glycoproteins are subsequently internalized and degraded. Human hepatoma cell lines have been also examined, with identical results. The Gal deficiency of *O*-glycans within the hinge region of IgA1 in patients with IgA nephropathy, as described above, should not profoundly decrease such binding to ASGP-R, because of the presence of terminal GalNAc. The apparent basis for the impaired clearance of nephritogenic IgA1 was experimentally elucidated by detailed analyses of circulating immune complexes. These studies demonstrated that in circulating immune complexes, GalNAc residues on IgA1 are covered by GalNAc-specific antibodies and, thus, this terminal GalNAc is inaccessible for interactions with the hepatic ASGP-R. Furthermore, the size of circulating immune complexes prevents their penetration through the fenestrae in endothelial cells of hepatic sinuses to enter the space of Disse and reach the ASGP-R for subsequent binding and catabolism. Because of the significantly larger size of fenestrae in endothelial cells of glomerular capillaries, circulating immune complexes are diverted from the normal hepatic catabolic pathway, are deposited in the mesangium, and induce glomerular injury. The validity of this concept has been reinforced in recent studies in mice. High-molecular-weight immune complexes composed of Gal-deficient IgA1 and anti-glycan IgG antibodies specifically displayed this deposition pattern, with ensuing pathogenic consequences.

29-5 Human IgA1 has a unique hinge region containing sites that are amenable to *O*-linked glycosylation.

The glycosylation of IgA1 is unique and provides some insights into the mechanisms underlying immune-complex formation and deposition in the glomerular mesangium of patients with IgA nephropathy. Heavy (α) chains of human IgA1 possess a unique hinge region, located between the C$_{\alpha}$1 and C$_{\alpha}$2 domains, containing a sequence rich with proline, serine, and threonine residues. There are nine serine and threonine amino acids, the potential sites of *O*-glycan attachment, in each hinge region (see Chapter 9 for details). However, usually

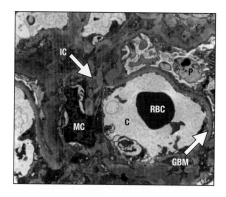

Figure 29.3 Electron microscopy image of an area of a glomerulus from a patient with IgA nephropathy. IC, electron-dense immunodeposits; C, capillary; MC, mesangial cell; P, podocyte; GBM, glomerular basement membrane; RBC, red blood cell. (Photograph courtesy of W. J. Cook.)

Figure 29.4 Hinge region of serum IgA1 and its *O*-glycans. Panel a shows that IgA1 has *O*-glycans attached in the hinge region between the first (C$_\alpha$1) and second (C$_\alpha$2) constant domains of the heavy chain. The hinge-region amino acid sequence is composed of two octapeptide repeats with as many as six of the nine potentially *O*-glycosylated sites occupied. Serine and threonine amino acids that can be theoretically glycosylated are in bold, and the five residues that have been shown to be most frequently glycosylated on circulatory IgA1 are numbered and shown in blue. Panel b depicts the variants of *O*-glycans on circulatory IgA1. Gal-deficient glycans that are found in elevated amounts in the circulation of patients with IgA nephropathy are in a large rectangle. The largest glycan on circulatory IgA1 is a GalNAc-Gal with two sialic acids, in other words a tetrasaccharide. Symbols: open rectangle, GalNAc; filled circle, Gal; filled diamond, sialic acid.

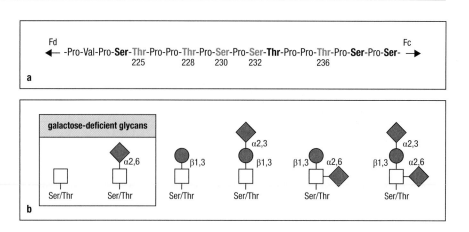

Figure 29.5 Biosynthesis of IgA1 *O*-glycans. *O*-linked glycans in the IgA1 hinge region are synthesized in a stepwise manner. The initial step is the attachment of GalNAc to the oxygen atom of the hydroxyl group of some of serine or threonine amino acids. The *O*-glycan chain is then extended by the sequential attachment of Gal and/or sialic acid residues to the GalNAc. The addition of Gal is mediated by core 1 β1,3-galactosyltransferase (C1GalT1), which transfers Gal from UDP-galactose to the GalNAc residue. The stability of this enzyme during biosynthesis depends on its interaction with its chaperone protein, Cosmc, which assists in protein folding. In the absence of Cosmc, the C1GalT1 protein is rapidly degraded. The glycan structure is completed by sialyltransferases that attach sialic acid to the Gal and/or GalNAc residues. If sialic acid is linked to GalNAc before the attachment of Gal, this 'premature' sialylation precludes the subsequent attachment of Gal. An imbalance in the activities or expression of specific glycosyltransferases accounts for the increased production of Gal-deficient *O*-linked glycans in the IgA1 hinge region with increased sialic acid residues: decreased C1GalT1 and Cosmc and increased ST6GalNAcII (marked by red arrows). Gal-deficient glycans that are found in elevated amounts in the circulation of patients with IgA nephropathy are shown in red rectangles. Symbols are as in Figure 29.4.

only up to six of these sites are glycosylated. Hinge-region glycoforms with four and five glycans are the most common.

The *O*-glycans of serum IgA1 are core 1 glycans in that they consist of GalNAc with a β1,3-linked Gal that are *O*-linked to either serine or threonine; both residues of the core 1 glycan may be sialylated. The carbohydrate composition of the *O*-linked glycans on normal serum IgA1 is variable; the prevailing forms include the Gal-GalNAc disaccharide, and its mono- and di-sialylated forms (Figure 29.4).

O-glycosylation is initiated by the attachment of GalNAc to seryl or threonyl residues by UDP-GalNAc:polypeptide *N*-acetylgalactosaminyltransferases (GalNAc-Ts). The sites to be *O*-glycosylated and their order are determined by the specific set of GalNAc-Ts expressed in a particular cell type. The IgA1 *O*-glycans are attached predominantly by GalNAc-T2, although other GalNAc-Ts (such as GalNAc-T1 and -T11, which can efficiently catalyze the same reaction) can use the IgA1 hinge region as an acceptor (Figure 29.5). GalNAc is then modified by the addition of β1,3-linked Gal in a reaction catalyzed by UDP-Gal:GalNAc-α-Ser/Thr β1,3-galactosyltransferase 1 (C1GalT1). This reaction results in a core 1 structure. Formation of the active C1GalT1 depends on core 1 β1,3-galactosyltransferase-specific chaperone (Cosmc).

The GalNAc-Gal structure in the hinge region of circulatory IgA1 can be further modified by attaching the sialic acid from cytidine monophosphate (CMP)-*N*-acetylneuraminic acid to the Gal residues in a reaction catalyzed by a Galβ1,3GalNAc α2,3-sialyltransferase (ST3Gal) and/or to the GalNAc residues catalyzed by a α2,6-sialyltransferase (ST6GalNAc). A ST3Gal, probably

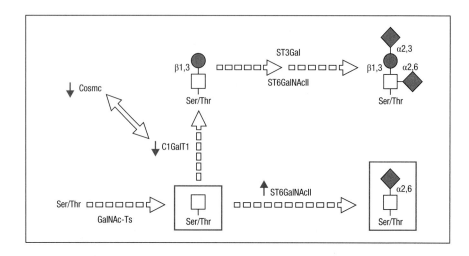

ST3GalI, is responsible for the sialylation of Gal on IgA1 *O*-glycans. Of the α2,6-sialyltransferases, IgA1-producing cells express exclusively ST6GalNAcII, which is apparently responsible for the sialylation of GalNAc in both the terminal and Gal-containing GalNAc on IgA1.

29-6 Circulating IgA1 in IgA nephropathy contains aberrant *O*-linked glycans.

The blood of patients with IgA nephropathy exhibits elevated levels of IgA1 molecules containing aberrantly glycosylated hinge-region *O*-glycans that are deficient in Gal. These aberrant molecules are present in circulating immune complexes. Furthermore, IgA1 in the mesangial deposits is enriched for this aberrantly glycosylated form of IgA1. IgA1-secreting cell lines from Epstein–Barr-virus-immortalized circulating B cells from patients with IgA nephropathy and from healthy controls have been used to characterize the glycosylation of the secreted immunoglobulins. The cells from the patients with IgA nephropathy secrete IgA1 that is mostly polymeric and contains Gal-deficient *O*-linked glycans, with terminal or sialylated GalNAc. These glycans also exhibit an increased content of sialic acid, as shown by an increased binding to *Sambucus nigra* agglutinin, a lectin specific for α2,6-bound sialic acid. In contrast, the IgA1 that is secreted by cells from healthy controls has a normal content of Gal. Thus, the cells from patients with IgA nephropathy secrete IgA1 that is Gal-deficient and contains terminal and sialylated GalNAc. These aberrances are due to the altered expression of specific glycosyltransferases: increased expression and activity of ST6GalNAcII and decreased expression and activity of C1GalT1 as well as decreased expression of Cosmc. Decreased expression of Cosmc further decreases the amount of C1GalT1 enzyme as a result of its greater degradation in the absence of the protein chaperone. The changes in enzyme activities mean that there will be fewer Gal but more sialic acid residues added to the GalNAc in the IgA1 hinge region. Furthermore, the overexpression of ST6GalNAcII may result in premature sialylation of GalNAc that would block other modifications, including galactosylation (see Figure 29.5).

29-7 Anti-glycan antibodies develop against aberrantly glycosylated IgA1 in IgA nephropathy and form immune complexes.

Antibodies specific for terminal GalNAc-containing glycoproteins have been reported in patients with IgA nephropathy and Henoch–Schoenlein purpura nephritis, and the rare Tn syndrome, also called 'permanent mixed-field polyagglutinability,' is a rare autoimmune disease in which subpopulations of blood cells in all lineages carry incompletely *O*-glycosylated cell-surface glycoproteins. These antibodies are also present in sera of healthy individuals, although at lower levels, and in sera of a diverse array of vertebrate species. The origin of such ubiquitous antibodies is unclear. However, many microorganisms, including bacteria and viruses, express glycan side chains with terminal GalNAc residues on their surfaces and may induce the synthesis of antibodies that cross-react with analogous structures in the hinge region of Gal-deficient IgA1 and other cell-associated glycoproteins.

In IgA nephropathy, Gal-deficient hinge-region glycans with terminal GalNAc on IgA1 represent neoepitopes that are recognized by the above-mentioned anti-glycan antibodies, resulting in the formation of circulating immune complexes. These complexes are relatively large, are not efficiently cleared from the circulation, tend to deposit in the renal mesangium, and bind to mesangial cells and activate them to induce glomerular injury.

Because the circulating lymphocytes from patients with IgA nephropathy produce IgG that is capable of binding to Gal-deficient IgA1, an approach has been developed to select Epstein–Barr-virus-immortalized lymphocytes

	CDR3 amino acid sequence	Binding to Gal-deficient IgA1
IgAN	Tyr-Cys-Ser (X)$_n$ n = 10–24	High
Controls	Tyr-Cys-**Ala** (X)$_n$ n = 6–9	Low

Figure 29.6 CDR3 of heavy chain of anti-glycan IgG antibodies binding to Gal-deficient IgA1. Sequence analyses of the cloned heavy- and light-chain antigen-binding domains of the glycan-specific IgG produced by immortalized lymphocytes from patients with IgA nephropathy (IgAN) identified unique features in CDR3 of the IgG heavy chain. Specifically, the third position in CDR3 in the heavy-chain variable region was serine in six of seven patients with IgA nephropathy (the first three amino acids were Tyr-Cys-Ser); in contrast, all six healthy controls had Ala in that position (the first three amino acids were Tyr-Cys-Ala). Mutagenesis experiments determined a critical role of serine for effective binding to Gal-deficient IgA1. Parenthesis shows the range of lengths of CDR3 regions. Note, they are longer in the CDR3 regions of IgAN. (Data from Suzuki, H. et al., *J. Clin. Invest.* 119:1668–1677, 2009. With permission from American Society for Clinical Investigation.)

that secrete IgG specific for Gal-deficient IgA1. Using this method, it has been shown that the IgG from the immortalized lymphocytes from patients with IgA nephropathy binds Gal-deficient IgA1 in a glycan-dependent manner, as does the patients' polyclonal serum IgG. Moreover, it has been demonstrated that the binding of these antibodies to Gal-deficient IgA1 is blocked by pre-incubation with the GalNAc-specific lectin (*Helix aspersa* agglutinin), suggesting that the binding is glycan-mediated. Sequence analyses of the cloned heavy- and light-chain antigen-binding domains of the glycan-specific antibodies produced by the immortalized lymphocytes from patients with IgA nephropathy have identified unique features in complementarity-determining region 3 (CDR3) of the IgG heavy chain. Specifically, the third position in CDR3 in the heavy-chain variable region is typically serine in IgA nephropathy (in which the first three amino acids are Tyr-Cys-Ser) as opposed to alanine in that position (in which the first three amino acids are Tyr-Cys-Ala) (Figure 29.6). Site-directed mutagenesis has proven that serine in the third position is necessary for efficient binding of IgG to Gal-deficient IgA1. These findings have revealed unique features on the IgG anti-glycan antibodies in patients with IgA nephropathy.

Elucidation of the mechanisms of immune-complex formation in which Gal-deficient IgA1 acts as an antigen reacting with ubiquitous naturally occurring GalNAc-specific antibodies provides a potential opportunity to interfere with the formation of pathogenic immune complexes. Generation of non-cross-linking, monovalent reagents (such as single-chain antibodies) with high affinity for GalNAc would theoretically prevent the binding of anti-glycan IgG and thus the formation of pathogenic complexes. This tactic may be one of the promising rational approaches to the development of a disease-specific treatment of IgA nephropathy.

29-8   Gal-deficient IgA1-containing immune complexes in IgA nephropathy are inflammatory.

Cultured human mesangial cells provide a convenient model for evaluating the biological activities of IgA1-containing complexes. Immune complexes from patients with IgA nephropathy containing Gal-deficient IgA1 bind to the cells more efficiently than do uncomplexed IgA1 or immune complexes from healthy controls. Moreover, the high-molecular-weight complexes associated with IgA nephropathy stimulate cellular proliferation and the production of cytokines (such as interleukin-6 (IL-6) and transforming growth factor-β (TGF-β)). IgA1-depleted fractions are devoid of such stimulatory activities. Consistent with this finding, when small quantities of desialylated polymeric IgA1 are added to sera of patients with IgA nephropathy, new immune complexes are formed and the amount of stimulatory complexes with molecular weights of 800–900 kDa is increased. These complexes contain IgG and IgA1. In contrast, uncomplexed IgA1 does not affect cellular proliferation. Complexes in the native sera of patients with IgA nephropathy enhance cellular proliferation more than do complexes of similar molecular weight from sera of healthy volunteers. In addition, immune complexes from sera of patients with IgA nephropathy collected during episodes of macroscopic hematuria stimulate more cellular proliferation than do complexes obtained during a later quiescent phase of disease. IgA1 complexes with high levels of Gal-deficient IgA1 induce proliferation of cultured human mesangial cells to a higher degree than complexes with a lower content of Gal-deficient IgA1.

Although the role of IgA1-containing immune complexes in IgA nephropathy is proven, the cellular mechanisms and receptors involved in the binding of IgA1 complexes and activation of mesangial cells are much less well understood. Several lines of evidence point to a specific IgA receptor expressed by mesangial cells that bind and at least partly internalize and catabolically degrade IgA.

IgA receptors reported on human cells include ASGP-R on hepatocytes; polymeric immunoglobulin receptor specific for J-chain-containing polymeric IgA and IgM on epithelial cells; FcαR (CD89) on monocytes, neutrophils, and eosinophils; and Fcα/μ receptor expressed by B cells and macrophages. However, none of the above receptors has been confirmed on human mesangial cells. In addition, a surface-bound Gal-transferase can also bind immunoglobulins via the glycans, but a role for this protein for IgA1 binding to mesangial cells has not been proven. More recently, it has been shown that a transferrin receptor (CD71) is able to bind polymeric IgA1 and that proliferating human mesangial cells (and other cell types) express CD71 on their surfaces. Moreover, CD71 on human mesangial cells effectively binds Gal-deficient IgA1 and immune complexes containing Gal-deficient IgA1, which further enhances the expression of CD71. This binding thus creates a positive-feedback loop that leads to overexpression of CD71 on proliferating mesangial cells in glomeruli. However, it is not clear whether CD71 is the only receptor involved in the binding of IgA1-containing immune complexes and whether it has a direct pathogenic role in IgA nephropathy.

Mesangial cells activated by immune complexes containing Gal-deficient IgA1 proliferate and overproduce extracellular matrix proteins and cytokines and chemokines. These processes, if they continue unchecked for substantial periods, may lead to expansion of the glomerular mesangium and, ultimately, glomerular fibrosis with the loss of glomerular filtration function. Furthermore, it has recently been shown that humoral factors (such as tumor necrosis factor and TGF-β) that are released from mesangial cells activated by IgA1-containing immune complexes alter podocyte gene expression and may thus alter glomerular permeability. This mesangio-podocyte communication may be a mechanism to explain the occurrence of proteinuria and tubulointerstitial injury in IgA nephropathy.

29-9 Urinary immunoglobulins are potential biomarkers of IgA nephropathy.

Currently, a definitive diagnosis of IgA nephropathy requires renal biopsy, because there is no valid noninvasive test that can be reliably used as a diagnostic alternative. Unfortunately, renal biopsy entails a risk for serious complications. Consequently, early detection is frequently impossible and monitoring of the disease is compromised. Some patients with IgA nephropathy will progress to end-stage renal disease even when their initial histological findings display relatively minor glomerular abnormalities. Therefore, a reliable noninvasive diagnostic test will be very useful if it can detect subclinical IgA nephropathy, estimate the degree of activity, monitor the progression or abatement of the renal damage, and assess the response to treatment without serial renal biopsies.

There have been numerous efforts to identify disease-specific markers, frequently using urine because it is easy to obtain and collection is noninvasive. The urinary concentrations of several cytokines involved in proliferation and repair have been evaluated as potential markers of histopathologic glomerular and tubulointerstitial changes and predictors of long-term prognosis. For example, urinary levels of IL-6 are elevated in patients with glomerulonephritis, but do not differentiate between different forms of the disease. However, urinary excretion of IL-6 is able to predict long-term renal outcome in patients with IgA nephropathy, and excretion of renal IL-6 and epidermal growth factor (EGF) has been shown to correlate with the degree of tubulointerstitial damage. Consequently, the ratio of urinary IL-6/EGF has been proposed as a prognostic marker for the progression of renal damage in IgA nephropathy. Among other potential urinary markers for renal disease, the excretion of monocyte chemotactic peptide-1 (MCP-1) and IL-8 correlates with tubulointerstitial damage, but also does not distinguish between the specific types of glomerulonephritis. Moreover, excretion of the membrane-attack complex is elevated in patients with membranous nephropathy, but not in patients with IgA nephropathy.

Urinary excretion of immunoglobulins has offered a promising new possibility in the search for surrogate markers of ongoing renal injury of IgA nephropathy. Urinary IgA and IgG concentrations have been shown to be higher in patients with IgA nephropathy than in normal healthy volunteers and patients with other renal diseases. Moreover, the amounts excreted correlate with the serum creatinine concentration and with proteinuria. Furthermore, a recent study has shown a correlation between glomerular filtration rate, urinary IgG excretion, and pathological grading of renal biopsies in patients with nephritis. Although these reports demonstrate the differential presence (levels) of proteins and protein complexes, including those of high molecular weight, in the urine of patients with IgA nephropathy, none of the tests has been developed into a clinically useful assay.

It is likely that the urine of patients with IgA nephropathy includes many polypeptides and/or their proteolytic fragments that are disease-specific. These peptides may be discovered by modern proteomic techniques and potentially refined into specific biomarkers. The potential markers include high-molecular-weight immune complexes and their components, immunoglobulin molecules, and smaller polypeptides. Enzyme-linked immunosorbent assay (ELISA) and Western blots are adequate for the analysis of large polypeptides and their complexes. Using capture ELISA, levels of urinary IgA:IgG-containing immune complexes in patients with IgA nephropathy are elevated compared with levels in patients with non-IgA nephropathy glomerulonephritis or in healthy controls. For the analysis of smaller polypeptides, modern separation techniques can be combined with mass-spectrometry-based identification and quantitation of the separated component.

Various mass-spectrometry-based technologies, such as surface-enhanced laser desorption ionization mass spectrometry (SELDI-MS) and capillary electrophoresis–mass spectrometry (CE-MS), have been used for proteomic analyses of clinical samples. Generally, CE-MS analysis displays a richer and more complex pattern of urinary polypeptides with higher resolution and higher mass accuracy than does an analysis with SELDI-MS; resolution and mass accuracy are functions of MS instruments that are linked with the two respective techniques. CE-MS has already been used for establishing markers of different renal diseases, including focal segmental glomerulosclerosis and diabetic nephropathy. Recently, CE-MS has been used to identify peptide markers of IgA nephropathy. Although proteomic urinary biomarkers are very appealing, more studies must be done before any assay will become a clinical application.

As the pathogenesis of IgA nephropathy is being uncovered and the role of IgA1 and IgA1-containing immune complexes has been identified, serum has become a promising source of potential biomarkers. Levels of Gal-deficient IgA1 and anti-glycan antibodies have been targeted as possible diagnostic markers of IgA nephropathy as well as markers of disease progression. An ELISA using *Helix aspersa* agglutinin lectin specific for GalNAc of the neoantigen of the nephritogenic IgA1 has shown that 74% of patients with IgA nephropathy have a serum level of Gal-deficient IgA1 that is above the 90th centile for healthy controls. Furthermore, serum levels of Gal-deficient IgA1 show significant heritability, suggesting a basis for the familial pattern of disease for some patients. A dot-blot test has been developed to semiquantitatively measure the amount of circulating IgG specific for Gal-deficient IgA1. As these methods are refined to improve their sensitivity and specificity, they may become lead candidates for development into clinical assays.

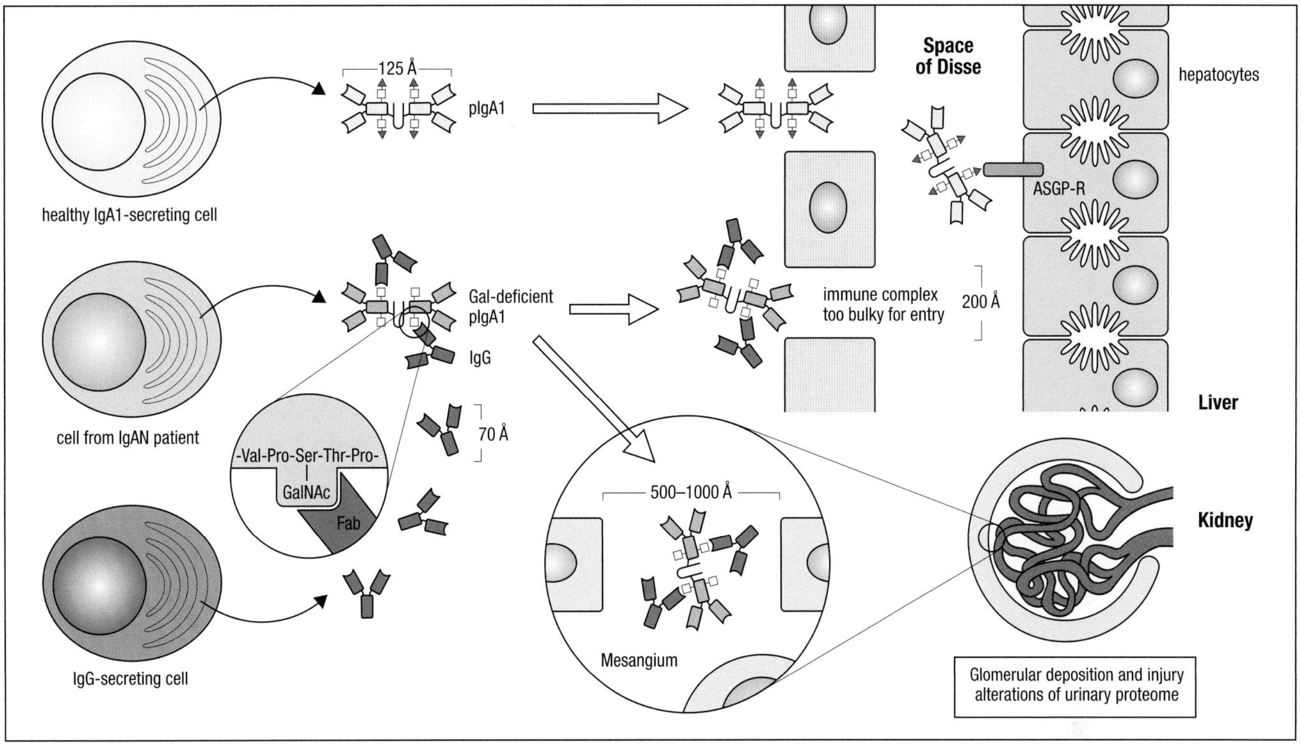

Figure 29.7 Model of IgA nephropathy pathogenesis. A fraction of polymeric IgA1 (pIgA1) produced by B cells and plasma cells in patients with IgA nephropathy (IgAN) is Gal-deficient and is recognized in the circulation by anti-glycan antibodies. The resultant immune complexes are too bulky to enter the space of Disse in the liver to reach the asialoglycoprotein receptor (ASGP-R) on hepatocytes, a major site of catabolism for normally glycosylated circulating IgA1, but are able to pass through the larger fenestrae in glomerular capillaries overlying the mesangium. These complexes bind to mesangial cells and activate them, thus initiating glomerular injury to induce alterations of the urinary proteome.

Summary.

In patients with IgA nephropathy, Gal-deficient IgA1 is produced at elevated levels and is recognized by unique anti-glycan antibodies. Immune complexes form, and on deposition in the mesangium they are capable of inducing a mesangioproliferative glomerular injury (Figure 29.7). Thus, formation of the Gal-deficient IgA1-containing circulating immune complexes is a critical factor in the pathogenesis of IgA nephropathy. The anti-glycan antibodies that bind to Gal-deficient IgA1 and the resultant immune complexes are key pathogenic factors contributing to the development of IgA nephropathy. On the basis of the understanding of the disease pathogenesis, biomarkers specific for IgA nephropathy can be identified and developed into clinical assays to aid in diagnosis, assessment of prognosis, and monitoring of disease progression. Furthermore, interference with specific pathogenetic pathways may be used in future disease-specific therapy.

Further Reading.

Barratt, J., and Feehally, J.: **IgA nephropathy.** *J. Am. Soc. Nephrol.* 2005, **16**:2088–2097.

Beerman, I., Novak, J., Wyatt, R.J., *et al.*: **Genetics of IgA nephropathy.** *Nat. Clin. Pract. Nephrol.* 2007, **3**:325–338.

Berger, J., and Hinglais, N.: **Les dépôts intercapillaires d'IgA-IgG. [Intercapillary deposits of IgA-IgG.]** *J. Urol. Nephrol.* 1968, **74**:694–695.

Cattran, D.C., Coppo, R., Cook, H.T., *et al.*: **The Oxford classification of IgA nephropathy: rationale, clinicopathological correlations, and classification.** *Kidney Int.* 2009, **76**:534–545.

Coppo, R., Troyanov, S., Camilla, R., *et al.*: **The Oxford IgA nephropathy clinicopathological classification is valid for children as well as adults.** *Kidney Int.* 2010, **77**:921–927.

Gharavi, A.G., Moldoveanu, Z., Wyatt, R.J., *et al.*: **Aberrant IgA1 glycosylation is inherited in familial and sporadic IgA nephropathy.** *J. Am. Soc. Nephrol.* 2008, **19**:1008–1014.

Jennette, J.C.: **The immunohistology of IgA nephropathy.** *Am. J. Kidney Dis.* 1988, **12**:348–352.

Julian, B.A.: **Treatment of IgA nephropathy.** *Semin. Nephrol.* 2000, **20**:277–285.

Julian, B.A., and Novak, J.: **IgA nephropathy: an update.** *Curr. Opin. Nephrol. Hypertens.* 2004, **13**:171–179.

Julian, B.A., Suzuki, H., Spasovski, G., *et al.*: **Application of proteomic analysis to renal disease in the clinic.** *Proteomics Clin. Appl.* 2009, **3**:1023–1028.

Julian, B.A., Suzuki, H., Suzuki, Y., *et al.*: **Sources of urinary proteins and their analysis by urinary proteomics for the detection of biomarkers of disease.** *Proteomics Clin. Appl.* 2009, **3**:1029–1043.

Lai, K.N., Leung, J.C., Chan, L.Y., *et al.*: **Podocyte injury induced by mesangial-derived cytokines in IgA nephropathy.** *Nephrol. Dial. Transplant.* 2009, **24**:62–72.

Mestecky, J., Moro, I., Kerr, M.A., *et al.*: **Mucosal immunoglobulins**, in Mestecky, J., Bienenstock, J., Lamm, M.E., *et al.* (eds): *Mucosal Immunology*, 3rd ed. Amsterdam, Elsevier Academic Press, 2005:153–181.

Moldoveanu, Z., Wyatt, R.J., Lee, J., *et al.*: **Patients with IgA nephropathy have increased serum galactose-deficient IgA1 levels.** *Kidney Int.* 2007, **71**:1148–1154.

Monteiro, R.C.: **Pathogenic role of IgA receptors in IgA nephropathy.** *Contrib. Nephrol.* 2007, **157**:64–69.

Moura, I.C., Arcos-Fajardo, M., Sadaka, C., *et al.*: **Glycosylation and size of IgA1 are essential for interaction with mesangial transferrin receptor in IgA nephropathy.** *J. Am. Soc. Nephrol.* 2004, **15**:622–634.

Moura, I.C., Centelles, M.N., Arcos-Fajardo, M., *et al.*: **Identification of the transferrin receptor as a novel immunoglobulin (Ig)A1 receptor and its enhanced expression on mesangial cells in IgA nephropathy.** *J. Exp. Med.* 2001, **194**:417–425.

Novak, J., Moldoveanu, Z., Renfrow, M.B., *et al.*: **IgA nephropathy and Henoch–Schoenlein purpura nephritis: aberrant glycosylation of IgA1, formation of IgA1-containing immune complexes, and activation of mesangial cells.** *Contrib. Nephrol.* 2007, **157**:134–138.

Suzuki, H., Fun, R., Zhang, Z., *et al.*: **Aberrantly glycosylated IgA1 in IgA nephropathy patients is recognized by IgG antibodies with restricted heterogeneity.** *J. Clin. Invest.* 2009, **119**:1668–1677.

Suzuki, H., Moldoveanu, Z., Hall, S., *et al.*: **IgA1-secreting cell lines from patients with IgA nephropathy produce aberrantly glycosylated IgA1.** *J. Clin. Invest.* 2008, **118**:629–639.

Mucosal Manifestations of Immunodeficiencies

30

For many years it has been considered almost axiomatic that the innate and acquired immune system must collectively defend against ingested and/or inhaled toxic foreign substances, and the ever-present potential assault of microbial pathogens, while simultaneously coexisting with the luminal commensal microflora. A critical role(s) for the innate and adaptive immune system in maintaining mucosal homeostasis has become evident through the study of humans with naturally occurring primary immunodeficiencies that result from dominant mutations in immune-related genes, and of murine strains with similar genetically defined mutations. Indeed, gastrointestinal pathology is a common clinical feature of many human primary immunodeficiencies that are inherited as simple Mendelian traits, with attendant implications for common polygenic disorders such as inflammatory bowel disease (IBD), celiac disease, and asthma. In order to highlight the specific roles of unique aspects of the mucosal immune system in maintaining mucosal homeostasis, this chapter will focus on those human immunodeficiencies where gastrointestinal manifestations usually occur. For this discussion, immunodeficiencies will be categorized based on those that present primarily as (1) antibody deficiencies; (2) combined (B cell and T cell) immunodeficiencies; (3) immunodeficiencies with immune dysregulation; and (4) immunodeficiencies with defects in innate immune cells (Table 30.1). Since recent large genome-wide association studies and other mapping approaches have implicated causative polymorphisms in several immune-related genes in the etiology of IBD, celiac disease, and asthma, an understanding of the specific role of the immune system in regulating mucosal immune function has acquired a much broader significance.

Table 30.1 Primary immunodeficiencies that most commonly present with mucosal manifestations

Antibody deficiencies	Combined or primary T-cell deficiencies	Immunodeficiency with immune dysregulation[1]	Immunodeficiencies with defects in innate immune cells
X-linked agammaglobulinemia (XLA)	Severe combined immunodeficiency (SCID)	Wiskott-Aldrich syndrome (WAS)	Chronic granulomatous disease (CGD)
Common variable immunodeficiency[2] (CVID)	CD40 ligand (CD40L) or CD40 deficiency	Immune dysregulation, polyendocrinopathy, enteropathy, X-linked (IPEX) syndrome	NEMO[3] (NFκB essential modulator) deficiency or X-linked immunodeficiency with ectodermal dystrophy
IgA deficiency		IL-10R/IL-10 deficiency IL-2Rα (CD25) deficiency	

[1] WAS and interleukin (IL)-10R/IL-10 deficiency affects both innate and adaptive immune function;
[2] T-cell dysfunction can also be an associated feature;
[3] NEMO also affects adaptive immune cell function.

Antibody deficiencies.

As discussed in Chapter 9, the presence and secretion of immunoglobulins are important to the maintenance of mucosal homeostasis and the proper architecture of microbial commensalism and defense against invading pathogens. It is not surprising therefore that the dysfunction in the ability to properly regulate the production of immunoglobulins represents an important class of immunodeficiencies. Such immunodeficiencies result from either primary defects in B cells or defects in T-cell regulation of B-cell function.

30-1   Congenital agammaglobulinemia, or X-linked agammaglobulinemia, is the prototypic disorder of genetically determined antibody deficiency.

X-linked agammaglobulinemia (XLA) results from mutations of the Bruton's tyrosine kinase (*BTK*) gene. *BTK* is located on the X chromosome and regulates signaling through the pre-B-cell receptor (pre-BCR) and the BCR. BTK deficiency results in a block at the pro-B to pre-B cell stage in B-cell differentiation in the bone marrow. Accordingly, patients with XLA have a severe reduction or absence of circulating B cells, associated with profound deficiency of all immunoglobulin isotypes. A similar phenotype can also be observed in patients with autosomal recessive forms of agammaglobulinemia, due to mutations of the mu (μ) heavy-chain gene; of the Igα, Igβ, and V pre-B components of the pre-BCR; or of the adaptor molecule B-cell linker protein, which is also involved in pre-BCR-mediated signaling.

The clinical phenotype of congenital agammaglobulinemia is mainly characterized by recurrent sinopulmonary infections that typically appear after 4 months of age, when levels of maternally derived IgG decrease, suggesting that among the immunoglobulin classes most associated with disease, the IgG class may be most relevant to the clinical manifestations of disease. The diagnosis is based upon demonstration of pan-hypogammaglobulinemia and severe reduction (<0.1% of the normal level of immunoglobulins) in the number of circulating B lymphocytes. Treatment consists of regular administration of intravenous or subcutaneous immunoglobulins, possibly associated with antimicrobial prophylaxis. Administration of live viral vaccines should be avoided.

Gastrointestinal infections are relatively common in patients with XLA and other forms of congenital agammaglobulinemia, although their frequency is lower compared with other forms of antibody deficiency that are also characterized by concomitant defective T-cell function, such as common variable immunodeficiency and combined immunodeficiency. Common causes of gastrointestinal infection and the resulting chronic diarrhea in patients with agammaglobulinemia include *Giardia lamblia*, *Salmonella* spp., *Campylobacter* spp., and rotavirus. In addition, enteroviral infections (Coxsackie virus, echovirus) may disseminate to the central nervous system and cause life-threatening encephalitis. Intestinal biopsies show lack of plasma cells in the lamina propria and no germinal centers in gut-associated lymphoid tissue, consistent with the block in B-cell differentiation.

Small-bowel Crohn's-like disease can occur in patients with XLA (Figure 30.1). However, IBD-like disorders are only an infrequent consequence of XLA, suggesting that a primary humoral abnormality alone is unlikely to explain IBD pathogenesis. Gastric adenocarcinoma and colorectal cancer may be more common in XLA than in the general population. A possible role for increased frequency of *Helicobacter pylori* infection has been hypothesized in the pathogenesis of the gastric tumors.

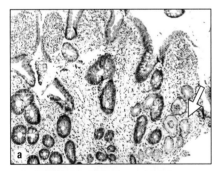

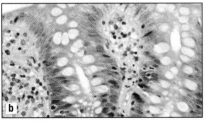

Figure 30.1 Intestinal inflammation in XLA. Panel a shows chronic changes and pyloric gland metaplasia (white arrow) in terminal ileum, ×100. Panel b shows the absence of plasma cells and mild acute inflammation in colonic lamina propria, ×400. (Courtesy of Thomas Walker and Jeffrey Goldsmith.)

30-2   Common variable immunodeficiency (CVID) is the most common form of clinically significant primary immunodeficiency.

CVID is defined by reduced levels of one or more immunoglobulin isotypes and impaired antibody production in response to immunization with antigens or to natural infections. It has been proposed to limit the diagnosis of CVID to individuals older than 4 years of age, with the idea that monogenic disorders of antibody production should be considered at a younger age. Patients with CVID are prone to recurrent respiratory infections, most often caused by *Haemophilus influenzae* and *Streptococcus pneumoniae*. In addition, autoimmune manifestations (cytopenias, inflammatory bowel disease), granulomas, lymphoid hyperplasia, and tumors (especially lymphoma) are also common.

The majority of patients with CVID have a normal number of B lymphocytes, but switched memory (CD27⁺IgD⁻) B lymphocytes are often significantly reduced and there is a low rate of somatic hypermutation of immunoglobulin genes. Plasma cells are absent on histological examination of lymphoid tissue and bone marrow. Other significant laboratory abnormalities in patients with CVID include defects of T-cell activation and of cytokine production, increased cellular apoptosis, and impaired B-cell activation following ligation of endosomal Toll-like receptors.

Although most cases of CVID represent sporadic forms of presentation, pedigrees consistent with autosomal dominant or autosomal recessive inheritance have been also described (Figure 30.2). Association with the human leukocyte antigen (HLA) region has been reported; however, the underlying gene defect has not been identified in most subjects with these disorders. Mutations of the transmembrane activator and calcium-modulator and cyclophilin ligand interactor (TACI) have been identified in approximately 10% of patients, most often in heterozygosity. TACI is expressed by B lymphocytes and interacts with the B cell-activating factor (BAFF) and with a proliferation inducing ligand (APRIL). In particular, APRIL–TACI interaction promotes B-cell activation and immunoglobulin class-switch recombination. Disruption of TACI in mice leads to lymphoid proliferation, resembling lymphoid hyperplasia that

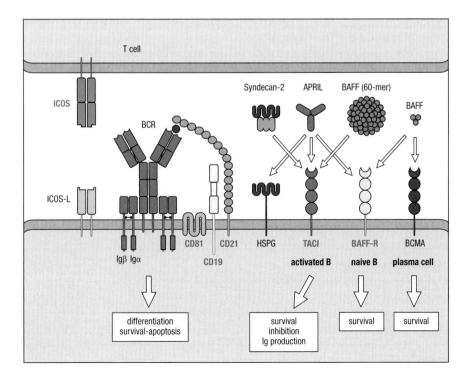

Figure 30.2 Genetic mutations associated with CVID. Depicted are proteins involved in B-cell survival and differentiation in peripheral lymphoid organs. CD21, CD19, and CD81 are part of a macromolecular complex that associates with B-cell receptor (BCR) and contributes to intracellular signaling. TACI and BAFF-R are expressed by mature B lymphocytes and share binding to APRIL and BAFF. ICOS is a cell-surface molecule expressed by activated T cells and binds to ICOS ligand (ICOS-L) expressed by B lymphocytes. Molecules identified in blue font have been associated with CVID. ICOS, inducible T cell co-stimulator; APRIL, a proliferation inducing ligand; BAFF, B cell-activating factor belonging to the tumor necrosis factor family; BAFF-R, BAFF receptor; TACI, transmembrane activator and calcium modulator cyclophilin ligand interactor; HSPG, heparan sulfate proteoglycans; BCMA, B cell maturation antigen; Ig, Immunoglobulin.

is often seen in humans with CVID. However, the significance of TACI mutations in CVID pathogenesis remains unclear; with few exceptions, it seems that they may represent susceptibility, rather than disease-causing mutations. A similar significance of CVID-predisposing factors has also been attributed to mutations of BAFF receptor (BAFF-R) and CD20 that have been reported in a few patients with CVID. In contrast, a clear pathogenic role has been demonstrated for mutations of ICOS, CD19, and CD81. ICOS is a co-stimulatory molecule, expressed by T cells, that interacts with ICOS-ligand (ICOS-L) expressed by B lymphocytes. CD19 and CD81 form a cell-surface protein complex that participates in BCR-mediated signaling. Patients with CVID are highly prone to a variety of gastrointestinal manifestations (Table 30.2), including infections, autoimmunity, inflammation, and tumors. In addition, abnormalities of liver function have also been reported.

Chronic diarrhea has been reported in 20–60% of CVID patients. It has been suggested that diarrhea is more common in CVID patients with undetectable IgA, arguing for a protective role of mucosal IgA antibodies. Consistent with this observation, diarrhea does not usually improve upon immunoglobulin substitution therapy that does not replace IgA. *G. lamblia* is a frequent cause of diarrhea in CVID. Upon ingestion of the protozoan cysts, the trophozoites may diffusely colonize the mucosa of the small intestine, causing abdominal cramps, profuse diarrhea, and malabsorption. Identification of the pathogen is based on culture and examination of the stool, but often requires duodenal aspiration and biopsies. Prolonged treatment with metronidazole is often required, but relapses are nonetheless common. Other common pathogens include *Salmonella*, *Campylobacter*, cytomegalovirus, and *Cryptosporidium*.

Chronic gastritis is seen in one-third of patients and is associated with *H. pylori* infection that may also predispose to gastric adenocarcinoma, a malignancy associated with increased frequency in CVID patients. Atrophic gastritis with achlorhydria and pernicious anemia is relatively common in CVID; autoantibodies are not detected, but there is significant lymphocytic infiltration in the gastric biopsies, suggesting an immune-mediated pathogenesis.

Diarrhea and malabsorption may also reflect immune-mediated damage to the small intestine, with flattening of villi and crypt hyperplasia. This complication is seen in up to 50% of patients, and may lead to significant weight loss, hypoproteinemia, and anemia. CD4 lymphopenia is also frequent. While these features are reminiscent of celiac sprue (disease), several pathological elements are distinctive. In particular, plasma cells are absent in intestinal

Table 30.2 Gastrointestinal manifestations of common variable immunodeficiencies

Infections	Autoimmune complications	Malignancies
Giardia lamblia	Diarrhea/malabsorption	Gastric adenocarcinoma
Salmonella	Atrophic gastritis	Intestinal lymphoma
Campylobacter	Pernicious anemia	Colonic adenocarcinoma
Cryptosporidium	Small and/or large bowel inflammation	
Cytomegalovirus	Nodular lymphoid hyperplasia	
Hepatitis C virus	Hepatitis	
	Nodular regenerative hyperplasia	

biopsies from CVID patients, whereas these are increased in patients with celiac disease. Furthermore, patients with CVID typically lack antibodies to gliadin, tissue transglutaminase (tTG), reticulin, and endomysium, which are common biomarkers of celiac disease. Finally, introduction of a gluten-free diet is not beneficial in patients with CVID. Treatment of the diarrhea and malabsorption associated with CVID is often challenging, and is based on correction of metabolic disturbances, generous support of water-soluble vitamins, and the use of anti-secretory agents. However, use of elemental diets and of parenteral nutrition may be necessary in severe cases. Judicious attempts with low-dose steroids are often considered.

IBD-like sequelae have been reported in up to 15% of CVID patients, and include both Crohn's disease- and ulcerative colitis-like forms. Clinical manifestations include diarrhea, abdominal pain, and rectal bleeding. Duodenal and colonic biopsies demonstrate villous flattening and lymphocytic infiltration of the lamina propria respectively (Figure 30.3). Apoptosis of epithelial cells is often prominent. In contrast to what is typically found in intestinal tissue in patients with Crohn's disease, there is often no increase in interleukin (IL)-17, IL-23, and tumor necrosis factor-α (TNF-α) production (but rather an increase in IL-12 and interferon-γ). Whether CVID and IBD have an overlapping genetic and environmental basis is unknown. Despite these distinct pathological features, treatment for the IBD-like disorder associated with CVID is similar to that used for Crohn's disease and ulcerative colitis and includes anti-inflammatory agents (e.g., 5-amino-salicylic acid compounds as well as topical and systemic steroids), immunomodulators (e.g., azathioprine, 6-mercaptopurine), and antibiotics (e.g., ciprofloxacin, metronidazole, and rifaximin). Beneficial effects have been reported with TNF-blocking agents, such as infliximab, but judicious surveillance for mycobacterial, fungal, and other infections is required.

Nodular lymphoid hyperplasia (NLH) is seen in up to 10% of patients with CVID in association with villous flattening and diarrhea, which may be associated with malabsorption; however, when nodules are particularly prominent, they may also cause obstruction. Histologically, large lymphoid follicles are found with germinal centers and an increase in B lymphocytes, but plasma cells are absent. It remains unclear whether NLH represents a lesion that precedes development of intestinal lymphoma. In any case, patients with CVID are at greater risk for non-Hodgkin's lymphoma of the intestine.

Finally, persistent elevation of liver enzymes can be seen in more than 50% of CVID patients. Most patients are found to have cholestasis without jaundice and, when liver biopsies have been available, have been found to exhibit a nodular regenerative hyperplasia. Nodular regenerative hyperplasia is characterized by the presence of regenerating liver cell plates alternating with atrophic areas with no significant portal fibrosis, bridging fibrosis, or definite cirrhosis. Non-caseating granulomas are often an associated feature. Portal vein endotheliitis, intrasinusoidal inflammation with prominent T-cell infiltration, and portal hypertension in the absence of cirrhosis are additional common features. The presence of nodular regenerative hyperplasia has been correlated with other autoimmune phenomena and it has been suggested that these liver modifications may represent the consequence of chronic gastrointestinal inflammation with excessive translocation of luminal antigens into the liver. Severe liver disease has been reported in the past in CVID patients who had acquired hepatitis C virus (HCV) infection through contaminated blood products (including immunoglobulin preparations). While this risk has been virtually eliminated with improvement in screening of blood products and the preparation of immunoglobulins, it appears that progression of liver disease is faster and more severe in patients with CVID who are infected with hepatotropic viruses.

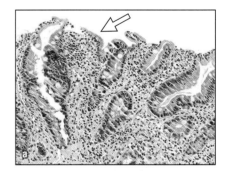

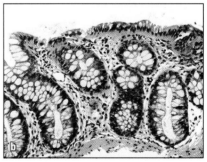

Figure 30.3 Intestinal findings in CVID. Panel a shows villous atrophy and acute inflammation (white arrow) in the duodenum, ×100. Panel b shows the absence of plasma cells in the colonic lamina propria. (Courtesy of Athos Bousvaros and Jeffrey Goldsmith.)

30-3 IgA deficiency is the most common primary immunodeficiency.

IgA deficiency, which is the most prevalent of the primary immunodeficiencies, affects approximately 1 in 700 individuals of European ancestry and significantly fewer of those of Asian ancestry. Both partial and complete forms of IgA deficiency are known. Association with HLA alleles and a higher frequency within families with CVID and celiac disease have been reported, but the pathophysiology remains unknown. Secretory IgA represents the most abundant mucosal immunoglobulin isotype; however, approximately two-thirds of IgA-deficient individuals are asymptomatic, while the remaining may suffer from recurrent infections, autoimmunity, or allergy. Respiratory and intestinal infections are more common when IgA deficiency is associated with defects of IgG subclasses. It is thought that IgM may compensate for the lack of IgA at the mucosal surface, because it can be efficiently transported by the polymeric immunoglobulin receptor.

Approximately one-third of adult individuals with IgA deficiency suffer from recurrent infections of the respiratory tract. Most of these infections are of viral origin and tend to be clinically mild. An increased incidence of pseudocroup (due to parainfluenza virus) has been observed in infants with IgA deficiency. Bacterial infections are less common, and are predominantly due to *Haemophilus influenzae* and *Streptococcus pneumoniae*. Bronchiectasis and chronic lung damage have been rarely observed.

Gastrointestinal manifestations of IgA deficiency are similar to those of patients with CVID, but their frequency is lower. Lack of secretory IgA is often associated with giardiasis, resulting in diarrhea with steatorrhea and malabsorption. Patients with IgA deficiency have a ten- to twentyfold increased risk of celiac disease, with typical villous flattening and an excellent response to a gluten-free diet. The biochemical diagnosis may be challenging, because of the lack of anti-tTG IgA antibodies; however, anti-tTG IgG are elevated. Interestingly, both IgA deficiency and celiac disease are associated with the extended haplotype HLA-A1, Cw7, B8, DR3, DQ2.

NLH has been also reported in IgA deficiency, but it is far less common than in patients with CVID. An important distinction is that the NLH of patients with IgA deficiency is characterized by accumulation of IgM-expressing B cells, possibly a compensatory mechanism for the lack of IgA in response to chronic stimulation. Otherwise, the clinical symptoms are the same as reported for NLH in CVID, and include diarrhea, abdominal pain, and even obstruction. NLH is also uniquely sensitive to oral steroids. Whenever an underlying infection is identified, treatment should be targeted to the specific pathogen. Lymphoma and gastric adenocarcinoma have also been reported in patients with IgA deficiency. On the other hand, IBD-like disease, pernicious anemia, and chronic gastritis are far less common in IgA deficiency than in CVID. Persistent elevation of liver enzymes and biliary cirrhosis have been described in IgA deficiency.

Combined immunodeficiencies.

Combined immunodeficiencies are those disorders that are genetically determined and lead to loss of both B- and T-cell function. As such, afflicted individuals exhibit a much more profound immune defect along their mucosal surfaces and are thus susceptible to a much broader combination of pathogens and associated clinical problems.

30-4 Severe combined immunodeficiencies (SCIDs) are a heterogeneous group of genetically determined disorders.

The diseases associated with SCID are characterized by profound defects in the development and function of T lymphocytes, and are variably associated with impaired development of B and/or natural killer (NK) lymphocytes. SCIDs have a prevalence of approximately 1 in 50,000–100,000 live births, and are more common in males, reflecting the fact that X-linked SCID is the most common form of SCID in humans. SCIDs are usually classified according to the immunological phenotype and are distinguished into: (1) SCID with absence of T lymphocytes, but presence of B lymphocytes (T⁻ B⁺ SCID); and (2) SCID with absence of both T and B lymphocytes (T⁻ B⁻ SCID). These distinct immunological phenotypes correspond to an even greater degree of genetic heterogeneity (Table 30.3). In addition, atypical forms of the disease exist, in which development of T lymphocytes is severely compromised, but not completely abrogated. In these 'leaky' variants, circulating T lymphocytes show an activated (CD45RO) phenotype, often display an oligoclonal T-cell receptor repertoire, and may infiltrate peripheral tissues, causing severe inflammatory manifestations. The prototype of such conditions is represented by Omenn syndrome. Patients with this variant present with significant skin rash and lymphadenopathy, and are prone to autoimmune manifestations. Another condition in which circulating T lymphocytes may be detected in patients with SCID is represented by transplacental passage of maternally derived T cells, that occurs in as many as 50% of SCID infants, and may cause graft-versus-host-like manifestations, including skin rash, severe diarrhea, elevation of liver enzymes, and bone marrow failure.

Irrespective of the nature of the genetic defect, patients with SCID present early in life with severe infections sustained by bacterial, viral, and fungal pathogens. Opportunistic respiratory infections (interstitial pneumonia due to *Pneumocystis jiroveci*, cytomegalovirus, adenovirus, and parainfluenza virus type 3) are particularly common. Infections of the gastrointestinal tract are a frequent cause of chronic diarrhea, leading to failure to thrive. Skin rash

Table 30.3 Genetic abnormalities associated with severe combined immunodeficiency

Cellular mechanism affected	Defective protein (gene)	Immunological phenotype	Inheritance
Lymphoid cell survival	Adenosine deaminase (*ADA*)	T⁻ B⁻ NK⁻ SCID	AR
	Adenylate kinase (*AK2*)	T⁻ B^low NK⁻ SCID	AR
Cytokine-mediated signaling	Common gamma chain (*IL2RG*)	T⁻ B⁺ NK⁻ SCID	XL
	Interleukin-7 receptor α chain (*IL7R*)	T⁻ B⁺ NK⁺ SCID	AR
	JAK3 (*JAK3*)	T⁻ B⁺ NK⁻ SCID	AR
V(D)J recombination	RAG proteins (*RAG1, RAG2*)	T⁻ B⁻ NK⁺ SCID	AR
	Artemis (*DCLRE1C*)	T⁻ B⁻ NK⁺ SCID	AR
	DNA-PKcs (*PRKDC*)	T⁻ B⁻ NK⁺ SCID	AR
	DNA ligase IV (*LIG4*)	T⁻ B⁻ NK⁺ SCID*	AR
	Cernunnos (*NHEJ1*)	T^low B^low NK⁺ SCID	AR
Signaling through pre-TCR and pre-BCR	CD3δ, ε, and ζ chains (*CD3D, CD3E, CD3Z*)	T⁻ B⁺ NK⁺ SCID	AR
	CD45 (*CD45*)	T⁻ B^{+/low} NK⁺ SCID	AR

TCR: T-cell receptor; BCR: B-cell receptor; JAK: Janus-associated kinase; RAG: recombination-activating gene; DNA-PKcs: DNA protein kinase catalytic subunit; AR: autosomal recessive; XL: X-linked. * DNA ligase IV associates with Cernunnos.

may be present, especially in infants with maternal T-cell engraftment or with atypical variants of the disease. Additional clinical features may be present in specific forms of SCID. For example, bone abnormalities are often observed in patients with adenosine deaminase deficiency, whereas microcephaly and growth failure are common in infants with SCID due to defects in genes encoding for proteins (DNA ligase IV, Cernunnos) involved in DNA repair.

The diagnosis of SCID is based on typical clinical and laboratory features. Severe lymphopenia is the hallmark of the disease. The number of circulating T lymphocytes is severely decreased in 85–90% of the patients, the exceptions being due to maternal T-cell engraftment or to atypical forms of SCID. When present, T cells are unable to proliferate *in vitro* in response to mitogens, and are often oligoclonal. Patients with SCID fail to produce specific antibodies upon immunization. However, total IgG serum levels may initially be within the normal range, reflecting transplacental passage of maternal IgG.

Unless immune reconstitution is provided by treatment, infants with SCID typically die at about 1–2 years of age, mostly due to infections. Recognition of SCID should prompt immediate and aggressive treatment of infections and use of broad antimicrobial prophylaxis, as well as immunoglobulin replacement therapy. Nutritional support is often necessary. The use of live virus immunizations should be strictly avoided, and all blood products should be irradiated and tested for cytomegalovirus before transfusion. Ultimately, treatment and immune reconstitution are based on hematopoietic cell transplantation (HCT). Survival is excellent (>90%) when HCT is performed from HLA-identical donors; however, this option is available only to a minority (<15%) of patients. Similarly good results are obtained when transplantation from HLA-mismatched related donors is performed early in life. Enzyme replacement therapy can be used in patients with adenosine deaminase deficiency and gene therapy has shown promising results in patients with adenosine deaminase deficiency and with X-linked SCID, although leukemic proliferation due to insertional mutagenesis has been observed in several patients with X-linked SCID following gene transfer.

Persistent oral candidiasis and chronic diarrhea are common findings in patients with SCID. Gastrointestinal infections may be sustained by a variety of microorganisms, including viruses (enteroviruses, but also adenovirus and cytomegalovirus), bacteria (*Salmonella* spp., *Enterococcus*), candida, and protozoa (*G. lamblia, Cryptosporidium parvum*). In addition, rotavirus vaccine (that has recently become available) may cause severe diarrhea and failure to thrive in infants with SCID. Enteroviral infections in SCID infants may lead to further complications, including encephalitis.

In patients with Omenn syndrome, leaky SCID, or SCID with maternal T-cell engraftment, diarrhea may be particularly pronounced and lead to significant protein loss and generalized edema. Moreover, SCID patients can present with clinical and pathological findings similar to IBD including colonic inflammation and fistulae. Pathological examination of small bowel biopsies of SCID patients frequently reveals prominent infiltrates with T lymphocytes and eosinophils, with villous flattening. Careful search for pathogens causing diarrhea, aggressive treatment of the infection (when possible), and nutritional support (often with use of parenteral nutrition) are the mainstay of treatment of diarrhea in infants with SCID. In addition, immunosuppression with steroids or cyclosporine A is often necessary to control the inflammatory reactions in the intestines of infants with Omenn syndrome. The association of intestinal inflammation with Omenn syndrome implicates a critical role for T-cell regulation in mucosal homeostasis.

Elevation of liver enzymes is also common in infants with SCID, and may reflect chronic infection in the intestine or in the liver. Inflammatory infiltrates of variable severity, with a predominance of mononuclear cells, may be observed in the lobules and in the periportal spaces in patients with Omenn syndrome or with SCID associated with maternal T-cell engraftment. Accumulation of toxic phosphorylated derivatives of adenosine and deoxyadenosine may cause increased apoptosis of hepatocytes in patients with adenosine deaminase deficiency.

30-5   CD40 ligand (CD40L) deficiency is a disorder with broad immunologic consequences due to the wide range of cell types that express its receptor, CD40.

CD40 ligand (CD40L, CD154) is a cell-surface molecule predominantly expressed by activated CD4$^+$ T lymphocytes. Interaction of CD40L with its counter-receptor CD40 on the surface of B cells is essential for germinal center formation and class-switch recombination. Furthermore, CD40 is also expressed on dendritic cells, macrophages, and activated endothelial and epithelial cells. Interaction of CD40L-expressing CD4$^+$ T cells with these cell types promotes B- and T-cell priming triggering protective responses against intracellular pathogens. Mutations in the *CD40LG* gene, mapping at Xq26, result in X-linked hyper-IgM syndrome (also known as type 1 hyper-IgM syndrome), a combined immunodeficiency characterized by an increased occurrence of bacterial and opportunistic infections, neutropenia, a high incidence of liver and biliary tract disease, increased risk of malignancies, and a high mortality rate.

Patients with CD40L deficiency show undetectable or very low levels of serum IgG and IgA, and normal to increased levels of IgM. Chronic or intermittent neutropenia is common, and is typically associated with a block in myeloid differentiation at the pro-myelocyte stage. Definitive diagnosis is based on the demonstration of defective expression of CD40L on the surface of T cells following *in vitro* activation, and is eventually confirmed by mutation analysis.

Regular immunoglobulin replacement therapy, prophylactic trimethoprim-sulfamethoxazole, and hygiene measures represent the mainstay of therapy. Severe neutropenia may be treated with recombinant granulocyte colony-stimulating factor. In spite of these measures, the long-term prognosis is poor because of severe infections and liver disease. The only curative approach is represented by HCT, and better results are achieved when transplantation is performed prior to development of lung problems or of *Cryptosporidium* infection.

Protracted or chronic diarrhea is very common in patients with CD40L deficiency, and is often due to *Cryptosporidium parvum* (leading to watery diarrhea), *G. lamblia*, and *Salmonella* spp. infections. In a European series, the frequency of *Cryptosporidium* infection was as high as 70%, but lower figures have been reported in the United States. Both *Cryptosporidium* and cytomegalovirus infection may lead to sclerosing cholangitis in patients with CD40L deficiency (Figure 30.4). This severe complication is more common in European than in North-American patients, presumably as a result of a different frequency of *C. parvum* infection in these populations. In any case, sclerosing cholangitis is a common cause of liver failure and death in patients with CD40L deficiency. The unique susceptibility to *Cryptosporidium*-related sclerosing cholangitis is also replicated in the mouse model of CD40L deficiency. It has been hypothesized that when infected with *C. parvum* or cytomegalovirus, bile duct cells may express CD40, and that CD40L-mediated signaling may induce apoptosis and block replication of the pathogen. Alternatively, because increased susceptibility to *C. parvum* infection is also seen in interferon-gamma (IFN-γ)-deficient mice, it is possible that CD40L deficiency may result in poor induction of IFN-γ, and hence impaired clearance of the pathogen. Persistent infection of the bile ducts may eventually cause dysplasia and neoplastic transformation.

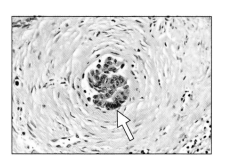

Figure 30.4 Sclerosing cholangitis in CD40L deficiency. Liver biopsy from a patient with CD40L deficiency and *Cryptosporidium parvum* infection, showing bile duct obliteration (white arrow) and periductular sclerosis; ×40. (Courtesy of Fabio Facchetti.)

Sclerosing cholangitis is often associated with dilated bile ducts. Careful monitoring of infections and of the liver and biliary tract should be included in the management of patients with CD40L deficiency. Search for *C. parvum* should be restricted to detection of oocysts in the stool, but should also be based on molecular detection by polymerase chain reaction. Approaches to cholangiography include endoscopic retrograde cholangiopancreatography (ERCP) and percutaneous transhepatic cholangiography (PTC). These also allow collection of bile fluid for microbiological and molecular analysis. Magnetic resonance cholangiopancreatography can also be used to visualize bile ducts, although it is not as sensitive as ERCP and PTC, and may miss minor abnormalities of the third and fourth order intrahepatic ducts. On the other hand, it does not carry the risk of ascending infection that is potentially associated with ERCP and PTC. Finally, liver histology is not as sensitive as cholangiography in detecting sclerosing cholangitis.

Patients with CD40L deficiency are at increased risk of malignancies that often involve the gastrointestinal tract. Cholangiocarcinoma, hepatocellular carcinoma, peripheral neuroectodermal tumors, and lymphoma have been frequently reported. Epithelial tumors and dysplasia of the liver and bile ducts in patients with CD40L deficiency are often associated with sclerosing cholangitis, and with infections due to *Cryptosporidium*, cytomegalovirus, and hepatitis B or C viruses. Sclerosing cholangitis and tumors of the gastrointestinal tract are associated with poor outcomes in patients with CD40L deficiency. In the absence of immune reconstitution, attempts to treat *Cryptosporidium*-related sclerosing cholangitis with liver transplantation are usually not successful in patients with CD40L deficiency, due to reinfection and relapse of the liver disease. Successful outcome has been reported following combined liver transplantation and HCT in some patients, but even using this approach, disseminated *C. parvum* infection and death may occur before immune reconstitution is achieved.

A variety of therapeutic and preventive approaches to *Cryptosporidium* infection have been proposed in patients with CD40L deficiency, including use of paromomycin, azithromycin, and nitazoxanide, but none of them are supported by controlled studies. However, azithromycin and standard hygiene measures are frequently recommended. Ursodeoxycholic acid may promote bile flow and reduce the risk of ascending infections in the biliary tree.

Although protracted diarrhea in patients with CD40L deficiency is typically due to infections, inflammatory bowel disease and intestinal nodular lymphoid hyperplasia have been also reported. Finally, patients with CD40L deficiency frequently suffer from oral and perirectal ulcers. These complications are typically associated with the neutropenia that is frequently seen in this condition.

30-6   CD40 deficiency is a disorder that reproduces the clinical phenotype of CD40L deficiency.

CD40 deficiency is inherited as an autosomal recessive trait and has been reported only in few patients, most of whom were of Turkish or Middle-Eastern origin. The clinical and laboratory manifestations of the disease are indistinguishable from those observed in patients with CD40L deficiency, including a higher risk of sclerosing cholangitis, *Cryptosporidium* infection, and oral ulcers associated with chronic neutropenia. HCT may provide permanent cure.

30-7   Major histocompatibility complex class II (MHC-II) deficiency leads to extensive deficiencies in humoral and cellular adaptive immune function.

MHC-II deficiency is a rare combined immunodeficiency characterized by markedly reduced or absent expression of MHC-II antigens, severe reduction of the number of circulating CD4$^+$ T lymphocytes, impaired antibody production (frequently associated with hypogammaglobulinemia), and defective cell-mediated immune responses. Patients with MHC-II deficiency are highly prone to bacterial, viral, and opportunistic infections. The disease is more common in selected populations (Northern Africa, Arabic countries). It is inherited as an autosomal recessive trait and is due to mutations in any of four genes that encode for transcription factors that regulate MHC-II antigen expression (e.g., class II transcriptional activator). Treatment is based on HCT, but results are far less satisfactory than in other cases of combined immunodeficiency.

Intestinal infections due to *Salmonella* spp., *E. coli*, *Proteus*, enteroviruses, cytomegalovirus, adenovirus, *Giardia*, *Cryptosporidium*, and *Candida* occur in nearly all patients with MHC-II deficiency, and are a frequent cause of chronic diarrhea, malabsorption, and failure to thrive. Histological examination shows various degrees of villous atrophy and increased intraepithelial lymphocytes, with a reduction of IgA plasmocytes.

Liver involvement is very common. Approximately one-third of patients develop biliary tract disease. Sclerosing cholangitis is frequently associated with *Cryptosporidium* or cytomegalovirus infection. Bacterial cholangitis has also been frequently described. The risk of biliary tract disease seems to increase with age. Viral infections (especially cytomegalovirus and adenovirus) are a cause of persistent elevation of liver enzymes. Drug-related toxicity and total parenteral nutrition may also lead to impaired liver function.

30-8   Deficiency of a nuclear protein, SP110, of unknown function leads to hepatic veno-occlusive disease with immunodeficiency.

Veno-occlusive disease with immunodeficiency is an autosomal recessive, rare disorder characterized by profound immunodeficiency and liver fibrosis and veno-occlusive disease. Patients present with severe symptoms within the first year of life. The disease has been reported predominantly in patients of Lebanese origin, and is due to mutations of the *SP110* gene that encodes for a nuclear protein of unclear function. Patients show profound hypogammaglobulinemia and a lack of germinal centers in the lymph nodes. T lymphocytes are present in normal numbers, but opportunistic infections (*P. jiroveci* pneumonia, cytomegalovirus infection, persistent candidiasis) are observed in most patients, indicative of a severe defect in cell-mediated immunity. Hepatomegaly and abnormalities of liver enzymes are often the first signs of severe liver disease that may progress to veno-occlusive disease and terminal liver failure. Liver biopsy shows sinusoidal congestion or obstruction, and narrowing of terminal hepatic venules. Ultrasonography demonstrates hepatosplenomegaly, increased portal vein diameter, reduced hepatic vein diameter, ascites, and re-canalization of the *ligamentum teres* due to shunting of blood-flow from the liver. Doppler ultrasonography shows reduced portal venous flow and increased resistance in the hepatic artery. Prognosis is dismal, and most patients die in the first years of life.

Immunodeficiencies affecting regulatory factors and populations of lymphocytes that produce these factors.

Regulatory T (T$_{reg}$) cells mediate suppression of immune responses in the periphery outside of the thymus. Generation of T$_{reg}$ cells in the thymus is controlled by the transcription factor Foxp3. Foxp3-expressing T$_{reg}$ cells can also be induced outside the thymus under the control of factors such as transforming growth factor-β (TGF-β) and retinoic acid (so-called induced T_{reg}). T_{reg} cells, which are discussed in more detail in Chapter 15, secrete a variety of immunoregulatory factors and cytokines such as IL-10, IL-35, and TGF-β. Such responses are important in preventing autoimmunity and restricting excessive immune responses to pathogens, thus delimiting the extent of inflammation and consequent tissue injury.

30-9   Genetically mediated Foxp3 deficiency leads to the immune dysregulation, polyendocrinopathy, enteropathy, X-linked (IPEX) syndrome.

Mutations of the *FOXP3* gene cause immune dysregulation, polyendocrinopathy, enteropathy, X-linked (IPEX) syndrome, with severe and early-onset autoimmune enteropathy, insulin-dependent diabetes, and eczema. There is hyperactivation of T lymphocytes, associated with elevated IgE, eosinophilia, and autoantibodies. In typical cases, the disease evolves rapidly, unless treated by HCT.

Intractable diarrhea is a key element of IPEX and is present in virtually all cases with typical early presentation, often being severe enough to cause significant malabsorption and cachexia. Intestinal villous flattening and lymphoid infiltration can be seen in patients with IPEX (Figure 30.5). A recent review of intestinal biopsies from a limited series of patients revealed that the majority of them exhibited abnormalities resembling acute graft-versus-host disease. Therapeutic attempts with elimination of presumed allergens from the diet were largely unsuccessful. Systemic autoimmunity with severe intestinal inflammation is also a hallmark of Foxp3-deficient mice.

30-10 Wiskott-Aldrich syndrome (WAS), due to deficiency in Wiskott-Aldrich protein (WASp), affects multiple lineages of hematopoietic cells and consequently diverse aspects of immune function.

Wiskott-Aldrich syndrome (WAS) is an X-linked disorder characterized by eczema, congenital thrombocytopenia with small-sized platelets, and immune deficiency. The responsible gene, named *WAS*, maps at Xp11.2 and encodes for a protein involved in cytoskeleton reorganization in hematopoietic cells. Most patients with classical WAS have mutations that impair expression and/or function of Wiskott-Aldrich protein (WASp). However, some missense mutations are associated with a milder phenotype (isolated X-linked thrombocytopenia).

The immune deficiency of WAS leads to recurrent bacterial and viral infections, autoimmune manifestations, and increased occurrence of tumors (leukemia, lymphoma). Autoimmune or inflammatory sequelae are seen in 40–70% of patients depending on the series, with a significant number of patients presenting with two or more autoimmune or inflammatory sequelae. The most common autoimmune disease or inflammatory manifestation includes hemolytic anemia, neutropenia, vasculitis, arthritis, IBD, and renal

Figure 30.5 Autoimmune enteropathy in the duodenum of an IPEX patient. Panel a shows villous flattening and acute inflammation (white arrow) in the duodenum, ×100. Panel b shows the lack of Paneth cells (white arrow) in the crypts, ×400. (Courtesy of Athos Bousvaros and Jeffrey Goldsmith.)

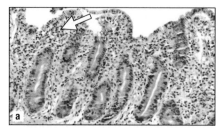
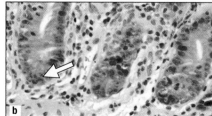

disease. Lymphopenia (particularly among CD8[+] T lymphocytes), impaired *in vitro* proliferation to immobilized anti-CD3, reduced serum IgM with increased levels of IgA and IgE, and inability to mount effective antibody responses, especially to T-independent antigens, are typical immunological findings. The basis for the autoimmunity in WAS is still unclear but may represent aberrant effector T cell, regulatory T cell, NK cell, dendritic cell, or other innate immune function.

Treatment of WAS is based on regular administration of intravenous immunoglobulins, antibiotic prophylaxis, topical steroids to control eczema, and use of vigorous immune suppression when autoimmune manifestations are present. Splenectomy may increase the platelet count in the majority of patients, but carries significant risks of invasive infection by encapsulated pathogens. The only curative approach to WAS is represented by HCT, which gives optimal results when performed from HLA-identical family donors or early in life from matched unrelated donors. Gene therapy trials have been initiated with promising results.

WAS patients can present with life-threatening gastrointestinal bleeding that is associated with severe thrombocytopenia (over 25% of patients presenting with hematemesis or melena in one series). Chronic diarrhea is also a common feature of classical WAS, most often associated with infectious etiologies. Moreover, a significant fraction of patients (nearly 10% in one series) can present with an inflammatory disorder of the intestine resembling inflammatory bowel disease. These patients can present with inflammatory lesions anywhere in the gastrointestinal tract with disease resembling both Crohn's disease and ulcerative colitis. Rarely, patients can present with a clinical picture consistent with necrotizing enterocolitis. Similar to WAS patients, mice deficient in WASp can also develop severe colitis associated with helper T (T_H) T_H2 cytokine skewing and defects in regulatory T cells.

30-11 Deficiency of the IL-2Rα (CD25) chain has major effects on T_{reg} function.

IL-2-mediated signaling is essential to maintain peripheral immune homeostasis. Deficiency of the α chain of the IL-2 receptor (IL-2Rα, CD25) is inherited as an autosomal recessive trait and results in immune dysregulation and lymphoproliferation. Patients are at higher risk for early-onset severe and recurrent viral and bacterial infections, oral thrush, and chronic diarrhea (that may lead to protein-losing enteropathy), associated with lymphadenopathy and hepatosplenomegaly. CD4[+] T-cell counts are low, and an *in vitro* proliferative response to mitogens is typically impaired. Treatment is based on HCT. Mice deficient in IL-2R (or IL-2) also develop severe intestinal inflammation which may result from defects in effector and/or regulatory T-cell homeostasis.

30-12 Genetically determined deficiency in the IL-10 receptor and IL-10 presents prominently with features associated with aberrant mucosal homeostasis.

Rare cases of patients presenting with mutations in IL-10 or in the IL-10 receptor alpha or beta (IL-10Rα or IL-10Rβ) chains have been reported in individuals presenting with very early onset IBD. These patients have been found to also suffer from severe infections and folliculitis. Perianal abscesses, enteric fistula, and inflammation in the colon (Figure 30.6) and to a lesser extent in the small bowel have been described in such patients. Bone marrow transplant has been effective therapy in such patients.

Genetic analyses of these rare patients have suggested an autosomal recessive inheritance pattern. Mutations of IL-10Rα or IL-10Rβ abrogate

Figure 30.6 Intestinal inflammation in IL-10RB deficiency. Lymphoplasmocytic infiltration of the colon; ×100. (Adapted from E.O. Glocker et al., *N. Engl. J. Med.* 361:2033–45, 2009. With permission from Massachusetts Medical Society.)

interleukin-10-induced signaling, as shown by deficient STAT3 (signal transducer and activator of transcription 3) phosphorylation. Moreover, since IL-10-dependent inhibition of lipopolysaccharide-induced secretion of TNF-α and other pro-inflammatory cytokines is defective in monocytes from these patients, it has been proposed that a block in IL-10-dependent negative feedback is responsible for the inflammatory gastrointestinal disorder. IL-10 and IL-10R knockout mice also develop IBD. Numerous cell types express IL-10R, including innate immune cells; and both naive and regulatory T cells and the IL-10Rβ chain can bind other cytokines, such as IL-22; however, the specific cell type(s) that require IL-10 and IL-10R signaling and the role of cytokines in maintaining mucosal homeostasis in these subjects remains unclear.

Immunodeficiencies affecting primarily innate immune cells.

Both parenchymal cells, such as epithelial cells, and hematopoietic cells, such as polymorphonuclear cells (or phagocytes) that include neutrophils, eosinophils, and basophils, are essential components of innate immune function within mucosal tissues. Consequently, environmentally mediated (e.g., drug-induced neutropenia) or genetically determined (e.g., chronic granulomatous disease) dysfunction of these have important and deleterious consequences for both the maintenance of homeostasis and protection from pathogenic exposures.

30-13 Chronic granulomatous disease is the archetypal example of a genetically specified disorder of phagocyte function.

Phagocytes play a key role in the defense against bacteria and fungi; accordingly, patients with defects of phagocytic cell number and/or function exhibit significant problems at their mucosal surfaces. Chronic granulomatous disease (CGD) represents the prototype of defects associated with phagocyte function, and is due to abnormalities in the NADPH oxidase complex. In 75% of the cases, CGD is inherited as an X-linked trait, due to mutations in the X-linked *CYBB* gene that encodes the gp91^{phox} component of NADPH oxidase. Autosomal recessive forms due to defects of the p22^{phox}, p47^{phox}, p67^{phox}, and p40^{phox} components are also known. The gp91^{phox} and p22^{phox} proteins are located in the phagosome membrane, where the cytosolic p47^{phox} and p67^{phox} isoforms are translocated to the membrane following phagocytic cell activation. Another subunit, p40^{phox}, plays an important role in phagocytosis-induced superoxide production. Assembly of the NADPH complex is regulated by two GTPases, Rac2 and Rap1, and defects of Rac2 have been identified in patients with CGD-like disease. Following phagocytosis, induction of the NADPH oxidase complex results in production of microbicidal compounds such as superoxide radicals and hydrogen peroxide and, even more importantly, activation of lytic enzymes (cathepsin G, elastase, myeloperoxidase) following potassium ion influx, ultimately permitting killing of bacteria and fungi. Patients with CGD suffer from recurrent and often severe infections of fungal or bacterial origin, leading to pneumonia, skin and perirectal abscesses (with frequent formation of fistulae), suppurative lymphadenitis, osteomyelitis, septicemia, and deep-seated abscesses in the liver and other organs. *Staphylococcus aureus*, *Serratia marcescens*, *Burkholderia cepacia*, and *Nocardia* spp. are responsible for most of the bacterial infections, whereas *Candida* spp. and *Aspergillus* spp. cause most of the fungal infections. Patients with CGD are also at higher risk for mycobacterial infections. The sustained inflammatory response observed in patients with CGD is also responsible for inflammatory manifestations (even in the absence of infection), presumably due to inappropriate responses to the commensal microbiota, that may cause obstruction in the gastrointestinal and/or urinary tract. Clinical symptoms tend to be more common, more severe, and to occur earlier in life in patients with X-linked versus autosomal recessive forms of the disease. The diagnosis of CGD is based on clinical history and laboratory demonstration of impaired NADPH function, as shown by impaired production of dihydrorhodamine 123 (DHR123) by *in vitro* activated phagocytes in a flow-cytometry-based assay.

CGD carries a high mortality risk, especially due to infections. However, significant progress in the management of the disease now permits almost half of the patients to survive into the third decade. Treatment is based upon prompt and aggressive investigation of any febrile episode in order to identify the responsible microorganism(s), and on continuous antimicrobial prophylaxis with trimethoprim-sulfamethoxazole (for bacterial infections) and itraconazole (for fungal infections). In addition, subcutaneous administration of IFN-γ has been shown to reduce occurrence of severe infections. However, the only definitive cure is represented by HCT, whose results are largely influenced by degree and duration of donor chimerism in the myeloid compartment, and by preexisting risk factors, such as lung aspergillosis. Results of HCT are better when the transplant is performed from an HLA-identical related donor. Use of a matched unrelated donor has been associated with increased risk of graft-versus-host disease, graft loss, and death; however, a successful outcome has been recently reported (including correction of inflammatory bowel disease) using reduced-intensity conditioning. Gene therapy has been attempted, but in the absence of conditioning only modest and transient presence of gene-corrected phagocytes is achieved. Furthermore, myelodysplasia has been observed due to insertional mutagenesis.

Gastrointestinal manifestations are observed in approximately one-third of CGD patients. Common symptoms include abdominal pain, fever, diarrhea, vomiting, and weight loss. Virtually any segment of the gastrointestinal tract may be involved; liver disease is also common. Gastrointestinal pathologies include infection, chronic inflammation with strictures, ulcerations, granuloma formation, and dysmotility.

Stomatitis and oral candidiasis are frequent. Involvement of the esophagus is less common but narrowing of the lumen, pre-stenotic dilatation, and dysmotility have been reported. Chronic gastritis and thickening of the stomach mucosa have been described more frequently, and may lead to gastric outlet obstruction. Inflammation in the small or large bowel may cause pathological features resembling IBD (Figure 30.7), and may also lead to obstruction. Histological examination of colonic biopsies reveals unique features compared with those typically seen in ulcerative colitis or Crohn's disease. In particular, there is prominent presence of eosinophils and pigmented macrophages in the basal mucosa or in the superficial submucosa, whereas neutrophils are very rare. Accumulation of eosinophils around the crypts is typically associated with damage to the fibroblastic pericryptal sheath, and may ultimately cause reduced absorption of nutrients. There is increased expression of MHC-II antigens in epithelial and vascular endothelial cells and of intercellular adhesion molecule 1 expression in the lamina propria.

Thorough microbiological evaluation, imaging studies, and empiric aggressive use of antibiotics are needed in CGD patients who develop intestinal manifestations of possible infectious origin. However, in patients with colitis, frequently no pathogens are isolated and antimicrobial treatment alone is not sufficient. In these cases, immunomodulation with steroids and sulfasalazine is often required. Cyclosporine has been also successfully used in some patients. Perirectal abscesses require surgical drainage; emergency ileostomy may be required in patients with severe colitis refractory to immunosuppressive treatment.

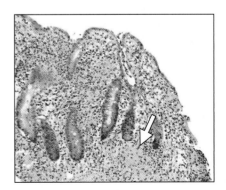

Figure 30.7 Active colitis and granulomas in CGD. Acute inflammation and granuloma (white arrow) in colon, ×400. (Courtesy of John Watkins and Jeffrey Goldsmith.)

Liver abscesses are very common. Often multiple, they have a tendency to persist or relapse. *Staphylococcus aureus* is the most common organism involved. A high index of suspicion should be used, with a low threshold for the performance of radiologic imaging studies. A generous use of aggressive surgical debridement, associated with prolonged use of antibiotics, is important to facilitate definitive treatment of the abscesses.

Ultimately, successful correction of the disease, including gastrointestinal and hepatic complications, is based on HCT. However, controversies exist whether this should be reserved for patients with a more severe phenotype and lack of adequate response to medical management.

30-14 NEMO (NFκB essential modulator) deficiency, or X-linked immunodeficiency with ectodermal dystrophy, is a disorder that typifies the consequences of innate defects which affect the epithelial barrier.

The nuclear factor kappa B (NFκB) is a multimeric transcription factor that regulates a variety of cellular processes, including innate and adaptive immune responses. Activation of NFκB is regulated by a complex consisting of two catalytic protein subunits with kinase activity (IKKα and IKKβ) and a regulatory component, IKKγ (also known as NFκB essential modulator, or NEMO), which function to phosphorylate and deactivate IκB, an inhibitor of NFκB. NEMO is encoded by a gene (*IKBKG*) located on the X chromosome. In humans, null *IKBKG* gene mutations lead to incontinentia pigmenti in heterozygous females, and result in embryonic lethality in males. In contrast, hypomorphic mutations in males result in X-linked immunodeficiency with ectodermal dystrophy. Typical manifestations of the disease include scanty hair, defective tooth formation with conical teeth, nail defects, hypohidrosis, increased susceptibility to recurrent infections (including mycobacterial disease), and aberrant inflammatory responses. Most patients show hypogammaglobulinemia, with reduced IgG levels and defective antibody responses. Impaired T-cell priming and defective cytolytic function of NK cells are also present. There is significant variability in the severity of the clinical phenotype. Treatment is based on immunoglobulin replacement therapy and prompt treatment of infections.

Gastrointestinal manifestations are common in patients with NEMO deficiency, and may reflect either infections or excessive inflammatory responses. Since NEMO is required for functional responses through Toll-like receptors, it is not unexpected that the functional response to infectious agents in the gastrointestinal tract is impaired. Accordingly, patients are more prone to bacterial and mycobacterial infections that may cause diarrhea, abdominal pain, fever, and malabsorption. However, NEMO is also critically involved in the generation and homeostasis of regulatory T cells and natural killer T (NKT) cells. Furthermore, NEMO functions in gut epithelial cells to control the integrity of the mucosal barrier and interaction of epithelial cells with the commensal microbiota. Consistent with this, NEMO deficiency also results in significant immune dysregulation and abnormal inflammatory responses. In the gastrointestinal tract, this aberrant pattern of response may lead to atypical enterocolitis (Figure 30.8), with abdominal pain and fever. Bloody stools and weight loss are common, and the erythrocyte sedimentation rate is markedly elevated. In such cases, no pathogens are cultured. Endoscopy reveals severe inflammation, especially in the colon, with exudates and mucosal bleeding. Histological examination shows superficial cryptitis, mucosal ulcerations, and edema, associated with abundance of neutrophils within the lamina propria juxtaposed to the mucosal surface. Absence of granuloma and deep cryptitis, and paucity of lymphocytic infiltrates, distinguishes these lesions from those typically observed in IBD. Treatment is based on the use of anti-inflammatory agents.

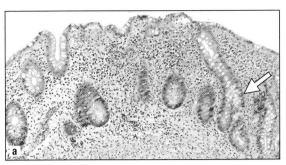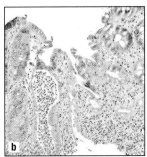

Figure 30.8 Enterocolitis in NEMO deficiency. Panel a shows acute inflammation in colon with evidence of chronic changes as evidenced by branched crypt (white arrow), ×100. Panel b shows moderately active colitis. (Courtesy of Athos Bousvaros and Jeffrey Goldsmith.)

Summary.

The common presentation of mucosal and particularly gastrointestinal symptomatology in the setting of primary immunodeficiencies underscores the role of the immune system in regulating mucosal homeostasis. Moreover, genome-wide association studies evaluating patients with inflammatory bowel diseases have also identified polymorphisms in genes that affect either innate (e.g., *NOD2*) and/or adaptive immune function (e.g., *IL23R*, *IL10*). Primary humoral disorders (e.g., XLA, selective IgA deficiency) are most often associated with diarrhea and malabsorption resulting from infectious consequences and only rarely aggressive inflammatory pathologies. Immunodeficiencies that affect both B- and T-cell signaling (e.g., CVID, SCID) present with gastrointestinal infections, aggressive inflammatory responses, and malignancies, thus emphasizing the role of appropriate T-cell function in maintaining mucosal function. Aberrant immunoregulatory circuits also result in severe intestinal pathology as evidenced by several immunodeficiencies (e.g., IPEX, WAS, IL-10R deficiency). In these disorders, regulatory T cell abnormalities may play a significant role in the underlying pathology. Finally, immunodeficiencies that affect innate immune cells (e.g., CGD, NEMO deficiency) also present with both intestinal infections and aggressive intestinal inflammatory sequelae.

Further Reading.

Agarwal, S., and Mayer, L.: **Gastrointestinal manifestations in primary immune disorders.** *Inflamm. Bowel Dis.* 2010, **16**:703–711.

Al-Muhsen, S.Z.: **Gastrointestinal and hepatic manifestations of primary immune deficiency diseases.** *Saudi J. Gastroenterol.* 2010, **16**:66–74.

Cannioto, Z., Berti, I., Martelossi, S., *et al.*: **IBD and IBD mimicking enterocolitis in children younger than 2 years of age.** *Eur. J. Pediatr.* 2009, **168**:149–155.

Cunningham-Rundles, C., and Bodian, C.: **Common variable immuno-deficiency: clinical and immunological features of 248 patients.** *Clin. Immunol.* 1999, **92**:34–48.

Daniels, J.A., Lederman, H.M., Maitra, A., *et al.*: **Gastrointestinal tract pathology in patients with common variable immunodeficiency (CVID): a clinicopathologic study and review.** *Am. J. Surg. Pathol.* 2007, **31**:1800–1812.

Hayward, A.R., Levy, J., Facchetti, F., *et al.*: **Cholangiopathy and tumors of the pancreas, liver and biliary tree in boys with X-linked immunodeficiency with hyper-IgM.** *J. Immunol.* 1997, **158**:977–983.

Huang, A., Abbasakoor, F., and Vaizey, C.J.: **Gastrointestinal manifestations of chronic granulomatous disease.** *Colorectal Dis.* 2006, **8**:637–644.

Khodadad, A., Aghamohammadi, A., Parvaneh, N., *et al.*: **Gastrointestinal manifestations in patients with common variable immunodeficiency.** *Dig. Dis. Sci.* 2007, **52**:2977–2983.

Malamut, G., Ziol, M., Suarez, F., *et al.*: **Nodular regenerative hyperplasia: the main liver disease in patients with primary hypogammaglobulinemia and hepatic abnormalities.** *J. Hepatol.* 2008, **48**:74–82.

Notarangelo, L.D.: **Primary immunodeficiencies.** *J. Allergy Clin. Immunol.* 2010, **125**:S182–194.

Ochs, H.D., and Thrasher, A.J.: **The Wiskott-Aldrich syndrome.** *J. Allergy Clin. Immunol.* 2006, **117**:725–738.

Patey-Mariaud de Serre, N., Canioni, D., Ganousse, S., *et al.*: **Digestive histopathological presentation of IPEX syndrome.** *Mod. Pathol.* 2009, **22**:95–102.

Picard, C., and Fischer, A.: **Hematopoietic stem cell transplantation and other management strategies for MHC class II deficiency.** *Immunol. Allergy Clin. North Am.* 2010, **30**:173–178.

Torgerson, T.R., and Ochs, H.D.: **Regulatory T cells in primary immunodeficiency diseases.** *Curr. Opin. Allergy Clin. Immunol.* 2007, **7**:515–521.

Westerberg, L.S., Klein, C., and Snapper, S.B.: **Breakdown of T cell tolerance and autoimmunity in primary immunodeficiency: lessons learned from monogenic disorders in mice and men.** *Curr. Opin. Immunol.* 2008, **20**:646–654.

Inflammatory Bowel Disease

31

Crohn's disease and ulcerative colitis are the major chronic inflammatory bowel diseases (IBD) in humans. IBD affects about 3 million individuals in the United States and Europe with an increasing incidence worldwide. Both forms of IBD have a peak age of onset in males and females in the second to third decades of life. Crohn's disease generally involves the ileum and colon, but it can present anywhere in the alimentary tract, from the mouth to the anus. Inflammation is often transmural and discontinuous. Histologically, gut lesions are characterized by the accumulation of chronic inflammatory cells, and by deep-fissuring ulcers (Figure 31.1). The presence of non-caseating granuloma, although highly characteristic of Crohn's disease, is neither unique to Crohn's disease nor universally found. Dense lymphoid aggregates can be seen in the submucosa (see Figure 31.1). Fibrosis in the deeper layers of the gut wall contributes to the development of strictures, as a result of the overproduction of collagen by fibroblasts and smooth muscle cells. Ulcerative colitis, in contrast, typically involves the rectum, with the inflammation being confined to the mucosa or submucosal layers. Disease progresses proximally in a continuous fashion, resulting in variable degrees of involvement of the colon, with the most extreme form of total inflammation presenting as a pancolitis. Inflammation is characterized by a marked infiltration of the lamina propria with neutrophils, lymphocytes, plasma cells, and macrophages (Figure 31.1). The neutrophils can invade the epithelium and cross into the lumen, particularly in the crypts, giving rise to cryptitis and crypt abscesses. Cryptitis is associated with a depletion of mucus from goblet cells and increased epithelial cell turnover. There is a marked increase in plasma cells in the lamina propria and an excess production of IgG, particularly IgG1 and IgG3; this contrasts markedly with Crohn's disease, in which there is also an increase in plasma cells, especially around ulcers, but in which the dominant isotype is IgG2.

Patients with IBD suffer from frequent and chronically relapsing flares, with rectal bleeding, abdominal pain, and/or diarrhea. In both types of IBD, there can also be local (for example cancer) and extraintestinal (for example primary sclerosing cholangitis, spondylitis, or iritis) complications that influence

Figure 31.1 Histology of Crohn's disease and ulcerative colitis.
Both Crohn's disease and ulcerative colitis are associated with massive inflammatory lesions in the bowel. The Crohn's disease image (left panel) shows a dense mononuclear infiltrate into the full thickness of the gut wall. Ectopic lymphoid follicles are obvious in the deeper layers, and a deep ulcer penetrates from the lumen into the submucosa. The ulcerative colitis lesion (right panel) shows the characteristic neutrophilic mucosal lesions, although in this case disease is so fulminant that there is some submucosal involvement. There are very obvious crypt abscesses where neutrophils have crossed the glandular epithelium. There is also extensive epithelial cell damage. Hematoxylin and eosin stain. Original magnification ×40.

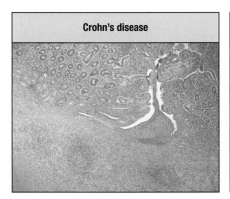

Crohn's disease

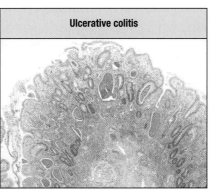

Ulcerative colitis

the natural history of IBD patients. Some of these complications directly reflect the extent and severity of the inflammatory activity of these disorders (for example cancer), whereas others are co-morbid conditions that follow their own course without regard necessarily to the intestinal inflammation (for example primary sclerosing cholangitis) and thus probably have a similar genetic and/or environmental foundation.

The etiology of both Crohn's disease and ulcerative colitis is unknown, but the rapid advancement of molecular techniques associated with genetics, microbiology, and immunology and the possibility of using a large number of animal models of intestinal inflammation have led to a better knowledge of factors that orchestrate the tissue-destructive inflammatory responses in these disorders. It is now well accepted that both forms of IBD develop in genetically predisposed individuals as a result of an exaggerated mucosal immune response directed against commensal microbiota that reside in the intestinal lumen. In this chapter we focus on recent advances in our understanding of the involvement of genetic, bacterial, and immune factors in the pathogenesis of Crohn's disease and ulcerative colitis.

Genetic basis of IBD.

Crohn's disease and ulcerative colitis have long been known to have a genetic basis; however, with the advent of population databases and modern molecular techniques for genetic analyses, the study of IBD has led the way in the discovery of gene polymorphisms in the development of complex disease.

31-1 Epidemiological observations support the idea of a genetic basis of IBD.

In the 1980s, several epidemiological studies confirmed the early findings of familial aggregation in IBD, with a positive family history ranging from 5 to 20%. IBD is thus considered to be familial in roughly 10% of afflicted individuals and sporadic in the remaining 90%. The relative risk to siblings of affected individuals has been estimated to be 30–40-fold for Crohn's disease and 10–20-fold for ulcerative colitis. Studies of twins have provided additional evidence for a genetic contribution in IBD. The concordance rate is significantly greater in monozygotic than dizygotic twins for both Crohn's disease (20–70% versus 0–10%) and ulcerative colitis (10–20% versus 0–5%). In families containing multiple affected individuals with IBD, 75% are concordant for disease type (either Crohn's disease or ulcerative colitis), with the remaining 25% being 'mixed' (both Crohn's disease and ulcerative colitis), which is consistent with an overlapping genetic and/or environmental basis as discussed further below. Overall, IBD is considered to be a polygenic disorder with complex inheritance except in a subset of individuals with early-onset IBD (in the first 5 years of life), which may be due to monogenic (Mendelian) inheritance.

Genetic factors influence not only disease onset but also disease phenotype and course, particularly in Crohn's disease. A population-based study of Scandinavian cohorts of twins with IBD revealed a pronounced genetic impact on clinical characteristics of Crohn's disease in comparison with ulcerative colitis. Monozygotic twins with Crohn's disease seem to be concordant for age at onset of symptoms, age at diagnosis, extent of inflammation, and disease behavior both at diagnosis and at 10 years after diagnosis. In contrast, monozygotic twins with ulcerative colitis are concordant for age at diagnosis and symptomatic onset only. Genetic factors seem also to influence the disease location, given that the percentage of familial cases is lower in Crohn's disease confined to the colon than in ileal disease.

31-2   IBD susceptibility genes were identified before the era of genome-wide association studies.

To determine the IBD-associated susceptibility genes, investigators originally used two main types of genetic studies, namely genetic linkage and candidate-gene association studies. Genetic linkage involves the study of the inheritance patterns of polymorphic genetic markers in families with multiple affected relatives. This approach led to the identification of at least nine disease loci designated IBD1–9 (Table 31.1); however, only five of these loci (IBD1, 2, and 4–6) showed significant linkage (LOD score > 3.6 or $P < 2.2 \times 10^{-5}$), with little replication across the scans. Such linkage studies supported the polygenic basis of these disorders and predicted future observations—to be discussed below—showing that the IBD probably results from the minor effects of a large number of genetic variants in a given individual. However, some genetic loci were unique and observed to contribute significant statistical risk for the development of IBD. One genetic locus that is worthy of being highlighted and was in fact the first genetic region to be identified was found within the pericentromeric

Table 31.1 Contribution of linkage and genome-wide association studies in the identification of Crohn's disease susceptibility loci

Association mapping of linkage region		Genome-wide association studies			
Locus	Candidate gene(s)	Chromosomal region	Relevant genes	Chromosomal region	Relevant genes
IBD1	NOD2/CARD15	1p13	PTPN22	9p24	JAK2
IBD2		1p31	IL23R	9q32	TNFSF15
IBD3	HLA/TNF	1p36	PLA2G2E	9q33	TLR4
IBD4		1q23	ITLN1	9q34	CARD9
IBD5	T_H2 cytokines	1q32	IL10	10p11	CCNY
IBD6	STAT6	1q44	NLRP3	10q21	ZNF365
IBD7		2q37	ATG16L1	10q23	DLG5
IBD8		2p16	PUS10	10q24	NKX2-3
IBD9		2q11	IL18RAP	10q26	DMBT1
		3p12	CADM2	11p15	NELL1
		3p21	MST1/BSN	11q13	C11orf30
		4p13	PHOX2B	12q12	LRRK2/MUC19
		5p13.1	PTGER4	12q15	IFNG
		5q13	S100Z	15q13	HERC2
		5q31	SLC22A5	16q12	NOD2
		5q33	IRGM/IL12B	16q24	FAM92B
		6p21	BTNL2/HLA	17q21.2	OMDL3/STAT3
		6p22	CDKAL1	18p11	PTPN2
		6q23	TNFAIP3	20q13	TNFRSF6B
		6q27	CCR6	21q22	ISOSLG
		7q32	IRF5	22q12	NCF4/XBP1

region of chromosome 16. This locus was named IBD1 and was later shown to contain *NOD2* (nucleotide-binding oligomerization domain-containing 2), also known as caspase recruitment domain protein 15 (*CARD15*), the first susceptibility gene for Crohn's disease, but not ulcerative colitis.

DNA sequencing of *NOD2/CARD15* in Crohn's disease demonstrated that three main polymorphisms situated in or close to the leucine-rich repeat (LRR) ligand-binding domain of the molecule contribute to 80% of the risk associated with this important genetic risk factor. A gene dosing effect is seen in most studies, such that carriage of one or two copies of the risk allele(s) increases the risk of developing Crohn's disease 2–4-fold or 20–40-fold respectively. Studies from France, Germany, the UK, and the United States have shown that up to 40% of patients with Crohn's disease (in contrast with 14% of controls) have one or more of these mutations. However, the allelic frequency is much lower in other European regions (for example Ireland, Scotland, Scandinavia, and Iceland), and NOD2 mutations are not found in patients with Crohn's disease in Asia, thus highlighting significant ethnic heterogeneity. Moreover, NOD2 mutations are associated with ileal Crohn's disease but not with colonic Crohn's disease or ulcerative colitis. Given that NOD2/CARD15 is an intracellular receptor for muramyl dipeptide (MDP) derived from peptidoglycan of Gram-negative and Gram-positive bacteria, and is a receptor for the glycolyl MDP of mycobacteria and single-stranded RNA of viruses and a member of the NOD-like pattern recognition receptors (NLR), these studies support a microbial basis for IBD, especially Crohn's disease.

Another instructive locus that has consistently been identified by linkage and association studies as important for both ulcerative colitis and Crohn's disease in many different populations is the major histocompatibility complex (MHC) region on chromosome 6p21.1–23 (IBD3 locus). The MHC locus is highly complex, with both human leukocyte antigen (HLA) and non-HLA alleles (for example tumor necrosis factor and complement genes), such that, despite extensive study, the specific genes and mechanism(s) involved in conferring genetic risk remain enigmatic. In ulcerative colitis, the most consistently replicated association is with the rare allele HLA-DRB1*0103. This variant is present in 0.2–3.2% of the population, in 6–10% of all ulcerative colitis cases, in 15.8% of severe and extensive ulcerative colitis cases, and in 14.1–25% of severe ulcerative colitis cases requiring colectomy. HLA-DRB1*0103 is also associated with colonic Crohn's disease. The common HLA allele HLA-DRB1*0701 is associated with Crohn's disease, and ileal Crohn's disease in particular. The alleles HLA-B27, HLA-B35, and HLA-DRB1*0103 are also associated with extraintestinal manifestations of disease. Together, these studies predict the importance of both MHC class I- and MHC class II-related immune responses and consequently CD8 and CD4 T cells in the immunopathogenesis of these disorders.

31-3   Genome-wide association studies reveal large numbers of genes associated with IBD.

The availability of the complete sequence of the human genome, the development of novel microarray technologies, and some limitations of linkage analysis (namely difficulties encountered in trying to identify genes of weak effect and the necessity of testing a large set of affected siblings) have favored the advent of genome-wide association studies, which seek statistical evidence for association between variants distributed across the genome and disease susceptibility in a non-biased manner. To provide adequate power, genome-wide association studies require large numbers of cases and controls, as well as the genotyping of several hundreds of thousands of markers to capture a majority of the common (more than 5% minor allele frequency) genomic variations, and replication in independent datasets on a different genotyping platform to confirm robust, true signals. So far, more than 100,000 single nucleotide polymorphisms (SNPs) have been used in the published genome-wide association studies. Overall these studies replicated the previously validated linkage and association studies (such as IBD1 and IBD3) and have identified new genetic risk factors. Up to now, nearly 100 risk-conferring loci have been identified for Crohn's disease and ulcerative colitis; some of these are described in Figure 31.2, which also highlights the complexity of these disorders. Genome-wide association studies and the associated deep DNA sequencing of candidate genes within identified loci and others have shown that (1) in the vast majority of identified examples the causal genetic variants remain unknown, because in only a minority of examples have nonsynonymous SNPs or other structural changes (for example variations in copy number) resulting in a change in the protein structure or promoter region been identified; (2) the genetic loci identified are also risk factors for many other immune-mediated disorders such as psoriasis, ankylosing spondylitis, celiac disease, and diabetes, demonstrating the similarities in disease pathogenesis with their attendant therapeutic implication; (3) the genetic loci so far identified by genome-wide association studies account for less than 20–30% of the heritable risk, suggesting that the remainder of risk is conferred by rare genetic variants with potentially important biological effects that may account for the so-called 'missing heritability'; (4) most genetic variations identified are common to both forms of IBD, whereas some of them are unique for Crohn's disease or ulcerative colitis (see Figure 31.2); and (5) each genetic variant contributes a very modest risk such that IBD is the cumulative result of the combined risk conferred by the overall genetic composition of the host that elicits this risk through convergence on a variety of functional pathways described in more detail below (see Figure 31.2). Overall, these pathways seem to influence the genetically determined relationship between the commensal microbiota, the intestinal epithelial cells, and the immune system.

31-4 Environmental factors are clearly involved in IBD pathogenesis.

Environmental factors, other than the commensal microbiota, are important in the pathogenesis of Crohn's disease and ulcerative colitis by modifying the genetic risk associated with disease. In particular, smoking is quite instructive in highlighting both the beneficial and the detrimental aspects of environmental factors. Smoking has been identified as a risk factor for an aggressive course of Crohn's disease, particularly after ileocecal resection, but seems to prevent the development of ulcerative colitis. Other studies indicate that environmental factors such as living in housing with hot water, fewer infections, and a low infant mortality rate coincide with a higher incidence of Crohn's disease; in contrast, early-life residence on a farm may be protective for Crohn's disease and ulcerative colitis. These findings support the notion of the 'hygiene hypothesis,' although many other variables associated with a westernized lifestyle may be more relevant and account for the increasing worldwide incidence of these disorders, especially Crohn's disease. It must be emphasized that whereas Crohn's disease is a disease of the developed world, and is increasing in the developing world as countries increase standards of living, ulcerative colitis shows a quite different distribution and is seen in developed as well as developing countries. In addition to these observations on the role of hygiene, further studies suggest that agents which disrupt mucosal barrier function and its relationship to the commensal microbiota such as nonsteroidal anti-inflammatory drugs (NSAIDs), antibiotics, and enteropathogenic exposures (viral, bacterial, and parasitic) that would be predicted to disrupt the normal composition of the commensal microbiota and function of epithelium and immune system may act as triggers of IBD in the genetically susceptible host. Finally, appendectomy before the age of 20 years is a protective factor for the future development of ulcerative colitis. Although the reasons for this finding are largely unknown, one can speculate that the appendix may

IBD	Ulcerative colitis	Crohn's disease
Adaptive immunity	**Adaptive immunity**	**Adaptive immunity**
CARD9 MST1 PRDM1 REL SMAD3 IL1R2 ICOSLG RTEL1 PTPN2 YDJC TNFSF15 IL10	IFNG/IL26 IL2/IL21 TNFRSF9 IL7R TNFRSF14 IRF5 LSP1 FCGR2A IL8RA / IL8RB	CCR6 PTPN22 IL27 IL2RA IL18RAP VAMP3 ITLN1 ERAP2 TNFS11 BACH2 CCL2/CCL7 TAGAP
HLA	**Gut epithelium**	**Innate immunity**
DRB03	HNF4A ECM1 LAMB1 GNA12 CDH1	LRRK2 IRGM NOD2 ATG16L1
T$_H$17	**Other**	**Other**
IL12B TYK2 STAT3 IL23R AK2	OTUD3/PLA2G2E CAPN10 PIM3 DAP	SP140 THADA ZPF36L1 GCKR ZMIZ1 PRDX5 CPEB4 FADS1 DENNDIB MUC1/SCAMP3 5q31 (IBD5) DNNT3A
Others		
C11orf30 CREM RTEL1 NKX2-3 ORMDL3 KIF21B ZNF365 CDKAL1 PTGER4		

Figure 31.2 The clustering of IBD susceptibility genes into groups according to disease and probable function. Genome-wide association studies have identified about 70 genes associated with IBD. Many of these genes are associated with innate immunity. Some are related to Crohn's disease, others to ulcerative colitis, and some are shared. It needs to be emphasized that, in the vast majority of cases, the causal variants are unknown and that the relative risk associated with these variants is extremely small, typically 1.2–1.3.

shape the composition of the commensal microbiota in childhood and/or the education of T and B cells within organized structures of the gut-associated lymphoid tissue, thereby modulating susceptibility to ulcerative colitis. In any case, different patient subpopulations seem to have different genetic predispositions and exposures to environmental factors that modify this risk.

31-5   Bacteria and potentially viruses are major drivers of pathogenic inflammation in the gut.

Studies in humans and animal models have identified important roles for the commensal bacteria in promoting the IBD-related immune response. Immunological tolerance exists toward resident intestinal flora in normal individuals but seems to be defective in IBD. Additionally, patients with IBD display aberrant T-cell activation, high levels of mucosal IgG antibodies, and cytokine responses to intestinal bacteria. Indirect evidence for a major role for the microflora also comes from clinical observations. For example, surgical reconstruction to divert the fecal stream is effective in ameliorating Crohn's disease, and antibiotics seem to provide some benefit in the treatment of Crohn's disease, especially disease in the distal colon. There is an increase in mucosa-associated bacteria in the neo-terminal ileum after ileocecal resection for Crohn's disease, which may be associated with postoperative relapse.

Studies in murine models of IBD, such as interleukin (IL)-10-null mice, have provided additional insights into the influence of intestinal bacteria on the pathogenesis of these diseases. IBD is seen if the mice are maintained under conventional (specific pathogen-free) conditions. In these circumstances, disease is often ameliorated by treatment of the animals with antibiotics. In comparison, germ-free mice are resistant to experimentally induced colitis or manifest few signs of inflammation in models in which the inflammation develops as a result of targeted changes in the immune system. Mice deficient in genes associated with IBD risk in humans (for example *ATG16L1*) only manifest the disease-associated phenotype (namely Paneth cell dysfunction) in the presence of specific viruses (such as norovirus), suggesting that infections may have a major role in the development of IBD in the susceptible host.

In normal subjects the intestinal wall is covered with mucus, which prevents most bacteria from contacting the mucosal surface, but in patients with IBD there is a dense coating of bacteria adjacent to the epithelial cell surface. Bacteria adhere to epithelial cells, enter crypts, and are occasionally found within cells. Of particular interest is an adherent and invasive *Escherichia coli* that seems particularly prominent in Crohn's disease. The use of next-generation DNA sequencing in the culture-independent rRNA sequence analysis of intestinal bacteria from patients with Crohn's disease or ulcerative colitis in comparison with non-IBD controls has shown that important differences can be detected between the microbiotas of patients with Crohn's disease or ulcerative colitis and those of non-IBD controls. Specifically, a subset of Crohn's disease and ulcerative colitis samples can be characterized by depletion of commensal bacteria, notably members of the phyla Firmicutes and Bacteroidetes. It remains unknown whether such bacterial alterations contribute to the development of IBD or represent a consequence of the ongoing inflammation. Genetically determined inflammation can induce dysbiosis that is inflammatory in its own right as a result of the unregulated presence of certain bacteria such as Proteobacteria (for example *Serratia* and *Klebsiella* sp.) that function as pathobionts in animals that lack T-bet and an adaptive immune system (*Rag2*$^{-/-}$ mice). In contrast, inflammation may result in the reduction of other organisms such as *Faecalibacterium prausnitzii* in the gut of patients with Crohn's disease. This organism is interesting because it induces high levels of IL-10 in mononuclear cells and absence of this organism from the gut is associated with a higher risk of postoperative recurrence of ileal Crohn's disease. Along similar lines, administration of *Bacteroides fragilis* or its polysaccharide A prevents colitis in IL-10-deficient mice or SCID/Rag 2-deficient mice colonized with *Helicobacter hepaticus*, a commensal bacterium with pathogenic potential in the setting of immune deficiency. Taken together, these findings suggest that changes in the composition of commensal bacteria (for example increased Proteobacteria or decreased numbers of *F. prausnitzii*) may contribute to the amplification of the ongoing inflammation in IBD. It remains unknown whether IBD is the result of an inappropriate immune response to an individual pathobiont that is a commensal microorganism with pathogenic potential under specific circumstances (for example enteroadherent *E. coli*) or even whether it is due to an immune response to a true pathogen that has yet to be defined. However, it is interesting that the genetic basis for Crohn's disease shares many of the same genetic underpinnings as a true infectious disease (such as leprosy), blurring the differences between disease caused by a pathogen and that caused by either a pathobiont or a benign member of the commensal flora.

Association of IBD with innate immunity abnormalities.

Although the role of the vast majority of genes associated with IBD remains to be elucidated, an emerging theme is that many of the gene associations found since NOD2, itself a molecule that functions as a bacterial sensor, are also involved in the innate response to microbes.

31-6 Genes involved in innate immunity, endoplasmic reticulum stress, and autophagy are important in IBD.

Autophagy is a cellular process that is classically responsible for the degradation of damaged organelles or long-lived proteins. It is activated by a variety of conditions associated with starvation and cellular stress such as that associated with the unfolded protein response as a consequence of stress caused by the accumulation of misfolded protein within the endoplasmic reticulum (ER) as well as the presence of cell-associated bacteria. Genes associated with bacterial sensing such as those encoding NOD2 (as discussed) and intelectin-1 have been associated with Crohn's disease, genes associated with autophagy such as autophagy-related gene 16 like-1 (*ATG16L1*) and that encoding immunity-related guanosine triphosphatase (*IRGM*) have been associated with Crohn's disease, and genes associated with ER stress such as those encoding X-box binding protein-1 (*XBP1*) and orsomucoid 1-like 3 (*ORMDL3*) have also been associated with both Crohn's disease and ulcerative colitis. Although the functional mechanisms for these susceptibilities are incompletely understood, as discussed below, they highlight the primary importance of innate immunity and the intestinal epithelium in the immunopathogenesis of IBD.

31-7 Autophagy-related proteins are linked to IBD.

ATG16L1 is expressed in intestinal epithelial cells, T and B lymphocytes, and antigen-presenting cells. Very little is known about the function of the disease-associated *ATG16L1* gene product. However, it has recently been shown that macrophages isolated from mice deficient in *Atg16l1* produce elevated levels of IL-1 in response to lipopolysaccharide (LPS) and commensal bacteria such as *E. coli*, and mice develop more severe experimental colitis when the hematopoietic system is Atg16l1-deficient. Mice deficient for *Atg16l1*, and humans

Figure 31.3 Autophagy and related genes seem to be important in Crohn's disease. A schematic view of the alterations that affect the NOD2–ATG16L1 protein interactions and xenophagy in Crohn's disease is shown. In the normal intestinal epithelium, Paneth cells secrete antimicrobial products into the intestinal lumen. In addition, resident macrophages mount a proficient xenophagic response, which is mediated by interaction between NOD2 and ATG16L1 and subsequent recruitment and activation of ATG5–ATG12 complexes. In Crohn's disease, Paneth cells have a reduced ability to produce and secrete antimicrobial peptides as a result of mutations of the genes encoding NOD2 and ATG16L1. The protein encoded by the T300A mutant of the *ATG16L1* gene retains the ability to recruit ATG5–ATG12 complexes, but it has less stability, which is responsible for a defective localization of the autophagic machinery to invading bacteria. Similarly, the xenophagic response to intracellular pathogens is highly impaired in macrophages displaying the NOD2 frameshift (FS) mutation, because NOD2FS prevents ATG16L1 from localizing at bacterial entry sites by retaining it in the cytoplasm.

with a disease-associated risk variant, show decreased autophagy in the ileum and abnormalities in Paneth-cell granule structure with a pro-inflammatory phenotype including an increased expression of genes involved in peroxisome proliferator-activated receptor pathways and adipokines, such as leptin and adiponectin, which are known to be increased in IBD. Furthermore, NOD2 engagement by peptidoglycans induces autophagy and bacterial clearance. This process is dependent on the ability of NOD2 to recruit the autophagy protein ATG16L1 to the plasma membrane at the site of bacterial entry. *IRGM*, the human homolog of the mouse *Irgm/Lrg47*, encodes a GTP-binding protein that induces autophagy and is involved in the elimination of intracellular bacteria, including *Mycobacterium tuberculosis*. Taken together, these findings suggest that reduced function and/or activity of NOD2/ATG16L1/IRGM proteins could alter the innate immune response to luminal bacteria, thereby leading to persistence of intracellular bacteria and the development of pathogenic intestinal inflammation (Figure 31.3).

31-8   Impairment of intestinal epithelial cell function may also be a key factor for the development and/or perpetuation of colitis.

In sharp contrast to the prominent pro-inflammatory role of nuclear factor κ B (NFκB) in myeloid cells, conditional ablation of IKKβ in epithelial cells results in increased inflammation in experimentally induced colitis; loss of IKKγ (NEMO (NFκB essential modulator)) in epithelial cells results in spontaneous colitis. This is consistent with an overall protective function of NFκB within the intestinal epithelium through either direct effects on epithelial cell survival or the ability to enhance the synthesis of chemokines, which may facilitate the mucosal recruitment of myeloid and T cells that in turn provide cytoprotective factors such as heat-shock proteins, IL-11, and IL-22 (Figure 31.4). Similarly, mice with an impaired ability to activate the inflammasome within epithelial cells, as a result of the lack of Nlrp3 (which in humans is a genetic risk factor for Crohn's disease), an adaptor protein, or caspase 1, are highly susceptible to experimentally induced colitis, which may be mediated by the ability of epithelial cells to secrete IL-18.

Studies in mice have shown that luminal bacteria stimulate epithelial cells to produce thymic stromal lymphopoietin (TSLP) and IL-25. These two proteins are directed to the lamina propria, where they exert anti-inflammatory effects

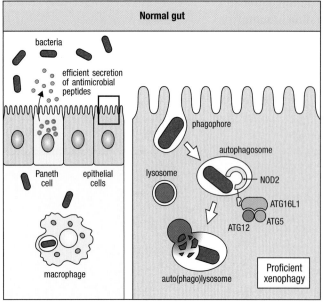
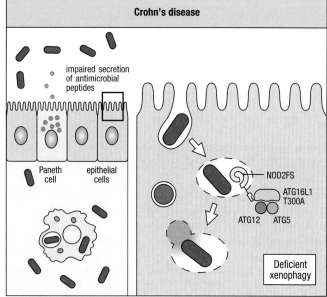

as they render intestinal dendritic cells (DCs) and macrophages unable to produce inflammatory cytokines such as IL-12 and IL-23 after bacterial stimulation. Production of both TSLP and IL-25 is diminished in IBD tissues, showing the relevance of this to human disease. Epithelial cells are also a major source of transforming growth factor (TGF)-β1, a powerful negative regulator of both macrophage and T-cell activation. In IBD, TGF-β1 is produced in excess but its activity is markedly decreased as a result of high levels of Smad7, an intracellular inhibitor of TGF-β1 signaling. Epithelial cells contribute to the amplification of the ongoing mucosal inflammation by making chemokines, which attract more inflammatory cells into the inflamed tissue, and cytokines, which enhance T-cell growth and survival.

An excellent example of how primary, genetically encoded dysfunction of the epithelial cell can phenotypically reveal itself as IBD is through studies of the unfolded protein response in response to ER stress. Genetic defects in proteins involved in the unfolded protein response itself (for example *XBP1*), chaperones involved in protein folding in the ER (for example *AGR2*), or highly secretory proteins themselves that make them susceptible to misfolding (for example *MUC2* and *HLAB27*) make the epithelium at risk for an unfolded protein response linked to the development of intestinal inflammation that may emanate directly from the epithelium. Mice with a gut epithelial deletion of *Xbp1*—a gene in the proximal effector pathway of the unfolded protein response and a genetic risk factor for Crohn's disease and ulcerative colitis—develop spontaneous enteritis. They also show increased susceptibility to experimental colitis and increased sensitivity of the epithelium to Toll-like receptor (TLR) and cytokine signaling, a marked depletion of Paneth cells, and a reduction in the goblet cells. In mice that are genetically engineered to possess a mutant Muc2 protein that exhibits aberrant oligomerization, this is associated with the induction of ER stress in goblet cells and the development of ulcerative colitis-like colitis. Along similar lines, mice that carry a missense mutation in a membrane-bound protease (site 1 protease) that generates transcriptionally active ATF6, which is important for promoting the unfolded protein response, have enhanced susceptibility to experimental colitis.

31-9 Excessive innate immune response toward the microbiota causes chronic inflammation in the gut.

The intestinal microbiota provides an abundant source of immunostimulatory organisms that can activate innate and adaptive immune responses. Indeed, in IBD tissues, innate myeloid cells overexpress various membrane-bound and endosomal receptors for microbial components; these can signal to the cell, resulting in the production of enormous quantities of pro-inflammatory molecules. Thus, increased levels of pro-inflammatory cytokines, including IL-1β, IL-6, IL-18, and tumor necrosis factor (TNF-α), are detected in active IBD and correlate with the severity of inflammation. These cytokines are probably important in the IBD-associated chronic inflammation, given their ability to alter tight junctions and intestinal permeability, to stimulate the secretion of chemokines and the expression of adhesion molecules on endothelial vessels, and to induce stromal cells to secrete extracellular matrix-degrading enzymes, such as matrix metalloproteinase 1 (MMP1) and MMP3, which digest away the extracellular matrix, leading to ulceration.

The synthesis and secretion of pro-inflammatory cytokines are governed by germline-encoded receptors such as the TLRs and nucleotide-binding-domain and LRR-containing (NLR) proteins that can be activated by microbe-associated molecular patterns (MAMPs). For example, the production of active (mature) IL-1β and IL-18 occurs through a two-step process initiated by the transcriptional induction of a pro-cytokine (for example a TLR stimulus) followed by caspase 1-mediated cleavage. In this process, NLRP3, also

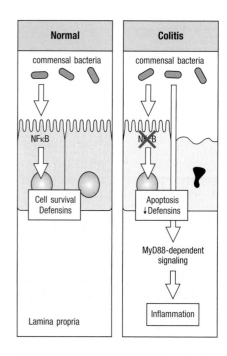

Figure 31.4 NFκB activation in colonic epithelial cells maintains epithelial integrity. Conditional knockout of NEMO in colonic epithelial cells results in a spontaneous colitis in mice. Mice lacking NEMO show increased epithelial apoptosis and reduced antibacterial defensin production, with penetration of commensal microbes into the lamina propria, triggering inflammation. Critically, in the absence of MyD88, the mice do not develop colitis, suggesting that Toll-like receptor signaling is important in driving colitis.

known as cryopyrin, associates with the NLR adaptor protein, apoptotic speck protein containing a CARD domain (ASC/PYCARD), to recruit pro-caspase 1. This complex is referred to as the inflammasome, and leads to the processing of pro-caspase 1 into active caspase 1. Caspase 1 is responsible for the subsequent cleavage of the IL-1β/IL-18 precursors into their functional forms. In addition to NLRP3, other NLRs, including NLRP1, NLRC4, and NAIP, also function in caspase 1 activation and IL-1β production through the formation of other inflammasomes in response to distinct sets of stimuli. Mice lacking NLRP3, ASC, or caspase 1 produce reduced levels of IL-1β and TNF-α and are protected from inflammation in acute, but not necessarily chronic, experimental colitis, suggesting that the NLRP inflammasome has a more critical role during the early phase of colitis.

TLR, NLR, and retinoic acid-inducible gene (RIG)-related molecules recognize 'alarm' signals provided by MAMPs from bacteria and viruses as well as damage-associated molecular patterns (DAMPs) associated with cellular injury such as uric acid. Most TLRs use the adaptor protein MyD88 to induce activation of the innate and adaptive immune systems. TLRs and MyD88 are critically involved in the pathogenesis of chronic intestinal inflammation on activation by the commensal microbiota. For instance, the absence of MyD88 prevents enterocolitis in IL-10-deficient mice in association with decreased activation of the T-cell dependent immune responses directed against the commensal microbiota. However, TLR- and NLR-related pathways are also important in maintaining intestinal homeostasis and thus preventing intestinal inflammation. Examples include the suppression of the early onset of colitis in multidrug-resistance-α-deficient mice by TLR2 agonists, the enhanced inflammation observed in MyD88-deficient mice when experimentally challenged with a barrier disruptive agent (dextran sodium sulfate), and the inhibition of experimental colitis by activation of NOD2 signaling. Innate recognition of MAMPs is thus important both for the maintenance of homeostasis and for the induction of inflammation under the appropriate environmental and probably genetic contexts.

Given the importance of the transcription factor NFκB in regulating TLR responses, it is important to consider the involvement of this transcription factor in the tissue-damaging inflammatory reaction associated with IBD. Increased NFκB activity is seen in intestinal epithelial cells and lamina propria mononuclear cells in IBD and is associated with an enhanced production of inflammatory cytokines such as IL-1β, IL-6, and TNF-α (Figure 31.5). In contrast, deletion of *Ikkb* in macrophages and neutrophils, inhibition of NFκB by antisense oligonucleotides directed against RelA/p65, and inhibition of IκB kinase by a small-molecule inhibitor (BMS-345541) can ameliorate disease in mouse models of intestinal inflammation. Although this highlights the important pro-inflammatory role of this transcription factor, NFκB clearly has important homeostatic properties, especially within the intestinal epithelium, such that deletion of *Ikkg* (NEMO) specifically within the intestinal epithelium results in spontaneous colitis, which may account in part for the beneficial effects of TLR-related signaling in maintaining homeostasis.

31-10 Innate immunity dysfunction is linked to Crohn's disease and Crohn's-like disease.

The pathophysiology of IBD, especially Crohn's disease, is currently undergoing a significant reassessment, because genetic studies have led to the identification of several susceptibility genes associated with innate immune dysfunction, suggesting that these may be central to disease pathogenesis. Another source of support for this idea comes from clinical observations showing that rare congenital disorders of innate immunity (and, in particular, phagocyte function) can associate with noninfectious bowel inflammation

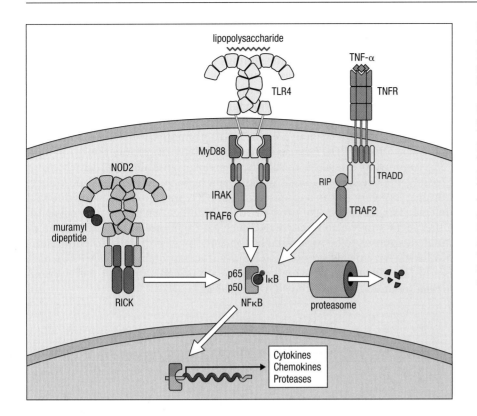

Figure 31.5 NFκB activation in myeloid cells is strongly pro-inflammatory in the gut. One of the features of IBD, especially Crohn's disease, is the influx of monocytes into diseased gut. In the inflamed tissues, monocytes potentially receive many pro-inflammatory signals, all of which can activate NFκB. Three examples are shown here: lipopolysaccharide can activate NFκB via TLR4; intracellular NOD2 can be activated by peptidoglycans; and cytokines such as TNF-α can signal via TNFR. In response to all these stimuli, monocytes produce extremely large amounts of TNF-α, IL-1β, IL-6, IL-8, and IL-12, as well as macrophage metalloelastase (MMP12), which serves to degrade the mucosa and maintain the continuing influx of white cells into the tissue.

that shows similarities to Crohn's disease, and that some patients with various neutropenias (namely leukocyte adhesion deficiency-1, glycogen storage disease type 1b, Hermansky–Pudlak syndrome, chronic granulomatous disease, and Chediak–Higashi syndrome) have been reported to develop chronic intestinal inflammation that looks similar to Crohn's disease. These results have been extrapolated to suggest that Crohn's disease is caused by a defect in macrophage function and that adaptive immune responses, as described below, are a secondary event. In reality the situation is more complex in that there is unlikely to be a single unifying problem with macrophages in Crohn's disease; instead, a combination of subtle effects in innate immunity, differing between individuals, may allow the accumulation of sufficient microbial antigens in the gut wall to trigger T-cell activation. Therefore, although the rare phagocyte defects show that an inability to handle antigens from the gut microflora properly can result in disease that resembles Crohn's disease, it is unlikely that this is the complete story and it is equally likely that some patients also have altered thresholds of T-cell activation.

31-11 NOD2/CARD15 is involved in the control of defensins in the gut.

NOD2 is strongly expressed in Paneth cells, a subset of epithelial cells at the base of intestinal crypts of the small intestine, predominantly in the terminal ileum, and in the cytoplasm of monocytes and tissue macrophages. Mutations in NOD2 have been reported to associate with a diminished activation of NFκB and decreased production of inflammatory cytokines and defensins in response to MDP. It is not clear how hypomorphic NOD2 function reconciles with intestinal inflammation and an absence of spontaneous intestinal inflammation in Nod2-deficient mice, although it is consistent with the absence of such inflammation in MyD88-deficient mice. The mechanisms by which hypomorphic NOD2 function predisposes to Crohn's disease are, however, likely to be multiple through effects on different cell types, which are in turn dependent on specific environmental factors such as enteropathogenic exposures. Such mechanisms potentially include decreased regulation of the inflammasome and consequently increased IL-1β production in response to the commensal microbiota; diminished TLR-associated tolerance to microbial stimuli with increased production of IL-12 in response to TLR2 ligands; and lessened IL-10 production through a pathway involving NOD2 association with active, phosphorylated p38 mitogen-activated protein kinase and the transcription factor heterogeneous nuclear ribonucleoprotein A1, which binds to and activates the IL-10 promoter. It has been suggested that mutations in NOD2 cause decreased α-defensin production by Paneth cells, leading to intestinal dysbiosis, decreased antimicrobial activity, and possibly decreased resistance to enteropathogens. As emphasized above, however, labeling Crohn's disease due to NOD2 mutations as being due to a defect in Paneth cells is likely to be an oversimplification of a complex disease. Other factors may also be involved, such as the failure of pro-inflammatory innate responses to be restrained by the zinc-finger protein A20, a ubiquitin-modifying enzyme also known as TNF-α-induced protein 3 (TNFAIP3), which is a negative regulator of signals downstream of TNF- and TLR-driven NFκB activation and MDP-induced NOD2 signaling.

Adaptive immunity in the pathogenesis of IBD.

31-12 Genes associated with T-cell activation and differentiation are also involved in IBD.

Gene variants related to specific pathways of inflammatory responses have been identified by genome-wide association studies in patients with IBD. One example is the signaling pathway associated with the IL-23 receptor, highlighting the importance of helper T (T_H) T_H17 cells, given the identification of polymorphisms in *IL23R* as well as IL-12p40 (*IL12B*), which forms a part of both IL-12 and IL-23, Janus-associated kinase 2 (*JAK2*), signal transducer of activation 3 (*STAT3*), and the *IL2/IL21* locus, which are associated with Crohn's disease and ulcerative colitis. Interestingly, both protective variants, such as the uncommon coding variant rs11209026 (Arg381Gln) of the IL-23R, and risk-associated variants have been identified, demonstrating that individual genes can contribute risk in quite opposite ways to others in an individual with these disorders (Figure 31.6).

31-13 Ongoing inflammation in IBD is driven by T cells.

In IBD, antigen-presenting cells such as dendritic cells (myeloid, plasmacytoid, mature DCs) and macrophages seem to have a key role in shaping the mucosal T-cell responses that lead to polarized T-cell responses associated with chronic intestinal inflammation (Figure 31.7). The number of antigen-presenting cells is increased in IBD, and these cells display signs of local cell activation in the lamina propria. In addition to these classical cell subsets, cell populations with an intermediate phenotype have been described. A CD14+ cell subset expressing both macrophage and dendritic cell markers has recently been identified in patients with Crohn's disease. This cell subset produces large quantities of pro-inflammatory cytokines such as IL-23, interferon-γ (IFN-γ), TNF-α, and IL-6 during active inflammation. In addition, antigen-presenting cells in IBD are known to produce cytokines such as IL-12 and Epstein–Barr virus-induced gene 3 (EBI3)-related cytokines such as IL-27 and IL-35 that control helper T cell polarization. IL-12 has been shown to have a key role in the induction of T_H1-cell differentiation by activating the signaling protein STAT4. The

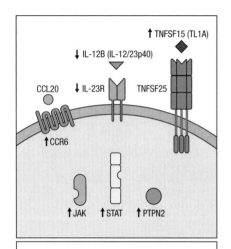

Figure 31.6 Several of the loci identified in genome-wide association studies in Crohn's disease are involved in T-cell activation. SNPs in *IL12B* (IL-12/23p40) and *IL23R* genes are associated with less risk of Crohn's disease, suggesting that the threshold of T_H1/T_H17 activation may be a determinant of disease. Similarly, TNFSF15 (TL1A) is a co-stimulatory molecule on antigen-presenting cells that, when it binds to its receptor on the T-cell surface (TNFSF25), increases interferon-γ production. *TL1A* SNPs are associated with increased risk of disease. *CCR6* SNPs are also associated with Crohn's disease; interestingly, this receptor binds CCL20, a chemokine made by epithelial cells to attract T cells into the gut. Finally, inside the cell, SNPs in the genes encoding the JAK/STAT pathways, which control cytokine signaling, and PTPN2, are associated with increased disease. The latter is interesting because the function of PTPN2 is to remove phosphates from signaling molecules and thus to dampen immune activation.

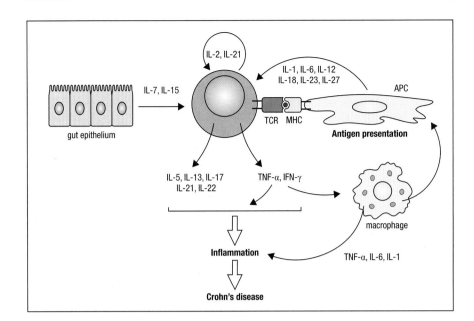

Figure 31.7 Chronic T-cell activation is a feature of IBD, especially Crohn's disease. In the inflamed environment, multiple factors are probably responsible for keeping T cells alive and may vary from patient to patient. There is very little evidence for antigen-driven T-cell activation in human IBD, because of the difficulty of identifying potential peptides from the myriad proteins made by gut bacteria. In contrast, there is a wealth of data to suggest that cytokines made by accessory cells are important in driving T-cell activation in IBD. APC, antigen-presenting cell.

increased production of IL-12 p35/p40 in patients with Crohn's disease in comparison with controls and patients with ulcerative colitis is therefore likely to contribute to the different helper T cell polarization between Crohn's disease and ulcerative colitis. Interestingly, blockade of the p40 subunit of IL-12 is associated with marked therapeutic effects in various murine models of IBD. Because anti-p40 antibodies neutralize the IL-23 (p19/p40) in addition to IL-12, the anti-p40 effects might be due to neutralization of IL-12, IL-23, or both cytokines. In fact, several studies show that IL-23 has a prominent role in murine models of chronic intestinal inflammation: blockade of IL-23 and p19 deficiency protects from intestinal inflammation. The relevance of these findings to human IBD is underlined by the observation that lamina propria cells in IBD produce augmented amounts of IL-23 and that IL-23R mutations are frequently found in patients with IBD as described earlier in the chapter.

The intestinal lamina propria contains many CD4 T lymphocytes that probably have a fundamental role in mucosal tolerance and host defense in the healthy intestine. Regulatory T cells in particular have been identified as a T-cell subset with anti-inflammatory properties in the gut. These cells express CD25 and the transcription factor Foxp3 and are known to counteract the function of mucosal effector T cells. The number of T cells with a normal regulatory function is increased in IBD. This is consistent with there being a marked expansion of effector T cells in IBD that can not be adequately controlled by these regulatory T cells (see Figure 31.7). Effector T lymphocytes in IBD are activated in IBD via antigen-presenting cells and luminal antigens. They show elevated cell proliferation and the activation of T-cell receptor-inducible transcription factors such as NFATc2. Furthermore, specific signature cytokines are produced by these cells in active IBD that permit their classification into helper T cell subsets. For instance, lamina propria T cells in Crohn's disease produce larger amounts of T_H1 cytokines such as IFN-γ on stimulation and express the T_H1 transcription factor T-bet. In ulcerative colitis, however, T cells produce increased amounts of some T_H2 cytokines such as IL-5 and IL-13. In contrast, IL-4 production by these cells is lower than in controls, suggesting the presence of an atypical T_H2 cytokine profile. In addition to prototypical T_H1 or T_H2 cytokines, large amounts of disease-perpetuating cytokines such as TNF-α and IL-6 are produced by lamina propria T cells in IBD as well as in cells associated with the innate immune system.

A new subset of T cells, designated T_H17 cells, has recently been described; the development of these cells requires the presence of specific commensal microbes that are yet to be defined in humans. T_H17 cells are generated via naive cells that are exposed to TGF-β plus IL-6 or TGF-β plus IL-21. In addition, it is known that these cells require IL-23 for stabilization of their phenotype. Various studies have analyzed the presence of this T_H17 cell subset in IBD. In patients with IBD, T_H17-inducing cytokines such as IL-6 and IL-21, and T_H17-related cytokines such as IL-17A/F and IL-22, are present at increased levels in the inflamed mucosa. Furthermore, T_H17-related transcription factors such as IRF4, STAT3, and RORA/C are upregulated in CD4 T cells isolated from IBD mucosa. Taken together, these results indicate the presence of active T_H17 cells in patients with IBD. Inactivation of cytokines that induce or augment T_H17 responses such as IL-6 and IL-21 has been shown to suppress T-cell dependent colitis. Together with the genetic information described showing an IL-23R/T_H17 genetic signature in IBD, T_H17 cells probably have an important role in the pathogenesis of chronic intestinal inflammation, but the precise role remains to be defined. There is, however, no doubt that IL-21 is strongly pro-inflammatory in the gut (Figure 31.8), but whether IL-17A is pathogenic or protective in the gut remains to be determined.

One of the major advances in IBD has been the translation of anti-cytokine therapy into the clinic, spectacularly demonstrated by the dramatic effects of anti-TNF therapy in both Crohn's disease and ulcerative colitis. However, subsequent studies blocking IFN-γ and IL-12/23 in Crohn's disease have not quite realized the same level of clinical efficacy, and many further studies are needed to identify which patients will respond best to individual therapies. In a similar fashion to the way in which the development of IBD in a single patient may be complex and even idiosyncratic and unique, it may well be that the disease-causing processes in individual patients may also be highly complex, and evolve as a result of therapy and disease duration, so that immune interventions will have to be tailored to the patient.

Figure 31.8 Newer cytokines such as IL-21 are strongly pro-inflammatory in the gut. IL-21 functions as an autocrine growth factor for both T_H1 and T_H17 cells but also induces epithelial cells to make CCL20 and fibroblasts to make matrix metalloproteinases (MMPs), which contribute to ulcer formation. IL-17A can also signal to many cell types and increase cytokine production. In the example shown here, IL-17A acts on fibroblasts to increase the production of neutrophil chemokines such as IL-8.

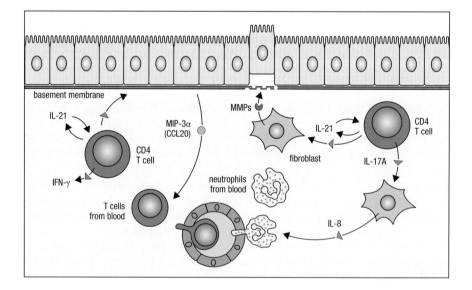

Summary.

Inflammatory bowel diseases are chronic conditions of unknown etiology. Recent data indicate that specific genetic factors predispose for disease development by modifying intestinal immune responses and barrier function. Changes in both innate and adaptive mucosal immune responses are associated with inflammatory bowel diseases and drive uncontrolled mucosal inflammation in the small and/or large bowel. In particular, activation of antigen-presenting cells and T cells in response to the commensal microflora is a key factor in disease perpetuation. Recent studies on mucosal immune responses have provided new insights into disease pathogenesis and will unequivocally result in new opportunities for optimized therapy.

Further Reading.

Fritz, T., Niederreiter, L., Adolph, T., *et al.*: **Crohn's disease: NOD2, autophagy and ER stress converge.** *Gut* 2011, **60**:1580–1588.

Kaser, A., Zeissig, S., and Blumberg, R.S.: **Inflammatory bowel disease.** *Annu. Rev. Immunol.* 2010, **28**:573–621.

Lees, C.W., Barrett, J.C., Parkes, M. *et al.*: **New IBD genetics: common pathways with other diseases.** *Gut* 2011, **60**:1739–1753.

Macdonald, T.T.: **Inside the microbial and immune labyrinth: totally gutted.** *Nat. Med.* 2010, **16**:1194–1195.

Macdonald, T.T., and Monteleone, G.: **Immunity, inflammation, and allergy in the gut.** *Science* 2005, **307**:1920–1925.

McCarroll, S.A., Huett, A., Kuballa, P., *et al.*: **Deletion polymorphism upstream of IRGM associated with altered IRGM expression and Crohn's disease.** *Nat. Genet.* 2008, **40**:1107–1112.

Monteleone, G., Pallone, F., and MacDonald, T.T.: **Interleukin-21: a critical regulator of the balance between effector and regulatory T-cell responses.** *Trends Immunol.* 2008, **29**:290–294.

Neurath, M.F., and Finotto, S.: **Translating inflammatory bowel disease research into clinical medicine.** *Immunity* 2009, **31**:357–361.

Ogura, Y., Bonen, D.K., Inohara, N., *et al.*: **A frameshift mutation in NOD2 associated with susceptibility to Crohn's disease.** *Nature* 2001, **411**:603–606.

Rioux, J.D., Goyette, P., Vyse, T.J., *et al.*: **Mapping of multiple susceptibility variants within the MHC region for 7 immune-mediated diseases.** *Proc. Natl Acad. Sci. USA* 2009, **106**:18680–18685.

Sartor, R.B.: **Genetics and environmental interactions shape the intestinal microbiome to promote inflammatory bowel disease versus mucosal homeostasis.** *Gastroenterology* 2010, **139**:1816–1819.

Sartor, R.B.: **Key questions to guide a better understanding of host-commensal microbiota interactions in intestinal inflammation.** *Mucosal Immunol.* 2011, **4**:127–132.

Sengupta, N., and MacDonald, T.T.: **The role of matrix metalloproteinases in stromal/epithelial interactions in the gut.** *Physiology (Bethesda)* 2007, **22**:401–409.

Strober, W., Zhang, F., Kitani, A., *et al.*: **Proinflammatory cytokines underlying the inflammation of Crohn's disease.** *Curr. Opin. Gastroenterol.* 2010, **26**:310–317.

Xavier, R.J., and Podolsky, D.K.: **Unravelling the pathogenesis of inflammatory bowel disease.** *Nature* 2007, **448**:427–434.

Food Sensitive and Eosinophilic Enteropathies

32

Food sensitive enteropathies are clinical response(s) to a given food (or foods) that are governed by an aberrant immune response. Inappropriate responses to innocuous food antigens can be IgE mediated, non-IgE mediated, or a mixed variety. Food allergies are increasingly common, especially in children; in the United States 3 million children (4 out of every 100) were reported to have an allergic response to food in 2007. Over the last decade, a new group of diseases associated with aberrant responses to food, the eosinophilic gastrointestinal diseases (EGIDs), has become increasingly recognized. These diseases are manifest by vague gastrointestinal complaints that occur in the setting of dense gastrointestinal mucosal eosinophilia. As for other allergic diseases there is a genetic basis for susceptibility to allergic responses to food. However, the rapidly increasing prevalence of food-related disease cannot be explained by genetic susceptibility alone and environmental, particularly microbial, influences have been implicated. No treatment other than strict avoidance of potential food antigens is currently available for any of the food-related diseases, making allergic reactions to food particularly problematic in school and childcare settings.

Basic principles of the immune response to foods.

It is estimated that in the USA, the average adult consumes 0.85 tons of food per year. Therefore the antigenic universe to which we are exposed in the food we eat is enormous in both its volume and its diversity, yet most of us remain non-responsive to this constant source of stimulation. Undoubtedly the process of digestion eliminates the vast majority of antigens from the diet, but the very fact that the immune system is designed to respond to very small amounts of antigen means that dose may not be a determinant of sensitivity.

32-1 Healthy individuals are tolerant to the great diversity of antigens present in food.

Our immune system does not *ignore* dietary antigens—the induction of non-responsiveness to food, often referred to as oral tolerance in animal models, is an active immune-mediated process. The integrity of the mucosal barrier and the production of secretory IgA also serve to exclude most food antigens from gaining access to the antigen-presenting cells of the mucosal immune system, but this barrier is not impenetrable. Small amounts of food proteins are detectable in the bloodstream and many healthy individuals have low levels of circulating food protein-specific IgG, IgA, and IgM in their serum.

Our understanding of the mechanism(s) by which non-responsiveness to dietary antigens is induced (and why these mechanisms fail in food sensitive enteropathies) remains incomplete. Tolerogenic antigen-presenting cell populations educated in the mucosal microenvironment are likely to present food antigens to naive T cells in the mesenteric lymph node to induce a predominantly immunoregulatory T- and B-cell response. The induction of anergy in these initially antigen-responsive clones and/or their deletion also help to maintain non-responsiveness to food antigen.

32-2 Aberrant immune responses to foods can be IgE and non-IgE mediated.

In some individuals, non-responsiveness to particular food antigens is either not induced, or it is abrogated. This failure to induce tolerance progresses to an aberrant or heightened immune response to particular food proteins. Food sensitive enteropathies can take a variety of forms and elicit responses that are predominantly allergic, inflammatory, or a mixture of the two (see Table 32.1).

IgE-mediated (allergic) disease is characterized by an acute, systemic response to oral food challenge that can include gastrointestinal symptoms (nausea, vomiting, abdominal pain, and diarrhea) as well as the symptoms common to allergic responses at other sites (urticaria, pruritis, anaphylaxis). The mechanisms involved are straightforward, involving classical type I hypersensitivity, IgE, and mast cells. Interestingly, classical food allergies produce very little morphological damage to the small bowel: the main finding is mucosal edema. However, local mediator release changes intestinal transit, increases epithelial permeability, and induces epithelial cells to become secretory instead of absorptive. The gut epithelium also expresses CD23, the low-affinity receptor for IgE, and there is compelling evidence that allergens coupled to trace amounts of IgE can be transported across the epithelium into the lamina propria, to propagate inflammation.

Non-IgE (cell-mediated) reactions are typically characterized by a delayed-onset inflammatory response that is confined to the gastrointestinal tract and include disorders such as cow's milk protein allergy, food protein induced enterocolitis, and celiac disease. It is often stated that this type of reaction

Table 32.1 Types of immune responses to food

Type	Key features	Disorders	Symptoms
IgE mediated	Acute onset Systemic	Food allergy	Nausea Abdominal pain Vomiting Diarrhea Pruritis Urticaria Anaphylaxis
Non-IgE (cell mediated)	Delayed onset Confined to gastrointestinal tract	Cow's milk allergy Food protein induced enterocolitis Celiac disease	Vomiting Diarrhea
Mixed reaction (both IgE and cell mediated)	Chronic	Eosinophilic esophagitis Eosinophilic gastritis Eosinophilic gastroenteritis Eosinophilic colitis	Feeding difficulty Vomiting Abdominal pain Dysphagia Food impaction

may be T-cell mediated, perhaps akin to celiac disease, which clearly involves a helper T (T_H) T_H1-mediated response to gluten. However, the extent of the mucosal flattening in non-IgE-mediated cow's milk hypersensitivity is minor compared with celiac disease, with only some villous blunting. It is entirely possible that late-onset reactions are T_H2 mediated, involving eosinophils and cytokines such as interleukin (IL)-13 that serve to open up tight junctions and make the gut leaky.

Mixed reactions have both an inflammatory and allergic component and are exemplified by the EGIDs.

32-3 Food sensitive enteropathies are diagnosed by laboratory and clinical parameters.

The diagnosis of food sensitive enteropathies relies on a combination of factors including a complete history, radioallergosorbent testing (RAST) to detect circulating food-specific IgE, skin-prick testing with allergen, and, in the case of the EGIDs, endoscopy with mucosal biopsy. The approach to making the diagnosis is dependent upon the type of food-induced immune response elicited and the symptoms at presentation. A detailed history is inherent to the diagnosis of any of these diseases. The patient must have a reproducible response to a specific food or group of foods. Historical elements that should be assessed include not only which food but also whether it is cooked or raw, the amount necessary to induce the response, timing of symptoms following the ingestion of foods, type, severity, and duration of the symptoms, and associated extraintestinal symptoms experienced. Because the food sensitive enteropathies are defined as an aberrant immune response to food, the elimination of other diseases related to the ingestion of foods (e.g., toxin-mediated reactions or inflammatory bowel diseases) or food intolerances is an important part of the diagnostic process. Food intolerances are non-immune-mediated physical responses to food products. An example of food intolerance is lactose intolerance, in which the patient does not have enough of the enzyme lactase present on the intestinal mucosa to digest lactose. This deficiency leads to maldigestion of the sugar with resultant gas production, diarrhea, and abdominal pain. Specific features important to the diagnosis of the different types of food sensitive enteropathies are discussed below.

While food allergies in IgE-mediated (allergic) disease can be manifested by gastrointestinal symptoms, other allergic symptoms related to the skin (urticaria) and lungs (wheezing and cardiovascular collapse) usually predominate. Typically, symptoms occur within minutes following the ingestion of the food. RAST and skin-prick testing with commercial food extracts are frequently positive for the suspected foods.

Non-IgE (cell-mediated) food sensitive diseases encompasses a wide variety of diseases including cow's milk protein allergy, food protein induced enterocolitis, and celiac disease, which is induced by an inappropriate response to dietary gluten (the reader is referred to Chapter 28 for a detailed description of celiac disease). Cow's milk protein allergy is a self-limited disease of infancy; the diagnosis is based on history and physical examination along with a prompt response to elimination of cow's milk. Food protein induced enterocolitis is an acute life-threatening disease; the diagnosis is based on history, physical, and biochemical analysis and does not require endoscopy with biopsy. Symptoms of food protein induced enterocolitis are profuse diarrhea leading to dehydration and acidosis that is typically caused by an immune-mediated response to milk protein. The diagnosis of celiac disease is based on a consistent history and laboratory testing including mucosal biopsy. Symptoms of celiac disease include diarrhea, abdominal pain, and slow growth. Serological testing for specific antibody titers (antitissue transglutaminase or endomysial antibodies) and an abnormal mucosal biopsy revealing villous disruption and lymphocytic inflammation are both required to make a diagnosis of celiac disease.

Mixed food sensitive enteropathies represent a broad range of diseases that include the EGIDs. EGIDs are characterized by various gastrointestinal symptoms that occur in the setting of mucosal eosinophilia. They can occur at any location in the gastrointestinal tract and are named accordingly; namely, eosinophilic esophagitis, eosinophilic gastritis, eosinophilic gastroenteritis, and eosinophilic colitis. The best defined of these is eosinophilic esophagitis or EoE. EoE presents with feeding difficulties, vomiting, or abdominal pain in the young child and dysphagia or food impaction in the adolescent/adult. Endoscopic appearances of EoE may reveal evidence of mucosal edema (linear furrows) and eosinophilic exudates (Figure 32.1), and in some circumstances luminal narrowing such as strictures. Mucosal sampling of the esophagus reveals dense eosinophilia numbering over 15 eosinophils per high-power field, basal zone hyperplasia, and in some circumstances microabscess formation (see Figure 32.2). Other diseases associated with esophageal eosinophilia, especially gastroesophageal reflux disease (GERD), must be ruled out before assigning a diagnosis of EoE.

32-4   Induction of food-specific IgE can lead to mast-cell degranulation and both local and systemic allergic symptoms.

In individuals with a genetic predisposition to allergic disease (atopy), non-responsiveness to certain food antigens is not induced. Instead, in the T_H2-biased mucosal microenvironment, exposure to food allergens induces an immune response biased to the production of the T_H2 cytokines IL-4 and IL-13, which support class switching to IgE.

Food-specific IgE binds to cells bearing high-affinity Fcε receptors (FcεRI) including mast cells and basophils (Figure 32.3). A myeloid cell subset which bridges innate and adaptive immunity, mast cells are characterized by large numbers of preformed granules which contain pro-inflammatory, vasoactive, and neuroactive mediators. Mast cells typically reside at host–environment interfaces in vascularized connective tissue (notably skin) and at mucosal surfaces of the respiratory and gastrointestinal tracts. Upon antigen ingestion, binding of allergen cross-links IgE bound to mast cell surface FcεRI, resulting in rapid degranulation. In the gastrointestinal tract, release of histamine by mucosal mast cells results in the contraction of smooth muscle and induces vomiting; transepithelial fluid loss into the gut lumen leads to diarrhea. Allergen rapidly absorbed from the gastrointestinal tract or introduced into the bloodstream also reaches the connective tissue mast cells associated

Figure 32.1 Endoscopic manifestations of eosinophilic esophagitis. The left panel shows normal esophageal mucosa. The mucosa is smooth and pink with visible vasculature underlying a translucent epithelial surface. The right panel shows eosinophilic esophagitis. The mucosal surface is irregular, dull, and has lost the appearance of the vascular pattern. Whitish patches are present that represent eosinophilic exudates (black arrows). The longitudinal furrows represent epithelial edema (white arrows).

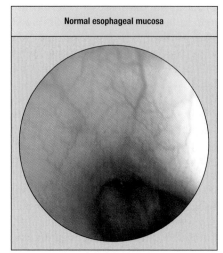
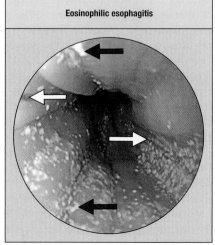

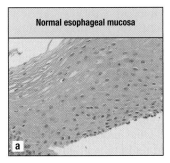

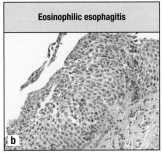

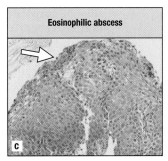

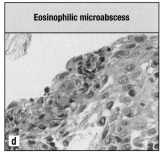

with blood vessels; degranulation at these sites results in urticaria (hives). Widespread degranulation of connective tissue mast cells and the concomitant release of histamine can also result in systemic anaphylaxis and potentially fatal anaphylactic shock. The histamine-induced increase in vascular permeability results in a precipitous loss of blood pressure and airway constriction. Swelling of the epiglottis further impairs breathing and can lead to asphyxiation. Rapid injection of epinephrine typically prevents fatal anaphylactic shock by inhibiting smooth muscle contraction and vasodilation. Food-allergic individuals are advised to carry an auto-injectable epinephrine-containing EpiPen® for protection against anaphylactic responses to accidental ingestion of food allergens.

Mast cells are also present in mucosal tissues of patients with EGIDs. Esophageal mast cells have not been well characterized but their increased numbers and evidence of their degranulation has become recognized as a potential key feature of EoE. Speculation as to their role in EoE, as with other IgE-mediated diseases, includes the stimulation of smooth muscle contraction leading to esophageal dysmotility. Mast cells are also increased in the mucosa in other EGIDs of the distal bowel. Murine models of intestinal eosinophilia support a role for mast cells in the generation of diarrhea and increase in intestinal permeability.

Figure 32.2 Histological representations of eosinophilic esophagitis. Panel a: normal esophageal mucosa. The normal esophageal mucosa is composed of stratified squamous epithelia layered on a 2–3-cell thick basal layer. A few scattered lymphocytes are present and neutrophils or eosinophils are absent. Panel b: in eosinophilic esophagitis, the epithelium is hyperplastic with an expanded basal layer. Numerous eosinophils are present throughout the epithelial surface. Panel c: in addition to a hyperplastic epithelium, an eosinophilic abscess (arrow) is present on the luminal surface. This abscess is the histological manifestation of a white exudate that is visualized endoscopically (see Figure 32.1). Panel d: magnified view of the eosinophilic microabscess on the luminal surface of the epithelia in esophagitis.

32-5 Eight types of food account for most allergic responses.

Although any food can elicit an allergic response, 90% of aberrant reactions are attributable to only eight major types of food: milk, eggs, fish, shellfish (particularly crustaceans), peanuts, soybeans, tree nuts, and wheat.

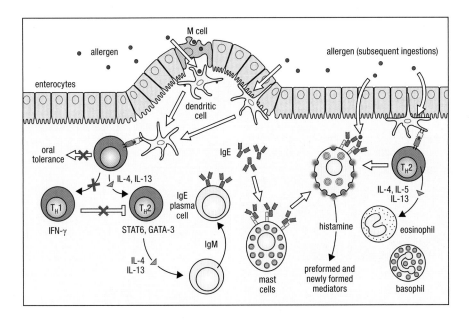

Figure 32.3 Cellular mechanisms of allergic responses to food. In healthy individuals ingestion of innocuous antigens leads to systemic non-responsiveness (oral tolerance). In allergic individuals oral tolerance is not induced. In the T$_H$2-biased mucosal microenvironment, antigen-specific T$_H$2 cells are generated when antigen fragments are presented to naive T cells by antigen-presenting cells, mainly dendritic cells. Once activated T$_H$2 cells produce IL-4 and IL-13, which promote IgE production by B cells. IgE binds to the Fcε receptor on allergic effector cells (notably mast cells). Upon subsequent allergen ingestion, antigen presentation leads to rapid T-cell activation and secretion of T$_H$2 cytokines, triggering mediator release by eosinophils and basophils. At the same time antigenic fragments may interact directly with receptor-bound IgE on mast cells and aggregation of receptors triggers the release of preformed and newly formed chemical mediators (particularly histamine) responsible for clinical symptoms. IFN-γ, interferon-γ.

Allergic responses to milk, soybeans, eggs, and wheat are typically transient and restricted to childhood while allergic responses to peanuts, tree nuts, fish, and shellfish often result in lifelong anaphylactic hypersensitivity (Figure 32.4). Most of the immunodominant protein allergens for these foods have been identified and share some features in common; they are water-soluble heat-stable glycoproteins that are generally resistant to digestion by gastric pepsin. Thus far, however, their biochemistry has not provided any other hints to explain their potent allergenicity. Marked geographic variations in the prevalence of allergic responses to particular foods may relate to diet, ethnic background, or both. Some evidence suggests that regional methods of food preparation might also impact allergenicity. For example, countries such as the USA, where the incidence of peanut allergy is high, tend to consume primarily roasted peanuts whereas in Asian countries (where the incidence is low) peanuts are typically boiled or fried.

32-6 Allergies initially manifest at peripheral sites such as the skin, but progress to the airways.

Food allergy is clinically linked to other allergic diseases, including asthma. It has been suggested that atopic dermatitis, food allergy, and asthma may be viewed as cutaneous, gastrointestinal, and pulmonary manifestations of atopy as a systemic allergic disorder.

All three diseases are characterized by eosinophilia and elevated IgE levels; pediatric patients who develop food allergy are at a higher risk of developing asthma over time.

The progression of allergic manifestations from cutaneous and gastrointestinal sites to the respiratory tract has been called the 'allergic march' and describes the natural history of atopy as a systemic disorder peaking in prevalence sequentially at different sites (Figure 32.5).

Atopic dermatitis and food allergy predominate in early childhood and begin to subside with age; some data suggest that bottle-feeding and introduction of solid foods during the first 4 months of life are risk factors for both diseases. In contrast, asthma, allergic rhinitis, and allergic responses to a subgroup of food allergens (notably peanuts) increase in prevalence with age. The

Figure 32.4 Most allergic responses are attributable to only eight foods. Allergic responses to some of these foods, including milk, eggs, wheat, and soy (blue shading), are transiently induced during childhood (typically ages 2 to 5 years) and outgrown. Other foods, including tree nuts, peanuts, fish, and shellfish (green shading), induce lifelong allergic disease with the risk of anaphylactic hyperreactivity.

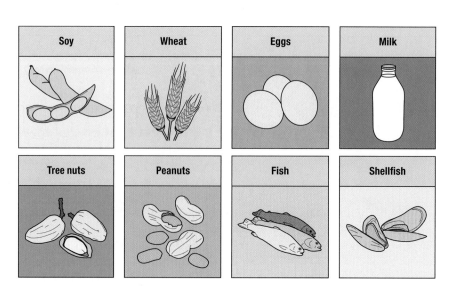

mechanism(s) responsible for the spread of allergic disease from the gut to the lung is not well understood. Many children with food allergies have a family history of allergy, which is often not restricted to food. Some aspects of the relationship between food allergies and asthma are intriguing. In a study in London, half of the children who were intubated for life-threatening asthma also had food allergies, compared with only 10% of controls, and fatal food-induced anaphylaxis occurred nearly exclusively in individuals with asthma. Peanuts or tree nuts accounted for 94% of these deaths.

32-7 Sensitization to food antigens may occur outside the gastrointestinal tract.

Respiratory sensitization to inhaled antigens in plant pollens that cross-react with allergens in food can lead to oral allergy syndrome with symptoms similar to those induced by allergen ingestion. In both animal models and human subjects, increasing evidence implicates sensitization to food antigens via the skin in allergic responses to ingested food. This may occur through breaks in the skin secondary to barrier dysfunction or through the use of oils or cream containing food allergens (particularly peanut oils).

32-8 Food allergy has a genetic component.

The heritability of the predisposition to allergic disease suggests a genetic basis, but susceptibility genes for food allergy are as yet largely unidentified. There may be both common (atopy) and unique (food allergy) genetic risk factors that contribute to allergic responses to food. Recent genome-wide association studies have identified intriguing candidates that may provide new insight into disease pathogenesis, particularly for EoE. The first genome-wide association study for EoE showed that eotaxin-3, a prominent eosinophil-recruiting chemokine, was the most upregulated gene in the esophageal epithelia of children with EoE. Further work has identified an epithelial cluster of EoE-associated genes including for filaggrin, an epidermal barrier protective protein. Loss-of-function variants in filaggrin are associated with atopic dermatitis in some patients. This finding is quite interesting when one considers its implications in EoE; in contrast to the rest of the gastrointestinal tract, the esophagus (like the skin) is lined by a squamous epithelium. If epithelial barrier function is impaired, as may be the case with loss of filaggrin function, luminal allergens could enter the lamina propria and initiate an allergic response. Finally, variants in thymic stromal lymphopoietin (TSLP), an IL-7-like cytokine secreted by the epithelial mucosa that educates dendritic cells to induce T_H2 responses, have been implicated in susceptibility to EoE. Thus, a model emerges of a series of genes related to epithelial barrier defects, eosinophil chemotaxis, and predisposition to a T_H2 phenotype (Figure 32.6).

32-9 The best time to introduce food antigens into the infant diet is unclear.

Pediatric guidelines had recommended that pregnant and lactating women with a family history of atopy avoid consumption of the major food allergens and delay introduction of these foods to their children. More recent work suggests that the reverse may be true; early oral food exposure may be necessary to induce oral tolerance and avoid allergic responses to food. Delayed introduction of foods may increase the chances of allergen sensitization via the skin (particularly in individuals with defective barrier function) before oral tolerance has had a chance to develop.

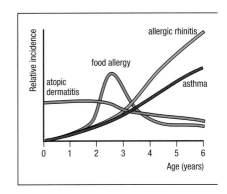

Figure 32.5 The allergic march. Atopic individuals tend to present with site-specific manifestations of allergic disease that occur in an ordered sequence. Atopic dermatitis appears first, in infancy. Transient food allergy peaks in early childhood; a subset of food allergic patients develops lifelong disease. Asthma and allergic rhinitis follow, and continue to increase into adulthood.

Figure 32.6 A model for the pathogenesis of EoE. Epithelial barrier dysfunction (possibly related to decreased expression of filaggrin) enhances allergen uptake. Antigen presentation by TSLP-primed dendritic cells induces an allergen-specific T$_H$2 response characterized by the production of IL-4, IL-13, and IL-5. IL-13 upregulates the expression of eotaxin-3 in the esophageal epithelium, which together with IL-5 mobilizes eosinophil recruitment. Their secretion of potent inflammatory mediators makes eosinophils key to both the initiation and perpetuation of pathogenesis in EoE. IL-4 and IL-13 are switch factors for the production of allergen-specific IgE.

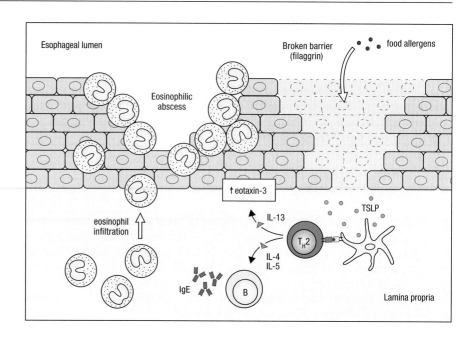

The eosinophilic gastrointestinal diseases (EGIDs).

The advent of flexible endoscopy and biopsy in the 1960s heralded a new era in mucosal immunology of the gastrointestinal tract. With the newfound ability to procure mucosal samples from the esophagus, stomach, small intestine, and colon, specific morphological and immunological features could be described and measured.

32-10 The definitions of mucosal eosinophilia are confusing.

With respect to the identification of eosinophils in the mucosa, several key features are critical. First, in healthy individuals, eosinophils are present at varying levels in all gastrointestinal organs, except the esophagus. Second, values for the 'normal' numbers of eosinophils in the mucosa are still being defined but clearly vary depending on which part of the gastrointestinal tract is examined. For instance, the cecum contains 40–50 eosinophils per high-power field whereas the stomach mucosa contains less than 10. Third, additional features regarding the association of this cell type with pathogenicity are under investigation. While current practice relies strictly on the numbers of eosinophils per high-power field, other features that may help identify inflammatory patterns include the location of eosinophils within the mucosa and their state of degranulation, as well as other associated histological abnormalities such as epithelial disruption, goblet cell hyperplasia, and evidence of tissue remodeling. Radiographic features of mucosal eosinophilia are nonspecific and reflect evidence of either mucosal edema (flocculation) or remodeling (luminal narrowing). If it is determined that pathological mucosal eosinophilia exists, the finding must be interpreted in the context of the clinical setting in which that biopsy was obtained. The differential diagnosis for gastrointestinal mucosal eosinophilia includes gastro-esophageal reflux disease, peptic disease, celiac disease, inflammatory bowel disease, the presence of bacterial, viral, or fungal infections, connective tissue diseases, and immunodeficiency (see Table 32.2).

32-11 Eosinophilic esophagitis (EoE) is the most common form of EGID.

EoE is, by far, the most common of the EGIDs. EoE occurs with a prevalence of 1–4 per 10,000 in the USA. Because of this, more is known regarding the pathophysiology of EoE than the other EGIDs, as will be summarized here. Over the course of the last decade, a series of basic and translational studies have identified potential mechanisms for the pathogenesis of EoE. The first murine model of eosinophilic esophagitis showed that sensitization and challenge by the ubiquitous aeroallergen *Aspergillus fumigatus* led to esophageal eosinophilia. The IL-5 dependency of this response was determined when IL-5-null mice were shown to be protected from disease in this model. Both murine studies and translational work with human samples have shown that the inflamed EoE mucosa has increased numbers of CD4$^+$ effector T cells and decreased numbers of regulatory T cells (T_{reg}). Further work determined that IL-5 was critical to the esophageal tissue remodeling associated with this eosinophilic inflammation. Together, these findings provided support for the development of therapeutics targeting IL-5 in the treatment of this disease. In this regard, initial studies have shown that anti-IL-5 antibody treatments are successful in significantly reducing mucosal eosinophilia but not symptoms. An increasing number of translational studies have brought new insights into the diagnostic features and pathogenesis of EoE. Traditionally, eosinophil number alone has been the major diagnostic criterion. Two recent studies showed that eosinophil degranulation is increased in EoE compared with GERD. The underlying esophageal immunological microenvironment contains increased T_H2 and T_H1 pro-inflammatory cytokines and infiltration with T cells, B cells, mast cells, and basophils. Although the role of regulatory T cells in EoE is uncertain, Foxp3$^+$ T_{reg} cells are increased in the esophageal mucosa in EoE. The downstream impact of chronic eosinophilic inflammation has focused on esophageal remodeling. Esophageal fibrosis is increased in children with EoE as shown by trichrome staining patterns.

32-12 The environmental allergens that contribute to EoE are dietary and airborne.

In light of basic work that focuses on the aeroallergen as an instigating cause of esophageal eosinophilia, case series studies support a role of environmental allergens as a potential antigenic etiology for EoE. Several reports note that some patients may develop both symptoms and esophageal inflammation during seasonal exposure to inhaled allergens. However, more evidence supports a role for food allergy in the pathogenesis of EoE. Over two-thirds of patients with EoE have evidence of an allergic diathesis including food allergic diseases. A number of studies have demonstrated that either the selected removal of suspected foods or the complete elimination of intact proteins and the use of an amino acid-based 'elemental' dietary formula leads to clinical and pathological remission of EoE. In this light, one successful strategy in the treatment of EoE has been selective exclusion of the major dietary protein allergens (see Figure 32.4).

32-13 Eosinophils play a critical role in the pathogenesis of EoE.

Although the precise functional role of eosinophils in EoE is uncertain it is derived from the impact of eosinophils in other allergic diseases. Eosinophils are biological powerhouses; they have both large numbers of granules containing preformed proteins and the capacity to synthesize and release newly generated mediators including cytokines and leukotrienes. Highly charged eosinophil granule proteins such as major basic protein can decrease

Table 32.2 Etiologies of eosinophilic gastrointestinal inflammation

Gastro-esophageal reflux disease
Peptic disease
Eosinophilic gastrointestinal diseases
Celiac disease
Inflammatory bowel diseases
Infections
Connective tissue diseases
Hypersensitivity reactions
Immunodeficiency
Cancer

epithelial barrier function and stimulate smooth muscle contraction. Major basic protein deposition in the esophageal mucosa may therefore facilitate transepithelial passage of luminal allergens to primed effector cells in the lamina propria and/or induce smooth muscle contraction leading to dysphagia or food impaction. Finally, eosinophils have been linked to tissue remodeling in the lung and recent evidence indicates that they may play a role in esophageal fibrosis through a transforming growth factor (TGF)-β-mediated mechanism.

32-14 Eosinophilic gastroenteritis is uncommon.

Eosinophilic gastroenteritis is a rare disease that is characterized by a wide variety of vague intestinal symptoms and histological evidence of gastrointestinal eosinophilia. The diagnosis is based on excluding other causes of intestinal eosinophilia. Since eosinophils reside in the normal gastrointestinal mucosa, most problematic in making the diagnosis of eosinophilic gastroenteritis is determining the exact number of eosinophils that defines a pathological state from a mucosal biopsy. Previously identified methods for categorizing eosinophilic gastroenteritis have differentiated patients in three categories: mucosal, muscular, and serosal disease. Most eosinophilic gastroenteritis patients have mucosal disease and typically have symptoms related to epithelial dysfunction including bleeding, pain, and diarrhea. A smaller number of patients with eosinophilic gastroenteritis have muscular disease characterized by vomiting and bloating, symptoms suggestive of intestinal obstruction. An even smaller fraction have serosal disease that presents with abdominal bloating and ascites. Pathophysiological mechanisms associated with eosinophilic gastroenteritis and these different phenotypes are unknown. Murine models of eosinophilic gastroenteritis have identified roles for eotaxin-1 and IL-5 in the generation of the mucosal response while others have shown that eosinophil granule proteins, eosinophil peroxidase, and major basic protein may impact inflammation and barrier function. Treatment of these diseases is based on either chronic steroid use or dietary elimination of suspected allergens.

32-15 Allergic reactions to foods sometimes manifest only in the colon.

IgE-mediated allergic reaction in the colon/rectum to foodstuffs, including cow's milk proteins and soy proteins, can result in serious disease. The condition most commonly occurs in children and can be seen even in exclusively breast-fed babies who become sensitized to dietary cow's milk in mothers' milk. The principal clinical features are vomiting, diarrhea, abdominal pain, rectal bleeding, and anemia. The condition may affect any part of the large intestine but the sigmoid colon and rectum are most commonly involved. Definitive diagnosis can only be achieved by elimination diet and challenge, but this is very rarely carried out since most infants thrive on a milk-free diet.

Colonoscopy reveals mucosal erythema and, in severe cases, ulceration. Colonic biopsy classically reveals intense eosinophilic infiltration of the lamina propria and muscularis mucosa. Features of chronicity (as seen in inflammatory bowel disease) such as glandular distortion are not seen. A peripheral eosinophilia is also commonly present. Serum IgE is often raised.

It is not clear why allergic disease manifests in the colon in some infants, especially breast-fed babies who are receiving minute amounts of allergen by mouth which is unlikely to reach the distal bowel and induce a local response. The immaturity of the intestinal mucosal barrier might allow allergens to enter the circulation to produce an IgE response, but there must be a major host genetic response since all infants show some deficit in barrier function and antigen penetrates into the blood in quite large amounts in all bottle-fed babies. Very low doses of antigen might also predispose toward an IgE-producing T_H2 response.

Changing to a hypoallergenic milk formula is often effective and usually produces a good clinical response. For infants who are sensitized through breast milk, the mother can commence a milk-free diet. Allergic colitis of infancy is a self-limiting condition that presents with symptoms of blood-streaked stools during infancy. Babies otherwise appear well, have no anemia, and respond to eliminating cow's milk or soy protein from the formula or breast milk. In rare circumstances, infants may need to be treated with an amino acid-based formula. Allergic colitis usually resolves spontaneously by age 6–12 months.

Prevalence of food sensitive enteropathies.

The Centers for Disease Control and Prevention has documented an 18% increase in the prevalence of reported food allergy for children under 18 in the USA between 1997 and 2007. Allergic responses to environmental allergens, including food, have become much more prevalent, particularly in developed countries. One explanation for the increased prevalence, modified from what has become known as the hygiene hypothesis, is that it is attributable to vaccination and improvements in hygiene. Reduced exposure to infectious diseases in childhood is thought to impair the development of the immunoregulatory networks that protect against both autoimmune inflammatory diseases and allergic hyperreactivity to environmental antigens. Recent work suggests that noninfectious microbial exposures may play an equally important, or greater, role. Very large numbers of bacteria are present in the gut (approximately 10^{12} microbes per milliliter of luminal contents at its highest point in the colon) and are estimated to outnumber the cells of the body by a factor of nearly 10:1. The luminal bacterial microbiome exists in a dynamic symbiotic relationship with its host and has many influences on host physiology (see Chapter 17). The commensal microbiota also has a profound impact on the development of both systemic and mucosal immunity; germ-free mice, which lack a colonizing microbiome, exhibit profound defects. Recent work has examined the influence of the enteric microbiota on intestinal inflammation, diabetes, and metabolic syndrome, but the mechanisms by which signals derived from a noninvasive commensal microbiota impact systemic immunity and regulate complex immune-mediated diseases are only beginning to be explored. Murine model work has shown that Toll-like receptor 4 (TLR4)-mediated signals from commensal bacteria control susceptibility to allergic responses to food. A recent study demonstrated that diet itself shapes the composition of the commensal microbiome. Microbial diversity and richness are significantly elevated in the fecal microbiota of children living in a rural African village when compared with that obtained from children living in the developed world and consuming a Western-style high-fat, low-fiber diet of processed foods. Decreasing exposure to infectious disease, increasing use of antibiotics, and other environmental factors, including diet, may be altering the composition of the bacterial microbiota in a manner that removes the bacterial populations that protect against allergic and inflammatory disease. This is an exciting area for future study.

Treatment for food sensitive enteropathies.

As mentioned above, food elimination diets are currently the most effective treatments for those suffering food sensitive enteropathies. Ongoing studies are examining whether sublingual (i.e., oral) administration of small doses of food allergens can desensitize atopic individuals to protect against severe anaphylactic responses to accidental food exposure. The nature of the anaphylactic response makes this approach inherently risky and requires that allergen dosing be confined to a carefully controlled medical setting. The realistic goal of this therapy is, however, not to make allergic individuals completely tolerant to food proteins; instead it is to use immunological tolerance as a strategy to raise the antigen threshold needed to elicit a response to a level beyond that which occurs as contaminants in poorly labeled foodstuffs. This is particularly relevant to peanut sensitization where raising the threshold for triggering anaphylaxis from trace amounts to gram amounts is potentially life-saving.

Omalizumab is a humanized anti-IgE monoclonal antibody licensed for use in moderate to severe allergic asthma. It is beginning to be used in patients, including children, with IgE-mediated food allergies with beneficial effects. It may be a useful adjunct to desensitization programs in the severely allergic child because it will help prevent anaphylaxis.

A better understanding of how microbiota-induced immunoregulatory signals impact susceptibility to food allergic disease will also inform strategies for the development of agonists that mimic these immunoregulatory signals. For example, repeated courses of antibiotics during infancy may be a contributing factor to the increasing incidence of allergic disease. Prophylactic coadministration of a bacterial TLR agonist during antibiotic treatment regimens in infancy might be able to maintain stable TLR-induced immunoregulatory signals under conditions that alter the composition of the microbiota and be an effective and novel strategy for preventing sensitization to food antigens. It remains a mystery, however, why the vast majority of food-allergic infants grow out of the disease when on an elimination diet. This supports the idea that the allergenicity is due to a passing phase of the immune response, perhaps due to delayed maturation of T_H1 responses. It is difficult to exclude trace amounts of the food proteins that may tolerize the infant during an elimination diet, and indeed it is not inconceivable to suggest that antigen-specific tolerization can occur best against the background of a maturing immune response. However, the epidemic of atopy, allergy, and asthma in the developing world is now seen as a major public health challenge that is best dealt with by public health measures such as determining the optimal time for weaning onto milk-based and cereal-based foods.

Summary.

Healthy individuals develop tolerance to the many potential antigens present in food. Food sensitive enteropathies can present in a variety of different forms. Allergic responses to food, in particular, appear to be increasing. New insights provided by genome-wide association studies point to dysregulated epithelial barrier function and the predisposition to a T_H2-biased response to food antigens as key factors in disease pathogenesis. The elucidation of how environmental factors, particularly alterations in the composition of the commensal microbiome, contribute to increasing disease prevalence will be an important area for future study that may provide insight into new strategies for preventing and treating inappropriate responses to food.

Further Reading.

Aceves, S.S., Newbury, R.O., Dohil, R., *et al.*: **Esophageal remodeling in pediatric eosinophilic esophagitis.** *J. Allergy Clin. Immunol.* 2007, **119**:206–212.

Atkins, D., and Furuta, G.T.: **Mucosal immunology, eosinophilic esophagitis, and other intestinal inflammatory diseases.** *J. Allergy Clin. Immunol.* 2010, **125**:S255–261.

Bashir, M.E., Louie, S., Shi, H.N., *et al.*: **Toll-like receptor 4 signaling by intestinal microbes influences susceptibility to food allergy.** *J. Immunol.* 2004, **172**:6978–6987.

Brandtzaeg, P.: **Food allergy: separating the science from the mythology.** *Nat. Rev. Gastroenterol. Hepatol.* 2010, **7**:380–400.

Branum, A.M., and Lukacs, S.L: **Food Allergy Among U.S Children: Trends in Prevalence and Hospitalizations.** National Center for Health Statistics, *Data Brief* No. 10, 2008.

De Filippo, C., Cavalieri, D., Di Paola, M., *et al.*: **Impact of diet in shaping gut microbiota revealed by a comparative study in children from Europe and rural Africa.** *Proc. Natl Acad. Sci. USA* 2010, **107**:14691–14696.

Du Toit, G., Katz, Y., Sasieni, P., *et al.*: **Early consumption of peanuts in infancy is associated with a low prevalence of peanut allergy.** *J. Allergy Clin. Immunol.* 2008, **122**:984–991.

Fuentebella, J., Patel, A., Nguyen, T., *et al.*: **Increased number of regulatory T cells in children with eosinophilic esophagitis.** *J. Pediatr. Gastroenterol. Nutr.* 2010, **51**:283–289.

Lack, G.: **Epidemiologic risks for food allergy.** *J. Allergy Clin. Immunol.* 2008, **121**:1331–1336.

Mishra, A., Wang, M., Pemmaraju, V.R., *et al.*: **Esophageal remodeling develops as a consequence of tissue specific IL-5-induced eosinophilia.** *Gastroenterology* 2008, **134**:204–214.

Prioult, G., and Nagler-Anderson, C.: **Mucosal immunity and allergic responses: lack of regulation and/or lack of microbial stimulation?** *Immunol. Rev.* 2005, **206**:204–218.

Rothenberg, M.E.: **Biology and treatment of eosinophilic esophagitis.** *Gastroenterology* 2009, **137**:1238–1249.

Sicherer, S.H., and Sampson, H.A.: **Food allergy.** *J. Allergy Clin. Immunol.* 2010, **125**:S116–125.

Turcanu, V., Maleki, S.J., and Lack, G.: **Characterization of lymphocyte responses to peanuts in normal children, peanut-allergic children, and allergic children who acquired tolerance to peanuts.** *J. Clin. Invest.* 2003, **111**:1065–1072.

Index

Note: page numbers followed by F refer to figures and those followed by T refer to tables. Prefixes are ignored in the alphabetical sequence – thus β-Catenin will be found after Caspase.

Keratoconjunctivitis, vernal 339
Keratoconjunctivitis sicca (dry eye) 336–338, 337F
Keyhole limpet hemocyanin (KLH), oral tolerance 226–227, 227F
Kidney
    biopsy in IgA nephropathy 446, 446F, 447F, 451
    head, of bony fish 23
KIR, natural IELs 82
c-Kit
    mast cells 181, 182F
    natural IELs 78
Klebsiella pneumonia, mast cell responses 185
Krause, accessory lacrimal glands of 331F
Kruppel-like factor 4 (Klf4) 40

L

L-selectin 206
    homing of lymphoid cells 14
    naive T-cell homing to lymph nodes 207–208, 208F
    respiratory tract 30
Lacrimal duct-associated lymphoid tissue (LDALT) *see* Tear-duct-associated lymphoid tissue
Lacrimal gland-associated lymphoid tissue 330, 330F, 331–332
Lacrimal glands 331–332, 331F
    plasma cells 28, 331–332
Lactobacillus 249F, 283
Lactobacillus acidophilus, protection against rotavirus 383
Lactoferrin 283
Lactose intolerance 491
Lamina propria 2
    B cells 105, 105F
        celiac disease 433–434
        homing mechanisms 116
        lineage activity 116–118
    dendritic cell populations 146F, 147
    distinction from MALT 105, 105F
    HIV-1 entry 389–390, 389F
    immune cells in celiac disease 433–434, 433F
    as immune effector site 3, 4F
    lymphocyte populations 87–101
    lymphoid cell homing 14–15, 15F
    macrophages *see* Macrophages, intestinal
    mast cells 180
    monocyte recruitment 161–162, 162F
    nasal 296, 297F
    ocular tissues 331, 331F
    plasma cells, celiac disease 433–434, 433F
    T cells
        celiac disease 433–434, 433F
        cytokine production 90–92
        IBD 485
        origin and phenotype 88–96
    transepithelial delivery of antigens 61–62, 61F
Laminin 5 (epiligrin) 338
Lampreys *see* Agnathans
Langerhans cells
    female genital mucosa 284
    genital tract infections 289
    HIV-1 capture and internalization 387–388, 387F, 388F
Langerin (CD207), HIV-1 uptake 387–388, 387F

Large intestine
    IELs 71
    immune inductive sites 30–31
    mucosal immunoglobulins 30
    see also Colon
Lectins, mucosal vaccine targeting 423
Legionnaire's disease, increased susceptibility 250
Lepidosaurs 21T
Leptin 258
Leucine-rich repeat (LRR) domains
    agnathan lymphocyte receptors 21, 22, 22F
    NOD-like protein receptors 236, 236F
Leukocytes
    female reproductive tract 266, 267F
    periodontal disease 397, 400
    see also specific types
Leukotriene B$_4$ (LTB$_4$) 404
Leukotrienes, biosynthesis 403F, 404
Lewis antigens, *Helicobacter pylori* 368, 368F, 373
Lewis^b antigen (Le^b) *see* MUC5AC
Leydig cells 268–269
Lgr5 (Gpr49) 38–39, 58
Linkage analysis, IBD 475–476, 475T
Lipid antigens, CD1d-restricted presentation 63
Lipopolysaccharide (LPS)
    B-cell activation 115
    gut bacteria interactions 358, 359F
    HIV-1 infection 393
    IgA-mediated internalization 135
    inhibition of IgD expression 13
    lipid A acylation 357
    NFκB activation 483F
    pattern-recognition receptors 235, 235F
    periodontal disease 400, 403
    preferential IgA2 responses 107, 108
    receptor *see* CD14
    regulation of pIgR expression 130
Lipoxins (LXA$_4$; LXB$_4$)
    biosynthesis 403F, 404
    metabolism 405F
    periodontal disease 404–405
5-Lipoxygenase (5-LO) 404
12-Lipoxygenase (12-LO) 404
15-Lipoxygenase (15-LO) 404
15R-Lipoxygenase 405F, 406
Lipoxygenases (LO) 403F, 404
    lipoxin biosynthesis 403F, 404
    omega-3 PUFA metabolism 405, 405F
Listeria monocytogenes 349T
    cell-to-cell spread 355
    epithelial adhesion/invasion 352–353, 353F, 355
    induced IEL responses 79, 82
    innate mucosal defenses 49
    NOD2-mediated clearance 240
Listeriolysin O 355
Liver abscess, chronic granulomatous disease 470
Liver cells *see* Hepatocytes
Liver disease
    CD40 ligand deficiency 463–464, 463F
    common variable immunodeficiency 459
    MHC class II deficiency 465
    severe combined immunodeficiency 462–463
    SP110 deficiency 465
LL-37 45, 45F

Lower respiratory tract 307–308
    antigen sampling 30–31, 308
    effector sites 308
    immune inductive sites 29–30, 307–308, 307F
    immunoglobulins 29, 308
    microbiota 248
    plasticity of Clara cells 41
LPS *see* Lipopolysaccharide
Lung
    antigen-presenting cells 313–314, 313F
    chronic inflammatory diseases 309–326
    epithelial cells 323
    interaction with environment 309
    see also Lower respiratory tract; Respiratory tract
Luteinizing hormone (LH) 263, 264, 265F
Ly6C-expressing monocytes 144, 144F
Ly49, natural IELs 82
Lymph nodes
    basophil migration to 174, 175
    cellular organization 210–211
    dendritic cell migration to 215
    effector/memory T cell migration to 214
    entry of naive T cells 207–208
    expression of homing molecules 209–210, 210F
    lymphocyte egress 211–212, 211F
    lymphocyte migration within 210–211, 211F
    organogenesis 333, 334T
    regulation of tissue-specific T-cell subsets 214–215
    respiratory tract 307, 307F, 308
    retention of activated T cells 211, 212
    see also Cervical lymph nodes; Mesenteric lymph nodes
Lymphatics, immune cell homing via 13–14
Lymphocyte function-associated antigen-1 (LFA-1) *see* α$_L$β$_2$ integrin
Lymphocytes
    attraction by FAE cells 192–193
    effector *see* Effector lymphocytes
    egress from lymph nodes 211–212, 211F
    eye-associated lymphoid tissue 331–332
    FAE induction 199
    intraepithelial *see* Intraepithelial lymphocytes
    lamina propria populations 87–101
    mast cell-mediated regulation 188–189
    migration 205–218
        different tissues 208–209, 208F
        homing to effector sites 13–17
        within lymph nodes 210–211, 211F
        molecules controling 206–207, 206F
        into mucosal tissues 212–217
        multi-step entry into tissues 207–208, 207F
        to sites of mucosal inflammation 217–218
        therapeutic opportunities 217–218
        vaccination implications 417–419, 418F
    phylogeny 20
    VLRA$^+$ and VLRB$^+$ 20, 21, 22
    see also B cell(s); T cell(s); *specific subsets*
Lymphocytic choriomeningitis virus (LCMV)
    induced autoimmune disease 226
    natural IEL function 81
Lymphoid aggregates
    bronchial mucosa 307–308, 308F
    Crohn's disease 473, 473F